CELL BIOLOGY
A Short Course

Companion Website

This book is accompanied by a companion website:

www.wileyshortcourse.com/cellbiology

The website includes:

- 5 Animations
- A thermocycler video
- All figures from the textbook as Power Point slides for downloading
- Additional textboxes
- Additional references and useful web links

CELL BIOLOGY
A Short Course

THIRD EDITION

Stephen R. Bolsover
Jeremy S. Hyams
Elizabeth A. Shephard
Hugh A. White

WILEY-BLACKWELL
A JOHN WILEY & SONS, INC., PUBLICATION

Published by John Wiley & Sons, Inc., Hoboken, New Jersey.
Published simultaneously in Canada

Library of Congress Cataloging-in-Publication Data:

Cell biology : a short course / Stephen R. Bolsover ... [et al.]. –3rd ed.
p. cm.
Includes index.
ISBN 978-0-470-52699-6 (pbk.)
1. Cytology. I. Bolsover, Stephen R., 1954–
QH581.2.C425 2011
571.6–dc22

2010040989

Printed in Singapore

10 9 8 7 6 5 4 3 2 1

CONTENTS IN BRIEF

CONTENTS

Companion Website

This book is accompanied by a companion website:

www.wileyshortcourse.com/cellbiology

The website includes:

- 5 Animations
- A thermocycler video
- All figures from the textbook as Power Point slides for downloading
- Additional textboxes
- Additional references and useful web links

PREFACE

Our aim in previous editions of *Cell Biology: A Short Course* was to cover a wide area of cell biology in a form especially suitable for first-year undergraduates, keeping the book to a manageable size so that neither the content, the cost, nor the weight were too daunting for the student. We have remained true to this aim while thoroughly revising the book and introducing new sections on cancer and the immune system. Many more medical examples are included, and these concentrate on topics covered in medical courses rather than obscure genetic diseases.

The overall theme for the book is the cell as the unit of life. We begin (Chapters 1–3) by describing the components of the cell as seen under the microscope. We then (Chapters 4–8) turn to the central dogma of molecular biology and describe how DNA is used to make RNA, which in turn is used to make protein. The next section (Chapters 9–11) describes how proteins are delivered to the appropriate location inside or outside the cell, and how proteins perform their many functions. We then (Chapters 12–14) describe how cells manipulate and use chemical and electrical energy. Chemical signaling within and between cells is covered in Chapters 15 and 16. The cytoskeleton is described in Chapter 17 and plays a major role in the phenomenon of cell division that, together with cell death, is the subject of Chapter 18. Chapter 19, new in this edition, describes the immune system. Finally Chapter 20 uses the example of the common and severe genetic disease cystic fibrosis to illustrate many of the themes discussed earlier in the book.

Boxed material throughout the book is divided into **examples** that illustrate the topics covered in the main text, explanations of the **medical relevance** of the material, and **in-depth** discussions that extend the coverage beyond the content of the main text. Extended matching questions at the end of each chapter help readers assess how well they have assimilated and understood the material, while each chapter also poses a thought question that tests concepts rather than facts.

A comprehensive website accompanies the book at www.wileyshortcourse.com/cellbiology. As well as giving additional self-test questions, the website includes a wealth of extra examples, in-depth explanations, and medical discussions for which there was no room in the printed book. The site also provides links to other Internet resources together with references to primary research publications to allow readers to trace the origin of statements in the text. The symbol 💻 in the book indicates that a corresponding entry in the website can be consulted for more information.

1

CELLS AND TISSUES

The **cell** is the basic unit of life. **Microorganisms** such as bacteria, yeast, and amoebae exist as single cells. By contrast, the adult human is made up of about 30 trillion cells (1 trillion = 10^{12}) which are mostly organized into collectives called **tissues**. Cells are, with a few notable exceptions, small (Fig. 1.1) with dimensions measured in micrometers (μm, 1μm = 1/1000 mm) and their discovery stemmed from the conviction of a small group of seventeenth-century microscope makers that a new and undiscovered world lay beyond the limits of resolution of the human eye. These pioneers set in motion a science and an industry that continues to the present day.

The first person to observe and record cells was Robert Hooke (1635–1703) who described the *cella* (open spaces) of plant tissues. But the colossus of this era of discovery was a Dutchman, Anton van Leeuwenhoek (1632–1723), a man with no scientific training but with unrivaled talents as both a microscope maker and as an observer and recorder of the microscopic living world. van Leeuwenhoek was a contemporary and friend of the Delft artist Johannes Vermeer (1632–1675) who pioneered the use of light and shade in art at the same time that van Leeuwenhoek was exploring the use of light to discover the microscopic world. Despite van Leeuwenhoek's efforts, which included the discovery of microorganisms and protozoa, red blood cells and spermatozoa, it was to be another 150 years before, in 1838, the botanist Matthias Schleiden and the zoologist Theodor Schwann formally proposed that all living organisms are composed of cells. Their "cell theory", which nowadays seems so obvious, was a milestone in the development of modern biology. Nevertheless general acceptance took many years, in large part because the **plasma membrane** (Fig. 1.2), the membrane surrounding the cell that divides the living inside from the nonliving **extracellular medium**, is too thin to be seen using a light microscope.

PRINCIPLES OF MICROSCOPY

Microscopes make small objects appear bigger. A light microscope will magnify an image up to 1500 times its original size. Electron microscopes can achieve magnifications up to several million times. However, bigger is only better when more details are revealed. The fineness of detail that a microscope can reveal is its **resolving power**. This is defined as the smallest distance that two objects can approach one another yet still be recognized as being separate. The resolution that a microscope achieves is mainly a function of the wavelength of the illumination source it employs. The smaller the wavelength, the smaller the object that will cause diffraction, and the better the resolving power. The light microscope, because it uses visible light of wavelength around 500 nanometers (nm; 1 nm = 1/1000 μm), can distinguish objects as small as about half this: 250 nm. It can therefore be used to visualize the smallest cells and the major intracellular structures or organelles. The microscopic study of cell structure and organization is known as **cytology**. An electron microscope is required to reveal the **ultrastructure** (the fine detail) of the organelles and other intracellular structures (Fig. 1.2). The wavelength

Cell Biology: A Short Course, Third Edition. Stephen R. Bolsover, Jeremy S. Hyams, Elizabeth A. Shephard and Hugh A. White.

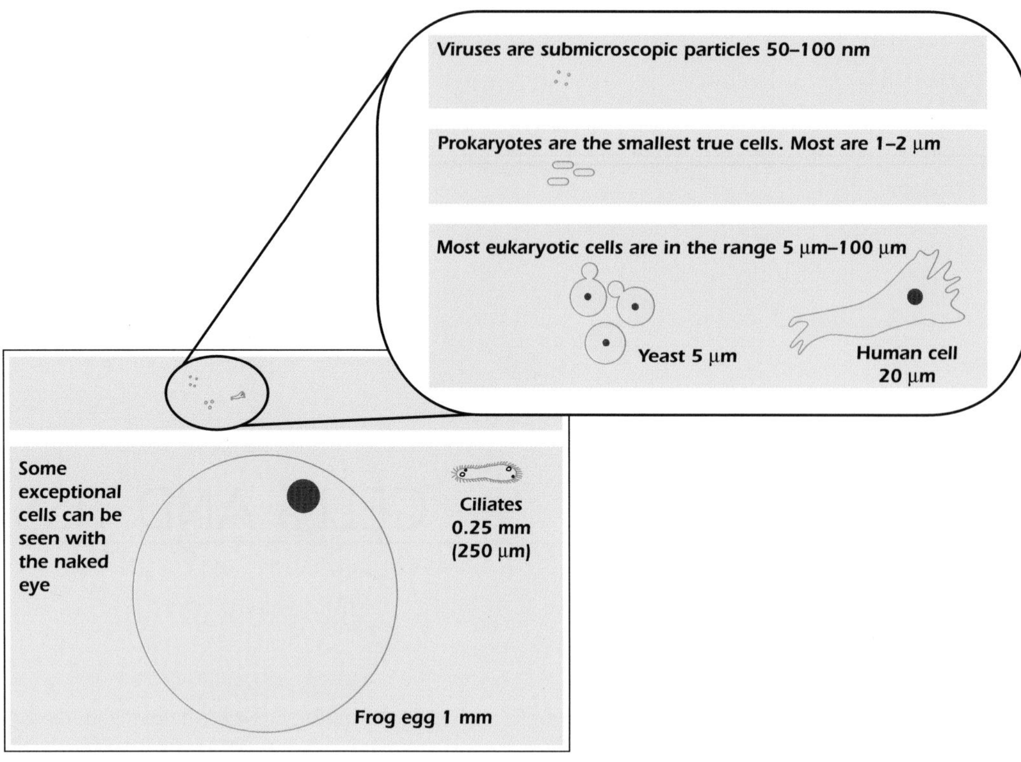

Figure 1.1. Dimensions of some example cells. 1 mm = 10^{-3} m; 1 μm = 10^{-6} m; 1 nm = 10^{-9} m.

of an electron beam is about 100,000 times less than that of white light. In theory, this should lead to a corresponding increase in resolution. In practice, the transmission type of electron microscope can distinguish structures about 1000 times smaller than is possible in the light microscope, that is, down to about 0.2 nm in size.

The Light Microscope

A light microscope (Figs. 1.3A and 1.4) consists of a light source, which may be the sun or an artificial light, plus three glass lenses: a **condenser lens** to focus light on the specimen, an **objective lens** to form the magnified image, and a **projector lens**, usually called the eyepiece, to convey the magnified image to the eye. Depending on the focal length of the various lenses and their arrangement, a given magnification is achieved. In **bright-field microscopy**, the image that reaches the eye consists of the colors of white light minus those absorbed by the cell. Most living cells have little color and are therefore largely transparent to transmitted light. This problem can be overcome by **cytochemistry**, the use of colored stains to selectively highlight particular structures and organelles. However, many of these compounds are highly toxic and to be effective they often require that the cell or tissue is subjected to a series of harsh chemical treatments. A different approach, and one that can be applied to living cells, is the use of **phase-contrast microscopy**. This relies on the fact that light travels at different speeds through regions of the cell that differ in composition. The phase-contrast microscope converts these differences in refractive index into differences in contrast, and considerably more detail is revealed (Fig. 1.5). Light microscopes come in a number of physical orientations (upright, inverted, etc.) but whatever the orientation of the microscope, the optical principles are the same.

Example 1.1 Sterilization by Filtration

Because even the smallest cells are larger than 1 μm, harmful bacteria and other organisms can be removed from drinking water by passing through a filter with 200 nm diameter holes. Filters can vary in size from huge, such as those used in various commercial processes, to small enough to be easily transportable by backpackers. Filtering drinking water greatly reduces the chances of bringing back an unwanted souvenir from your camping trip!

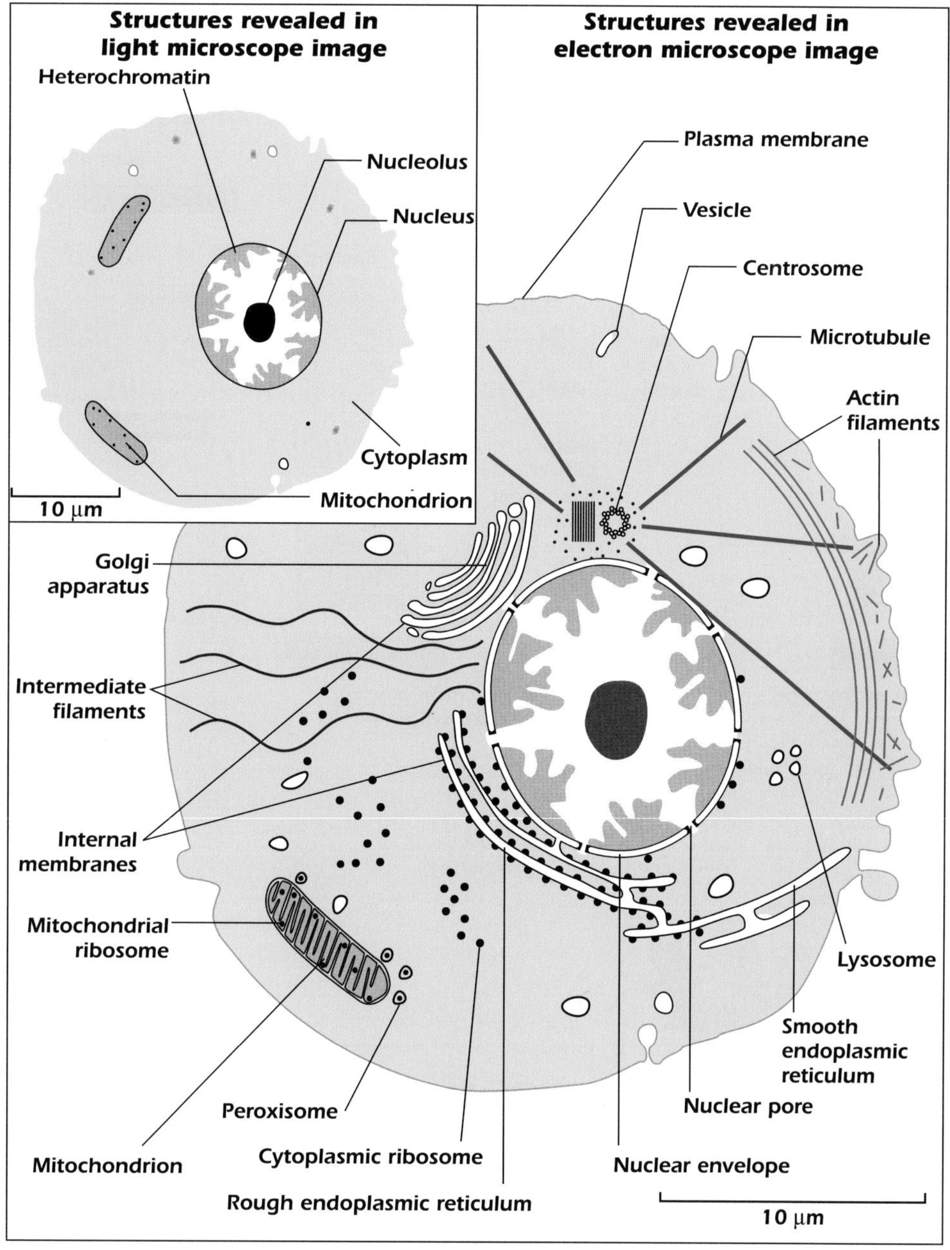

Figure 1.2. Cell structure as seen through the light and transmission electron microscopes.

The Electron Microscope

The most commonly used type of electron microscope in biology is called the **transmission electron microscope** because electrons are transmitted through the specimen to the observer. The transmission electron microscope has essentially the same design as a light microscope, but the lenses, rather than being glass, are electromagnets that bend beams of electrons (Fig. 1.3B). An electron gun generates a beam of electrons by heating a thin, V-shaped piece of tungsten wire to 3000°C. A large voltage accelerates the beam down the microscope column, which is under vacuum because the electrons are slowed and scattered if they collide with air molecules. The magnified image can be viewed on a fluorescent screen that emits light when struck by electrons. While the electron microscope offers great improvements in resolution, electron beams are potentially highly destructive, and biological material must be subjected to a complex processing schedule before it can be examined. The preparation of cells for electron microscopy is summarized in Figure 1.6. The transmission electron microscope produces a detailed image but one that is static, two-dimensional, and highly processed (Fig. 1.7). Often, only a small region of what was once a dynamic, living, three-dimensional cell is revealed. Moreover, the

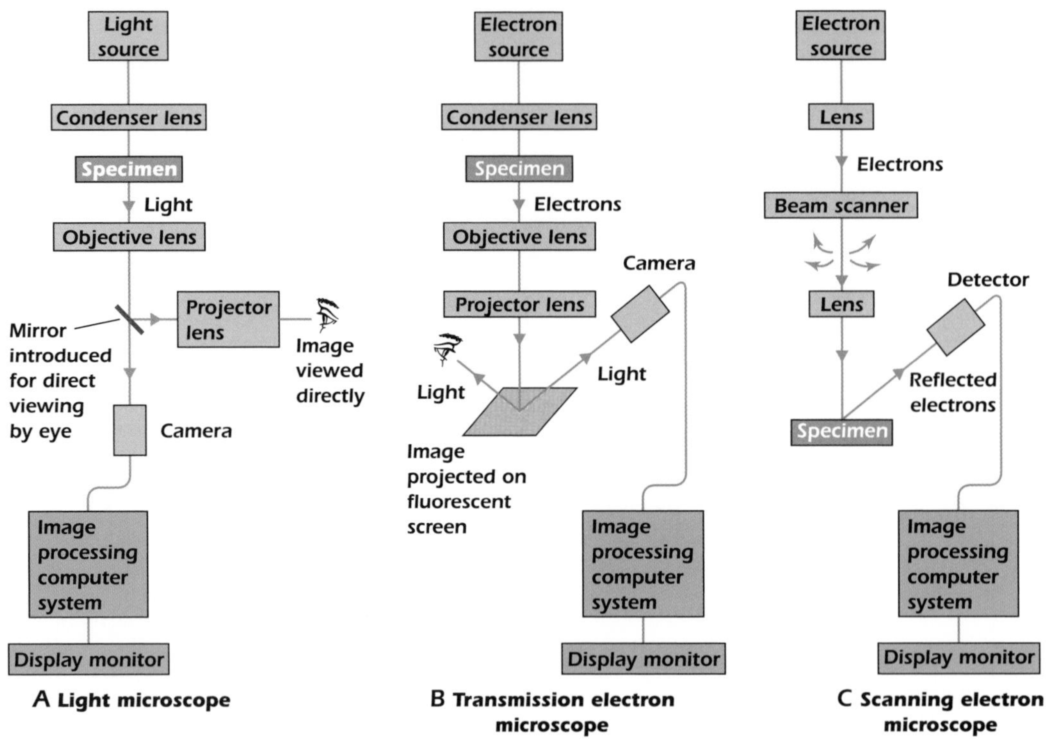

Figure 1.3. Basic design of light and electron microscopes.

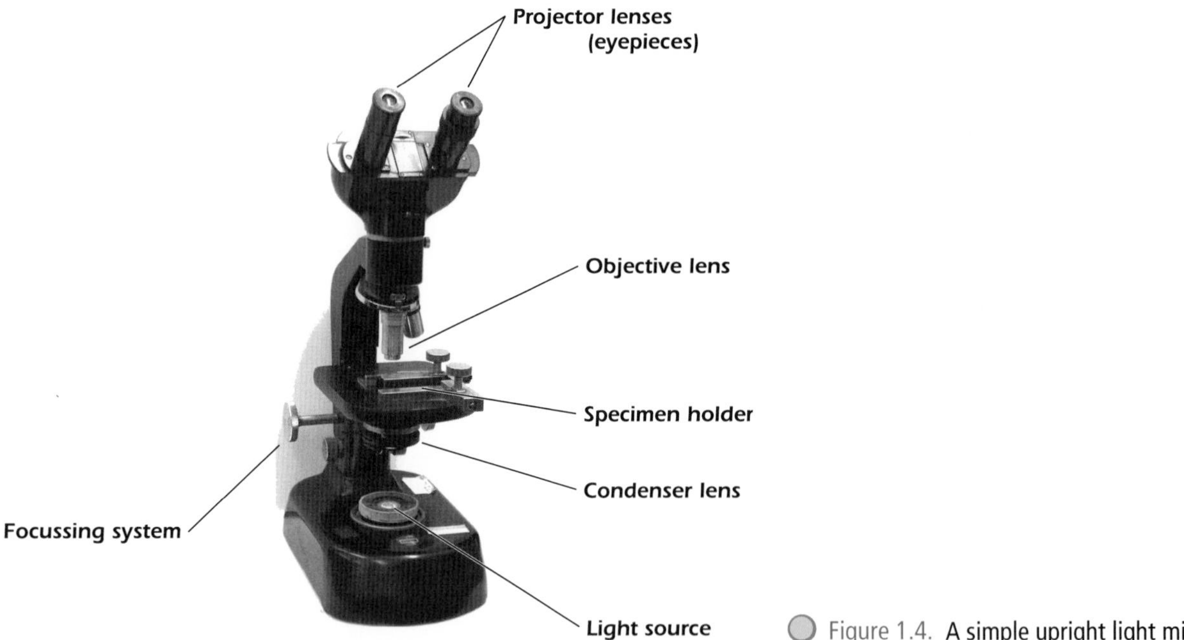

Figure 1.4. A simple upright light microscope.

picture revealed is essentially a snapshot taken at the particular instant that the cell was killed. Clearly, such images must be interpreted with great care. Also, electron microscopes are large, expensive and require a skilled operator. Nevertheless, they are the main source of information on the ultrastructure of the cell at the nanometer scale.

The Scanning Electron Microscope

Whereas the image in a transmission electron microscope is formed by electrons transmitted through the specimen, in the **scanning electron microscope** it is formed from electrons that are reflected back from the surface of a specimen as the electron beam scans rapidly back and forth

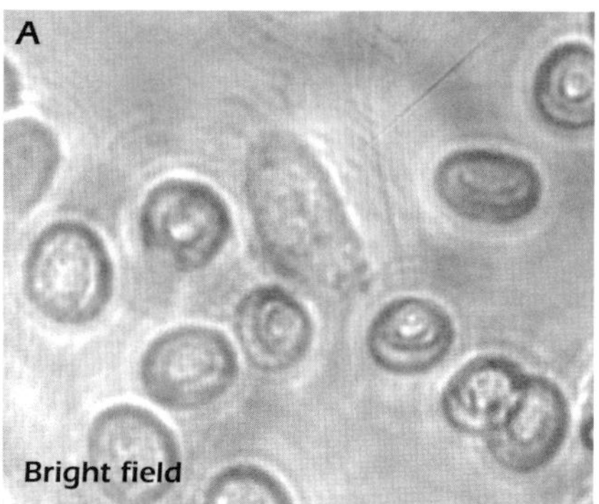

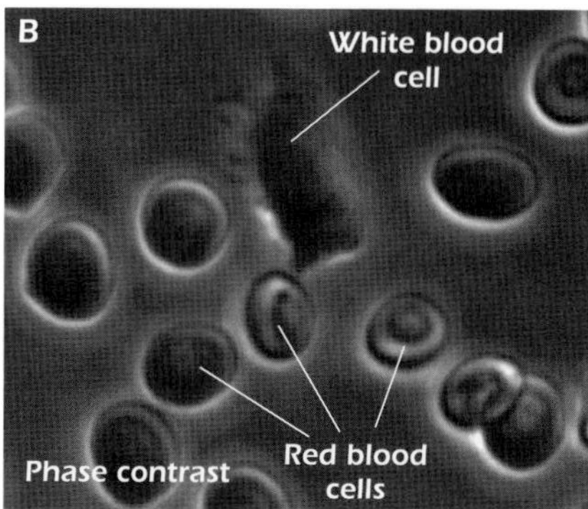

After addition of a formyl-methionine peptide

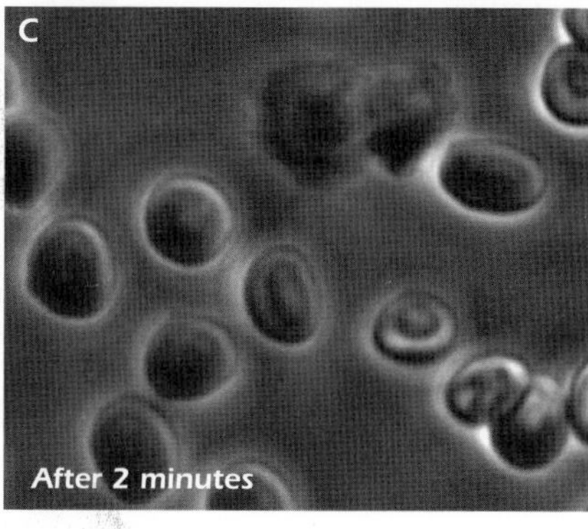

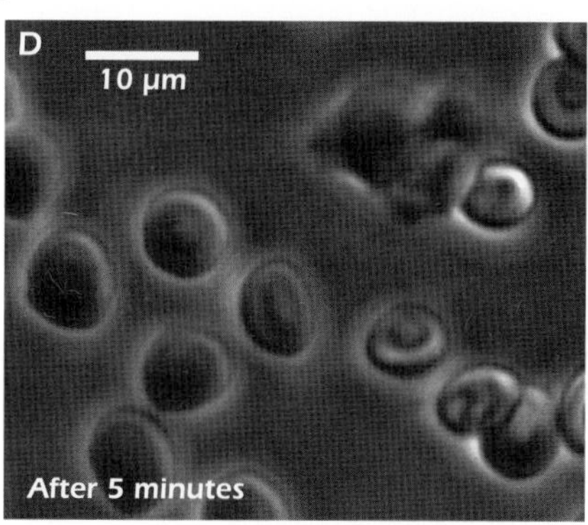

Figure 1.5. Human blood cells viewed by bright-field (A) and phase-contrast (B) light microscopy. Thin extensions of the white blood cell are clear in the phase contrast image but invisible in the bright field image. (C) and (D) are phase contrast images acquired 2 and 5 minutes after addition of a formyl methionine peptide (see page 128). The white blood cell is activated and begins crawling to the right.

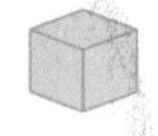

A small piece of tissue (~1 mm^3) is immersed in glutaraldehyde and osmium tetroxide. These chemicals bind all the component parts of the cells together; the tissue is said to be **fixed**. It is then washed thoroughly.

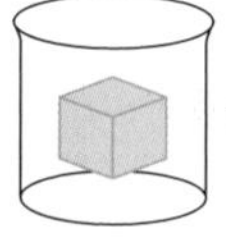

The tissue is **dehydrated** by soaking in acetone or ethanol.

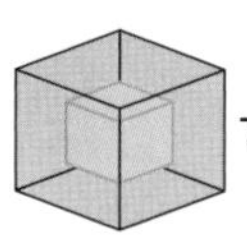

The tissue is **embedded** in resin which is then baked hard.

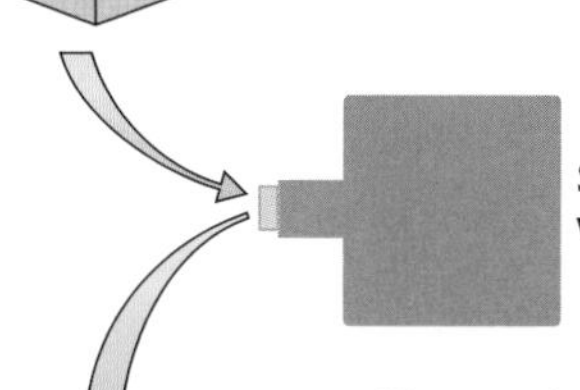

Sections (thin slices less than 100 nm thick) are cut with a machine called an ultramicrotome.

The sections are placed on a small copper grid and **stained** with uranyl acetate and lead citrate. When viewed in the electron microscope, regions that have bound lots of uranium and lead will appear dark because they are a barrier to the electron beam.

Figure 1.6. Preparation of tissue for electron microscopy.

over it (Fig. 1.3C). These reflected electrons are detected and used to generate a picture on a display monitor. The scanning electron microscope operates over a wide magnification range, from 10 times to 100,000 times, and has a wide **depth of focus**. The images created give an excellent impression of the three-dimensional shape of objects (Fig. 1.8). The scanning electron microscope is therefore particularly useful for providing topographical information on the surfaces of cells or tissues. Modern instruments have a resolution of about 1 nm.

IN DEPTH 1.1 FLUORESCENCE MICROSCOPY

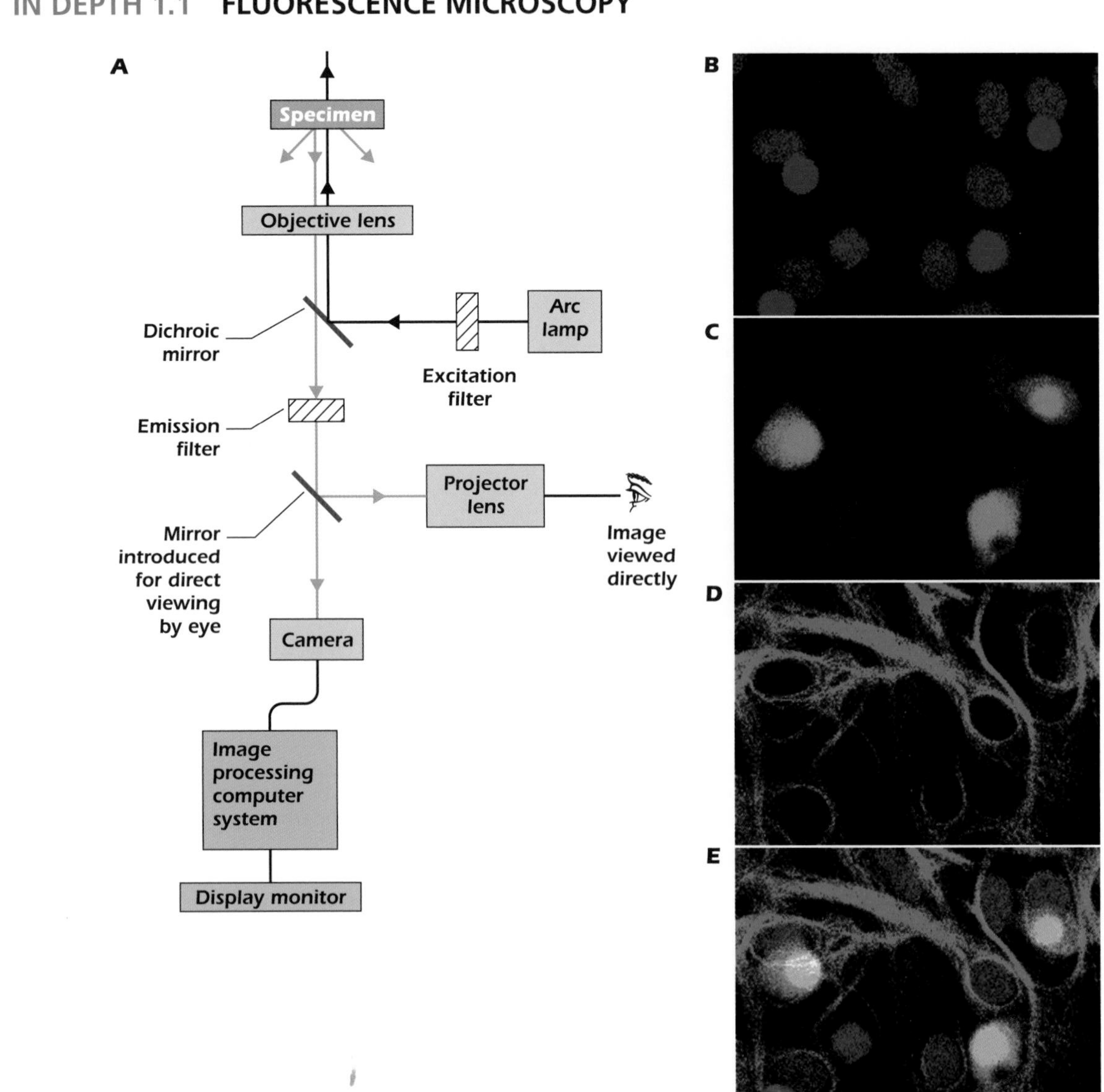

Fluorescent molecules emit light when they are illuminated with light of a shorter wavelength. Familiar examples are the hidden signature in bank passbooks, which is written in fluorescent ink that glows blue (wavelength about 450 nm) when illuminated with ultraviolet light (UV) (wavelength about 360 nm), and the whitener in fabric detergents that causes your white shirt to glow blue when illuminated by the ultraviolet light in a club. The fluorescent dye Hoechst 33342 has a similar wavelength dependence: it is excited by UV light and emits blue light. However, it differs from the dyes used in ink or detergent in that it binds tightly to the DNA in the nucleus and only fluoresces when so bound. Diagram A shows the optical path through a microscope set up so as to look at a preparation stained with Hoechst. White light from an arc lamp passes through an excitation filter that allows only UV light to pass. This light then strikes the heart of the fluorescent microscope: a special mirror called a dichroic mirror that reflects light of wavelengths shorter than a designed cutoff but transmits light of longer wavelength. To view Hoechst, we use a dichroic mirror of cutoff wavelength 400 nm, which therefore reflects the UV excitation light up through the objective lens and onto the specimen. Any Hoechst bound to DNA in the preparation will emit blue light. Some of this will be captured by the objective lens and, because its wavelength is greater than 400 nm, will not be reflected by the dichroic mirror but will instead pass through. An emission filter, set to pass only blue light, cuts out any scattered UV light.

The blue light now passes to the eye or camera in the usual way. Image B shows a field of cells cultured from rat brain after staining with Hoechst. Only the nuclei are seen, as bright ovals.

Although some of the structures and chemicals found in cells can be selectively stained by specific fluorescent dyes, others are most conveniently revealed by using antibodies. In this technique an animal (usually a mouse, rabbit, or goat) is injected with a protein or other chemical of interest. The animal's immune system recognizes the chemical as foreign and generates antibodies that bind to (and therefore help neutralize) the chemical. Some blood is then taken from the animal and the antibodies purified. The antibodies can then be labeled by attaching a fluorescent dye. Images C and D show the same field of brain cells but with the excitation filter, dichroic mirror, and emission filter changed so as to reveal in C a protein called ELAV that is found only in nerve cells; and in D, an intermediate filament protein (page 291) found only in glial cells. The antibody that binds to ELAV is labeled with a fluorescent dye that is excited by blue light and emits green light. The antibody that binds to the glial filaments is labeled with a dye that is excited by green light and emits red light. Because these wavelength characteristics are different, the location of the three chemicals—DNA, ELAV, and intermediate filament—can be revealed independently in the same specimen. Image E shows the previous three images superimposed.

The technique just described is **primary immunofluorescence** and requires that the antibody to the chemical of interest be labeled with a dye. Only antibodies to chemicals that many laboratories study are so labeled. In order to reveal other chemicals, scientists use **secondary immunofluorescence**. In this approach, a commercial company injects an animal (e.g., a goat) with an antibody from another animal (e.g., a rabbit). The goat then makes "goat anti-rabbit" antibody. This, the **secondary antibody**, is purified and labeled with a dye. All the scientist has to do is make or buy a rabbit antibody that binds to the chemical of interest. No further modification of this specialized **primary antibody** is necessary. Once the primary antibody has bound to the specimen and excess antibody rinsed off, the specimen is then exposed to the fluorescent secondary antibody that binds selectively to the primary antibody. Viewing the stained preparation in a fluorescence microscope then reveals the location of the chemical of interest. The same dye-labeled secondary antibody can be used in other laboratories or at other times to reveal the location of many different chemicals because the specificity is determined by the unlabeled primary antibody.

A completely different approach uses genetically encoded fluorescent molecules. The first to be used was **green fluorescent protein** from the jellyfish *Aequorea victoria*. Cells can be induced to make this protein and the living cell viewed by fluorescence microscopy. If the cell is induced to make a chimaeric protein comprising green fluorescent protein fused to some or all of a protein of interest, then fluorescence microscopy can be used to follow the changing location of this protein within a living cell (e.g., Fig. 18.13 on page 310). Green fluorescent protein is described in more detail later (pages 115, 148).

ONLY TWO TYPES OF CELL

Superficially at least, cells exhibit a staggering diversity. Some lead a solitary existence; others live in communities; some have defined, geometric shapes; others have flexible boundaries; some swim, some crawl, and some are sedentary. Given these differences, it is perhaps surprising that there are only two types of cell (Fig. 1.9). **Prokaryotic** (Greek for "before nucleus") cells have very little visible internal organization so that, for instance, the genetic material, stored in the molecule deoxyribonucleic acid (DNA),

IN DEPTH 1.2 MICROSCOPY REWARDED

Such has been the importance of microscopy to developments in biology that two scientists have been awarded the Nobel prize for their contributions to microscopy. Frits Zernike was awarded the Nobel prize for physics in 1953 for the development of phase-contrast microscopy whilst Ernst Ruska received the same award in 1986 for the invention of the transmission electron microscope. Ruska's prize marks one of the longest gaps between a discovery (in the 1930s, in the research labs of the Siemens Corporation in Berlin) and the award of a Nobel prize. Anton van Leeuwenhoek died almost two centuries before the Nobel prizes were introduced in 1901 and the prize is not awarded posthumously.

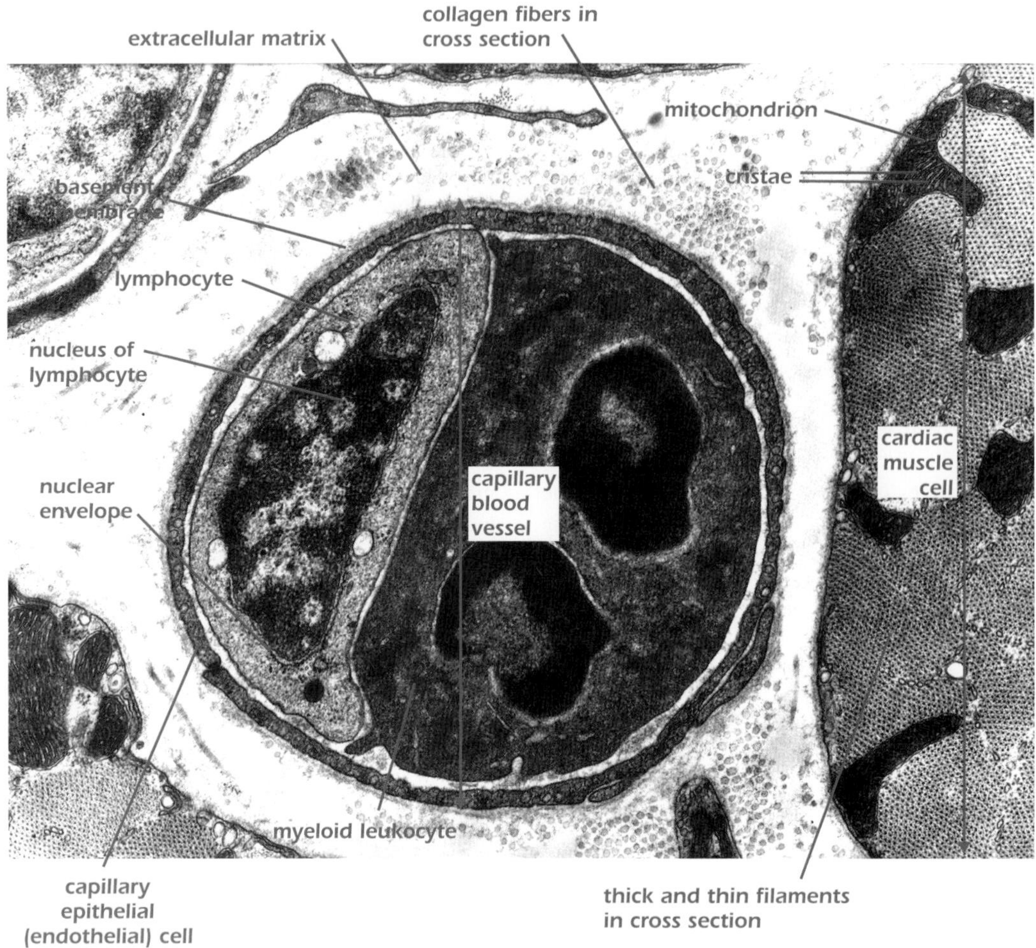

Figure 1.7. Transmission electron micrograph of a capillary blood vessel running between heart muscle cells. Image by Giorgio Gabella, Department of Cell and Developmental Biology, University College London. Reproduced by permission.

is free within the cell. They are especially small, the vast majority being 1–2 μm in length. The prokaryotes are made up of two broad groups of organisms, the bacteria and the archaea (Fig. 1.10). The archaea were originally thought to be an unusual group of bacteria but we now know that they are a distinct group of prokaryotes with an independent evolutionary history. The cells of all other organisms, from yeasts to plants to worms to humans, are **eukaryotic** (Greek for "with a nucleus"). These are generally larger (5–100 μm, although some eukaryotic cells are large enough to be seen with the naked eye; see Fig. 1.1) and structurally more complex. Eukaryotic cells contain a variety of specialized structures known collectively as organelles, surrounded by a viscous substance called **cytosol**. Their DNA is held within the largest organelle, the nucleus. The structure and function of organelles will be described in detail in subsequent chapters. Table 1.1 summarizes the differences between prokaryotic and eukaryotic cells.

Cell Division

One of the major distinctions between prokaryotic and eukaryotic cells is their mode of division. In prokaryotes the circular chromosome is duplicated from a single replication origin by a group of enzymes that reside on the inside of the plasma membrane. At the completion of replication the old and new copies of the chromosome lie side by side on the plasma membrane, which then pinches inwards between them. This process, which generates two equal, or roughly equal, progeny cells is described as **binary fission**. In eukaryotes the large, linear chromosomes, housed in the nucleus, are duplicated from multiple origins of replication by enzymes located in the nucleus. Some time later the **nuclear envelope** breaks down and the replicated chromosomes are compacted so that they can be segregated without damage during **mitosis**. We will deal with mitosis in detail in Chapter 18. For the moment we should be aware that although it is primarily about changes to the nucleus, mitosis is accompanied by

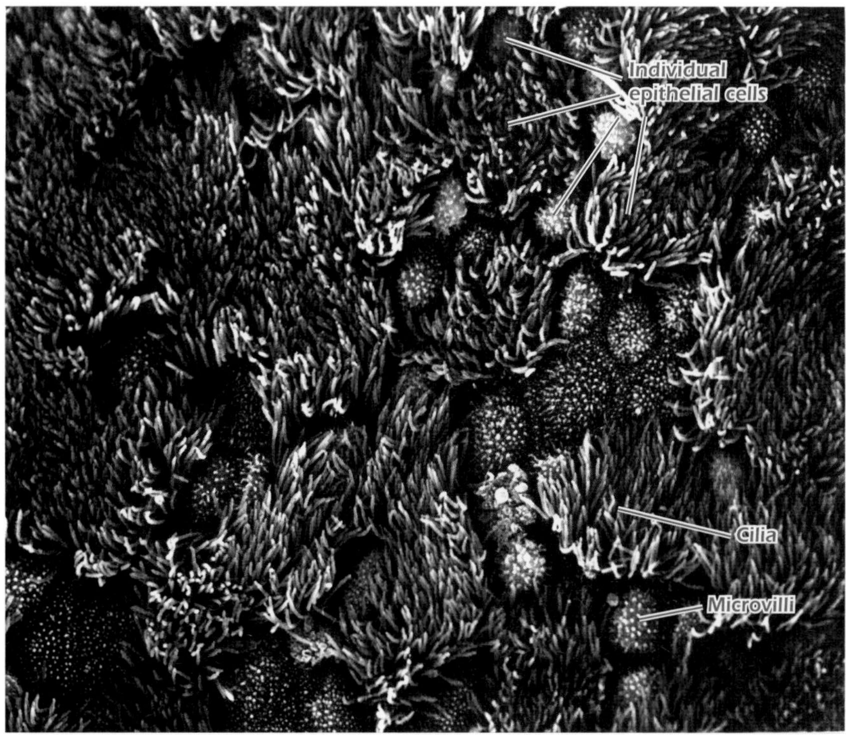

Figure 1.8. Scanning electron micrograph of airway epithelium. Image by Giorgio Gabella, Department of Cell and Developmental Biology, University College London. Reproduced by permission.

TABLE 1.1. Differences between Prokaryotic and Eukaryotic Cells

	Prokaryotes	Eukaryotes
Size	Usually 1–2 μm	Usually 5–100 μm
Nucleus	Absent	Present, bounded by nuclear envelope
DNA	Usually a single circular molecule (= chromosome)	Multiple linear molecules (chromosomes)[a]
Cell division	Simple fission	Mitosis or meiosis
Internal membranes	Rare	Complex (nuclear envelope, Golgi apparatus, endoplasmic reticulum, etc.)
Ribosomes	70S[b]	80S (70S in mitochondria and chloroplasts)
Cytoskeleton	Rudimentary	Microtubules, microfilaments, intermediate filaments
Motility	Rotary motor (drives bacterial flagellum)	Dynein (drives cilia and flagella); kinesin, myosin
First appeared	3.5×10^9 years ago	1.5×10^9 years ago

[a]The tiny chromosomes of mitochondria and chloroplasts are exceptions; like prokaryotic chromosomes they are often circular.
[b]The S value, or Svedberg unit, is a sedimentation rate. It is a measure of how fast a molecule moves in a gravitational field, and therefore in an ultracentrifuge.

dramatic changes to the organization of the rest of the cell. A new structure, the **mitotic spindle**, is assembled specifically to move the chromosomes apart while other structures such as the **Golgi apparatus** and endoplasmic reticulum are dismantled so that their components can be divided among the two progeny cells following cell division.

VIRUSES

Viruses occupy a unique position between the living and nonliving worlds. On the one hand they are made of the same molecules as living cells. On the other they are incapable of independent existence, being completely dependent

A Bacterium, prokaryotic

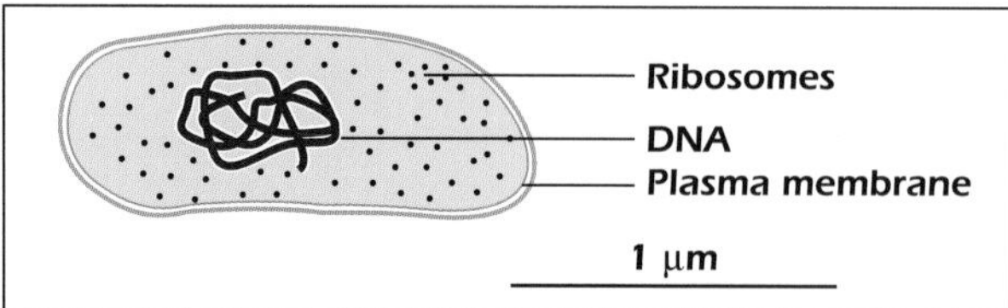

B Animal cell, eukaryotic

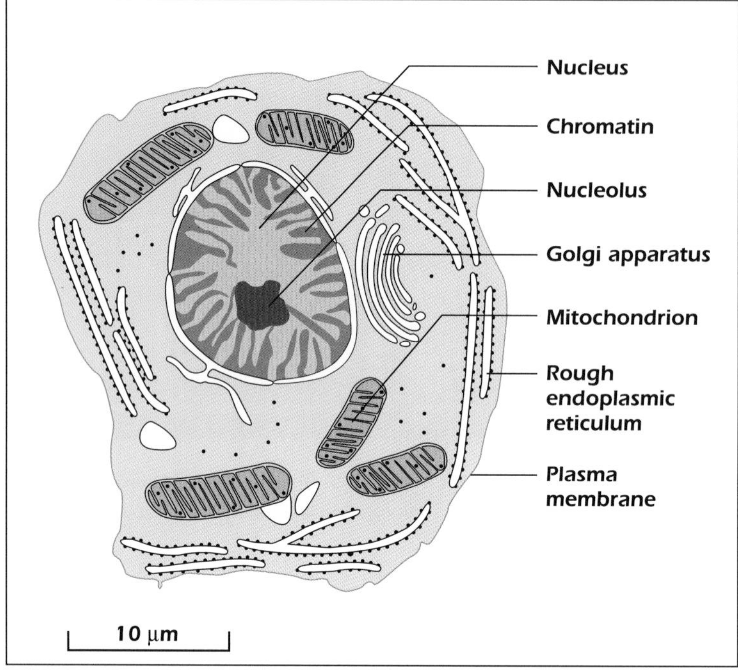

Figure 1.9. Organization of prokaryotic and eukaryotic cells.

on a host cell for reproduction. Almost all living organisms have viruses that infect them. Human viruses include polio, influenza, herpes, rabies, ebola, smallpox, chickenpox, and HIV (human immunodeficiency virus, the causative agent of AIDS). Viruses are submicroscopic particles consisting of a core of genetic material enclosed within a protein coat called the capsid. Some have an extra membrane layer called the **envelope**. Viruses are metabolically inert until they enter a host cell, whereupon their genetic material directs the host cell machinery to produce viral protein and viral genetic material. Viruses often insert their genome into that of the host, an ability that is widely made use of in molecular biology research (Chapter 7). Bacterial viruses, **bacteriophages**, are used by scientists to transfer genes between bacterial strains. Human viruses are used as vehicles for gene therapy. By exploiting the natural infection cycle of a virus such as adenovirus, it is possible to introduce a functional copy of a human gene into the cells of a patient suffering from a genetic disease such as Leber congenital amaurosis (page 333).

ORIGIN OF EUKARYOTIC CELLS

Prokaryotic cells are simpler in their organization than eukaryotic cells and are assumed to be more primitive. According to the fossil record, prokaryotic organisms antedate, by at least 2 billion years, the first eukaryotes that appeared some 1.5 billion years ago. It seems highly likely that eukaryotes evolved from prokaryotes, and the most likely explanation of this process is the **endosymbiotic theory**. The basis of this theory is that some eukaryotic organelles originated as free-living bacteria that were engulfed by larger cells in which they established a mutually beneficial relationship. For example, **mitochondria** would have originated as free-living aerobic bacteria and **chloroplasts** as cyanobacteria, which are photosynthetic prokaryotes formerly known as blue-green algae. The endosymbiotic theory provides an attractive explanation for the fact that mitochondria and chloroplasts contain their own DNA and ribosomes, both of which are more closely related to those of bacteria than to all the other DNA and ribosomes in the same cell. The case for

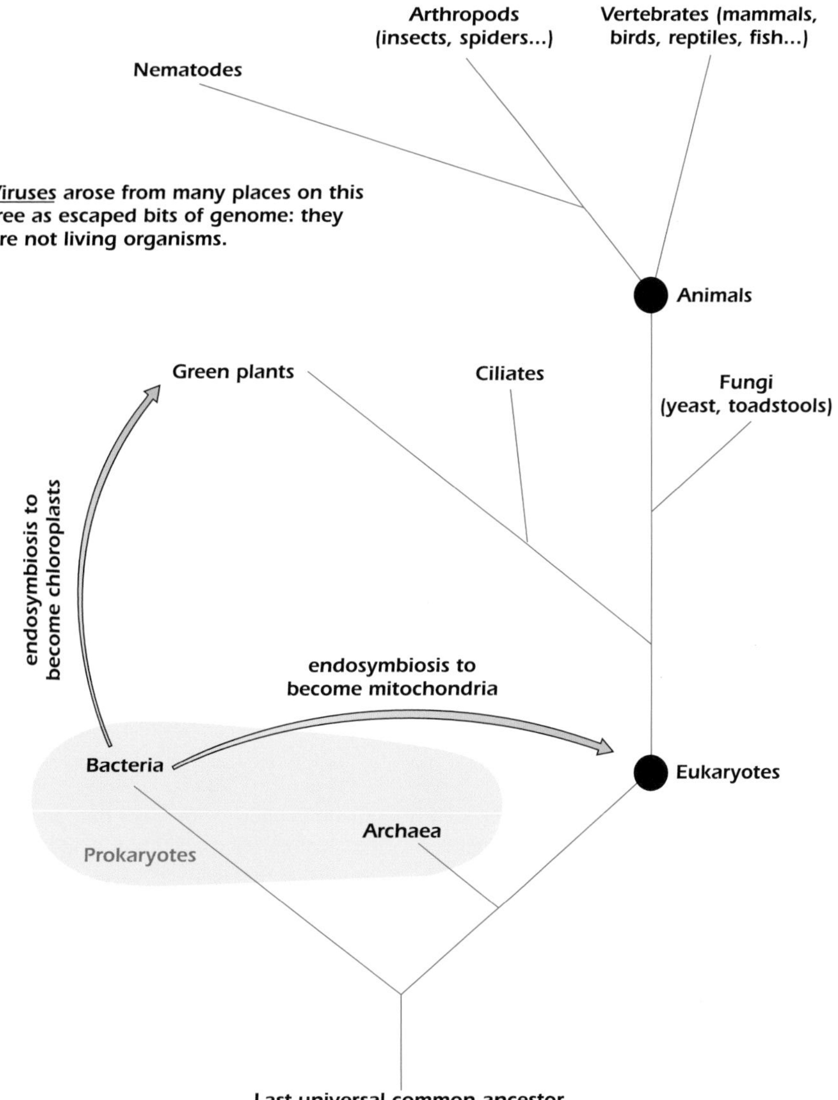

Figure 1.10. The tree of life. The diagram shows the currently accepted view of how the different types of organism arose from a common ancestor. Many minor groups have been omitted. Distance up the page should not be taken as indicating complexity or how "advanced" the organisms are. All organisms living today represent lineages that have had the same amount of time to evolve and change from the last universal common ancestor.

the origin of other eukaryotic organelles is less persuasive. Nevertheless, while it is clearly not perfect, most biologists are now prepared to accept that the endosymbiotic theory provides at least a partial explanation for the evolution of the eukaryotic cell from prokaryotic ancestors.

CELL SPECIALIZATION IN ANIMALS

Animals are multicellular communities of individual cells. Lying between and supporting the cells is the **extracellular matrix** (Fig. 1.7) of different types of fiber around which the fluids and solute of the **interstitial fluid** can easily pass. All the body cells that comprise a single organism share the same set of genetic instructions in their nuclei (with the single exception of lymphocytes, page 317). Nevertheless, the cells are not all identical. Rather, they form a variety of **tissues**, groups of cells that are specialized to carry out a common function. This specialization occurs because different cell types read out different parts of the DNA blueprint and therefore make different proteins. In animals there are four major tissue types: epithelium, connective tissue, nervous tissue, and muscle. Some examples of the cells that make up these tissues are shown in Figure 1.11.

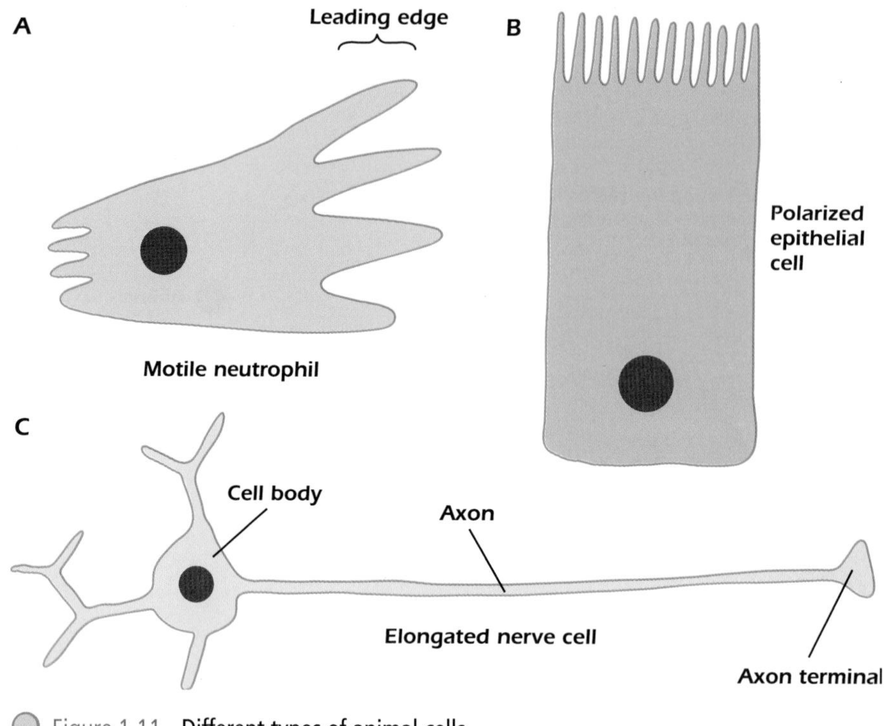

Figure 1.11. Different types of animal cells.

Epithelia are sheets of cells that cover the surface of the body and line its internal cavities such as the lungs and intestine. The cells may be **columnar**, taller than they are broad (Fig. 1.11B), or **squamous**, meaning flat (e.g., the capillary cell in Fig. 1.7). They are often **polarized**, meaning that one surface of the cell is distinct in its organization, composition and appearance from the other. In the intestine, the single layer of columnar cells lining the inside, or **lumen**, has an absorptive function that is increased by the folding of the surface into villi (Fig. 1.12). The luminal surfaces of these polarized cells have **microvilli** that increase the surface area even further. The basal (bottom) surface sits on a thin planar sheet of specialized extracellular matrix called the **basement membrane** or **basal lamina**. Many of the epithelial cells of the airways, for instance, those lining the trachea and bronchioles, have **cilia** on their surfaces (Fig. 1.8). These are hairlike appendages that actively beat back and forth, moving a layer of mucus away from the lungs (Chapter 17). Particles and bacteria are trapped in the mucus layer, preventing them from reaching the delicate air exchange membranes in the lung. In the case of the skin, the epithelium is said to be **stratified** because it is composed of several layers.

Connective tissues provide essential support for the other tissues of the body. They include bone, cartilage, and adipose (fat) tissue. Unlike other tissues, connective tissue contains relatively few cells within a large volume of extracellular matrix that consists of different types of fiber embedded in **amorphous** ground substance (Fig. 1.12). The most abundant of the fibers is **collagen**, a protein with the tensile properties of steel that accounts for about a third of the protein of the human body. Other fibers have elastic properties that permit the supported tissues to be displaced and then to return to their original position. The amorphous ground substance absorbs large quantities of water, facilitating the diffusion of metabolites, oxygen, and carbon dioxide to and from the cells in other tissues and organs. Of the many cell types found in connective tissue, two of the most important are **fibroblasts**, which make and secrete the ground substance and fibers, and **macrophages**, which remove foreign, dead, and defective material. A number of inherited diseases are associated with defects in connective tissue. Marfan's syndrome, for example, is characterized by long arms, legs, and torso and by a weakness of the cardiovascular system and eyes. These characteristics result from a defect in the organization of the collagen fibers.

Nervous tissue is a highly modified epithelium that is composed of several cell types. Principal among these are the **nerve cells**, also called **neurons** (Fig. 1.11C), along with a variety of supporting cells that help maintain them. Neurons extend processes called **axons**, which can be over a meter in length. Neurons constantly monitor what is occurring inside and outside the body. They integrate and summarize this information and mount appropriate responses to it (Chapters 14–16). Another type of cell, **glia**, has other roles in nervous tissue including forming the electrical insulation around axons.

Muscle tissue can be of two types, **smooth** or **striated**. Smooth muscle cells are long and slender and are usually found in the walls of tubular organs such as the intestine and

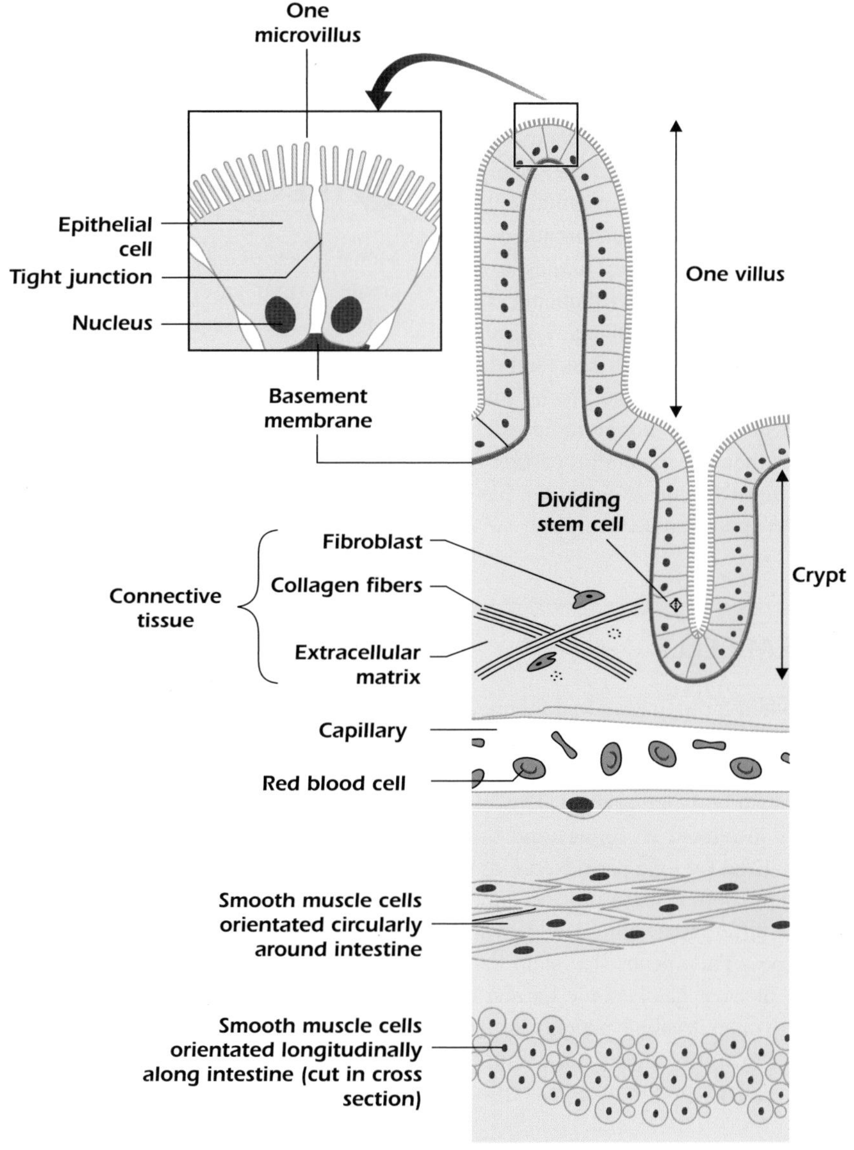

Figure 1.12. Tissues and structures of the intestine wall.

many blood vessels. In general, smooth muscle cells contract slowly and can maintain the contracted state for a long period of time. There are two classes of striated muscle: **cardiac** and **skeletal**. Cardiac muscle cells (Fig. 1.7) make up the walls of the heart chambers. These are branched cells that are connected electrically by **gap junctions** (page 46), and their automatic rhythmical contraction powers the beating of the heart. Each skeletal muscle is a bundle of hundreds to thousands of fibers, each fiber being a giant single cell with many nuclei. This rather unusual situation is the result of an event that occurs in the embryo when the cells that give rise to the fibers fuse together, pooling their nuclei in a common **cytoplasm** (the term cytoplasm is historically a crude term meaning the semi-viscous ground substance of cells; we use the term to mean everything inside the plasma membrane except the nucleus). The mechanism of muscle contraction will be described in Chapter 17.

STEM CELLS AND TISSUE REPLACEMENT

Cells multiply by division. In the human body an estimated 25 million cell divisions occur every second! These provide new cells for the blood and immune systems, for the repair of wounds and the replacement of dead cells. In complex tissues such as those described above, division is restricted to a small number of **stem cells** from which all of the other cells of the tissue derive. In the case of the intestine, folds in the surface epithelium form crypts, each of which contains approximately 250 cells (Fig. 1.12). Mature cells at the top die and must be replaced by the division of between four and six stem cells near the base of the crypt. Each stem cell divides roughly twice a day, the resulting cells moving up the crypt to replace those lost at the surface. Benign (noncancerous) polyps can be formed in the intestine if this normal balance between birth and death is disturbed.

As in the intestine, stem cells in other tissues exist in specific locales, called niches, with environments that support their special and vital functions. In many tissues the requirement to replace dead cells is much less than it is in the intestine and in such cases the stem cell niche must maintain its occupants in a quiescent (nondividing) state until needed. Like stem cells themselves, the properties of niches remain deeply mysterious. For the moment we have few markers that allow us to specifically distinguish stem cells from the cells around them and hence unambiguously identify the territories they occupy. The resolution of such questions is vital if the potential of stem cells in cell therapy is to be realized.

THE CELL WALL

Many types of cell, particularly bacteria and plant cells, create a rigid case around themselves called a **cell wall**. For cells that live in an extracellular medium more dilute than their own cytosol, the cell wall is critical in preventing the cell bursting. For example, penicillin and many other **antibiotics** block the synthesis of bacterial cell walls with the result that the bacteria burst. Within trees, plant cells modify the cell wall to generate the woody trunk. Animal cells do not have cell walls.

IN DEPTH 1.3 STEM CELLS

Few developments in modern biology have had the medical and social impact that has followed the discovery of stem cells. These are unspecialized cells from which all specialized cells in the human body are derived. They are present in small numbers in almost all tissues and they can either remain unspecialized for long periods, dividing at intervals to produce more stem cells (a process called self-renewal) or they can differentiate into a wide variety of specialized cell types. This means that stem cells have enormous potential in what has become known as stem cell therapy. This involves treating patients suffering from spinal cord injury or from a stroke, for example, with stem cells that have been preprogrammed to replace the damaged tissue.

There are two kinds of stem cells: adult stem cells and embryonic stem cells. It is the latter that has caused much of the controversy about stem cell therapy since they are derived from human embryos that are left over after in vitro fertilization treatment. Cells from a four- to five-day-old human embryo are pluripotent, that is, they can be induced to create just about any cell type. There are fewer ethical objections to the use of adult stem cells, but as yet, these have nothing like the versatility of their embryonic counterparts. Eventually it should be possible to isolate stem cells from a patient and program these appropriately before reintroducing them into the damaged tissue. Because they are the patient's own cells, questions of rejection and other complications do not arise.

In some respects, stem cell therapy is not really new. Bone marrow transplantation has been familiar in the treatment of a variety of blood diseases for more than 40 years. Patients suffering from cancers such as leukemia and lymphoma, for example, receive doses of chemotherapy that kill the stem cells that are located in the bone marrow. These give rise to the different cells in the blood; the red cells that carry oxygen around the body, the white cells that help fight infection, and the platelets that help blood clot. Replacement bone marrow, usually from a healthy relative, is therefore essential to restock these stem cells and allow the patient to make a full recovery.

Answer to Thought Question: Transmission electron microscopy. If the appropriate fluorescent dyes are used light microscopes can reveal the location of the Golgi complex within a cell, but only the electron microscope has sufficient resolution to show the structure of the organelle and hence whether it is malformed. Malformation of the endoplasmic reticulum and Golgi apparatus is thought to underlie one type of inherited spastic paraplegia.

SUMMARY

1. All living organisms are made of cells.
2. Our understanding of cell structure and function has gone hand in hand with developments in microscopy and its associated techniques.
3. Light microscopy revealed the diversity of cell types and the existence of the major organelles such as the nucleus and mitochondria.
4. The electron microscope reveals the detailed structure of the larger organelles and resolves the cell ultrastructure, the fine detail, at the nanometer scale.
5. There are only two types of cells, prokaryotic and eukaryotic.
6. Prokaryotic cells have little visible internal organization. They are usually 1–2 μm in size.
7. Eukaryotic cells usually measure 5–100 μm. They contain a variety of specialized internal organelles, the largest of which, the nucleus, contains the genetic material.
8. The endosymbiotic theory proposes that some eukaryotic organelles, such as mitochondria and chloroplasts, originated as free-living prokaryotes.
9. In multicellular organisms, cells are organized into tissues. In animals there are four tissue types: epithelium, connective tissue, nervous tissue, and muscle.
10. The extracellular matrix is found on the outside of animal cells.
11. In tissues, specialized cells arise from unspecialized stem cells.

FURTHER READING

Booth, C., and Potten, C. S. (2000) Gut instincts: thoughts on intestinal epithelial stem cells. *Journal of Clinical Investigation* **105**, 1493–1499.

Gest, H. (2004) The discovery of microorganisms by Robert Hooke and Antoni van Leeuwenhoek, Fellows of the Royal Society. *Notes and Records of the Royal Society of London*, **58**, 187–201.

Harris, H. (1999) *The Birth of the Cell*, Yale University Press, New Haven, Connecticut.

REVIEW QUESTIONS

We use the same format of review questions throughout the book. For each of the numbered questions choose the best response from the lettered list. The same response may apply to more than one numbered question. Unless specifically told to do so, you should not refer back to the chapter text or figures in answering the questions. Answers are at the back of the book, starting on page 381.

1.1 Theme: Dimensions in Cell Biology

A 0.025 nm
B 0.2 nm
C 20 nm
D 250 nm
E 2,000 nm
F 20,000 nm
G 200,000 nm
H 5,000,000 nm
I 1,000,000,000 nm
J 20,000,000,000 nm

From the above list, select the dimension most appropriate for each of the descriptions below.

1. A typical bacterium
2. A typical eukaryotic cell
3. The longest cell in the human body
4. Resolution of a light microscope
5. Resolution of a transmission electron microscope

1.2 Theme: Types of Cell

A bacterium
B epithelial cell
C fibroblast
D macrophage
E glial cell
F skeletal muscle cell
G stem cell

From the above list of cell types, select the cell corresponding to each of the descriptions below.

1. A cell that synthesizes collagen
2. A cell type found in nervous tissue
3. A cell type that forms sheets, e.g., to separate different spaces in the body
4. A cell whose role is to remove dead and foreign material
5. A cell with no nuclear envelope
6. Large cells with multiple nuclei
7. Undifferentiated cells capable of multiple rounds of cell division

1.3 Theme: Some Basic Components of the Eukaryotic Cell

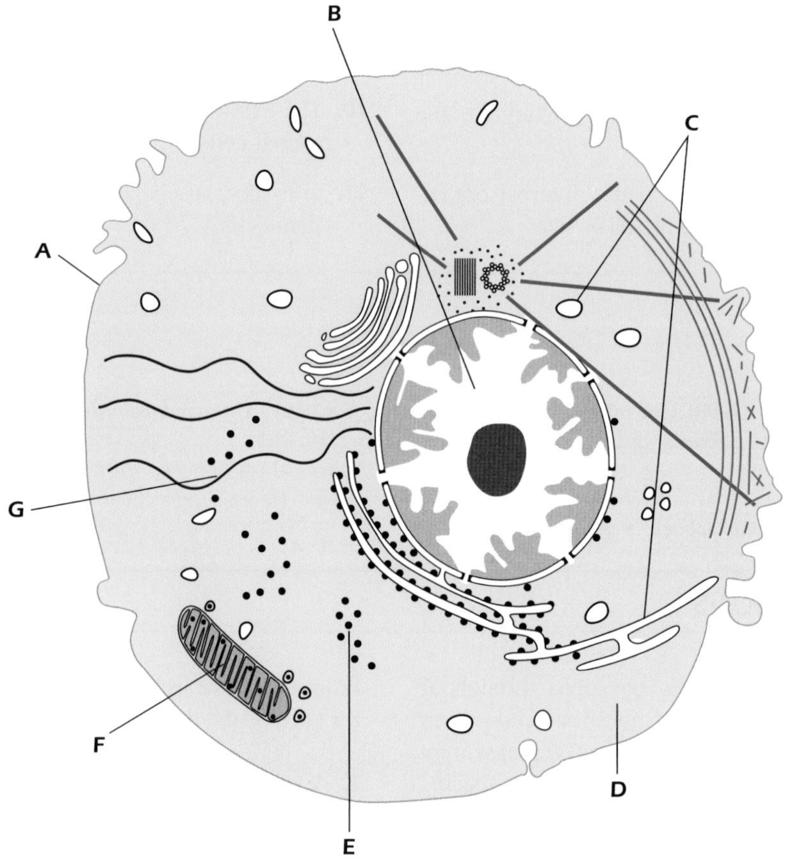

Identify each of the cellular components below from the figure above.

1. Cytosol
2. Internal membranes
3. Mitochondrion
4. Nucleus
5. Plasma membrane

THOUGHT QUESTION

Each chapter has a thought question. For these you are encouraged to refer back to the text and diagrams within the chapter to formulate your response. Answers appear earlier in the relevant chapter, printed upside down.

You wish to test the hypothesis that a particular inherited human disease is characterized by malformation of the Golgi apparatus. What type of microscopy would you use to examine a tissue biopsy from a patient?

2

WATER AND MACROMOLECULES

Organisms are made up of a lot of different chemicals. These vary in size, from small molecules like water to large molecules like DNA. In this chapter we will introduce some basic concepts of how these chemicals are made and how they interact to generate the processes of life. We will then describe the most important of these chemicals: water, carbohydrates, nucleotides, amino acids, and lipids.

THE CHEMICAL BOND: SHARING ELECTRONS

Water is the most abundant substance in living cells. Cytoplasm consists of organelles floating in a watery cytosol that also contains proteins. The situation is not so different outside our cells. Although we are land animals living in air, most of our cells are bathed in a watery extracellular medium. We will therefore start by considering water itself.

Figure 2.1A shows a molecule of water, consisting of one atom of oxygen and two hydrogen atoms, joined to form an open V shape. The lines represent covalent bonds formed when atoms share electrons, each seeking the most stable structure. Oxygen has a greater affinity for electrons than does hydrogen so the electrons are not distributed equally. The oxygen grabs a greater share of the available negative charge than do the hydrogen atoms. The water molecule is polarized, with partial negative charge on the oxygen and partial positive charges on the two hydrogens. We write the charge on each hydrogen as δ^+ to indicate that it is smaller than the charge on a single hydrogen nucleus. The oxygen atom has the small net negative charge $2\delta^-$. Molecules that, like water, have positive regions sticking out one side and negative regions sticking out the other are called **polar**. Two linked atoms with an unequal distribution of charge form a **dipole**; water therefore has two dipoles.

Figure 2.1B shows a molecule of chlorine gas. It consists of two chlorine atoms, each of which consists of a positively charged nucleus surrounded by negatively charged electrons. Like oxygen, chlorine atoms tend to accept electrons when they become available, but the battle is equal in the chlorine molecule: the two atoms share their electrons equally and the molecule is **nonpolar**.

Figure 2.2A shows what happens when a chlorine molecule is allowed to react with the metal sodium. Each atom of chlorine captures an electron from a sodium atom. This leaves each sodium atom with a single positive charge because it now has one less electron than there are positive charges in its nucleus. Similarly, each chlorine atom now has a single negative charge because it now has one more electron than there are positive charges in its nucleus. Chemical species that have either gained or lost electrons, and therefore bear an overall charge, are **ions**. The reaction of chlorine and sodium has produced sodium ions and chloride ions. Positively charged ions like sodium are **cations** while negatively charged ones like chloride are **anions**. The positively charged sodium ions and the negatively charged chloride ions now attract each other strongly. If there are no other chemicals around, the ions

Cell Biology: A Short Course, Third Edition. Stephen R. Bolsover, Jeremy S. Hyams, Elizabeth A. Shephard and Hugh A. White.

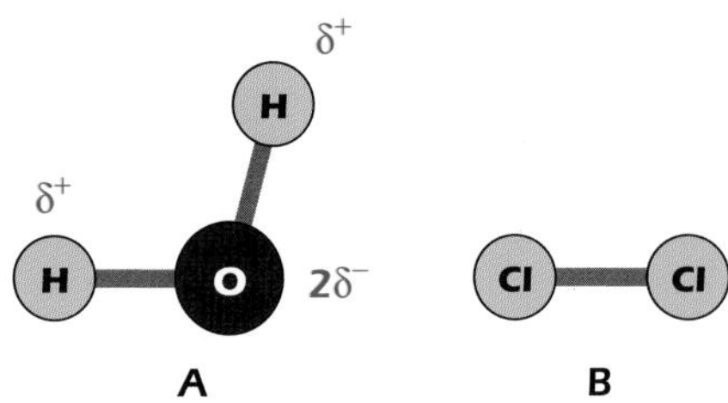

Figure 2.1. Water is a polar molecule, while the chlorine molecule is nonpolar.

will arrange themselves to minimize the distance between sodium and chloride, and the resulting well-packed array of ions is a crystal of sodium chloride, shown in Figure 2.2B.

INTERACTIONS WITH WATER: SOLUTIONS

Ionic Compounds Will Dissolve Only in Polar Solvents

Figure 2.3A shows one molecule of octane, the main constituent of gasoline. Octane is an example of a nonpolar solvent. Electrons are shared equally between carbon and hydrogen, and the component atoms do not bear a net charge.

Figure 2.3B shows a small crystal of sodium chloride immersed in octane. At the edge of the crystal, positively charged sodium ions are being pulled in toward the center of the crystal by the negative charge on chloride ions, and negatively charged chloride ions are being pulled in toward the center of the crystal by the positive charge on sodium ions. The sodium and chloride ions will not leave the crystal. Sodium chloride is insoluble in octane. However, sodium chloride will dissolve in water, and Figure 2.4A shows why. The chloride ion at the top left is being pulled into the crystal by the positive charge on its sodium ion neighbors, but at the same time it is being pulled out of the crystal by the positive charge on the hydrogen atoms of nearby water molecules. Similarly, the sodium ion at the bottom left is being pulled into the crystal by the negative charge on its chloride ion neighbors, but at the same time it is being pulled out of the crystal by the negative charge on the oxygen atoms of nearby water molecules. The ions are not held in the crystal so tightly and can leave. Once the ions have left the crystal, they become surrounded by a **hydration shell** of water molecules, all

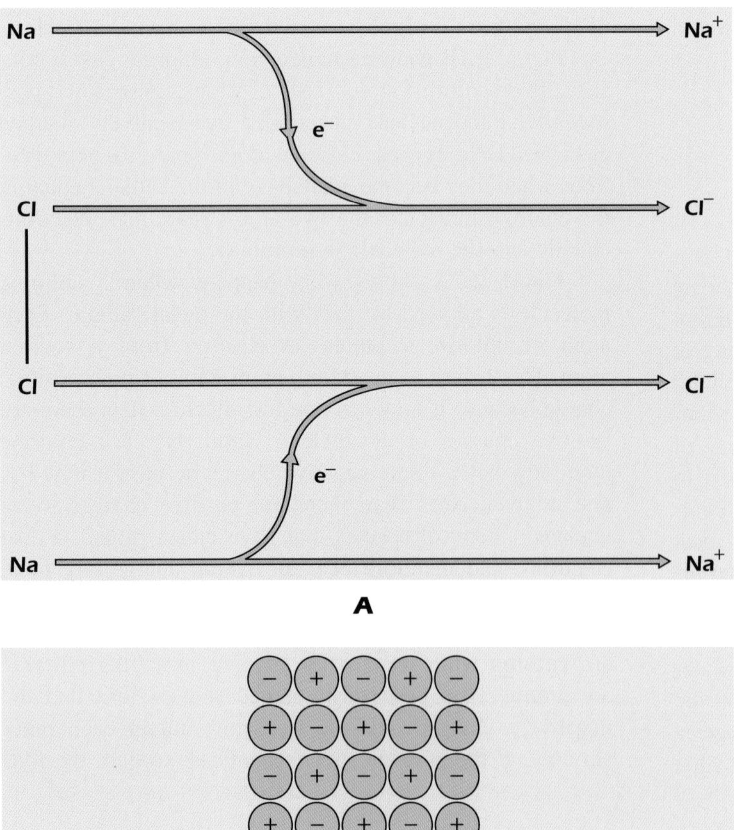

Figure 2.2. Formation of sodium chloride, an ionic compound.

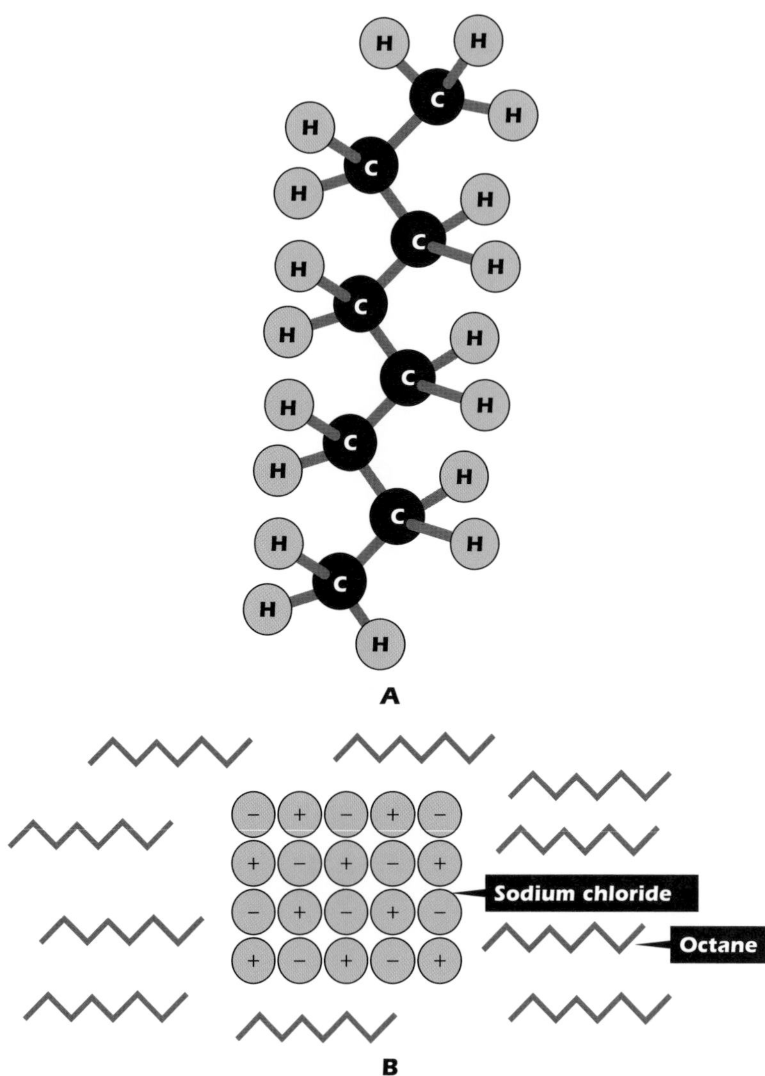

Figure 2.3. (A) Structure of the nonpolar compound octane. (B) Ionic compounds are insoluble in nonpolar solvents.

oriented in the appropriate direction (Fig. 2.4B)—oxygen inward for a positive ion like sodium, hydrogen inward for a negative ion like chloride. A chemical species in solution, whether in water or in any other solvent, is called a **solute**. Liquids whose main constituent is water are **aqueous**.

Example 2.1 Salad Dressing Is a Mixture of Solvents

Vinegar is a dilute solution of acetic acid in water. Acetic acid gives up an H^+ to water to leave the negatively charged acetate ion. The simplest salad dressing is a shaken-up mixture of olive oil and vinegar. Olive oil is hydrophobic and so does not dissolve in the water; the two liquids remain as separate droplets and soon separate again after shaking. However much one shakes, all the acetate will remain in the water because it is an ion. If salt (Na^+Cl^-) is added to the dressing, it also dissolves in the water, with none dissolving in the oil. In contrast, if you add chilies to the salad dressing, the active chemical component capsaicin will dissolve in the oil because it is nonpolar. Only by shaking up the mixture can you get all the tastes together.

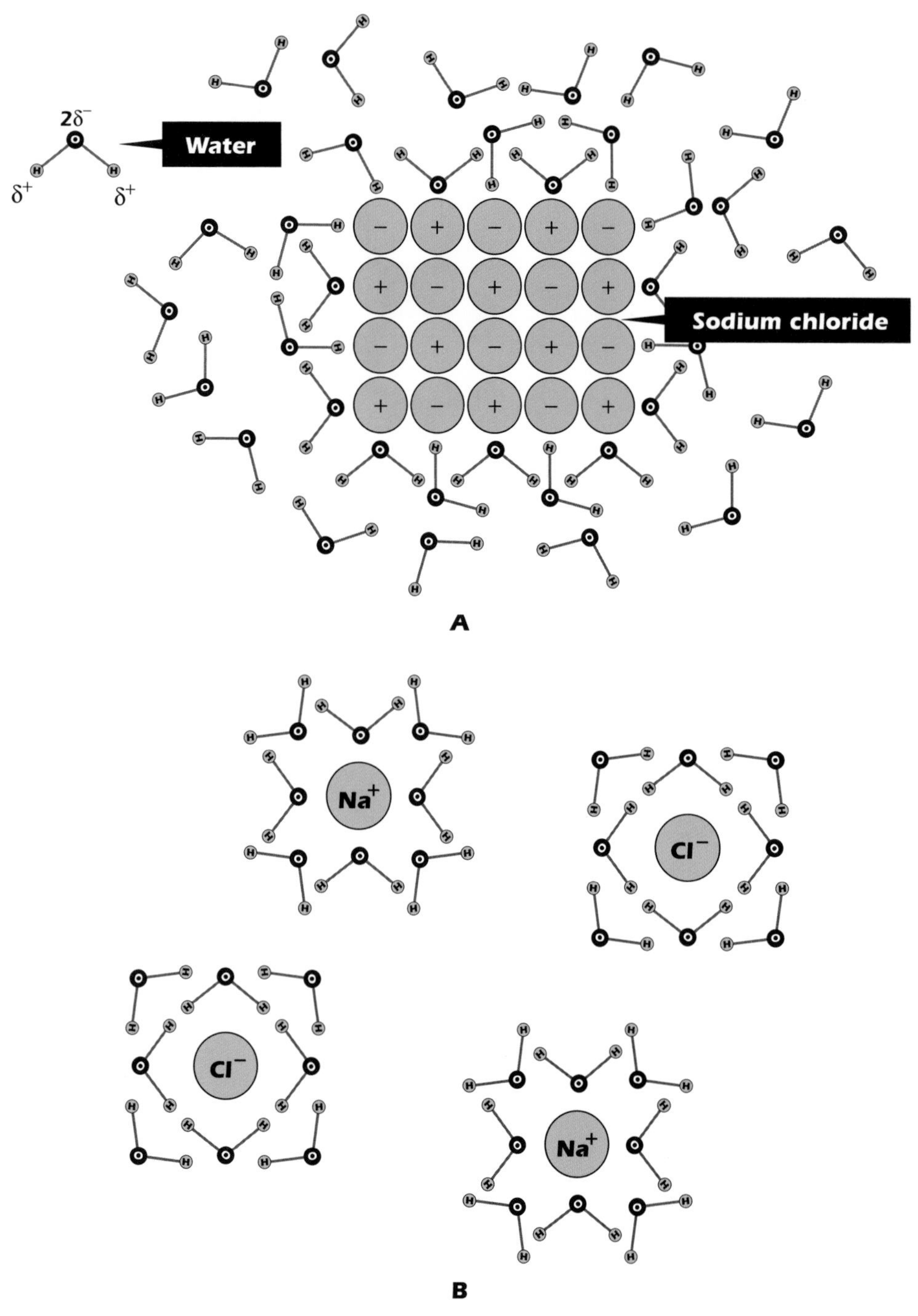

Figure 2.4. Ionic compounds dissolve readily in water, forming hydrated ions.

Acids Are Molecules That Give H^+ to Water

When we exercise, our muscle cells can become acidic, and this is what creates the pain of cramping muscles and the heart pain of angina. Acidity is important in all areas of biology, from the acidity gradient that drives our mitochondria (page 193) to the failure of coral reefs resulting from CO_2 buildup in the oceans.

Acidic solutions contain a high concentration of hydrogen ions. The hydrogen atom is unusual in that it only has one electron while, in its most common isotope, its nucleus comprises a single proton. In gases at very low pressure it is possible for bare protons to exist alone and be manipulated, for example, in linear accelerators. However in water protons never exist alone but always associate with another molecule, for example with water to create the H_3O^+ ion. Acidic solutions are those with an H_3O^+ concentration higher than 100 nanomoles per liter (nmol liter^{-1}).

Sour cream contains lactic acid. Pure lactic acid has the structure shown at the left of Figure 2.5A. The —COOH part in the box is a **carboxyl group**. Both oxygens have a tendency to pull electrons away from the hydrogen and,

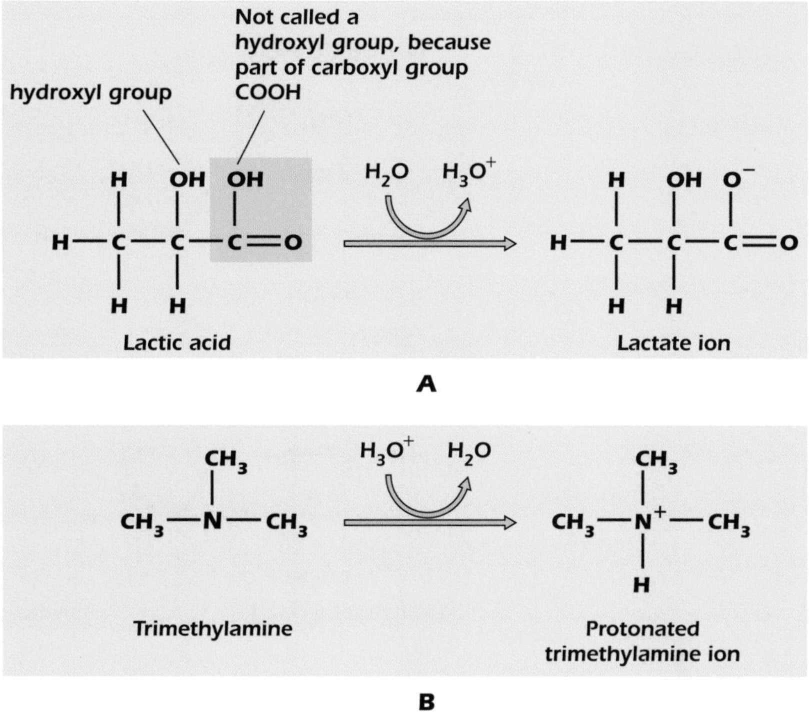

Figure 2.5. Acids and bases, respectively, give up and accept H^+ when dissolved in water.

in aqueous solution, the hydrogen is donated with a full positive charge to a molecule of water. The electron is left behind on the now negatively charged lactate ion:

$$CH_3CH(OH)COOH + H_2O \rightarrow CH_3CH(OH)COO^- + H_3O^+$$

For convenience we often forget about the water and write this as

$$CH_3CH(OH)COOH \rightarrow CH_3CH(OH)COO^- + H^+$$

Here we are using H^+ as a convenient symbol to denote H_3O^+. We do not mean that there are real H^+ ions, that is, bare hydrogen nuclei, in aqueous solutions.

The equilibrium constant K_a for the dissociation of lactic acid is defined as

$$K_a = \frac{[CH_3CH(OH)COO^-]_e[H^+]_e}{[CH_3CH(OH)COOH]_e}$$

where the square brackets refer, by convention, to concentrations and the subscripts, e, denote that these are the concentrations of each species at equilibrium, that is, when the rate of the forward reaction

$$CH_3CH(OH)COOH \rightarrow CH_3CH(OH)COO^- + H^+$$

and the backward reaction

$$CH_3CH(OH)COOH \leftarrow CH_3CH(OH)COO^- + H^+$$

are the same.

Dissolving lots of lactic acid in water produces an acidic solution, that is, one with a high concentration of H^+ (really H_3O^+) ions. For historical reasons, the acidity of a solution is given as the pH, defined thus:

$$pH = -\log_{10}[H^+]$$

where $[H^+]$ is measured in moles per liter. Pure water has a pH of 7, corresponding to $[H^+] = 100$ nmol liter^{-1}. This is said to be neutral as regards pH. If the pH is lower than this, then there is more H^+ about and the solution is acidic. Cytosol has a pH that lies very slightly on the alkaline side of neutrality, at about 7.2.

The pH of a solution determines the ratio of lactate to undissociated lactic acid. As the H^+ concentration rises and pH falls, the equilibrium is pushed over from lactate toward lactic acid. At pH 3.9, the concentrations of lactate and lactic acid become equal. Looking at the equation above, we can therefore see that when this happens $K_a = [H^+]$. Just as acidity is given on the logarithmic pH scale, so is the scale of strengths of different acids, pK_a being defined as $-\log_{10} K_a$. The pK_a is the pH at which the concentration of the dissociated acid is equal to the concentration of the undissociated acid. For an acid that is weaker than lactic acid, the pK_a is higher than 3.9, meaning that the acid is less readily dissociated and needs a lower concentration of hydrogen ions before it will give up its H^+. For an acid that is stronger than lactic acid, the pK_a is less than 3.9: such an acid is more readily dissociated and needs a higher H^+ concentration before it will accept H^+ and form undissociated acid.

Medical Relevance 2.1 Diabetic Acidosis

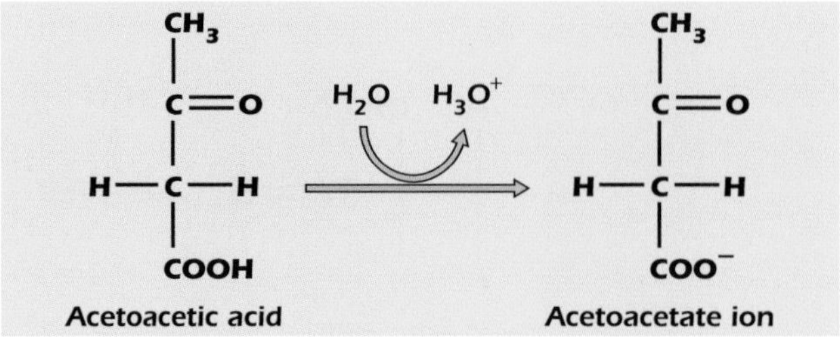

CH_3–C(=O)–CH_2–COOH + H_2O → CH_3–C(=O)–CH_2–COO^- + H_3O^+

Acetoacetic acid → Acetoacetate ion

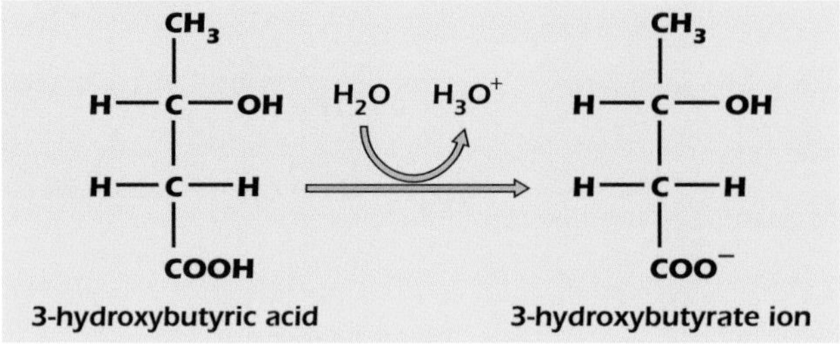

CH_3–CH(OH)–CH_2–COOH + H_2O → CH_3–CH(OH)–CH_2–COO^- + H_3O^+

3-hydroxybutyric acid → 3-hydroxybutyrate ion

During starvation, the body mobilizes fat reserves by converting fat into the small, soluble compounds acetoacetic acid and 3-hydroxy butyric acid, collectively called ketone bodies (page 215). These componds have pK_a values of 4 and 5, respectively, and therefore give H^+ to water at the near-neutral pH of body fluids.

In untreated type 1 diabetes the body activates fat mobilization to an extreme degree, generating large amounts of ketone bodies. These in turn release large amounts of H^+, making the body dangerously acidic in a condition known as diabetic acidosis.

Bases are Molecules that Take H^+ from Water

Trimethylamine is the compound that gives rotting fish its unpleasant smell. Pure trimethylamine has the structure shown at the left of Figure 2.5B. When trimethylamine is dissolved in water, it accepts an H^+ to become the positively charged trimethylamine ion shown on the right of the figure. We refer to molecules that have accepted H^+ ions as **protonated**, using "proton" as a short way of saying "hydrogen nucleus." Dissolving lots of trimethylamine in water produces an alkaline solution, that is, one with a low concentration of H^+ ions and hence a pH greater than 7. The solution never runs out of H^+ completely because new H^+ are formed from water:

$$H_2O \rightarrow OH^- + H^+$$

Thus if we keep adding trimethylamine to water and using up H^+, we end up with an alkaline solution: one with a low concentration of H^+, but lots of OH^-.

The pH of a solution determines the position of equilibrium between protonated and deprotonated trimethylamine, and as before we define the pK_a as the pH at which the concentration of protonated and deprotonated base are the same. The pK_a of trimethylamine is 9.7, meaning that the concentration of H^+ must fall to the low level of $10^{-9.7}$ mol liter^{-1}, that is, 0.2 nmol liter^{-1}, before half of the trimethylamines will give up their H^+s.

Isoelectric Point

The large molecules called proteins (page 33) have many acidic and basic sites that will give up or accept an H^+ as the pH changes. In alkaline solutions proteins will tend to have an overall negative charge because the acidic sites have lost H^+ and bear negative charge. As the pH falls, the acidic sites accept H^+ and become uncharged, and basic sites also accept H^+ to gain positive charge. Thus as the pH falls from an initial high value, the overall charge on the protein becomes less and less negative and then more and more positive. The pH at which the protein has no overall charge is the **isoelectric point**. The isoelectric points of different proteins are different, and this property is useful in separating them during analysis (see "In Depth 8.3" on page 133). The majority of intracellular proteins have an isoelectric point that is less than 7.2, so that at normal intracellular pH they bear a net negative charge.

A Hydrogen Bond Forms When a Hydrogen Atom Is Shared

We have seen how oxygen tends to grab electrons from hydrogen, forming a polar bond. Nitrogen and sulfur are similarly electron-grabbing. If a hydrogen attached to an oxygen, nitrogen or sulfur by a covalent bond gets close to a second electron-grabbing atom; then that second atom also grabs a small share of the electrons to form what is known as a **hydrogen bond** (Fig. 2.6). The atom to which the hydrogen is covalently bound is called the **donor** because it is losing some of its share of electrons; the other electron-grabbing atom is the **acceptor**. For a strong hydrogen bond to form, the donor and acceptor must be within a fixed distance of one another (typically 0.3 nm) with the hydrogen on a straight line between them.

Liquid water is so stable because the individual molecules can hydrogen bond (Fig. 2.6A). Hydrogen bonding also plays a critical role in allowing DNA to store and replicate genetic information. Figure 2.6B shows how the base pairs (page 54) of DNA form hydrogen bonds in which hydrogen atoms are shared between nitrogen and oxygen and between nitrogen and nitrogen.

 BIOLOGICAL MACROMOLECULES

Very large molecules, or **macromolecules**, are central to the working of cells. Large biological molecules are **polymers**: they are assembled by joining together small, simpler molecules, which are therefore called **monomers**. Chemical technology has mimicked nature by producing many important polymers—polyethylene is a polymer of ethylene monomers. Cells make a number of macromolecules that we will introduce, together with their monomer building blocks, in this chapter.

 CARBOHYDRATES: CANDY AND CANES

Carbohydrates—sugars and the macromolecules built from them—have many different roles in cells and organisms.

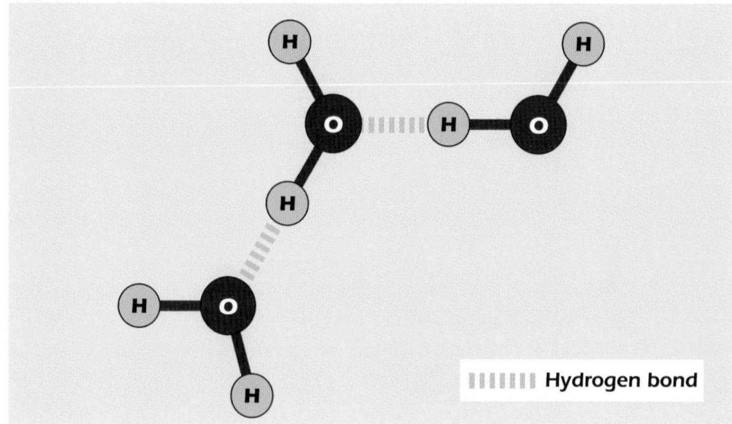

A

A hydrogen bond cannot form here because the N H N atoms are not in a straight line

Deoxyribose

Deoxyribose

Thymine (T)

Adenine (A)

B

Figure 2.6. The hydrogen bond.

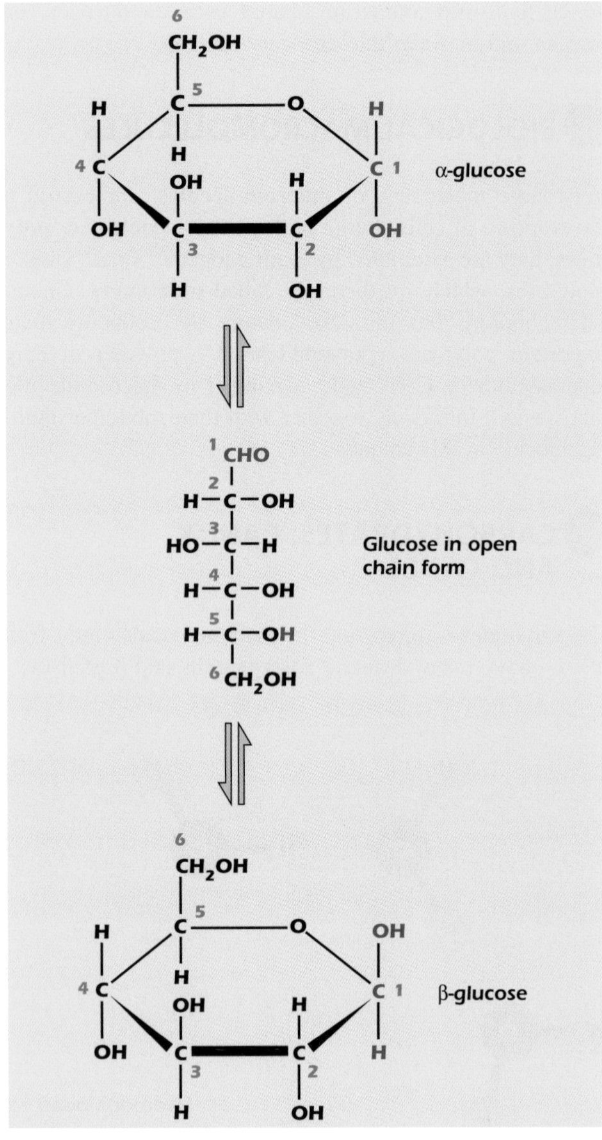

Figure 2.7. Glucose, a monosaccharide, easily switches between three isomers.

An Assortment of Sweets

All carbohydrates are formed from the simple sugars called **monosaccharides**. Figure 2.7 shows the monosaccharide glucose. The form shown at the top has five of its carbons, plus an oxygen atom, arranged in a ring. In addition to the oxygen in the ring, glucose has five other oxygens, each in an —OH (hydroxyl) group. Glucose easily switches between the three forms, or **isomers**, illustrated in Figure 2.7. The two ring structures are **stereo isomers**: although they comprise the same atoms connected by the same bonds, they represent two different ways of arranging the atoms in space. The two stereo isomers, named α and β, continually interconvert in solution via the open-chain form.

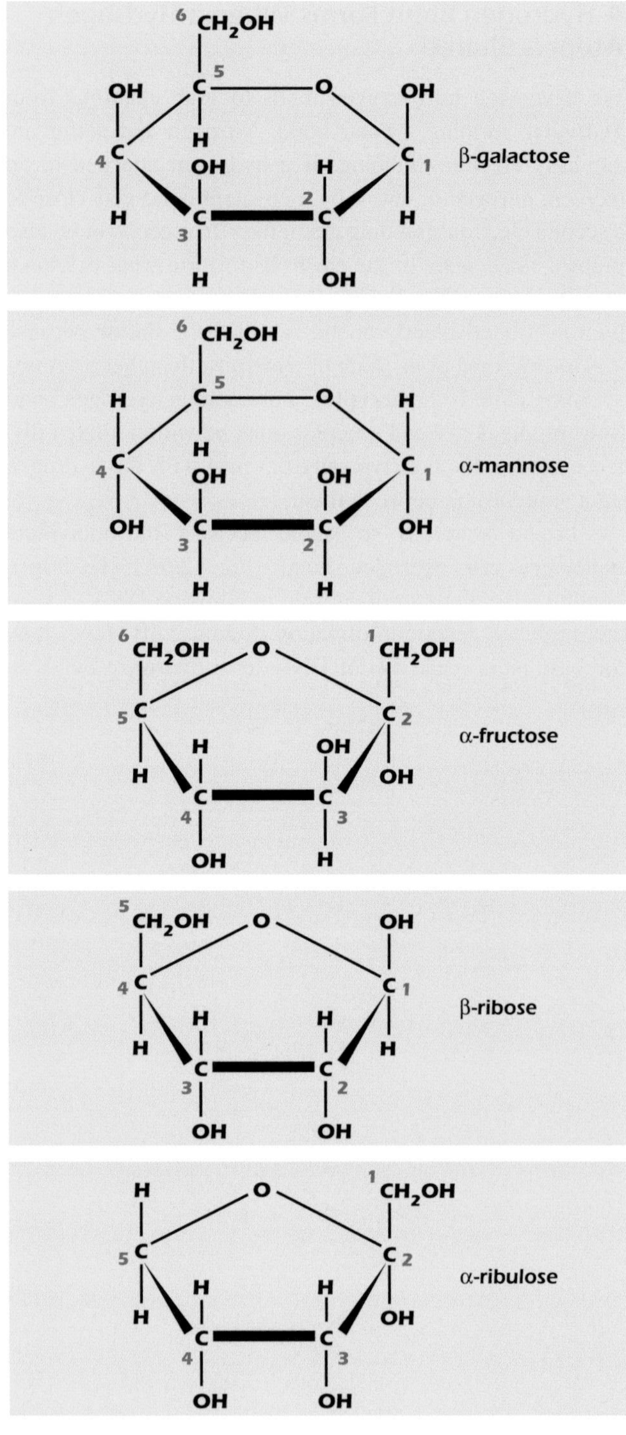

Figure 2.8. Some monosaccharides, each shown as a common isomer.

Figure 2.8 shows five other monosaccharides that we will meet again in this book. Like glucose, each of these monosaccharides can adopt an open-chain form and a number of ring structures. In Figure 2.8 we show each sugar in a form that it adopts quite often. These sugars share with glucose the two characteristics of monosaccharides: they can adopt a form in which an oxygen atom completes a ring of carbons, and they have many hydroxyl groups. The generic names for monosaccharides are derived from the Greek for the number of their carbon atoms, so glucose, galactose, mannose, and fructose are hexoses (six carbons) while ribose and ribulose are pentoses (five carbons). Classically, a monosaccharide has the general formula $C_n(H_2O)_n$, hence the name carbohydrate. All the monosaccharides shown in Figures 2.7 and 2.8 fit this rule—the four hexoses can be written as $C_6(H_2O)_6$ and the two pentoses as $C_5(H_2O)_5$.

It is worth noting that although both sugars and organic acids contain an —OH group, the behavior of an —OH group that forms part of a carboxyl group is very different from one that is not next to a double-bonded oxygen. In general the —OH group in a carboxyl group will readily give up an H^+ to water or other acceptors; this is not true of —OH groups in general. We therefore reserve the term **hydroxyl group** for —OH groups on carbon atoms that are not also double-bonded to oxygen (Fig. 2.5A).

Disaccharides

Monosaccharides can be easily joined by **glycosidic bonds** in which the carbon backbones are linked through oxygen and a water molecule is lost. The bond is identified by the carbons linked. For example, Figure 2.9 shows lactose, a sugar found in milk that is formed when galactose and glucose are linked by a $(1 \rightarrow 4)$ glycosidic bond. Sugars formed from two monosaccharides are called **disaccharides**. Whenever a more complicated molecule is made from simpler building blocks we call the remnants of the individual building blocks **residues**. Thus lactose is made of one galactose residue and one glucose residue.

A complication arises here. Although free monosaccharides can easily switch between the α and β forms, formation of the glycosidic bond locks the shape. Thus, although galactose in solution spends time in the α and β forms, the galactose residue in lactose is locked into the β form, so the full specification of the bond in lactose is $\beta(1 \rightarrow 4)$.

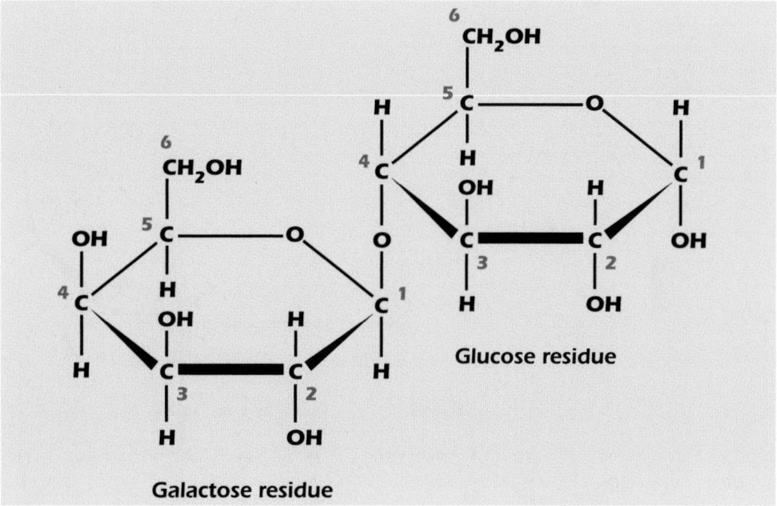

Figure 2.9. The disaccharide lactose.

Out of the Sweet Comes Forth Strength

Formation of glycosidic bonds can continue almost indefinitely. A chain of up to 100 or so monosaccharides is called an **oligosaccharide**, from the Greek word oligo meaning few. Longer polymers are **polysaccharides** and have characteristics very different from those of their monosaccharide building blocks. Figure 2.10A shows part of a molecule of **glycogen**, a polymer made exclusively of glucose monomers organized in long chains with $\alpha(1 \rightarrow 4)$ links. The glycogen chain branches at intervals, each branch being an $\alpha(1 \rightarrow 4)$ linked chain of glucose linked to the main chain with an $\alpha(1 \rightarrow 6)$ bond. Solid lumps of glycogen are found in the cytoplasm of muscle, liver, and some other cells. These glycogen granules are 10–40 nm in diameter with up to 120,000 glucose residues. Glycogen is broken down to release glucose when the cell needs energy (Chapter 13).

Cellulose makes up the cell wall of plants and is the world's most abundant macromolecule. Like glycogen, cellulose is a polymer of glucose, but this time the links are $\beta(1 \rightarrow 4)$. It is shown in Figure 2.10B. Drawn on the flat page, it looks very like glycogen. However, the difference in bond type is critical. Glucoses linked by $\alpha(1 \rightarrow 4)$ links arrange themselves in a floppy helix, while glucoses linked by $\beta(1 \rightarrow 4)$ links form extended chains, ideal for building the rigid plant cell wall. Animals have enzymes (protein catalysts) that can break down the $\alpha(1 \rightarrow 4)$ bond in glycogen, but only certain bacteria and fungi can break the $\beta(1 \rightarrow 4)$ link in cellulose. All animals that eat plants rely on bacteria in their intestines to provide the enzymes to digest cellulose.

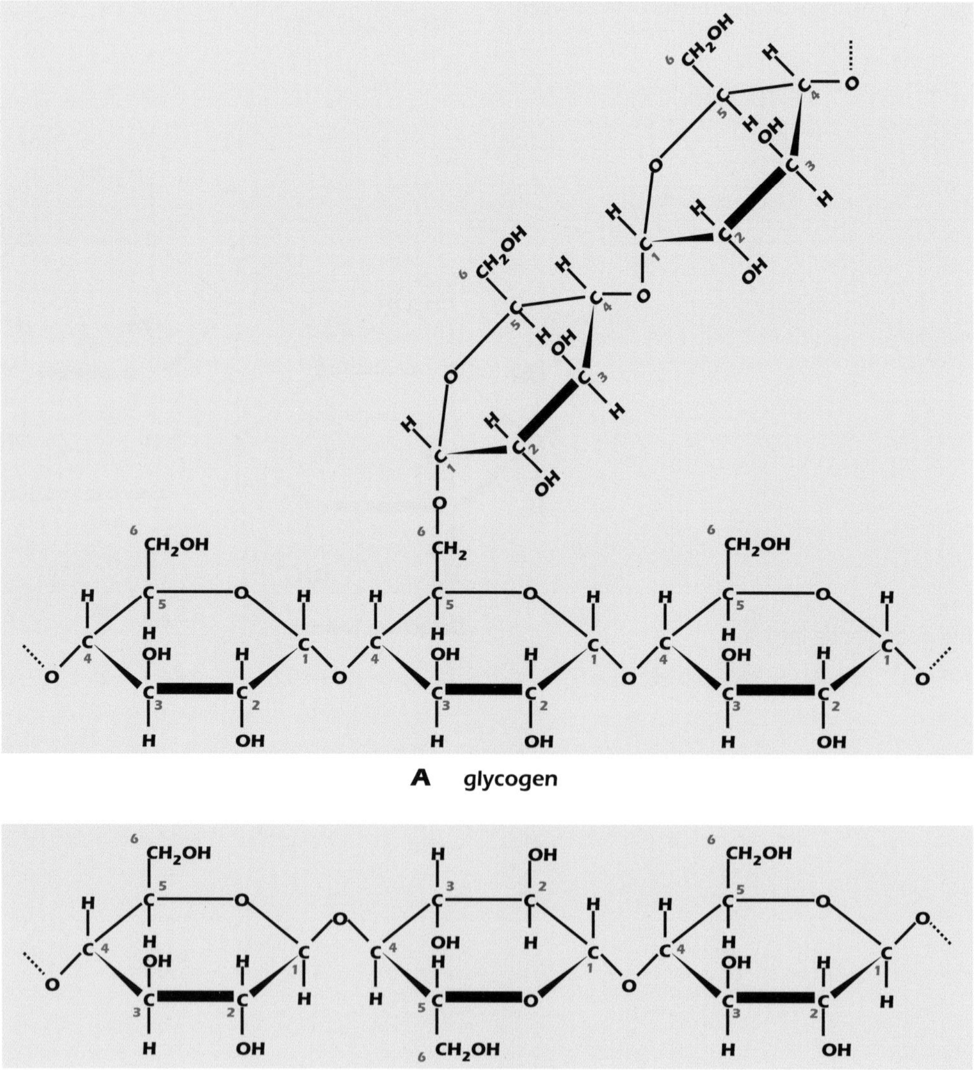

A glycogen

B cellulose

Figure 2.10. The polysaccharides glycogen and cellulose.

Modified Sugars

A number of chemically modified sugars are important in biology. Figure 2.11 shows three. **Deoxyribose** is a ribose that is missing the —OH group on carbon number two. Its main use is in making deoxyribonucleic acid—DNA. N-acetyl glucosamine is a glucose in which an —NHCOCH$_3$ group replaces the —OH on carbon two. It is used in a number of oligosaccharides and polysaccharides, for example, in chitin, which forms the hard parts of insects. **Mannose-6-phosphate** is a mannose that has been **phosphorylated**, that is, had a **phosphate** group attached, on carbon number six. We will meet more phosphorylated sugars in the next section.

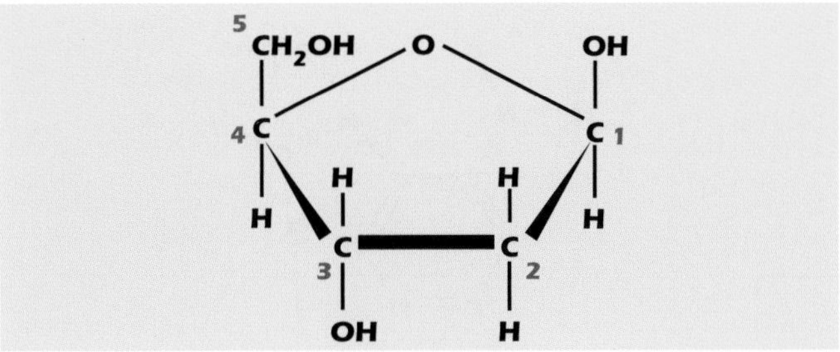

Deoxyribose

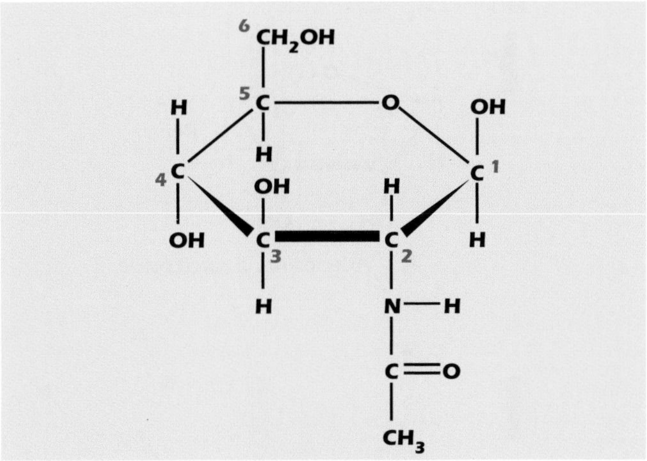

N-acetyl glucosamine

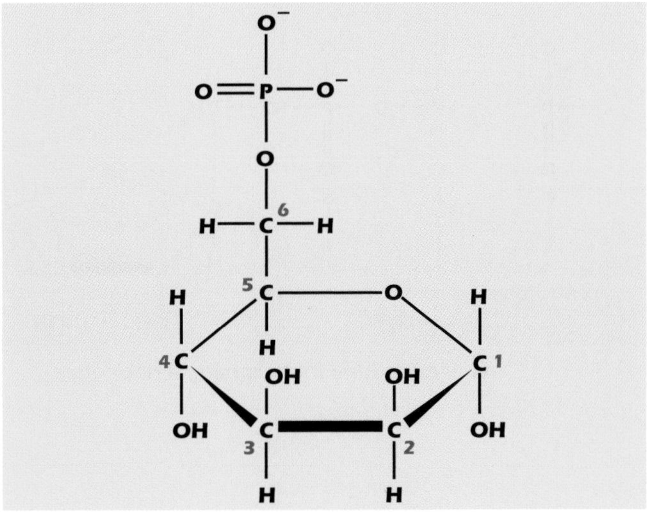

Mannose-6-phosphate

Figure 2.11. Deoxyribose, N-acetyl glucosamine and mannose-6-phosphate are modified sugars.

Figure 2.12A shows a **nucleoside** called **adenosine**. It is composed of ribose coupled to a nitrogen-rich compound called **adenine**. The numbers on the sugar—1′, 2′; and so forth—are the same numbering system we have seen before; the ′symbol is pronounced "prime" and is there to indicate that we are identifying the atoms of the sugar, not the atoms of the adenine. The name nucleoside reflects the fact that phosphorylated nucleosides are the building blocks of the **nucleic acids** that form the genetic material in the nucleus. However, nucleosides also play important roles in other places inside and outside the cell. Seven different compounds are commonly used to generate nucleosides (Fig. 2.13). All seven contain many nitrogen atoms and one or more ring structures.

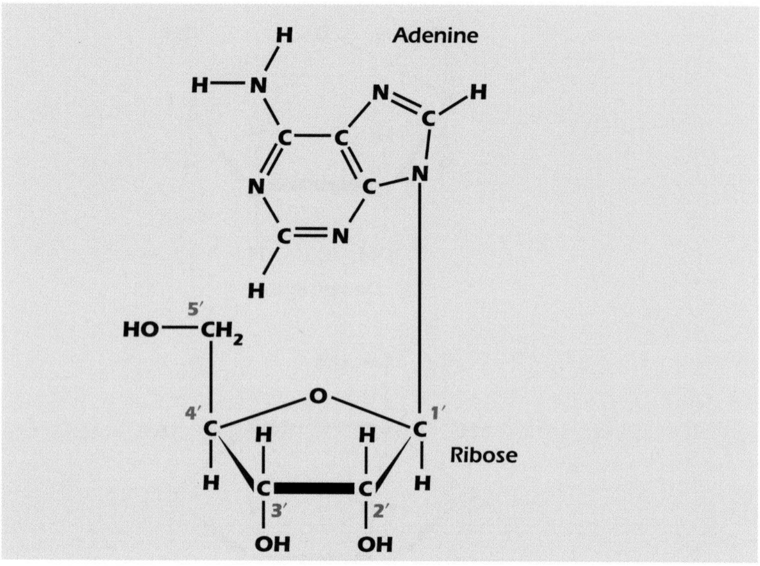

A Adenosine, a nucleoside

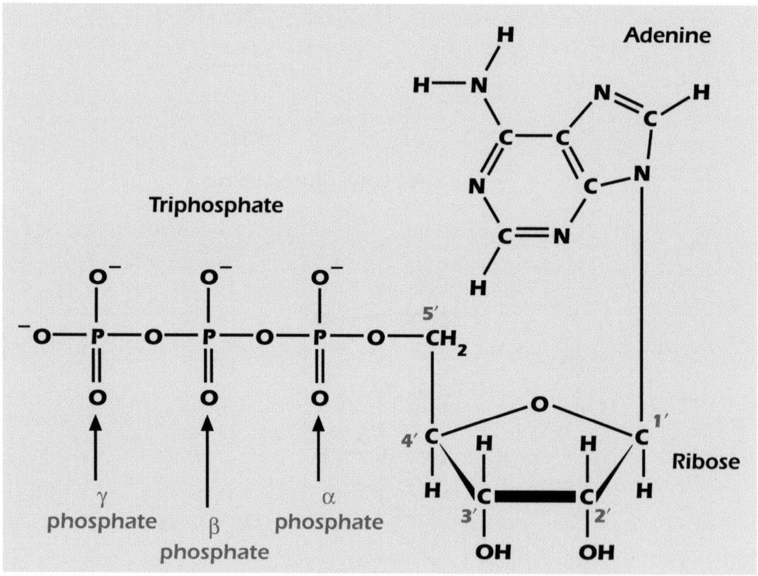

B Adenosine triphosphate, a nucleotide

Figure 2.12. A nucleoside comprises a base and a ribose; a nucleotide comprises a nucleoside plus phosphate.

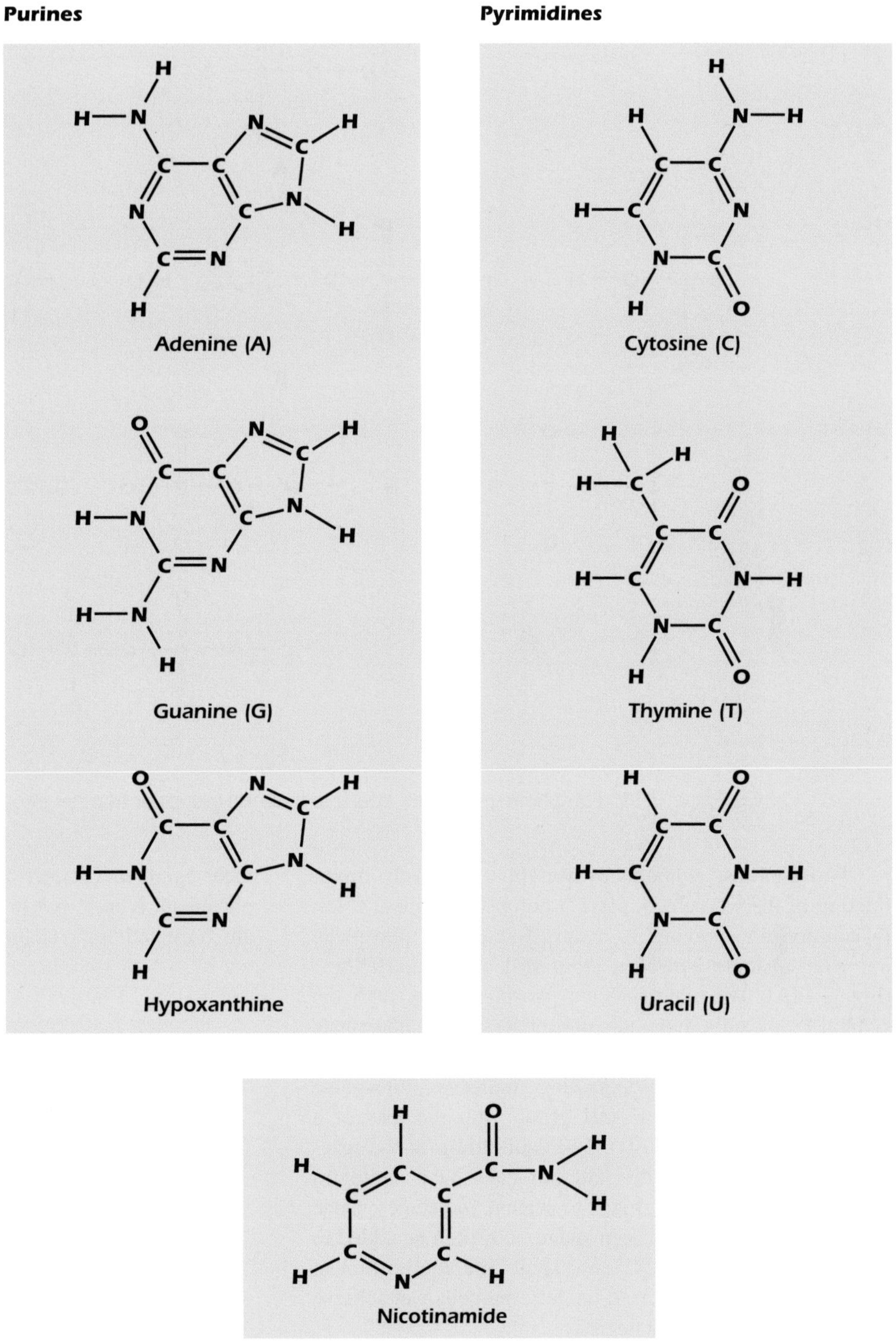

Purines

Adenine (A)

Guanine (G)

Hypoxanthine

Pyrimidines

Cytosine (C)

Thymine (T)

Uracil (U)

Nicotinamide

Figure 2.13. Seven bases found in nucleosides.

They are the three **purines** adenine, guanine and hypoxanthine; the three **pyrimidines** cytosine, thymine, and uracil; and an odd man out called **nicotinamide**. These ring compounds are called **bases**. Historically, the name arose because the compounds are indeed bases in the sense used earlier in this chapter—they will accept an H^+ from water. The roots of the name are now forgotten by most molecular biologists, who now use the word base to mean a purine, a pyrimidine, or nicotinamide. In general, a nucleoside is formed by attaching one of these bases to the 1′-carbon atom of ribose.

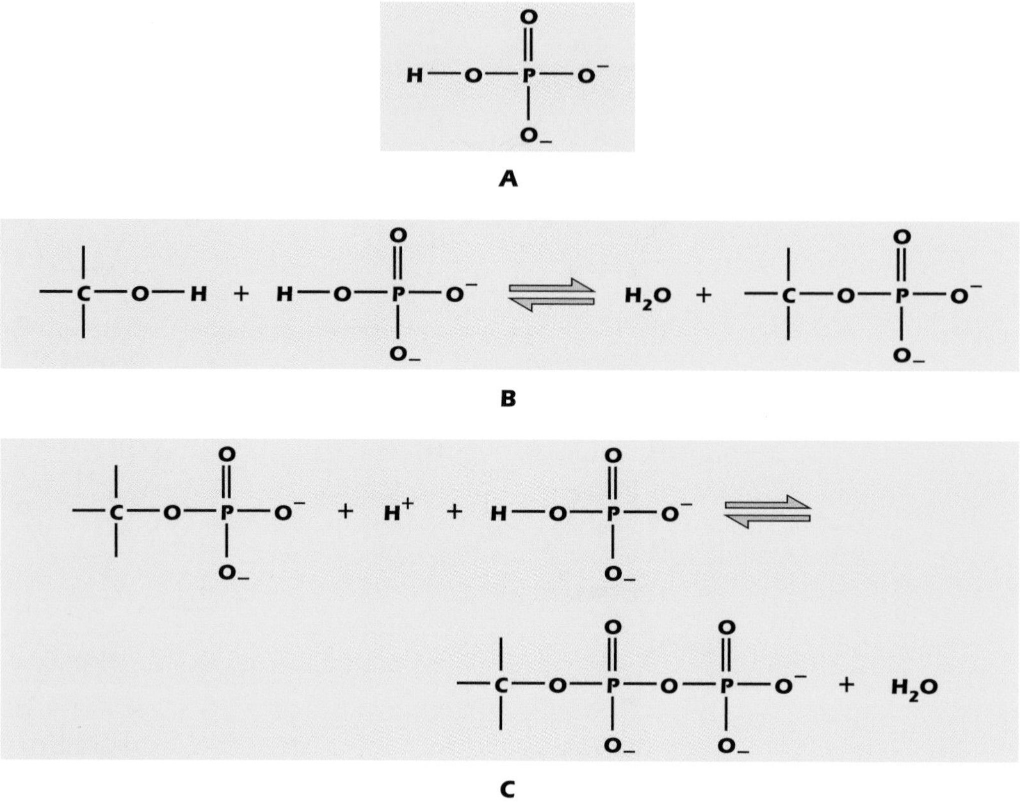

A

B

C

Figure 2.14. Phosphate groups can attach to hydroxyl groups or to other phosphate groups.

Phosphorous, although by weight a relatively minor fraction of the whole cell, plays a number of critical roles. In solution phosphorous is mainly found as a phosphate ion with a single hydrogen atom still attached, HPO_4^{2-} (Fig. 2.14A). We often indicate phosphate ions with the symbol Pi, meaning inorganic phosphate. Where phosphate becomes important, however, is when it is attached to an **organic**, that is, carbon-containing, molecule. Phosphate can substitute into any C–OH group with the loss of a water molecule (Fig. 2.14B). The equilibrium in the reaction shown lies far to the left, but cells have strategies for attaching phosphate groups to organic molecules. Once one phosphate group has been added, more can be added to form a chain (Fig. 2.14C). Once again, the equilibrium in the reaction shown lies far to the left, but cells can achieve this result using other strategies.

Figure 2.12B shows adenosine with a chain of three phosphate groups attached to the 5′ carbon of the ribose. This important molecule is **adenosine triphosphate (ATP)**, and we will meet it many times in the course of this book. The three phosphate groups are denoted by the Greek letters α, β, and γ. Phosphorylated nucleosides are called **nucleotides**.

Molecules with several OH groups can become multiply phosphorylated. Figure 2.15 shows **inositol trisphosphate (IP_3)**, an important messenger molecule we will meet again in Chapter 15. Both ATP and IP_3 have three phosphate groups, but to indicate the fact that in ATP these are arranged in a chain, while in IP_3 they are attached

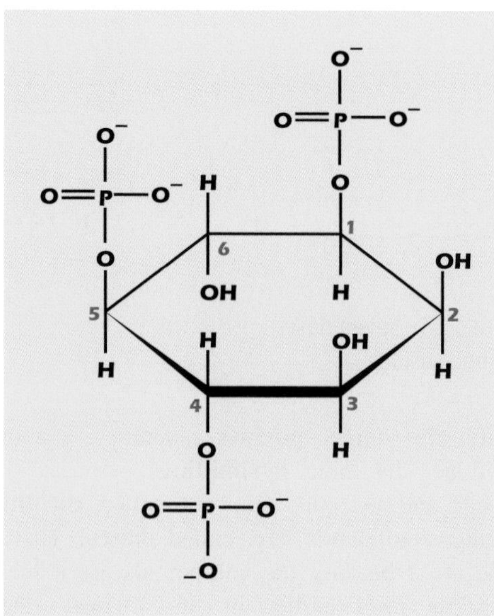

Figure 2.15. Inositol trisphosphate, a multiply phosphorylated polyalcohol.

to different carbons, we use the prefix *tri* in adenosine *tri*phosphate and the prefix *tris* in inositol *tris*phosphate. Similarly, a compound with two phosphates in a chain is called a *di*phosphate, while one with one phosphate on each of two different carbons is a *bis*phosphate.

OXIDATION AND REDUCTION INVOLVE THE MOVEMENT OF ELECTRONS

Figure 2.16A shows another important molecule formed from adenosine. Here adenosine and a nicotinamide nucleotide are joined through their phosphate groups. The resulting molecule, reduced **nicotinamide adenine dinucleotide** (NADH), is a strong **reducing agent**.

Addition of hydrogen atoms to molecules, or the removal of oxygen atoms, is called **reduction**. The opposite of reduction is **oxidation**, the addition of oxygen or the removal of hydrogen atoms. When they form bonds, hydrogen atoms give up a generous share of their electrons, while oxygen atoms tend to take more than their fair share of electrons in any bonds they make. Thus addition of oxygen to a compound means removal of electrons, and vice versa, so the most general definition of oxidation is the loss of electrons, with reduction being defined as the gain of electrons (a silly but useful mnemonic is "$L_{oss}E_{lectrons}O_{xidation}$ the lion says $G_{ain}E_{lectrons}R_{eduction}$"). Oxidation and reduction occur together, so that one reactant is oxidized while the other is reduced. Figure 2.17 shows the interconversion of pyruvate and lactate, an important reaction in metabolism (page 211). In the forward reaction pyruvate gains two hydrogens (and two electrons), so it is reduced. At the same time NADH loses one electron plus a hydrogen atom to become NAD^+ (Fig. 2.16B). The NADH lost two electrons, so it has been oxidized.

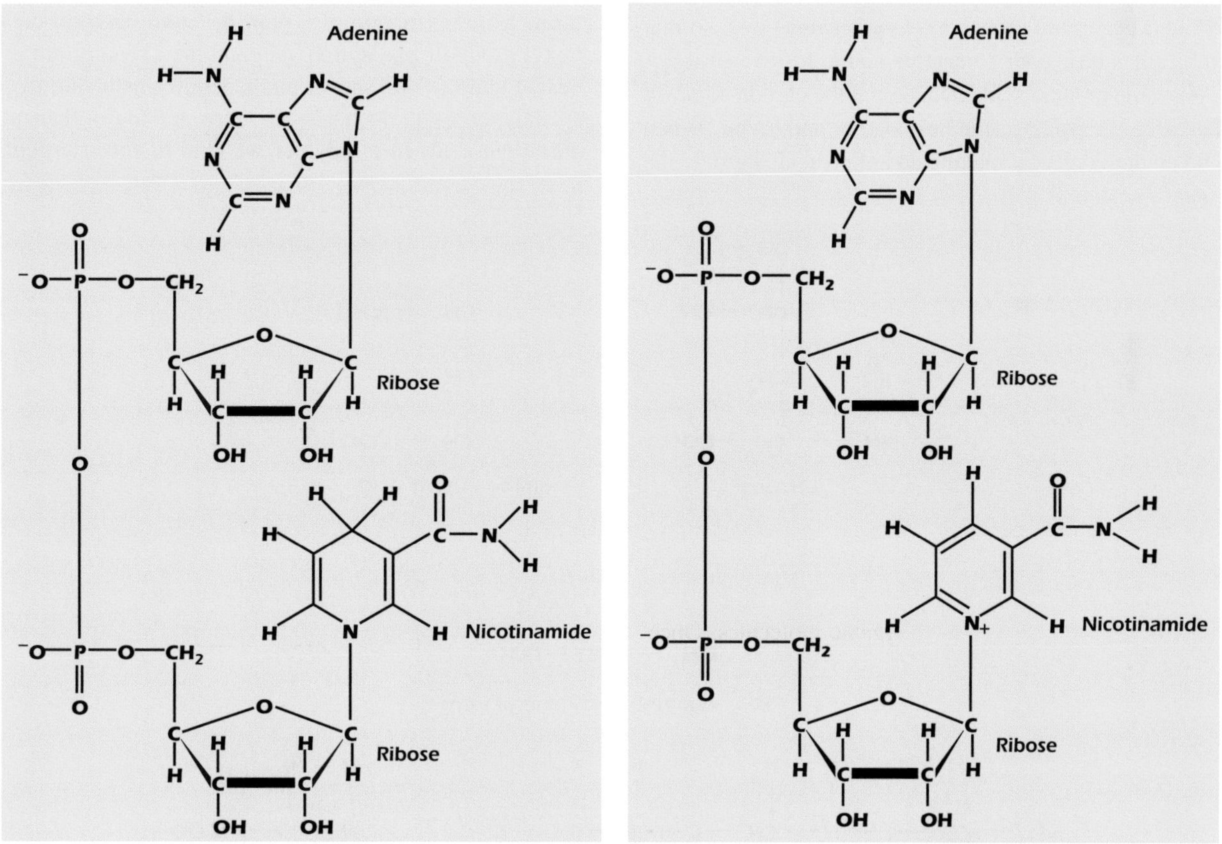

A NADH

B NAD^+

Figure 2.16. Nicotinamide adenine dinucleotide is formed of two nucleotides joined via their phosphate groups. The reduced form, NADH (A), is a strong reducing agent and energy currency. The oxidized form, NAD^+, is shown in (B).

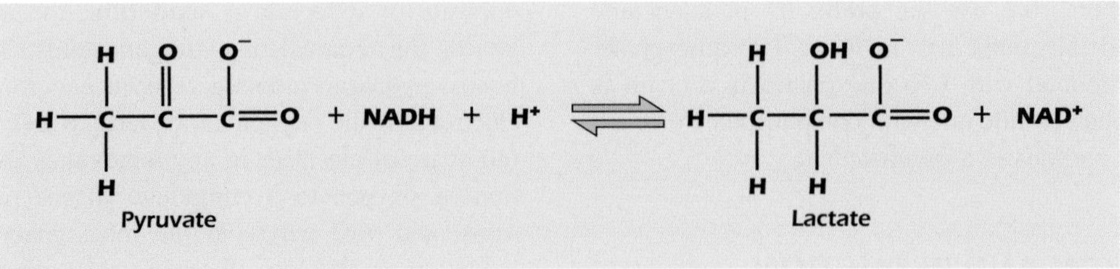

Figure 2.17. The reversible reduction of pyruvate to lactate.

AMINO ACIDS, POLYPEPTIDES, AND PROTEINS

Amino acids contain both a carboxyl group, which readily gives an H^+ to water and is therefore acidic, and a basic **amino group** (NH_2), which readily accepts H^+ to become $NH_3{}^+$. Figure 2.18A shows two amino acids, leucine and **γ-amino butyric acid** (GABA), in the form in which they are found at normal pH: the carboxyl groups have each lost an H^+ and the amino groups have each gained one, so that the molecules bear both a negative and a positive charge.

We name organic acids by labeling the carbon adjacent to the carboxyl group α, the next one β, and so on. When we add an amino group, making an amino acid, we state the letter of the carbon to which the amino group is attached. Hence leucine is an **α-amino acid** while GABA stands for *gamma*-amino butyric acid. α-Amino acids are the building blocks of proteins. They have the general structure shown in Figure 2.18B where R is the **side chain**. Leucine has a simple side chain of carbon and hydrogen. Other amino acids have different side chains and so have different properties. It is the diversity of amino side chains that give proteins their characteristic properties (page 138).

α-Amino acids can link together to form long chains through the formation of a **peptide bond** between the carboxyl group of one amino acid and the amino group of the next. Figure 2.18C shows the generalized stucture of such a chain of α-amino acids. If there are fewer than about 50 amino acids in a polymer we tend to call it a **peptide**.

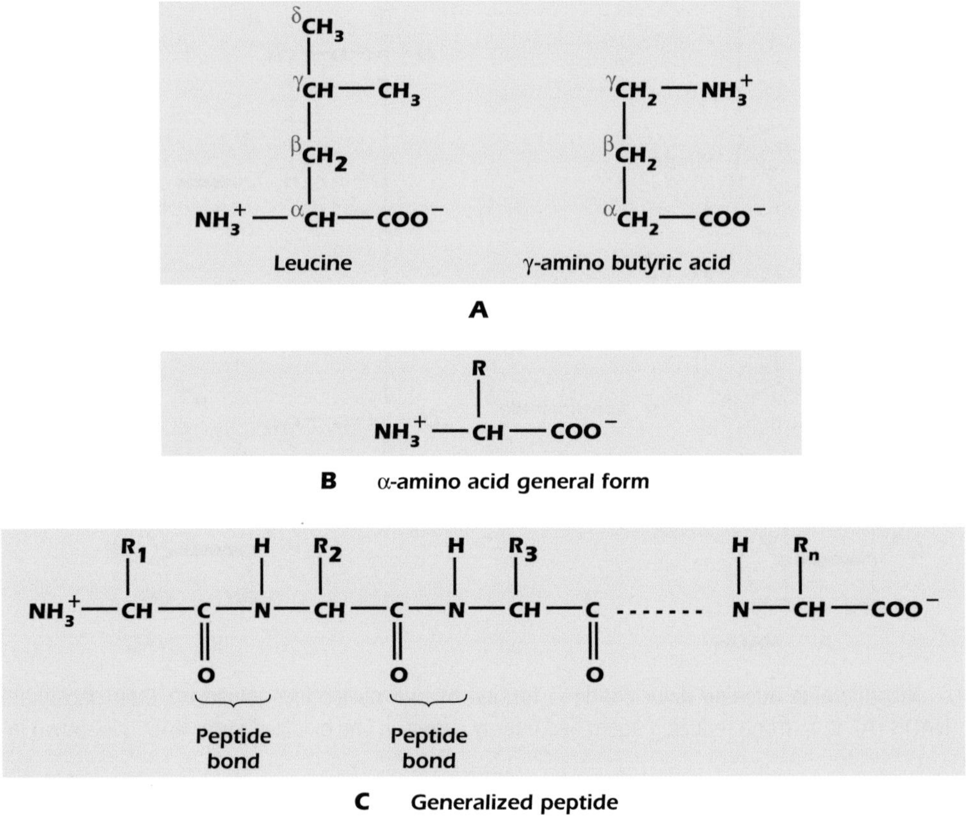

Figure 2.18. Amino acids and the peptide bond.

IN DEPTH 2.1 NOMENCLATURE

T. S. Eliot wrote "I tell you a cat must have three different names. First there's the name that the family use daily..." Chemicals that we use daily often have traditional, easy-to-say names such as butyric acid for $CH_3CH_2CH_2$-COOH. There is no clue in the name as to the structure, but once learned, the name is convenient.

For many years biologists have used a naming system for organic acids in which the carbon atoms are given Greek letters. The α carbon is the one *next to* the COOH group, the β carbon the next one along, and the γ carbon the third one along and so on. For long chain fatty acids the last letter in the Greek alphabet (omega) is used to denote the carbon farthest from the carboxyl group. Thus the compound at the top right of Figure 2.18 is γ-amino butyric acid, GABA for short.

Like Eliot's cats, chemicals have a third name. Chemists need a nomenclature that will allow the unequivocal naming of all compounds, so the International Union of Pure and Applied Chemistry (IUPAC) generated such a system. IUPAC has a different convention for organic acids: the carbons are numbered 1, 2, 3, and so on, and number one is the carbon in the COOH group. Thus in the IUPAC system, GABA is 4-aminobutanoic acid.

In this book we will alternate between nomenclatures, using the most common name for each compound.

More and it is a **polypeptide**. Polypeptides that fold into a specific shape are **proteins**. Peptides and polypeptides are formed inside cells by adding specific amino acids to the end of a growing polymer. As we will see in Chapter 4, the particular amino acid to be added at each step is defined by the instructions on a molecule called **messenger ribonucleic acid (mRNA)**, which in turn contains a copy of the information recorded in the cell's DNA that is stored in the nucleus. This is the **central dogma** of molecular biology: DNA makes RNA makes protein.

Most of the components of the machinery of the cell—the chemicals that carry out the processes of life—are proteins. From Chapter 8 onwards this book will largely be describing proteins and structures made out of them. However the central dogma dictates that we should concentrate on DNA and RNA first, as we do in Chapters 4 through 7.

LIPIDS

Unlike the chemical groups we have so far mentioned, the term lipid is not a single chemical class. Instead it refers to all of the water-insoluble chemicals that can be extracted from a cell using a solvent such as octane. Included under lipids are **fats** (triacylglycerols), phospholipids, and sterols such as cholesterol.

Figure 2.19A shows the **fatty acid** oleic acid. It comprises a carboxyl group plus a long tail of carbons and hydrogens. At neutral pH oleic acid gives up its H^+ to become the oleate ion (Fig. 2.19B). The two ends of the oleate ion are very different. The carboxyl group is negatively charged and hence will associate readily with water molecules. The **hydrocarbon tail** is nonpolar and does not readily associate with water. The molecule is said to be **amphipathic**, from the Greek for "hating both," meaning that one half does not like to be in water while the other half does not like to be in a nonpolar environment like octane. An alternative word with the same meaning is **amphiphilic** ("loving both").

Figure 2.19C shows the small molecule **glycerol**. Like sugars, glycerol has many hydroxyl groups and, like sugars, tastes sweet, as its name suggests. However, it is not a sugar because it cannot adopt an oxygen-containing ring structure. Rather, because compounds containing hydroxyl groups are called alcohols, glycerol is a polyalcohol.

Cells can join fatty acids and glycerol to make **glycerides**. The bond is formed by removing the elements of water between the carboxyl group of the fatty acid and a

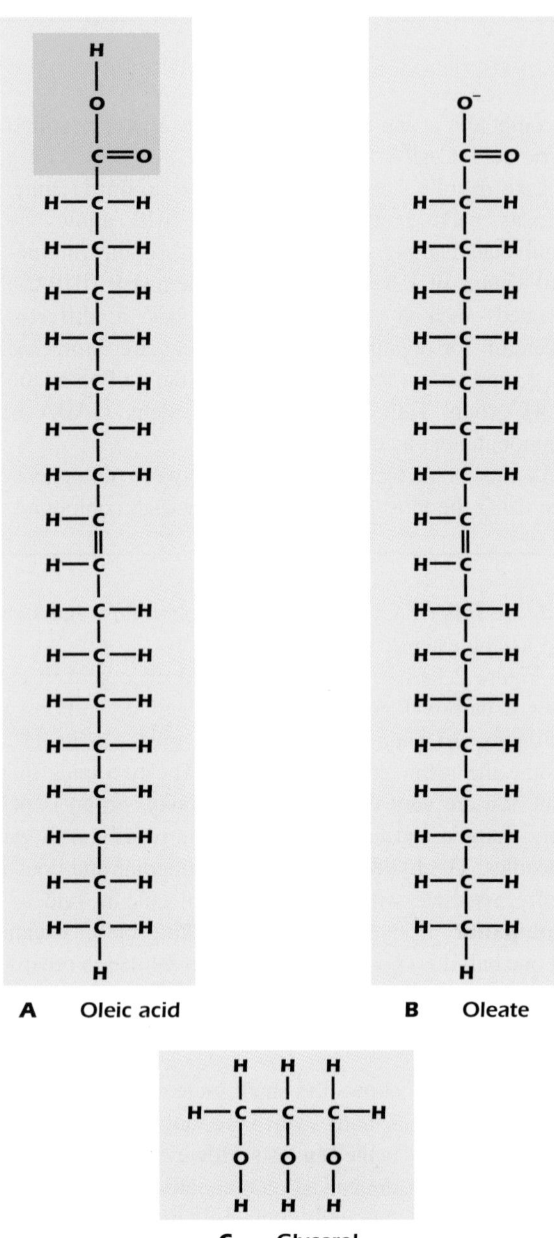

Figure 2.19. Oleic acid, oleate, and glycerol.

hydroxyl group of glycerol. Any bond of this type, between a carboxyl group and a hydroxyl group, is called an **ester bond**. Figure 2.20 shows two glycerides. In Figure 2.20A is trioleoylglycerol, the main component of olive oil. It is a **triacylglycerol** (or **triglyceride**) formed from three molecules of oleic acid and one molecule of glycerol. The fatty acid residues are called **acyl groups**. Formation of the ester bonds has removed the charged carboxyl groups that rendered the oleate amphipathic. Almost all of the triacylglycerol molecule is simple hydrocarbon chains that cannot hydrogen bond with water molecules. For this reason, olive oil and other triacylglycerols do not mix with water. They are therefore said to be **hydrophobic**. Triacylglycerols that are liquid at room temperature are called oils, those that are solid are called fats, but they are all the same sort of molecule.

Cells contain droplets of triacylglycerol within their cytoplasm. These are usually small, but in fat cells, which are specialized as fat stores, the droplets coalesce into a single large globule so that the cell's cytoplasm is squeezed into a thin layer surrounding the fat globule. During fasting triacylglycerols are broken down into free fatty acids and glycerol. The fatty acids and the glycerol then enter the circulation for use by tissues.

Figure 2.20B shows phosphatidylcholine, an example of the **phospholipids** that make the plasma membrane and other cell membranes. Like triacylglycerols, phospholipids have fatty acids attached to glycerol, but only two of them. In place of the third fatty acid is a polar, often electrically charged **head group**. The head group is joined to the glycerol through phosphate in a structure called a **phosphodiester link**. The combination of head group and negatively charged phosphate is able to associate strongly with water—it is said to be **hydrophilic**. The two acyl groups, on the other hand, form a tail that, like olive oil, is hydrophobic. Phospholipids can therefore neither dissolve in water (because of their hydrophobic tails) nor remain completely separate, like olive oil (because then the head group could not associate with water). Phospholipid molecules therefore spontaneously form **lipid bilayers** (Fig. 2.20C) in which each part of the molecule is in its preferred environment. Cell membranes are lipid bilayers plus some added protein.

HYDROLYSIS

Many of the macromolecules of which cells are made are generated from their individual building blocks by the removal of the elements of water. Equally, macromolecules can be broken into their individual building blocks by **hydrolysis**—breakage by the addition of water. Figure 2.21 shows three examples. Lactose is hydrolyzed to the monosaccharides galactose and glucose by the addition of one water molecule. Next, we show a dipeptide being broken into individual amino acids by hydrolysis. Here, the reaction would just be called hydrolysis, but when the peptide bonds in a protein are hydrolyzed, breaking it into separate fragments, we call this **proteolysis**. Enzymes catalyse these hydrolysis reactions all the time in our intestines, though some people lose the ability to hydrolyze lactose as they get older: they are **lactose intolerant**. Third, we show the **dimeric** (formed of two parts) inorganic phosphate ion **pyrophosphate** being hydrolyzed to regular phosphate ions. The hydrolysis of pyrophosphate is catalyzed by enzymes found throughout the body, both inside cells and

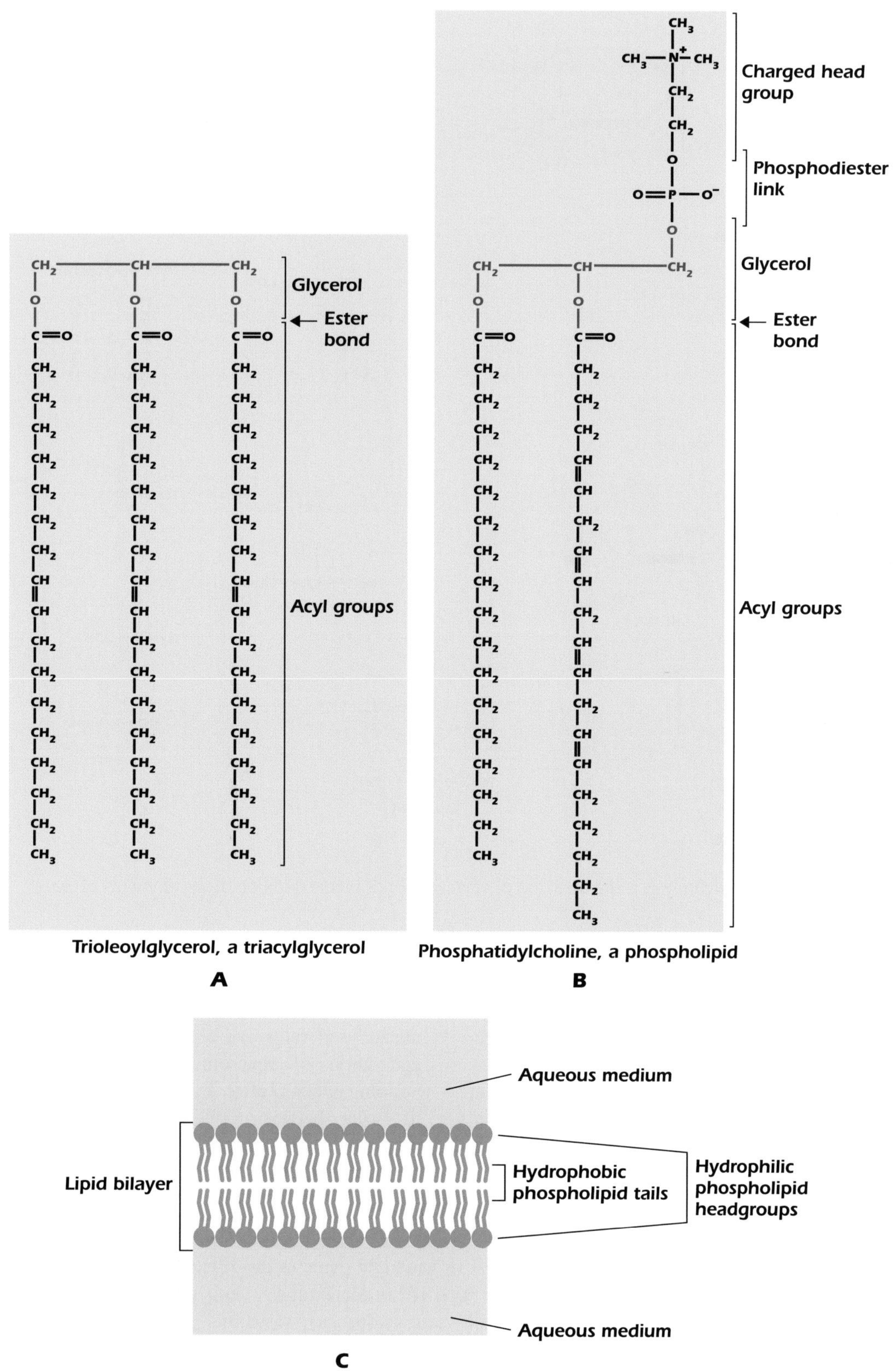

Figure 2.20. Glycerides and the lipid bilayer.

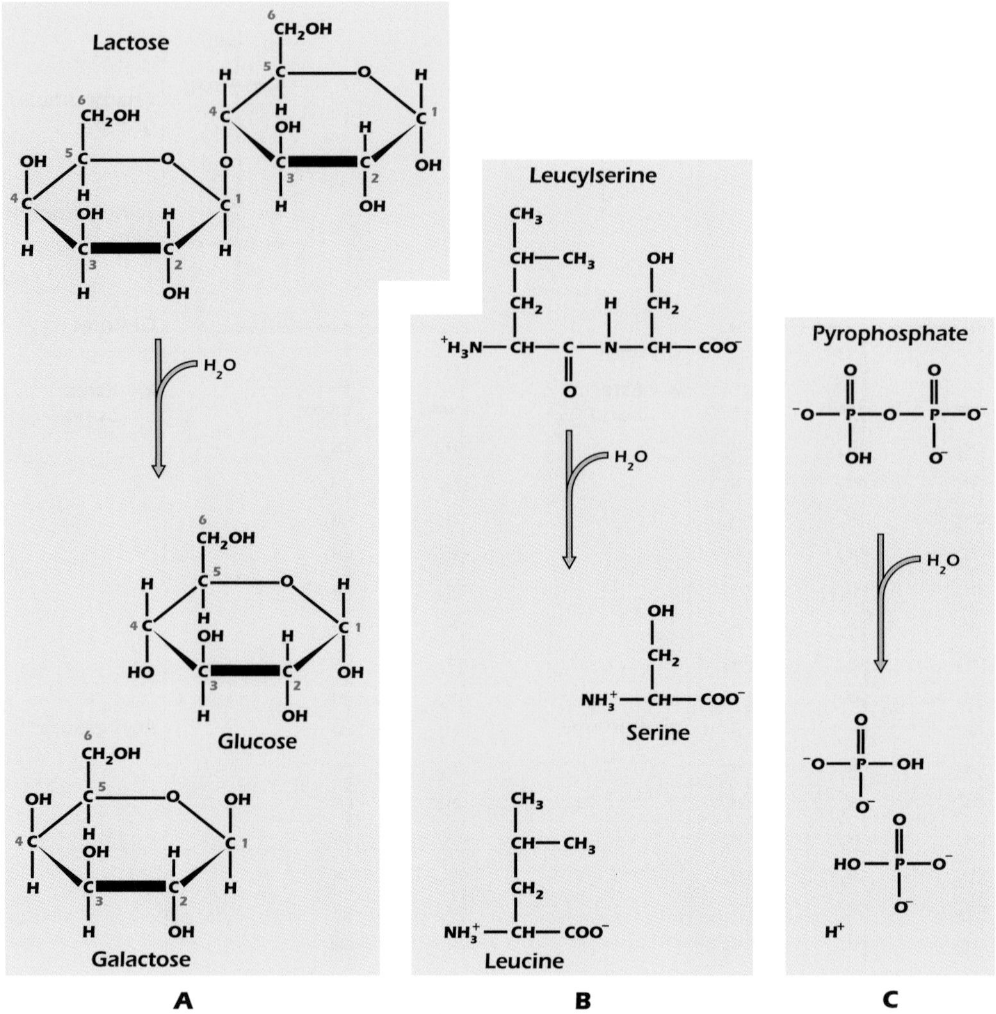

Figure 2.21. Hydrolysis is the breaking of covalent bonds by the addition of the elements of water.

out. Later in the book we will meet a number of instances where a reaction creates pyrophosphate, but as soon as the pyrophosphate is produced it will be hydrolyzed to regular phosphate. In Figure 2.22 we show the complete hydrolysis of the phospholipid phosphatidylcholine. Five molecules result: one phosphate ion together with choline, glycerol, and two fatty acids. This hydrolysis also occurs in our intestines, where it is a multistage process.

Hydrolysis reactions usually proceed rapidly in living tissue when the appropriate catalyst is present, because water is present at high concentration, although the uncatalysed rates are usually very slow. In contrast the generation of macromolecules from their constituent building blocks, often by the elimination of the elements of water between constituent building blocks, usually requires the expenditure of energy by the cell. We will see how these reactions are performed in later chapters.

Answer to Thought Question: (i) Pure water has a pH of 7, that is, the H^+ concentration is 10^{-7} moles liter^{-1}. Adding HCl to a final concentration of 10^{-4} moles liter^{-1} releases 10^{-4} moles liter^{-1} of extra H^+, so the final H^+ concentration is $10^{-7} + 10^{-4}$ moles liter^{-1} = 1.001×10^{-4} moles liter^{-1}. To three significant figures, the pH is 4. In fact, we could have ignored the H^+ already present in the pure water and just considered the H^+ from the 10^{-4} moles liter^{-1} of HCl. (ii) In this case we cannot ignore the 10^{-7} moles liter^{-1} of H^+ already present in the pure water. Adding HCl to a final concentration of 10^{-8} moles liter^{-1} releases 10^{-8} moles liter^{-1} of extra H^+, so the final H^+ concentration is $10^{-7} + 10^{-8}$ moles liter^{-1} = 1.1×10^{-7} moles liter^{-1}, equivalent to pH 6.96. If you answered pH 8, then you are not thinking clearly: how could adding hydrochloric acid, even in small amounts, to water make the solution *less* acid than it was to start with?

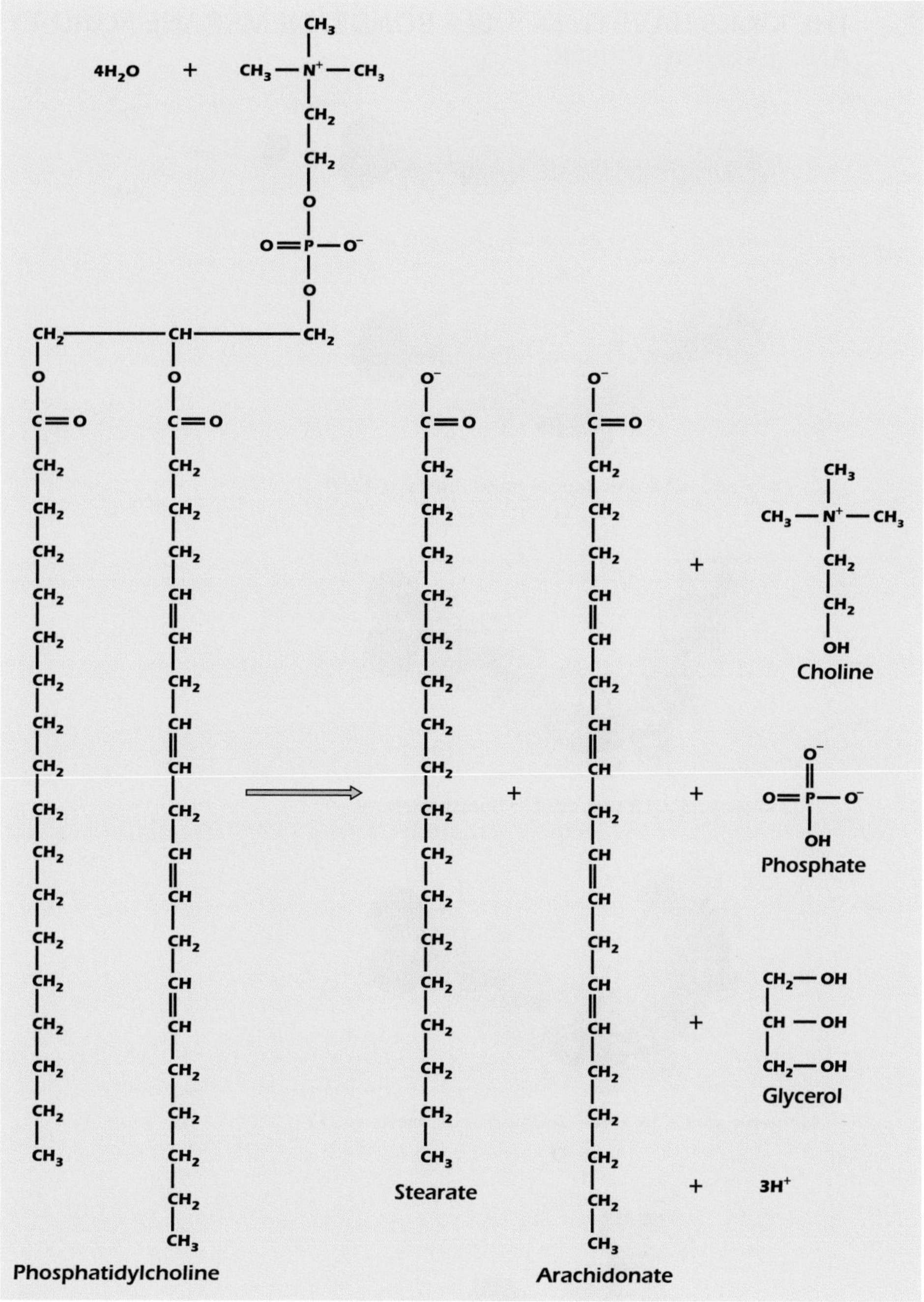

$4H_2O$ + Phosphatidylcholine $\Longrightarrow$ Stearate + Arachidonate + Choline + Phosphate + Glycerol + $3H^+$

Phosphatidylcholine

Stearate

Arachidonate

Choline

Phosphate

Glycerol

Figure 2.22. Hydrolysis of a phospholipid.

IN DEPTH 2.2 THE KINKS HAVE IT: DOUBLE BONDS, MEMBRANE FLUIDITY AND EVENING PRIMROSES

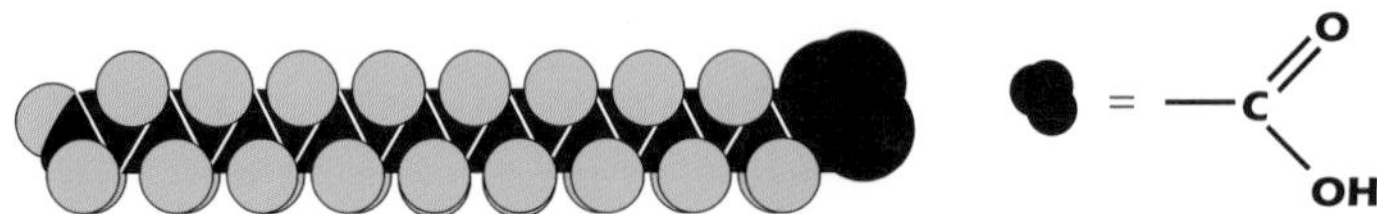

Stearic acid, C18 no double bonds, melts at 69.6°

A

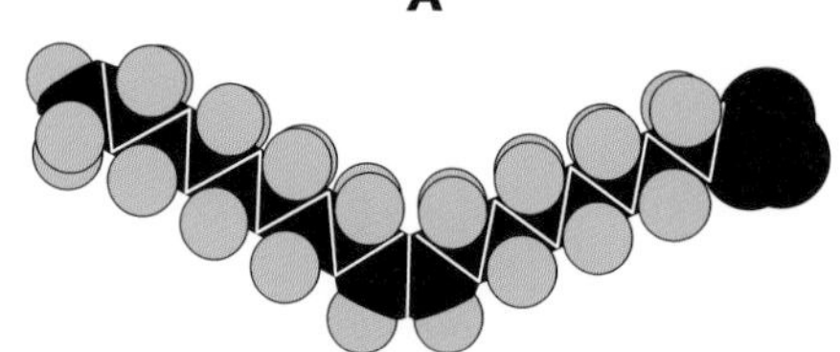

Oleic acid, C18 one double bond, melts at 13.4°

B

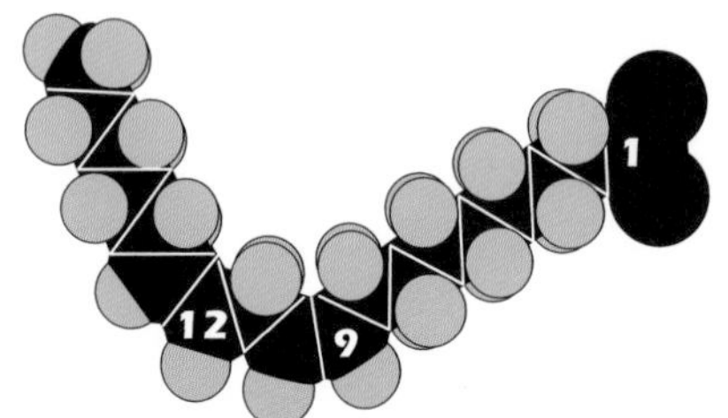

Linoleic acid, C18 two double bonds, melts at −9°

C

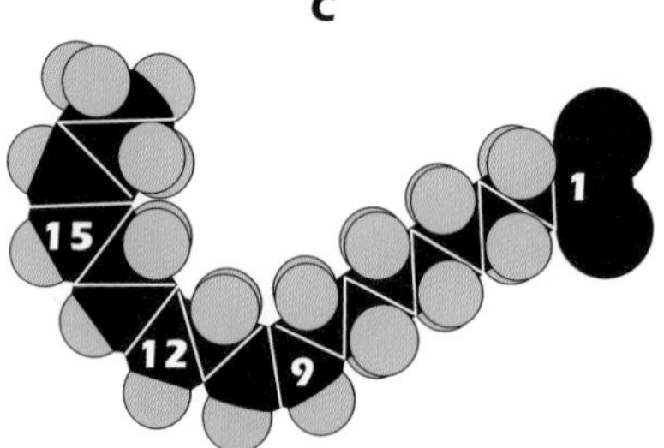

Linolenic acid, C18 three double bonds, melts at −17°

D

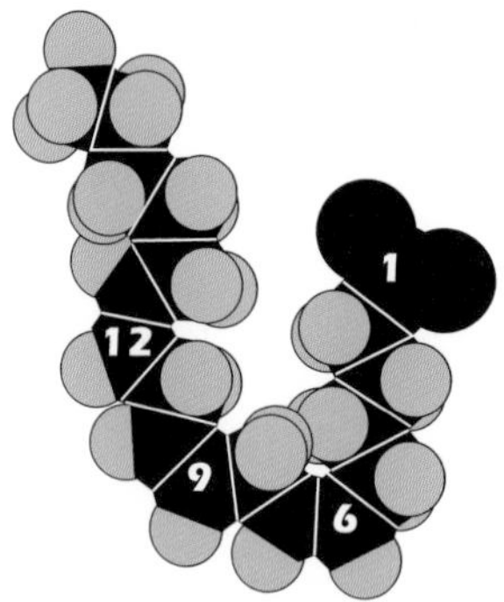

γ-Linolenic acid, C18 three double bonds

E

One of the fatty acids commonly found as a component of animal fat is stearic acid, illustrated as A in the diagram. Its 18 carbon atoms are joined by single bonds, making a long straight molecule. In contrast, oleic acid (B) has a double bond between the ninth and tenth carbons. This introduces a kink in the chain. A fatty acid with kinks is less able to solidify because it is less able to pack in a regular fashion. The more double bonds in a fatty acid, the lower its melting point. Thus stearic acid melts at 70°C, while oleic acid melts at 13°C and is therefore liquid at room temperature. Fatty acids containing double bonds between the carbon atoms are said to be **unsaturated**. **Polyunsaturated** fatty acids have more than one double bond, more than one bend, and therefore even lower melting points. Linoleic acid (C) has two, and melts at −9°C; linolenic acid (D) has three and a melting point of −17°C.

For the triacylglycerols inside our cells, and the phospholipids in our membranes, the same rule applies: the more double bonds in the acyl groups, the lower the melting point. The large masses of triacylglycerols containing stearic acid in animal bodies are liquid at body temperature but form solid lumps of fat at room temperature, while trioleoylglycerol is liquid at room temperature (but should not be kept in the fridge!). Membranes must not solidify: if they did, then they would crack and the cell contents would leak out each time the cell was flexed. Unsaturated fatty acids play an essential role in maintaining membrane liquidity.

Mammals are unable to introduce double bonds beyond carbon 9 in the fatty acid chain. This means that linoleic and linolenic acids must be present in the diet. They are known as **essential fatty acids**. Fortunately, the biochemical abilities of plants are not so restricted, and plant oils form a valuable source of unsaturated fatty acids.

The normal form of linolenic acid, shown in D, is α-linolenic acid, which has its double bonds between carbons 9 and 10, 12 and 13, and 15 and 16. Some plant seed oils contain an isomer of linolenic acid with double bonds between carbons 6 and 7, 9 and 10, and 12 and 13. This is γ-linolenic acid (E). The attractive, yellow-flowered evening primrose (*Oenethera perennis*) has seeds that contain an oil with γ-linolenic in its triacyglycerols. The 6 to 7 double bond introduces a kink closer to the glycerol. This is thought to increase fluidity when incorporated into membrane lipids. No one really knows if it really has the marvelous health effects that some claim for it, or if it does, how it works. This ignorance has not stopped people from making lots of money from evening primrose oil. It is included in cosmetics and in alternative medicines for internal and external application. γ-Linolenic acid occurs in other plants—the seeds of borage (*Borago officinalis*) are one of the richest sources. It is even found in some fungi. However, these lack the romance of the evening primrose!

Medical Relevance 2.2 Protease Inhibitors as AIDS Therapy

HIV, the virus that causes AIDS, uses a transcription factor (page 92) called NFκB that is already present in the host cell. NFκB triggers the generation of mRNA from the viral DNA genome. Most of the NFκB in the host cell is present in a large inactive form. Proteolysis cleaves this large protein to release the smaller, active transcription factor. The virus itself contains an enzyme called a protease that does this job, and because this enzyme is distinct from the host cell's enzymes, it is possible to find drugs that inhibit it without dramatically affecting the host cell. Protease inhibitors are used in antiviral therapy in the treatment of AIDS.

SUMMARY

1. When two atoms interact, electrons may be shared between the two to form a covalent bond, or may pass completely from one atom to another, forming ions.
2. In water the electrons are not shared equally but are displaced toward the oxygen atom, so that the molecule is polar.
3. Ionic compounds will only dissolve in polar solvents.
4. Acids are molecules that give an H^+ to water, forming H_3O^+. Dissolving an acid in water produces a solution of pH less than 7.
5. Bases are molecules that accept an H^+ from water, leaving OH^-. Dissolving a base in water produces a solution of pH greater than 7.
6. Pure water has the neutral pH of 7.0.
7. A hydrogen bond can form when a hydrogen atom takes up a position between two electron-grabbing atoms (oxygen, nitrogen, or sulfur), the three forming a straight line.
8. Monosaccharides are compounds with a central skeleton of carbon to which many OH groups are attached. They can switch between two oxygen-containing ring structures via an open-chain configuration.
9. Monosaccharides can join together to form disaccharides such as lactose and to form long chains such as glycogen and cellulose.
10. Adenine, guanine, hypoxanthine, cytosine, thymine, uracil, and nicotinamide are nitrogen-rich ring-shaped molecules called bases. Bases combine with ribose or deoxyribose to form nucleosides. Phosphorylation of a nucleoside on the 5′ carbon produces a nucleotide.
11. Amino acids are compounds with a carboxyl acidic group and an amino basic group. At neutral pH these groups will, respectively, lose and gain an H^+ to become —COO^- and —NH_3^+.
12. Polypeptides are polymers of α-amino acids linked by peptide bonds. The sequence of amino acids is determined by instructions on the cell's DNA. Proteins are polypeptides that fold into a specific shape.
13. Many lipids are formed by the attachment of hydrophobic long-chain fatty acids to a glycerol backbone. Phospholipids also have a hydrophilic head group and spontaneously form lipid bilayers.
14. Hydrolysis is the breakage of covalent bonds by the addition of the elements of water.

FURTHER READING

Voet, D., and Voet, J. (2011). *Biochemistry*, 4th edition, John Wiley & Sons, Hoboken.

REVIEW QUESTIONS

2.1 Theme: Types of Organic Chemicals

A adenine
B adenosine
C adenosine triphosphate
D galactose
E lactic acid
F lactose
G leucine
H oleic acid
I ribose
J trimethylamine

From the above list of specific chemicals, select the chemical that belongs to each of the broader classes below.

1. A disaccharide
2. A fatty acid
3. A nucleoside
4. A nucleotide
5. a pentose sugar

2.2 Theme: Chemical Groups and Bonds

Diagrams i and ii show two small organic compounds, while diagram iii shows part of a macromolecule. The bonds indicated −−− are connections to other parts of the macromolecule. The red letters A through I indicate particular groups and bonds.

i

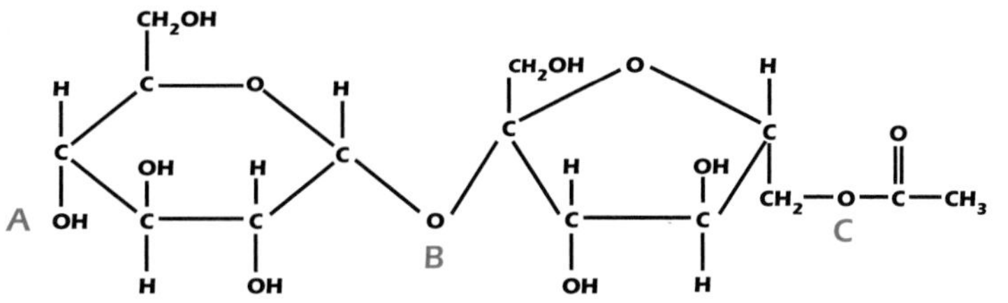

ii

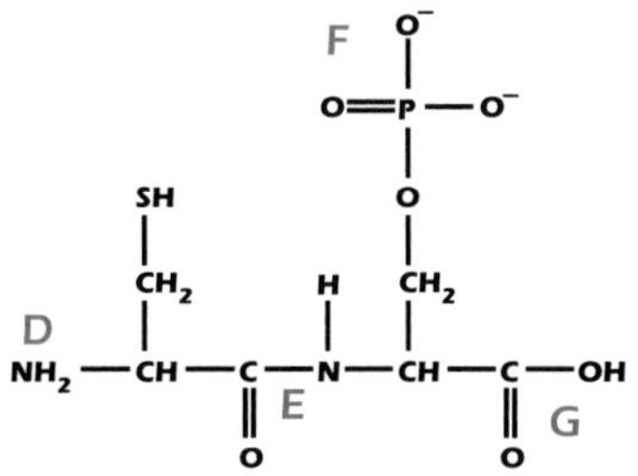

iii

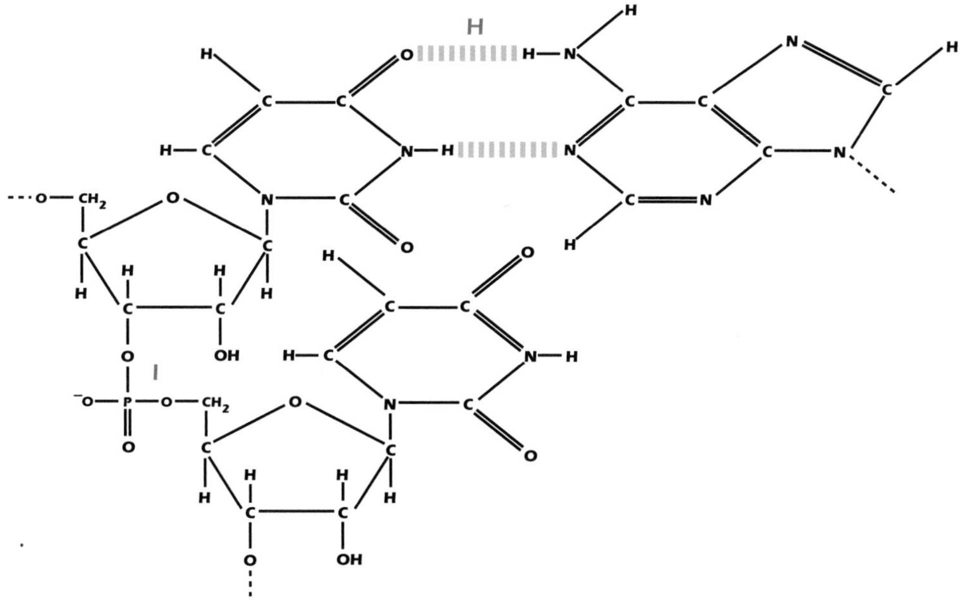

Use the letters to identify examples of each of the groups and bonds listed below.

1. amino group
2. carboxyl group
3. hydroxyl group
4. phosphate group
5. ester bond
6. glycosidic bond
7. hydrogen bond
8. peptide bond
9. phosphodiester link

2.3 Theme: Acids and Bases

Acetic acid (ethanoic acid), CH_3COOH, is a simple organic acid that is the main ingredient in vinegar. The pK_a of the reaction $CH_3COOH \rightleftarrows CH_3COO^- + H^+$ is 4.8. The gas ammonia dissolves in water and the dissolved NH_3 is a base which accepts a proton to become the ammonium ion NH_4^+. The pK_a of the reaction $NH_3 + H^+ \rightleftarrows NH_4^+$ is 9.2.

Consider a solution of ammonium acetate, that is, a solution that initially contains both ammonium ions and acetate ions. By adding acid or alkali (for example, HCl or NaOH) the pH can be changed.

A pH = 3
B pH = 5
C pH = 7
D pH = 9
E pH = 11
F the described condition is impossible

From the above list, choose the pH at which each of the conditions described below applies. Note that at least one of the conditions is impossible: for this/these, select answer F.

1. Considering the acetate, both the protonated form CH_3COOH and the deprotonated form CH_3COO^- are present at significant concentrations.
2. Considering the ammonium, both the protonated form NH_4^+ and the unprotonated form NH_3 are present at significant concentrations.
3. The vast majority of both the acetate and the ammonium are in their protonated forms, CH_3COOH and NH_4^+ respectively.
4. The vast majority of both the acetate and the ammonium are in their unprotonated forms, CH_3COO^- and NH_3 respectively.
5. The vast majority of both the acetate and the ammonium are in their ionic forms, CH_3COO^- and NH_4^+ respectively.
6. The vast majority of both the acetate and the ammonium are in their uncharged forms, CH_3COOH and NH_3 respectively.

THOUGHT QUESTION

If you add HCl (hydrochloric acid) to water it dissociates completely into H^+ and Cl^-. What is the pH of the solution generated when HCl is dissolved in pure water to a final concentration of (i) 10^{-4} mole liter^{-1} (ii) 10^{-8} mole liter^{-1}?

3

MEMBRANES AND ORGANELLES

The vast majority of the reactions that cells carry out take place in water. Eukaryotic cells are, at any one time, carrying out an enormous range of such chemical manipulations which are collectively referred to as **metabolism** (Chapter 13). In much the same way that our homes are divided into rooms that are adapted for particular activities, so eukaryotic cells contain distinct compartments or **organelles** to house specific functions. The term organelle is used rather loosely. At one extreme some scientists use it to mean any distinct cellular structure that has a more or less well-defined job to do; at the other are those who would reserve the name organelle for those cellular compartments that contain their own DNA and have some limited genetic autonomy. In this book, we will define organelles as those cellular components whose limits, like those of the cell itself, are defined by membranes. It is first necessary to consider some of the fundamental properties of cell membranes.

BASIC PROPERTIES OF CELL MEMBRANES

It is difficult to overstate the importance of membranes to living cells; without them life as we know it could not exist. The **plasma membrane**, also known as the **cell membrane** or **plasmalemma**, defines the boundary of the cell. It regulates the movement of materials into and out of the cell and facilitates electrical and chemical signaling between cells. Other membranes define the boundaries of organelles and provide a matrix upon which complex chemical reactions can occur. Some of these themes will be developed in subsequent chapters. In the following section the basic structure of the cell membrane will be outlined.

The basic structure of a biological membrane is shown in Figure 3.1. Approximately half the mass is phospholipid, which spontaneously organizes to form a bilayer about 4 nm thick. All the membranes of the cell, including the plasma membrane, also contain proteins. These may be tightly associated with the membrane and extracted from it only with great difficulty, in which case they are **integral proteins** (e.g., connexin, Fig. 3.3); or they may be separated with relative ease, in which case they are **peripheral membrane proteins** (e.g., clathrin adaptor protein, page 168). Membrane proteins are free to move laterally, within the plane of the membrane. Integral plasma membrane proteins are often **glycosylated**: they have sugar residues attached on the side facing the extracellular medium. In addition to phospholipids and proteins, eukaryotic cell membranes also contain **cholesterol**. This is a bulky molecule, most of which is embedded in the lipid bilayer, a single hydroxyl group at one end interacting with the polar head groups of the membrane phospholipids. Cholesterol makes the membrane less permeable to small hydrophilic molecules and makes the membrane more fluid. Cholesterol is essential for life but excess cholesterol in the bloodstream is strongly implicated in the development of atherosclerosis and heart disease.

Cell Biology: A Short Course, Third Edition. Stephen R. Bolsover, Jeremy S. Hyams, Elizabeth A. Shephard and Hugh A. White.

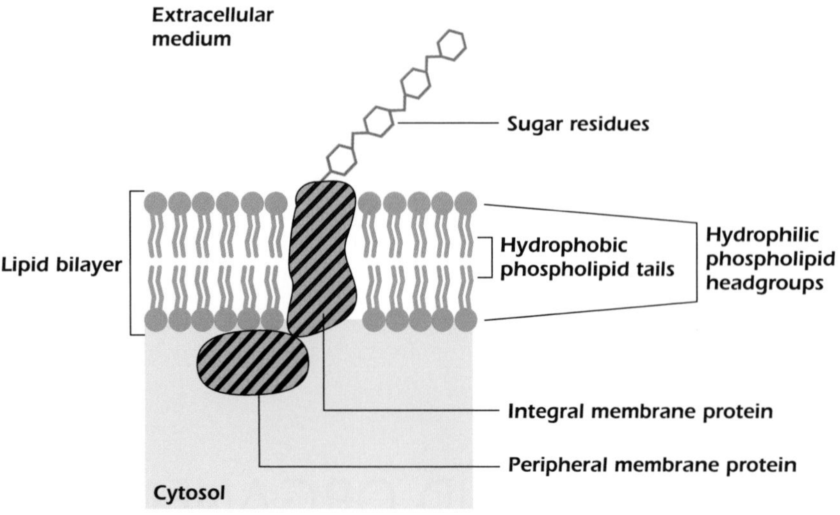

Figure 3.1. Membranes comprise a lipid bilayer plus integral and peripheral proteins.

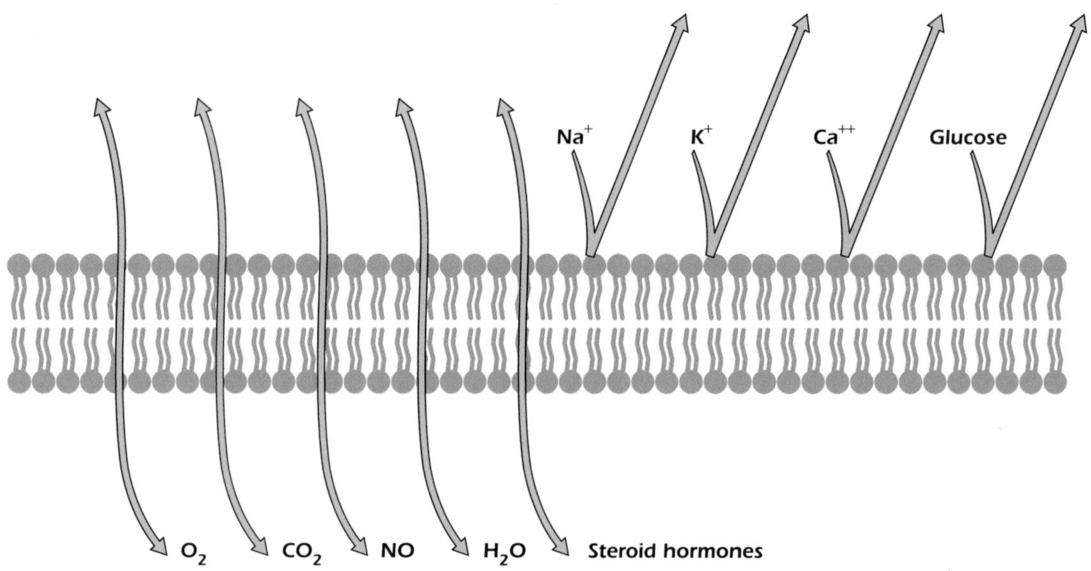

Figure 3.2. Small uncharged molecules can pass through membranes by simple diffusion, but hydrophilic solute cannot.

Straight Through the Membrane: Diffusion Through the Bilayer

Molecules of oxygen are uncharged. Although they dissolve readily in water, they are also able to dissolve in the hydrophobic interior of lipid bilayers. Oxygen molecules can therefore pass from the extracellular medium into the interior of the plasma membrane, and from there on into the cytoplasm, in a simple diffusion process (Fig. 3.2). Three other small molecules with important roles in biology—carbon dioxide, nitric oxide, and water itself—also pass across the plasma membrane by simple diffusion, as do the uncharged hormones of the steroid family. In contrast, ions cannot dissolve in hydrophobic regions (page 18) and therefore cannot cross membranes by simple diffusion.

Cell Junctions

In multicellular organisms, and particularly in epithelia, it is often necessary for neighboring cells within a tissue to be connected. This function is provided by cell junctions. In animal cells there are three types of junction. Those that form a tight seal between adjacent cells are known as **tight junctions**; those that allow communication between cells are known as **gap junctions**. A third class

Example 3.1 Rapid Diffusion in the Lungs

Blood comprises red and white cells in plasma, a solution of sodium chloride, other ions, various organic molecules, and proteins. Red blood cells are very simple. They have no nucleus and their plasma membrane encloses a cytosol packed with the oxygen-carrying protein hemoglobin, bathed in a salt solution. The sodium concentration in the red blood cell cytosol is much lower than in plasma. It is important that the sodium concentration in the cytosol remain low: if it were to increase, the red blood cells would swell and burst as water rushed in. As red blood cells pass through the lungs, they quickly gain oxygen because oxygen molecules can pass rapidly across the plasma membrane by simple diffusion. The cells do not, however, gain sodium ions from the plasma as they pass around the body because the lipid bilayer is impermeable to these ions. Red blood cells therefore remain the right size to pass though even the tiniest blood vessels in our bodies.

of cell junction that anchors cells together, allowing the tissue to be stretched without tearing, are called **anchoring junctions**.

Tight junctions are found wherever flow of extracellular medium is to be restricted and are particularly common in epithelial cells such as those lining the small

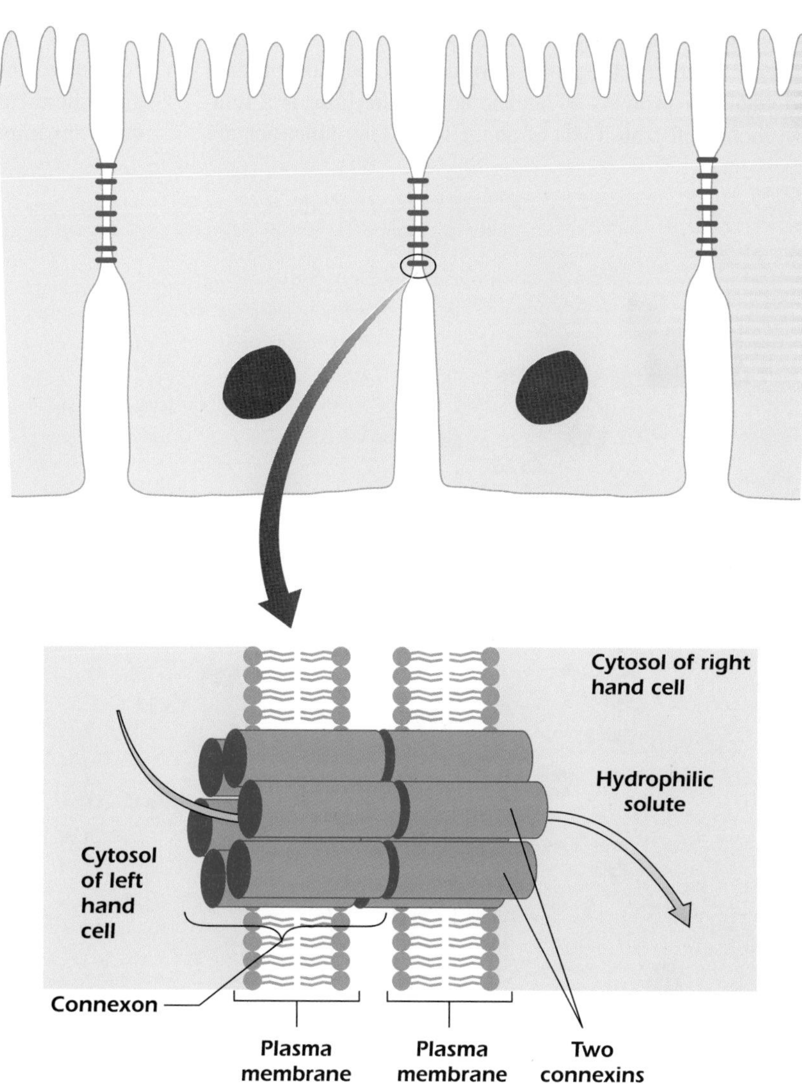

Figure 3.3. Gap junctions allow solute and electrical current to pass from the cytosol of one cell to the cytosol of its neighbor.

intestine. The plasma membranes of adjacent cells are pressed together so tightly that no intercellular space exists between them (Fig. 1.12 on page 13). Tight junctions between the epithelial cells of the intestine ensure that the only way that molecules can get from the lumen of the intestine to the blood supply that lies beneath is by passing through the cells, a route that can be selective.

Gap junctions are specialized structures that allow cell-to-cell communication in animals (Fig. 3.3). When two cells form a gap junction, ions and small molecules can pass directly from the cytosol of one cell to the cytosol of the other cell without going into the extracellular medium. Since ions can move through the junction, changes in electrical voltage are also rapidly transmitted from cell to cell by this route. In vertebrates the structure that makes this possible is the **connexon**. When two compatible connexons meet, they form a tube, about 1.5 nm in diameter, that runs through the plasma membrane of the first cell, across the small gap between the cells, and through the plasma membrane of the second cell. This hole is large enough to allow through small ions (and therefore to pass electrical current) together with amino acids and nucleotides, but it is too small for proteins or nucleic acids. The limit is a relative molecular mass (Mr) of about 1000. Gap junctions are especially important in the heart, where they allow an electrical signal to pass rapidly between all the cardiac muscle cells, ensuring that they all contract at the appropriate time. Each connexon is composed of six protein subunits called connexins that can twist against each other to open and close the central channel in a process called **gating** (page 233). This allows the cell to control the degree to which it shares solutes with its neighbor.

There are over 20 different members of the connexin gene family in humans. Each can, if made in a population of cells, generate complete, working gap junction channels of 12 identical connexins. However, not all connexin **isoforms** are compatible. A cell with connexin 26 cannot form gap junctions with another cell that makes connexin 43 (Fig. 3.4). However a connexon made from connexin 26 is able to form a good working channel if it meets a connexon made from connexin 32, and a connexon made from connexin 32 is in turn compatible with connexin 43. The interaction of connexins shows the importance of protein shape and binding. As the connexons from adjacent cells brush across each other they will bind and form a complete channel, or just touch and move on, depending on the three-dimensional shape and chemical makeup of their surfaces. We will meet many examples of protein-protein interaction throughout

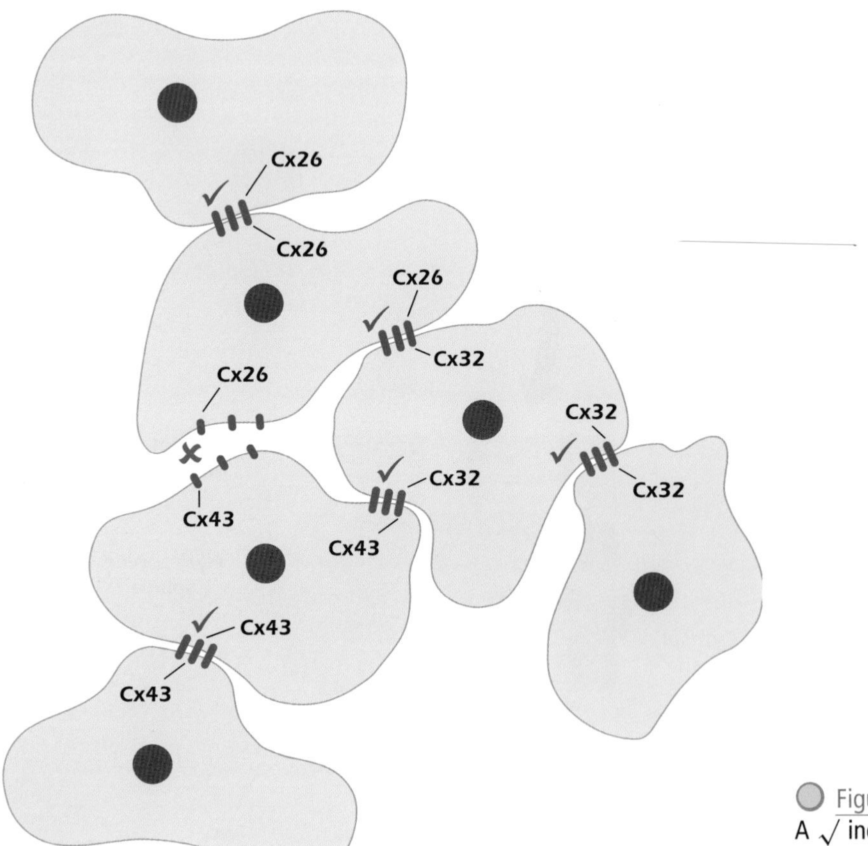

Figure 3.4. Not all connexins are compatible. A √ indicates a working gap junction, x indicates that gap junction channels cannot form.

Example 3.2 Gap Junctions Keep Eggs Ready

In the days leading up to ovulation, oocytes that are to be released from the ovary are kept in a state of suspended development by a chemical in their cytosol called cyclic AMP, or cAMP (page 257). The oocytes themselves do not make cAMP. Rather, the follicle cells that surround them make cAMP, which then passes through gap junctions into the oocyte. Only when the oocyte is released and begins its passage down the Fallopian tube does it complete its maturation into a gamete ready to be fertilized by a spermatozoon.

this book. The chemical composition of proteins and the generation of their three-dimensional shape is discussed in Chapter 9.

Anchoring junctions bind cells tightly together and are found in tissues such as the skin and heart that are subjected to mechanical stress. They are described laterbreak (page 292).

ORGANELLES BOUNDED BY DOUBLE-MEMBRANE ENVELOPES

Two of the major cell organelles, the nucleus and mitochondrion (plus chloroplasts in plants), share two distinctive features: they are enclosed within an envelope consisting of two parallel membranes 'and they contain the genetic material DNA.

The Nucleus

The nucleus is often the most prominent cell organelle (e.g., Fig. 1.7 on page 8). It contains the **genome**, the cell's database, which is encoded in molecules of the nucleic acid, DNA. The nucleus is bounded by a nuclear envelope composed of two membranes separated by an intermembrane space (Fig. 3.5). The inner membrane of the nuclear envelope is lined by the **nuclear lamina**, a meshwork of **lamin** proteins which provide rigidity to the nucleus and anchorage for the DNA. A two-way traffic of proteins and nucleic acids between the nucleus and the cytoplasm passes through holes in the nuclear envelope called **nuclear pores**. The nucleus of a cell that is synthesizing proteins at a low level will have few nuclear pores. In cells that are undergoing active protein synthesis, however, virtually the whole nuclear surface is perforated.

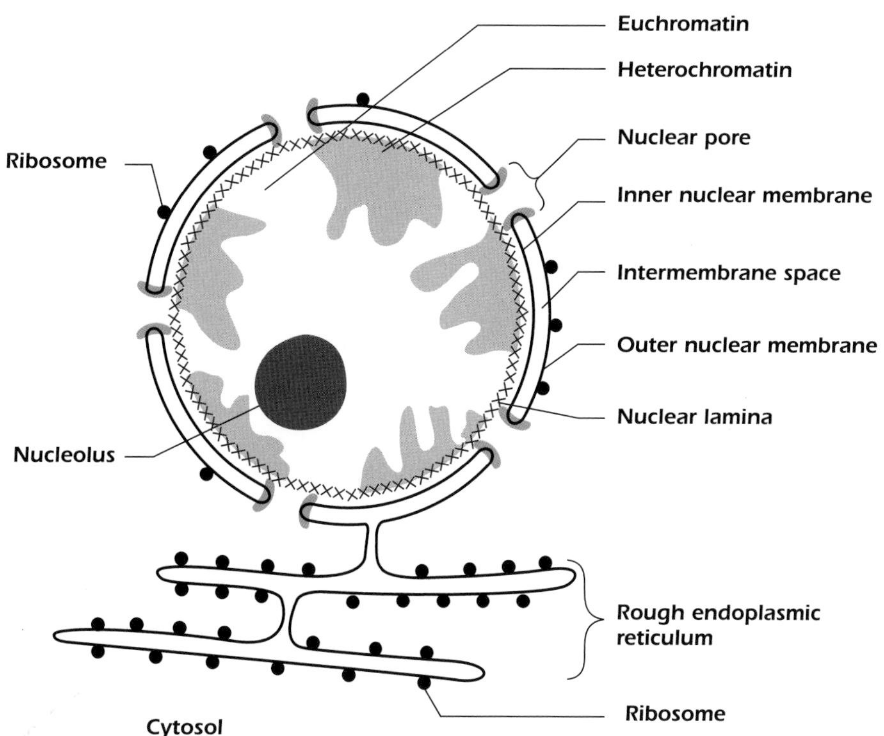

Figure 3.5. The nucleus and the relationship of its membranes to those of the endoplasmic reticulum.

Within the nucleus it is usually possible to recognize discrete areas. Much of it is occupied by chromatin, a complex of DNA and certain DNA-binding proteins such as **histones** (page 57). In most cells it is possible to recognize two types of chromatin. A central region of lightly staining **euchromatin** is that portion of the cell's DNA that is being actively transcribed into RNA while peripheral, darkly staining **heterochromatin** is the inactive portion of the genome where no RNA synthesis is occurring. The DNA in heterochromatin is densely packed, leading to its dark appearance in both the light and electron microscopes.

Unlike DNA, RNA is not confined to the nucleus and other organelles but is also found in the cytoplasm associated with particles called ribosomes whose function is to make proteins. Ribosomes are made in the nucleus, in specialized regions called **nucleoli** that form at specific **nucleolar organizer** regions sites on the DNA. These contain blocks of genes that code for the ribosomal RNA. Nuclear pores allow ribosomes to exit the nucleus.

It should be stressed that the appearance of the nucleus we have described thus far relates to the cell in **interphase**, the period between successive rounds of cell division. As the cell enters mitosis (Chapter 18) the organization of the nucleus changes dramatically. The DNA becomes more and more tightly packed and is revealed as a number of separate rods called **chromosomes**, of which there are 46 in human cells. The nucleolus disperses, and the nuclear envelope fragments. Upon completion of mitosis, these structural rearrangements are reversed and the nucleus resumes its typical interphase organization.

Mitochondria

Although most of the genetic information of a eukaryotic cell resides in nuclear DNA, some of the information necessary for the function of mitochondria is stored within the organelle itself. Mitochondria (Fig. 3.6) contain many small circular DNA molecules that are very different from the long, linear DNA molecules in the nucleus. This is strong evidence for the endosymbiotic theory of the origin of mitochondria (page 10), which proposes that the small circular DNA molecules found in mitochondria are all that is left of the chromosomes of the original symbiotic bacteria. Mitochondria also contain ribosomes (again, more like those of bacteria than the ribosomes in the cytoplasm of their own cell), and synthesize a small subset of their own proteins. The great majority of proteins of the mitochondrion, however, are encoded by nuclear genes and synthesized in the cytoplasm. Perhaps the most distinctive feature of mitochondria is that the inner of their two membranes is markedly elaborated and folded to increase its surface area. These shelf-like projections, named **cristae**, make mitochondria among the most easily recognizable organelles (e.g., Fig. 1.7 on page 8). The number of cristae, like the number of mitochondria themselves, depends upon the energy budget of the cell in which they are found. In muscle cells, which must contract and relax repeatedly over long periods of time, there are many mitochondria that contain numerous cristae; in fat cells, which generate little energy, there are few mitochondria and their cristae are less well developed. This gives a clue as to the function of mitochondria: they are the cell's power stations. Mitochondria produce the molecule adenosine triphosphate (ATP) (page 30), one of the cell's energy currencies that provide the energy to drive a host of cellular reactions and mechanisms (Chapter 12). The double-membrane structure of the mitochondrion provides four distinct domains: the

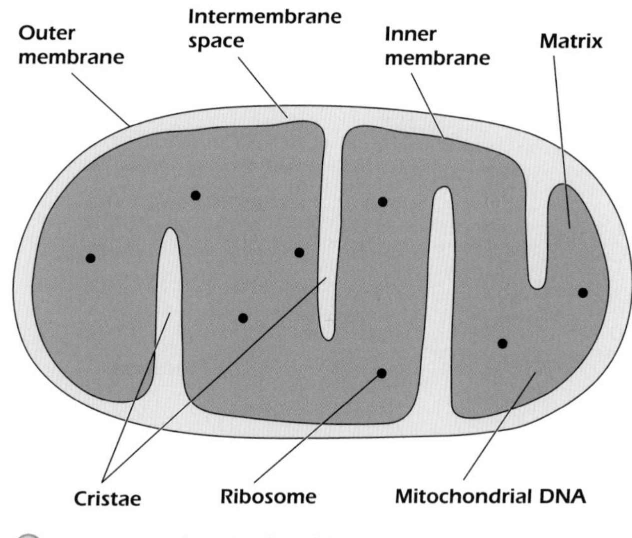

Figure 3.6. The mitochondrion.

Example 3.3 DNA Destruction in the Cytosol

An animal cell's own DNA should remain in the nucleus, except for the tiny amount that is within mitochondria. DNA in the cytoplasm will likely belong to a pathogen such as an invading virus. Cells therefore contain active DNAses in the cytoplasm that rapidly destroy DNA, while leaving RNA intact. It is to evade this defense mechanism that many viruses use RNA as their genetic material, even though RNA is a much less stable molecule than is DNA.

IN DEPTH 3.1 NOBEL PRIZES FOR ORGANELLE RESEARCHERS

In addition to Camillo Golgi's Nobel prize in 1906, modern studies of cell organelles have also been recognized by the Nobel committee. George Palade, Albert Claude, and Christian de Duve shared the 1974 Nobel prize in physiology or medicine for their work on the identification and isolation of cell organelles. Gunter Blobel, a former student of Palade's, was awarded the same prize in 1999 for the discovery that proteins carry so-called targeting sequences that determine where in the cell they should reside (page 158).

outer membrane, the inner membrane, the intermembrane space, and the matrix. Each domain houses a distinct set of functions, details of which are given in Chapters 12 and 13.

ORGANELLES BOUNDED BY SINGLE MEMBRANES

Eukaryotic cells contain many sacs and tubes bounded by a single membrane. Although these are often rather similar in appearance, they can be subdivided into different types specialized to carry out distinct functions.

Peroxisomes

Mitochondria are frequently found close to another membrane-bound organelle, the peroxisome (Fig 1.2 on page 3). In human cells peroxisomes have a diameter of about 500 nm, and their dense matrix contains a heterogeneous collection of proteins concerned with a variety of metabolic functions, some of which are only now beginning to be understood. Peroxisomes are so named because they are frequently responsible for the conversion of the highly reactive molecule hydrogen peroxide (H_2O_2), which is formed as a by-product of the reactions in the mitochondrion, into water:

$$2H_2O_2 \rightarrow 2H_2O + O_2$$

This reaction is carried out by a protein called catalase, which sometimes forms an obvious crystal within the peroxisome. Catalase is an **enzyme**—a protein catalyst that increases the rate of a chemical reaction (page 179). In fact, it was one of the first enzymes to be discovered. In humans, peroxisomes are primarily associated with lipid metabolism. Understanding peroxisome function is important for a number of inherited human diseases such as X-linked adrenoleukodystrophy, where peroxisome malfunction and the consequent inability to metabolise lipid properly typically leads to death in childhood or early adulthood unless dietary lipid is extremely restricted.

Endoplasmic Reticulum

The endoplasmic reticulum is a network of membrane-enclosed channels that run throughout the cell, forming a continuous network whose lumen (interior) is at all points separated from the cytosol by a single membrane. The membrane of the endoplasmic reticulum is continuous with the outer nuclear membrane (Fig. 3.5). Two regions can be recognized in most cells, known as smooth endoplasmic reticulum and rough endoplasmic reticulum (Fig. 1.2 on page 3). The basic difference is that the rough endoplasmic reticulum is covered in ribosomes, which give it its rough appearance in the electron microscope.

The function of the smooth endoplasmic reticulum varies from tissue to tissue. In the ovaries, testes, and the adrenal gland it is where steroid hormones are made; in the liver it is the site of detoxication of foreign chemicals, including drugs. Probably the most universal role of the smooth endoplasmic reticulum is the storage and sudden release of calcium ions. Calcium ions are pumped from the cytosol into the lumen of the smooth endoplasmic reticulum to more than 100 times the concentration found in the cytosol. Many stimuli can cause this calcium to be released back into the cytosol, where it activates many cell processes (Chapter 15).

The rough endoplasmic reticulum is where cells make the proteins that will end up as integral membrane proteins in the plasma membrane, and proteins that the cells will

Answer to Thought Question: This can only be speculation, but one obvious effect of cells having incompatible connexin isoforms is that the intracellular route for cell-cell communication is lost. Consider a tissue whose cells make connexin 26 that lies adjacent to another tissue whose cells use connexin 43. The cells of each tissue can coordinate their activity by passing signals via gap junctions, for example as small soluble chemicals (such as cAMP, Example 3.2) or as voltage changes. However, the signals will remain private to each tissue and will not be shared with the neighboring tissue. The two tissues can still communicate when necessary, for instance by transmitter chemicals that the cells release into the extracellular medium (see Chapter 16).

export to the extracellular medium (such as the proteins of the extracellular matrix, page 11).

Golgi Apparatus

The Golgi apparatus, named after its discoverer, 1906 Nobel prize winner Camillo Golgi, is a distinctive stack of flattened sacks called **cisternae**. The Golgi apparatus is the distribution point of the cell where proteins made within the rough endoplasmic reticulum are further processed and then directed to their final destination, the interior of the cell or the cell surface. Appropriately, given this central role, the Golgi apparatus is situated at the so-called cell center, a point immediately adjacent to the nucleus that is also occupied by a structure called the **centrosome** (Fig. 1.2 on page 3). The centrosome helps to organize the cytoskeleton (Chapter 17).

Lysosomes

Lysosomes are sometimes called "cell stomachs" because they contain a battery of enzymes that digest cellular components. Sequestering such potentially destructive enzymes in the lysosome protects the cytosol and provides the enzymes with the acidic conditions that allows them to function efficiently. Lysosomes are roughly spherical and usually 250–500 nm in diameter. They are particularly plentiful in cells that digest and destroy other cells, such as macrophages. They ensure the turnover of other organelles by the process of autophagy (self-eating). Organelles that are to be digested are engulfed in a membrane sac called an autophagosome which then fuses with one or more lysosomes to form an autophagolysosome. Exposed to lysosomal enzymes, engulfed organelles are rapidly digested and their components recycled. We shall return to lysosomes in Chapter 10.

Medical Relevance 3.1 Lysosomal Storage Disorders

A large number of inherited diseases are characterized by cells filled with very large lysosomes. Many of these diseases involve abnormalities of development of the skeleton and connective tissues as well as of the nervous system. The severity varies with the particular disease but they often lead to death in infancy. In the majority of these diseases one lysosomal enzyme is missing or defective. Lysosomes work to degrade cellular components that are damaged or no longer needed and their hydrolytic enzymes function under the acidic conditions in the lysosome. If one of these hydrolytic enzymes is defective then the substrate will accumulate, filling the lysosome. The distended lysosomes eventually fill and damage the cell. Some of the best understood lysosomal storage diseases involve deficiencies in one or the other of the enzymes required to degrade the complex glycosylated proteins and lipids found on cell surfaces.

Tay-Sachs disease involves severe mental retardation and blindness with death by the age of three. In this case an enzyme required to break down a particular complex membrane lipid called a ganglioside is missing and undegraded ganglioside accumulates, swelling the lysosomes. Gangliosides are particularly important in nerve cell membranes so the nerve cells are particularly damaged.

SUMMARY

1. Membranes are made of phospholipids, protein and cholesterol.
2. Cells are bounded by membranes while cell functions are compartmentalized into membrane-bound organelles.
3. Solutes with a significant solubility in hydrophobic solvents can pass across biological membranes by simple diffusion.
4. Tight junctions prevent the passage of extracellular water or solute between the cells of an epithelium.
5. Gap junctions allow solute and electrical current to pass from the cytosol of one cell to the cytosol of its neighbor.
6. The nucleus and mitochondrion are bounded by double-membrane envelopes. In the case of the nucleus, this is perforated by nuclear pores. Both organelles contain DNA.
7. Mitochondria produce the energy currency adenosine triphosphate (ATP).
8. Peroxisomes carry out a number of reactions including the destruction of hydrogen peroxide.
9. The endoplasmic reticulum is the site of much protein synthesis. Cell stimulation can often cause calcium ions stored in the endoplasmic reticulum to be released into the cytosol.
10. The Golgi apparatus is concerned with the modification of proteins after they have been synthesized.
11. Lysosomes contain powerful degradative enzymes.

FURTHER READING

Balda, M. S., and Matter, K. (2008) Tight junctions at a glance. *Journal of Cell Science* **121**, 3677–3682.

Duchen, M. R. (2004) Mitochondria in health and disease: Perspectives on a new mitochondrial biology. *Molecular Aspects of Medicine* **25**, 365–451.

Dundr, M., and Misteli, T. (2001) Functional architecture in the cell nucleus. *Biochemical Journal* **356**, 297–310.

Gall, J. G., and McIntosh, J. R. (2001) *Landmark Papers in Cell Biology*. Cold Spring Harbor Laboratory Press, New York.

Lamond, A. I., and Earnshaw, W. C. (1998) Structure and function in the nucleus. *Science* **280**, 547–553

Short, B., and Barr, F. A. (2000) The Golgi apparatus. *Current Biology* **10**, R583–585.

van der Klei, I., and Veenhuis, M. (2002) Peroxisomes: Flexible and dynamic organelles. *Current Opinion in Cell Biology* **14**, 500–505.

REVIEW QUESTIONS

3.1 Theme: Membranes

A The membrane or set of membranes surrounding the endoplasmic reticulum
B The membrane or set of membranes surrounding the Golgi apparatus
C The membrane or set of membranes surrounding the lysosome
D The membrane or set of membranes surrounding the mitochondrion
E The membrane or set of membranes surrounding the peroxisome
F The plasma membrane

From the above list, select the membrane that best fits the descriptions below.

1. A double layer comprising an inner and an outer membrane
2. Is continuous with a membrane surrounding the nucleus
3. Forms a set of flattened sacks called cisternae
4. Often contains connexons as an integral membrane protein
5. Surrounds a space containing calcium ions at a much higher concentration than in the cytosol. This store of calcium ions can be released into the cytosol upon cell stimulation. (Other such releasable calcium stores may exist in cells, but this is the largest)

3.2 Theme: Organelles in Eukaryotic Cells

A endoplasmic reticulum
B Golgi apparatus
C lysosome
D mitochondrion
E nucleus
F peroxisome

From the above list of organelles, select the organelle described by each of the descriptions below.

1. A site of protein synthesis
2. Contains many powerful digestive enzymes
3. Contains small circular chromosomes
4. Contains the enzyme catalase
5. Filled with chromatin
6. Made up of flattened sacks called cisternae
7. Most of the cell's ATP is made here
8. Usually found at the cell center

3.3 Theme: Transport Across Membranes

A Can move from the cytosol of one cell to the cytosol of a neighboring cell, crossing the lipid bilayer component of the plasma membrane as it does so
B Cannot cross lipid bilayers, but can move from the cytosol of one cell to the cytosol of a neighboring cell via gap junctions
C Cannot move to a neighboring cell either by crossing the lipid bilayer component of the plasma membrane or via gap junctions

Above we list three different possible constraints on the movement of a cytosolic solute. For each of the molecules below, state which of the three conditions apply.

1. An RNA molecule of Mr = 10,000
2. Inositol trisphosphate (Mr = 649)
3. K^+ (atomic weight = 39)
4. Nitric oxide (NO) (Mr = 30)

THOUGHT QUESTION

Why might it be useful for the genome to code for many connexin isoforms, some of which are incompatible with each other?

4

DNA STRUCTURE AND THE GENETIC CODE

Our genes are made of **deoxyribonucleic acid (DNA)**. This remarkable molecule contains all the information necessary to make a cell, and to pass on this information when a cell divides. This chapter describes the structure and properties of DNA molecules, the way in which our DNA is packaged into chromosomes, and how the information stored within DNA is retrieved via the genetic code.

THE STRUCTURE OF DNA

Deoxyribonucleic acid is an extremely long polymer made from units called deoxyribonucleotides, which are often simply called **nucleotides**. These nucleotides differ from those described in Chapter 2 in one respect: the sugar is deoxyribose, not ribose. Figure 4.1 shows one deoxyribonucleotide, deoxyadenosine triphosphate. Note that deoxyribose, unlike ribose (page 24), has no OH group on its 2′ carbon. Four bases are found in DNA; they are the two purines **adenine** (A) and **guanine** (G) and the two pyrimidines **cytosine** (C) and **thymine** (T) (Fig. 4.2). The combined base and sugar is known as a nucleoside to distinguish it from the phosphorylated form, which is called a nucleotide. Four different nucleotides are used to make DNA. They are 2′-deoxyadenosine-5′-triphosphate (dATP), 2′-deoxyguanosine-5′-triphosphate (dGTP), 2′-deoxycytidine-5′-triphosphate (dCTP), and 2′-deoxythymidine-5′-triphosphate (dTTP).

DNA molecules are very large. The single chromosome of the bacterium *Escherichia coli* is made up of two strands of DNA that are hydrogen-bonded together to form a single circular molecule comprising 9 million nucleotides. Humans have 46 DNA molecules in each cell, each forming one chromosome. We inherit 23 chromosomes from each parent. Each set of 23 chromosomes encodes a complete copy of our **genome** and is made up of 6×10^9 nucleotides (or 3×10^9 **base pairs**—see below).

Figure 4.3 illustrates the structure of the DNA chain. As nucleotides are added to the chain by the enzyme DNA polymerase (Chapter 5), they lose two phosphate groups. The last (the α phosphate) remains and forms a phosphodiester link between successive deoxyribose residues. The bond forms between the hydroxyl group on the 3′ carbon of the deoxyribose of one nucleotide and the α-phosphate group attached to the 5′ carbon of the next nucleotide. Adjacent nucleotides are hence joined by a 3′–5′ phosphodiester link. The linkage gives rise to the sugar-phosphate backbone of a DNA molecule. A DNA chain has polarity because its two ends are different. In the first nucleotide in the chain, the 5′ carbon of the deoxyribose is phosphorylated but otherwise free. This is called the 5′ end of the DNA chain. At the other end is a deoxyribose with a free hydroxyl group on its 3′ carbon. This is the 3′ end.

The DNA Molecule Is a Double Helix

In 1953 Rosalind Franklin used X-ray diffraction to show that DNA was a helical (i.e., twisted) polymer. James Watson and Francis Crick demonstrated, by building three-dimensional models, that the molecule is a double helix

Cell Biology: A Short Course, Third Edition. Stephen R. Bolsover, Jeremy S. Hyams, Elizabeth A. Shephard and Hugh A. White.

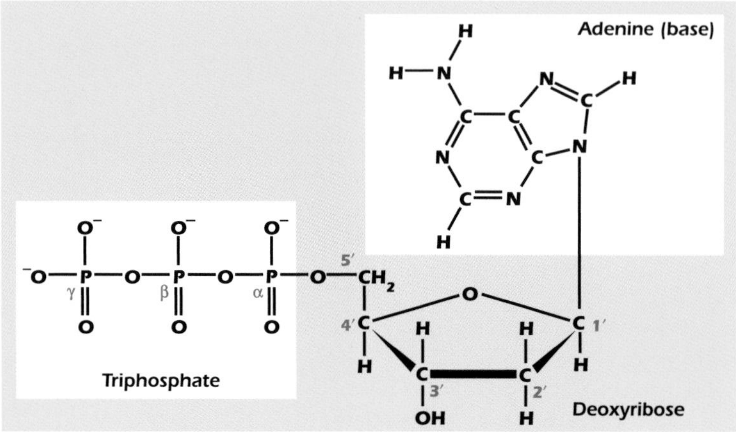

Adenine (base)

Triphosphate

Deoxyribose

Figure 4.1. 2′-deoxyadenosine-5′-triphosphate.

Purines

Adenine (A)

Guanine (G)

Pyrimidines

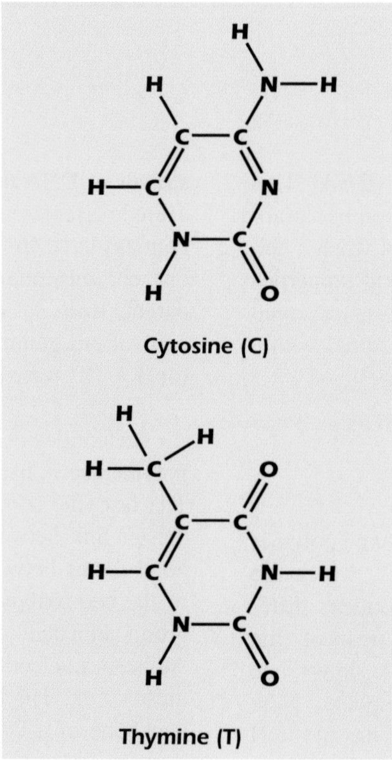

Figure 4.2. The four bases found in DNA.

(Fig. 4.4). Two hydrophilic sugar-phosphate backbones lie on the outside of the molecule, and the purines and pyrimidines lie on the inside of the molecule. There is just enough space for one purine and one pyrimidine in the center of the double helix. The Watson-Crick model showed that the purine guanine (G) would fit nicely with the pyrimidine cytosine (C), forming three hydrogen bonds. The purine adenine (A) would fit nicely with the pyrimidine thymine (T), forming two hydrogen bonds. Thus A always pairs with T, and G always pairs with C. The three hydrogen bonds formed between G and C produce a relatively strong base pair. Because only two hydrogen bonds are formed between A and T, this weaker base pair is more easily broken. The difference in strengths between a G-C and an A-T base pair is important in the initiation and termination of RNA synthesis (page 82). The two chains of DNA are said to be antiparallel because they lie in the opposite orientation with respect to one another, with the 3′-hydroxyl terminus of one strand opposite the 5′-phosphate terminus of the second strand. The sugar-phosphate backbones do not completely conceal the bases inside. There are two grooves along the surface of the DNA molecule. One is wide and deep—the major groove—and the other is narrow and shallow—the minor groove (Fig. 4.4). Proteins can use the grooves to gain access to the bases (page 149).

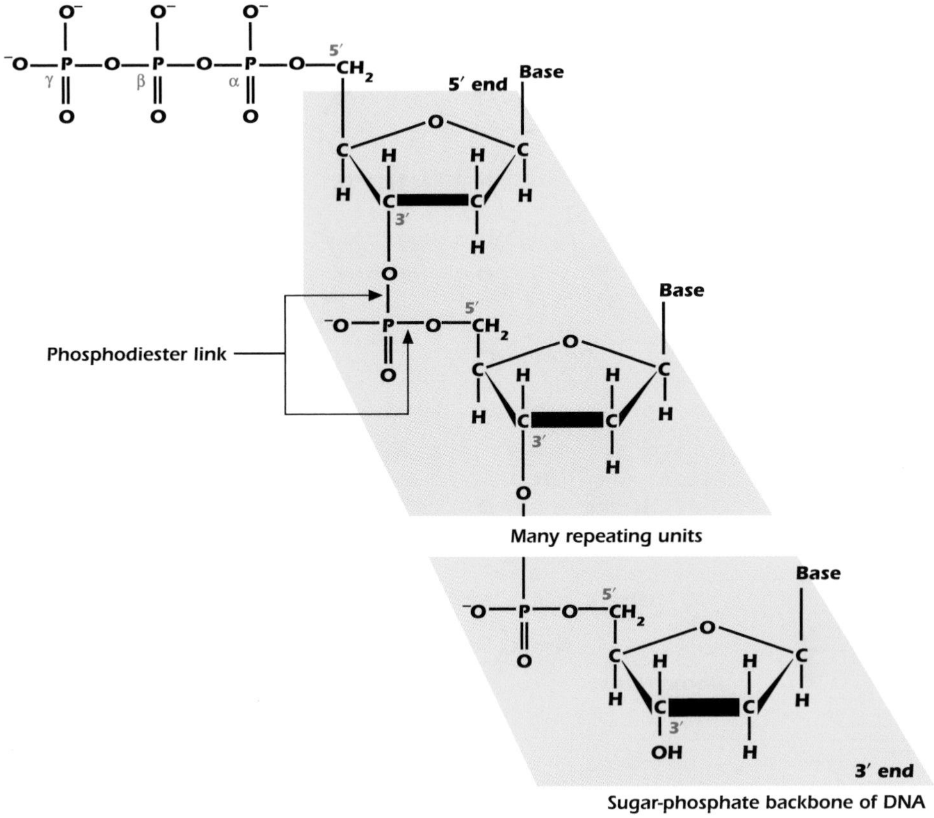

Figure 4.3. The phosphodiester link and the sugar-phosphate backbone of DNA.

Example 4.1 Erwin Chargaff's Puzzling Data

In a key discovery of the 1950s, Erwin Chargaff analyzed the purine and pyrimidine content of DNA isolated from many different organisms and found that the amounts of A and T were always the same, as were the amounts of G and C. Such an identity was inexplicable at the time, but helped James Watson and Francis Crick build their double-helix model in which every A on one strand of the DNA helix has a matching T on the other strand, and every G on one strand has a matching C on the other.

The Two DNA Chains Are Complementary

A consequence of the base pairs formed between the two strands of DNA is that if the base sequence of one strand is known, then that of its partner can be inferred. A G in one strand will always be paired with a C in the other. Similarly an A will always pair with a T. The two strands are therefore said to be **complementary**.

Different Forms of DNA

The original Watson-Crick model of DNA is now called the B-form. In this form, the two strands of DNA form a right-handed helix. If viewed from either end, it turns in a clockwise direction. B-DNA is the predominant form in which DNA is found. Our genome, however, also contains several variations of the B-form double helix. One of these, Z-DNA, so called because its backbone has a zigzag shape, forms a left-handed helix and occurs when the DNA sequence is made of alternating purines and pyrimidines. Thus the structure adopted by DNA is a function of its base sequence.

DNA AS THE GENETIC MATERIAL

Deoxyribonucleic acid carries the genetic information encoded in the sequence of the four bases—adenine, guanine, cytosine, and thymine. The information in DNA is transferred to its daughter molecules through **replication** (the duplication of DNA molecules) and subsequent cell division. DNA directs the synthesis of proteins through

Cytosine-guanine base pair (C≡G)

Hydrogen bond
Deoxyribose
Deoxyribose
Cytosine (C)
Guanine (G)

Thymine-adenine base pair (T═A)

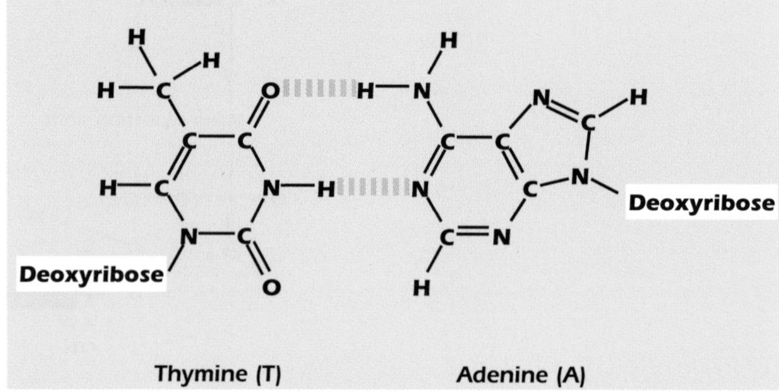

Schematic representation of the double-helical structure of DNA

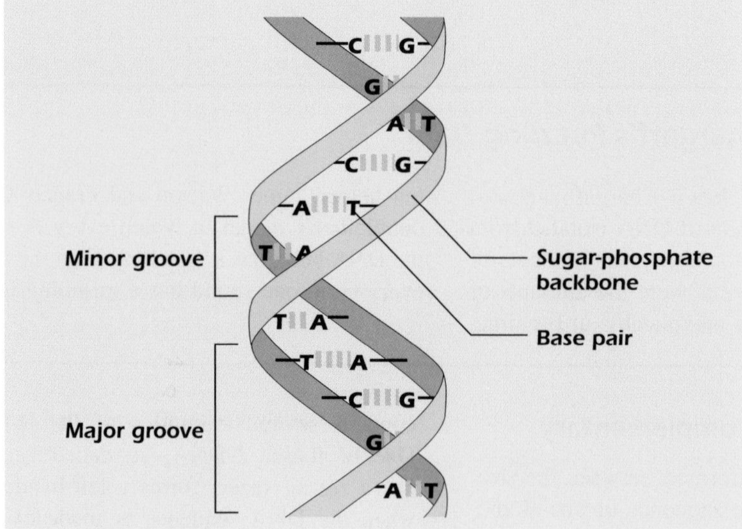

Figure 4.4. The DNA double helix is held together by hydrogen bonds.

the intermediary molecule messenger RNA (mRNA). The DNA code is transferred to mRNA by a process known as transcription (Chapter 6). The mRNA code is then translated into a sequence of amino acids during protein synthesis (Chapter 8). This is the central dogma of molecular biology: DNA makes RNA makes protein.

Retroviruses such as human immunodeficiency virus, the cause of AIDS, are an exception to this rule. As their name suggests, they reverse the normal order of data transfer. Inside the virus coat is a molecule of RNA plus an enzyme that can make DNA from an RNA template by the process known as **reverse transcription**.

We do not yet know the exact number of genes that encode messenger RNA in the human genome. The current estimate is 20,500. Table 4.1 compares the number of predicted messenger RNA genes in the genomes of different organisms. In each organism, there are also a small number of genes (about 100 in humans) that code for ribosomal RNAs and transfer RNAs. The roles these three types of RNA play in protein synthesis is described in Chapter 8.

Medical Relevance 4.1 Antiviral Drugs

Because viruses use the host cell's synthetic machinery, it is hard to find drugs that will inhibit their replication without damaging the host cell. However, the first act of RNA viruses is to make DNA from their RNA genome. Often the only enzyme contained within the viral envelope is the one that does this, called reverse transcriptase. Azidothymidine (AZT), the most widely used anti-AIDS drug, inhibits this enzyme.

IN DEPTH 4.1 DNA—A GORDIAN KNOT

At the start of his career Alexander the Great was shown the Gordian Knot, a tangled ball of knotted rope, and told that whoever untied the knot would conquer Asia. Alexander cut through the knot with his sword. A similar problem occurs in the nucleus, where the 46 chromosomes form 2 m of tangled, knotted DNA. How does the DNA ever untangle at mitosis? The cell adopts Alexander's solution—it cuts the rope. At any place where the DNA helix is under strain, for instance, where two chromosomes press against each other, an enzyme called **topoisomerase II** cuts one chromosome double helix so that the other can pass through the gap. Then, surpassing Alexander, the enzyme rejoins the cut ends. Topoisomerases are active all the time in the nucleus, relieving any strain that develops in the tangled mass of DNA.

Concerns that a terrorist organization might release large amounts of anthrax spores have caused several governments to stockpile large amounts of the antibiotic Cipro. This works by inhibiting the prokaryotic form of topoisomerase II (sometimes called gyrase), hence preventing cell replication.

PACKAGING OF DNA MOLECULES INTO CHROMOSOMES

Eukaryotic Chromosomes and Chromatin Structure

A human cell contains 46 chromosomes (23 pairs), each of which is a single DNA molecule bundled up with various proteins. On average, each human chromosome contains about 1.3×10^8 base pairs (bp) of DNA. If the DNA in a human chromosome were stretched as far as it would go without breaking it would be about 5 cm long, so the 46 chromosomes in all represent about 2 m of DNA. The nucleus in which this DNA must be contained has a diameter of only about 10 μm, so large amounts of DNA must be packaged into a small space. This represents a formidable problem that is dealt with by binding the DNA to proteins to form chromatin. As shown in Figure 4.5, the DNA double helix is packaged at both small and larger scales. In the first stage, shown on the right of the figure, the DNA double helix with a diameter of 2 nm is bound to proteins known as **histones**. Histones are positively charged because they contain large amounts of the amino acids arginine and lysine (page 138) and bind tightly to the negatively charged phosphates on DNA. A 146 bp length of DNA is wound around a protein complex composed of two molecules each of four different histones—H2A, H2B, H3, and H4—to form a **nucleosome**. Because each nucleosome is separated from its neighbor by about 50 bp of linker DNA, this unfolded chromatin state looks like beads on a string when viewed in an electron microscope. Nucleosomes undergo further packaging. A fifth type of histone, H1, binds to the linker

TABLE 4.1. Numbers of Predicted Genes in Various Organisms

Organism	Number of Predicted Genes
Bacterium: *Escherichia coli*	**4,377**
Yeast: *Saccharomyces cerevisiae*	**5,770**
Fruit fly: *Drosophila melanogaster*	**13,379**
Worm: *Caenorhabdites elegans*	**19,427**
Plant: *Arabidopsis thaliana*	**~28,000**
Human: *Homo sapiens*	**~20,500**

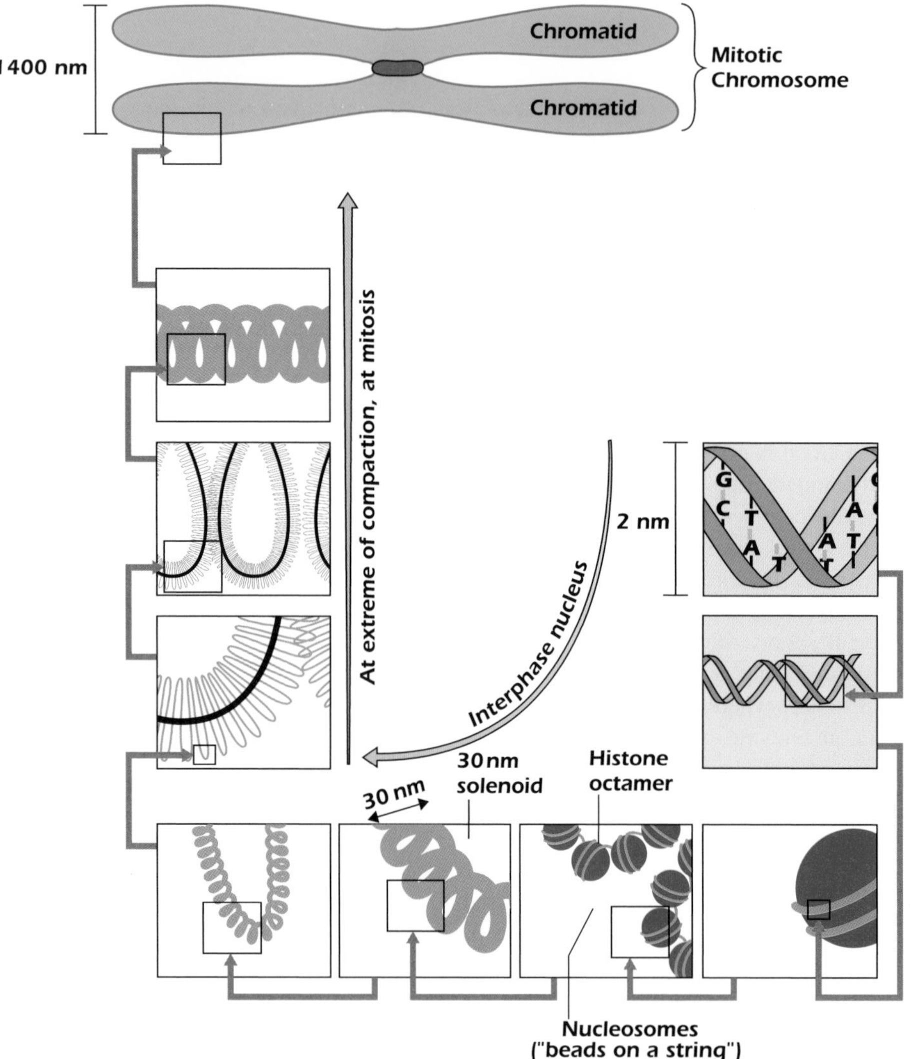

Figure 4.5. How DNA is packaged into chromosomes.

DNA and pulls the nucleosomes together helping to further coil the DNA into chromatin fibers 30 nm in diameter, which are referred to as 30-nm solenoids. The fibers then form loops with the help of a class of proteins known as nonhistones, and this further condenses the DNA (panels on left-hand side of Fig. 4.5) into a higher order set of coils in a process of **supercoiling**.

In a normal interphase cell about 10% of the chromatin is highly compacted and visible under the light microscope (page 48). This form of chromatin is called heterochromatin and is the portion of the genome where no RNA synthesis is occurring. Euchromatin, that is, chromatin that is being transcribed into RNA, is wholly or partly unpacked from the histones to allow it to be read and has a less dense appearance in the microscope. Chromatin is in its most compacted form when the cell is preparing for mitosis, as shown at the top left of Figure 4.5. The chromatin folds and condenses further to form the 1400 nm-wide chromosomes we see under the light microscope. Because the cell is to divide, the DNA has been replicated, so that each chromosome is now formed by two chromatids, each one a DNA double helix. This means the progeny cell, produced by division of the **progenitor** cell, will receive a full set of 46 chromosomes. Figure 4.6 is a photograph of human chromosomes as they appear at cell division.

Prokaryotic Chromosomes

The chromosome of the bacterium *E. coli* is a single circular DNA molecule of about 4.5×10^6 base pairs. It has a circumference of 1 mm, yet must fit into the 1 μm cell, so like eukaryotic chromosomes it is coiled, supercoiled, and packaged with basic proteins that are similar to eukaryotic histones. However, an ordered nucleosome structure similar to the "beads on a string" seen in eukaryotic cells is not observed in prokaryotes. Prokaryotes do not have nuclear envelopes so the condensed chromosome, together with its associated proteins, lies free in the cytoplasm, forming a mass that is called the **nucleoid** to emphasize its functional equivalence to the eukaryotic nucleus.

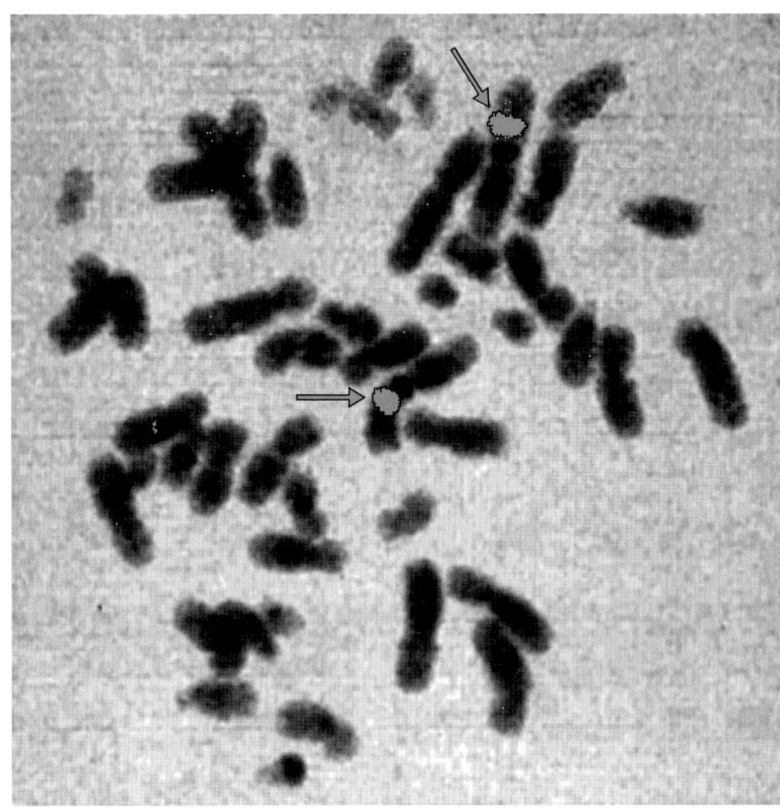

Figure 4.6. A spread of human chromosomes (at metaphase-see page 298). The orange signal reveals the gene called *FMO3*, which when mutated causes trimethylaminuria (fish odor syndrome). There are two copies of the gene, one inherited from each parent, indicated by the arrows. The *FMO3* gene is located on the long arm of chromosome 1, the longest human chromosome.

Plasmids

Plasmids are small circular minichromosomes found in bacteria and some eukaryotes. They are several thousand base pairs long and are probably tightly coiled and supercoiled inside the cell. Plasmids often code for proteins that confer resistance to a particular antibiotic. In Chapter 7 we describe how plasmids are used by scientists to artificially introduce foreign DNA molecules into bacterial cells.

Viruses

Viruses (page 9) rely on the host cell to make more virus. Once viruses have entered cells, the cells' machinery is used to copy the viral genome. Depending on the virus type, the genome may be single- or double-stranded DNA, or even RNA. A viral genome is packaged within a protective protein coat. Viruses that infect bacteria are called bacteriophages. One of these, lambda, has a fixed-size DNA molecule of 4.5×10^4 base pairs. In contrast, the bacteriophage M13 can change its chromosome size, its protein coat expanding in parallel to accommodate the chromosome.

THE GENETIC CODE

Proteins are linear polymers of individual amino acid building blocks (page 32). The sequence of bases along the DNA strand determines the sequence of the amino acids in proteins. There are 20 different amino acids in proteins but only 4 different bases in DNA (A, T, C, and G). Each amino acid is specified by a **codon**, a group of three bases. Because there are 4 bases in DNA, a three-letter code gives 64 ($4 \times 4 \times 4$) possible codons. These 64 codons form the **genetic code**—the set of instructions that tells a cell the order in which amino acids are to be joined to form a protein (Fig. 4.7). Despite the fact that the sequence of codons on DNA determines the sequence of amino acids in proteins, the DNA helix does not itself play a role in protein synthesis. The **translation** of the sequence from codons into amino acids occurs through the intervention of members of a third class of molecule—messenger RNA (mRNA). Messenger RNA acts as a template, guiding the assembly of amino acids into a polypeptide chain. Messenger RNA uses the same code as the one used in DNA with one difference: in mRNA the base uracil (U) (Fig. 2.13 on page 29) is used in place of thymine (T). When we write the genetic code we usually use the RNA format, that is, we use U instead of T.

The code is read in sequential groups of three, codon by codon. Adjacent codons do not overlap, and each triplet of bases specifies one particular amino acid. This discovery was made by Sydney Brenner, Francis Crick, and their colleagues by studying the effect of various mutations (changes in the DNA sequence) on the bacteriophage T4, which infects the common bacterium *E. coli*. If a mutation

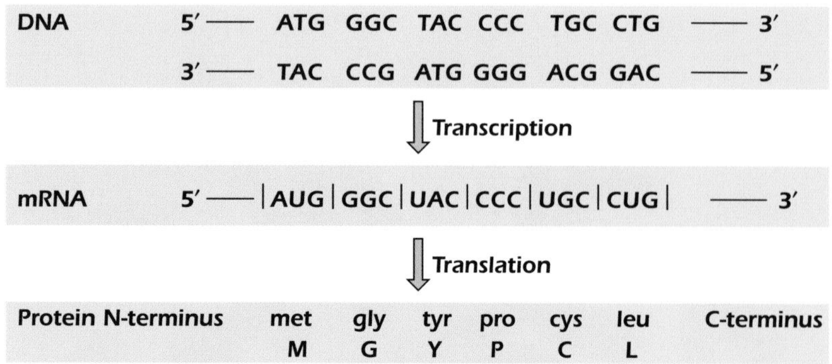

Figure 4.7. DNA makes RNA makes protein: the central dogma of molecular biology.

caused either one or two nucleotides to be added or deleted from one end of the T4 DNA, then a defective polypeptide was produced, with a completely different sequence of amino acids. However, if three bases were added or deleted, then the protein made often retained its normal function. These proteins were found to be identical to the original protein, except for the addition or loss of one amino acid.

The identification of the triplets encoding each amino acid began in 1961. This was made possible by using a cell-free protein synthesis system prepared by breaking open *E. coli* cells. Synthetic RNA polymers, of known sequence, were added to the cell-free system together with the 20 amino acids. When the RNA template contained only uridine residues (poly-U) the polypeptide produced contained only phenylalanine. The codon specifying this amino acid must therefore be UUU. A poly-A template produced a polypeptide of lysine, and poly-C one of proline: AAA and CCC must therefore specify lysine and proline, respectively. Synthetic RNA polymers containing all possible combinations of the bases A, C, G, and U were added to the cell-free system to determine the codons for the other amino acids. A template made of the repeating unit CU gave a polypeptide with the alternating sequence leucine–serine. Because the first amino acid in the chain was found to be leucine, its codon must be CUC and that for serine UCU. Although much of the genetic code was read in this way, the amino acids defined by some codons were particularly hard to determine. Only when specific **transfer RNA** molecules (page 123) were used was it possible to demonstrate that GUU codes for valine. The genetic code was finally solved by the combined efforts of several research teams. The leaders of two of these, Marshall Nirenberg and Har Gobind Khorana, received the Nobel prize in 1968 for their part in cracking the code.

Amino Acid Names Are Abbreviated

To save time we usually write an amino acid as either a three-letter abbreviation (for example, glycine is written as gly and leucine as leu), or as a one-letter code (for example, glycine is G and leucine is L). Figure 4.8 shows the full name, and the three- and one-letter abbreviations, used for each of the 20 amino acids found in proteins.

The Code Is Degenerate But Unambiguous

To introduce the terms *degenerate* and *ambiguous*, consider the English language. English shows considerable degeneracy, meaning that the same concept can be indicated using a number of different words—think, for example, of *lockup, cell, pen, pound, brig*, and *dungeon*. English also shows ambiguity, so that it is only by context that one can tell whether *cell* means *a lockup* or *the basic unit of life*. Like the English language the genetic code shows degeneracy but, unlike language, the code is unambiguous.

Medical Relevance 4.2 A Premature STOP

We described earlier (Medical Relevance 3.1 on page 50) how the failure to produce one of a number of lysosomal enzymes causes severe disease. A majority of cases of one such disease, Hurler (or Scheie) syndrome, are caused by one of two mutations in the gene coding for the lysosomal enzyme α-L-iduronidase. Each is a single base substitution: one has UAG in place of the CAG that normally codes for a glutamine, the other has a UAG in place of the UGG that normally codes for tryptophan (these are therefore nonsense mutations—page 63). In each case protein synthesis stops at the UAG and the resulting abbreviated polypeptide has no catalytic activity. Although relatively uncommon in the general population (1 in every 26,000 births) the syndrome is as common as one in every 400 births in some populations.

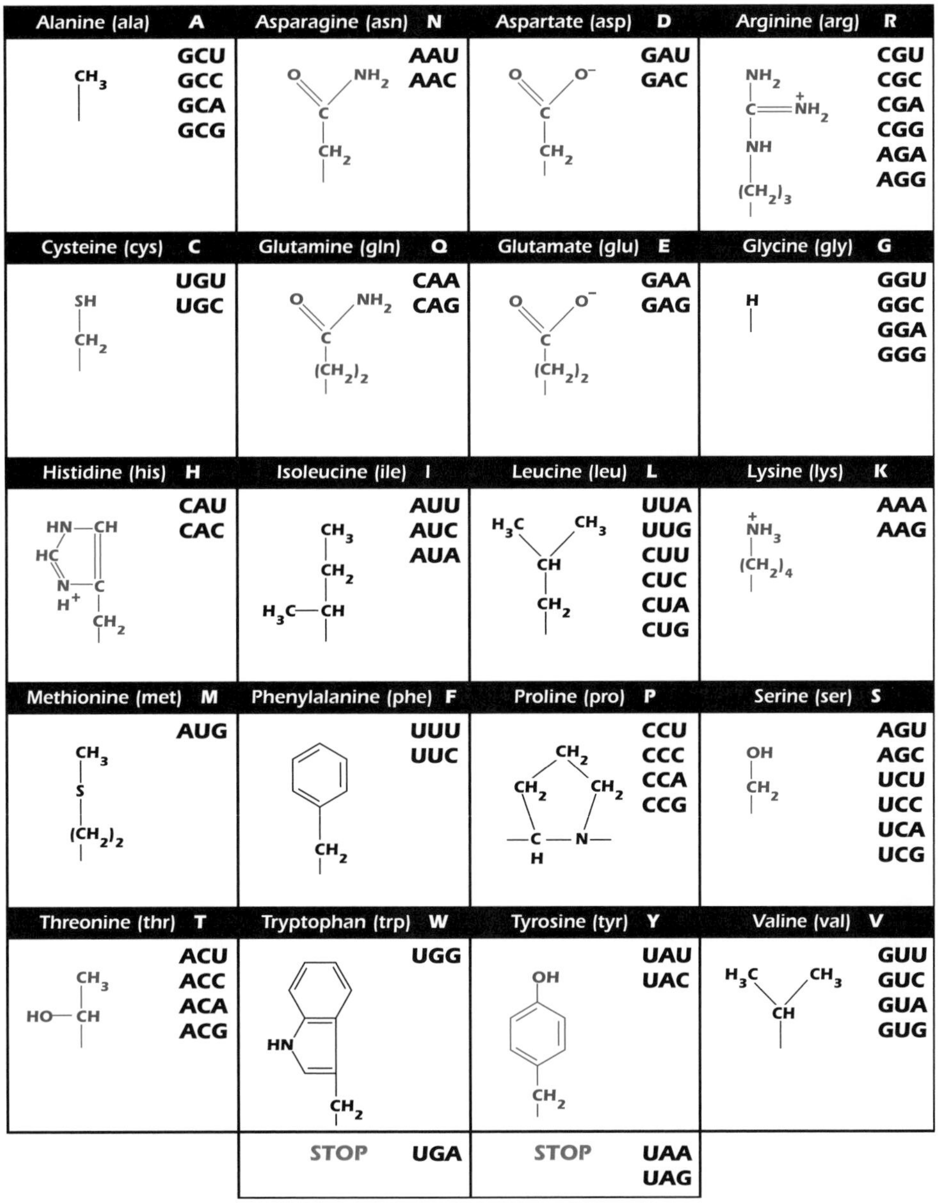

Alanine (ala) A	Asparagine (asn) N	Aspartate (asp) D	Arginine (arg) R
GCU GCC GCA GCG	AAU AAC	GAU GAC	CGU CGC CGA CGG AGA AGG
Cysteine (cys) C	**Glutamine (gln) Q**	**Glutamate (glu) E**	**Glycine (gly) G**
UGU UGC	CAA CAG	GAA GAG	GGU GGC GGA GGG
Histidine (his) H	**Isoleucine (ile) I**	**Leucine (leu) L**	**Lysine (lys) K**
CAU CAC	AUU AUC AUA	UUA UUG CUU CUC CUA CUG	AAA AAG
Methionine (met) M	**Phenylalanine (phe) F**	**Proline (pro) P**	**Serine (ser) S**
AUG	UUU UUC	CCU CCC CCA CCG	AGU AGC UCU UCC UCA UCG
Threonine (thr) T	**Tryptophan (trp) W**	**Tyrosine (tyr) Y**	**Valine (val) V**
ACU ACC ACA ACG	UGG	UAU UAC	GUU GUC GUA GUG
	STOP UGA	STOP UAA UAG	

Figure 4.8. The genetic code. Amino acid side chains are shown in alphabetical order together with the three- and one-letter amino acid abbreviations. Hydrophilic side chains are shown in green, hydrophobic side chains in black; the significance of this distinction is discussed in Chapter 9. To the right of each amino acid we show the corresponding mRNA codons.

The 64 codons of the genetic code are shown in Figure 4.8 together with the side chains of the amino acids for which each codes. Amino acids with hydrophilic side chains are shown in green while those with hydrophobic side chains are in black. The importance of this distinction will be discussed in Chapter 9. Sixty-one codons specify an amino acid, and the remaining three act as **stop signals** for protein synthesis. Methionine and tryptophan are the only amino acids coded for by single codons. The other 18 amino acids are encoded by either 2, 3, 4, or 6 codons and so the code is degenerate. No triplet codes for more than one amino acid and so the code is unambiguous. Notice that when two or more codons specify the same amino acid, they usually only differ in the third base of the triplet. Thus mutations can arise in this position of the codon without altering amino acid sequences. Perhaps degeneracy evolved in the triplet system to avoid a situation in which 20 codons each meant one amino acid, and 44 specified none. If this were the case, then most mutations would stop protein synthesis dead.

Medical Relevance 4.3 Trimethylaminuria

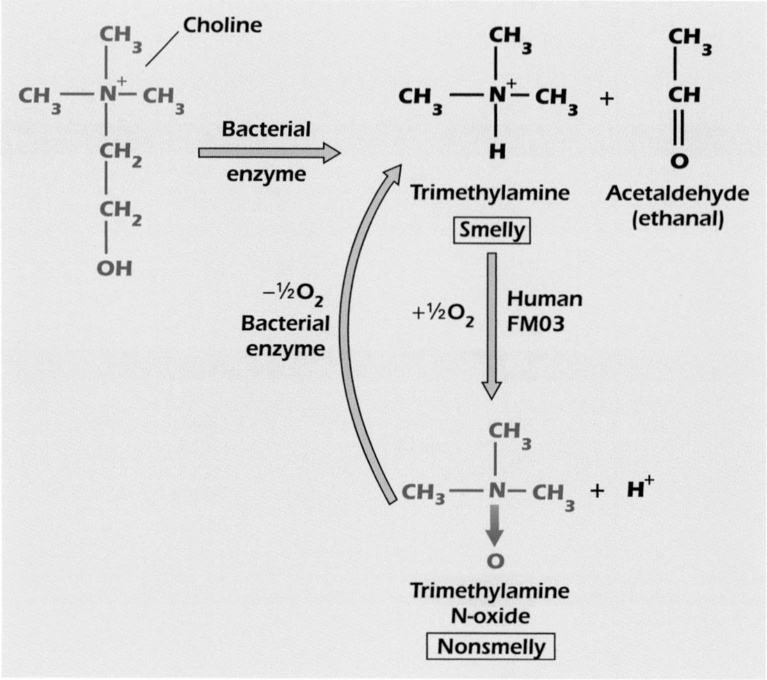

Trimethylaminuria is an inherited disorder characterized by an unpleasant body odor similar to rotting fish. Mutations in the gene *FMO3*, located on human chromosome 1 (Figure 4.6), produce a defective version of a protein called flavin-containing monooxygenase 3 (FMO3). One of the functions of this protein, which is found in the liver, is to convert the chemical trimethylamine (very smelly) to trimethylamine *N*-oxide (nonsmelly). FMO3 carries out this reaction by adding an oxygen onto the nitrogen of trimethylamine. Defective FMO3 protein cannot carry out this reaction. Affected individuals therefore excrete large amounts of trimethylamine in their breath, sweat, and urine. Trimethylamine is the chemical that gives rotting fish its distinctive and unpleasant smell, and this gives rise to the alternative name for trimethylaminuria—fish odor syndrome.

Trimethylamine is derived from our diet in two ways. Fish contain lots of trimethylamine *N*-oxide (page 158). The bacteria that live in our intestines remove the oxygen from the nitrogen of the *N*-oxide to produce trimethylamine. However, one cannot avoid trimethylamine simply by avoiding fish. The phospholipids in the cell membranes of food include phosphatidylcholine, and this is hydrolyzed by our digestive enzymes to release free choline (page 35). Gut bacteria then break down the choline further, producing trimethylamine together with acetaldehyde (also called ethanal).

Trimethylaminuria is a difficult disorder to live with. Because of their body odor, sufferers often experience rejection and social isolation. A diet very low in choline and other chemicals that contain a trimethylamine group can help to reduce the problem. Analysis of the faulty genes in different families has revealed different mutations in the *FMO3* gene. For example, some people have a missense mutation in which a CCC that codes for a proline at amino acid 153 in the protein is mutated to CTC (leu), a change that turns out to be critical for the operation of the protein. Other families carry a nonsense mutation in which the GAA (glutamate) at amino acid 305 in the normal protein is mutated to TAA, a stop codon. This results in a shortened protein with no enzyme activity. In yet other families an insertion or deletion of a base causes frameshift mutations. Because over thirty different mutations can give rise to trimethylaminuria, the development of a general DNA diagnostic test for carriers of the faulty gene (who also have one good copy and are therefore carriers who do not themselves have the disorder; page 302) is not possible.

In contrast, a disorder such as sickle cell anemia is caused, in all individuals, by the same missense mutation: a GAG (glu) becomes GTG (val) in the hemoglobin β chain (page 64). One test will diagnose the defective gene and prospective parents can be advised and counseled.

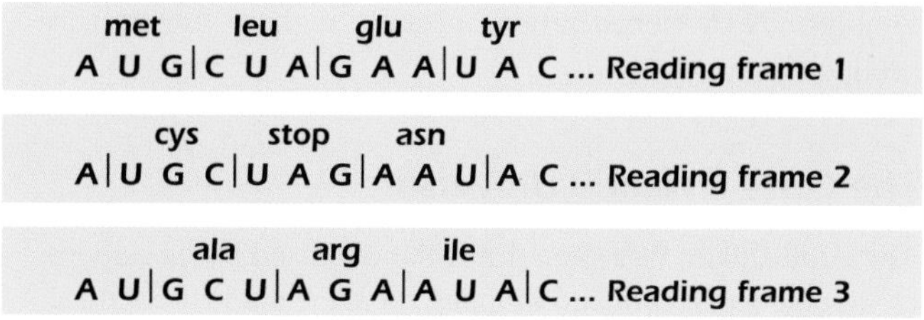

met	leu	glu	tyr	
A U G	C U A	G A A	U A C ...	Reading frame 1

	cys	stop	asn		
A	U G C	U A G	A A U	A C ...	Reading frame 2

	ala	arg	ile		
A U	G C U	A G A	A U A	C ...	Reading frame 3

Figure 4.9. Reading frames. The genetic code is read in blocks of three.

Frameshift mutation

met	gln	trp	val	glu	
AUG	CAA	(U)GG	GUC	GAG A.....	Normal

U deleted

met	gln	gly	ser	arg	
AUG	CAA	GGG	UCG	AGA......	Mutant

Nonsense mutation

met	gln	trp	val	glu	
AUG	CAA	UG(G)	GUC	GAG......	Normal

changed to **A**

met	gln	STOP	val	glu	
AUG	CAA	UGA	GUC	GAG......	Mutant

Missense mutation

met	gln	trp	val	glu	
AUG	CA(A)	UGG	GUC	GAG......	Normal

changed to **U**

met	his	trp	val	glu	
AUG	CAU	UGG	GUC	GAG......	Mutant

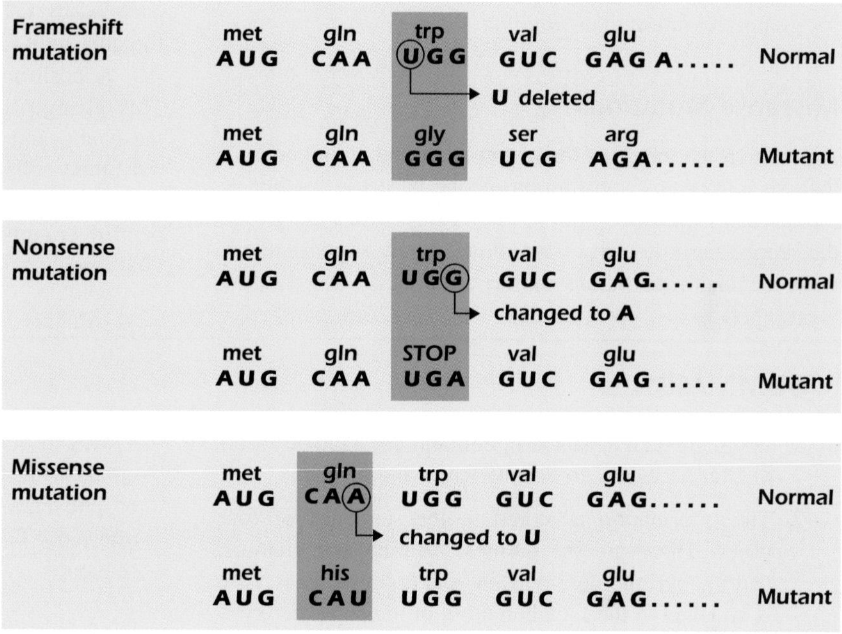

Figure 4.10. Mutations that alter the sequence of bases.

Start and Stop Codons and the Reading Frame

The order of the codons in DNA and the amino acid sequence of a protein are colinear. The **start signal** for protein synthesis is the codon AUG specifying the incorporation of methionine. Because the genetic code is read in blocks of three, there are three potential **reading frames** in any mRNA. Figure 4.9 shows that only one of these results in the synthesis of the correct protein. When we look at a sequence of bases, it is not obvious which of the reading frames should be used to code for protein. As we shall see later (page 127), the ribosome scans along the mRNA until it encounters an AUG. This both defines the first amino acid of the protein and the reading frame used from that point on. A mutation that inserts or deletes a nucleotide will change the normal reading frame and is called a **frameshift mutation** (Fig. 4.10).

The codons UAA, UAG, and UGA are stop signals for protein synthesis. A base change that causes an amino acid codon to become a **stop codon** is known as a **nonsense mutation** (Fig. 4.10). If, for example, the codon for tryptophan UGG changes to UGA, then a premature stop signal will have been introduced into the messenger RNA

Answer to Thought Question: The molecule they created, when translated into the single letter amino acid code, spelled out a Christmas poem by Viktor Rydberg. The part of the poem we show spells "Stars send a sparkling light." For more on this project see www.dna20.com/files/PDF/Christmas2005.pdf.

template. A shortened protein, usually without function, is produced.

The Code Is Nearly Universal

The code shown in Figure 4.8 is the one used by organisms as diverse as *E. coli* and humans for their nuclear-encoded proteins. It was originally thought that the code would be universal. However, several mitochondrial genes use UGA to mean tryptophan rather than *stop*. The nuclear code for some unicellular eukaryotes uses UAA and UAG to code for glutamine rather than *stop*.

Missense Mutations

A mutation that changes the codon from one amino acid to that for another by substitution of one base for another is a **missense mutation** (Fig. 4.10). As shown in Figure 4.8, the second base of each codon shows the most consistency with the chemical nature of the amino acid it encodes. Amino acids with charged, hydrophilic side chains usually have A or G—a purine—in the second position. Those with hydrophobic side chains usually have C or U—a pyrimidine—in that position. This has implications for mutations of the second base. Substitution of a purine for a pyrimidine is very likely to change the chemical nature of the amino acid side chain significantly and can therefore seriously affect the protein. Sickle cell anemia is an example of such a mutation. At position 6 in the β-globin chain of hemoglobin the mutation in DNA changes a glutamate residue encoded by GAG to a valine residue encoded by GTG (GUG in RNA). The shorthand notation for this mutation is E6V, meaning that the glutamate (E) at position 6 of the protein becomes a valine (V). This change in amino acid alters the overall charge of the chain, and the hemoglobin tends to precipitate in the red blood cells of those affected. The cells adopt a sickle shape and therefore tend to block blood vessels, causing painful cramplike symptoms and progressive damage to vital organs.

SUMMARY

1. DNA, the cell's database, contains the genetic information necessary to encode RNA and protein.
2. The information is stored in the sequence of four bases. These are the purines, adenine and guanine, and the pyrimidines, cytosine and thymine. Each base is attached to the 1′-carbon atom of the sugar deoxyribose. A phosphate group is attached to the 5′-carbon atom of the sugar. The base + sugar + phosphate is called a nucleotide.
3. The enzyme DNA polymerase joins nucleotides together by forming a phosphodiester link between the hydroxyl group on the 3′-carbon of deoxyribose of one nucleotide and the 5′-phosphate group of another. This gives rise to the sugar–phosphate backbone structure of DNA.
4. The two strands of DNA are held together in an antiparallel double-helical structure because guanine hydrogen bonds with cytosine and adenine hydrogen bonds with thymine. This means that if the sequence of one strand is known, that of the other can be inferred. The two strands are complementary in sequence.
5. DNA binds to histone and nonhistone proteins to form chromatin. DNA is wrapped around histones to form a nucleosome structure. This is then folded again and again. This packaging compresses the DNA molecule to a size that fits into the cell.
6. The genetic code specifies the sequence of amino acids in a polypeptide. The code is transferred from DNA to mRNA and is read in groups of three bases (a codon) during protein synthesis. There are 64 codons; 61 specify an amino acid and 3 are the stop signals for protein synthesis.

FURTHER READING

Annunziato, A.T. (2008) DNA packaging:Nucleosomes and chromatin. *Nature Education* **1**(1). www.nature.com/scitable/topicpage/DNA-Packaging-Nucleosomes-and-Chromatin-310.

DiGuilo, M. (1997) The origin of the genetic code. *Trends Biochem. Sci*. **22**, 49–50.

Dolphin, C. T., Janmohamed, A., Smith, R. L., Shephard, E. A., and Phillips, I. R. (1997) Missense mutation in flavin-containing monooxygenase 3 gene, FMO3, underlies fish-odour syndrome. *Nature Genetics* **17**, 491–494.

Maddox, B. (2002) *Rosalind Franklin: The Dark Lady of DNA*. Harper Collins, New York.

Roca, J. (1995) The mechanisms of DNA topoisomerases. *Trends Biochem. Sci*. **20**, 156–160.

Watson, J. D., and Crick, F. H. C. (1953) A structure for deoxyribose nucleic acid. *Nature* **171**, 737.

REVIEW QUESTIONS

4.1 Theme: Mutations

A frameshift
B missense
C nonsense
D none of the above

Consider the mRNA strand 5′ACU AUC UGU AUU AUG UUA CAC CCA3′ coding for the amino acid sequence TICIMLHP. For each of the errors listed below, choose the appropriate description from the list above. Refer to Figure 4.8 on page 61 while answering this question.

1. A change of a U to a A in the 6th codon of the sequence, generating the sequence 5′ACUAUCUGUAUUAUGUAACACCCA3′
2. A change of a U to a C in the 6th codon of the sequence, generating the sequence 5′ACUAUCUGUAUUAUGCUACACCCA3′
3. A change of a U to a G in the 2nd codon of the sequence, generating the sequence 5′ACUAGCUGUAUUAUGUUACACCCA3′
4. Deletion of a U in the 3rd codon, generating the sequence 5′ACUAUCUGAUUAUGUUACACCCA3′
5. Deletion of an A in the 4th codon, generating the sequence 5′ACUAUCUGUUUAUGUUACACCCA3′

4.2 Theme: Bases and Amino Acids

A adenine
B alanine
C arginine
D aspartate
E cytosine
F glutamate
G glycine
H guanine
I thymine
J uracil
K valine

From the above list of compounds, select the one that fits each of the statements or questions below.

1. A nitrogen-rich base that is not a component of DNA
2. A positively charged amino acid that is found in large amounts in chromatin, where it neutralizes the negative charge on the phosphodiester links of DNA
3. A protein is described as having the mutation G5E. Which amino acid is present in this protein in place of the amino acid present in the normal protein?
4. The base that pairs with guanine in double stranded DNA
5. The base that pairs with thymine in double stranded DNA

4.3 Theme: Structures Associated with DNA

A 30 nm solenoid
B codon
C euchromatin
D gene
E heterochromatin
F nucleoid
G nucleosome

From the above list of structures, select the one that fits each of the descriptions below.

1. A highly compacted, darkly staining substance comprising DNA and protein found at the nuclear periphery
2. A mass of DNA and associated proteins lying free in the cytoplasm
3. A structure formed when a 146 base pair length of DNA winds around a complex of histone proteins
4. The form adopted by those parts of chromosomes that are being transcribed into RNA

THOUGHT QUESTION

As a Christmas joke, a company called DNA2.0 of Menlo Park, California, created an artificial DNA sequence part of which read:

5′TCTACTGCGCGCTCTAGCGAAAAC
GACGCATCTCCGGCGCGTAAACTG
ATCAACGGCCTGATCGGTCATACG3′

Why is this amusing?

5

DNA AS A DATA STORAGE MEDIUM

The genetic material DNA must be faithfully replicated every time a cell divides to ensure that the information encoded in it is passed unaltered to the progeny cells. DNA molecules have to last a long time compared to RNA and protein. The sugar-phosphate backbone of DNA is a very stable structure because there are no free hydroxyl groups on the sugar—they are all used up in bonds, either to the base or to phosphate. The bases themselves are protected from chemical attack because they are hidden within the DNA double helix. Nevertheless, chemical changes—**mutations**—do occur in the DNA molecule, and cells have had to evolve mechanisms to ensure that mutation is kept to a minimum. Repair systems are essential for both cell survival and to ensure that the correct DNA sequence is passed on to progeny cells. This chapter describes how new DNA molecules are made during chromosome duplication and how the cell acts to correct base changes in DNA.

DNA REPLICATION

During replication the two strands of the double helix unwind. Each then acts as a template for the synthesis of a new strand. This process generates two double-stranded daughter DNA molecules, each of which is identical to the parent molecule. The base sequences of the new strands are complementary in sequence to the template strands upon which they were built. This means that G, A, C, and T in the old strand cause C, T, G, and A, respectively, to be placed in the new strand.

The DNA Replication Fork

Replication of a new DNA strand starts at specific sequences known as **origins of replication**. The small circular chromosome of *Escherichia coli* has only one of these, whereas eukaryotic chromosomes, which are much larger, have many. At each origin of replication, the parental strands of DNA untwist to give rise to a structure known as the **replication fork** (Fig. 5.1). This unwinding permits each parental strand to act as a template for the synthesis of a new strand. The structure of the double helix and the nature of DNA replication pose a mechanical problem. How do the two strands unwind and how do they stay unwound so that each can act as a template for a new strand?

PROTEINS OPEN UP THE DNA DOUBLE HELIX DURING REPLICATION

The DNA molecule must be opened up before replication can proceed. The helix is a very stable structure, and in a test tube the two strands separate only when the temperature reaches about 90°C. In the cell the combined actions of several proteins help to separate the two strands. Much of our knowledge of replication comes from studying *E. coli*, but similar systems operate in all organisms, prokaryote and eukaryote. The proteins *E. coli* uses to open up the double helix during replication include DnaA, DnaB, DnaC, and single-strand binding proteins.

Cell Biology: A Short Course, Third Edition. Stephen R. Bolsover, Jeremy S. Hyams, Elizabeth A. Shephard and Hugh A. White.

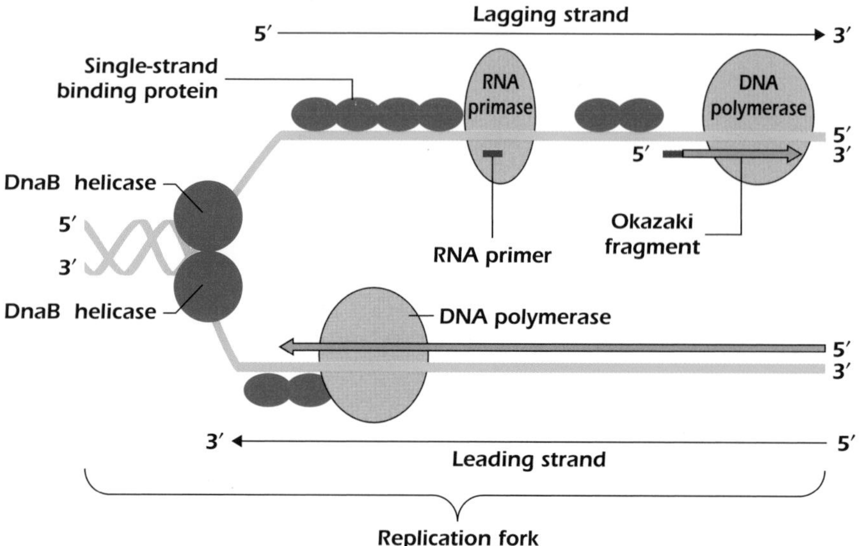

Figure 5.1. DNA replication. The helicases, and the replication fork, are moving to the left.

DnaA Protein

Several copies of the protein DnaA, which is activated by a molecule of ATP, bind to four sequences of nine base pairs within the *E. coli* origin of replication (*ori C*). This causes the two strands to begin to separate (or "melt") because the hydrogen bonds in DNA are broken near to where the DnaA protein binds. The DNA is now in the **open complex** formation and has been prepared for the next stage in replication, which is to open up the helix even further.

DnaB and DnaC Proteins

DnaB is a **helicase**. It moves along a DNA strand, breaking hydrogen bonds, and in the process unwinds the helix (Fig. 5.1). Two molecules of DnaB are needed, one for each strand of DNA. One DnaB attaches to one of the template strands and moves in the 5′ to 3′ direction; the second DnaB attaches to the other strand and moves in the 3′ to 5′ direction. The unwinding of the DNA double helix by DnaB is an ATP-dependent process. DnaB is escorted to the DNA strands by another protein, DnaC. However, having delivered DnaB to its destination, DnaC plays no further role in replication.

Single-Strand Binding Proteins

As soon as DnaB unwinds the two parental strands, they are engulfed by single-strand binding proteins. These proteins bind to adjacent groups of 32 nucleotides. DNA covered by single-strand binding proteins is rigid, without bends or kinks. It is therefore an excellent template for DNA synthesis. Single-strand binding proteins are sometimes called helix-destabilizing proteins.

BIOCHEMISTRY OF DNA REPLICATION

In prokaryotes the synthesis of a new DNA molecule is catalyzed by **DNA polymerase** III. Its substrates are the four deoxyribonucleoside triphosphates, dATP, dCTP, dGTP, and dTTP. DNA polymerase III catalyzes the formation of a phosphodiester link (Fig. 4.3 on page 55) between the 3′-hydroxyl group of one sugar residue and the 5′-phosphate group of a second sugar residue (Fig. 5.2A). The base sequence of a newly synthesized DNA strand is dictated by the base sequence of its parental strand. If the sequence of the template strand is 3′ CATCGA 5′, then that of the daughter strand is 5′ GTAGCT 3′. In eukaryotes, DNA replication is performed by three isoforms, DNA polymerases α, δ and ε, but the mechanism is much the same.

DNA polymerase III can only add a nucleotide to a free 3′-hydroxyl group and therefore synthesizes DNA in the 5′ to 3′ direction. The template strand is read in the 3′ to 5′ direction. However, the two strands of the double helix are antiparallel. They cannot be synthesized in the same direction because only one has a free 3′-hydroxyl group, the other has a free 5′-phosphate group. No DNA polymerase has been found that can synthesize DNA in the 3′ to 5′ direction, that is, by attaching a nucleotide to a 5′ phosphate, so the synthesis of the two daughter strands must differ. One strand, the **leading strand**, is synthesized continuously while the other, the **lagging strand**, is synthesized discontinuously. DNA polymerase III can synthesize both daughter strands, but must make the lagging strand as a series of short 5′ to 3′ sections (Fig. 5.1). The fragments of DNA, called **Okazaki fragments** after Reiji Okazaki

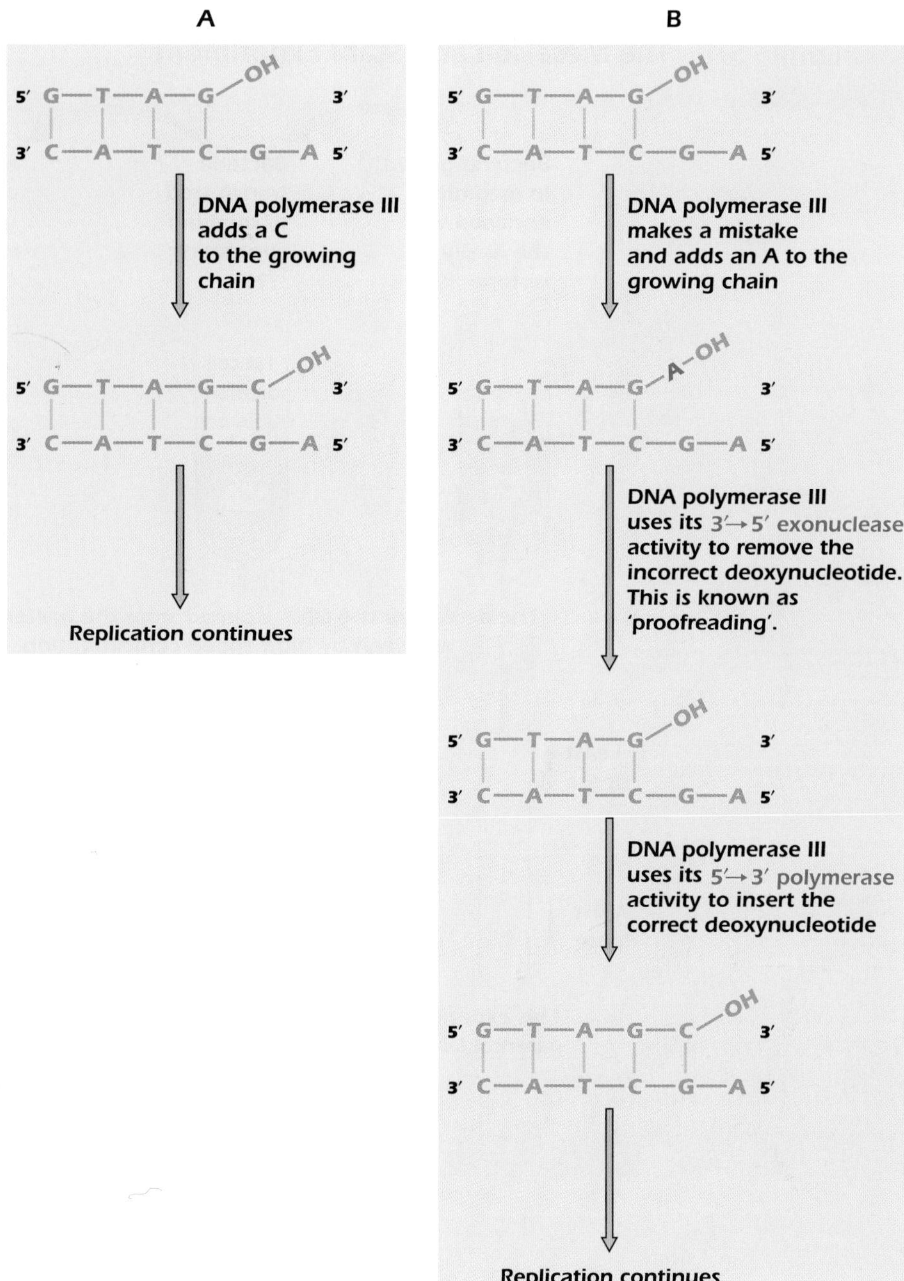

Figure 5.2. DNA polymerase III can correct its own mistakes.

who discovered them in 1968, are then joined together by another DNA polymerase and **DNA ligase**.

DNA Synthesis Requires an RNA Primer

DNA polymerase III cannot itself initiate the synthesis of DNA. The enzyme **primase** is needed to catalyze the formation of a short stretch of RNA complementary in sequence to the DNA template strand (Fig. 5.1). This RNA chain, the primer, is needed to prime (or start) the synthesis of the new DNA strand. DNA polymerase III catalyzes the formation of a phosphodiester link between the 3′-hydroxyl group of the RNA primer and the 5′-phosphate group of the appropriate deoxyribonucleotide. Several RNA primers are made along the length of the lagging strand template. Each is extended in the 5′ to 3′ direction by DNA polymerase III until it reaches the 5′ end of the next RNA primer. In prokaryotes the lagging strand is primed about every 1000 nucleotides whereas in eukaryotes this takes place each 200 nucleotides.

Example 5.1 The Meselson and Stahl Experiment

Bacteria grown in medium enriched with the heavy isotope ^{15}N

Bacteria transferred to medium containing ^{14}N

Bacteria continue to grow in medium containing ^{14}N

1st cell division

2nd cell division

The density of the DNA isolated from the bacteria was analyzed by high speed centrifugation

Least dense

Most dense

This experiment demonstrated that the two strands of the parental DNA unwind and each acts as template for a new daughter strand

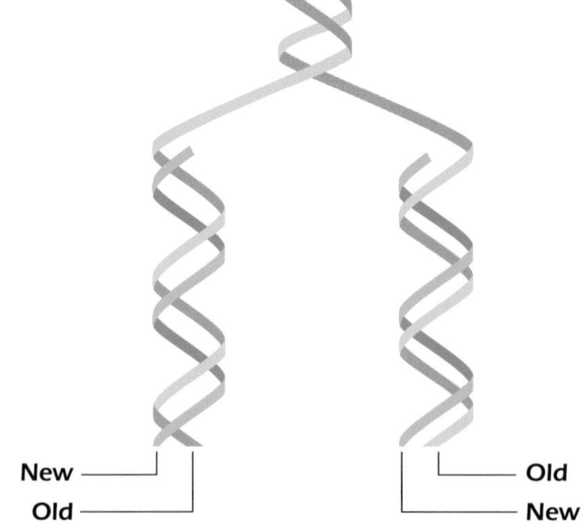

In 1958 Matthew Meselson and Franklin Stahl designed an ingenious experiment to test whether each strand of the double helix does indeed act as a template for the synthesis of a new strand. They grew the bacterium *E. coli* in a medium containing the heavy isotope ^{15}N that could be incorporated into new DNA molecules. After several cell divisions they transferred the bacteria, now containing "heavy" DNA, to a medium containing only the lighter, normal, isotope ^{14}N. Any newly synthesized DNA molecules would therefore be lighter than the original parent DNA molecules containing ^{15}N. The difference in density between the heavy and light DNAs allows their separation using very high speed centrifugation. The results of this experiment are illustrated in the figure. DNA isolated from cells grown in the ^{15}N medium had the highest density and migrated the furthest during centrifugation. The lightest DNA was found in cells grown in the ^{14}N medium for two generations, whereas DNA from bacteria grown for only one generation in the lighter ^{14}N medium had a density halfway between these two. This is exactly the pattern expected if each strand of the double helix acts as a template for the synthesis of a new strand. The two heavy parental strands separated during replication, with each acting as a template for a newly synthesized light strand, which remained bound to the heavy strand in a double helix. The resulting DNA was therefore of intermediate density. Only in the second round of DNA replication, when the light strands created during the first round of replication were allowed to act as templates for the construction of complementary light strands, did DNA double helices composed entirely of ^{14}N-containing building blocks appear.

The two scientists were awarded the Nobel prize for this discovery that DNA replication is "semiconservative," meaning that the results are not completely new but are half new and half old.

Medical Relevance 5.1 Inhibiting DNA Polymerase Fights Cancer

Drugs that inhibit DNA replication prevent cells dividing. Cytosine arabinoside, sold as Cytosar-U, Tarabine PFS and Depocyt, inhibits eukaryote DNA polymerases α, δ and ε, and is therefore widely used to treat cancers.

RNA Primers Are Removed

Once the synthesis of the DNA fragment is complete, the RNA primers must be replaced by deoxyribonucleotides. In prokaryotes the enzyme DNA polymerase I removes ribonucleotides using its 5′ to 3′ **exonuclease** activity and then uses its 5′ to 3′ polymerizing activity to incorporate deoxyribonucleotides. In this way, the entire RNA primer gets replaced by DNA. Synthesis of the lagging strand is completed by the enzyme DNA ligase, which joins the DNA fragments together by catalyzing formation of phosphodiester links between adjacent fragments.

Eukaryotic organisms probably use an enzyme called **ribonuclease H** to remove their RNA primers. This enzyme breaks phosphodiester links in an RNA strand that is hydrogen-bonded to a DNA strand.

The Self-Correcting DNA Polymerase

The genome of *E. coli* consists of about 4.5×10^6 base pairs of DNA. DNA polymerase III makes a mistake about every 1 in 10^4 bases and joins an incorrect deoxyribonucleotide to the growing chain. If unchecked, these mistakes would lead to a catastrophic mutation rate. Fortunately, DNA polymerase III has a built-in proofreading mechanism that corrects its own errors. If an incorrect base is inserted into the newly synthesized daughter strand, the enzyme recognizes the change in shape of the double-stranded molecule, which arises through incorrect base pairing, and DNA synthesis stops (Fig. 5.2B). DNA polymerase III then uses its 3′ to 5′ exonuclease activity to remove the incorrect deoxyribonucleotide and replace it with the correct one. DNA synthesis then proceeds. DNA polymerase III hence functions as a self-correcting enzyme.

Mismatch Repair Backs Up the Proofreading Mechanism

The proofreading function of DNA polymerase III improves the accuracy of DNA replication about a hundredfold. However, sometimes the enzyme does miss a nucleotide that has been incorrectly inserted into the newly synthesized DNA strand. Cells have evolved a back-up mechanism, **mismatch repair**, that detects when an incorrect nucleotide has been inserted into the daughter strand (Fig. 5.3). The repair mechanism relies on the cell being able to distinguish, within the double helix, between the template strand

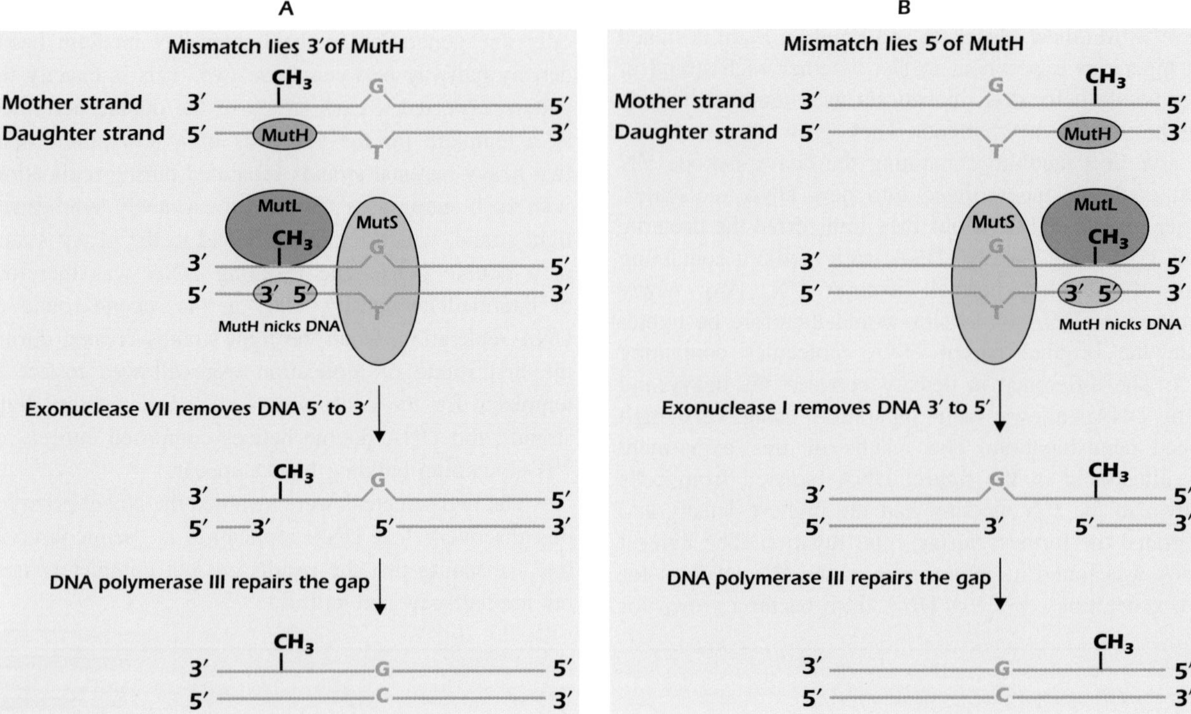

Figure 5.3. Mismatch repair of DNA.

(mother strand) and the newly synthesized strand (daughter strand).

We best understand this repair process in *E. coli*. The bacterium has an enzyme called Dam methylase that adds a methyl group on the A of the sequence 5′ GATC 3′. This sequence occurs very frequently in DNA, about once every 256 bp. The methylation of DNA happens very soon after a DNA strand has been replicated. However, for a short time during replication the double-stranded DNA molecule will have one strand methylated (mother strand) and one strand not methylated (daughter strand). The DNA molecule is said to be hemi-methylated (half methylated). Because the daughter strand has not yet been methylated the cell knows that if a mismatch in base pairing has occurred between the two strands then it is the nonmethylated, newly synthesized, strand that must carry the mistake.

A protein called MutH binds on the daughter strand at a site opposite a methylated A in the mother strand. If there is no mismatched base pair nearby then MutH does nothing. However, if two other proteins called MutL and MutS have detected a mismatched base pair then MutH, which is an **endonuclease**, is activated and nicks (cleaves a phosphodiester link in) the unmethylated daughter strand. This allows a stretch of DNA containing the mismatched base pair to be removed. Two different proteins are involved in removing the stretch of DNA. If MutH nicks the DNA 5′ to the mismatch (Fig 5.3A), then exonuclease VII degrades the DNA strand in the 5′ to 3′ direction. However, if MutH nicks the DNA 3′ to the mismatch (Fig 5.3B), then the DNA strand is removed by exonuclease I in the 3′ to 5′ direction. In either case, the gap in the daughter strand is then replaced by DNA polymerase III.

DNA REPAIR AFTER REPLICATION

Deoxyribonucleic acid can be damaged by a number of agents, which include oxygen, water, naturally occurring chemicals in our diet, and radiation. Because damage to DNA can change the sequence of bases, a cell must be able to repair alterations in the DNA code if it is to survive and pass on the DNA database unaltered to its progeny cells.

Spontaneous and Chemically Induced Base Changes

The most common damage suffered by a DNA molecule is **depurination**—the loss of an adenine or guanine because the bond between the purine base and the deoxyribose sugar to which it is attached spontaneously hydrolyzes (Fig. 5.4). Within each human cell about 5000–10,000 depurinations occur every day.

Deamination is a less frequent event; it happens about 100 times a day in every human cell. Collision of H_3O^+ ions with the bond linking the amino group to cytosine sets off a spontaneous deamination that produces uracil (Fig. 5.4). Cytosine base pairs with guanine, whereas uracil

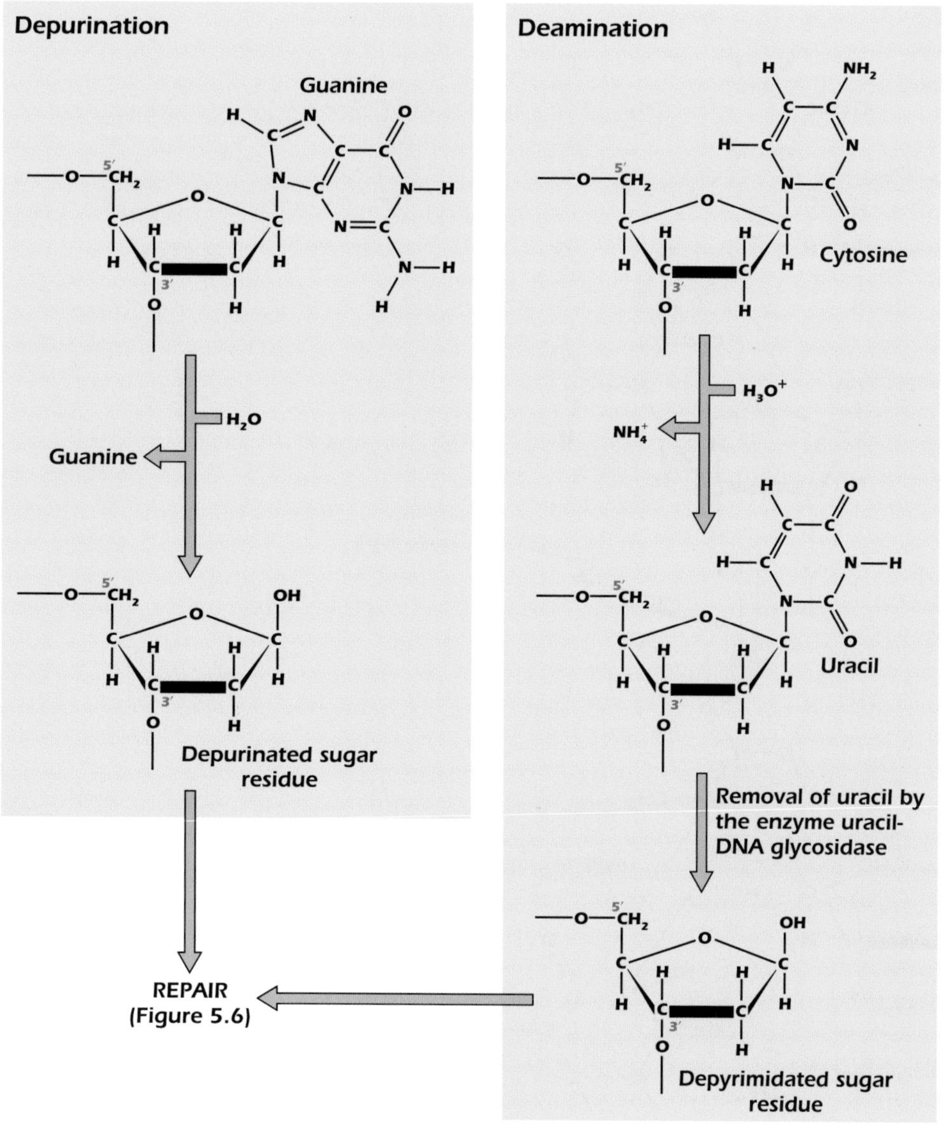

Figure 5.4. Spontaneous reactions corrupt the DNA database.

pairs with adenine. If this change were not corrected, then a CG base pair would mutate to a UA base pair the next time the DNA strand was replicated.

Ultraviolet light or chemical carcinogens such as benzopyrene, found in cigarette smoke, can also disrupt the structure of DNA. The absorption of ultraviolet light can cause two adjacent thymine residues to link and form a thymine dimer (Fig. 5.5). If uncorrected, thymine dimers create a distortion in the DNA helix known as a **bulky lesion**. This inhibits normal base pairing between the two strands of the double helix and blocks the replication process. Ultraviolet light has a powerful germicidal action and is widely used to sterilize equipment. One of the reasons why bacteria are killed by this treatment is because the formation of large numbers of thymine dimers prevents replication.

Repair Processes

If there were no way to correct altered DNA, the rate of mutation would be intolerable. **DNA excision** and **DNA repair enzymes** have evolved to detect and to repair altered DNA. The role of the repair enzymes is to cut out (excise) the damaged portion of DNA and then to repair the base sequence. Much of our knowledge on DNA repair has been derived from studies on *E. coli*, but the general principles apply to other organisms such as ourselves. Repair is possible because DNA comprises two complementary strands. If the repair mechanisms can identify which of the two strands is the damaged one, it can then be repaired as good as new by rebuilding it to be complementary to the undamaged one.

Two types of excision repair are described in this section: base excision repair and nucleotide excision repair.

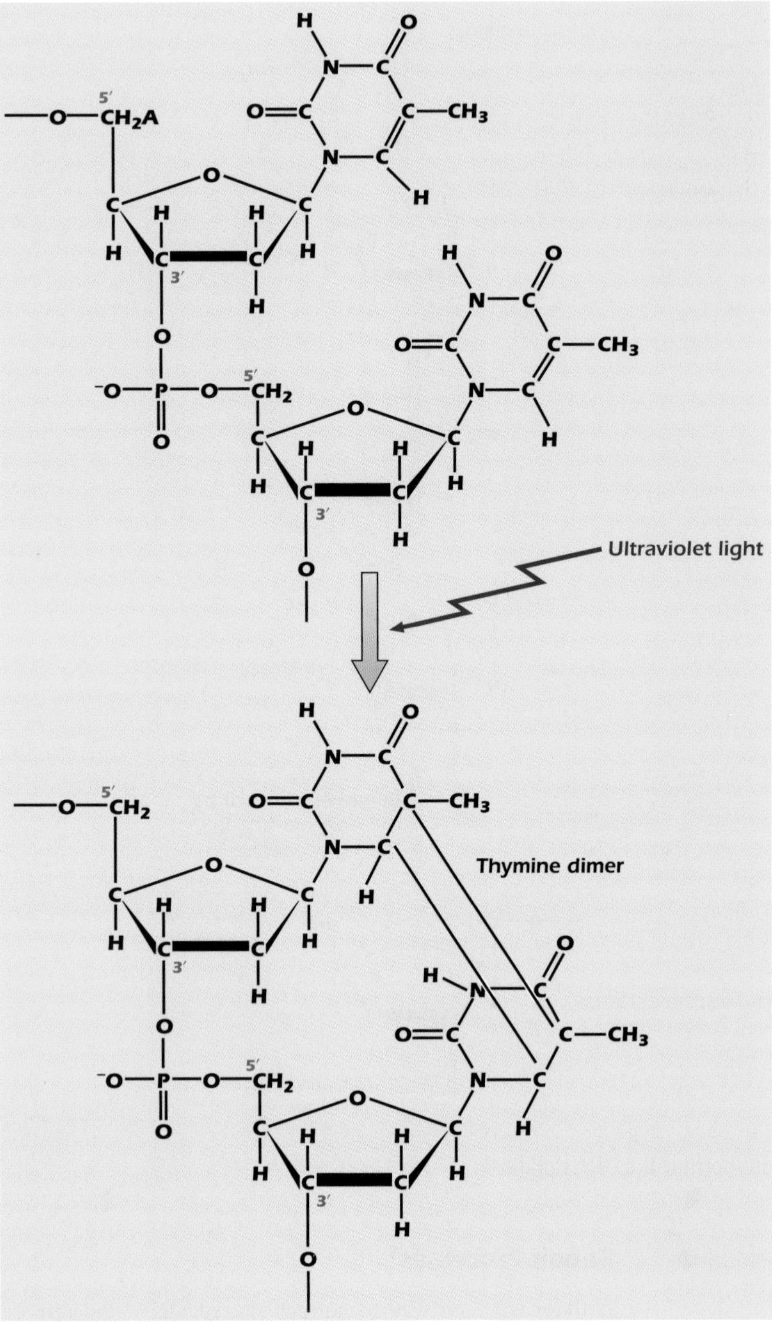

Figure 5.5. Formation of a thymine dimer in DNA.

The common themes of each of these repair mechanisms are: (1) An enzyme recognizes the damaged DNA, (2) the damaged portion is removed, (3) DNA polymerase inserts the correct nucleotide(s) into position (according to the base sequence of the second DNA strand), and (4) DNA ligase joins the newly repaired section to the remainder of the DNA strand.

Base excision repair is needed to repair DNA that has lost a purine (depurination), or where a cytosine has been deaminated to uracil (U). Although uracil is a normal constituent of RNA, it does not form part of undamaged DNA and is recognized and removed by the repair enzyme **uracil-DNA glycosidase** (Fig. 5.4). This leaves a gap in the DNA where the base had been attached to deoxyribose. There is no enzyme that can simply reattach a C into the vacant space on the sugar. Instead, an enzyme called **AP endonuclease** recognizes the gap and removes the sugar by breaking the phosphodiester links on either side (Fig. 5.6). When DNA has been damaged by the loss of a purine (Fig. 5.4), AP endonuclease also removes the sugar that has lost its base. The AP in the enzyme's name means apyrimidinic (without a pyrimidine) or apurinic (without a purine).

Medical Relevance 5.2 **Bloom's Syndrome and Xeroderma Pigmentosum**

DNA helicases are essential proteins required to open up the DNA helix during replication. In Bloom's syndrome, mutations give rise to a defective helicase. The result is excessive chromosome breakage, and affected people are predisposed to many different types of cancers when they are young.

People who suffer from the genetic disorder xeroderma pigmentosum are deficient in one of the enzymes for excision repair. As a result, they are very sensitive to ultraviolet light. They contract skin cancer even when they have been exposed to sunlight for very short periods because thymine dimers produced by ultraviolet light are not excised from their genomes.

The repair process for reinserting a purine, or a pyrimidine, into DNA is now the same (Fig. 5.6). DNA polymerase I replaces the appropriate deoxyribonucleotide into position. DNA ligase then seals the strand by catalyzing the reformation of a phosphodiester link.

Nucleotide excision repair is required to correct a thymine dimer. The thymine dimer, together with some 30 surrounding nucleotides, is excised from the DNA. Repairing damage of this bulky type requires several proteins because the exposed, undamaged, DNA strand must be protected from nuclease attack while the damaged strand is repaired by the actions of DNA polymerase I and DNA ligase.

Even with all these protection systems in place, the cell divisions that create and repair our bodies generate errors, so that the adult human contains many cells with **somatic mutations**. Most are irrelevant to the specialized operation of that cell, or merely reduce its ability to function. Some, however, can cause cancer (page 308).

GENE STRUCTURE AND ORGANIZATION IN EUKARYOTES

Introns and Exons: Additional Complexity in Eukaryotic Genes

Genes that code for proteins should be simple things: DNA makes RNA makes protein, and a gene codes for the amino acids of a protein by the three-base genetic code. In prokaryotes, indeed, a gene is a continuous series of bases that, read in threes, code for the protein. This simple and

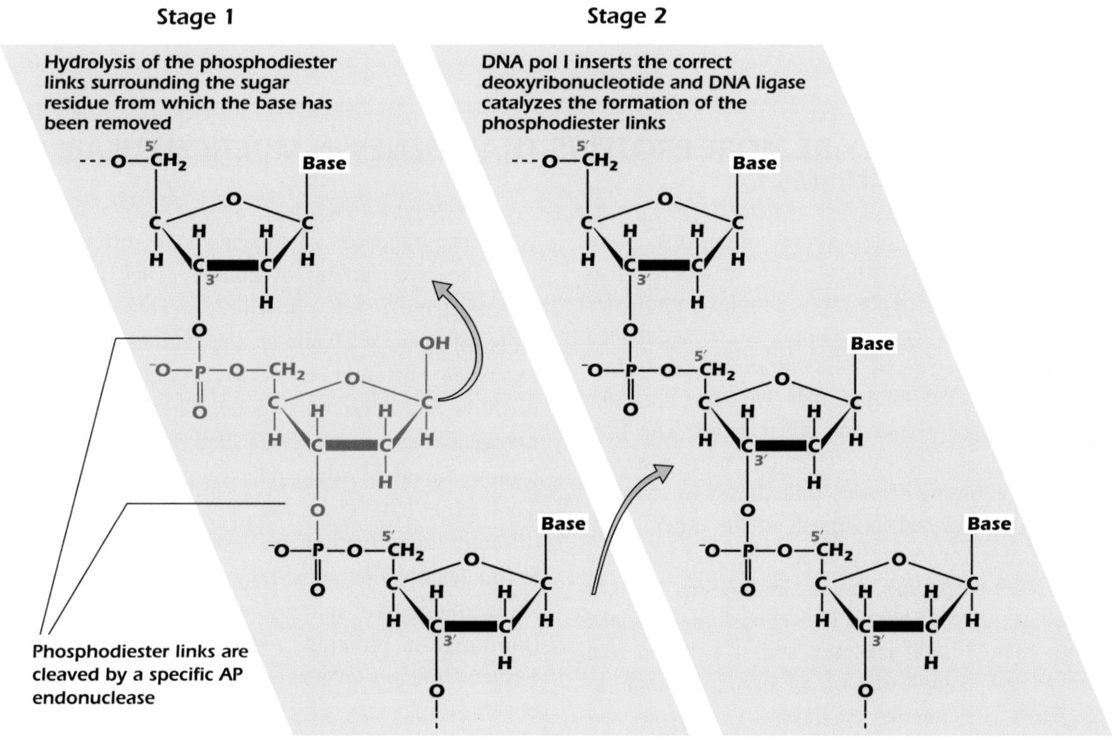

Stage 1

Hydrolysis of the phosphodiester links surrounding the sugar residue from which the base has been removed

Phosphodiester links are cleaved by a specific AP endonuclease

Stage 2

DNA pol I inserts the correct deoxyribonucleotide and DNA ligase catalyzes the formation of the phosphodiester links

Figure 5.6. Base excision repair.

apparently sensible system does not apply in eukaryotes. Instead, the protein-coding regions of almost all eukaryotic genes are organized as a series of separate bits interspersed with noncoding regions. The protein-coding regions of the split genes are **exons**. The regions between are called **introns**, short for intervening sequences. At the bottom of Figure 5.7 we show the structure of the β-globin gene, which contains three exons and two introns. Introns are often very long compared to exons. As happens in prokaryotes, messenger RNA complementary to the DNA is synthesized, but then the introns are spliced out before the mRNA leaves the nucleus (page 91). This means that a gene is much longer than the mRNA that ultimately codes for the protein. The name exon derives from the fact that these are the regions of the gene that, when transcribed into mRNA, exit from the nucleus.

In fact there is an evolutionary rationale to this apparently perverse arrangement. As we will see (page 147), a single protein is often composed of a series of **domains**, with each domain performing a different role. The breaks between exons usually correspond to domain boundaries. During evolution, reordering of exons has created new genes that have some of the exons of one gene, and some of the exons of another, and hence generates novel proteins composed of new arrangements of domains, each of which still does its job.

The Major Classes of Eukaryotic DNA

We do not yet fully understand the construction of our nuclear genome. Only about 1.5% of the human genome codes for exonic sequences (i.e., makes protein) with about 23.5% coding for introns, tRNA genes, and ribosomal RNA (rRNA) genes. Most protein-coding genes occur only once in the genome and are called single-copy genes.

Many genes have been duplicated at some time during their evolution. Mutation over the succeeding generations causes the initially identical copies to diverge in sequence and produce what is known as a gene family. Members of a gene family usually have a related function—we have already met the connexin family (page 46) whose products make gap junctions. These genes then generate related proteins or **isoforms**, which are often distinguished by placing a Greek letter after the protein name, for example, hemoglobin α and hemoglobin β. Different members of a family sometimes encode proteins that carry out the same specialized function but at different times during development. The α- and β-globin gene families, illustrated in Figure 5.7, are an example. The α-globin gene cluster is on human chromosome 16 while the the β-globin gene cluster is on human chromosome 11. Hemoglobin is composed of two α-globins and two β-globins (page 152). The gene clusters encode proteins produced at specific times during development, from embryo to fetus to adult. The different globin proteins are produced at different stages of gestation to cope with the different oxygen transport requirements during development (Medical Relevance 11.1 on page 176). The duplication of genes and their subsequent divergence allows the expansion of the gene repertoire, the

IN DEPTH 5.1 THERE ARE MORE PROTEINS THAN GENES IN MULTICELLULAR ORGANISMS

As genomes of more and more organisms were sequenced, the most surprising feature to emerge was just how few genes supposedly complex organisms possess. The first eukaryotic genome to be sequenced was that of the budding yeast, *Saccharomyces cerevisiae*, the simple unicellular fungus that we use to make bread and beer. *S. cerevisiae* has 5770 genes. The fruit fly, *Drosophila melanogaster*, a much more complex organism with a brain, nervous and digestive systems and the ability to fly and navigate, on the other hand, has 13,379 genes, or roughly twice the number in a yeast. Even more surprising was the finding that humans have only about 20,500 genes. However, humans make many more than 20,500 proteins and it is these that contribute to the complexity of an organism such as ourselves. How is it possible to have so few genes and yet make hundreds of thousands of different proteins? It is the arrangement of human genes into exons and introns (page 76) that provides the solution. **Alternative splicing** (page 92) allows the cell to "cut and paste" exons in different ways to produce many different mRNAs from the same gene. The most extreme case known is the human gene called *SLO*, which encodes a protein found in some potassium channels (page 229). This gene has 35 exons, which can produce 40,320 different combinations of exons from a single gene. Estimates are that something like 50% of human genes show alternative splicing with the pattern of splicing (the range of proteins produced) varying from tissue to tissue. *Drosophila* genes also show alternative splicing but those of yeast, which contain few introns, do not.

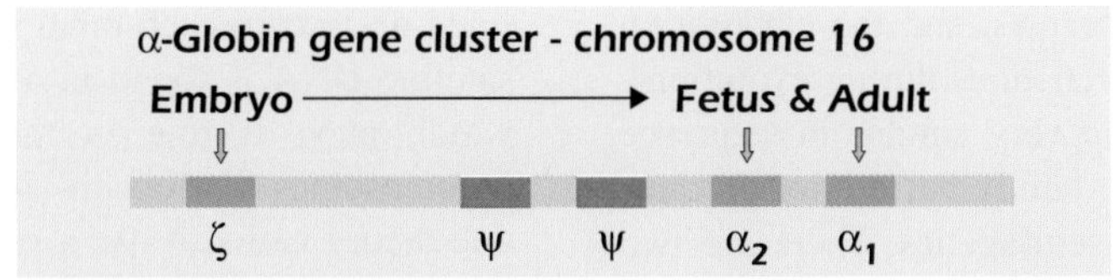

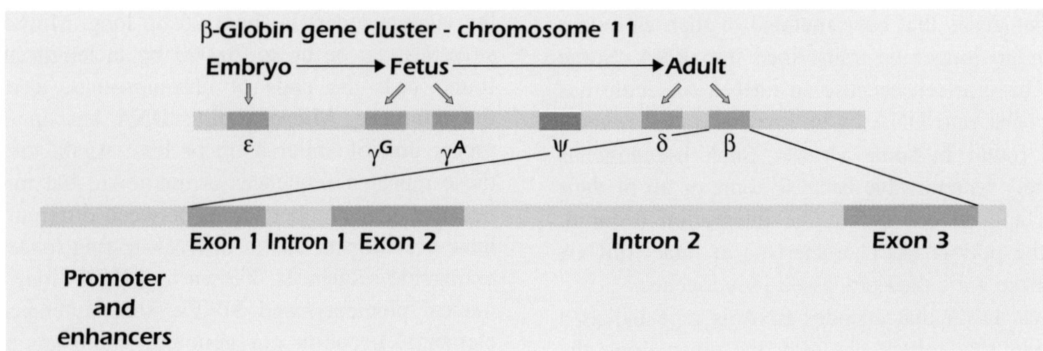

Figure 5.7. The human α- and β-globin gene family clusters. Ψ indicates a pseudogene. Adults only express α, β, and δ, and of these the expression of δ is very low. The exon/intron boundaries of the β-globin gene are indicated at the bottom.

IN DEPTH 5.2 GENOME PROJECTS

The publication in 1996 of the sequence of the 5,770 genes that make up the genome of the single celled yeast *Saccharomyces cerevisiae* was a milestone in biology. Not only did scientists have before them the complete genetic blueprint of a eukaryotic organism, but the technology for obtaining and curating huge amounts of genetic data was now established. The genomes of other simple organisms such as the tiny nematode worm *Caenorhabditis elegans*, with just 959 body cells and 19,427 genes, and the fruit fly *Drosophila melanogaster* with 13,379 genes, were published soon after, followed by more complex organisms such as the mouse and, of course, humans (both around 20,000 genes). Genomes from every branch of the tree of life are now available for study, including the platypus, our most distant mammalian relative.

Sophisticated databases have been created to store and analyze base sequence information from the various genome projects. Computer programs analyze the data for exon sequences and compare the sequence of one genome to that of another. In this way sequences encoding related proteins (proteins that share stretches of similar amino acids) can be identified. Some important programs, which can be easily accessed through the Internet are BLASTN for the comparison of a nucleotide sequence to other sequences stored in a nucleotide database, and BLASTP, which compares an amino acid sequence to protein sequence databases.

The next goal is to determine what all the genes do. This is a difficult task, which is made easier by a comparative genomics approach. The genome of our closest relative, the chimpanzee, is more than 98% identical to our own. Do complex human traits such as language lie in the 2% of genes that are different or are common genes employed differently? The genome of man's best friend, the dog, may help to answer other fundamental biological questions. Dogs have been selected for specific, homogeneous features and each breed shows little genetic diversity. This may be key to tracking down the causes of diseases such as bone and skin cancers which affect both species. The genome of the cow may not be of such direct utility but may be vital in reducing the production of greenhouse gases by our livestock.

The Human Genome Project, completed in 2003, was a 13-year international effort that was described at the time as the biological equivalent of putting a man on the moon. As more and more genomes were sequenced, the technology became quicker and, more importantly, cheaper. Within the next two or three years any of us will be able to have our genomes sequenced for a few hundred dollars. Quite what we would do with this information is not clear, but eventually clinicians will be able to tailor treatment for a range of diseases to our own unique genetic makeup.

production of new protein molecules, and the elaboration of ever-more specialized gene functions during evolution.

Some sections of DNA are very similar in sequence to other members of their gene family but do not produce mRNA. These are known as **pseudogenes**. There are two in the α-globin gene cluster (Ψ in Fig. 5.7). Pseudogenes may be former genes that have mutated to such an extent that they can no longer be transcribed into RNA. Some pseudogenes have arisen because an mRNA molecule has been copied back into DNA by an enzyme called reverse transcriptase found in some viruses. Such pseudogenes are immediately recognizable because some or all of their introns were spliced out before the integration occured. Some have the poly-A tail characteristic of intact mRNA (page 91). These are called processed pseudogenes.

Sometimes DNA that encodes RNA is repeated as a series of copies that follow one after the other along the chromosome. Such genes are said to be **tandemly repeated** and include the genes that code for ribosomal RNAs (about 250 copies/cell), transfer RNAs (50 copies/cell), and histone proteins (20–50 copies/cell). The products of these genes are required in large amounts.

This still leaves about 75% of our genome with no very clearly understood function. A large proportion of this **extragenic DNA** is made up of **repetitious DNA** sequences that are repeated many times in the genome. Some sequences are repeated more than a million times and are called **satellite DNA**. The repeating unit is usually several hundred base pairs long, and many copies are often lined up next to each other in tandem repeats. Most of the satellite DNA is found in a region called the **centromere**, which plays a role in the physical movement of the chromosomes that occurs at cell division (page 298), and one theory is that it has a structural function.

Our genome also contains **minisatellite DNA** where the tandem repeat is about 25 bp long. Minisatellite DNA stretches can be up to 20,000 bp in length and are often found near the ends of chromosomes, a region called the **telomere**. **Microsatellite DNA** has an even smaller repeat unit of about 4 bp or less. Again the function of these repeated sequences is unknown, but microsatellites, because their number varies between different individuals, have proved very useful in DNA testing (page 111). Other extragenic sequences, known as LINEs (long interspersed nuclear elements) and SINEs (short interspersed nuclear elements) occur in our genome. There are about 50,000 copies of LINEs in a mammalian genome and they make up about 17% of the human genome.

 GENE NOMENCLATURE

One of the great difficulties that has arisen out of genome-sequencing projects is how to name the genes and the proteins they encode. This has not been easy and a number of committees have been set up to deal with this problem. In general, each gene is designated by an abbreviation, written in capitalized italics. For example, the human gene

Answer to Thought Question: Guanine cannot base pair with uracil, so there is certainly a mismatch in the DNA. However, mismatch repair cannot correct the error, because the mutation has occurred in a mature chromosome in which both DNA strands are methylated.

When the bacterium replicates its DNA, the strands will separate and each will act as a template for the synthesis of a new strand. The unmodified strand 5′ TGAA 3′ will have the matching strand 3′ ACTT 5′ synthesized on it, and the resulting daughter strand will be unmutated, as will its own daughters.

However the modified strand 3′ AUTT 5′ will have the matching strand 5′ TAAA 3′ synthesized on it, since adenine base pairs with uracil. The progeny cell that inherits this chromosome, assuming that it is still infected with PBS2 and therefore allows the uracil to remain, will now have a chromosome with the structure

5′TAAA3′
3′AUTT5′

with each base pair now nicely hydrogen bonded to its partner. The presence of uracil in the DNA molecule betrays the fact that a deamination event has occurred.

When this cell replicates its DNA the strands will separate and each will act as a template for the synthesis of a new strand. The lower strand, 3′AUTT5′ will as before have the matching strand 5′TAAA3′ synthesized on it. The upper strand 5′TAAA3′ will generate a chromosome with the structure

5′TAAA3′
3′ATTT5′

and the progeny cell that inherits this, and all its progeny in turn, will have a mutation that can no longer be easily identified as arising from a deamination.

This answer neglects the fact that a bacterium infected by a bacteriophage will likely die without generating any progeny.

for flavin containing monooxygenase is designated *FMO*. Because there is more than one *FMO* gene, we assign a number to identify the specific gene to which we are referring. The gene which when mutated gives rise to trimethylaminuria (Medical Relevance 4.3 on page 62) is called *FMO3*. The protein encoded by the *FMO3* gene is written in normal capitals, as FMO3. Similarly, cytochrome P-450 genes (In Depth 11.4 on page 185) are abbreviated to *CYP*. *CYP3A4* (Medical Relevance 6.1 on page 95) is a gene that belongs to the *CYP3* family. This family has several members so we must include additional information in the gene name to specify precisely the member of the *CYP3* gene family we are referring to. The protein name is written as CYP3A4.

SUMMARY

1. During replication each parent DNA strand acts as the template for the synthesis of a new daughter strand. The base sequence of the newly synthesized strand is complementary to that of the template strand.
2. Replication starts at specific sequences called origins of replication. The two strands untwist and form the replication fork. Helicase enzymes unwind the double helix, and single-strand binding proteins keep it unwound during replication. In prokaryotes DNA polymerase III synthesizes the leading strand continuously in the 5′ to 3′ direction. The lagging strand is made discontinuously in short pieces in the 5′ to 3′ direction. These are joined together by DNA ligase. DNA polymerase is a self-correcting enzyme. It can remove an incorrect base using its 3′- to 5′-exonuclease activity and then replace it.
3. DNA repair enzymes can correct mutations. Uracil in DNA, resulting from the spontaneous deamination of cytosine, is removed by uracil-DNA glycosidase. The depyrimidinated sugar is cleaved from the sugar-phosphate backbone by AP endonuclease, and DNA polymerase then inserts the correct nucleotide. The phosphodiester link is reformed by DNA ligase.
4. In eukaryotes protein-coding genes are split into exons and introns. Only exons code for protein. The human genome has a large amount of DNA whose function is not obvious. This includes much repetitious DNA, whose sequence is multiplied many times.
5. Protein-coding genes may be found in repeated groups of slightly diverging structure called gene families, either close together or scattered over the genome. Some of the family members have lost the ability to operate—they are pseudogenes.

FURTHER READING

Brenner, S., Elgar, G., Sandford, R., Macrae, A., Venkatesh, B., and Aparicio, S. (1993) Characterization of the pufferfish (Fugu) genome as a compact model vertebrate genome. *Nature* **366**, 265–268.

Friedberg, E. C. (2001) How nucleotide excision repair protects against cancer. *Nature Rev. Cancer* **1**, 22–33.

Radman, M., and Wagner, R. (1988) The high fidelity of DNA duplication. *Sci. Am.* **259**(2), 40–46.

Scharer, O. D., and Jiricny, J. (2001) Recent progress in the biology, chemistry and structural biology of DNA glycosylases. *Bioessays* **23**, 270–281.

REVIEW QUESTIONS

5.1 Theme: Synthesis on a DNA Template

A 5′ TACGACTTCGC 3′
B 5′ UACGACUUGCG 3′
C 5′ ATGCTGAAGCG 3′
D 3′ GCGAAGTCGTA 5′
E 3′ UACGACUUGCG 5′
F 5′ CGCUUCAGCAU 3′
G 5′ CGCTTCAGCAT 3′

From the above list of compounds, select the one described by each of the descriptions below.

1. The sequence that is generated from 3′GCGAAGTCGTA5′ during the process of transcription
2. The sequence that is generated from 3′GCGAAGTCGTA5′ during the process of replication
3. The sequence that is generated from 5′ATGCTGAAGCG3′ during the process of transcription
4. The sequence that is generated from 5′ATGCTGAAGCG3′ during the process of replication

5.2 Theme: DNA Replication

A Breaking of base:base hydrogen bonds
B Connection of adjacent Okazaki fragments
C Formation of an RNA sequence complimentary to the DNA chain
D Hydrolysis and removal of a DNA strand
E Incorporation of deoxyribonucleotide monomers into both the leading and lagging strand
F Incorporation of deoxyribonucleotide monomers into the lagging strand but not into the leading strand
G In eukaryotes, removal of an RNA sequence complimentary to the DNA chain

From the above list of processes, select the one perfomed by each of the enzymes below.

1. DNA ligase
2. DNA polymerase I
3. DNA polymerase III
4. exonuclease I
5. exonuclease VII
6. helicase
7. primase
8. ribonuclease H

5.3 Theme: Regions within Eukaryotic Chromosomes

A exon
B gene family
C intron
D long interspersed nuclear element
E pseudogene
F satellite DNA
G tandemly repeated DNA

From the above list of DNA regions within eukaryotic chromosomes, select the region described by each of the descriptions below.

1. A section of DNA that, read in triplets of bases, encodes successive amino acids in a polypeptide chain
2. A section of DNA with a sequence similar to a working gene but which no longer encodes a functional protein
3. A section of DNA within a gene that is transcribed into RNA but which does not encode amino acids and which must be removed from the RNA before it leaves the nucleus
4. A series of identical or almost identical genes, all of which are transcribed so as to generate identical or almost identical RNA products and, in the case of protein coding genes, identical proteins
5. A type of DNA with a presumed structural role that makes up much of the chromosome in the centromere region
6. An extragenic DNA sequence that is repeated more than a million times

THOUGHT QUESTION

If the bacterium *Bacillus subtilis* is infected by the bacteriophage PBS2, its uracil DNA glycosidase is turned off. Consider a section of the bacterial chromosome with the structure

5′TGAA3′
3′ACTT5′

and suppose that a deamination converts the cytosine to a uracil. Given that with uracil-DNA glycosidase deactivated the bacterium cannot repair the error by base excision repair, can it repair the error using mismatch repair enzymes? If the error is not corrected, suggest what will happen to the DNA sequence (i) when the bacterium replicates its DNA and generates two progeny cells, and (ii) when these cells replicate their DNA to generate four second generation progeny?

6

TRANSCRIPTION AND THE CONTROL OF GENE EXPRESSION

Transcription (or RNA synthesis) is the process whereby the information held in the nucleotide sequence of DNA is transferred to RNA. The three major classes of RNA are **ribosomal RNA (rRNA)**, **transfer RNA (tRNA)**, and **messenger RNA (mRNA)**. All play key roles in protein synthesis. Genes encoding mRNAs are known as protein-coding genes. A gene is said to be **expressed** when its genetic information is transferred to mRNA and then to protein. Two important questions are addressed in this chapter: how is RNA synthesized, and what factors control how much mRNA is made?

 STRUCTURE OF RNA

Ribonucleic acid is a polymer made up of monomeric nucleotide units. RNA has a chemical structure similar to that of DNA, but there are two major differences. First, the sugar in RNA is ribose instead of deoxyribose (Fig. 6.1). Second, although RNA contains the two purine bases adenine and guanine and the pyrimidine cytosine, the fourth base is different. The pyrimidine uracil (U) replaces thymine (Fig. 6.1). The building blocks of RNA are therefore the four ribonucleoside triphosphates adenosine 5′-triphosphate, guanosine 5′-triphosphate, cytidine 5′-triphosphate, and uridine 5′-triphosphate. These four nucleotides are joined together by phosphodiester links (Fig. 6.2). Like DNA, the RNA chain has direction. In the first nucleotide in the chain, the 5′ carbon of the ribose is phosphorylated. This is the 5′ end of the RNA chain. At the other end is a ribose with a free hydroxyl group on its 3′ carbon. This is the 3′ end. As in DNA, phosphodiester links are made between the free 3′-hydroxyl group and the α phosphate group on the 5′ carbon of ribose. RNA molecules are single stranded along much of their length, although they often contain regions that are double stranded due to intramolecular base pairing (see for example Fig. 6.5B).

 RNA POLYMERASE

In any gene only one DNA strand acts as the template for transcription. The sequence of nucleotides in RNA depends on their sequence in the DNA template. The bases T, A, G, and C in the DNA template will specify the bases A, U, C, and G, respectively, in RNA. DNA is transcribed into RNA by the enzyme **RNA polymerase**. Transcription requires that this enzyme recognize the beginning of the gene to be transcribed and catalyze the formation of phosphodiester links between nucleotides that have been selected according to the sequence within the DNA template.

 GENE NOTATION

Figure 6.3 shows the notation used in describing the positions of nucleotides within and adjacent to a gene. The nucleotide in the template strand at which transcription begins is designated with the number +1. Transcription proceeds in the **downstream** direction, and nucleotides in the

Cell Biology: A Short Course, Third Edition. Stephen R. Bolsover, Jeremy S. Hyams, Elizabeth A. Shephard and Hugh A. White.

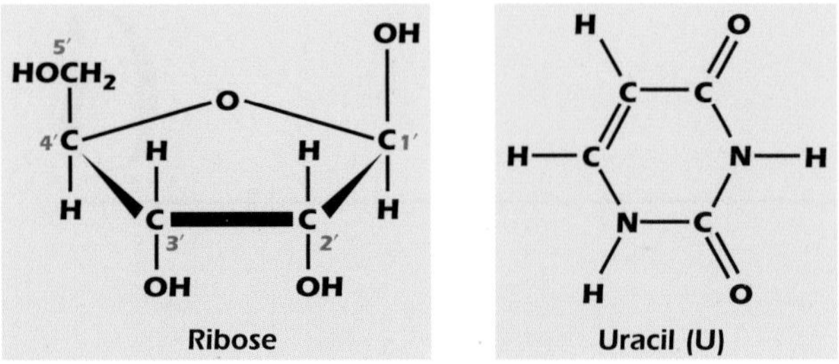

Figure 6.1. RNA contains the sugar ribose and the base uracil in place of deoxyribose and thymine.

transcribed DNA are given successive positive numbers. Downstream sequences are drawn, by convention, to the right of the transcription start site. Nucleotides that lie to the left of this site are the **upstream** sequences and are identified by negative numbers.

BACTERIAL RNA SYNTHESIS

Escherichia coli genes are all transcribed by the same RNA polymerase. This enzyme is made up of five subunits (polypeptide chains). The subunits are named α (there are two of these), β, β', and σ. Each of the subunits has its own job to do in transcription. The role of the sigma (σ) factor is to recognize a specific DNA sequence called the **promoter**, which lies just upstream of the gene to be transcribed (Fig. 6.3). *E. coli* promoters contain two

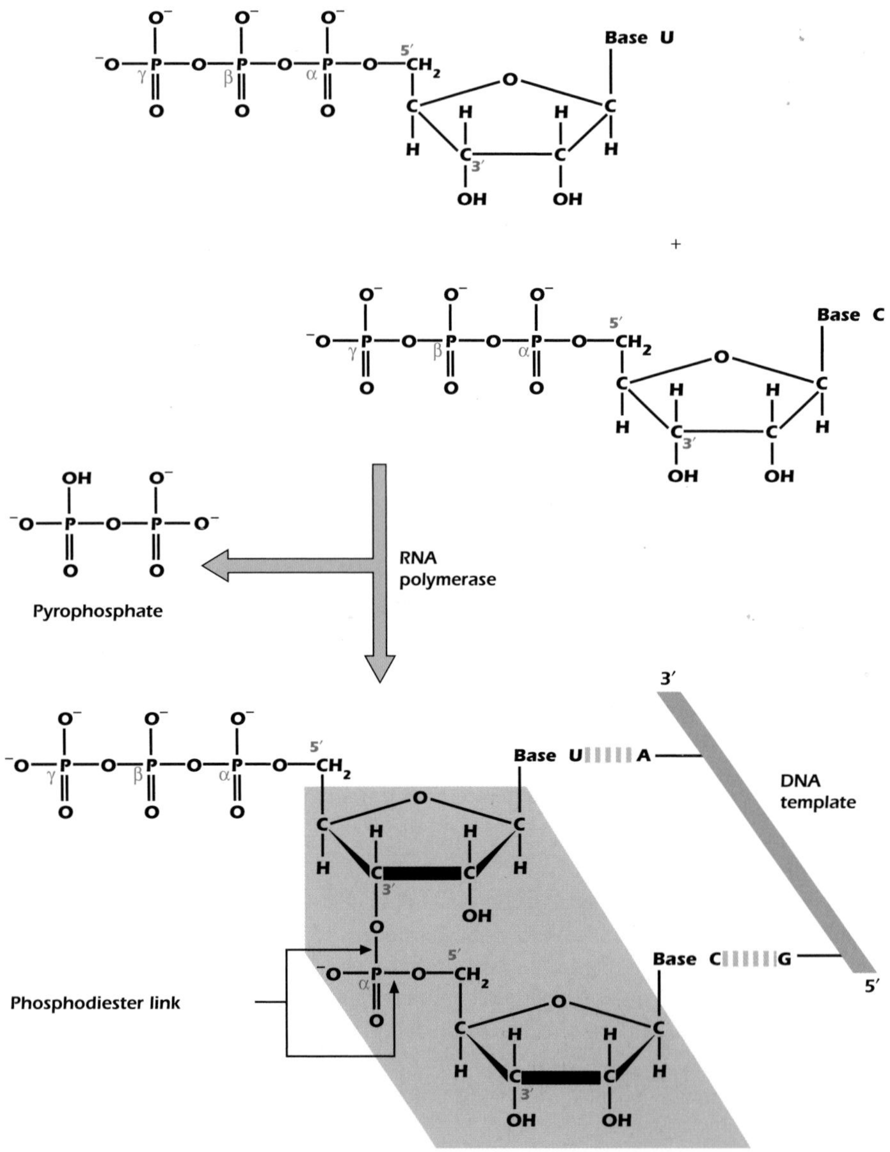

Figure 6.2. Synthesis of an RNA strand.

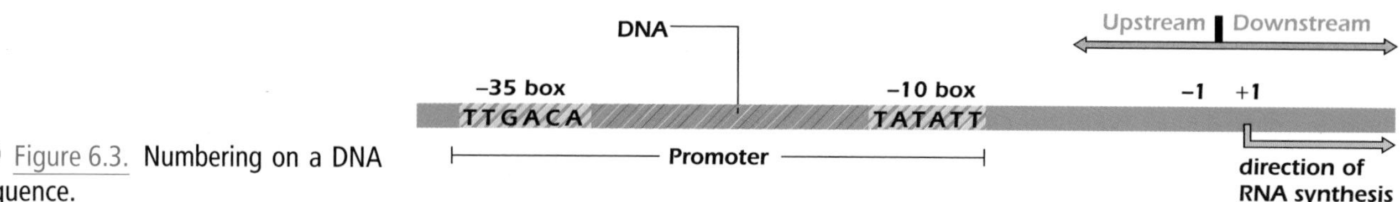

Figure 6.3. Numbering on a DNA sequence.

important regions. One centered around nucleotide −10 usually has the sequence TATATT. This sequence is called the −10 box (or the Pribnow box). The second, centered near nucleotide −35 often has the sequence TTGACA. This is the −35 box.

On binding to the promoter sequence (Fig. 6.4A), the σ factor brings the other subunits (two of α plus one each of β and β′) of RNA polymerase into contact with the DNA to be transcribed. This forms the **closed promoter complex**. For transcription to begin, the two strands of DNA must separate, enabling one strand to act as the template for the synthesis of an RNA molecule. This formation is the **open promoter complex**. The separation of the two DNA strands is helped by the AT-rich sequence of the −10 box. There are only two hydrogen bonds between the bases adenine and thymine; thus it is relatively easy to separate the two strands at this point. DNA unwinds and rewinds as RNA polymerase advances along the double helix, synthesizing an RNA chain as it goes. This produces a **transcription bubble** (Fig. 6.4B). The RNA chain grows in the 5′ to 3′ direction, and the template strand is read in the 3′ to 5′ direction.

When the RNA chain is about 10 bases long, the σ factor is released from RNA polymerase and plays no further role in transcription. The β subunit of RNA polymerase binds ribonucleotides and joins them together by catalyzing the formation of phosphodiester links as it moves along the DNA template. The β′ subunit helps to keep the RNA polymerase attached to DNA. The two α subunits are important as they help RNA polymerase to assemble on the promoter.

RNA polymerase must recognize when it has reached the end of a gene. *E. coli* has specific sequences, called **terminators**, at the ends of its genes that cause RNA polymerase to stop transcribing DNA. A terminator sequence

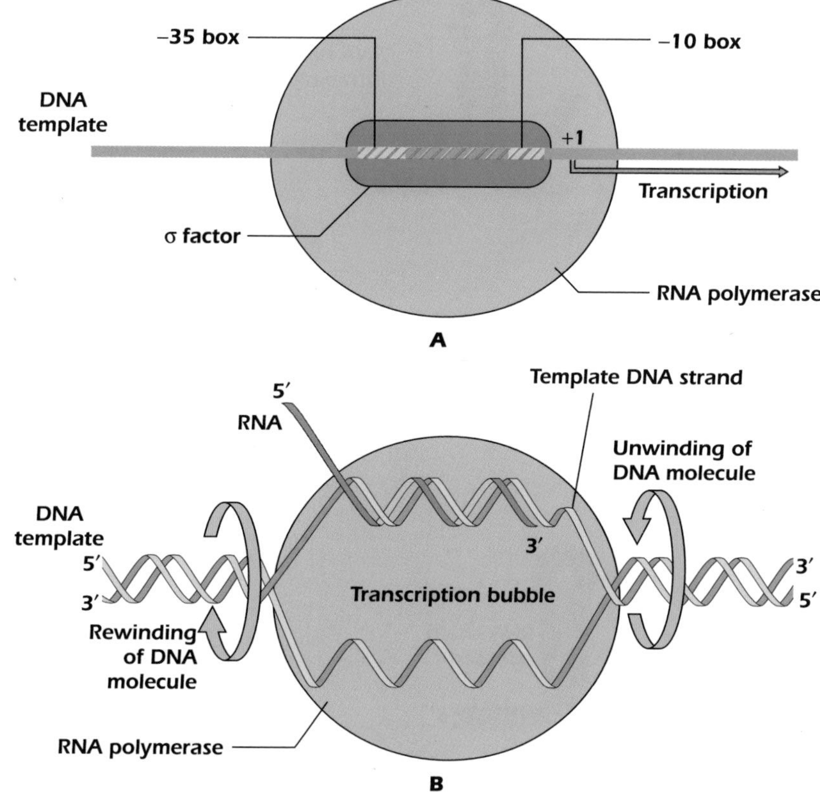

Figure 6.4. (A) RNA polymerase binds to the promoter to form the closed promoter complex. (B) The open promoter complex: The DNA helix unwinds and RNA polymerase synthesizes an RNA molecule.

consists of two regions rich in the bases G and C that are separated by about 4bp. This sequence is followed by a stretch of A bases. Figure 6.5 shows how the terminator halts transcription. When the GC-rich regions are transcribed, a hairpin loop forms in the RNA with the first and second GC-rich regions aligning and pairing up. Formation of this structure within the RNA molecule causes the transcription bubble to shrink because where the template DNA strand can no longer bind to the RNA molecule it reconnects to its sister DNA strand. The remaining interactions between the adenines in the DNA template and the uracils in the RNA chain have only two hydrogen bonds per base pair and are therefore too weak to maintain the transcription bubble. The RNA molecule is then released, transcription terminates, and the double helix reforms. This type of transcription termination is known as rho-independent termination.

Some *E. coli* genes contain a different type of terminator site. These are recognized by a protein, known as rho, which frees the RNA from the DNA. In this case transcription is terminated by a process known as rho-dependent termination.

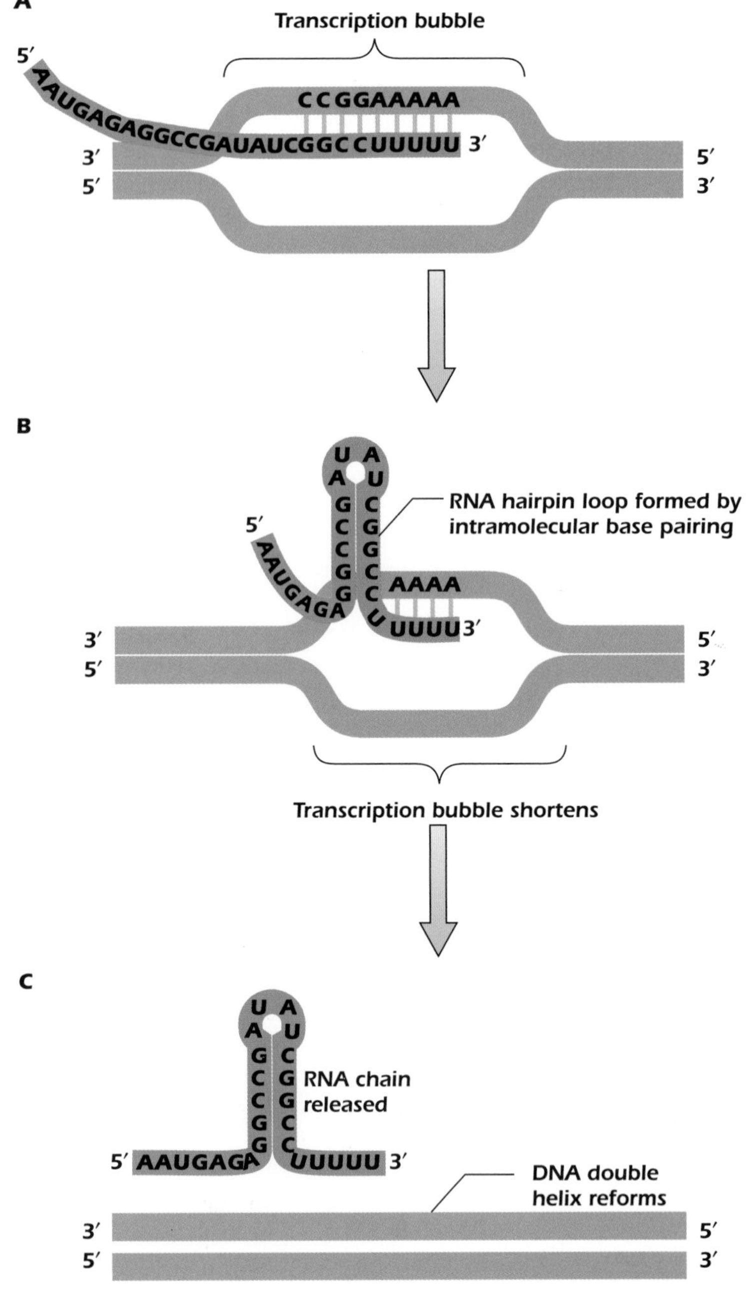

Figure 6.5. Rho-independent trancription termination in *Escherichia coli*.

CONTROL OF BACTERIAL GENE EXPRESSION

Many bacterial proteins are always present in the cell in a constant amount. However, the amount of other proteins is regulated by the presence or absence of a particular nutrient. To grow and divide and make the most efficient use of the available nutrients, bacteria have to adjust quickly to changes in their environment. They do this by regulating the production of proteins required for either breakdown or synthesis of a particular compound. Gene expression in bacteria is controlled mainly at the level of transcription. This is because bacterial cells have no nuclear envelope, and RNA synthesis and protein synthesis are not separate but occur simultaneously. This is one reason why bacteria lack the more sophisticated control mechanisms that regulate gene expression in eukaryotes.

Each bacterial promoter usually controls the transcription of a cluster of genes coding for proteins that work together on a particular task. This collection of related genes is called an **operon** and is transcribed as a single mRNA molecule called a **polycistronic mRNA**. As shown in Figure 6.6, translation of this mRNA produces the required proteins because there are several start and stop codons for protein synthesis along its length. Each start and stop codon (page 63) specifies a region of RNA that will be translated into one particular protein. The organization of genes into operons ensures that all the proteins necessary to metabolize a particular compound are made at the same time and hence helps bacteria to respond quickly to environmental changes.

The three major factors involved in regulating how much RNA is made are (1) nucleotide sequences within or flanking a gene, (2) proteins that bind to these sequences, and (3) the environment. The human intestine contains many millions of *E. coli* cells that must respond very quickly to the sudden appearance of a particular nutrient. For instance, most foods do not contain the disaccharide lactose, but milk contains large amounts. Within minutes of our drinking a glass of milk, *E. coli* in our intestines start to produce the enzyme β-galactosidase that cleaves lactose to glucose and galactose (Fig. 6.7). In general, the substrates of β-galactosidase are compounds like lactose that contain a β-galactoside linkage and are therefore called β-galactosides.

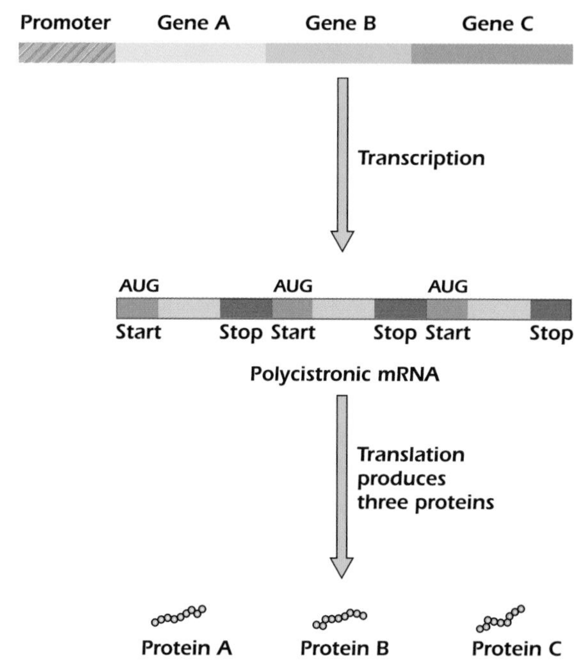

Figure 6.6. A bacterial operon is transcribed into a polycistronic mRNA.

lac, an Inducible Operon

β-galactosidase is encoded by one of the genes that make up the lactose (*lac*) operon, which is shown in Figure 6.8. The operon contains three protein-coding genes: *lac z, lac y*, and *lac a*. β-galactosidase is encoded by the *lac z* gene. As noted before, gene names are always italicized, while the protein product is always in standard type. *Lac y* encodes β-galactoside permease, a carrier (page 236) that helps lactose get into the cell. The *lac a* gene codes for transacetylase. This protein is thought to remove compounds that have a structure similar to lactose but that are not useful to the cell.

Answer to Thought Question: Exons are the parts of the newly synthesized mRNA that are spliced together to produce the mRNA that will be translated into protein. Exons can contain sequences that code for proteins (protein-coding exons) but exons also contain non-coding sequences (i.e., sequences found at the 5′ and 3′ ends of the mature mRNA). It is quite possible therefore for alternative RNA splicing to produce two mRNAs that have different 5′ leader sequences, or different so-called 3′ untranslated regions (3′UTRs), but for the protein product to be identical because all the protein-coding exons are common to both mature mRNAs. This phenomenon is often cell-specific, so one mature mRNA is produced in one cell type and an alternative mRNA product is produced in another cell type. Different 5′ non-coding sequences may help to regulate how efficiently the mRNA is translated in different tissues. So, two cells may still produce the same protein, but alternative splicing, by modifying the 5′ sequence of the mRNA may help to determine how much of a protein is made in each cell. 3′UTRs are thought to contain sequences that allow the mRNA to be picked up by intracellular transport systems (page 287) so that one form of the mRNA might remain in the cell body and generate protein there, while the other might be carried out to distant cell processes so that the same protein is now preferentially made at the remote location.

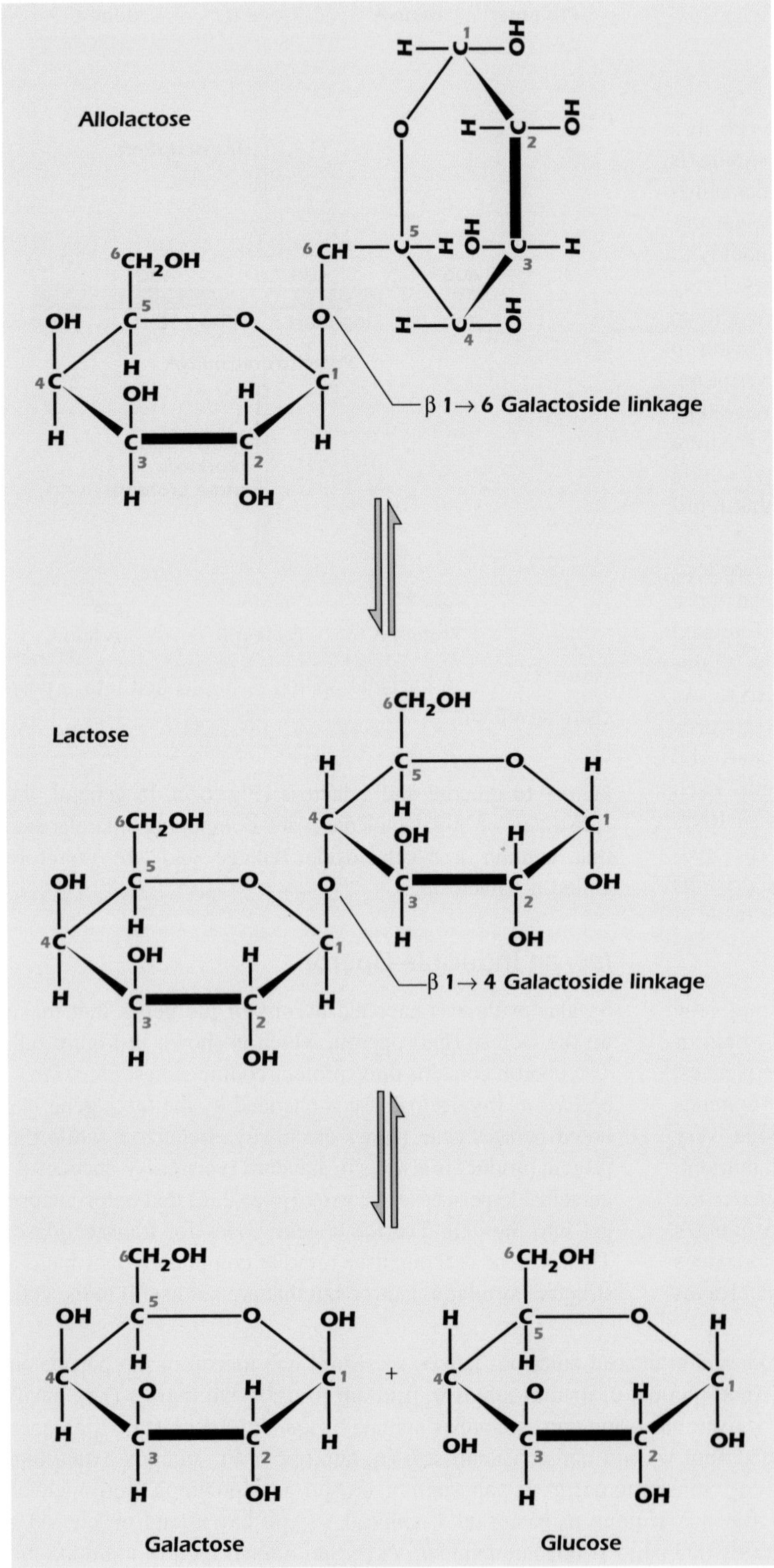

Allolactose

β 1→ 6 Galactoside linkage

Lactose

β 1→ 4 Galactoside linkage

Galactose

Glucose

Figure 6.7. Reactions catalyzed by β-galactosidase.

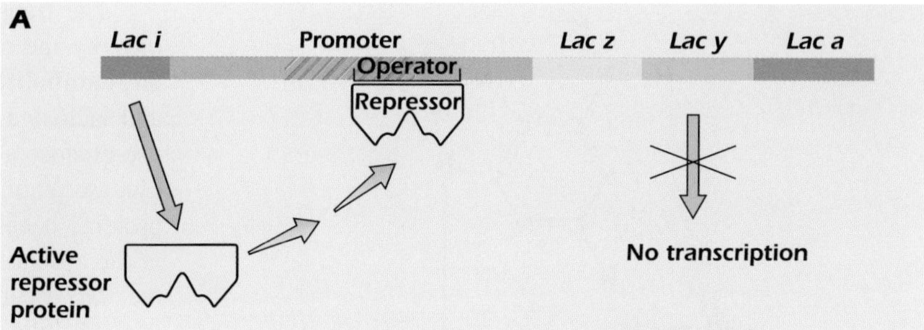

No β-galactoside sugars present—operon repressed

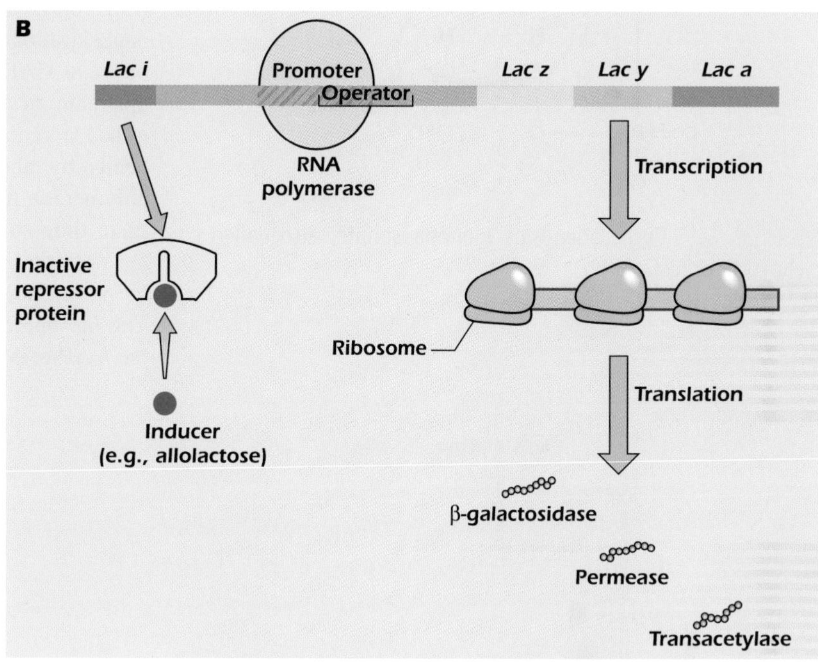

β-galactoside sugars present—operon derepressed

Figure 6.8. Transcription of the *lac* operon requires the presence of an inducer.

In the absence of β-galactoside compounds like lactose, there is no need for *E. coli* to produce β-galactosidase or β-galactoside permease, and the cell contains only a few molecules of these proteins. The *lac* operon is said to be inducible because the rate of transcription into RNA increases greatly when a β-galactoside is present. How is the transcription of the *lac z, lac y*, and *lac a* genes switched on and off? A repressor protein (the product of the *lac i* gene) binds to a sequence in the *lac* operon known as the operator. The operator lies next to the promoter so that, when the repressor is bound, RNA polymerase is unable to bind to the promoter. In the absence of a β-galactoside, the *lac* operon spends most of its time in the state shown in Figure 6.8A. The repressor is bound to the operator, RNA polymerase cannot bind, and no transcription occurs. Only for the small fraction of time that the operator is unoccupied by the repressor can RNA polymerase bind and generate mRNA. Thus in the absence of a β-galactoside, only very small amounts of β-galactosidase, β-galactoside permease, and transacetylase are synthesized.

If lactose appears, it is converted to an isomer called allolactose. This conversion is carried out by β-galactosidase (Fig. 6.7); as we have seen, a small amount of β-galactosidase is made even when β-galactoside is absent. The repressor protein has a binding site for allolactose and undergoes a conformational change when bound to this compound (Fig. 6.8B). This means that the repressor is no longer able to bind to the operator. The way is then clear for RNA polymerase to bind to the promoter and to transcribe the operon. Thus in a short time the bacteria produce the proteins necessary for utilizing the new food source. The concentration of the substrate (lactose in this case) determines whether or not mRNA is synthesized. The *lac* operon is said to be under negative regulation by the repressor protein.

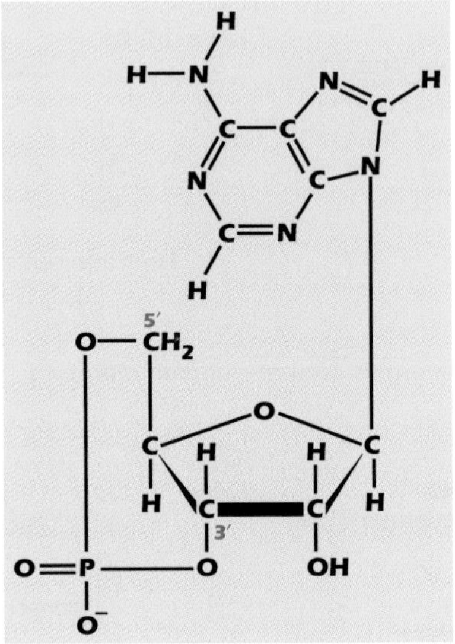

Figure 6.9. Cyclic adenosine monophosphate, also called cyclic AMP or just cAMP.

The transcription of the *lac* operon is controlled not only by the repressor protein but also by another protein, the **catabolite activator protein (CAP)**. If both glucose and lactose are present, it is more efficient for the cell to use glucose as the carbon source because the utilization of glucose requires no new RNA and protein synthesis, all the proteins necessary being already present in the cell. Only in the absence of glucose, therefore, does *E. coli* transcribe the *lac* operon at a high rate. This control operates through the intracellular messenger molecule **cyclic adenosine monophosphate (cyclic AMP or cAMP)** (Fig. 6.9). When glucose concentrations are low, the concentration of cyclic AMP increases. Cyclic AMP binds to CAP, and the complex then binds to a sequence upstream of the *lac* operon promoter (Fig. 6.10) where it has a remarkable effect. The DNA surrounding CAP bends by about 90 degrees. Both α subunits of RNA polymerase are now able to make contact with CAP at the same time so that the affinity of RNA polymerase for the *lac* promoter is increased. The result is that now the *lac z*, *lac y*, and *lac a* genes can be transcribed very efficiently. The *lac* operon is said to be under **positive regulation** by the CAP-cAMP complex.

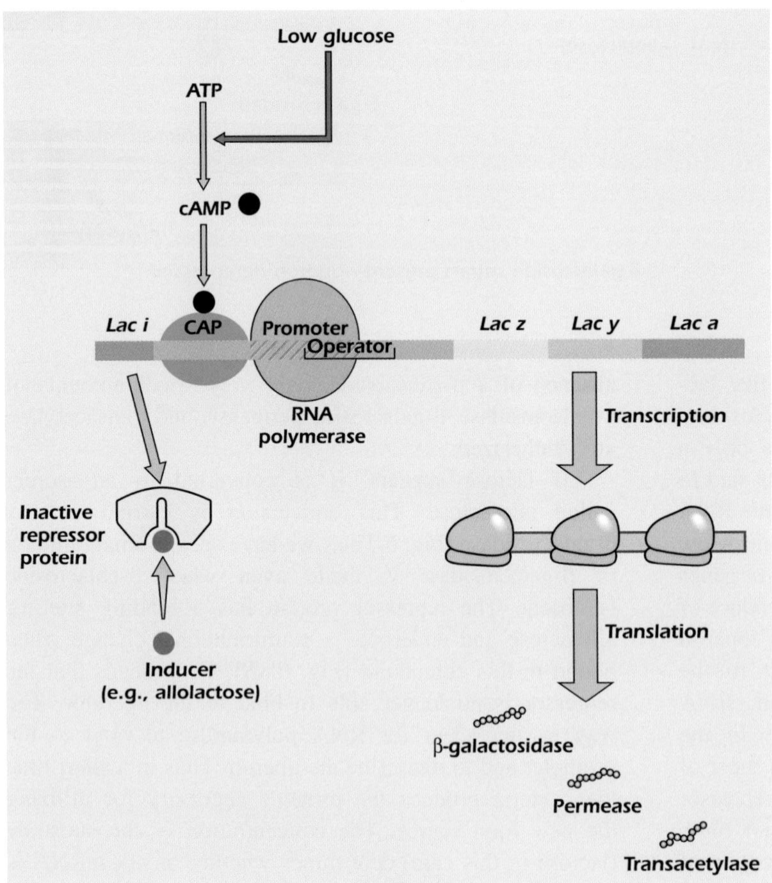

Figure 6.10. For efficient transcription of the *lac* operon, both cAMP and a β-galactoside sugar must be present.

To recap, the control of the *lac* operon is not simple. Several requirements need to be met before it can be transcribed. The repressor must not be bound to the operator, and the CAP–cyclic AMP complex and RNA polymerase must be bound to their respective DNA binding sites. These requirements are only met when glucose is absent and a β-galactoside, such as the sugar lactose, is present.

Other compounds such as isopropylthio-ß-D-galactoside (IPTG) (Fig. 6.11) can bind to the repressor but are not metabolized. These **gratuitous inducers** are very useful in DNA research and in biotechnology. Chapter 7 deals with this and with some of the industrial applications of the *lac* operon.

Example 6.1 Quorum Sensing: Squids That Glow in the Dark

The bacterium *Vibrio fischeri* lives free in seawater but is also found at high densities in the light-emitting organs of the nocturnal squid *Euprymna scolopes*, where it synthesizes an enzyme called luciferase that generates light. This phenomenon is called bioluminescence. When living free in seawater *V. fischeri* synthesizes almost no luciferase; indeed, there would be little point in doing so because the light emitted by a single bacterium would be too dim for anything to see. *V. fischeri* only begins making lots of luciferase when the density of bacteria is high—just as it is in the squid light organs. The word *quorum* is defined in Webster's Dictionary as "the number of... members of a body that when duly assembled is legally competent to transact business." Thus *quorum sensing* is a good description of *V. fischeri's* behavior. How does it work?

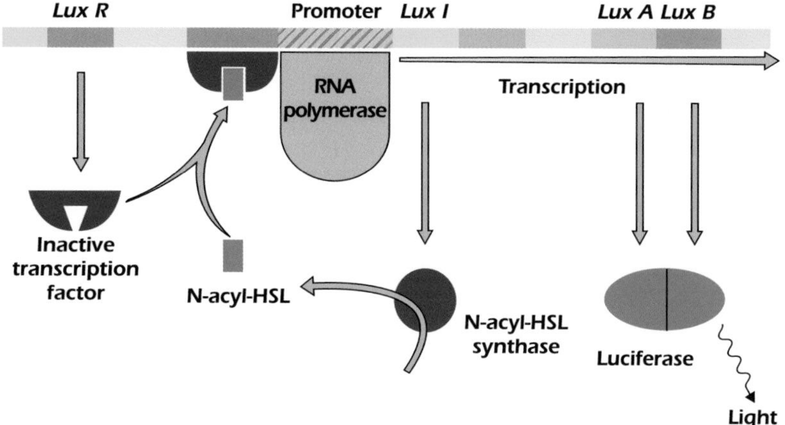

Luciferase is a product of the *lux a* and *lux b* genes in a bacterial operon called the *lux* operon. A region in the *lux* operon promoter binds a transcription factor, LuxR, which is only active when it has bound the small, uncharged molecule N-acyl-HSL (also called VAI for *V. fischeri* autoinducer). N-acyl-HSL in turn is made by the enzyme N-acyl-HSL synthase (or VAI synthase) that is encoded by the gene *lux i*, which is part of the *lux* operon. In free-living *V. fischeri* the *lux* operon is transcribed at a low level. Small amounts of N-acyl-HSL are made, which immediately leak out of the cell into the open sea without binding to LuxR. When the bacteria are concentrated in the squid's light organs, then some N-acyl-HSL binds to LuxR, increasing transcription of the *lux* operon. This makes more luciferase—but it also makes more N-acyl-HSL synthase. The concentration of N-acyl-HSL therefore rises—so transcription of the *lux* operon increases further. This means that the genes for N-acyl-HSL synthase, luciferase, and the enzymes that produce the substrate for luciferase are now transcribed at a high rate. This **autoinduction** of the *lux* operon by N-acyl-HSL is a form of positive feedback (page 223). When luciferase carries out its reaction, the squid will luminesce based on the intensity of moonlight so that as it glides at night over the coral reefs of Hawaii it does not cast a dark shadow; thus its prey are less likely to notice its presence.

IPTG (isopropylthio-β-D-galactoside)

Figure 6.11. Isopropylthio-β-D-galactoside (IPTG), which can bind to the *lac* repressor protein but which is not metabolized.

trp, a Repressible Operon

Operons that code for proteins that synthesize amino acids are regulated in a different way from the *lac* operon. These operons are only transcribed if the amino acid is not present, and transcription is switched off if there is already enough of the amino acid around. In this way the cell carefully controls the concentration of free amino acids. The tryptophan (*trp*) operon is made up of five structural genes encoding enzymes that synthesize the amino acid tryptophan (Fig. 6.12). This is a **repressible operon**. The cell regulates the amount of tryptophan produced by preventing transcription of the *trp* operon mRNA when there is sufficient tryptophan about. As with the *lac* operon, the transcription of the *trp* operon is controlled by a regulatory protein. The gene *trp r* encodes an inactive repressor protein that is called an **aporepressor**. Tryptophan binds to this to produce an active repressor complex. The active repressor complex binds to the operator sequence of the *trp* operon and prevents the attachment of RNA polymerase to the *trp* promoter sequence. Therefore, when the concentration of tryptophan in the cell is high, the active repressor complex will form, and transcription of the *trp* operon is prevented. However, when the amount of tryptophan in the cell decreases, the active repressor complex cannot be formed. RNA polymerase binds to the promoter, transcription of the *trp* operon proceeds, and the enzymes needed to synthesize tryptophan are produced. This is an example of **negative feedback** (page 223).

Many other operons are regulated by similar mechanisms in which specific regulatory proteins interact with specific small molecules.

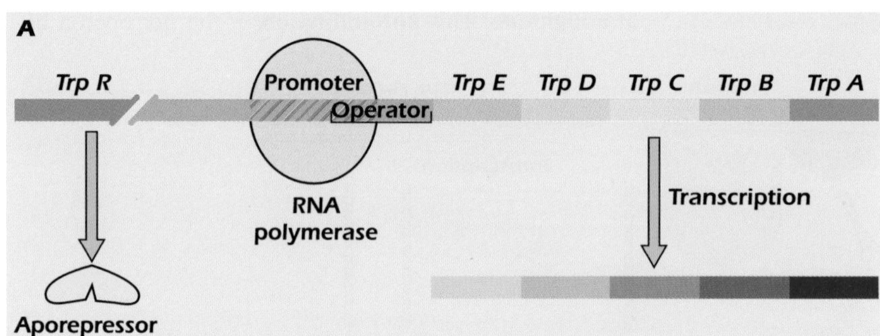

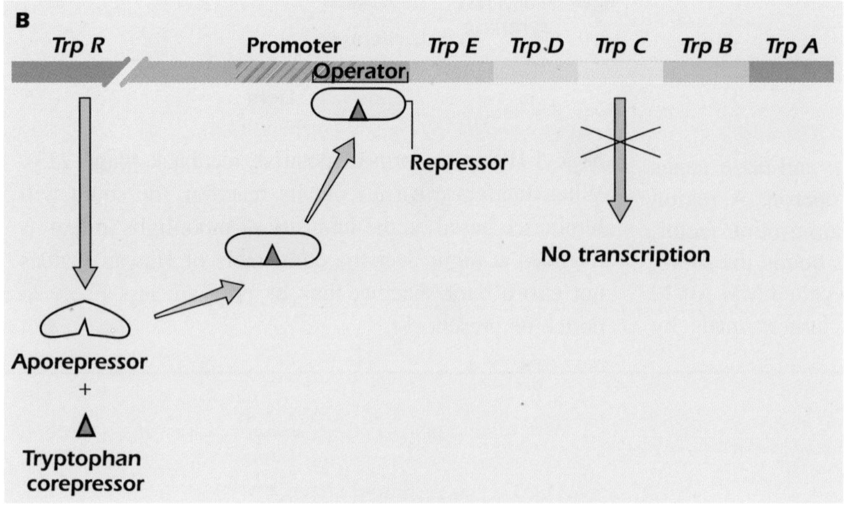

Figure 6.12. Transcription of the *trp* operon is controlled by the concentration of the amino acid tryptophan.

EUKARYOTIC RNA SYNTHESIS

Eukaryotes have three types of RNA polymerase. **RNA polymerase I** transcribes the genes that code for most of the ribosomal RNAs. All messenger RNAs are synthesized using **RNA polymerase II**. Transfer RNA genes are transcribed by **RNA polymerase III**. The chemical reaction catalyzed by these three RNA polymerases, the formation of phosphodiester links between nucleotides, is the same in eukaryotes and bacteria.

Messenger RNA Processing in Eukaryotes

A newly synthesized eukaryotic mRNA undergoes several modifications before it leaves the nucleus (Fig. 6.13). The first is known as capping. Very early in transcription the 5′-terminal triphosphate group is modified by the addition of a guanosine via a 5′-5′-phosphodiester link. The guanosine is subsequently methylated to form the **7-methyl guanosine cap**. The 3′ ends of nearly all eukaryotic mRNAs are modified by the addition of a long stretch of adenosine residues, the **poly-A tail**. A sequence 5′ AAUAAA 3′ is found in most eukaryotic mRNAs about 20 bases from where the poly-A tail is added and is probably a signal for the enzyme poly-A polymerase to bind and to begin the polyadenylation process. The length of the poly-A tail varies; it can be as long as 250 nucleotides. Unlike DNA, RNA is an unstable molecule, and the capping of eukaryotic mRNAs at their 5′ ends and the addition of a poly-A tail to their 3′ ends increases the lifetime of mRNA molecules by protecting them from digestion by nucleases.

Many eukaryotic protein-coding genes are split into exon and intron sequences. Both the exons and introns are transcribed into mRNA. The introns have to be removed and the exons joined together by a process known as **RNA splicing** before the mRNA can be used to make protein. Removal of introns takes place within the nucleus. Splicing is complex and not yet fully understood. It has, however, certain rules. Within an mRNA the first two bases following an exon are always GU and the last two bases of the intron are AG. Several **small nuclear RNAs** (snRNAs) are involved in splicing. These are complexed with a number of proteins to form a structure known as the **spliceosome**. One of the snRNAs is complementary in sequence to either end of the intron sequence. It is thought that binding of this snRNA to the intron, by complementary base pairing, brings the two exon sequences together, which causes the intron to loop out (Fig. 6.13). The proteins in the spliceosome remove the intron and join the exons together. Splicing is the final modification made to the mRNA in the nucleus. The mRNA is now transported to the cytoplasm for protein synthesis.

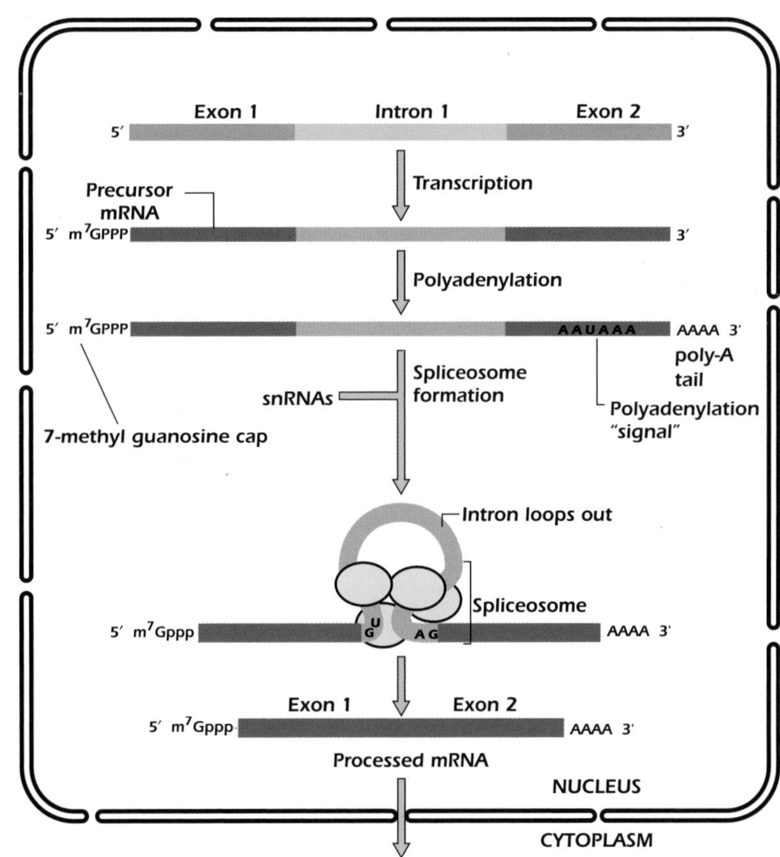

Figure 6.13. mRNA processing in eukaryotes.

As well as removing introns, splicing can sometimes remove exons in the process of **alternative splicing**. This allows the same gene to give rise to different proteins at different times or in different cells. Alternative splicing is a powerful mechanism that allows the approximately 20,500 protein-coding genes in the human genome to code for hundreds of thousands of proteins (In Depth 5.1 on page 76). For example, alternative splicing of the gene for the molecular motor dynein produces motors that transport different types of cargo (page 288).

CONTROL OF EUKARYOTIC GENE EXPRESSION

Since most eukaryotes are multicellular organisms with many cell types, gene expression must be controlled so that different cell lineages develop differently and remain different. A brain cell is quite different from a liver cell because it contains different proteins even though the DNA in the two cell types is identical. During development and differentiation, different sets of genes are switched on and off. Hemoglobin, for example, is only expressed in developing red blood cells even though the globin genes are present in all types of cell. Genetic engineering technology (Chapter 7) has made the isolation and manipulation of eukaryotic genes possible. This has given us some insight into the extraordinarily complex processes that regulate transcription of eukaryotic genes and allow a fertilized egg to develop into a multicellular, multitissue adult.

Unlike the situation in bacteria, the eukaryotic cell is divided by the nuclear envelope into nucleus and cytoplasm. Transcription and translation are therefore separated in space and in time. This means that the expression of eukaryotic genes can be regulated at more than one place in the cell. Although gene expression in eukaryotes is controlled primarily by regulating transcription in the nucleus, there are many instances in which expression is controlled at the level of translation in the cytoplasm or by altering the way in which the primary mRNA transcript is processed.

The interaction of RNA polymerase with its promoter is far more complex in eukaryotes than it is in bacteria (Fig. 6.14). This section describes how the transcription of a gene, encoding mRNA, is transcribed by RNA polymerase II. In contrast to bacterial RNA polymerase, RNA polymerase II cannot recognize a promoter sequence. Instead, other proteins known as **transcription factors** bind to the promoter and guide RNA polymerase II to the beginning of the gene to be transcribed.

The promoter sequence of most eukaryotic genes encoding mRNAs contains an AT-rich region about 25bp upstream of the transcription start site. This sequence, called the **TATA box**, binds a protein called transcription factor IID (TFIID), one of whose subunits is the TATA-binding protein, or TBP (Fig. 6.14A). Several other transcription factors (TFIIA, TFIIB, TFIIE, TFIIF,

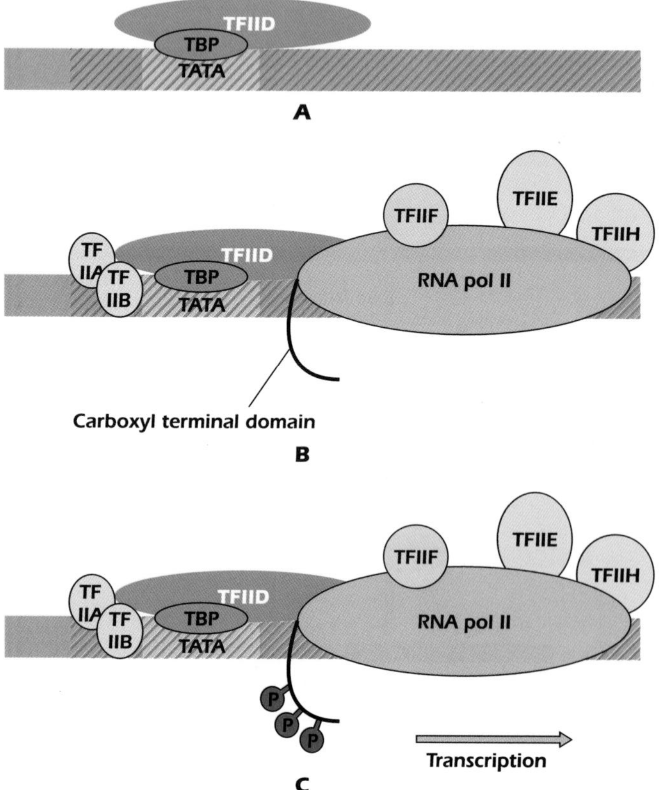

Figure 6.14. In eukaryotes, RNA polymerase II is guided to the promoter by TFII accessory proteins. (A) TBP binds to the TATA box. (B) The complete transcription preinitiation complex. (C) Phosphorylated RNA polymerase is active.

and TFIIH) then bind to TFIID and to the promoter region (Fig. 6.14B). TFIIF is the protein that guides RNA polymerase II to the beginning of the gene to be transcribed. The complex formed between the TATA box, TFIID, the other transcription factors, and RNA polymerase is known as the transcription preinitiation complex. Note that the proteins with the prefix TFII are so named because they are transcription factors that help RNA polymerase II to bind to promoter sequences.

Although many gene promoters contain a TATA box, some do not. These TATA-less genes usually encode proteins that are needed in every cell and are hence called housekeeping genes. The promoters of these genes contain the sequence 5′ GGGGCGGGGC 3′, called the GC box. A protein called Sp1 binds to the GC box and is then able to recruit TATA-binding protein to the DNA even though there is no TATA box for the latter to bind to. TATA-binding protein then recruits the rest of the transcription preinitiation complex so that transcription can proceeed.

In either case, transcription begins when the carboxy-terminal domain of RNA polymerase II is phosphorylated. This region is rich in the amino acids serine and threonine which contain OH groups in their side chains. When these OH groups are phosphorylated (pages 142, 178), RNA polymerase II breaks away from the preinitiation complex and proceeds to transcribe DNA into mRNA (Fig. 6.14C).

Although the formation of the transcription preinitiation complex is sometimes enough to produce a few molecules of RNA, the binding of other proteins to sequences next to the gene greatly increases the rate of transcription producing much more mRNA. These proteins are also called transcription factors, and the DNA sequences to which they bind are **enhancers**, so named because their presence enhances transcription. Enhancer sequences often lie upstream of a promoter, but they have also been found downstream. Enhancer sequences and the proteins that bind to them play an important role in determining whether a particular gene is to be transcribed. Some transcription factors bind to a gene to ensure that it is transcribed at the right stage of development or in the right tissue. Figure 6.15 shows how one gene can be transcribed in skeletal muscle but not in the liver simply because of the presence or absence of proteins that bind to enhancer sequences.

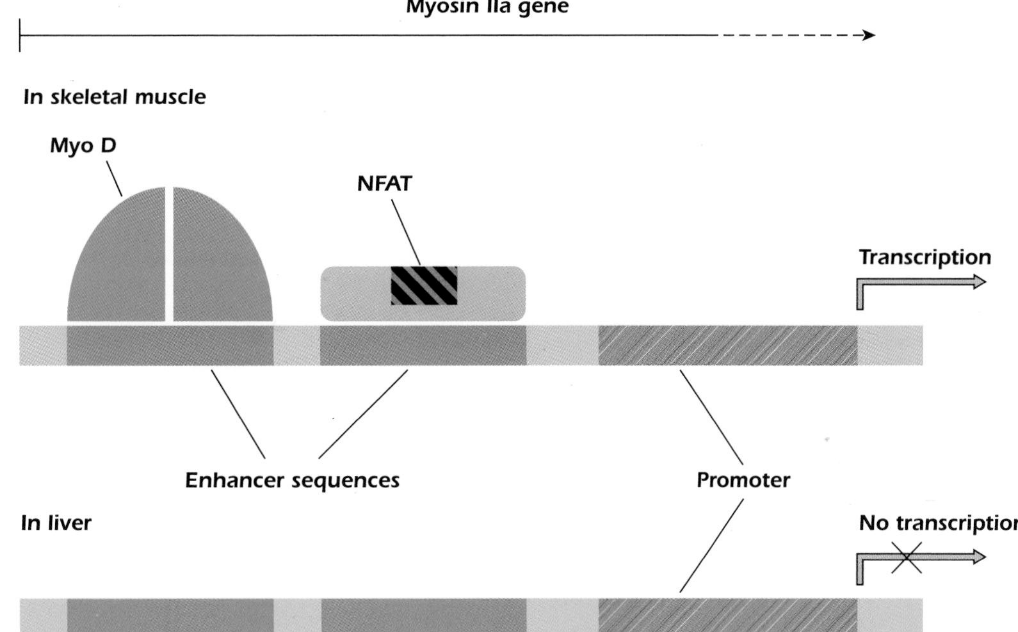

Figure 6.15. Tissue-specific transcription. The myosin IIa gene is not transcribed in liver cells, which do not contain the transcription factors Myo D and NFAT.

IN DEPTH 6.1 RNA MOLECULES SILENCE GENE EXPRESSION

A special group of RNA molecules called microRNAs (miRNAs) silence (switch off) the expression of certain genes. miRNAs are themselves encoded in the genome and are transcribed as longer RNA molecules that are then cleaved by a ribonuclease to produce short molecules, usually about 21 to 23 nucleotides in length. MicroRNAs silence gene expression by binding to other RNAs through complementary base pairing. An RNA that is bonded to an miRNA is not translated into protein. For example, we will later describe how the transcription factor p53 supresses cell division. One way it does this is by causing expression of human microRNA 34a, which then blocks translation of Cdk4 and Cdc25, two enzymes that are necessary for entry into mitosis (page 305).

Production of miRNAs is a normal function of a cell. However, when increased amounts of miRNAs are produced this can lead to disease because important genes are silenced.

Glucocorticoids Cross the Plasma Membrane to Activate Transcription

Glucocorticoids are steroid hormones produced by the adrenal cortex that increase the transcription of several genes important in carbohydrate and protein metabolism. Because they are uncharged and relatively nonpolar, steroid hormones can pass through the plasma membrane by simple diffusion to enter the cytosol. Here they encounter a class of transcription factors that have a binding site for steroid hormones and are therefore called **steroid hormone receptors** (Fig. 6.16). In the absence of glucocorticoid hormone, its receptor remains in the cytosol and is inactive because it is complexed to two molecules of an inhibitor protein known as Hsp90. However, when the glucocorticoid hormone enters the cell and binds to its receptor, the Hsp90 protein is displaced. The targeting sequence (page 158) that targets the receptor to the nucleus is uncovered, and the glucocorticoid receptor:hormone complex can now move into the nucleus. Here, two molecules of the complex bind to a 15-bp sequence known as the **hormone response element** (HRE) that lies upstream of the TATA box. The HRE is an enhancer sequence. The glucocorticoid receptor–hormone complex interacts with the preinitiation complex bound to the TATA box, and the rate at which RNA polymerase transcribes genes containing the HRE is increased.

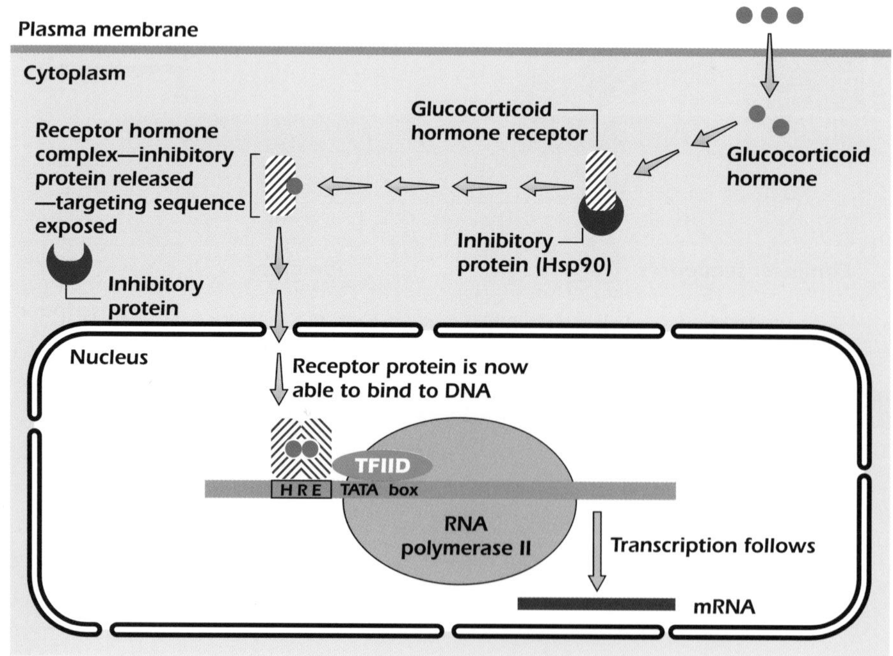

Figure 6.16. The glucocorticoid hormone receptor acts to increase gene transcription in the presence of hormone.

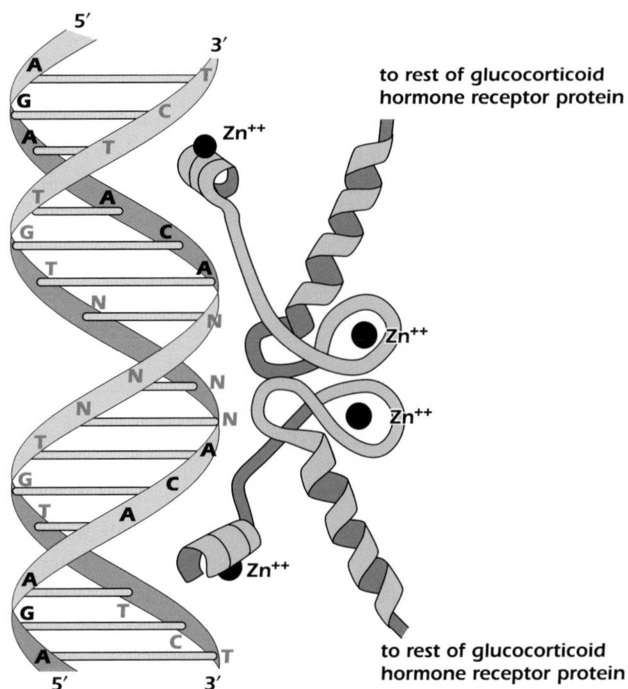

Figure 6.17. The dimerized glucocorticoid hormone receptor binds to a palindromic HRE.

Figure 6.17 shows why the glucorticoid hormone receptor binds to DNA as a dimer. The HRE is a palindrome—the sequence on both the top and bottom strands is the same when read in the 5′ to 3′ direction. Each strand of the HRE has the 6-bp sequence 5′ AGAACA 3′ that is known as the core recognition motif. This is the sequence to which a single receptor molecule binds. Because the HRE contains two recognition motifs, it binds two molecules of glucocorticoid receptor. The two 6-bp sequences are separated by three base pairs shown as N, which presumably are there to provide sufficient space for the receptor homodimer to fit snugly on the double helix. The glucocorticoid receptor is unaffected by the identity of these three base pairs.

Chemicals that are released from one cell and that alter the behavior of other cells are in general called **transmitters**. Glucocorticoids are an example of transmitters that alter gene transcription. The cells of multicellular organisms turn the transcription of genes on and off in response to many extracellular chemicals. Unlike steroid hormones, most of these transmitters cannot enter the cell and must activate **intracellular messenger** systems inside the cell that in turn carry the signal onward from the plasma membrane to the nucleus (Chapter 15).

Medical Relevance 6.1 St. John's Wort, Grapefruit and a Natural Detox

When we ingest plant products and other chemicals foreign to us, the body responds by increasing the amounts of a group of proteins known as the cytochromes P450 (CYPs). This remarkable response, found largely in the liver, is our built-in detoxication system. The foreign chemical signals the appropriate *CYP* gene to activate, transcription takes place, and more CYP protein is produced. The CYP protein then metabolizes the foreign chemical so that it will be cleared from the body quickly and efficiently through the urine or feces.

St. John's wort is a herbal remedy, taken by many as a natural antidepressant. One of the active components is the chemical hyperforin, which upregulates transcription of the gene for one particular cytochrome P450 called CYP3A4. The transcription of the *CYP3A4* gene is increased because hyperforin passes to the nucleus and binds to a receptor called the pregnane X receptor (PXR). The receptor-hyperforin complex is then able to dimerize with another receptor called RXR. The dimer binds to an enhancer sequence of the *CYP3A4* gene, interacts with the transcription preinitiation complex, and triggers RNA polymerase II to begin transcription of CYP3A4 mRNA. The extra CYP3A4 will help clear toxins and other foreign chemicals from the body faster.

However, among the foreign chemicals destroyed by CYP3A4 are about 50% of currently prescribed medicinal drugs. By increasing the rate at which drugs are removed from the body, St. John's wort will reduce the concentration achieved for a standard dose. For example, CYP3A4 helps destroy the progesterone analogues in birth control pills, so if one takes St. John's wort while on the pill one can get pregnant.

In contrast several ingredients of grapefruit juice inhibit the enzymic activity of the CYP3A4 protein and as a result the drugs CYP3A4 destroys can build up to a higher, dangerously toxic, concentration in the body. For this reason patients on many medicines are advised not to eat grapefruit.

Medical Relevance 6.2 Aldosterone in the Kidney

We will learn on page 247 how the kidney uses a channel called ENaC to recover sodium that would otherwise be lost in the urine. When body sodium is low, the adrenal gland produces the steroid hormone aldosterone. This binds to glucocorticoid hormone receptors in kidney cells. The receptor proteins then dimerize and bind in turn to a hormone response element in the enhancer region of the *ENaCα* gene. The result is that more ENaC protein is made and more sodium is reabsorbed.

Medical Relevance 6.3 Glucocorticoid Hormones Can Repress Transcription: Rheumatoid Arthritis

Glucocorticoid treatment has been shown to give some relief to individuals who suffer from the debilitating disorder rheumatoid arthritis. Collagenase, an enzyme that digests collagen, is generated in the joints of these patients, causing destruction of the extracellular matrix and therefore chronic inflammation. Transcription of the collagenase gene is controlled by an enhancer sequence called the AP1 site. For transcription to occur, the enhancer must be occupied by a transcription factor called AP1, which, like the active glucocorticoid receptor, is a dimer. Glucocorticoid hormones inhibit transcription of the collagenase gene by an ingenious mechanism. The glucocorticoid hormone, on entering the cell, interacts with the glucocorticoid receptor as shown in Figure 6.16. The receptor–hormone complex moves to the nucleus and binds to the proteins that would otherwise dimerize to form AP1. The heterodimer is unable to activate transcription of the collagenase gene. By depleting the pool of AP1 subunits, the glucocorticoid receptor–hormone complex prevents transcription of the collagenase gene.

SUMMARY

1. DNA is transcribed into RNA by the enzyme RNA polymerase. The three types of RNA are ribosomal RNA (rRNA), transfer RNA (tRNA), and messenger RNA (mRNA). Uracil, adenine, cytosine, and guanine are the four bases in RNA.

2. In bacteria, RNA polymerase binds to the promoter sequence just upstream of the start site of transcription. The enzyme moves down the DNA template and synthesizes an RNA molecule. RNA synthesis stops once the enzyme has transcribed a terminator sequence.

3. Bacterial genes encoding proteins for the same metabolic pathway are often clustered into operons. Some operons are induced in the presence of the substrate of the pathway, for example, the *lac* operon. Others are repressed in the presence of the product of the pathway, for example, the *trp* operon.

4. Eukaryotic mRNAs are modified by the addition of a 7-methyl-guanosine cap at their 5′ end. A poly-A tail is added to their 3′ end. Intron sequences are removed, and the exon sequences joined together by the process known as splicing. The fully processed mRNA is then ready for transport to the cytoplasm and protein synthesis.

5. In eukaryotes, there are three RNA polymerases—RNA polymerases I, II, and III. RNA polymerase II needs the help of the TATA-binding protein and other transcription factors to become bound to a promoter. This group of proteins is called the transcription preinitiation complex, and this is sufficient to make a small number of RNA molecules. However, to make a lot of RNA in response to a signal, such as a hormone, other proteins bind to sequences called enhancers. These proteins interact with the preinitiation complex and increase the rate of RNA synthesis.

FURTHER READING

Lloyd, G., Landini, P., and Busby, S. (2001) Activation and repression of transcription initiation in bacteria. *Essays Biochem*. **37**, 17–31.

Roberts, G. C., and Smith, C. W. (2000) Alternative splicing: Combinatorial output from the genome. *Curr. Opin. Chem. Biol*. **6**, 375–383.

Tjian, R. (1995) Molecular machines that control genes. *Sci. Am*. **272**, 38–45.

REVIEW QUESTIONS

6.1 Theme: Codes within the Base Sequence

A Bending of the DNA double helix through a right angle
B Formation of a 7-methyl guanosine cap on eukaryotic mRNA
C Polyadenylation of eukaryotic mRNA
D Removal of introns in eukaryotic mRNA
E Transcription initiation in eukaryotes
F Transcription initiation in prokaryotes
G Transcription termination in eukaryotes
H Transcription termination in prokaryotes

From the above list of processes, identify the process triggered by or associated with each of the base sequences below.

1. A stretch of DNA rich in adenine and thymine, called the TATA box
2. A stretch of DNA rich in guanine and cytosine, followed by a string of adenines
3. The DNA sequence GGGGCGGGGC, called the GC box
4. The DNA sequence TATATT, called the −10 or Pribnow box
5. The RNA motif GU . . . . AG, where the . . . . indicates a long sequence of bases
6. The RNA sequence AAUAAA

6.2 Theme: The Control of Transcription

A A protein that binds to a regulatory site on DNA only in the absence of a small soluble molecule. Binding of the protein to the DNA increases transcription of the relevant gene. If the small soluble molecule appears, the protein adopts a conformation that cannot bind to the regulatory site, so the transcription promoting effect disappears. (Increases transcription when small soluble molecule absent)
B A protein that binds to a regulatory site on DNA only in the absence of a small soluble molecule. Binding of the protein to the DNA reduces transcription of the relevant gene. If the small soluble molecule appears, the protein adopts a conformation that cannot bind to the regulatory site, so the blocking effect on transcription disappears. (Inhibits transcription when small soluble molecule absent)
C A protein that binds to a regulatory site on DNA only in the presence of a small soluble molecule. Binding of the protein to the DNA increases transcription of the relevant gene. If the small soluble molecule disappears, the protein adopts a conformation that cannot bind to the regulatory site, so the transcription promoting effect disappears. (Increases transcription when small soluble molecule present)
D A protein that binds to a regulatory site on DNA only in the presence of a small soluble molecule. Binding of the protein to the DNA reduces transcription of the relevant gene. If the small soluble molecule disappears, the protein adopts a conformation that cannot bind to the regulatory site, so the blocking effect on transcription disappears. (Inhibits transcription when small soluble molecule present)

From the above list of descriptions, select the description applicable to each of the regulatory proteins listed below.

1. Catabolite activator protein
2. Glucocorticoid hormone receptor
3. *Lac* repressor protein
4. *Trp* aporepressor protein

6.3 Theme: Events That Occur after Transcription in Eukaryotes

A Capping: the addition of a methylated guanosine to the end of the RNA chain
B Digestion by nucleases
C Export from the nucleus
D Polyadenylation
E RNA splicing

From the above list of processes, select the one described by each of the descriptions below.

1. A chemical modification of the $3'$ end of the RNA molecule
2. A chemical modification of the $5'$ end of the RNA molecule
3. A process that allows two or more polypeptide chains of different amino acid sequence to be generated from the same mRNA transcript
4. A process that reduces, often dramatically, the length of the RNA molecule prior to its subsequent translation into protein

THOUGHT QUESTION

How is it that alternative splicing of a primary mRNA transcript sometimes generates two different mature mRNA molecules, both of which are translated, yet the proteins produced when the two mature mRNAs are translated are identical?

7

RECOMBINANT DNA AND GENETIC ENGINEERING

Deoxyribonucleic acid is the cell's database. Within the base sequence is all the information necessary to encode RNA and protein. A number of biological and chemical methods now give us the ability to isolate DNA molecules and to determine their base sequence. Once we have the DNA and know the sequence, many possibilities open up. We can identify mutations that cause disease, make a human vaccine in a bacterial cell, or alter a sequence and hence the protein it encodes. The knowledge of the entire base sequence (In Depth 5.2 on page 77) of the human genome, and of the genomes of many other organisms, such as bacteria that cause disease, is revolutionizing medicine and biology. In future years the power of genetic engineering is likely to impact ever more strongly on industry and on the way we live. This chapter describes some of the important methods involved in recombinant DNA technology at the heart of which is DNA cloning.

DNA CLONING

DNA cloning has had an enormous impact on our understanding of the information stored within cells. This is because the technology used in DNA cloning allows us to fragment large DNA molecules (e.g., a chromosome) into smaller ones and to separate these from each other. Cloning is a way to make many copies of selected DNA molecules and to store particular DNA sequences for later copying. It is very difficult to work with just one copy of a molecule of DNA: cloning provides the investigator with many copies of an identical DNA sequence that are then amenable to analysis.

Since all DNA molecules have very similar chemical properties, it is extremely difficult to purify individual species of DNA by classical biochemical techniques similar to those used successfully for the purification of proteins. However, we can use DNA cloning to help us to separate DNA molecules from each other. A **clone** is a population of cells that arose from one original cell and, in the absence of mutation, all members of a clone will be genetically identical. If a foreign gene or gene fragment is introduced into a cell and the cell then grows and divides repeatedly, many copies of the foreign gene can be produced, and the gene is then said to have been cloned. A DNA fragment can be cloned from any organism. The basic approach to cloning a gene is to take the genetic material from the cell of interest, which in the examples we will describe is a human cell, and to introduce this DNA into bacterial cells. Clones of bacteria are then generated, each of which contains and replicates one fragment of the human genetic material. The clones that contain the gene we are interested in are then identified and grown separately. We therefore use a biological approach to isolate DNA molecules rather than physical or chemical techniques.

Cell Biology: A Short Course, Third Edition. Stephen R. Bolsover, Jeremy S. Hyams, Elizabeth A. Shephard and Hugh A. White.

CREATING THE CLONE

How do we clone a human DNA sequence? The human genome has 3×10^9 base pairs of DNA, and with the exception of gametes (Chapter 18) and lymphocytes (Chapter 19) the DNA content of each cell is identical. However, each cell expresses only a fraction of its genes. Different types of cells express different sets of genes and thus their mRNA content is not the same. In addition, processed mRNA is shorter than its parent DNA sequence and contains no introns (page 91). Consequently, it is much easier to isolate a DNA sequence by starting with its mRNA. We therefore start the cloning process by isolating mRNA from the cells of interest. The mRNA is then copied into DNA by an enzyme called **reverse transcriptase** that is found in some viruses. As the newly synthesized DNA is complementary in sequence to the mRNA template, it is known as **complementary DNA**, or **cDNA**. The sample of cDNAs, produced from the mRNA, will represent the products of many different genes.

The way in which a cDNA molecule is synthesized from mRNA is shown in Figure 7.1. Most eukaryotic mRNA molecules have a string of As at their 3′ end, the poly-A tail (page 91). A short run of T residues can therefore be used to prime the synthesis of DNA from an mRNA template using reverse transcriptase. The resulting double-stranded molecule is a hybrid containing one strand of DNA and one of RNA. The RNA strand is removed by digestion with the enzyme ribonuclease H. This enzyme cleaves phosphodiester links in the RNA strand of the paired RNA-DNA complex, making a series of nicks down the length of the RNA. DNA polymerase (page 68) is then added. This homes in on the nicks and then moves along replacing ribonucleotides with deoxyribonucleotides. Lastly, DNA ligase is used to reform any missing phosphodiester links. In this way a double-stranded DNA molecule is generated by the replacement of the RNA strand with a DNA strand. If the starting point had been mRNA isolated from liver cells, then a collection of cDNA molecules representative of all the mRNA molecules within the liver will have been produced. These DNA molecules now have to be introduced into bacteria.

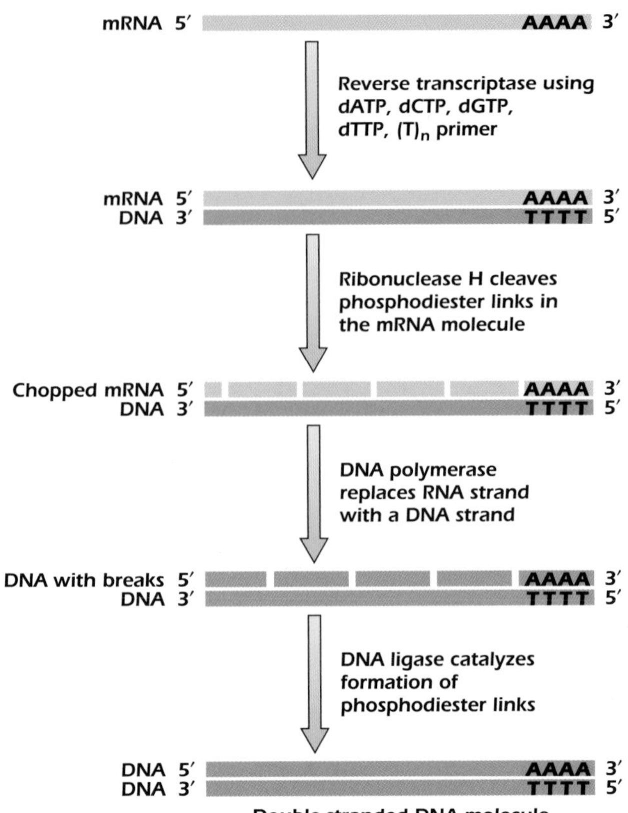

Figure 7.1. Synthesis of a double-stranded DNA molecule.

Introduction of Foreign DNA Molecules into Bacteria

Cloning Vectors. To ensure the survival and propagation of foreign DNAs, they must be inserted into a vector that can replicate inside bacterial cells and be passed on to subsequent generations of the bacteria. The vectors used for cloning are derived from naturally occurring bacterial plasmids or bacteriophages. Plasmids (page 59) are small circular DNA molecules found within bacteria. Each contains an origin of replication (page 67) and thus can replicate independently of the bacterial chromosome and produce many copies of itself. Plasmids often carry genes that confer antibiotic resistance on the host bacterium. The advantage of this to the scientist is that bacteria containing the plasmid can be selected for in a population of other bacteria simply by applying the antibiotic. Those bacteria with the antibiotic resistance gene will survive, whereas those without it will die. Figure 7.2 shows the basic components of a typical plasmid cloning vector: an antibiotic resistance gene, a restriction endonuclease site (see next section) at which foreign DNA can be inserted, and an origin of replication so the plasmid can copy itself many times inside the bacterial cell.

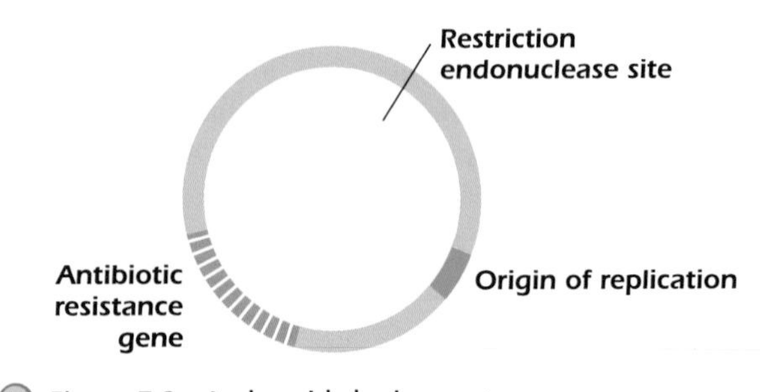

Figure 7.2. A plasmid cloning vector.

Bacteriophages are viruses that infect bacteria and utilize the host cell's components for their own replication. The bacteriophage genome is, like a plasmid, circular, although many viruses use RNA rather than DNA as their genetic material. If human DNA is inserted into a bacteriophage, the bacteriophage will do the job of introducing it into a bacterium.

Joining Foreign DNAs to a Cloning Vector. Enzymes known as **restriction endonucleases** are used to insert foreign DNA into a cloning vector. Each restriction endonuclease recognizes a particular DNA sequence of (usually) 4 or 6 bp. The enzyme binds to this sequence and then cuts both strands of the double helix. Many restriction endonucleases have been isolated from bacteria. The names and recognition sequences of a few of the common ones are shown in Figure 7.3. Restriction endonuclease names are conventionally written in italics because they are derived from the Latin name for the bacterium in which the protein occurs.

Example 7.1 The Cloning Vector pBluescript

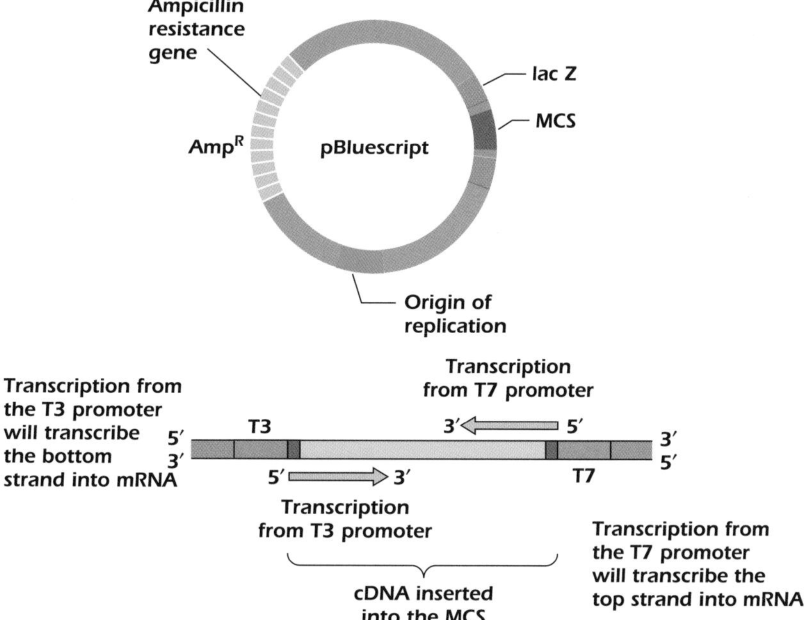

The plasmid pBluescript is based on a naturally occurring plasmid that has been engineered to include several valuable features. pBluescript has an origin of replication, and the ampicillin resistance gene for selecting bacterial cells that have taken up the plasmid. The multiple cloning site (MCS) allows the scientist to cut the plasmid with the most appropriate restriction endonuclease for the task at hand. The MCS lies within the *lac Z* gene, which codes for the enzyme β-galactosidase (page 85). β-galactosidase converts a substrate known as X-gal to a bright blue product. Cells containing pBluescript without a foreign DNA in the MCS will produce blue colonies when grown on agar plates. However, when the *lac Z* gene is disrupted by insertion of a foreign DNA in the MCS, the cells containing the recombinant plasmids will grow to produce a colony with the normal color of white. This is because the function of the *lac Z* gene is destroyed and β-galactosidase is not produced. This is the basis of a test, called the blue/white assay, to identify colonies containing recombinant plasmids.

Another important feature of pBluescript is the presence of the bacteriophage T7 and T3 promoter sequences, which flank the MCS. These promoter sequences are used to transcribe mRNA from a cDNA cloned into one of the sites within the MCS. By selecting the promoter sequence, and adding the appropriate RNA polymerase (T3 or T7 RNA polymerase), either the sense or the antisense mRNA can be synthesized in a test tube. The mRNAs produced can then be used for a number of different techniques. For example, antisense RNAs produced in this way are used for *in situ* hybridization (page 112) to detect cells producing a specific mRNA.

Bacterial species/strain	Enzyme name	Recognition sequences and cleavage sites
***Bacillus amyloliquefaciens* H**	*Bam* H1	5′ G|GATCC 3′ 3′ CCTAG|G 5′
***Escherichia coli* Ry13**	*Eco* R1	5′ G|AATTC 3′ 3′ CTTAA|G 5′
***Providencia stuartii* 164**	*Pst* 1	5′ CTGCA|G 3′ 3′ G|ACGTC 5′
***Serratia marcescens* SB**	*Sma* H1	5′ CCC|GGG 3′ 3′ GGG|CCC 5′
Rhodopseudomonas sphaeroides	*Rsa* 1	5′ GT|AC 3′ 3′ CA|TG 5′

Figure 7.3. Recognition sites of some common restriction endonucleases.

Some enzymes such as *Bam* H1, *Eco* R1, and *Pst* 1 make staggered cuts on each strand. The resultant DNA molecules are said to have **sticky ends** (Fig. 7.4) because such fragments can associate by complementary base pairing to any other fragment of DNA generated by the same enzyme. Other enzymes such as *Sma* H1 cleave the DNA smoothly to produce **blunt ends** (Fig. 7.4). DNA fragments produced in this way can be joined to any other blunt-ended fragment.

Figure 7.5 illustrates how human DNA is inserted into a plasmid that contains a *Bam* H1 restriction endonuclease site. A short length of synthetic DNA (an **oligonucleotide**) that includes a *Bam* H1 recognition site is added to each end of the human DNA fragment. Both the human DNA and the cloning vector are cut with *Bam* H1. The cut ends are now complementary and will anneal together by hydrogen bonding. DNA ligase then catalyzes the formation of phosphodiester links between the vector and the human DNA. The resultant molecule is known as a **recombinant plasmid**. If our starting material was mRNA from a sample of liver, we would now have a collection of plasmids each carrying a cDNA from one of the genes that was being transcribed in this organ.

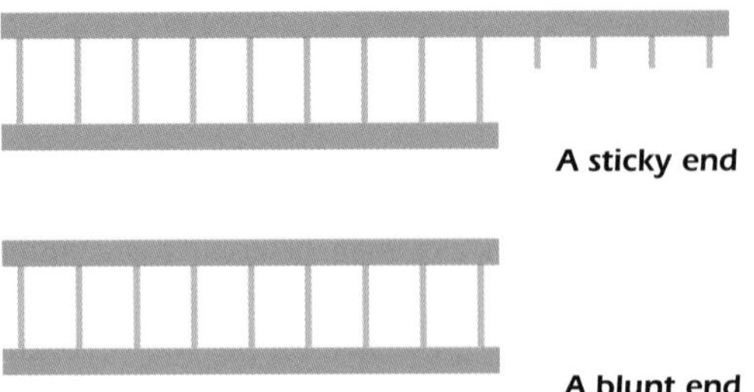

Figure 7.4. Restriction endonucleases generate two types of cut ends in double-stranded DNA.

Introduction of Recombinant Plasmids into Bacteria. Figure 7.6 summarizes how recombinant plasmids are introduced into bacteria such as *Escherichia coli*. Bacteria are first treated with concentrated calcium chloride to make the cell wall more permeable to DNA. DNA can now enter these cells, which are said to have been made **competent**. Cells that take up DNA in this way are said to be **transformed**. The transformation process is very inefficient, and only a small percentage of cells actually take up the recombinant molecules. This means that it is extremely unlikely that any one bacterium has taken up two plasmids. The presence of an antibiotic resistance gene in the cloning vector makes it possible to select those bacteria that have taken up a molecule of foreign DNA, since only the transformed cells can survive in the presence of the antibiotic. The collection of bacterial colonies produced after this selection process is a **clone library**. All the cells of a single colony harbor identical recombinant molecules that began as one mRNA molecule in the original cell sample. Other colonies in the same clone library contain plasmids carrying different DNA inserts. Isolating individual bacterial colonies will produce different clones of foreign DNA. In the example we have described, where the starting DNA material used to produce these clones was a population of cDNA molecules, the collection of clones is called a **cDNA library**.

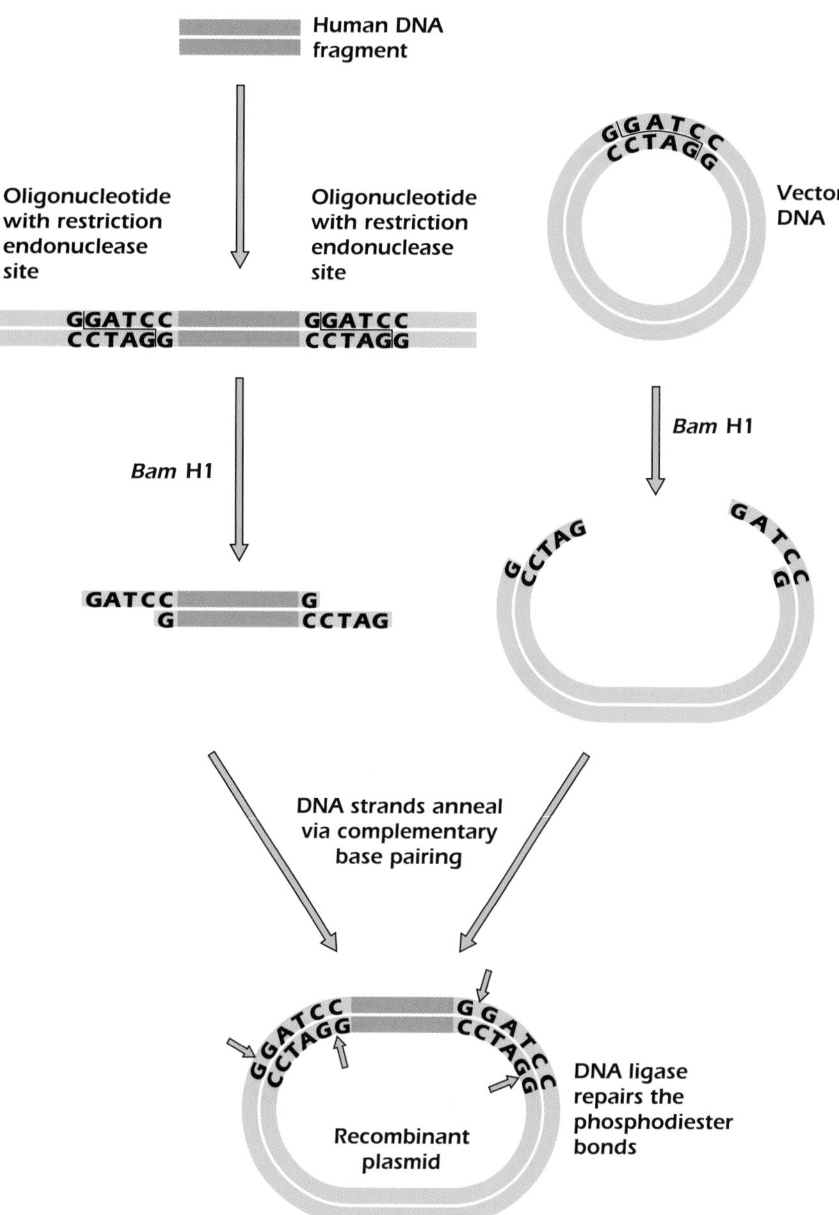

Figure 7.5. Generation of a recombinant plasmid.

Selection of cDNA Clones

Having constructed a cDNA library—which may contain many thousands of different clones—the next step is to identify the clones that contain the cDNA of interest. There are many ingenious ways of doing this. We will describe two ways of selecting cDNA clones from a library. One method simply detects the presence of the foreign DNA attached to the plasmid vector, and the second detects the protein encoded for by the foreign DNA. We call the process of selecting specific clones "screening the library."

Preparation of the cDNA Library for Screening. Bacterial colonies are plated onto agar plates, and the colonies are replica-plated onto a nylon membrane, which is then treated with detergent to burst (or lyse) the bound cells (Fig. 7.7). If the clone is to be selected by virtue of its DNA sequence, the nylon membrane is processed with sodium hydroxide. This is necessary to break all hydrogen bonds between the DNA strands bound to the nylon membrane and ensures that the DNA is single-stranded. The processed nylon membrane is an exact replica of the DNA contained within each bacterial colony on the agar plate. If the clone is to be selected from the library by detecting the protein encoded by the foreign DNA, then colonies are again replica-plated on to a nylon membrane. This time, however, the nylon membrane is processed to produce an exact copy of the proteins synthesized by each bacterial colony.

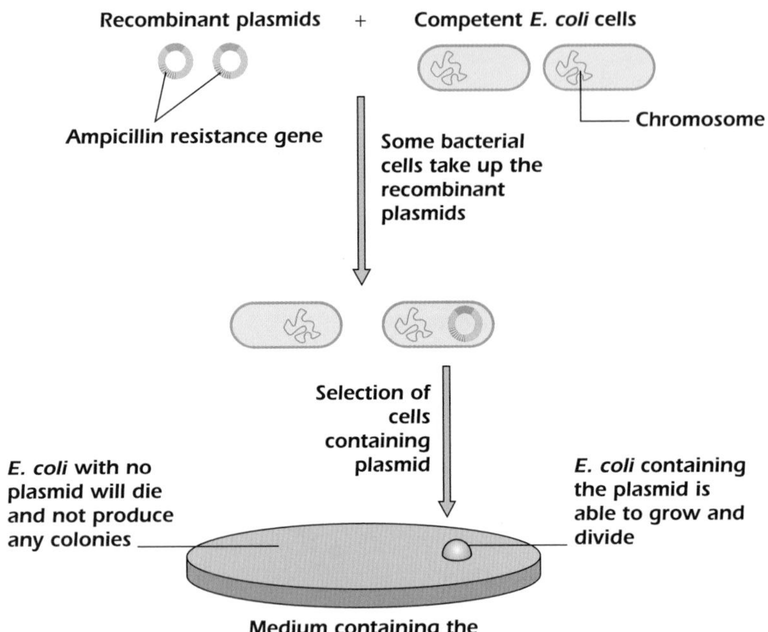

Figure 7.6. Introduction of recombinant plasmids into bacteria.

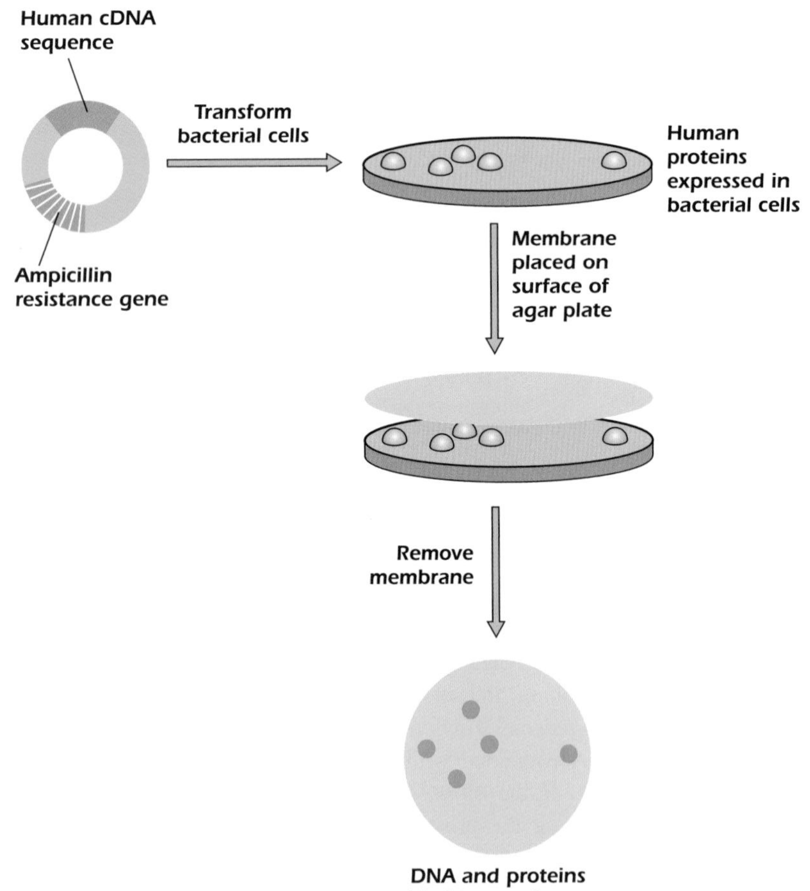

Figure 7.7. A membrane lift takes images of DNA and protein from colonies.

Oligonucleotide Probes for cDNA Clones. If some of the amino acid sequence is known for the protein whose cDNA is to be cloned, an oligonucleotide can be synthesized *in vitro* (i.e., in a machine rather than in a cell) that has a sequence complementary to one of the strands of the cDNA. The first step is to use the genetic code (page 61) to predict all possible DNA sequences that could code for a short stretch of amino acids within the protein. This strategy is shown in Figure 7.8. The sequence—met gln lys phe asn—can be coded for by 16 possible sequences, because of the degeneracy of the genetic code (page 61). All 16 oligonucleotide sequences are synthesized. One of the 16 sequences will be complementary to the cDNA we want to select from the library. The oligonucleotides are tagged with a radioactive phosphate group (^{32}P) at their 5′ ends using the enzyme polynucleotide kinase (PNK) and the substrate [$\gamma - {}^{32}$P]ATP, that is, ATP whose γ phosphate is the radioactive isotope ^{32}P. PNK removes the 5′-phosphate group from each oligonucleotide, leaving a 5′-hydroxyl group. The enzyme then transfers the γ(^{32}P) phosphate group of [$\gamma - {}^{32}$P]ATP to the 5′-hydroxyl.

The nylon membrane to which the library DNA is attached is incubated together with the mixture of radiolabeled oligonucleotides. This process is called **hybridization**, a word used whenever two nucleic acid strands associate together by hydrogen bonding. In this case the oligonucleotide complementary in sequence to the clone we want to select will hydrogen bond to the single-stranded DNA on the nylon membrane. Once hybridization is complete, excess oligonucleotide is washed from the nylon membrane, which is now covered with a sheet of X-ray film and placed in a light-tight

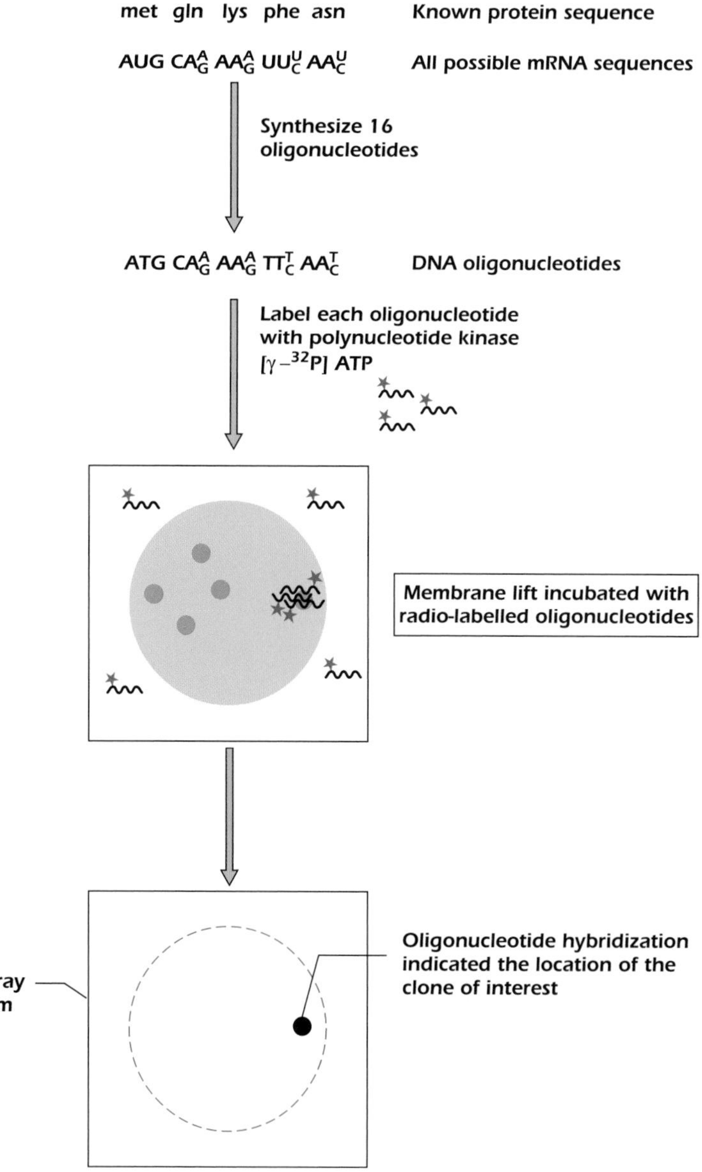

Figure 7.8. Use of a radioactive oligonucleotide probe to select a cDNA clone.

cassette. The radioactivity in the oligonucleotide will darken the silver grains on the X-ray film—a process known as **autoradiography**. A positive clone will show up as a black spot on the film. Superimposing the X-ray film back onto the original bacterial plate will identify the living bacterial colony that contains the desired foreign DNA clone.

Antibody Probes for cDNA Clones. This method makes use of specific antibodies to detect bacteria expressing the protein product of the DNA to be cloned. For this to work, the foreign DNA must be expressed in the bacterial cells; that is to say, its information must be copied first into mRNA and then into protein. To ensure efficient expression, the plasmid vector contains a bacterial promoter sequence that is used to control transcription of foreign DNA. Such cloning vectors are known as **expression vectors**. The promoter of the *lac* operon is commonly used in this way. The clone library is plated onto agar plates containing an inducer of the *lac* operon such as IPTG (page 89) to ensure that lots of mRNA and in turn lots of protein is synthesized. Figure 7.9 shows how an antibody, linked to an enzyme (usually alkaline phosphatase), can detect a positive clone by generating a colored product. The pattern of colored spots on the nylon membrane is used to identify the bacterial clones of interest on the original agar plate.

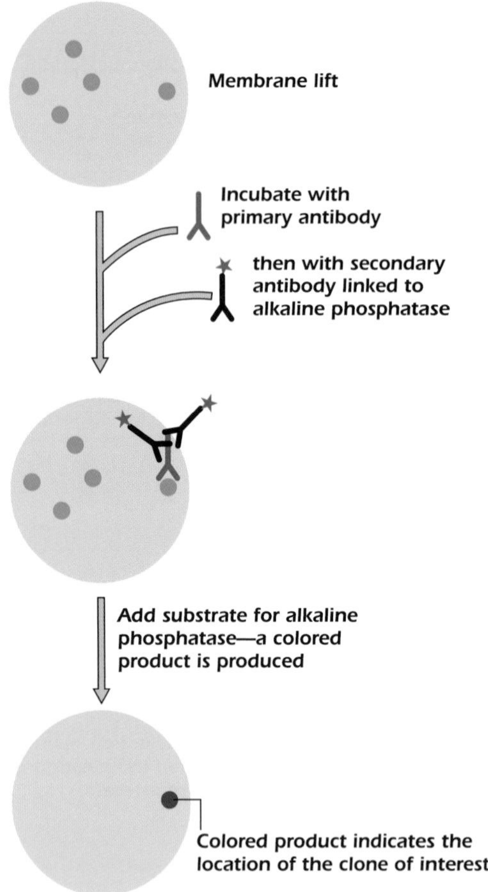
Membrane lift
Incubate with primary antibody
then with secondary antibody linked to alkaline phosphatase
Add substrate for alkaline phosphatase—a colored product is produced
Colored product indicates the location of the clone of interest

Figure 7.9. Use of an antibody probe to select a cDNA clone.

Genomic DNA Clones

The approach described in the previous section permits the isolation of cDNA clones. Complementary DNA clones have many important uses, some of which are described below. However, as a cDNA is a copy of mRNA only, when we want to isolate a gene to investigate its structure and function, we need to create genomic DNA clones. Genes contain exons and introns and have regulatory sequences at their 5′ and 3′ ends and are therefore much larger than cDNAs. The vectors used to clone genes must therefore be able to hold long stretches of DNA. Plasmids used for cDNA cloning cannot do this. A selection of vectors used to clone genes is shown in Table 7.1. Vectors such as the P1 artificial chromosomes (PACs)—based on the bacteriophage P1—can hold about 150,000 bp of DNA. Vectors called BACs (bacterial artificial chromosome) can hold up to 300,000 bp. PACs and BACs have been used very successfully in the Human Genome Project and in the sequencing of the genomes of other organisms such as the mouse (page 77). A yeast artificial chromosome (YAC) can hold between 200,000 and 600,000 base pairs of foreign DNA. The choice of genomic vector is governed by the size of DNA the scientist needs to clone. PAC, BAC and YAC vectors are needed if an entire gene sequence is to be represented in a single clone. They contain all the sequences needed to produce a minichromosome in the appropriate cell type: PACs and BACs in bacterial cells, and YACs to replicate foreign sequences in a yeast cell. The YAC vector therefore contains sequences that will allow the DNA to be processed by the host yeast cell as if it were a normal chromosome and to allow replication alongside the other chromosomes in the cell. Thus a YAC vector has sequences that specify the yeast centromere (page 298), telomeres (the ends of the new chromosome), and the yeast origin of replication. A YAC vector also contains selectable marker genes so that only those cells that have been transformed by a correctly constructed YAC chromosome will survive.

TABLE 7.1. Vectors Used for Cloning Genomic DNA

Genomic DNA Cloning Vector	Size of Insert (kb)
Bacteriophage	9–23
Cosmid	30–44
PAC (P1 artificial chromosome)	130–150
BAC (bacterial artificial chromosome)	Up to 300
YAC (yeast artificial chromosome)	200–600

Example 7.2 Cloning a Receptor Protein cDNA

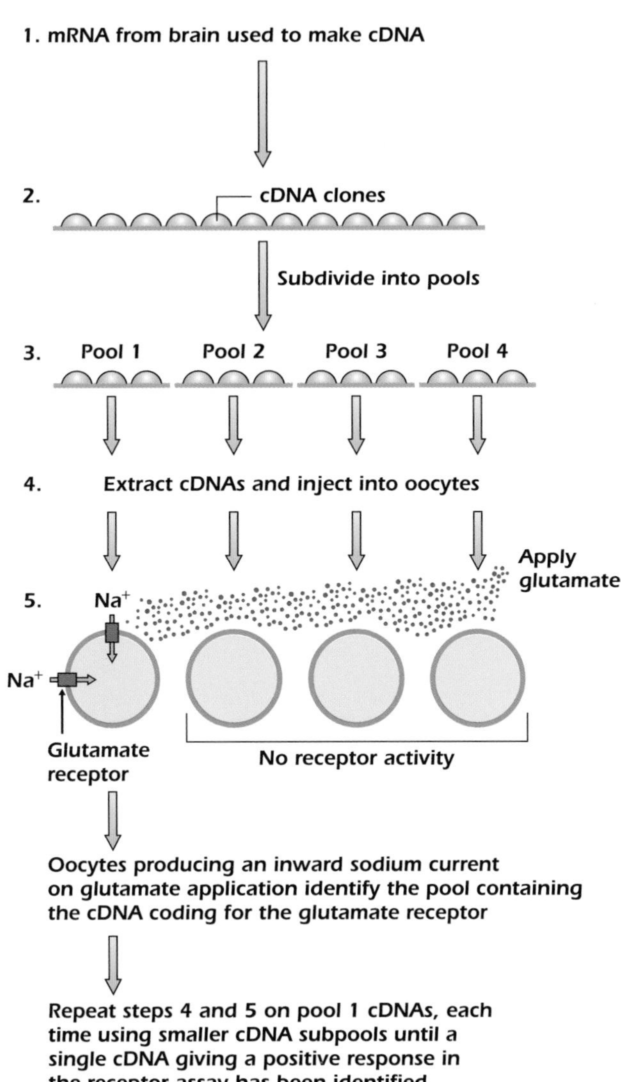
1. mRNA from brain used to make cDNA
2. cDNA clones
Subdivide into pools
3. Pool 1 Pool 2 Pool 3 Pool 4
4. Extract cDNAs and inject into oocytes
Apply glutamate
5. Na+
Na+
Glutamate receptor
No receptor activity
Oocytes producing an inward sodium current on glutamate application identify the pool containing the cDNA coding for the glutamate receptor
Repeat steps 4 and 5 on pool 1 cDNAs, each time using smaller cDNA subpools until a single cDNA giving a positive response in the receptor assay has been identified

Glutamate is one of the most important transmitters in the brain. The gene coding for the ionotropic glutamate receptor, the protein on the surface of nerve cells that, upon binding glutamate, allows an influx of sodium ions (page 269), remained uncloned for a number of years. Success came with the use of a very clever cloning strategy, based on the function of the receptor. mRNA was isolated from brain cells and used as the template for the production of cDNA molecules. These were inserted into a plasmid expression vector. Following the introduction of these cDNAs into bacteria, a cDNA library representative of all the mRNAs in the brain was produced. The many thousands of cDNA clones in the library were then divided into pools. Each of the many pools was then injected into a frog egg (Xenopus oocyte), which transcribed the cDNAs into RNA and translated the RNA into protein. To see which of the oocytes had been injected with the cDNA for the glutamate receptor, these cells were whole-cell patch clamped (page 231). Glutamate was applied to the oocytes, and the oocyte whose injected pool had included the cDNA for the glutamate receptor responded with an inward current of sodium ions indicating the presence of glutamate receptors in the plasma membrane.

The pool of cDNAs giving this response was further divided into smaller pools. Each of these was rescreened for the presence of glutamate receptor activity. This was followed by several rounds of rescreening. For each round a further subdivision was made of the cDNAs into pools containing fewer and fewer cDNA molecules. Eventually each pool contained only a single cDNA so that the cDNA for the glutamate receptor could be identified. A number of other receptors have now been isolated using the same strategy in which a functional assay is used to identify the cDNA encoding the receptor.

To generate the large DNA fragments needed for genomic cloning, the chromosomal DNA is incubated with a restriction endonuclease for a very short time. Not all the recognition sites for that enzyme are cleaved, and large fragments of DNA are hence produced by what is called "partial digestion." Genomic DNA fragments are joined to genomic cloning vectors in the same way we join cDNAs to cDNA cloning vectors. In the example shown in Figure 7.10, human DNA has been introduced into the genome of a bacteriophage known as lambda (λ). This particular vector can accommodate up to 23,000 bp of foreign DNA in its genome. Bacteria are then infected, generating a collection of bacteria called a **genomic DNA library**.

To select the genomic DNA sequence of interest, the library is plated onto a layer (or lawn) of cultured bacteria so that many copies of the recombinant bacteriophage can be produced. A single λ bacteriophage infects a single *E. coli*. The recombinant bacteriophages then multiply inside the host cells. The cells die and lyse, and the bacteriophages spread to the surrounding layer of bacteria and infect them. These cells lyse, in turn, and the process is repeated. The dead cells give rise to a clear area on the bacterial lawn called a **plaque**. Each plaque contains many copies of a recombinant bacteriophage that can be transferred to a nylon membrane (Fig. 7.10). Specific DNA clones are selected by incubating the nylon membrane with

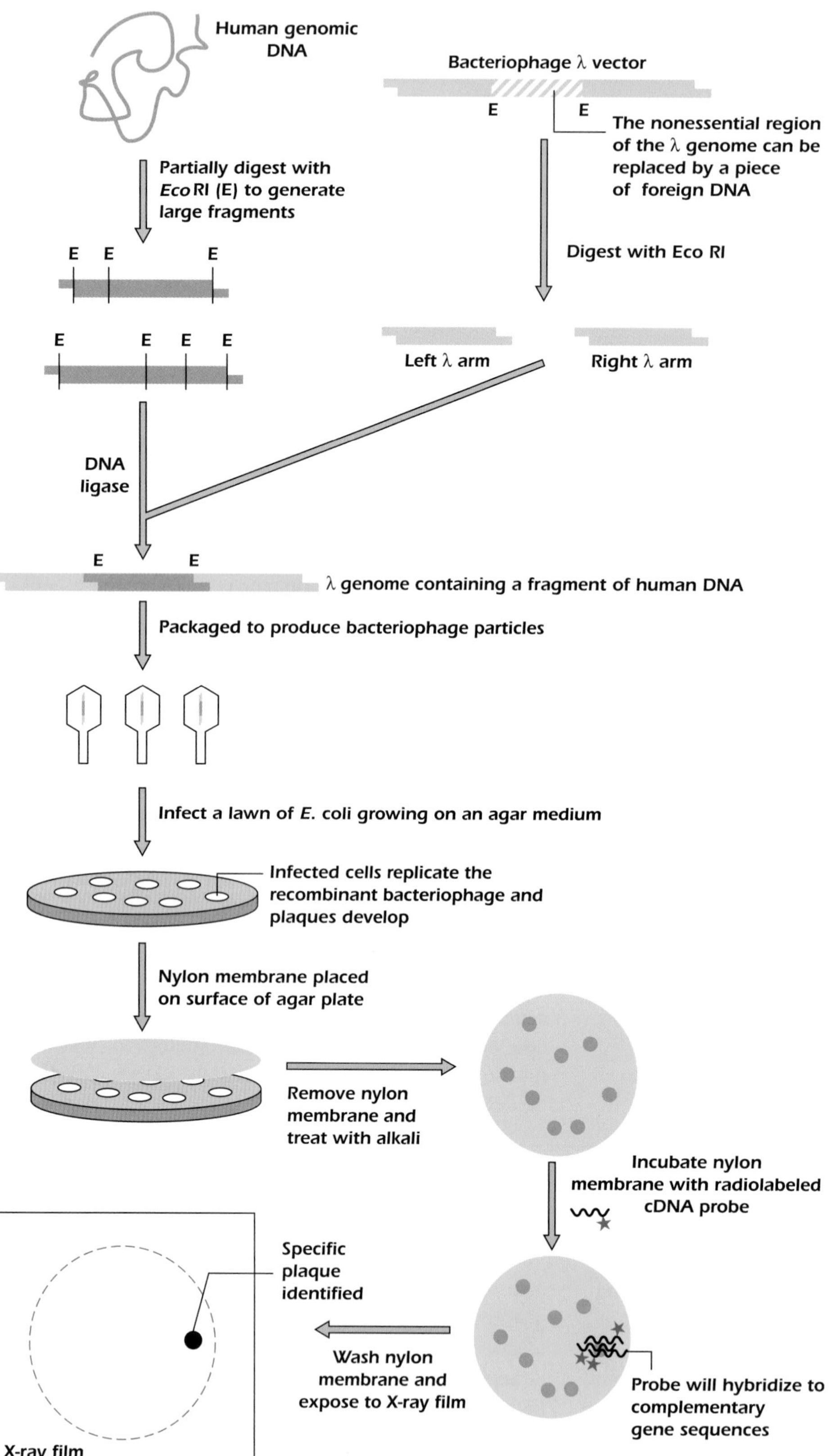

Figure 7.10. Generation and selection of genomic DNA clones.

a radiolabeled cDNA probe complementary to the genomic sequence being searched for. This produces a radioactive area on the nylon membrane that is identified by autoradiography. The use of a cDNA sequence as a gene probe makes the task of isolating the corresponding genomic sequence much easier.

A radioactive cDNA probe can be synthesized using the method called random priming. The cDNA clone is heated so that the two strands will separate. Each strand will act as the template for the synthesis of a new DNA strand. A mixture of random hexamers, six nucleotide sequences, containing all possible combinations of the four bases (A, T, C, G) is added to the denatured cDNA along with DNA polymerase and the four deoxynucleotides dATP, dTTP, dCTP, and dGTP. The hexamers will hydrogen bond (anneal) to their corresponding sequences on the cDNA templates and prime the synthesis of new DNA strands. If, for example, radiolabeled [$\alpha - {}^{32}$P]dATP is included in the reaction, the newly synthesized DNA strands will be radioactive.

USES OF DNA CLONES

The following techniques need large amounts of identical DNA and therefore can only be performed if one has cloned the gene and can therefore grow up the bacteria containing it in large numbers.

DNA Sequencing

The ability to determine the order of the bases within a DNA molecule has been one of the greatest technical contributions to molecular biology. DNA is made by the polymerization of the four deoxynucleotides dATP, dGTP, dCTP, and dTTP. These are joined together when DNA polymerase catalyzes the formation of a phosphodiester link between a free 3′-hydroxyl on the deoxyribose sugar moiety of one nucleotide and a free 5′-phosphate group on the sugar residue of a second nucleotide. However, the artificial dideoxynucleotides ddATP, ddGTP, ddCTP, and ddTTP have no 3′-hydroxyl on their sugar residue (Fig. 7.11), and so if they are incorporated into a growing DNA chain, synthesis will stop. This is the basis of the dideoxy chain termination DNA sequencing technique devised by Frederick Sanger, for which he was awarded a Nobel prize in 1980.

This technique is illustrated in Figure 7.12. A cloned piece of DNA of unknown sequence is first joined to a short oligonucleotide whose sequence is known. The DNA is then made single-stranded so that it can act as the template for the synthesis of a new DNA strand. All DNA synthesis requires a primer (page 69); in this case a primer is provided that is complementary in sequence to the oligonucleotide attached to the template DNA. Four

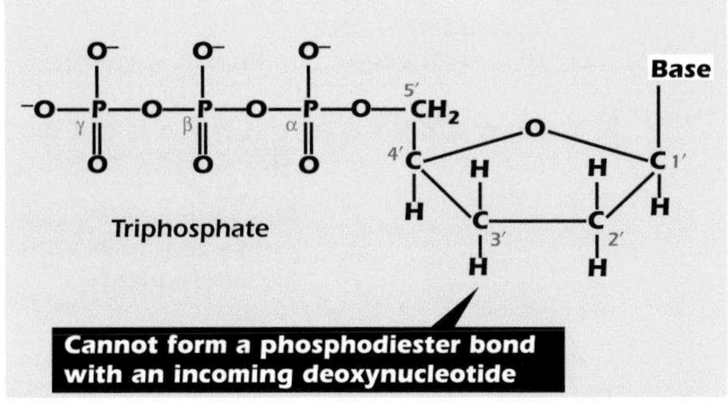

Figure 7.11. General structure of a dideoxynucleotide.

separate mixtures are prepared. Each contains the DNA template, the primer (which has been radiolabeled), DNA polymerase, and the four deoxynucleotides. The mixtures differ in that each also contains a low concentration of one of the four dideoxynucleotides ddATP, ddGTP, ddCTP, or ddTTP. When a molecule of dideoxynucleotide is joined to the newly synthesized chain, DNA synthesis will stop.

Let us follow what happens in the tube containing ddTTP. The first base that DNA polymerase encounters in the DNA template to be sequenced is an A. Since the tube contains much more dTTP than ddTTP, DNA polymerase will add a dTTP to most of the primer molecules. However, a small fraction of the primers will have a ddTTP added to them instead of dTTP since DNA polymerase can use either nucleotide as a substrate. The next base encountered is a G. DNA polymerase is unable to attach dCTP to the ddTTP since there is no OH group on the 3′ carbon of the sugar, and so DNA synthesis is terminated. The majority of strands, however, had a dTTP added, and for these DNA polymerase can proceed, building the growing strand. No problems are encountered with the next six bases. However, the eighth base in the template strand is another A, and once again a small fraction of the growing strands will have ddTTP added instead of dTTP. In the same way as before, these strands can grow no further. This process will be repeated each time an A occurs on the template strand. When the reaction is over, the tube will contain a mixture of DNA fragments of different length, each of which ends in a ddTTP. Similarly, each of the other three tubes will contain a mixture of DNA chains of different length, each of which ends in either ddCTP, ddATP, or ddGTP.

To determine the sequence of the newly synthesized chains, each of the four samples is loaded onto a polyacrylamide gel. The monomeric form, acrylamide, is poured into a mold. A solid but porous gel forms as the acrylamide polymerizes. The shape of the mold is such that wells are formed at the top of the gel into which the samples can be loaded. The samples are then subjected to **electrophoresis**, that is, a voltage is applied across the gel and the DNA strands move, with the smallest ones moving the quickest. The polyacrylamide gel used for DNA sequencing has

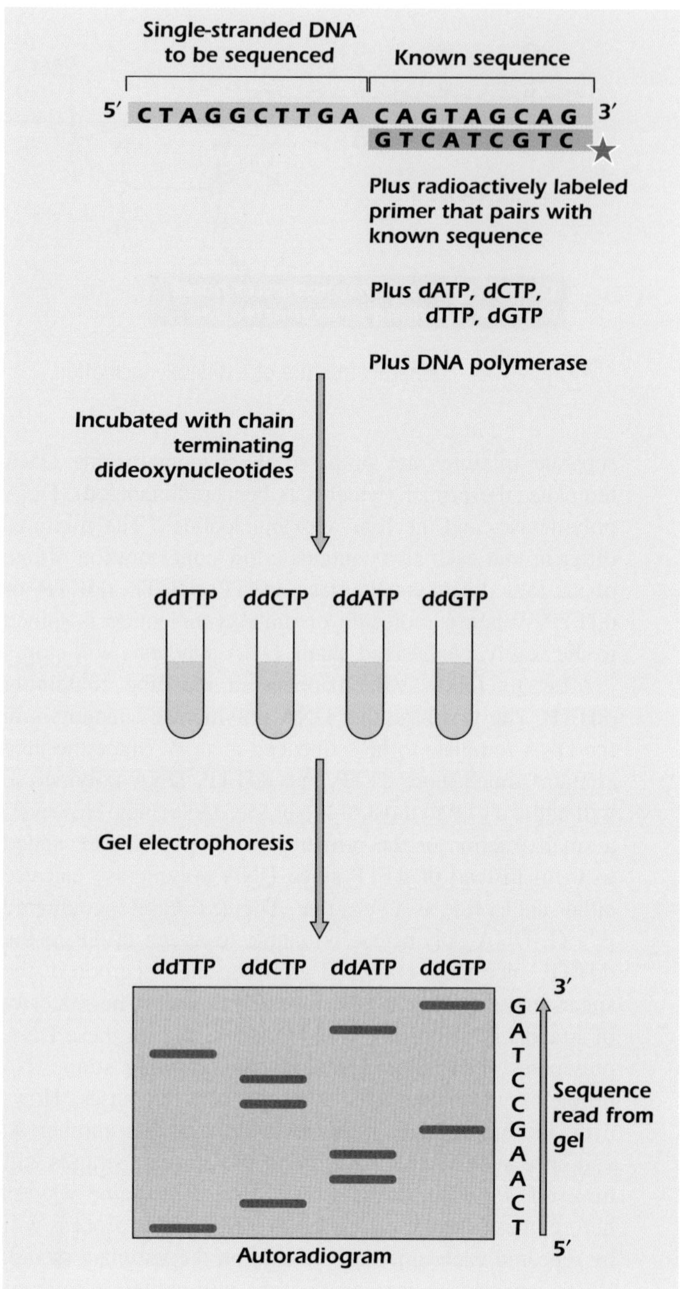

Figure 7.12. DNA sequencing by the dideoxy chain termination method.

a resolving power of one nucleotide. This is an important feature, because it is the ability to separate out strands terminated at each nucleotide along a strand of DNA that makes it possible to read a DNA sequence and determine the order of the As, Ts, Cs and Gs. Because the primer used was radiolabeled, all the new DNA chains will carry a radioactive tag so that, after electrophoresis, the pattern of DNA fragments on the gel can be detected by autoradiography. Each terminated reaction will show up as a black band on the X-ray film. The smallest DNA molecules are that fraction in the tube containing ddTTP whose growth was blocked after the first base, T. These move farthest and produce the band at the bottom of the T lane. DNA molecules one nucleotide larger were produced in the tube containing ddCTP—their growth was blocked after the second base, C. These move almost as far, but not quite, producing the band at the bottom of the C lane. Reading bands up from the bottom of the gel therefore tells us the sequence in which bases were added to the unknown strand: T, C, and so on. Because the new chain is complementary in sequence to its template strand, the sequence of the template strand can be inferred.

The Human Genome Project required the process of DNA sequencing to be automated. Instead of using radioactivity and one reaction tube for each dideoxynucleotide, we now use dideoxynucleotides tagged with fluorescent dyes. Each of the four dideoxynucleotides ddATP, ddGTP, ddCTP, and ddTTP is tagged with a different fluorescent dye. This means that all four of the termination reactions can be carried out in a single reaction tube and loaded into the same well on the polyacrylamide gel. As the reaction product drips from the bottom of the gel, the fluorescence intensity of each of the four colors corresponding to the four dideoxynucleotides is monitored, and this information is transferred straight to a computer where the data are analyzed. An example of a DNA trace produced using fluorescently tagged dideoxynucleotides is shown in Figure 7.13. Each peak represents a terminated DNA product, so by reading the order of the peaks, the DNA sequence is determined. In this example G is yellow, A is green, T is red, and C is blue. To determine the base sequence of the entire human genome, the DNA was cut by partial restriction endonuclease digestion (page 101) to give large fragments of about 150,000 bp. These were cloned into a vector such as PAC (Table 7.1). The aim was to create a library of clones that overlapped one another. Each clone was sequenced, as described above, and their base sequences compared. Because the clones overlapped, it was possible, by comparing their sequences, to line up the position of each clone with respect to its neighbors. This required the development of sophisticated computer programs, and the creation of a large database of information to order the 3×10^9 base pairs that comprise the human genome. Examples of the databases constructed to handle the information from the human genome project and many other genome projects that are now underway can be viewed at the Sanger Institute website (www.sanger.ac.uk/) and that of the National Institutes of Health (www.ncbi.nlm.nih.gov/).

Southern Blotting

In 1975 Ed Southern developed an ingenious technique, now known as **Southern blotting**, which can be used to detect specific genes (Fig. 7.14). Genomic DNA is isolated and digested with one or more restriction endonucleases. The resultant fragments are separated according to size by agarose gel electrophoresis. The gel is soaked in alkali to break hydrogen bonds between the two DNA strands and then transferred to a nylon membrane. This produces an exact replica of the pattern of DNA fragments in the agarose gel. The nylon membrane is incubated with a cloned DNA fragment tagged with a radioactive label. The gene probe is heated before being adding to the nylon membrane to make it single-stranded so it will base pair, or hybridize, to its complementary sequences on the nylon membrane. As the gene probe is radiolabeled, the sequences to which it has hybridized can be detected by autoradiography.

Mutations that change the pattern of DNA fragments—for instance, by altering a restriction endonuclease recognition site or deleting a large section of the gene—can easily be detected by Southern blotting. This technique is therefore useful in determining whether an individual carries a certain genetic defect. All that is needed is a small DNA sample from white blood cells or, in the case of a fetus, from the amniotic fluid in which it is bathed, or by removing a small amount of tissue from the chorion villus that surrounds the fetus in the early stages of pregnancy.

Forensic laboratories use Southern blotting to generate DNA fingerprints from samples of blood or semen left at the scene of a crime. A DNA fingerprint is a person-specific Southern blot. The gene probe used in the test is a sequence that is repeated very many times within the human genome—a microsatellite sequence (page 78). Everyone carries a different number of these repeated sequences, and because they lie adjacent to each other on the chromosome they are called **VNTRs** (**variable number tandem repeats**). When genomic DNA is digested with a restriction endonuclease and then analyzed by Southern blotting, a DNA pattern of its VNTRs is produced. Unless they are identical twins, it is extremely unlikely that two individuals will have the same DNA fingerprint profile. It has been estimated that if eight restriction endonucleases are used,

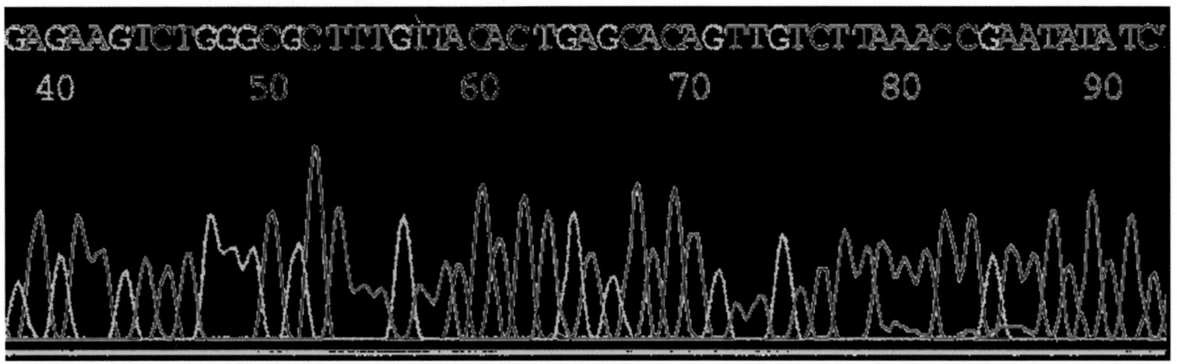

Figure 7.13. Typical output of an automated DNA sequencer.

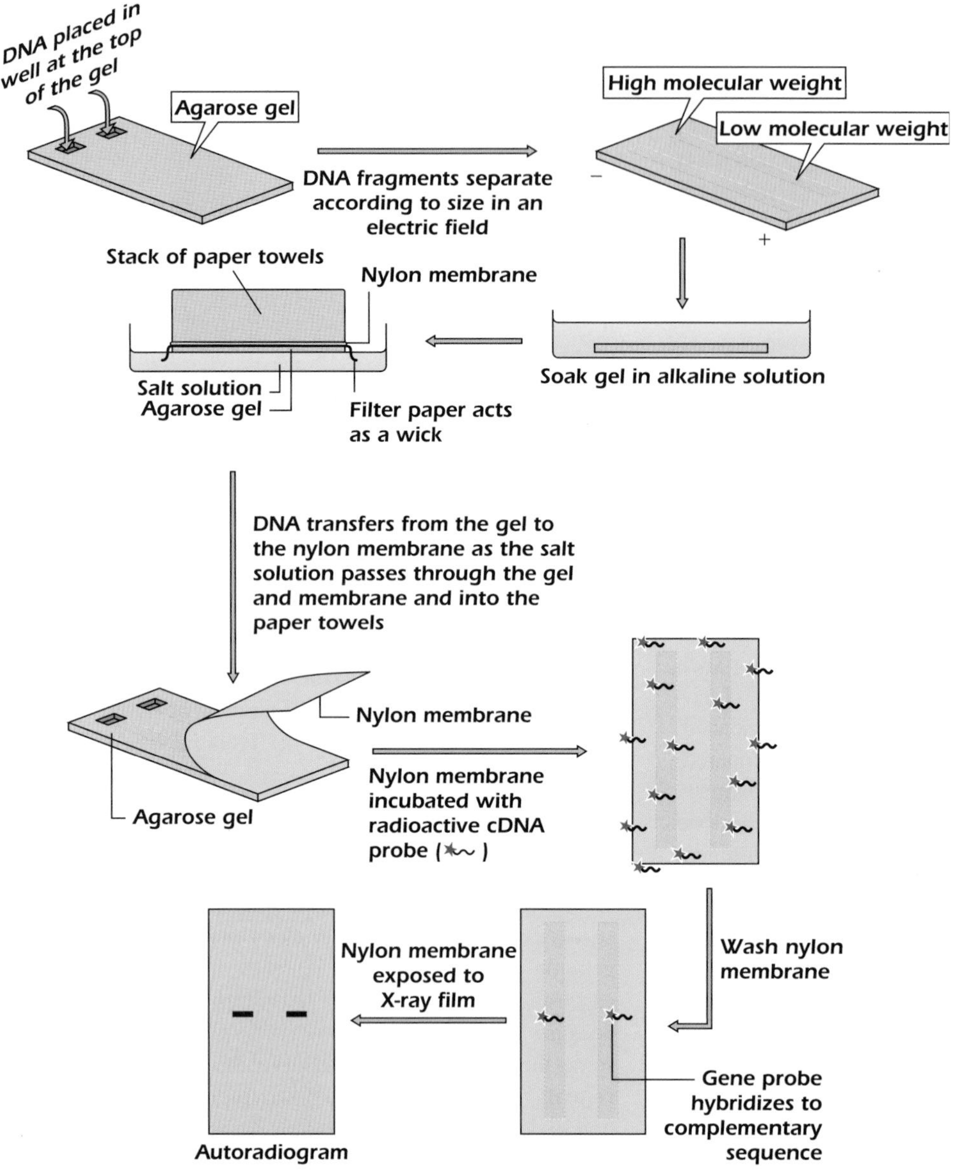

Figure 7.14. The technique of Southern blotting.

the probability of two people who are not identical twins generating the same pattern is one in 10^{30}.

A special type of Southern blot, called a zoo blot, can be used to reveal genes that are similar in different species. Such genes, which have been conserved through evolution, are likely to code for crucial proteins. A probe generated from a genomic DNA library is used to probe genomic DNA from many different species. A probe that hybridizes with DNA from a number of species is likely to represent all or part of such a conserved gene.

In situ Hybridization

It is possible, using the technique of ***in situ* hybridization**, to identify individual cells expressing a particular mRNA. To do this we need to synthesize an antisense RNA molecule—an RNA that is complementary in sequence to the mRNA of interest. In a test tube, the appropriate strand of the cloned cDNA is copied into the antisense RNA using RNA polymerase. The RNA is then labeled with a modified nucleotide that can subsequently be detected using an antibody and a color reaction. The cDNA must first be cloned into an expression vector that contains a promoter sequence to which RNA polymerase can bind. A thin tissue section, attached to a glass microscope slide, is incubated with the antisense RNA. The antisense RNA will hybridize, in the cell, to its complementary mRNA partner. Excess antisense RNA is washed off the slide, leaving only the hybridized probe. The color reaction is now carried out so that the cells expressing the mRNA of interest can be identified by bright-field microscopy (page 2).

Medical Relevance 7.1 Microarrays and Cancer Classification

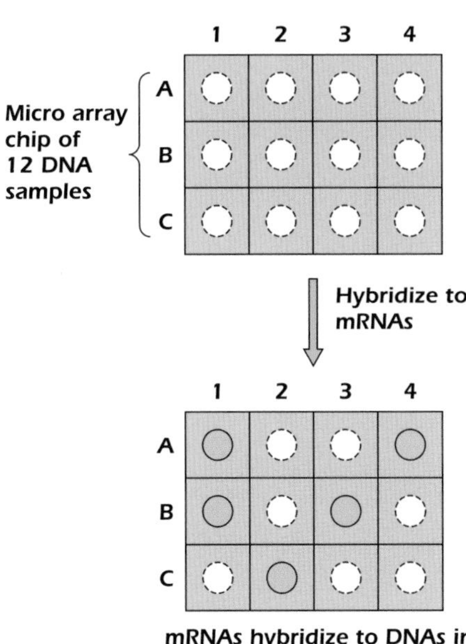

Microarrays or gene chips are tiny glass wafers to which cloned DNAs are attached. The principle of microarray technology is to isolate mRNA from a particular cell and to hybridize this to the DNAs on the chip. Because the mRNAs are tagged with a fluorescent dye, the DNAs to which they hybridize on the chip can be detected. Excess mRNA is removed, and fluorescent areas are viewed using a special scanner and microscope. Computer algorithms have been written to analyze the hybridization patterns seen for a particular microarray. The number and type of DNAs used to make a microarray is dependent on the question to be answered.

One of the spinoffs of the Human Genome Project is to identify sets of genes involved in disease. Microarrays are being used to type the mRNAs produced by different cancers in the hope that this will lead to better diagnosis and therefore better treatment. Leukemia, a cancer affecting the blood, is not a single type of disease. Microarray analysis is helping to classify different types of leukemia more precisely by cataloging the mRNAs expressed in different patients. Preliminary studies are encouraging. A microarray of 6817 cDNAs was used to compare the blood mRNAs from patients with acute lymphoblastic leukemias or acute myeloid leukemias. The mRNAs of each patient were passed over the microarray and complementary sequences hybridized. Fifty cDNA/mRNA hybrids were found that could be used to refine the classification of lymphoblastic and myeloid leukemias. In addition, the patterns allowed the classification of lymphoblastic leukemias into T-cell or B-cell classes (Chapter 19). This separation of lymphoblastic leukemias into two classes is an important distinction when it comes to deciding on the best treatment for a patient. As more patients with leukemia are investigated using microarrays, it should be possible to design smaller chips, with fewer cDNAs, but with greater prognosis value.

A recent study has shown that the fate of women with breast cancer can be predicted by microarray analysis of their cancer tissue. Several genes were shown to be important for this prediction. If the women had a certain gene expression pattern, then the disease was likely to recur within a five-year period. However, if the gene pattern was of a second type, the cancer almost never returned. Women that fall into this second class also show no additional benefit from chemotherapy or radiotherapy. Microarray analysis not only offers a survival prognosis for women with breast cancer but has also shown that unpleasant treatment for the disease is unnecessary for those possessing the second type of gene pattern.

Northern Blotting

Figure 7.15A shows a blotting technique that can determine the size of an mRNA and tell us about its expression patterns. RNA is denatured by heating to remove any intramolecular double-stranded regions and then electrophoresed on a denaturing agarose gel. The RNA is transferred to a nylon membrane (as described in Figure 7.14 for the transfer of DNA). The nylon membrane is incubated with a radiolabeled, single-stranded cDNA probe, or an antisense RNA probe (page 112). Following hybridization, excess probe is washed off and the nylon membrane exposed to X-ray film. The mRNA is visualized on the autoradiogram because it hybridized to the radioactive probe. By analogy with Southern blotting, this technique is called **northern blotting** (Table 7.2). Figure 7.15B shows a northern blot for a cytochrome P450 (pages 94, 185) mRNA known as CYP2B1. The gene for this protein is activated in the liver by the barbiturate phenobarbital and hence a lot of CYP2B1 mRNA is made.

Production of Mammalian Proteins in Bacteria

The large-scale production of proteins using cDNA-based expression systems has wide applications for medicine and industry. It is increasingly being used to produce

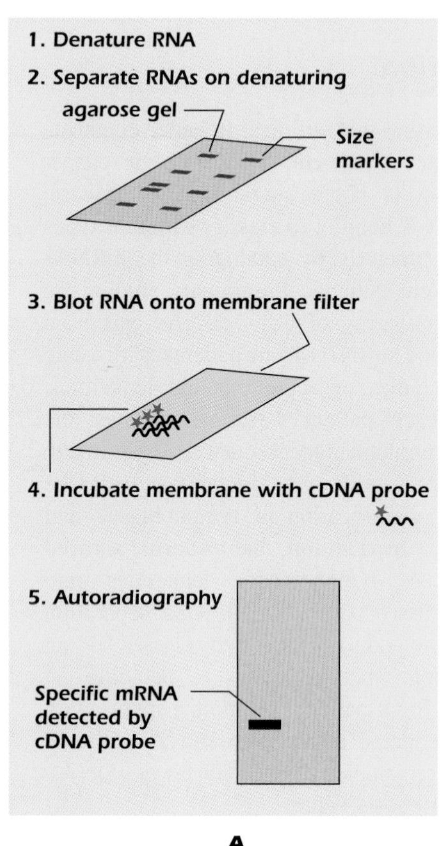

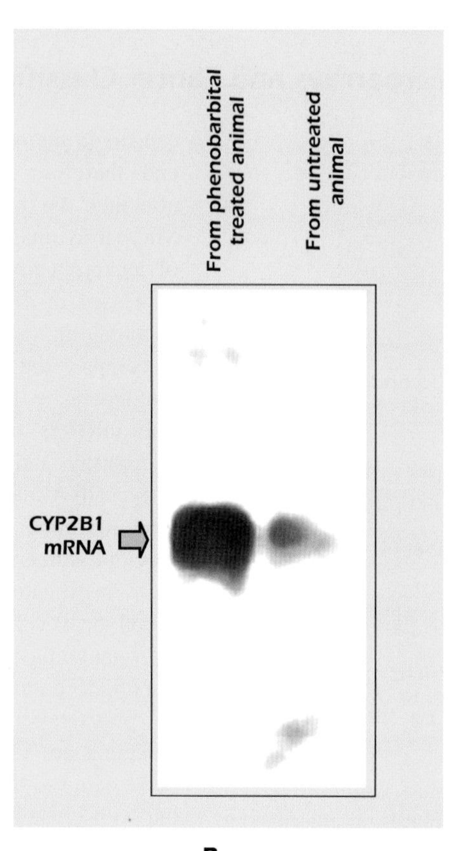

Figure 7.15. (A) The technique of northern blotting. (B) A northern blot reveals that transcription of the *CYP2B1* gene is increased in animals given phenobarbital.

TABLE 7.2. Blotting Techniques

	What Is Probed	Nature of the Probe	Book Page for Description
Southern blot	DNA	DNA	111
Northern blot	RNA	cDNA or RNA	113
Western blot	Protein	Antibody	127

polypeptide-based drugs, vaccines, and antibodies. Such protein products are called **recombinant** because they are produced from a recombinant plasmid. For a mammalian protein to be synthesized in bacteria its cDNA must be cloned into an expression vector (as described on page 106). Insulin was the first human protein to be expressed from a plasmid introduced into bacterial cells and has now largely replaced insulin from pigs and cattle for the treatment of diabetes. Other products of recombinant DNA technology include growth hormone and factor VIII, a protein used in the treatment of the blood clotting disorder hemophilia. Factor VIII was previously isolated from donor human blood. However, because of the danger of infection from viruses such as HIV, it is much safer to treat hemophiliacs with recombinant factor VIII. It should, in theory, be possible to express any human protein via its cDNA.

Protein Engineering

The ability to change the amino acid sequence of a protein by altering the sequence of its cDNA is known as **protein engineering**. This is achieved through the use of a technique known as **site-directed mutagenesis**. A new cDNA is created that is identical to the natural one except for changes designed into it by the scientist. This DNA can then be used to generate protein in bacteria, yeast, or other eukaryotic cell lines.

The first use of protein engineering is to study the protein itself. A comparison of the catalytic properties of the normal and mutated form of an enzyme helps to identify amino acid residues important for substrate and cofactor binding sites (Chapter 11). This technique was also used to identify the particular charged amino acid residues responsible for the selectivity of ion channels (page 233). Now scientists are using protein engineering to generate new proteins as tools, not only for scientific research but for wider medical and industrial purposes.

Subtilisin is a protease and is one of the enzymes used in biological washing powder. The natural source of this enzyme is *Bacillus subtilis*, an organism that grows on pig feces. To produce, from this source, the 6000 tons of subtilisin used per year by the soap powder industry is a difficult and presumably unpleasant task. The cDNA for subtilisin has been isolated and is now used by industry to synthesize the protein on a large-scale in *E. coli*. The wild-type (natural) form of subtilisin is, however, prone to oxidation because of a methionine present at position 222 in the protein. Its susceptibility to oxidation makes it an unsuitable enzyme for a washing powder that must have a long shelf life and be robust enough to withstand the rigors of a washing machine with all its temperature cycles. Scientists therefore changed the codon for methionine (AUG) to the codon for alanine (GCG). When the modified cDNA was expressed in *E. coli*, the resulting enzyme was found to be active and not susceptible to oxidation. This was excellent news for the makers of soap powder. However, it is always necessary to check the kinetic parameters of a new protein produced from a modified cDNA. For subtilisin (met222) the K_M (page 183) is 1.4×10^{-4} moles liter^{-1} while for subtilisin (ala222) the K_M is 7.3×10^{-4} moles liter^{-1}. This means that at micromolar concentrations of dirt the modified enzyme will bind less dirt than the wild-type one, but the dirt concentrations caked onto our clothes are well above micromolar. The turnover number, k_{cat} (page 182), is 50 s^{-1} for subtilisin (met222) and 40 s^{-1} for subtilisin (ala222): The mutant enzyme is slightly slower, but not by much. By changing a met to an ala, a new enzyme has been produced that can do a reasonable job and is stable during storage and in our washing machines.

Green fluorescent protein is found naturally in certain jellyfish. Protein engineering has now created a palette of proteins with different colors. However, the great advantage of these proteins to biologists is that **chimeric proteins** (proteins composed of two parts, each derived from a different protein) incorporating a fluorescent protein are intrinsically fluorescent. This means that our protein of interest can be imaged inside a living cell using a fluorescence microscope (In Depth 1.1 on page 6). The fluorescent part of the chimeric protein tells us exactly where our protein is targeted in the cell and if this location changes in response to signals.

Figure 7.16 illustrates how this approach can be used to determine what concentration of glucocorticoid drug is required to cause the glucocorticoid receptor to move to the nucleus. The plasmid, like many plasmids designed for convenience of use, contains a multiple cloning site (MCS), sometimes called a polylinker, which is a stretch of DNA that contains several restriction endonuclease recognition sites. A convenient restriction endonuclease is used to cut the plasmid (which already contains the sequence that codes for green fluorescent protein) and the cDNA for the glucocorticoid receptor is inserted. The plasmid also contains a promoter sequence, derived from a virus, that will drive the expression of the DNA into

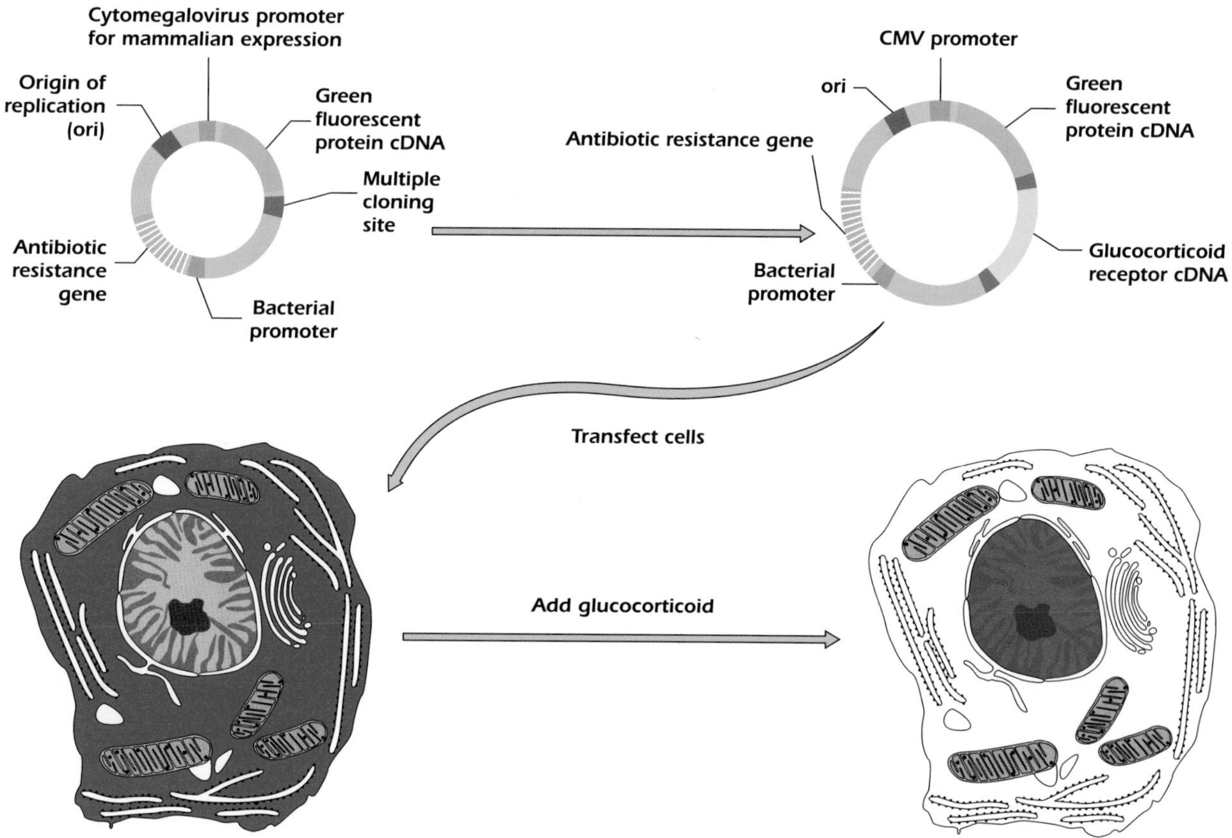

Figure 7.16. A chimera of green fluorescent protein and the glucocorticoid receptor reveals its location in living cells.

mRNA in mammalian cells that have been infected with the plasmid (or **transfected**). The plasmid is grown up in bacteria and then used to transfect mammalian cells. The chimeric protein, part green fluorescent protein and part glucocorticoid receptor, is synthesized in the cells from the mRNA. In the absence of glucocorticoid the protein, and therefore the green fluorescence, is in the cytosol. When enough glucocorticoid is added, it binds to the chimeric protein, which then moves rapidly to the nucleus.

Polymerase Chain Reaction

The **polymerase chain reaction (PCR)** is a technique that has revolutionized recombinant DNA technology. It can amplify DNA from as little material as a single cell and from very old tissue such as that isolated from Egyptian mummies, a frozen mammoth, and insects trapped in ancient amber. A simple mouth swab can yield enough cheek cell DNA to determine carriers of a particular recessive genetic disorder. PCR is used to amplify DNA from fetal cells or from small amounts of tissue found at the scene of a crime. The tool that makes PCR possible is a thermostable DNA polymerase, an enzyme that can function at extremely high temperatures that would denature (page 151) most enzymes. Thermostable DNA polymerases are isolated from prokaryotes that live in extremely hot deep-sea volcanic environments.

Figure 7.17 shows how PCR uses a thermostable DNA polymerase and two short oligonucleotide DNA sequences called primers. Each primer is complementary in sequence to a short length of one of the two strands of DNA to be amplified. The DNA duplex is heated to 90°C to separate the two strands (step 1). The mixture is cooled to 60°C to allow the primers to anneal to their complementary sequences (step 2). At 72°C the primers direct the thermostable DNA polymerase to copy each of the template strands (step 3). These three steps, which together constitute one cycle of the PCR, produce twice the number of original templates. The process of template denaturation, primer annealing, and DNA synthesis is repeated many times in a tube in an automated heater block to yield many thousands of copies of the original target sequence.

Identifying the Gene Responsible for a Disease

Until recently, the starting point for an identification of the gene responsible for a particular inherited disease was a pattern of inheritance in particular families plus a knowledge of the tissues affected. It is very difficult to find the gene responsible for a disease when the identity of the

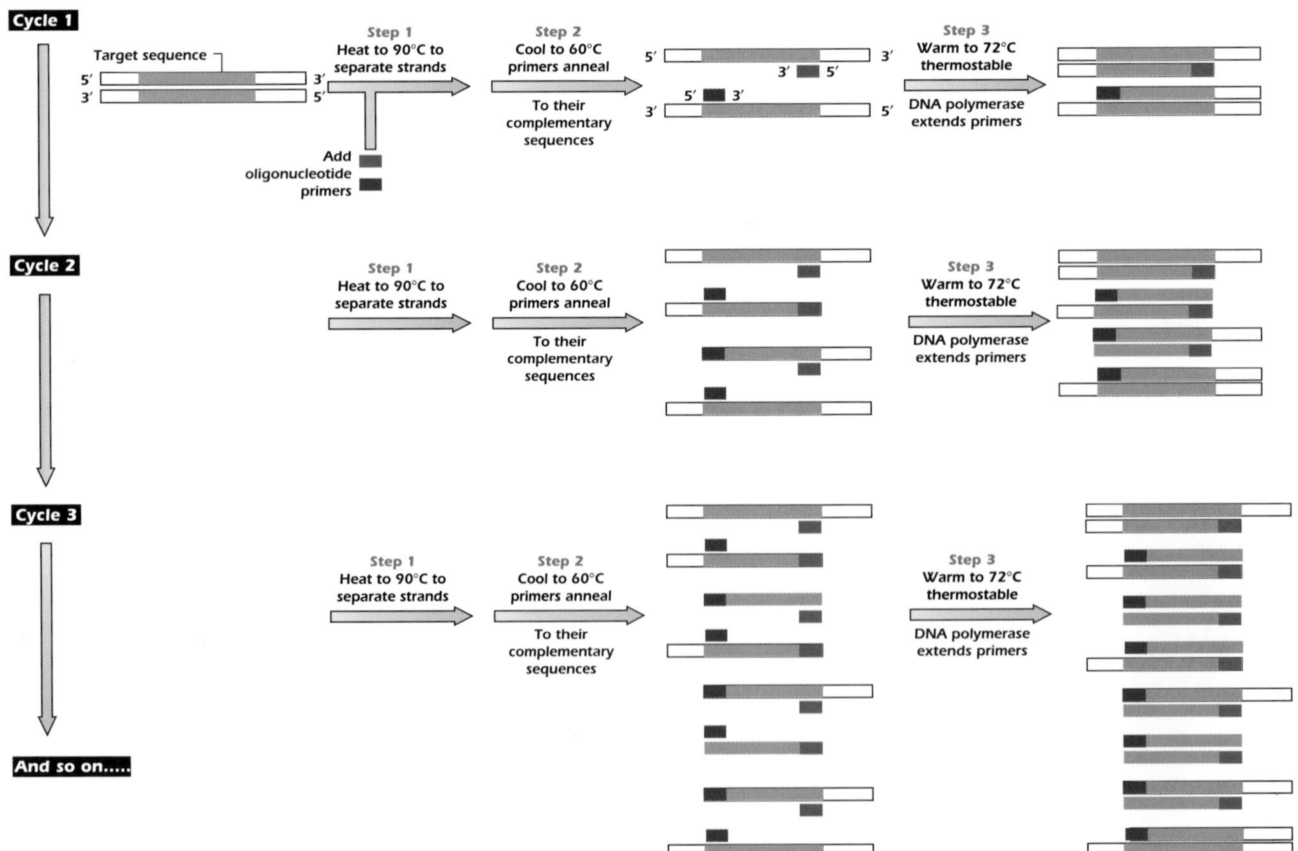

Figure 7.17. Amplification of a DNA sequence using the polymerase chain reaction.

normal protein is unknown. Very often, the first clue is the identification of other genes that are inherited along with the malfunctioning gene and that are therefore likely to lie close on the same chromosome (this is **linkage**, page 302). In the past, **chromosome walking** was then used to identify the disease gene. This is a slow and tedious process that involves isolating a series of overlapping clones from a genomic clone library. One starts by isolating a single clone and then uses this as a probe to find the next clone along the chromosome. The second clone is used to find the third, and this is used to find a fourth clone, and so on. Each successive clone is tested to find out if it might include all or part of the gene of interest (e.g., by using northern blotting to see if the gene is expressed in a tissue known to be affected in the disease). Once a candidate gene is identified, one can find out if it was the gene of interest by sequencing it in unaffected individuals and in disease sufferers; if it is the disease gene, its sequence will be different in the affected people. With the publication of the entire human genome, chromosome walking to generate a series of overlapping clones for testing is unnecessary, but identifying the gene that is responsible for a particular inherited condition is even now a time-consuming task.

Reverse Genetics

Because beginning with an inherited defect in function and working toward identification of a gene is even today a time-consuming task, more and more scientists are now working the other way: they take a gene with a known sequence but unknown function and deduce its role. Since we know the complete genome of a number of species, we can sit at a computer and identify genes that look interesting—for example, because their sequence is similar to a gene of known function. The gene of interest can be mutated and reinserted into cells or organisms, and the cells or organisms tested for any altered function. This approach is called **reverse genetics**.

Transgenic and Knockout Mice

A transgenic animal is produced by introducing a foreign gene into the nucleus of a fertilized egg (Fig. 7.18A). The egg is then implanted into a foster mother and the offspring are tested to determine whether they carry the foreign gene. If they do, a transgenic animal has been produced. The first transgenic mice ever made were used to identify an enhancer sequence that activates the metallothionein gene when an animal is exposed to metal ions in its diet. The 5′ flanking sequence of the metallothionein gene was fused to the rat growth hormone gene (Fig. 7.18B). This DNA construct, the transgene, was injected into fertilized eggs. When the mice were a few weeks old, they were given drinking water containing zinc. Mice carrying the transgene grew to twice the size of their litter mates because the metallothionein enhancer sequence, stimulated by zinc, had increased growth hormone production.

Transgenic farm animals—such as sheep synthesizing human factor VIII in their milk—have been created. This is an alternative to producing human proteins in bacteria.

IN DEPTH 7.1 GENETICALLY MODIFIED (GM) PLANTS—CAN THEY HELP TO FEED THE WORLD?

The arguments about the value of GM crops and the damage they cause the environment will continue for a long time. These have largely concerned plants that produce insecticide and plants that are resistant to herbicide. Almost unnoticed in the maelstrom of claim and counterclaim has been the use of genetic engineering to produce nutritionally enhanced crops. Rice is a staple food in many countries but lacks many vital nutrients. The World Health Organization estimates that 250,000 to 500,000 children go blind each year because of vitamin A deficiency and millions of children in the developing world suffer from a weakened immune system because their diets do not contain sufficient quantities of this vitamin. In an attempt to overcome this severe nutritional deficiency, a group of Swiss scientists have engineered the rice endosperm (the part we eat) to produce provitamin A (beta-carotene). This is converted in the body to vitamin A. Rice does produce beta-carotene in the green tissue but some of the genes needed to produce this chemical are switched off in the endosperm. The Swiss scientists inserted two of the genes that had been switched off in the endosperm back into the rice genome. The genetically modified rice plant with the inserted genes produces an endosperm that is golden in colour because beta-carotene is produced. Hence the name golden rice for this crop. The more yellow the crop, the more beta-carotene is produced.

The creators of golden rice have donated their technology to the developing world and the paperwork necessary to allow its use in Bangladesh, India, Indonesia, and the Philippines is at the time of writing being funded by the Rockefeller Foundation in New York. The hope is that the required daily dose of beta-carotene will be delivered in about 100–200g of rice, which corresponds to the average daily rice consumption in countries where this crop is a major part of the diet.

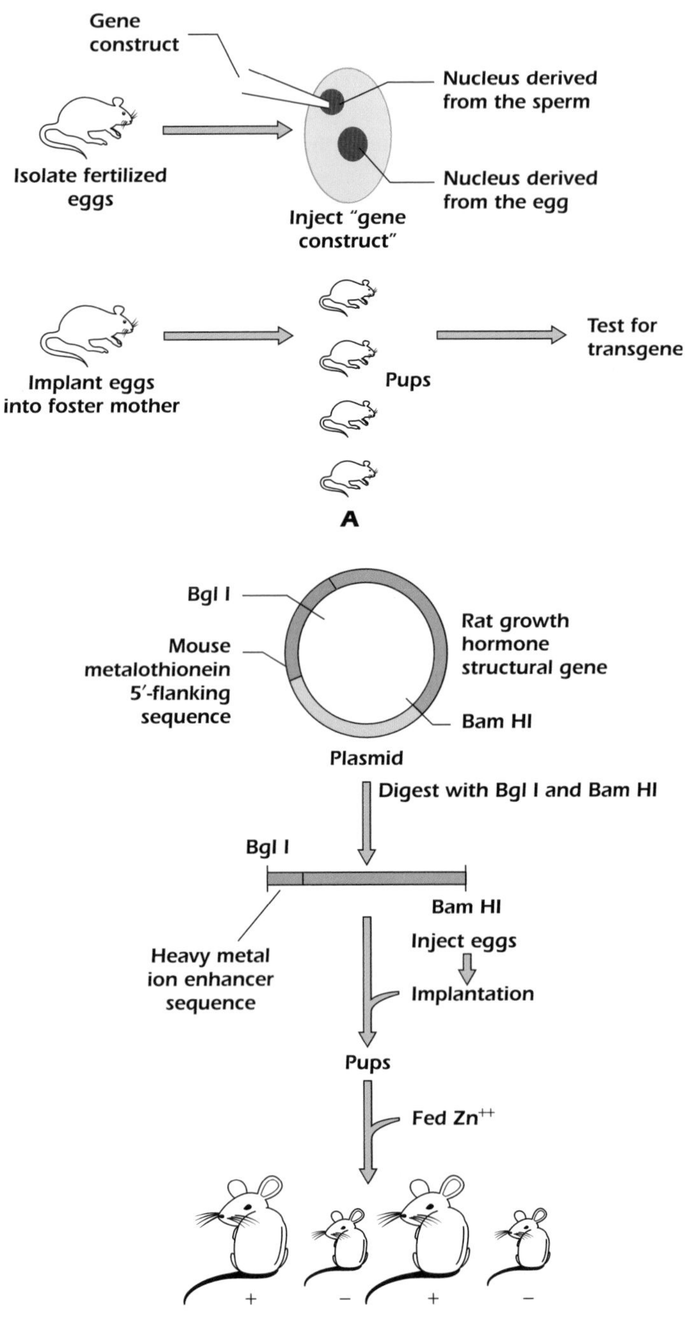

Figure 7.18. (A) Transgenic mouse carrying a foreign gene. (B) The metallothionein gene contains a heavy-metal ion enhancer sequence. The + mice carry the transgene while littermates without the transgene are indicated by −.

Genetically modified mice are increasingly being used to prove a protein's function. To do this the gene sequence is modified so that protein function is knocked out. In this case a **knockout mouse** is produced. This is done by either inserting a piece of foreign DNA into the gene, or by deleting the gene from the mouse genome. The consequences of the protein's absence are then established. Figure 7.19 describes the method called **insertional mutagenesis** for knocking out a gene's function. The first step is to isolate a genomic clone containing the gene to be knocked out. A marker gene, such as the drug resistance gene *neo* (for resistance to the antibiotic G418, an analogue of the antibiotic neomycin), is then inserted into the genomic clone, usually in exon 2 of the gene. This means that the normal, functional product of the gene cannot be synthesized. This modified piece of DNA is then introduced into embryonic stem (ES) cells. These are cells derived from the inner mass of a mouse **blastocyst**—that is, a very early embryo. The mouse strain from which the ES cells are derived has a butterscotch coat color. Homologous recombination (see

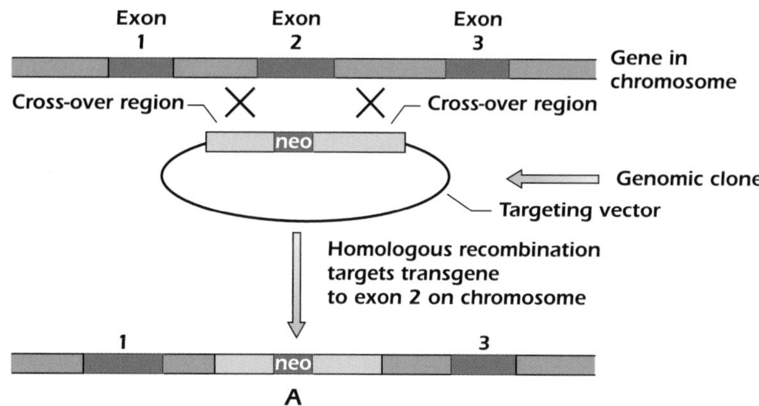

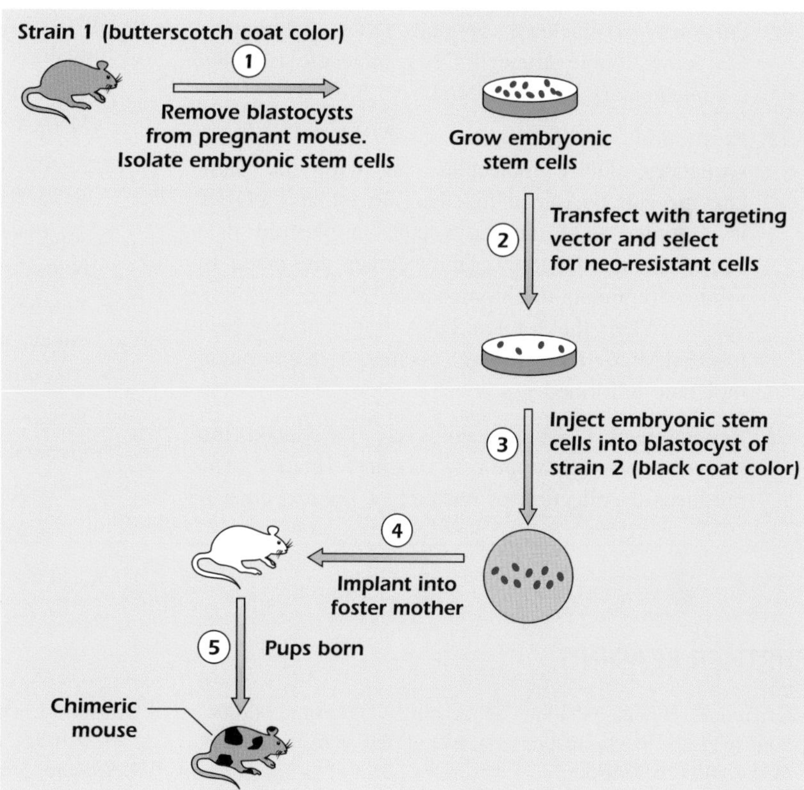

Figure 7.19. Knockout mice. (A) A region of the targeting vector is incorporated into the genome of embryonic stem cells by homologous recombination. (B) Genetically modified embryonic stem cells are injected into a blastocyst, which is implanted into a foster mother.

page 302) inside the embryonic stem cells will replace the normal gene with the modified piece of DNA carried on the plasmid. Cells in which this rare event happens will survive when grown on G418 while embryonic stem cells not containing the antibiotic-resistance gene will die. The genetically modified embryonic stem cells are inserted into the blastocyst cavity of a black mouse and the blastocyst implanted into a foster mother. Knockout mice will be chimeric and have a mixed color coat because the cells derived from the genetically modified embryonic stem cells will give a butterscotch-color coat while the cells from the blastocyst will give a black-color coat. Subsequent breeding of the chimeric mouse to mice with a black-coat color will produce a pure black mouse with both copies of the gene nonfunctional. The effect of knocking out the gene can then be analyzed.

ETHICS OF DNA TESTING FOR INHERITED DISEASE

The applications of recombinant DNA technology are exciting and far-reaching. However, the ability to examine the base sequence of an individual raises important ethical questions. Would you want to know that you had inherited a gene that will cause you to die prematurely? Some of you might feel fine about this and decide to live life to the full. We suspect most people would not want to know their fate. But what if you have no choice and DNA testing becomes obligatory should you wish to take out life insurance? In the United Kingdom insurance companies are now able to ask for the results of the test for Huntington's disease. This is a fatal degenerative brain disorder that strikes people in their forties. From the insurance company's point of view DNA testing could mean higher premiums according to life expectancy or at worst refusal of insurance cover. There is much ongoing debate on this issue.

SUMMARY

1. DNA sequences can be cloned using reverse transcriptase, which copies mRNA into DNA to make a hybrid mRNA:DNA double-stranded molecule. The mRNA strand is then converted into DNA by the enzymes ribonuclease H and DNA polymerase. The new double-stranded DNA molecule is called complementary DNA (cDNA).

2. Restriction endonucleases cut DNA at specific sequences. DNA molecules cut with the same enzyme can be joined together. To clone a cDNA, it is joined to a cloning vector—a plasmid or a bacteriophage. Genomic DNA clones are made by joining fragments of chromosomal DNA to a cloning vector. When the foreign DNA fragment has been inserted into the cloning vector, a recombinant molecule is formed.

3. Recombinant DNA molecules are introduced into bacterial cells by the process of transformation. This produces a collection of bacteria (a library) each of which contains a different DNA molecule. The DNA molecule of interest is then selected from the library using either an antibody or a nucleic acid probe.

4. There are many important medical, forensic and industrial uses for DNA clones. These include:
   - Determination of the base sequence of the cloned DNA fragment
   - *In situ* hybridization to detect specific cells making RNA complementary to the clone
   - Southern blotting and genetic fingerprinting to analyze an individual's DNA pattern
   - Synthesis of mammalian proteins in bacteria or eukaryotic cells
   - Changing the DNA sequence to produce a new protein
   - Generation of fluorescent protein chimeras for subsequent microscopy on live cells

FURTHER READING

International Human Genome Sequencing Consortium. (2004) Finishing the euchromatic sequence of the human genome. *Nature* **431**, 931–945.

Mullis, K. B. (1990) The unusual origin of the polymerase chain reaction. *Scientific American* **262**, 56–65.

Watson, J. D., Caudy, A. A., Myers, R. M. and Witkowski, J. A. (2007) *Recombinant DNA: Genes and Genomes—a Short Course*, 3rd edition, W. H. Freeman.

Answer to Thought Question: Transform *E. coli* and plate the bacteria on agar containing ampicillin, IPTG (gratuitous inducer of the *lac* operon) and X-gal. Cells transformed with the plasmid containing the glucocorticoid receptor cDNA will appear white because insertion of DNA into pBluescript disrupts the *lac Z* sequence. These bacterial cells will not be able to produce β-galactosidase and are unable to convert X-gal to its blue product. However, the pBluescript plasmid has an intact *lac Z* sequence. Cells transformed with this plasmid will appear blue, because these cells can produce β-galactosidase. By selecting white bacterial cells and using these to isolate plasmid DNA you will produce a solution in which all the plasmids contain the glucocorticoid cDNA.

REVIEW QUESTIONS

7.1 Theme: A Mammalian Expression Plasmid

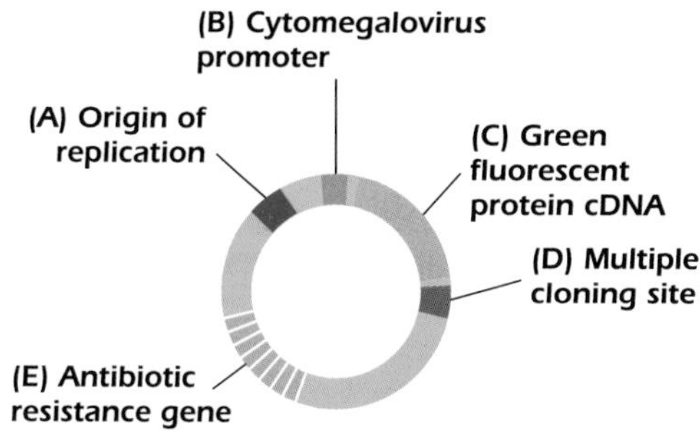

Figure 18.13 on page 310 shows fluorescent images of human cells containing a chimera of green fluorescent protein and cytochrome *c*. We tabulate below the steps that the experimenters went through to generate these cells. Identify the elements in the plasmid shown above that allow each step to be performed.

1. DNA encoding cytochrome *c* is inserted into the plasmid. Which element in the plasmid makes this possible?
2. To generate large amounts of the recombinant plasmid, the plasmid is grown up in bacteria. The plasmid is used to transform competent *E. coli* which are then cultured in such a way that only bacteria containing the plasmid survive. Which element of the plasmid allows the survival of host bacteria when sister bacteria are dying?
3. The transformed bacteria divide repeatedly, producing colonies each derived from a single transformed progenitor. What element in the plasmid allows it to be copied in parallel with the host bacterium's DNA?
4. Some clones will contain nonrecombinant plasmid, without the cytochrome *c* insert. However, the recombinant plasmid can be recognised by its higher relative molecular mass. Clones containing the recombinant plasmid are grown up further and then lysed, allowing purification of large amounts of recombinant plasmid. The purified plasmid is then used to transfect human cells, which synthesize the green fluorescent protein:cytochrome *c* chimaera. Which element in the plasmid allows the chimaeric protein to be expressed in the HeLa cells, even though it was not expressed in the bacteria?

7.2 Theme: Choosing an Oligonucleotide for a Specific Task

The first two questions refer to the DNA sequence shown at the bottom of the page. Note that we show only the sequence at the ends of the double-stranded DNA molecule.

A 5′ TTTTTTTTTTTTTTT 3′
B 5′ TGCCTACTGCAGCGTCTGCA 3′
C 5′ TACGGATCCCTTTGCAGGATGAATTC 3′
D 5′ TTCTGCAGACGCTGCAGTAG 3′
E 5′ GAATTCTACGGATCCCTTTGCAGGAT 3′
F 5′ GTGCATCTGACTCCTGTGGAGAAGTCT 3′
G 5′ GACTGCCATCGTAAGCTGAC 3′

From the above table of DNA sequences, select the one that best fits each of the descriptions below.

1. In the polymerase chain reaction: indicate the oligonucleotide that should be used together with the oligonucleotide 5′ TACGGATCCCTTTGCAGGAT 3′ to amplify the double stranded DNA molecule shown at the bottom of the page.
2. You wish to use the polymerase chain reaction to create, using the double stranded DNA molecule shown at the bottom of the page, a DNA product that can then be cloned into the EcoR1 site of a plasmid. Indicate the oligonucleotide that would you use in place of 5′ TACGGATCCCTTTGCAGGAT 3′ in the PCR reaction mix.
3. An oligonucleotide that could be used to prime the synthesis of DNA from most of the mRNAs present in a tissue, in order to generate a cDNA library.
4. An oligonucleotide that could be used in a Southern blot to identify carriers of sickle cell anemia. Note that this disease is caused because an A in the sequence 5′ GTGCATCTGACTCCTG**A**GGAGAAGTCT 3′ in the normal β globin gene is mutated to a T to generate the sequence 5′ GTGCATCTGACTCCTG**T**GGAGAAG TCT 3′.
5. An oligonucleotide that could be used for northern blotting to detect mRNA containing the sequence 5′ GUCAGCUUACGAUGGCAGUC 3′.

5′ TACGGATCCCTTTGCAGGATCCAG——TTCTGCAGACGCTGCAGTAGGCA 3′
3′ ATGCCTAGGGAAACGTCCTAGGTC——AAGACGTCTGCGACGTCATCCGT 5′

7.3 Theme: Uses of cDNA Clones

A A pair of oligonucleotides, one complementary to a short length of one strand of a DNA molecule, the other complementary to a short length, up to about 4,000 base pairs distant, of the *other* strand

B A pair of oligonucleotides, one complementary to a short length of one strand of a DNA molecule, the other complementary to a short length, up to about 4,000 base pairs distant, of the *same* strand

C A radiolabelled oligonucleotide complementary to a known sequence within the molecule of interest

D An oligonucleotide complementary to the bases at the 3′ end of a partially known, but largely unknown, DNA sequence

E An oligonucleotide complementary to the bases at the 5′ end of a partially known, but largely unknown, DNA sequence

From the above list of oligonucleotides, select the one appropriate for each of the techniques described below.

1. Amplifying a known or partially known DNA sequence using the polymerase chain reaction

2. Automated DNA sequencing by the dideoxy chain termination method

3. Detection of specific DNA sequences by Southern blotting, for example to differentiate DNA from two human subjects

4. Investigation, by northern blotting, of the degree to which a gene of interest is transcribed in a particular tissue

THOUGHT QUESTION

You have been provided with a plasmid mixture. Most of the mixture contains the pBluescript plasmid (page 101). A small proportion of the mix contains pBluescript into which you have cloned the cDNA for the glucocorticoid receptor cDNA. What can you do to separate from your mix a plasmid that contains the glucocorticoid receptor cDNA?

8

MANUFACTURING PROTEIN

The genetic code (page 59) dictates the sequence of amino acids in a protein molecule. The synthesis of proteins is quite complex, requiring three types of RNA. Messenger RNA (mRNA) contains the code and is the template for protein synthesis. Transfer RNAs (tRNAs) are adapter molecules that carry amino acids to the mRNA. Ribosomal RNAs (rRNAs) form part of the ribosome that brings together all the components necessary for protein synthesis. Several enzymes also help in the construction of new protein molecules. This chapter describes how the nucleotide sequence of an mRNA molecule is translated into the amino acid sequence of a protein.

Figure 8.1 shows the basic mechanism of protein synthesis, also called translation. In the first step, free amino acids are attached to tRNA molecules. In the second step, a ribosome assembles on the mRNA strand to initiate synthesis. In the third step, the ribosome travels along the mRNA. At each codon on the RNA a tRNA binds, bringing the amino acid defined by that codon to be added to the growing polypeptide chain. In the last, fourth, step the ribosome encounters a stop codon and protein synthesis is terminated.

ATTACHMENT OF AN AMINO ACID TO ITS tRNA

Amino acids are not directly incorporated into protein on a messenger RNA template. An amino acid is carried to the mRNA chain by a tRNA molecule. tRNAs are small, about 7000 nucleotides in length, and are folded into precise three-dimensional structures because of hydrogen bonding between bases in particular stretches of the molecule. This gives rise to four double-stranded regions, and it is these that give tRNA its characteristic cloverleaf structure when drawn in two dimensions as in Figure 8.2.

Each tRNA molecule has an amino acid attachment site at its 3′ end and an **anticodon**, three bases that are complementary in sequence to a codon on the mRNA. The tRNA binds to the mRNA molecule because hydrogen bonds form between the anticodon and codon. For example, the codon for methionine is 5′ AUG 3′ which will base pair with the anticodon 3′ UAC 5′.

Transfer RNA, the Anticodon, and the Wobble

Although 61 codons specify the 20 different amino acids, there are not 61 tRNAs; instead the cell economizes. The codons for some amino acids differ only in the third position of the codon. Figure 4.8 on page 61 shows that when an amino acid is encoded by only two different triplets the third bases will be either U and C, or A and G. For example aspartate is coded by GAU and GAC and glutamine by CAA and CAG. The **wobble** hypothesis suggests that the pairing of the first two bases in the codon and anticodon follows the standard rules—G bonds with C and A bonds with U—but the base pairing in the third position is not as restricted and can wobble. If the pyrimidine uracil (U) is in

Cell Biology: A Short Course, Third Edition. Stephen R. Bolsover, Jeremy S. Hyams, Elizabeth A. Shephard and Hugh A. White.

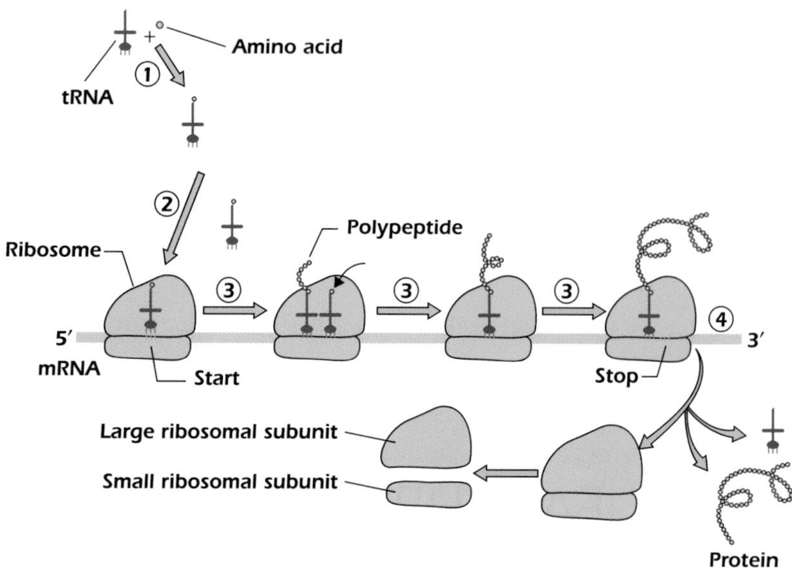

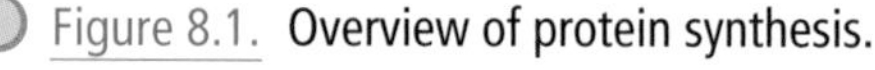
Figure 8.1. Overview of protein synthesis.

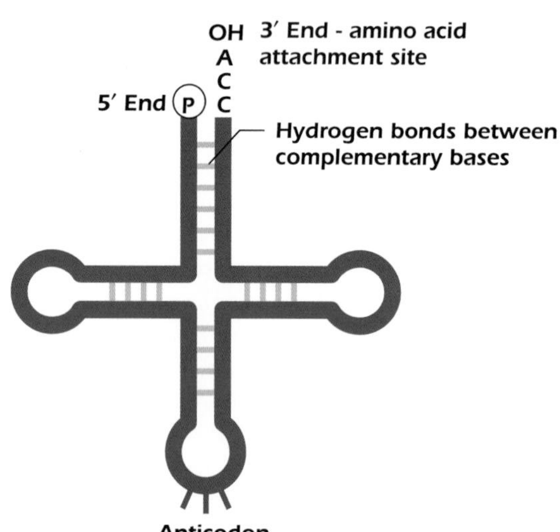

Figure 8.2. Transfer RNA (tRNA).

the third position of the codon, it can fit with any purine, G or A, in the 5′ position of the anticodon. Thus, only one tRNA molecule is required for two codon sequences. The anticodon of some tRNAs contains the unusual nucleoside inosine (I), whose base is the purine hypoxanthine (Fig. 2.13, page 29). Inosine can base pair with any of U, C, or A in the third position of the codon. Some tRNA molecules can therefore base pair with as many as three different codons provided the first two bases of the codon are the same. For example, the tRNA for isoleucine has the anticodon UAI and can therefore base pair with any of AUU, AUC, or AUA.

The attachment of an amino acid to its correct tRNA molecule is illustrated in Figure 8.3. This process occurs in two stages, both catalyzed by the enzyme **aminoacyl tRNA synthase**. During the first reaction (A in the figure), the amino acid is joined, via its carboxyl group, to an adenosine monophosphate (AMP) and remains bound to the enzyme. All tRNA molecules have at their 3′ end the nucleotide sequence CCA. In the second reaction (B in the figure) aminoacyl tRNA synthase transfers the amino acid from AMP to the tRNA, forming an ester bond between its carboxyl group and either the 2′- or 3′-hydroxyl group of the ribose of the terminal adenosine (A) on the tRNA to form an **aminoacyl tRNA**. This step is often referred to as amino acid activation because the energy of the ester bond can be used in the formation of a lower energy peptide bond between two amino acids. A tRNA that is attached to an amino acid is known as a **charged tRNA**. There are at least 20 aminoacyl tRNA synthases, one for each amino acid and its specific tRNA.

THE RIBOSOME

The ribosome is the cell's factory for protein synthesis. Each ribosome consists of two subunits, one large and one small, each of which is made up of rRNA plus a large number of proteins. The ribosomal subunits and their RNAs are named using a parameter, called the **S value**. The S value, or **Svedberg unit**, is a sedimentation rate. It is a measure of how fast a molecule moves in a gravitational field. For example, the bigger a ribosomal subunit, the quicker it will sediment and the larger the S value. Prokaryotic ribosomes, and those found inside mitochondria and chloroplasts, are 70S when fully assembled and comprise a larger, 50S subunit and a smaller 30S one (Fig. 8.4A). Because the

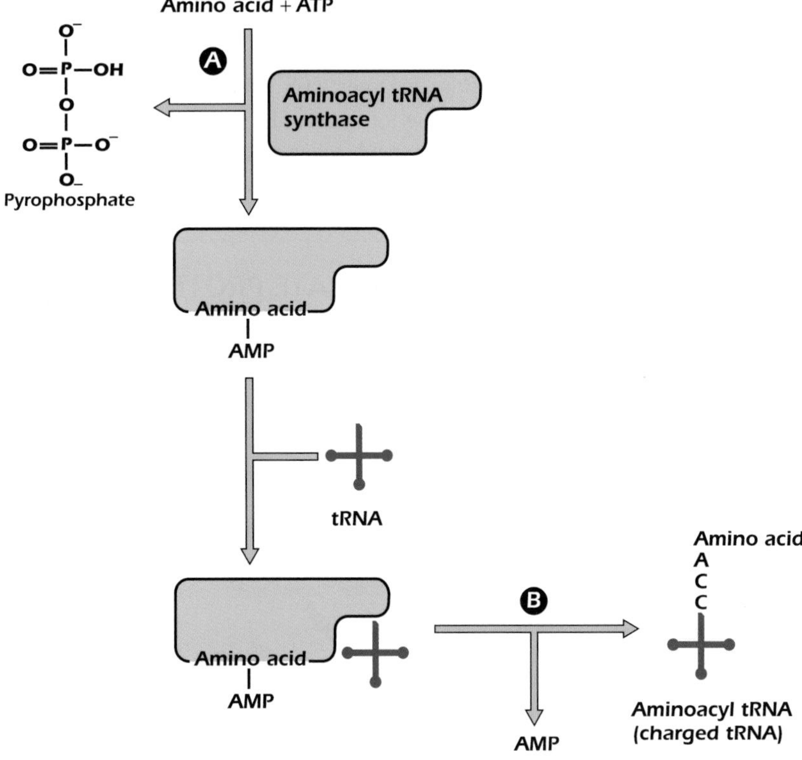

Figure 8.3. Attachment of an amino acid to its tRNA.

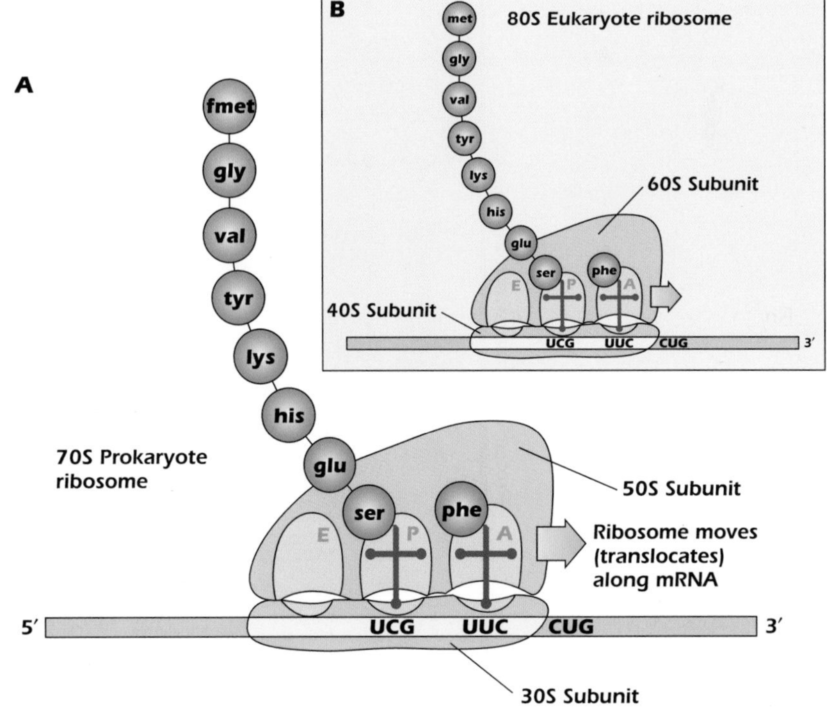

Figure 8.4. Prokaryote and eukaryote ribosomes.

S value refers to the sedimentation rate (and not the molecular mass) of the ribosome or its subunits, 70S is less than the sum of 50S and 30S. Eukaryotic ribosomes are 80S when fully assembled and comprise a larger, 60S subunit and a smaller 40S one (Fig. 8.4B). The formation of a peptide bond (page 32) between two amino acids takes place on the ribosome. The ribosome has binding sites for the mRNA template and for two charged tRNAs. An incoming tRNA with its linked amino acid occupies the **aminoacyl site (A site)**, and the tRNA attached to the growing polypeptide chain occupies the **peptidyl site (P site)**. The ribosome has a third binding site for tRNA, the **exit site (E site)**. This is the site to which a tRNA moves before it leaves the ribosome.

IN DEPTH 8.1 HOW WE SEPARATE PROTEINS IN ONE DIMENSION

The technique known as **SDS-PAGE** is widely used to analyze the spectrum of proteins made by a particular tissue, cell type, or organelle. It is also invaluable for assessing the purity of isolated proteins. SDS stands for sodium dodecyl sulfate and PAGE for polyacrylamide gel electrophoresis.

The aim of the technique is to denature the proteins to be analyzed and then to separate them according to their size in an electrical field. To do this we first add the chemical 2-mercaptoethanol to the protein sample. This will break any **disulfide bonds** (page 138) within a protein or between protein subunits. Next SDS,

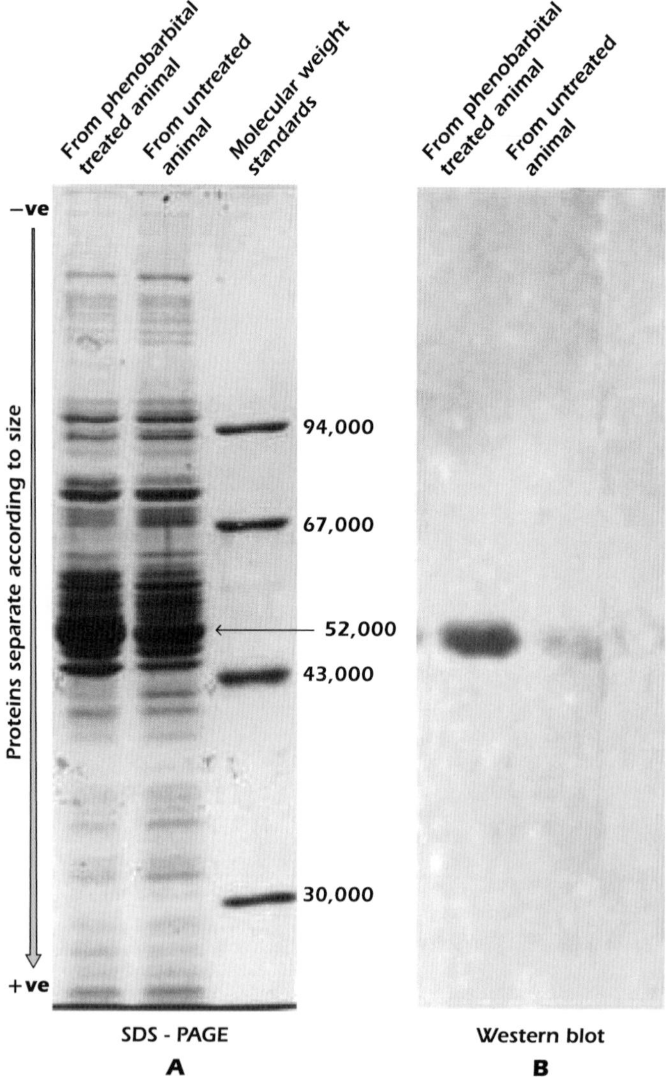

which is an anionic detergent, is added and the protein sample boiled. SDS coats each protein chain with negative charge. Each individual polypeptide in the sample becomes covered with an overall net negative charge. This means that when subjected to polyacrylamide gel electrophoresis (page 109) the SDS-coated proteins will separate according to their size because the smaller proteins move most quickly toward the positive electrode or **anode**.

When electrophoresis is complete, the proteins are stained by incubating the gel in a solution of Coomassie brilliant blue. Each protein band stains blue and is detectable by eye. However, if the amount of protein is very low, a more sensitive detection system is needed, such as a silver stain. Proteins of known molecular mass are also electrophoresed on the gel. By comparison with the standard proteins, the mass of an unknown protein can be determined.

If we want to follow the fate of a single protein in a complex mixture of proteins, we combine SDS-PAGE with a technique called western blotting. The name western blotting is, like northern blotting, a play on the name of Dr. Ed Southern who devised the technique of Southern blotting to analyze DNA (Table 7.2 on page 114).

The protein mixture is separated by SDS-PAGE. A nylon membrane is then placed up against the polyacrylamide gel and picks up the proteins, so that the pattern of protein spots on the original polyacrylamide gel is preserved on the nylon membrane. The nylon membrane is then incubated with an antibody specific for the protein of interest. This antibody, the primary antibody, will seek out and bind to its target protein on the nylon membrane. A second antibody is added that will bind to the primary antibody. To be able to detect the specific protein of interest on the membrane, the secondary antibody is attached to an enzyme. In the figure shown, the enzyme used was horseradish peroxidase. A substrate is added and is converted by the enzyme into a colored product. The protein of interest is seen as a colored band on the nylon membrane. The same enzyme-linked secondary antibody can be used in other laboratories or at other times for western blotting of many different proteins because the specificity is determined by the unlabeled primary antibody.

Part A of the figure shows the Coomassie brilliant-blue-stained pattern of proteins isolated from the endoplasmic reticulum of liver. The leftmost lane is from a phenobarbital-treated animal while the middle lane is from an untreated control animal. The dark bands indicate the presence of protein. The spectrum of proteins is very similar in the two samples, except that a band with a relative molecular mass (Mr) of about 52,000 is darker in the sample from the treated animal. This tells us that drug treatment has caused an increase in the production of a protein with this relative molecular mass. Western blotting (part B) using an antiCYP2B1 antibody confirms that the induced protein is the cytochrome P450 protein known as CYP2B1. The CYP2B1 gene is activated by phenobarbital to produce more CYP2B1 protein to metabolize and clear the drug from the body (page 94).

We already showed how northern blotting revealed that transcription of the *CYP2B1* gene is increased after phenobarbital treatment (Fig. 7.15 on page 114). The western blot shown here demonstrates that, as expected, the amount of CYP2B1 protein is increased as well.

Sodium dodecyl sulfate (SDS) is a major constituent of hair shampoo, where it is usually called by its alternative name of sodium lauryl sulfate.

BACTERIAL PROTEIN SYNTHESIS

Ribosome-Binding Site

For protein synthesis to take place, a ribosome must first attach to the mRNA template. AUG is not only the start codon for protein synthesis; it is used to code for all the other methionines in the protein. How does the ribosome recognize the correct AUG at which to begin protein synthesis? All bacterial mRNAs have at their 5′ end a stretch of nucleotides called the **untranslated** (or leader) **sequence**. These nucleotides do not code for the protein but are nevertheless essential for the correct placing of the ribosome on the mRNA. A nucleotide sequence 5′ GGAGG 3′ (or similar) is usually found with its center about 8 to 13 nucleotides upstream of (5′ of) the AUG start codon (Fig. 8.5). This sequence is complementary to a short stretch of sequence, 3′ CCUCC 5′, found at the 3′ end of the rRNA molecule within the 30S ribosomal subunit. The mRNA and the rRNA interact by complementary base pairing to place the 30S ribosomal subunit in the correct position to start protein synthesis. The sequence on the mRNA molecule is called the ribosome-binding site. This is sometimes referred to as the **Shine–Dalgarno sequence** after the two scientists who found it.

Because the genetic code is read in triplets of three bases, there are three possible reading frames (page 63). The reading frame that is actually used by the cell is defined by the first AUG that the ribosome encounters downstream of the ribosome-binding site.

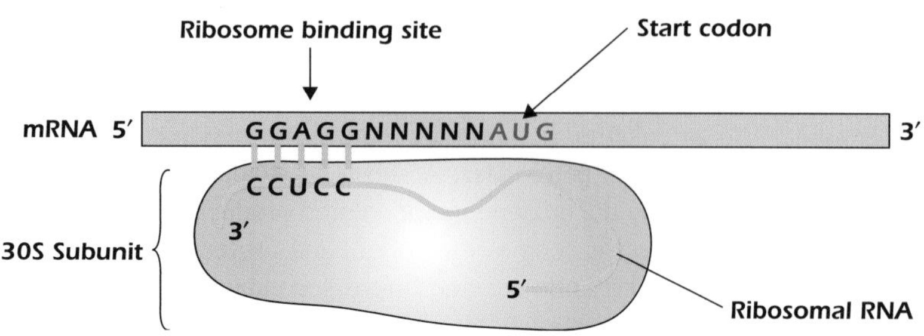

Figure 8.5. The initial binding of the prokaryote 30S subunit to mRNA. N indicates "any nucleotide."

Chain Initiation

The first amino acid incorporated into a new bacterial polypeptide is always a modified methionine, formyl methionine (fmet) (Fig. 8.6). Methionine first attaches to a specific tRNA molecule, $tRNA^{fmet}$, and is then modified by the addition of a formyl group that attaches to its amino group. $tRNA^{fmet}$ has the anticodon sequence 5′ CAU 3′ that binds to its complementary codon, the universal start codon AUG.

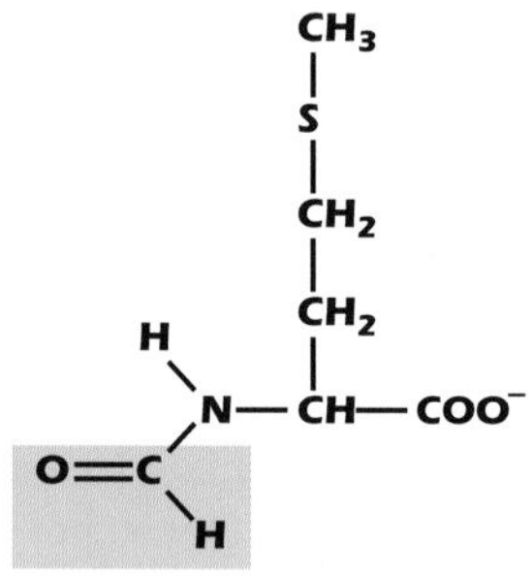

Formyl group

Figure 8.6. Formyl methionine.

The 70S Initiation Complex

The initiation phase of protein synthesis involves the formation of a complex between the ribosomal subunits, an mRNA template, and $tRNA^{fmet}$. Three proteins called **initiation factors (IF)**, IF1, IF2, and IF3, together with the nucleotide guanosine triphosphate (GTP), are needed to help the 70S initiation complex form. First, the initiation factors bind to the 30S subunit (Fig. 8.7A). IF1 and IF3 attach to the 30S subunit. IF1 binds near to the site on the 30S subunit that will become part of the A site of the ribosome. IF2 is an enzyme that breaks down GTP to GDP and inorganic phosphate (Pi). IF3 prevents the 50S subunit binding to the 30S subunit. This is necessary to allow $tRNA^{fmet}$ to bind to an mRNA and a 30S subunit and to form the 30S initiation complex (Fig. 8.7B). The 50S subunit can now bind to 30S subunit; this is accompanied by the release of IF3 (Fig. 8.7C). Lastly, IF1 and IF2 are released and the GTP in IF2 is hydrolyzed, losing its γ phosphate to become guanosine diphosphate. The ribosome is now complete (Fig. 8.7D), and the first tRNA and its amino acid are in place in the P site of the ribosome. A 70S initiation complex has been formed, and protein synthesis can begin. The ribosome is orientated so that it will move along the mRNA in the 5′ to 3′ direction, the direction in which the information encoded in the mRNA molecule is read.

Elongation of the Protein Chain in Bacteria

The synthesis of a protein begins when an aminoacyl tRNA enters the A site of the ribosome (Fig. 8.8). The identity of the incoming aminoacyl tRNA is determined by the codon on the mRNA. If, for example, the second codon is 5′ AAA 3′, then lysyl $tRNA^{Lys}$, whose anticodon is 5′ UUU 3′, will occupy the A site. The P site has, of course, already been occupied by $tRNA^{fmet}$ during the formation of the initiation complex.

Example 8.1 The Irritating Formyl Methionine

White blood cells are strongly attracted by any peptide that begins with formyl methionine: they assume an amoeboid shape (Fig. 1.5 on page 5) and begin to crawl toward the source of the peptide. To a white blood cell, the presence of a peptide beginning with formyl methionine means that there is an infection nearby that needs to be fought. This is because the body's own proteins do not contain formyl methionine: only prokaryotes begin protein synthesis with this modified amino acid.

In mice, scent-sensitive nerve cells in a specialized region of the nose respond strongly to formyl methionine peptides. The hypothesis, as yet unproven, is that this allows the mice to smell the presence of bacteria and therefore avoid spoiled food.

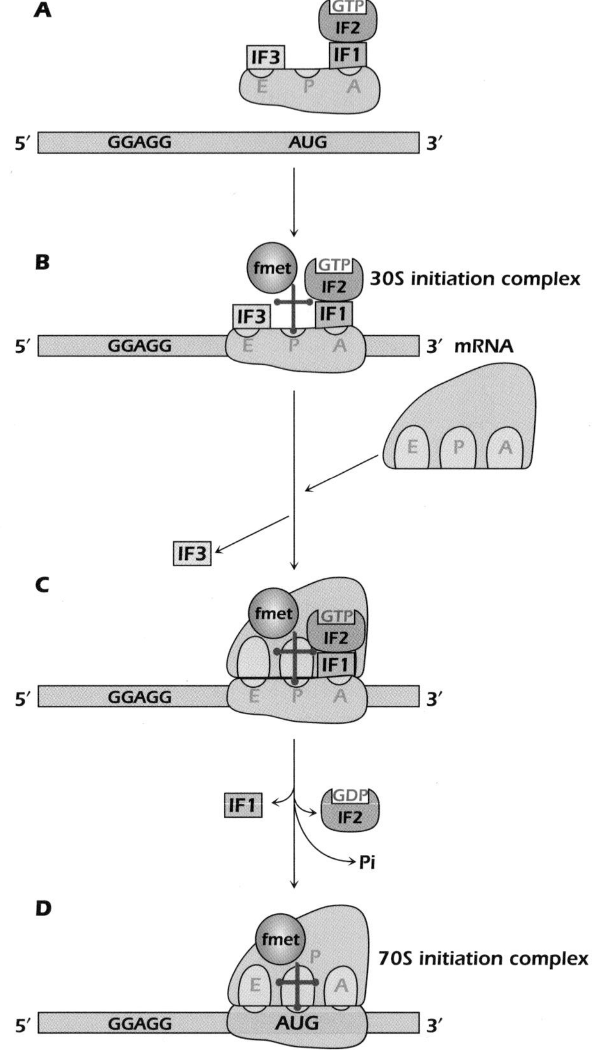

Figure 8.7. Formation of the prokaryote 70S initiation complex.

Elongation of a polypeptide chain needs the help of proteins called **elongation factors (EF)**. The aminoacyl tRNA cannot bind on its own to the A site. Instead the aminoacyl tRNA must first form a complex with EF-Tu and a molecule of guanosine triphosphate (GTP) (Fig. 8.8A). The aminoacyl tRNA is now able to enter the A site (Fig. 8.8B). The presence of EF-Tu is important because it both guides and helps to correctly place the aminoacyl tRNA in the A site. However, for the peptide bond to form between two amino acids, EF-Tu must first be released from the ribosome. This happens when GTP is hydrolysed to GDP plus an inorganic phosphate ion (Pi). Now that both the A and P sites are occupied, the enzyme **peptidyl transferase** catalyzes the formation of a peptide bond between the two amino acids (fmet and lys in this example) (Fig. 8.8C). The dipeptide is attached to the tRNA occupying the A site.

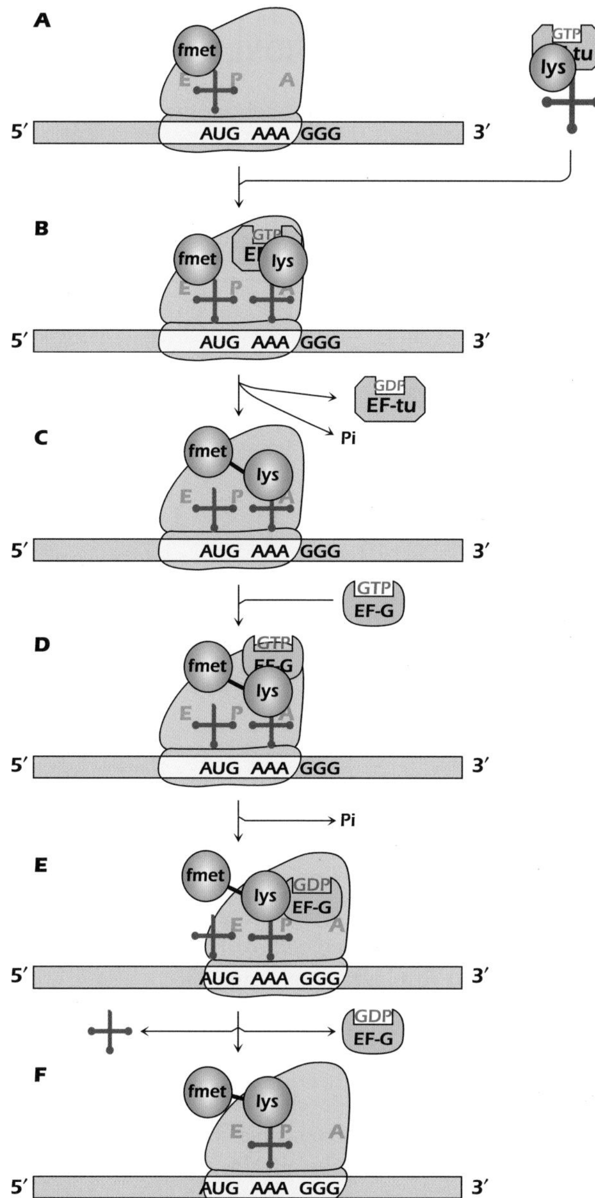

Figure 8.8. Elongation of the protein chain.

For the polypeptide chain to grow, the ribosome must move down the mRNA, this is known as **translocation**. For this to happen the protein EF-G, bound to a molecule of GTP, enters the A site (Fig. 8.8D). When the GTP molecule is hydrolysed to GDP and inorganic phosphate (Pi) the ribosome moves (Fig. 8.8E). This causes the tRNA in the P site, which is no longer attached to an amino acid, to move to the E site (Fig. 8.8E) from where it is released from the ribosome (Fig. 8.8F). The tRNA and attached peptide chain move to occupy the P site. The movement of the ribosome releases EF-G from the A site (Fig. 8.8F) which is once again available for an incoming aminoacyl tRNA. The process of peptide bond

IN DEPTH 8.2 PEPTIDYL TRANSFERASE IS A RIBOZYME

Peptidyl transferase, which catalyses the formation of the peptide bond between two amino acids, is not a protein molecule. In *Escherichia coli* it is the rRNA molecule within the large ribosomal subunit that is responsible for peptide bond formation and this is therefore called a **ribozyme**. This came as a surprise to scientists who discovered that the large subunit of the ribosome had peptidyl transferase activity even when all its protein had been removed.

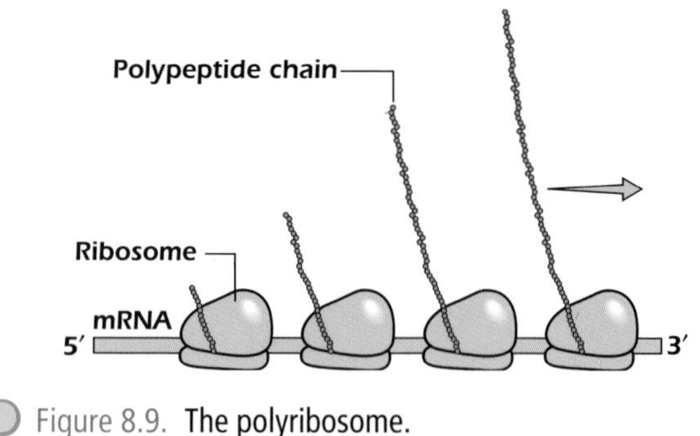

Figure 8.9. The polyribosome.

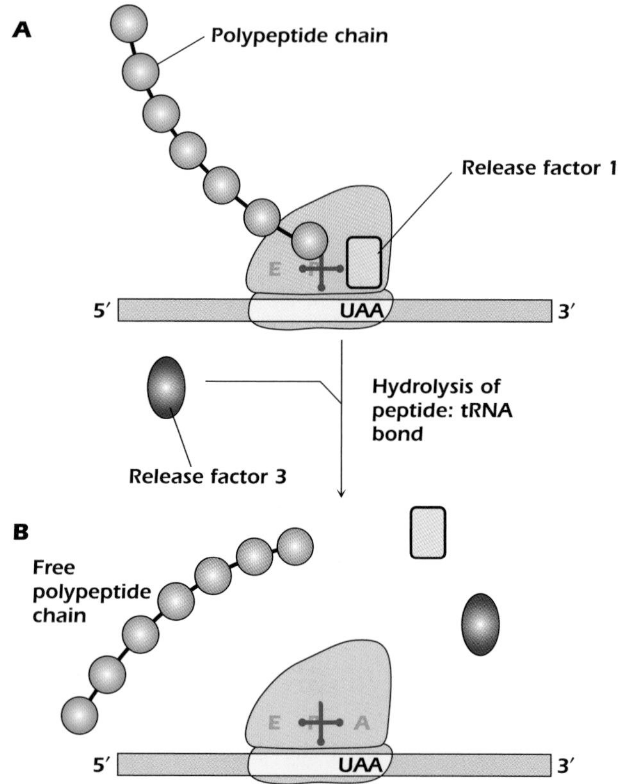

Figure 8.10. Termination of protein synthesis.

formation, followed by translocation, is repeated until the ribosome reaches a stop signal and protein synthesis terminates. Proteins are synthesized beginning at their **amino** or **N terminus** (page 138). The first amino acid hence has a free (although formylated) amino group. The last amino acid in the chain has a free carboxyl group and is known as the **carboxyl** or **C terminus** (page 138).

The Polyribosome

More than one polypeptide chain is synthesized from an mRNA molecule at any given time. Once a ribosome has begun translocating along the mRNA, the start AUG codon is free, and another ribosome can bind. A second 70S initiation complex forms. Once this ribosome has moved away, a third ribosome can attach to the start codon. This process is repeated until the mRNA is covered with ribosomes. Each of these spans about 80 nucleotides. The resultant structure, the **polyribosome** or **polysome** (Fig. 8.9), is visible under the electron microscope. This mechanism allows many protein molecules to be made at the same time on one mRNA.

Termination of Protein Synthesis

There are three codons, UAG, UAA, and UGA, that have no corresponding tRNA molecule. These are stop codons. Instead of interacting with tRNAs, the A site occupied by one of these codons is filled by proteins known as **release factors (RF)** (Fig. 8.10). In the presence of these factors the newly synthesized polypeptide chain is freed from the ribosome. RF1 causes polypeptide chain release from UAA and UAG, and RF2 terminates chains with UAA and UGA. The RF proteins mimic the structure of tRNA. Despite being made up of amino acids instead of nucleotides, RF1 and RF2 have a very similar three-dimensional structure to tRNA. When the A site is occupied by a release factor (Fig. 8.10A), the enzyme peptidyl transferase is unable to add an amino acid to the growing polypeptide chain and instead catalyzes the hydrolysis of the bond joining the polypeptide chain to the tRNA. The carboxyl (COOH) end of the protein is therefore freed from the tRNA (Fig. 8.10B), and the protein is released. The release factors themselves must be removed from the ribosome. A third protein, RF3, helps remove RF1 from the ribosome.

Medical Relevance 8.1 Reading through a STOP

We described earlier (Medical Relevance 4.2 on page 60) how two nonsense mutations are responsible for the majority of cases of Hurler/Scheie syndrome. Remarkably, some antibiotics originally used to block translation in prokaryotes, such as gentamycin, allow eukaryotic ribosomes to travel on through a STOP codon, introducing an amino acid as they do so. Even better, the amino acid introduced at UAG is glutamine (normally encoded by CAA or CAG) meaning that the protein generated in cells from patients with one of the commonest mutations is completely normal. To date there have been no clinical trials of these drugs in Hurler/Scheie patients but the science looks very promising.

The Ribosome Is Recycled

The end of protein synthesis sees the polypeptide and release factor proteins dissociate from the ribosome. At this stage the ribosome is still bound to the mRNA and the P site is still occupied by a tRNA. Bacterial cells contain a protein called **ribosome recycling factor (RRF)**. RRF, like the release factors RF1 and RF2, mimics the structure of a tRNA and binds to the empty A site (Fig. 8.11). The RRF protein is then able to recruit the EF-G protein, the same protein used in ribosome translocation. In a similar way to that described above, EF-G helps remove the tRNA from the P site. To dissociate the two ribosomal subunits, the initiation factor protein IF3 binds to the 30S subunit. The 30S subunit has been prepared to begin the assembly of the 30S initiation complex and a new round of protein synthesis.

EUKARYOTIC PROTEIN SYNTHESIS IS A LITTLE MORE COMPLEX

Elongation of the polypeptide chain and the termination of protein synthesis in eukaryotes does not differ very much from that described for bacteria. The elongation factors and release factors found in eukaryotes have different names to those found in bacteria, but the equivalent proteins do a very similar job. However, the initiation of protein synthesis is more complex in eukaryotes. Their proteins always start with methionine instead of the formyl methionine used in bacterial protein synthesis. A special transfer RNA, $tRNA_i^{met}$, is used to initiate protein synthesis from the AUG start codon. The methionine is often removed from the protein after synthesis. Eukaryotic mRNAs do not contain the bacterial Shine-Dalgarno sequence for ribosome binding. Eukaryotic mRNAs have at their 5′ end a 7-methyl guanosine cap (page 91). The cap is a key feature in the assembly of the ribosome on mRNA. The initiator $tRNA_i^{met}$ binds directly to the P site on the small ribosome subunit (the 40S subunit) along with several proteins called **eukaryotic initiation factors (eIF)**. One of the eIF proteins binds directly to the cap and so positions the small subunit at the 5′ end of the mRNA. The small subunit then has to move forward to find the AUG start codon. It is helped to do this by other eIF proteins, which bind ATP that provides the power for the subunit to move. One of the eIFs is a helicase which unwraps any kinks in the mRNA produced by intramolecular hydrogen bonds so that the ribosome is able to slide along the mRNA. All eukaryotic mRNAs have a sequence very similar to 5′ CCACC 3′ adjacent to the initiating AUG codon. This sequence (known as the Kozak sequence after the scientist who noted it) tells the ribosome that it has reached the start AUG codon. The recognition of the AUG codon that specifies the start site for translation also requires the help of at least nine eIF proteins. Once the large 60S ribosomal subunit has attached to form the 80S initiation complex protein synthesis can begin.

Example 8.2 The Diptheria Bacterium Inhibits Protein Synthesis

Some bacteria cause disease because they inhibit eukaryotic protein synthesis. Diphtheria was once a widespread and often fatal disease caused by infection with the bacterium *Corynebacterium diphtheriae*. This organism produces an enzyme (diphtheria toxin) that inactivates eukaryotic elongation factor 2 (the equivalent of the bacterial elongation factor G). Diphtheria toxin splits the bond between ribose and nicotinamide in NAD^+ (Fig. 2.16 on page 31), releasing free nicotinamide and attaching the remainder, ADP-ribose, to elongation factor 2, a process known as ADP ribosylation. The protein is now inactive and is unable to assist in the movement of the ribosome along the mRNA template. Protein synthesis therefore stops in the affected human cells. All the amino acids that the host was using to make its own protein are now available for the bacterium's use.

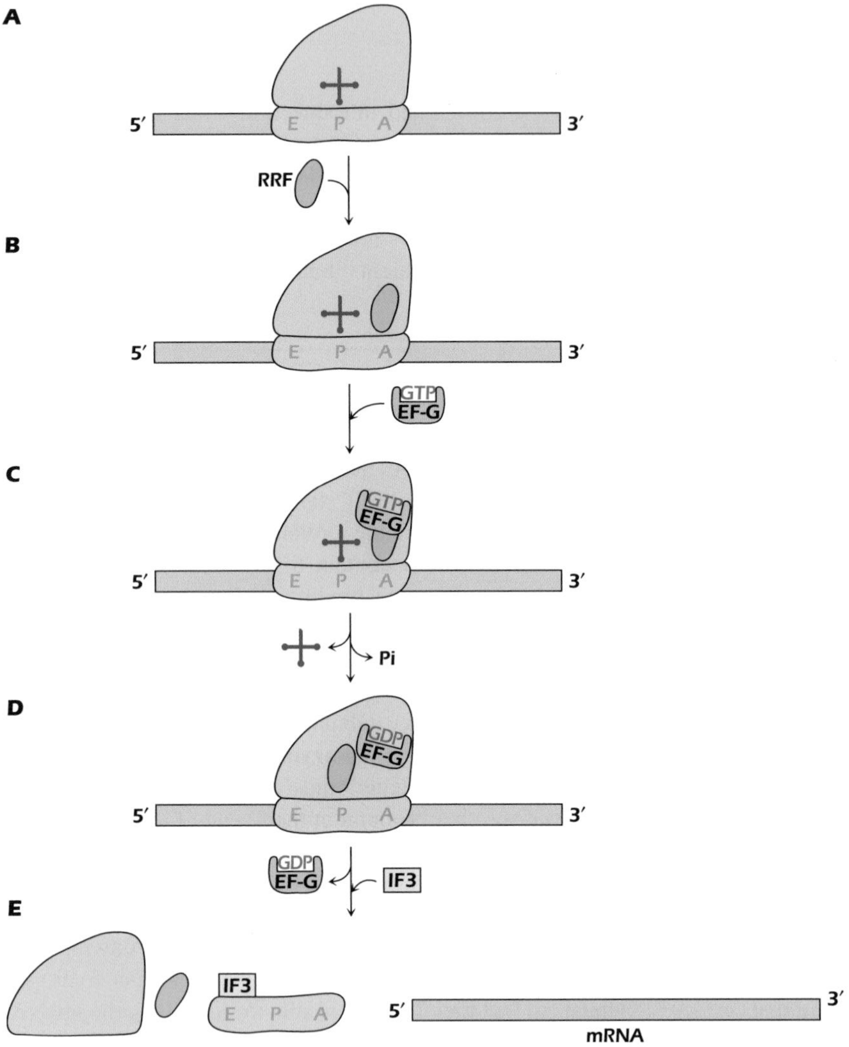

Figure 8.11. The ribosome is recycled.

Antibiotics and Protein Synthesis

Many antibiotics work by blocking protein synthesis, a property that is extensively exploited in research and medicine. Many antibiotics only inhibit protein synthesis in bacteria and not in eukaryotes. They are therefore extremely useful in the treatment of infections because the invading bacteria will die but protein synthesis in the host organism remains unaffected. Examples are chloramphenicol, which blocks the peptidyl transferase reaction, and tetracycline, which inhibits the binding of an aminoacyl tRNA to the A site of the ribosome. Both of these antibiotics therefore block chain elongation. Streptomycin, on the other hand, inhibits the formation of the 70S initiation complex because it prevents $tRNA^{fmet}$ from binding to the P site of the ribosome.

Puromycin causes the premature release of polypeptide chains from the ribosome and acts on both bacterial and eukaryotic cells. This antibiotic has been widely used in the study of protein synthesis. Puromycin can occupy the A site of the ribosome because its structure resembles the 3′ end of an aminoacyl-tRNA (Fig. 8.12). However, puromycin does not bind to the mRNA. Puromycin blocks protein synthesis because peptidyl transferase uses it as a substrate and forms a peptide bond between the growing polypeptide and the antibiotic. Once translocation has occurred, the growing polypeptide has no strong attachment to the mRNA and is therefore released from the ribosome.

Answer to Thought Question: SDS-PAGE separates multi-subunit proteins into their component polypeptides. *E. coli* RNA polymerase is made of five subunits, one each of β, β′, and σ plus two identical α subunits. Four bands will therefore be visible on the stained polyacrylamide gel. The two α subunits are identical and will run together down the gel to form one band.

IN DEPTH 8.3 PROTEOMICS

Proteomics is the study of the **proteome**—the complete protein content of a cell. It is the proteins a cell makes that allow it to carry out its specialized functions. Although, for example, a liver and a kidney cell have many proteins in common, they also each possess a unique subset of proteins, which gives to them their own characteristics. Similarly, a cell will need to make different proteins, and proteins in different amounts, according to its metabolic state. The goal of proteomics is to identify all of the proteins produced by different cells and how a particular disease changes a cell's protein profile.

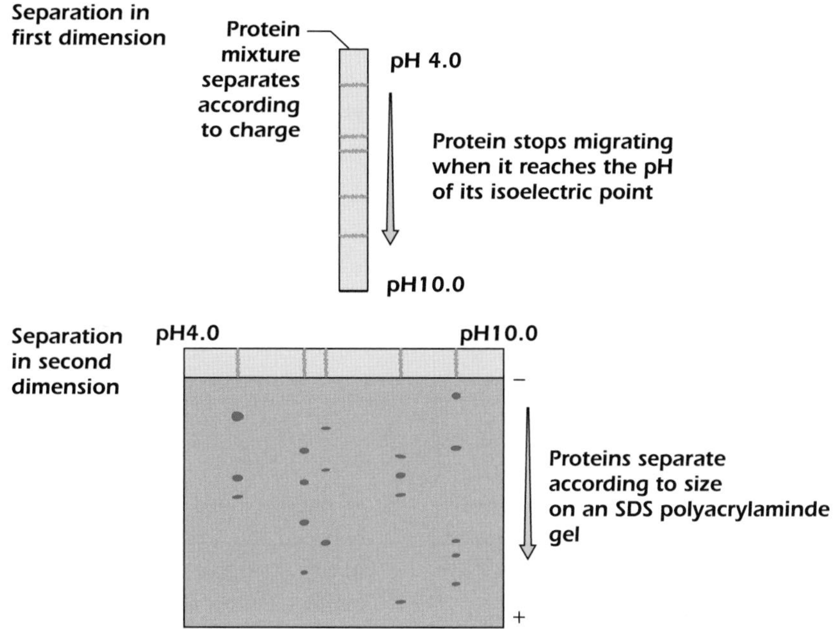

The separation of a cell's protein mixture into individual components is tackled using a technique known as two-dimensional polyacrylamide gel electrophoresis. This produces a pattern of protein spots. These patterns are recorded and serve as templates for the comparison of the proteomes of different cells. Spots that change during a cell's development, or changes that occur in disease, can be easily identified. A protein spot of interest is excised from the polyacrylamide gel and the protein broken into small, overlapping peptide fragments by proteolytic enzymes. The fragments are fed into a mass spectrometer, and their peptide mass fingerprint determined. The fingerprint identifies the protein.

The dream of many scientists, now that we know the base sequence of the human genome, is to define the human proteome and to determine all of its variations. However, the genome project used a single automated method to sequence the order of the four bases in DNA, and data was obtained at a relatively rapid pace. In contrast it is very labor intensive to identify proteins and much effort is being made to devise ways of speeding things up. The aim of proteomic research centers is to increase their throughput from about 40 to 100 peptide samples an hour, to the daunting number of 1 million peptides a day.

Protein Destruction

Different proteins have enormously different lifetimes. The **keratin** (page 291) in hair lasts months until it is cut or falls off, while the α subunit of ATP synthase, the enzyme that makes ATP (page 200) has a half life of 2 hours. Cytosolic proteins usually end their lives in the cell's equivalent of an office shredder, the **proteasome**. This is a barrel-shaped proteolytic machine that chops old or damaged proteins up into short lengths of peptide which are then in turn hydrolysed by cytosolic peptidases into their constituent amino acids, ready to be used again for protein synthesis.

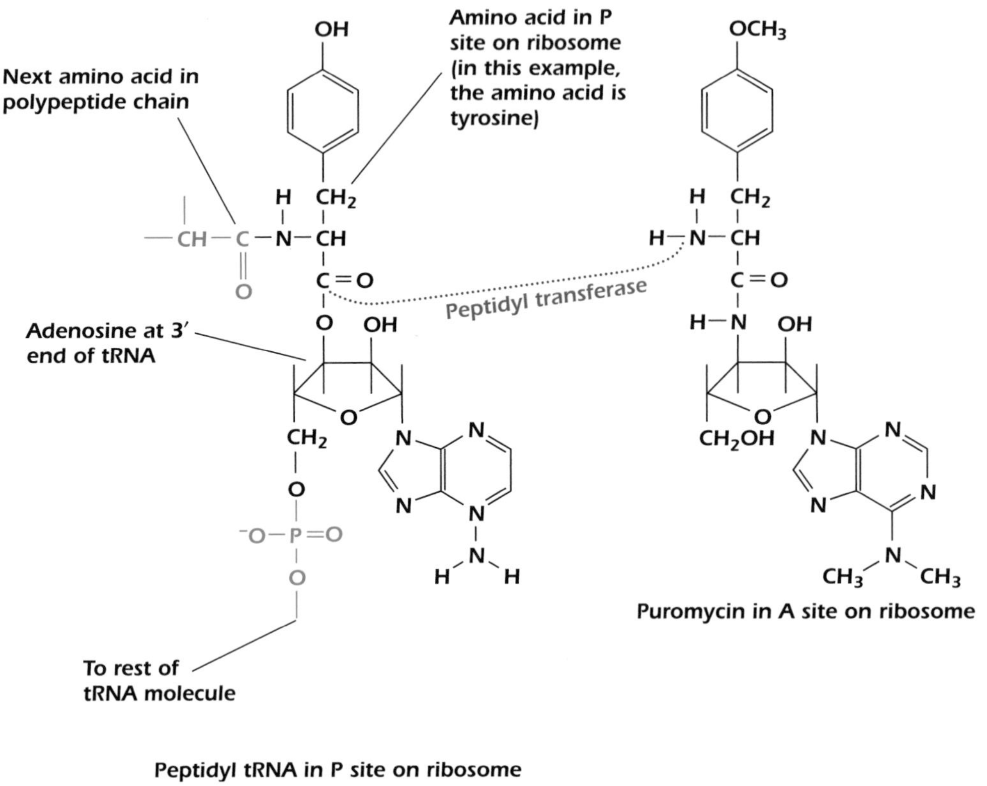

Next amino acid in polypeptide chain

Amino acid in P site on ribosome (in this example, the amino acid is tyrosine)

Adenosine at 3′ end of tRNA

Peptidyl transferase

To rest of tRNA molecule

Peptidyl tRNA in P site on ribosome

Puromycin in A site on ribosome

Figure 8.12. Puromycin can occupy the ribosome A site.

SUMMARY

1. Amino acids are prepared for use in protein synthesis by being attached to the 3′ end of a tRNA to form an aminoacyl tRNA.
2. The genetic code in an mRNA is translated into a sequence of amino acids on the ribosome, which comprises a small and large subunit and has two aminoacyl tRNA binding sites, the P site and the A site, plus an E site from which tRNAs are lost.
3. Initiation of protein synthesis involves the binding of the small ribosomal subunit to the mRNA. A special tRNA ($tRNA^{fmet}$ in prokaryotes, $tRNA_i^{met}$ in eukaryotes) binds to the initiation codon, and then the large ribosomal subunit attaches and the initiation complex is formed.
4. Protein synthesis begins when a second aminoacyl tRNA occupies the A site. Each incoming amino acid is specified by the codon on the mRNA. The anticodon on the tRNA hydrogen bonds to the codon, thus positioning the amino acid on the ribosome.
5. A peptide bond is formed, by peptidyl transferase, between the amino acids in the P and A sites. The newly synthesized peptide occupies the P site, and another amino acid is brought into the A site. This process of elongation requires a number of proteins (elongation factors); as it continues, the peptide chain grows.
6. When a stop codon is reached, the polypeptide chain is released with the help of proteins known as release factors.
7. More than one ribosome can attach to an mRNA. This forms a polyribosome, and many protein molecules can be made simultaneously from the same mRNA.
8. Many antibiotics fight disease because they inhibit particular steps in protein synthesis.

FURTHER READING

Arnez, J. G., and Moras, D. (1997) Structural and functional considerations of the aminoacylation reaction. *Trends Biochem. Sci*. **22**, 211–216.

Moore, P. B., and Steitz, T. A. (2002) The involvement of RNA in ribosome function. *Nature* **418**, 229–235.

Ribas de Pouplana, L., and Schimmel, P. (2001) Aminoacyl-tRNA synthetases: Potential markers of genetic code development. *Trends Biochem. Sci*. **26**, 591–596.

Shaw, K. (2008) The role of ribosomes in protein synthesis. *Nature Education* **1**(1). www.nature.com/scitable/topicpage/The-Role-of-Ribosomes-in-Protein-Synthesis-1021.

REVIEW QUESTIONS

The figure above shows a ribosome during translation to provide a reminder of the basic structure.

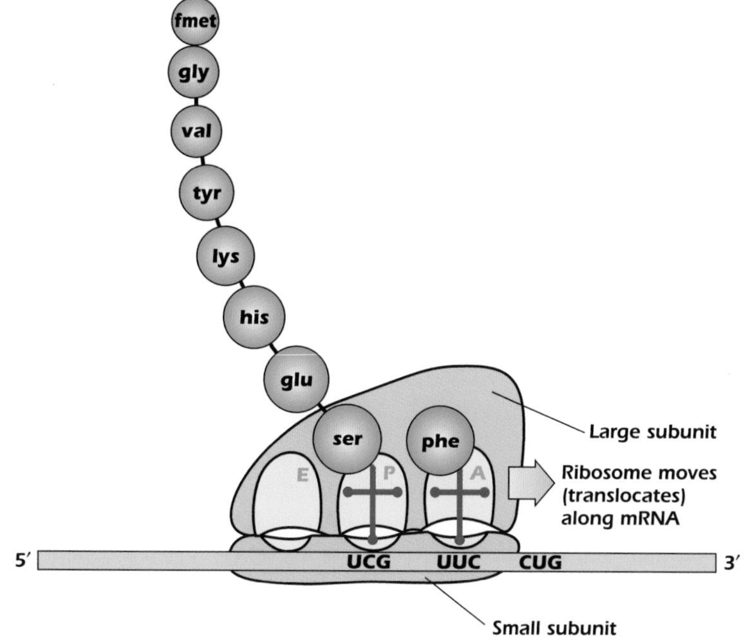

8.1 Theme: Translation Initiation

A A site
B E site
C P site
D 5′ AUG 3′
E 5′ CCACC 3′
F 5′ CCUCC 3′
G 5′ CUG 3′
H 5′ GGAGG 3′
I 5′ UGCUUC 3′
J Formyl methionine
K Methionine
L Poly-A tail
M 7-methyl guanosine cap

From the above list of base sequences and other components, select the one relevant to each of the steps in translation initiation which are described below in sequence.

1. In the early stages of translation initiation in prokaryotes, the small ribosomal subunit attaches by complimentary base pairing to this sequence at the 5′ end of the mRNA, sometimes called the Shine-Dalgano sequence.
2. In contrast in the early stages of translation initiation in eukaryotes, the small ribosomal subunit attaches to this group at the extreme 5′ end of the mRNA molecule.
3. The eukaryotic small subunit then slides along the mRNA until it encounters this sequence, known as the Kozak sequence.
4. The subsequent steps are similar in prokaryotes and eukaryotes. The small subunit slides a few more bases along until it encounters this sequence, the start codon for translation.
5. Initiation factors then act to catalyze the assembly of the complete ribosome. The first tRNA, with its formyl methionine (in prokaryotes) or methionine (in eukaryotes) attached, locates in one of the three tRNA binding sites on the ribosome: state which one.

8.2 Theme: Translation Elongation and Termination

A A site
B E site
C P site
D EF-G
E EF-Tu
F IF-3
G IF-Tu
H puromycin
I release factor 1 or 2
J $tRNA^{fmet}$ in prokaryotes, $tRNA_i^{met}$ in eukaryotes

From the above list of proteins, compounds and components, select the one relevant to each of the steps in translation initiation which are described below in sequence.

1. Following translocation of the ribosome three bases along the mRNA, this site on the ribosome is empty and can be occupied by a charged tRNA whose anticodon is complimentary to the corresponding codon on the mRNA.
2. Peptidyl transferase now catalyzes the formation of a peptide bond between the new amino acid and the existing polypeptide chain. Immediately following this, the polypeptide chain is only attached to the mRNA via the tRNA in which of the three sites on the ribosome.
3. The next step is translocation, the physical movement of the ribosome three bases along the mRNA. Energy from this is provided by the hydrolysis of GTP by which enzyme, which occupies the A site on the ribosome.
4. As a result of translocation the uncharged tRNA, which gave up its amino acid during the formation of the peptide bond, moves to this site on the ribosome, from where it is released.
5. When translocation brings a stop codon UGA, UAA or UAG into the position facing the A site on the ribosome, the A site becomes occupied not by a normal charged tRNA molecule but rather by this molecule.
6. Finally the ribosome splits into two subunits as a result of energy released in GTP hydrolysis by EF-G. The released small ribosomal subunit already has one initiation factor attached, ready to accept the other initiation factors and the first charged amino acid to initiate synthesis of a new polypeptide. Name this pre-attached initiation factor.

8.3 Theme: The Wobble

You should refer to Figure 4.8 on page 61 while answering this question.

A 5′ AAG 3′
B 5′ CAU 3′
C 5′ GAA 3′
D 5′ GUU 3′
E 5′ IAU 3′
F 5′ UAI 3′
G 5′ UUG 3′
H 5′UAC 3′

From the above list of base sequences, select the one that is likely to be the anticodon on a tRNA specific for each of the amino acids below.

1. Methionine
2. Asparagine
3. Phenylalanine
4. Isoleucine

THOUGHT QUESTION

Predict the appearance of an SDS-PAGE gel used to study a purified sample of *E. coli* RNA polymerase (see page 82).

9

PROTEIN STRUCTURE

Virtually everything cells do depends upon proteins. We are all, regardless of build, made up of water plus more or less equal amounts of fat and protein. Although the DNA in our cells contains the information necessary to make our bodies, DNA itself is not a significant part of our body mass. Nor is it a chemically interesting molecule, in the sense that one length of DNA is much the same as another in terms of shape and chemical reactivity. The simplicity of DNA arises because it is a polymer made up of only four fairly similar monomers, and this is appropriate because the function of DNA is simply to remain as a record and to be read during transcription. In contrast, proteins made using the instructions in DNA vary enormously in physical characteristics and function and can be considered as constituting the much more complex proteome (In Depth 8.3 on page 133). Silk, hair, the lens of an eye, an immunoassay (such as found in a pregnancy test kit), and cottage cheese are all just protein plus more or less water, but they have very different properties because the proteins they contain are different. Proteins carry out almost all of the functions of the living cell including, of course, the synthesis of new DNA. Neither growth nor development would be possible without proteins.

Most proteins have functions that depend on their ability to recognize other molecules by binding. This recognition depends on specific three-dimensional **binding sites** that make multiple interactions with the **ligand**, the molecule being bound. To do this a protein must itself have a specific three-dimensional structure. Each of the huge number of protein functions demands its own protein structure. Evolution has produced this diversity by using a palette of 20 amino acid monomers, each with its own unique shape and chemical properties, as the building blocks of proteins. Huge numbers of very different structures are therefore possible.

 NAMING PROTEINS

To be able to discuss proteins, we need to give them names. Naming conventions vary between different areas of biology. We have already seen how enzymes such as DNA polymerase are named for the reaction they catalyze, and their names usually end in *-ase*. Many proteins have names that describe their structure or their role in cells, such as hemoglobin and connexin. However, the pace with which new proteins are being discovered at the moment means that many are not given proper names but are simply referred to by their size: p38 (page 311) and p53 (page 306) have relative molecular masses of about 38,000 and about 53,000, respectively. Clearly this could cause confusion, so we sometimes add the name of the gene as a superscript: $p16^{INK4a}$ is a protein of relative molecular mass (Mr) of about 16,000 that is the product of the *INK4a* gene.

 POLYMERS OF AMINO ACIDS

Translation produces linear polymers of α-amino acids. If there are fewer than around 50 amino acids in a polymer, we tend to call it a **peptide**. More and it is a **polypeptide**.

Cell Biology: A Short Course, Third Edition. Stephen R. Bolsover, Jeremy S. Hyams, Elizabeth A. Shephard and Hugh A. White.

Proteins are polypeptides, and most have dimensions of a few nanometers (nm), although structural proteins like keratin in hair are much bigger. The relative molecular masses of proteins can range from 5,000 to hundreds of thousands.

The Amino Acid Building Blocks

The general structure of **α-amino acids**, the building blocks of polypeptides and proteins, is shown in Figure 9.1A. R is the side chain. It is the side chain that gives each amino acid its unique properties.

During the process of translation (page 129), peptidyl transferase joins the amino group of one amino acid to the carboxyl of the next to generate a peptide bond. A generalized polypeptide is shown in Figure 9.1B. The backbone, a series of peptide bonds separated by the α carbons, is shown in red. At the left-hand end is a free amino group; this is known as the N terminus or amino terminus (in a freshly synthesized prokaryote polypeptide, the amino group would be masked by a formyl group, page 128). At the right-hand end is a free carboxyl group; this is known as the C terminus or carboxy terminus. Peptides are normally written this way, with the C terminus to the right.

The properties of individual polypeptides are conferred by the side chains of their constituent amino acids. Many different properties are important—size, electrical charge, the ability to participate in particular reactions—but the most important is the affinity of the side chain for water. Side chains that interact strongly with water are hydrophilic. Those that do not are hydrophobic. We have already encountered the 20 amino acids coded for by the genetic code (page 61), but we will now describe each in turn, beginning with the most hydrophilic and ending with the most hydrophobic. Each amino acid has both a three-letter abbreviation and a one-letter code (Fig. 9.2). In the following section we refer to each amino acid by its full name and give the three- and one-letter abbreviations. This will help you to familiarize yourself with the amino acid abbreviations, which are used in other sections of the book.

Several amino acids have hydrophilic side chains. Of these, four amino acids have side chains that are either acid groups or derived from them. Aspartate (Asp, D) and glutamate (Glu, E) have acidic carboxyl side chains. At the pH of the cytoplasm, these side chains bear negative charges, which interact strongly with water molecules. As they are usually charged, we generally name them as their ionized forms: aspartate and glutamate rather than aspartic acid and glutamic acid. A polypeptide made entirely of these amino acids is very soluble in water. Asparagine (Asn, N) and glutamine (Gln, Q) are the amides of aspartate and glutamate. They are hydrophilic but not charged.

Lysine (Lys, K) and arginine (Arg, R) have basic side chains. At the pH of the cytoplasm, these side chains bear positive charges, which again interact strongly with water molecules.

Figure 9.3 shows histidine (His, H), whose side chain is weakly basic with a pKa (page 21) of 7. At neutral pH about half the histidine side chains, therefore, bear a positive charge. The fact that histidine is equally balanced between protonated and unprotonated forms gives it important roles in enzyme catalysis. At neutral pH a polypeptide made entirely of histidine will be very soluble in water.

Figure 9.3 also shows cysteine (Cys, C) whose thiol (–SH) group is weakly acidic with a pKa of about 8. At neutral pH most (about 90%) of cysteine side chains will have their hydrogen attached. Even so the charge on the remaining 10% means that a polypeptide made entirely of cysteine will be soluble in water. Cysteine's thiol group is chemically reactive and has important roles in some enzyme active sites. Under oxidizing conditions the thiol groups of two cysteine residues can link together to form a **disulfide bond** (or **disulfide bridge**) (Fig. 9.4). Proteins made for use outside the cell often have disulfide bonds that confer additional rigidity on the protein molecule. If all the peptide bonds in a polypeptide are hydrolyzed, any cysteines that were linked by disulfide bonds remain connected in dimers called—very confusingly—cystine molecules.

The three amino acids serine (Ser, S), threonine (Thr, T), and tyrosine (Tyr, Y) are hydrophilic as their side chains

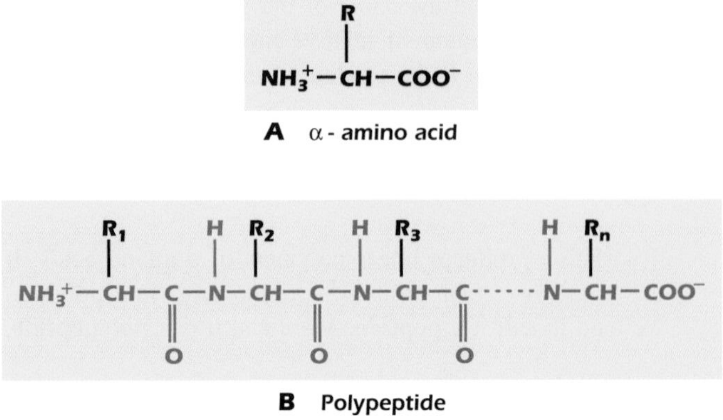

$$\begin{array}{c} R \\ | \\ NH_3^+-CH-COO^- \end{array}$$

A α - amino acid

$$\begin{array}{ccccccccccccc} & R_1 & & H & R_2 & & H & R_3 & & H & R_n & \\ & | & & | & | & & | & | & & | & | & \\ NH_3^+ - & CH & - C - & N & - CH - & C - & N & - CH - & C \cdots\cdots & N & - CH - & COO^- \\ & & \| & & & \| & & & \| & & & \\ & & O & & & O & & & O & & & \end{array}$$

B Polypeptide

Figure 9.1. α-amino acids and the peptide bond.

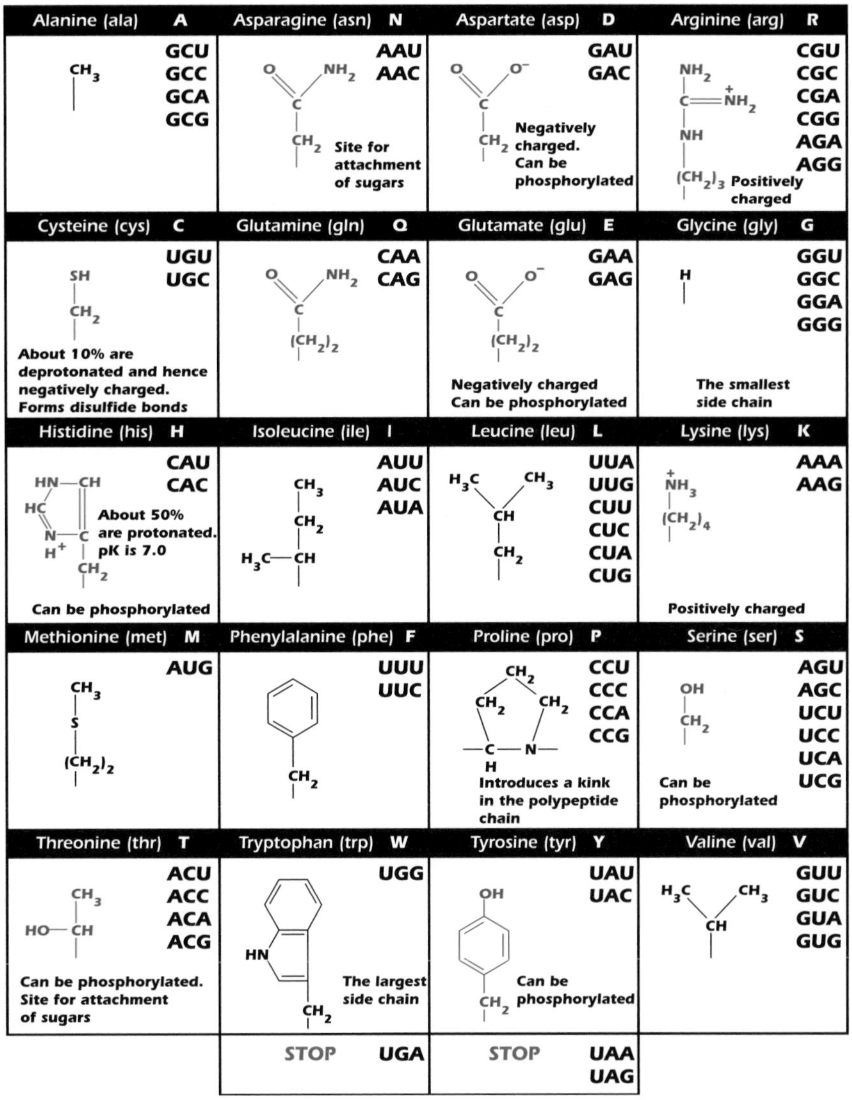

Amino acid	Codons	Side chain	Notes
Alanine (ala) A	GCU, GCC, GCA, GCG	CH_3	
Asparagine (asn) N	AAU, AAC	CH_2–C(=O)NH_2	Site for attachment of sugars
Aspartate (asp) D	GAU, GAC	CH_2–C(=O)O^-	Negatively charged. Can be phosphorylated
Arginine (arg) R	CGU, CGC, CGA, CGG, AGA, AGG	$(CH_2)_3$–NH–C(=$\overset{+}{N}H_2$)NH_2	Positively charged
Cysteine (cys) C	UGU, UGC	CH_2–SH	About 10% are deprotonated and hence negatively charged. Forms disulfide bonds
Glutamine (gln) Q	CAA, CAG	$(CH_2)_2$–C(=O)NH_2	
Glutamate (glu) E	GAA, GAG	$(CH_2)_2$–C(=O)O^-	Negatively charged. Can be phosphorylated
Glycine (gly) G	GGU, GGC, GGA, GGG	H	The smallest side chain
Histidine (his) H	CAU, CAC	CH_2–imidazole (HN–CH, HC, N–C, H^+)	About 50% are protonated. pK is 7.0. Can be phosphorylated
Isoleucine (ile) I	AUU, AUC, AUA	H_3C–CH–CH_2–CH_3	
Leucine (leu) L	UUA, UUG, CUU, CUC, CUA, CUG	CH_2–CH(CH_3)CH_3	
Lysine (lys) K	AAA, AAG	$(CH_2)_4$–$\overset{+}{N}H_3$	Positively charged
Methionine (met) M	AUG	$(CH_2)_2$–S–CH_3	
Phenylalanine (phe) F	UUU, UUC	CH_2–phenyl	
Proline (pro) P	CCU, CCC, CCA, CCG	CH_2–CH_2–CH_2 ring with C–N (H)	Introduces a kink in the polypeptide chain
Serine (ser) S	AGU, AGC, UCU, UCC, UCA, UCG	CH_2–OH	Can be phosphorylated
Threonine (thr) T	ACU, ACC, ACA, ACG	HO–CH–CH_3	Can be phosphorylated. Site for attachment of sugars
Tryptophan (trp) W	UGG	CH_2–indole (HN)	The largest side chain
Tyrosine (tyr) Y	UAU, UAC	CH_2–phenyl–OH	Can be phosphorylated
Valine (val) V	GUU, GUC, GUA, GUG	CH(CH_3)CH_3	
STOP	UGA		
STOP	UAA, UAG		

Figure 9.2. The genetic code. Amino acid side chains are shown in alphabetical order together with the three- and one-letter amino acid abbreviations. Hydrophilic side chains are shown in green, hydrophobic side chains in black. Other important characteristics of each side chains are noted. To the right of each amino acid we show the corresponding mRNA codons.

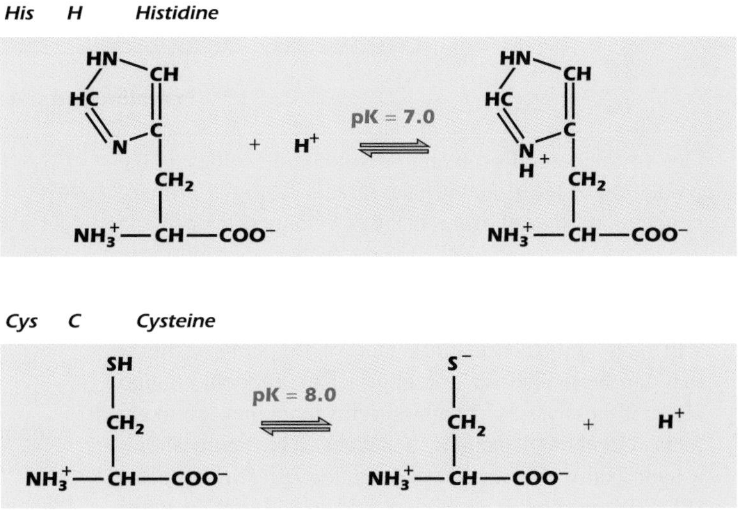

His H Histidine

Histidine (imidazole HN–CH, HC, N–C; CH_2; NH_3^+—CH—COO^-) + H^+ ⇌ (pK = 7.0) protonated histidine (imidazole HN–CH, HC, N^+–H, C; CH_2; NH_3^+—CH—COO^-)

Cys C Cysteine

SH–CH_2–CH(NH_3^+)–COO^- ⇌ (pK = 8.0) S^-–CH_2–CH(NH_3^+)–COO^- + H^+

Figure 9.3. Histidine and cysteine have pK values in the physiological range.

Disulfide bond between 2 cysteines

The double amino acid cystine

Figure 9.4. Oxidation of adjoining cysteine residues produces a disulfide bond. Proteolysis of the polypeptide releases cystine.

have hydroxyl (–OH) groups that can hydrogen bond with water molecules. A polypeptide composed of these amino acids is soluble in water.

Glycine (Gly, G) has nothing but a hydrogen atom for its side chain. It is relatively indifferent to its surroundings.

Five amino acids have hydrophobic side chains of carbon and hydrogen only. These are alanine (Ala, A), valine, (Val, V), leucine (Leu, L), isoleucine (Ile, I), and phenylalanine (Phe, F). The side chains cannot interact with water so they are hydrophobic. A polypeptide composed entirely of these amino acids does not dissolve in water but will dissolve in olive oil.

Tryptophan (Trp, W) is the largest of the amino acids. Its double ring side chain is mainly hydrophobic. Methionine (Met, M) is also hydrophobic: its sulfur atom is in the middle of the chain so cannot interact with water. Last comes proline (Pro, P). Proline is not really an amino acid at all—it is an **imino acid**, but biologists give it honorary amino acid status. Because the side chain (which is hydrophobic) is connected to the nitrogen atom and therefore to the peptide bond, including proline in the polypeptide chain introduces a kink.

IN DEPTH 9.1 HYDROPATHY PLOTTING—THE PDGF RECEPTOR

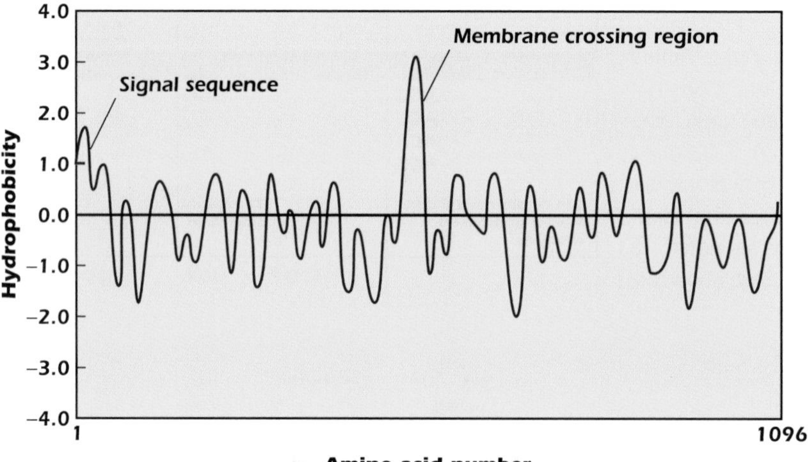

One of the central problems in structural biology is the prediction of the structure of a protein from the primary structure or indeed from the DNA sequence once the gene has been found. Although we have come some way in understanding protein folding, the problem remains unsolved. It is possible, however, to identify proteins that will have similar structures. One of the simpler things that can be looked for is regions of hydrophobic amino acids: if these are 21–22 amino acids long, they are likely to be membrane-spanning α helices. The figure shows a hydropathy plot for the platelet-derived growth factor (PDGF) receptor. Each amino acid in the protein has a hydrophobicity allotted to it, so that ionized groups like those on aspartate and arginine get a big negative score while groups like those on phenylalanine and leucine get a big positive score. The plot is a running average of the hydrophobicity along the polypeptide chain. The protein begins with a somewhat hydrophobic region: this is the signal sequence (page 158) that directs the growing protein to the endoplasmic reticulum. The rest of the protein is neutral or somewhat hydrophilic except for a prominent short hydrophobic region in the center. We can therefore predict that this protein will cross the membrane once at this location. Figure 9.6 shows the full predicted structure of this protein.

The Unique Properties of Each Amino Acid

Although we have classified side chains on the basis of their affinity for water, their other properties are important.

Charge. If a side chain is charged it will be hydrophilic. But charge has other effects too. A positively charged residue such as a lysine will attract a negatively charged residue such as glutamate. If the two residues are buried deep within a folded polypeptide, where neither can interact with water, then it will be very difficult to pull them apart. We call such an electrostatic bond inside a protein a **salt bridge**. Negatively charged residues will attract positively charged ions out of solution, so a pocket on the surface of a protein lined with negatively charged residues will, if it is the right size, form a binding site for a particular positively charged ion like sodium or calcium.

Post-translational Modifications. After a polypeptide has been synthesized some of the amino acid side chains may be modified. A good example of this is glycosylation, in which chains of sugars may be added to asparagine or threonine side chains (page 165). A very specific post-translational modification occurs in the connective tissue protein collagen (page 12). After the polypeptide chain has been synthesized, specific prolines and lysines are hydroxylated to hydroxyproline and hydroxylysine. Because hydroxyl groups can hydrogen bond with water, this helps the extracellular matrix to form a hydrated gel.

Example 9.1 The Salt Bridges in ROMK Hold the Channel Open

A potassium channel called ROMK plays a critical role in allowing the kidney to recover sodium that would otherwise be lost in the urine (Medical Relevance 14.3 on page 246). The complete channel is formed from four identical protein subunits. The pore through which potassium ions pass lies at the centre where all four subunits meet.

The illustration shows this critical pore region. The quaternary structure is stabilized by salt bridges that form between positively charged arginine residues and negatively charged glutamate residues on the neighboring subunit. A rare genetic disease called Bartter's syndrome is caused by the absence of one or other of the arginine or glutamate partners. In the absence of the salt bridge the four subunits slump inwards, occluding the central pore. Without a functional ROMK, patients lose considerable sodium in the urine and must take salt supplements.

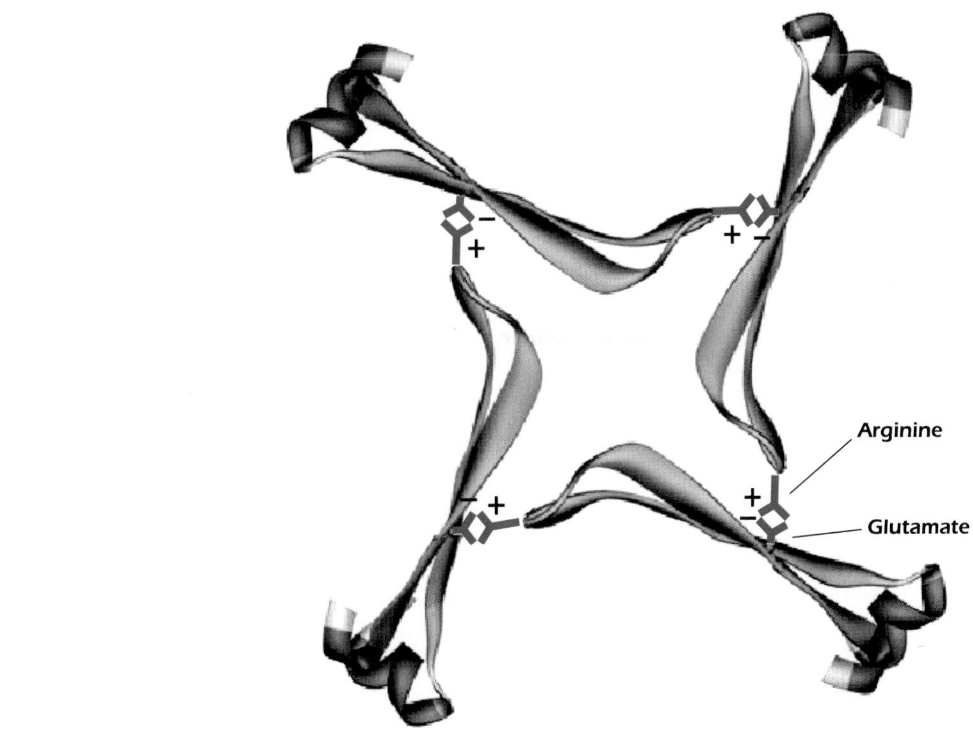

Not all post-transational modifications are so permanent. One very important modification is **phosphorylation**, the attachment of a phosphate group. We have already seen how sugars can be phosphorylated (page 30), but so can the side chains of the six amino acids serine, threonine, tyrosine, aspartate, glutamate, and histidine (Fig. 9.5). Usually the phosphate group comes from ATP, in which case the enzyme that does the job is called a **kinase**. Phosphate groups carry two negative charges and can therefore markedly alter the balance of electrical forces within a protein. Another group of enzymes called **phosphatases** remove the phosphate groups. Since phosphorylation is readily reversible, it is often used to turn protein activity off and on. For example, the transcription factor NFAT (page 161) is only active in the dephosphorylated state, while eukaryote RNA polymerase II (page 93) is activated by phosphorylation. In this book the majority of the protein kinases we meet will fall into two classes: **serine-threonine kinases**, which can phosphorylate on either of these two side chains, and **tyrosine kinases**.

Medical Relevance 9.1 Adding a Hydrophobic Group to Ras

Ras is an important protein involved in control pathways. It will be described in Chapter 15. Ras as made at the ribosome is a small, soluble protein. It is post-translationally modified by the attachment of a long hydrophobic group. This targets it to the inside of the plasma membrane where it participates in processes leading to cell survival and cell division. Mutations in Ras can disrupt these pathways and lead to some types of cancer. The anti-cancer drug Zarnestra (also called Tipifarnib) inhibits the enzyme that carries out the addition of the hydrophobic group.

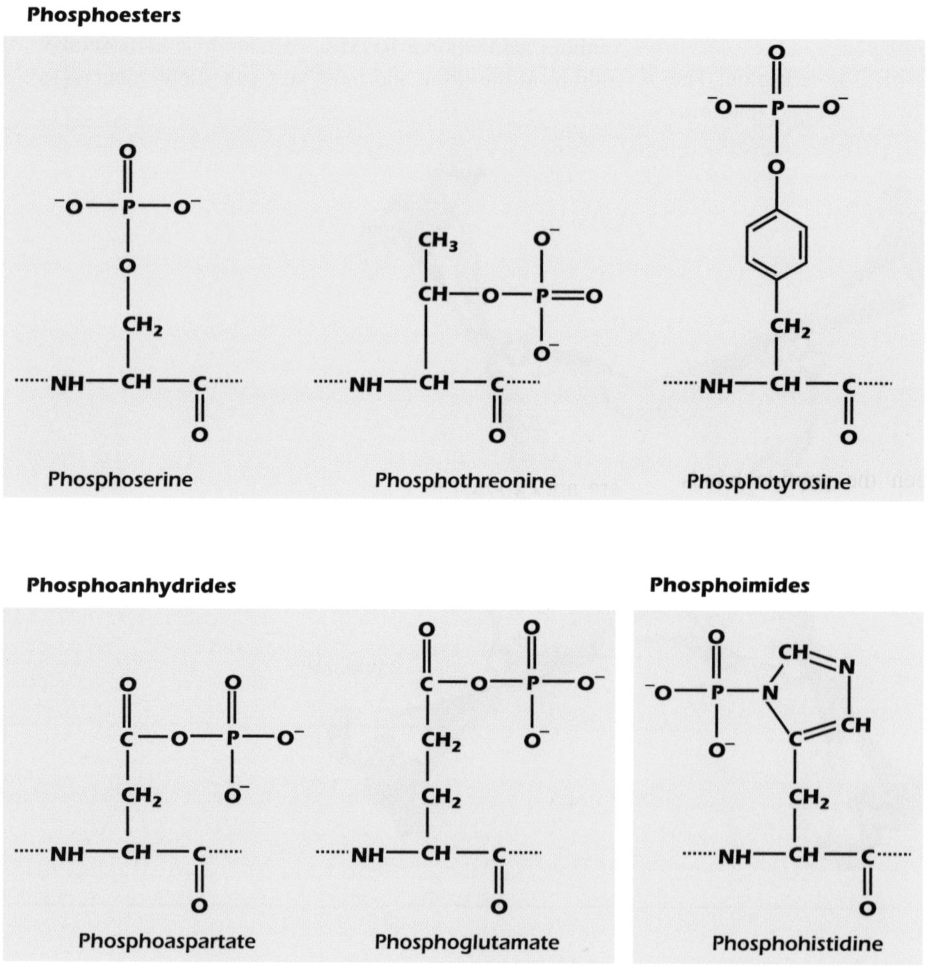

Phosphoesters

Phosphoserine

Phosphothreonine

Phosphotyrosine

Phosphoanhydrides

Phosphoaspartate

Phosphoglutamate

Phosphoimides

Phosphohistidine

Figure 9.5. Six amino acid side chains can be phosphorylated.

OTHER AMINO ACIDS ARE FOUND IN NATURE

Ornithine and citrulline are α-amino acids that play a vital role in the body—they are used as part of the urea cycle to get rid of ammonium ions, which would otherwise poison us (page 218). However, they are not used as building blocks in the synthesis of polypeptides.

There are many other nonprotein amino acids. Very early in the evolution of life, the palette of amino acids used to make polypeptides became fixed at the 20 that we have described. Almost the entire substance of all living things on earth is either polypeptide, composed of these 20 monomers, or other molecules synthesized by enzymes that are themselves proteins made of these 20 monomers. Natural selection has directed evolution within the constraints imposed by the palette.

THE THREE-DIMENSIONAL STRUCTURES OF PROTEINS

Proteins are polypeptides folded into specific, complex three-dimensional shapes. Generally, hydrophobic amino acids pack into the interior of the protein while hydrophilic amino acid side chains end up on the surface where they can interact with water. The amino acid side chains pack together tightly in the interior: there are no spaces. It is the three-dimensional structure of a particular protein that allows it to carry out its role in the cell. The shape can be fully defined by stating the position and orientation of each amino acid, and such knowledge lets us produce representations of the protein.

Protein structures are held together by a large number of relatively weak interactions between the amino acid side chains interacting with one another, with the peptide bonds of the polypeptide backbone, and with other bound molecules; and between the peptide bonds themselves. These interactions are often between amino acids far apart in the linear sequence because the chain is folded in three dimensions. Although individually weak, collectively these interactions are sufficient to produce protein molecules that are stable in their environments.

The forces that stabilize the three-dimensional structures are hydrogen bonds, electrostatic interactions, van der Waals forces, hydrophobic interactions and, in some proteins only, the covalent disulphide bond.

Hydrogen Bonds

As we will see, hydrogen bonds (page 23) are important in all of the higher levels of protein structure.

Electrostatic Interactions

If positive and negatively charged amino acid residues are buried in the hydrophobic interior of a protein, where neither can interact with water, then they will attract each other and the force between them will be stronger than it would be in water. Such an electrostatic bond inside a protein is called a salt bridge.

Polar groups such as hydroxyl and amide groups are dipoles. They have an excess of electrons at one atom and a corresponding deficiency at another, as we have already seen for the water molecule (page 17). The partial charges of dipoles will be attracted to other dipoles and to fully charged groups.

van der Waals Forces

van der Waals forces are relatively weak close-range interactions between atoms. Imagine two atoms sitting close together. At a given instant more of the electrons on one atom may be on one side, and this exposes the positive charge on the nucleus. This positive charge attracts the electrons of the adjacent atom thus exposing its nuclear charge, which would attract the electrons of another atom and so on. The next instant the electrons will have moved so we have a situation of fluctuating attractions between atoms. These forces are important in the close-packed interiors of proteins and membranes and in the specific binding of a ligand to its binding site.

Hydrophobic Interactions

Molecules that do not interact with water (and that are therefore classed as hydrophobic) force the water around them to become more organized: hydrogen-bonded "cages" form. This organization is thermodynamically undesirable and is minimized by the clustering together of such molecules. This is called the **hydrophobic effect**. A polypeptide with hydrophilic and hydrophobic residues will spontaneously adopt a configuration in which the hydrophobic residues are not exposed to water. They can achieve this either by sitting in a lipid bilayer (Fig. 9.6) or by adopting a globular shape in which the hydrophobic residues are clustered in the center of the protein.

Disulfide Bonds

Extracellular proteins often have disulfide bonds between specific cysteine residues. These are strong covalent bonds, and they tend to lock the molecule into its conformation. Although relatively few proteins contain disulfide bonds, those that do are more stable and are therefore easy to purify and study. For this reason many of the first proteins studied in detail, such as the digestive enzymes chymotrypsin and ribonuclease, and the bacterial cell wall degrading enzyme lysozyme, have disulfide bonds.

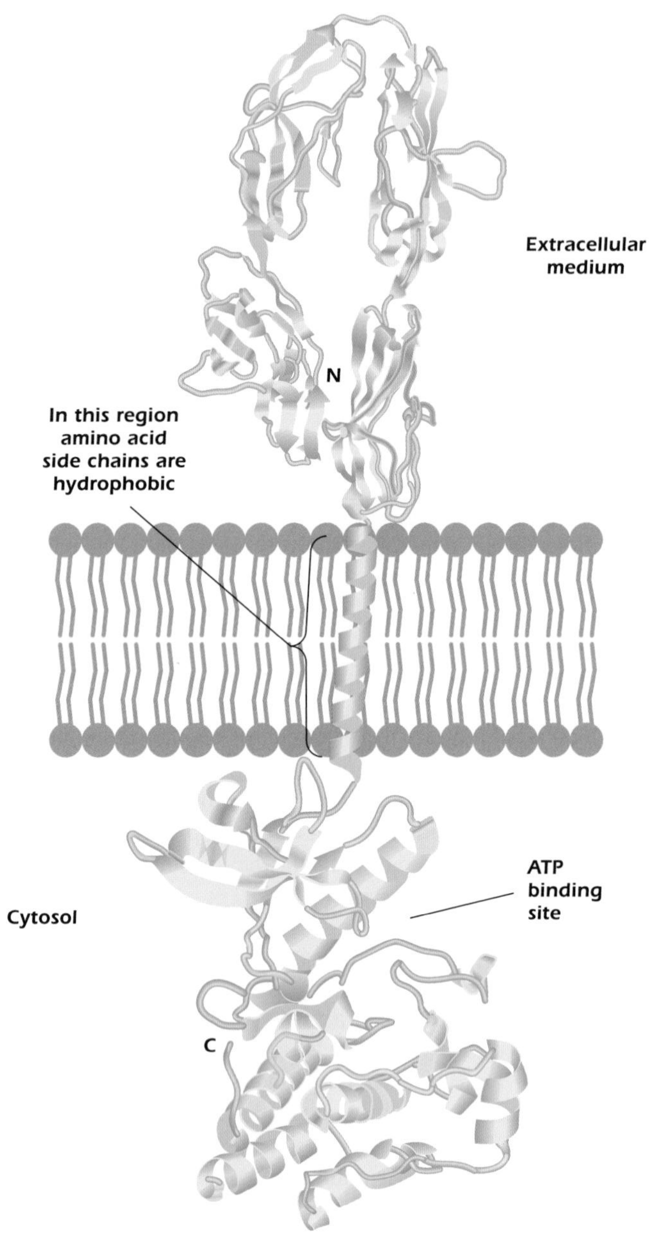

Figure 9.6. Drawing of the platelet-derived growth factor receptor.

LEVELS OF COMPLEXITY

When discussing protein structure, it is helpful to think in terms of a series of different levels of complexity. These levels are (unimaginatively) called primary, secondary and tertiary structures. Some proteins are made up of individually folded, tertiary structured subunits and such proteins are said to have quaternary structures.

The Primary Structure

The **primary structure** of a protein is the sequence of amino acids. For example, lysozyme is an enzyme that attacks bacterial cell walls. It is found in secretions such as tears and in the white of eggs. Lysozyme has the following primary structure:

(NH_2)KVFGRCELAAAMKRHGLDNYRGYSLGNWVC
AAKFESNFNTQATNRNTDGSTDYGILQINSRWWCD
NGRTPGSRNLCNIPCSALLSSDITASVNCAKKIVSDG
DGMNAWVAWRNRCKGTDVQAWIRGCRL(COOH)

Numbering is always from the amino terminal end where synthesis of the protein begins on the ribosome. Lysozyme has four disulfide bonds between four pairs of cysteines. The 129 amino acids of hen egg white lysozyme are shown in linear order in Figure 9.7A with the disulfide bonds indicated. Lysozyme was the first enzyme to have

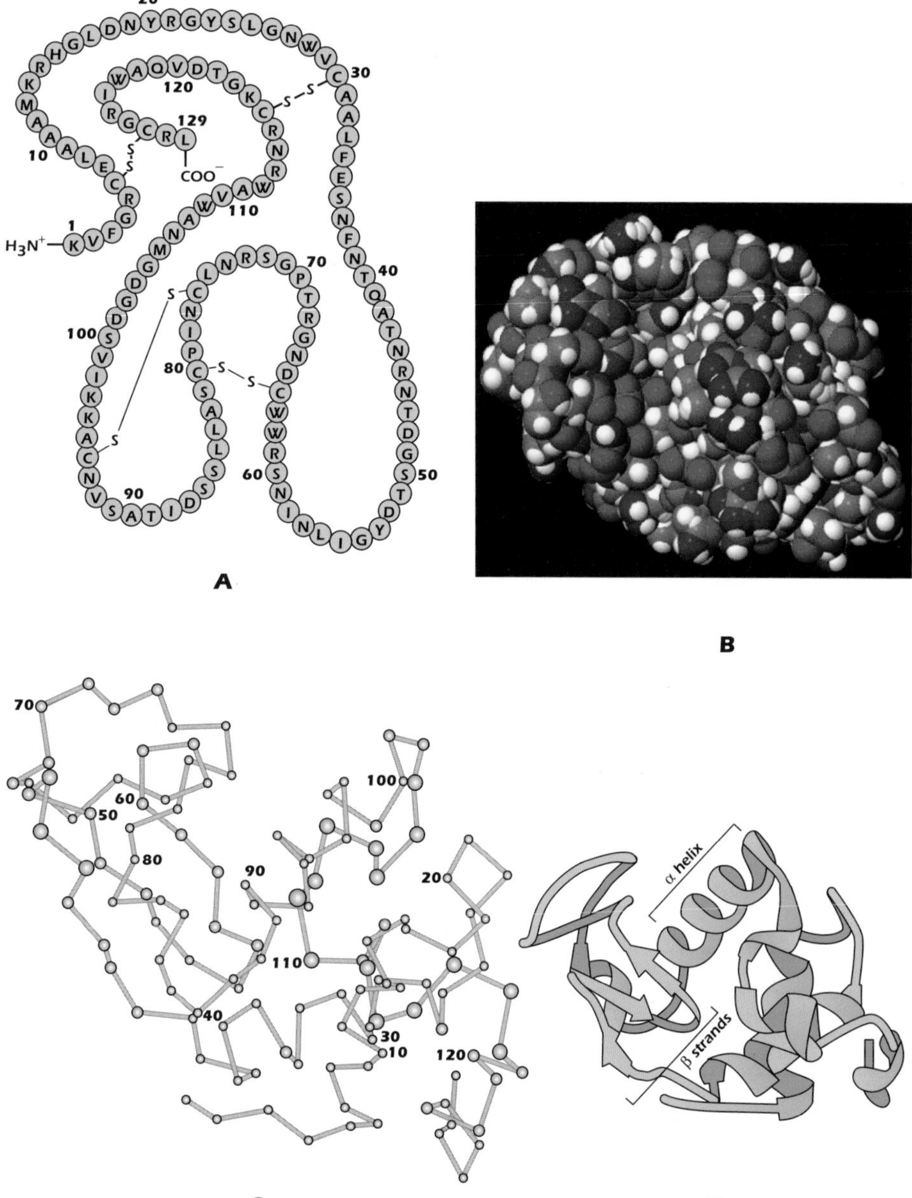

Figure 9.7. Lysozyme: (A) Linear map. (B) Space-filling model in which carbon atoms appear gray, oxygen red, nitrogen blue and hydrogen white. None of the sulfur atoms are visible on the surface. (C) Backbone representation. (D) Cartoon. Panel B uses data from Wang et al. (2007) *Acta Crystallogr. Sect. D* **63**, p. 1254 viewed via the Research Collaboratory for Structural Bioinformatics (www.rcsb.org) database.

its three-dimensional structure fully determined (in 1965). Figure 9.7B shows that structure, with all of the atoms that form the molecule displayed. We see little except an irregular surface. However, if the amino acid side chains are stripped away and the path of the peptide-bonded backbone drawn (Fig. 9.7C), we see that some regions of the protein backbone are ordered in a repeating pattern.

The Secondary Structure

Two types of protein backbone organization are common to many proteins. These are named the **α helix** and the **β sheet**. Figure 9.7D redraws the peptide backbone of lysozyme to emphasize these patterns, with the lengths of peptide participating in β sheets represented as arrows. These repeating patterns are known as **secondary structures**. There are other regions of the protein that do not have any such ordered pattern.

In an α helix the polypeptide chain twists around in a spiral, each turn of the helix taking 3.6 amino acid residues. This allows the nitrogen atom in each peptide bond to form a hydrogen bond with the oxygen four residues ahead of it in the polypeptide chain (Fig. 9.8A, B). All the peptide bonds in the helix are able to form such hydrogen bonds (Fig. 9.8C), producing a rod in which the amino acid side chains point outward (Fig. 9.8D). Because it introduces a kink into the polypeptide chain and has no hydrogen to donate to a hydrogen bond, proline cannot be within an α helix.

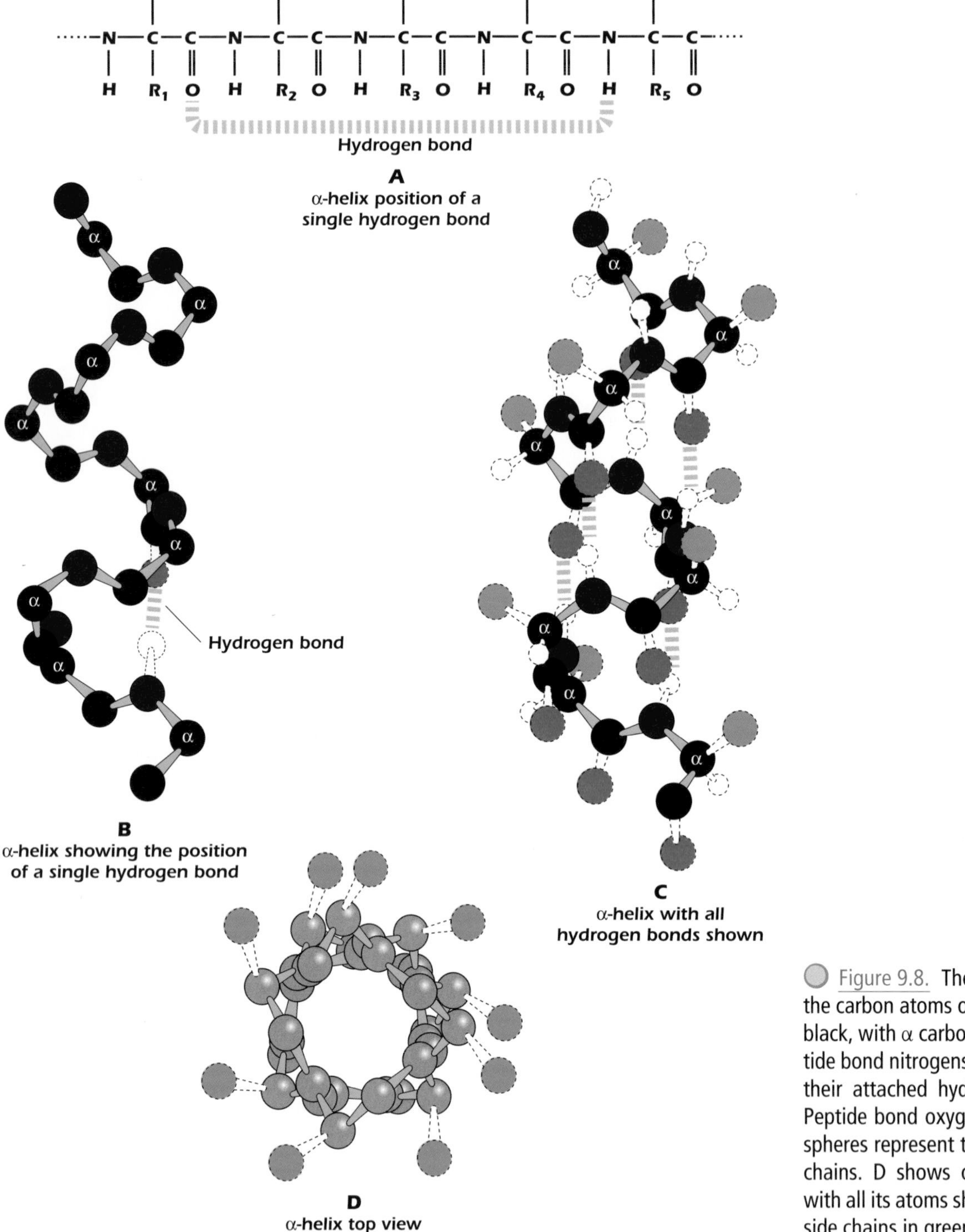

Figure 9.8. The α helix. In B and C the carbon atoms of the backbone are black, with α carbons so marked. Peptide bond nitrogens are deep blue and their attached hydrogens are white. Peptide bond oxygens are red. Green spheres represent the amino acid side chains. D shows only the backbone with all its atoms shown gray, plus the side chains in green.

In a β sheet lengths of polypeptide run alongside each other, and hydrogen bonds form between the peptide bonds of the strands. This generates a sheet that has the side chains protruding above and below it (Fig. 9.9). Along a single strand the side chains alternate up then down, up then down. Because the actual geometry prevents them from being completely flat, they are sometimes called β pleated sheets. A polypeptide chain can form two types of β sheet: Either all of the strands in the sheet are running in the same direction (Fig. 9.9A) forming a **parallel β sheet** or they can alternate in direction (Fig. 9.9B) making an **antiparallel β sheet**. The polypeptide chains in sheets are fully extended, unlike the chain in an α helix. Again, proline is unlikely to be found within a β sheet strand.

IN DEPTH 9.2 CHIRALITY AND AMINO ACIDS

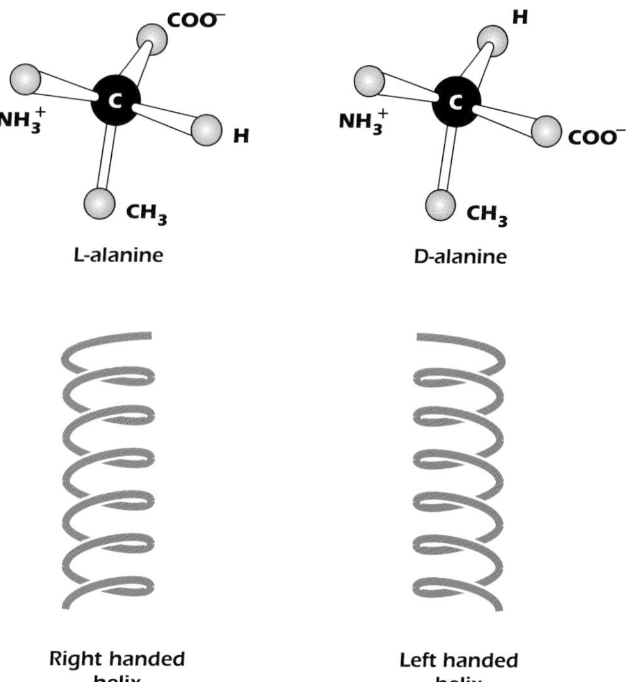

A **chiral** structure is one in which the mirror image of the structure cannot be superimposed on the structure. The shape of your whole body is not chiral. Your mirror image could be rotated through 180°, so that it faces into the mirror, then step back and be superimposed on you. Your right hand, however, is chiral: Its mirror image is not a right hand any more but a left hand. Molecules are often chiral—this arises if a carbon atom has four different groups attached to it and is therefore asymmetric. The α carbon of α-amino acids is asymmetric for all except glycine. It is possible for there to be more than one asymmetric carbon in a molecule. If there is an asymmetric carbon in a molecule, there will be two different ways that the groups on that carbon can be arranged and so there will be two **optical isomers**. These isomers interact with polarized light differently. Although a different system is used in chemistry, optical isomers of amino acids (and sugars) are denoted L and D. L amino acids are exclusively used in proteins and predominate in the metabolism of amino acids generally. However D amino acids are found in nature. D alanine occurs in bacterial cell walls while some antibiotics such as valinomycin and gramicidin A contain D amino acids. These molecules are synthesized in entirely different ways from proteins. Among sugars it is the D forms that are mainly used by cells.

Helices are chiral too. The α helix found in proteins is right handed, like a regular screw thread. Reflect a length of right-handed α helix in a mirror, and you will get a left-handed α helix composed of D amino acids. Because of the actual structures, it turns out that L amino acids fit nicely in a right-handed α helix, while D amino acids fit nicely in a left-handed α helix.

In structural proteins like **keratin** in hair or fibroin in silk, the whole polypeptide chain is ordered into one of these secondary structures. Such fibrous proteins have relatively simple repeating shapes and do not have binding sites for other molecules. In contrast most proteins have regions without secondary structures, the precise folding and packing of the amino acids being unique to the protein, side by side with regions of secondary structure.

Tertiary Structure: Domains and Motifs

The three-dimensional protein structure often has protrusions, clefts, or grooves on the surface where particular amino acids are positioned to form sites that bind ligands and, in the case of enzymes, catalyze reactions within or between ligands. The whole three-dimensional arrangement of the amino acids in the protein is called the **tertiary structure**. The tertiary structure is stabilized by all the

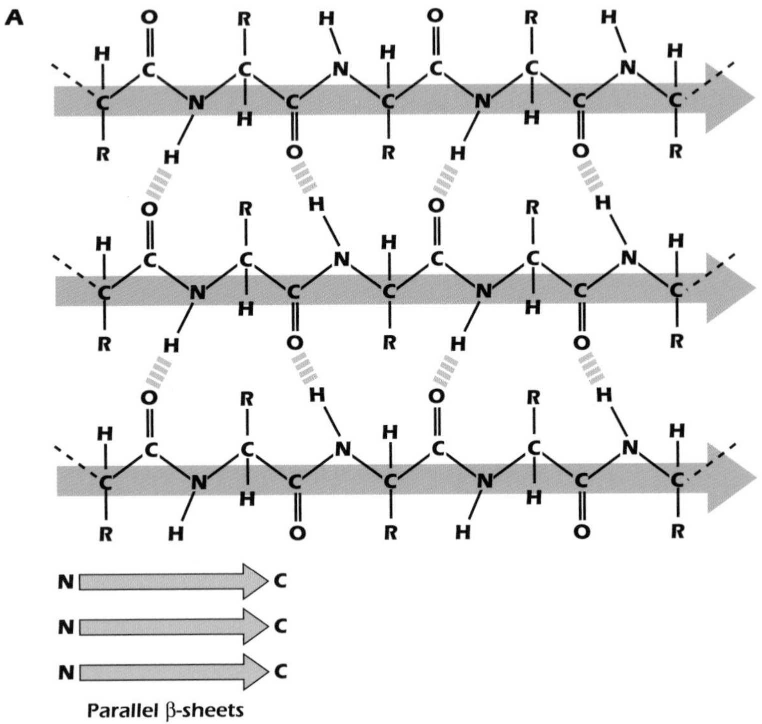

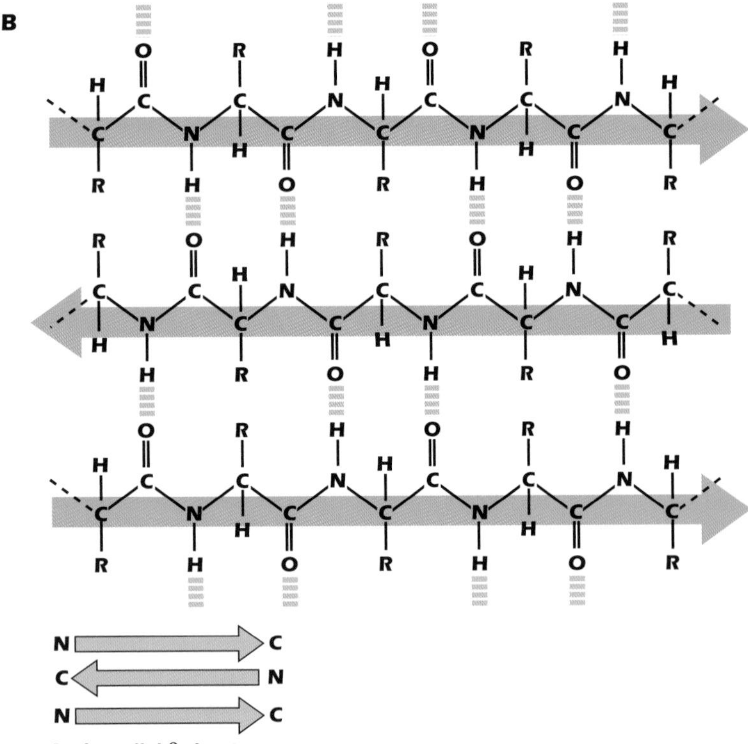

Figure 9.9. β sheets.

interactions we have listed above. In particular, hydrogen bonds, which we have already seen to stabilize the α helix and β sheet, also mold the tertiary structure by linking pairs of amino acid side chains, or one side chain to a physically adjacent section of the peptide backbone.

A tertiary structure is unique to a particular protein. However, common patterns or **motifs** occur in tertiary structures. For example, many proteins with different functions show a β-barrel structure where β sheet is rolled up to form a tube. Green fluorescent protein (page 115)

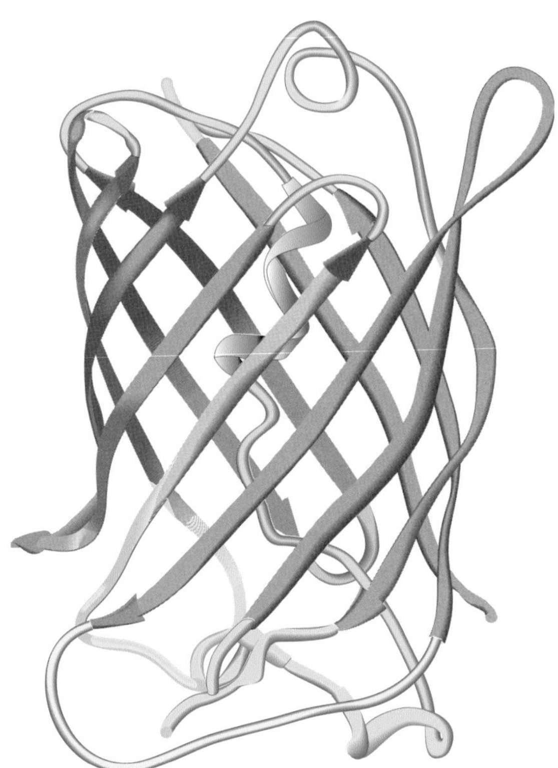

Figure 9.10. The green fluorescent protein molecule comprises a β barrel (shown gray) and a central α helix.

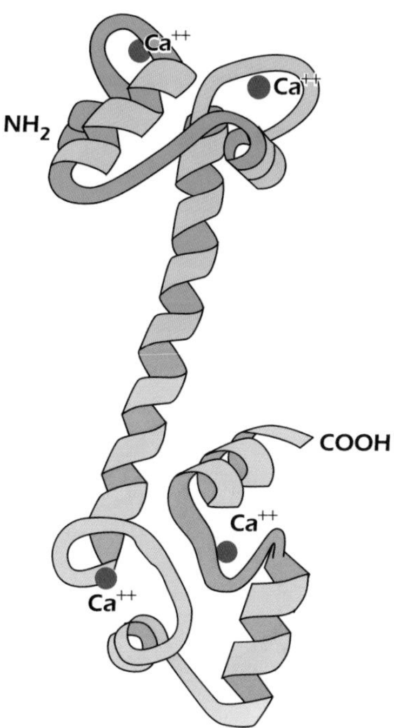

Figure 9.11. Calmodulin is composed of two very similar domains.

is one example (Fig. 9.10). Often the tertiary structure is seen to divide into discrete regions. The calcium-binding protein **calmodulin** shows this clearly (Fig. 9.11). Its single chain is organized into two **domains**, one shown in green, the other in pink, joined by only one strand of the polypeptide chain. In calmodulin the two domains are very similar, and the modern gene probably arose through duplication of an ancestral gene that was half as big. Domains are easier to see than to define. "A separately folded region of a single polypeptide chain" is as good a definition as any.

Domains may be similar or different in both structure and function. The catabolite activating protein (CAP) of *Escherichia coli* (page 88) binds to a specific sequence of bases on DNA, assisting RNA polymerase to bind to its promoter and initiate transcription of the *lac* operon. It does this only when it has bound cAMP (which in turn only happens when the intracellular glucose concentration is low). One of the domains of CAP has the job of binding cAMP. Another recognizes DNA sequences using a **helix-turn-helix** motif (Fig. 9.12). One of these helices fits into the major groove of the DNA where it can make specific interactions with the exposed edges of the bases.

Proteins that interact with DNA often do so via **zinc fingers**. The DNA binding domain of the glucocorticoid receptor (page 94) contains two zinc finger motifs (Fig. 9.13A). Each finger is generated by the binding of four cysteine residues with a Zn^{2+} ion. The domain contains two α helices (Fig. 9.13B), one in each finger. The first helix, also known as the recognition helix, fits into the major groove of the DNA and makes contacts with specific bases. The second helix is bent at right angles to the first and helps to stabilize the receptor homodimer on DNA by promoting dimerization between the two receptor monomers. This finger can alternatively interact with other proteins that help to activate or repress transcription, such as AP1 subunits (page 96).

Domains usually correspond to exons (page 76) in the gene. It is therefore relatively easy for evolution to create new proteins by mixing and matching domains from existing proteins. The calcium-dependent transcription factor NFAT (page 161) is thought to have arisen in the ancestor of the vertebrates when a mutation occurred that spliced the DNA-binding domain from the transcription factor NFκB or a related protein into a preexisting protein that was a target for the calcium-activated phosphatase calcineurin.

Integral membrane proteins have special tertiary structures. The polypeptide chain may cross the membrane once or many times, but each time it does so it has a region of hydrophobic side chains that can be in a hydrophobic environment such as the interior of a lipid bilayer. The commonest membrane-spanning structure is a 22-amino-acid α helix. Figure 9.6 shows an integral membrane protein called the platelet-derived growth factor

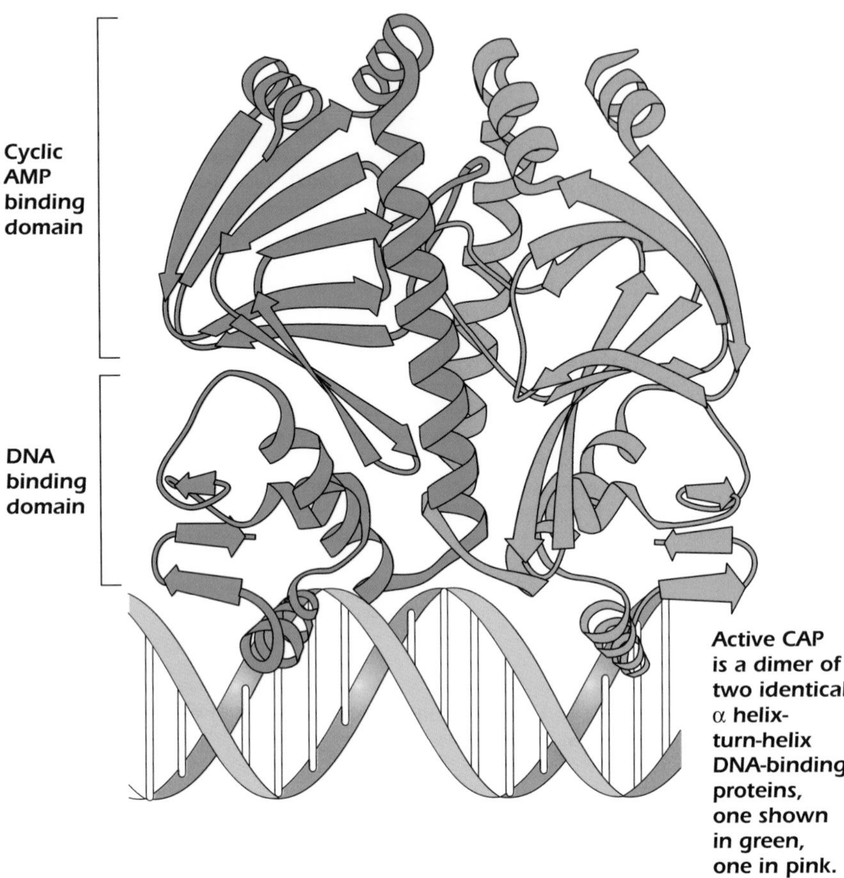
Cyclic
AMP
binding
domain
DNA
binding
domain
Active CAP
is a dimer of
two identical
α helix-
turn-helix
DNA-binding
proteins,
one shown
in green,
one in pink.

Figure 9.12. Active catabolite activator protein is a dimer.

receptor (page 260). Hydrophobic side chains on the transmembrane α helix interact with the hydrophobic interior of the membrane. The protein is held tightly in the membrane because for it to leave would expose these hydrophobic side chains to water.

Quaternary Structure: Assemblies of Protein Subunits

Many proteins associate to form multiple molecular structures that are held together tightly but not covalently by the same interactions that stabilize tertiary structures. For example, connexons (page 46) are formed from six identical connexin monomers. CAP is only active when it dimerizes (Fig. 9.12). Hemoglobin, the protein that carries oxygen in our red blood cells, is formed from four individual polypeptide chains, two α chains and two β chains (Fig. 9.14). In all these cases, we call the three-dimensional arrangement of the protein subunits as they fit together to form the multimolecular structure the **quaternary structure** of the protein.

PROSTHETIC GROUPS

Even with the enormous variety of structures available, there are some functions that proteins need help with because the 20 protein amino acid side chains do not cover the properties required. Moving electrons in oxidation and reduction reactions and binding oxygen are good examples. Proteins therefore associate with other chemical species that have the required chemical properties. Hemoglobin uses iron-containing heme groups to carry oxygen molecules (Fig. 9.15). It is worth noting that the iron remains as Fe^{2+} throughout: although it binds the oxygen molecule it is not oxidized to Fe^{3+} when hemoglobin is working normally. The general name for a nonprotein species that is tightly bound to a protein and helps it perform its function is **prosthetic group**.

Some proteins have tightly bound metal ions that are essential to their function. We have already met the zinc fingers of the glucocorticoid receptor and the iron in hemoglobin. Other proteins use molybdenum, manganese, or copper.

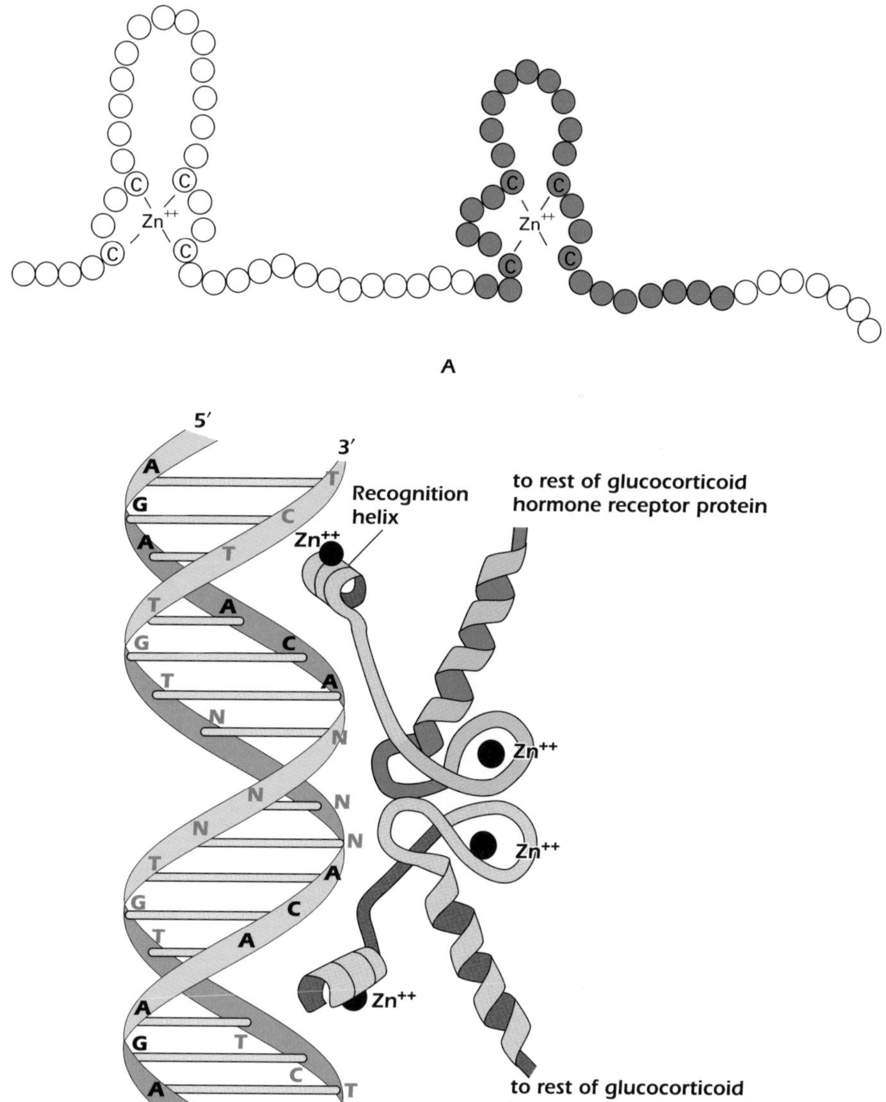

Figure 9.13. (A) Two zinc finger motifs in the glucocorticoid receptor. Each circle is one amino acid residue. Four cysteine residues, indicated c, bind each zinc ion. (B) Drawing of the zinc finger domains of a dimerized pair of glucocorticoid hormone receptors interacting with DNA. One zinc finger is indicated in green and corresponds to the amino acids colored green in (A).

THE PRIMARY STRUCTURE CONTAINS ALL THE INFORMATION NECESSARY TO SPECIFY HIGHER-LEVEL STRUCTURES

Protein structures are stable and functional over a small range of environmental conditions. Outside this range the pattern of interactions that stabilizes the tertiary structure is disrupted and the molecule **denatures**—activity disappears as the molecule loses its structure. Denaturation may be caused by many factors, which include excessive temperature, change of pH, and detergents. Concentrated solutions of urea (8 mol per liter) have long been used by biochemists to denature proteins. Unlike heat and pH, urea

Answer to Thought Question: In addition to any charges on the side chain, all free amino acids bear two charges, a positive charge on the amino group and a negative charge on the carboxyl group at cellular pH values. Therefore like all small ions they are soluble in water. Formation of the peptide bond removes these charges leaving only any present on the side chain.

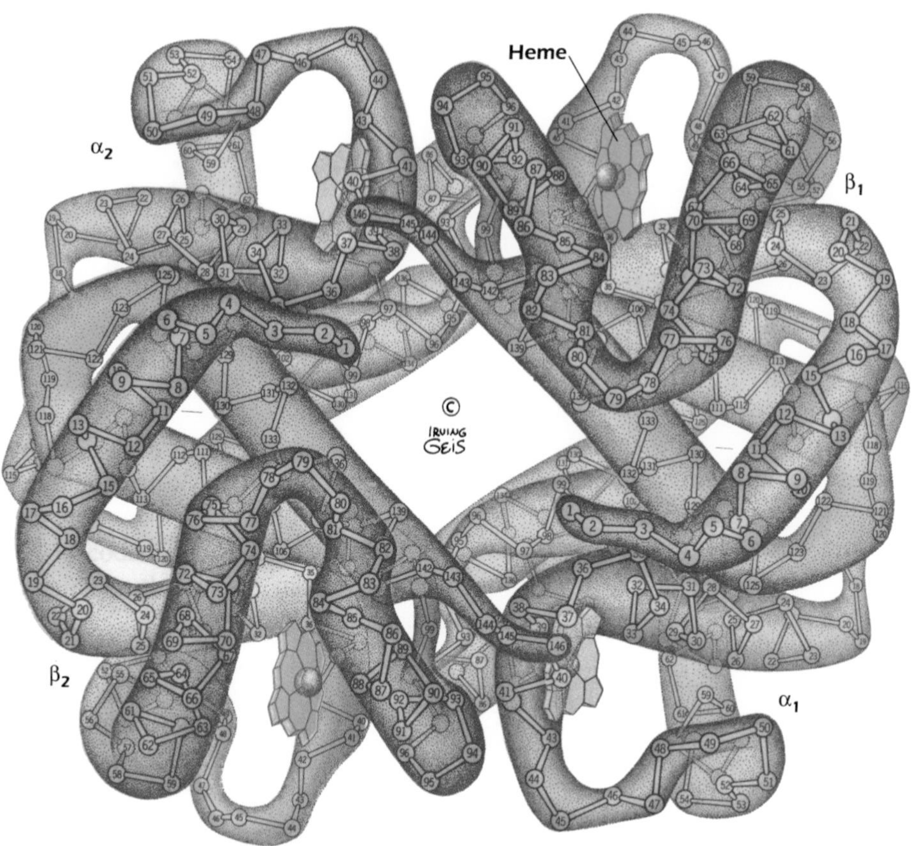

Figure 9.14. Hemoglobin, a tetrameric protein with heme prosthetic groups. Illustration by Irving Geis. Rights owned by Howard Hughes Medical Institute. Reproduced by permission.

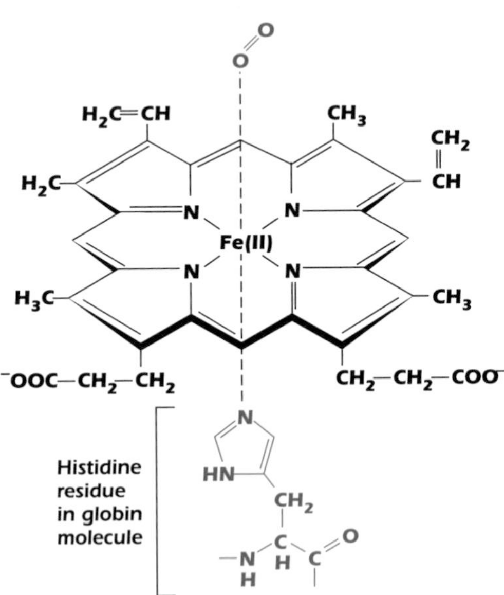

Figure 9.15. The iron-containing prosthetic group heme in the form in which it is found in oxygenated hemoglobin. The oxygen molecule is at the top. From D. Voet and J. D. Voet, *Biochemistry*, 2nd ed., p. 216. © 1995 John Wiley & Sons, Inc. Used by permission of John Wiley & Sons, Inc.

does not cause the protein to precipitate. Physicochemical techniques have shown that in urea solution all of the higher levels of structure are lost and that polypeptide chains adopt random, changing conformations. Reagents such as urea that do this are called **chaotropic reagents**. If the urea is removed (by dialysis or simply by dilution), the protein refolds, regaining its structure and biological activity. This shows that the sequence of amino acids contains all of the information necessary to specify the final structure. The refolding of a urea-denatured protein cannot be random. Even a small protein with 100 amino acids would take some 10^{50} years to try out all of the structural conformations available. The fact that refolding does happen and happens on a time scale of seconds tells us that there must be a folding pathway, and the process is not random. Secondary structures may form first and act as folding units. In the cell folding is assisted by proteins called **chaperones** (page 164).

Medical Relevance 9.2 Protein Folding Gone Awry: Mad Cow Disease

Some years ago one of the dogmas of biology was overthrown—it was long believed that diseases could only be transmitted by structures that contained nucleic acids, that is, by viruses or by microorganisms. It is now clear that a group of brain diseases can be transmitted by a protein. These are the diseases called spongiform encephalopathies: bovine spongiform encephalopathy in cows (mad cow disease), scrapie in sheep, and Creutzfeldt–Jacob disease and kuru in humans. These diseases are rare in humans but have recently increased in farm animals.

The infectious agent is called a prion and is a protein. It is coded for by a gene the animals have as part of their genome. In healthy individuals the gene is expressed in the cells of the nervous system and generates an innocuous protein called PrP^{c} (prion-related protein, cellular). PrP^{c} has a globular region at the C-terminal end but the N-terminal region seems to be unstructured.

Sometimes the same polypeptide chain folds up differently with the disordered N-terminal section instead folding into a structure rich in β sheet. This is called PrP^{sc} (prion-related protein, scrapie). The disease arises because small aggregates of PrP^{sc} can cause normal PrP^{c} protein molecules to fold up into the abnormal form. This keeps happening and lumps of PrP^{sc} form that damage nerve cells. The disease spread in cattle herds through food that contained recycled meat from infected animals, and seems to have been triggered by a change in the late 1970s in the way such material was processed.

Before the problem was fully comprehended, infected animals were used for human food and caused what became known as new variant Creutzfeldt–Jacob disease. Deaths from this in the United Kingdom peaked at 28 in 2000 and have fallen steadily since then (to one in 2008) due to stringent controls on animal feed and meat preparation.

IN DEPTH 9.3 CURING MAD MICE WITH SMELLY FISH

If a cell is immersed in a solution more dilute than its own cytosol, it tends to swell. On the other hand, cells immersed in concentrated solutions shrink. The overall strength of a solution is its osmolarity, and the movement of water into or out of cells because of solution strength is called osmosis. Many fish are exposed to changes in the salt concentration of the water around them, either because they swim from salt water to fresh and back (salmon, eels) or because they live in estuaries. The cells of these fish adjust the osmolarity of their cytosol to match the surrounding water by increasing the concentration of various small, easily synthesized molecules, one of which is urea. However, this introduces a further problem because high concentrations of urea are chaotropic—they cause cellular proteins to unfold.

The small molecule trimethylamine *N*-oxide, in contrast, helps proteins to fold into their correct shape. One of the most dramatic examples of this is that trimethylamine *N*-oxide, added to mouse nerve cells in culture along with the scrapie agent PrP^{sc} (Medical Relevance 9.2), protects the endogenous PrP^{c} prion protein from being incorrectly refolded to generate PrP^{sc}. Fish that produce urea, therefore, also upregulate a gene encoding a flavin-containing monooxygenase (page 62) and produce trimethylamine *N*-oxide. A battle of opposing forces then ensues, with urea tending to denature the protein and trimethylamine *N*-oxide counteracting this to maintain the protein in its correctly folded state. When fish die and rot, bacterial action converts trimethylamine *N*-oxide to trimethylamine, which gives rotting fish its distinctive smell (Medical Relevance 4.3 on page 62).

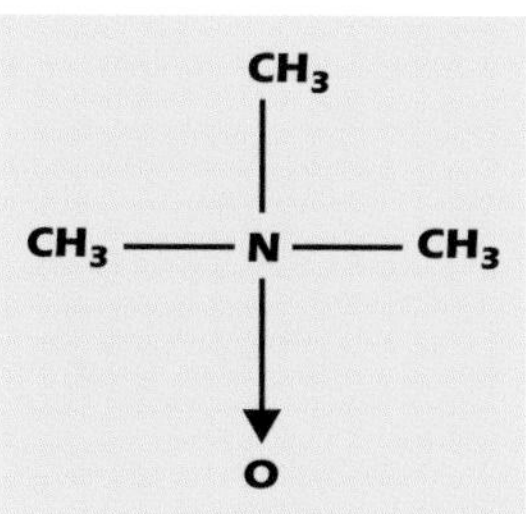

Trimethylamine N-oxide

SUMMARY

1. Polypeptides are linear polymers of α-amino acids linked by peptide bonds. There are 20 amino acids coded for by the genetic code. They differ in the properties of their side chains, which range from hydrophobic groups to charged and uncharged hydrophilic groups.
2. In addition to their hydrophilicity, side chains of individual amino acids have specific properties of which the most important are charge, the ability to form disulfide bonds, and the ability to undergo posttranslational modification.
3. Phosphorylation (on S, T, Y, D, E, or H) is a post-translational modification that allows the balance of electrical charges on a protein to be dramatically and reversibly altered.
4. Proteins are polypeptides that have a complex three-dimensional shape.
5. It is convenient to consider protein structure as having three levels. The primary structure is the linear sequence of the amino acid monomers.
6. In some parts of a protein regular, repeated foldings of the polypeptide chain can be seen: these are secondary structures. Secondary structures are held together by hydrogen bonds between the carboxyl oxygens and the hydrogens on the nitrogens of the peptide bonds.
7. There are two common types of secondary structure. In the α helix the chain coils upon itself making a spiral with hydrogen bonds running parallel to the length of the spiral. In β sheets the hydrogen bonds are between extended strands of polypeptide that run alongside one another. The hydrogen bonds are at right angles to the strands and in the plane of the sheet. There are two types—parallel and antiparallel.
8. The final, complex folding of a protein is its tertiary structure. Interactions between side chains stabilize the tertiary structure.
9. Some proteins have a quaternary structure. This is an association of subunits, each of which has a tertiary structure.

FURTHER READING

Branden, C., and Tooze. J. (1999) *Introduction to Protein Structure*, 2nd edition, New York: Garland.

Creighton, T. (1992) *Proteins, Structures and Molecular Properties*, 2nd edition, New York: W. H. Freeman.

Fersht, A. (1999) *Structure and Mechanism in Protein Science*, New York: W. H. Freeman.

McGee, H. (2004) *On Food and Cooking*, London: Unwin.

Perutz, M. (1992) *Protein Structure: New Approaches to Disease and Therapy*. New York: W. H. Freeman.

Tanford, C., and Reynolds, J. (2001) *Nature's Robots—A History of Proteins*, Oxford: Oxford University Press.

Voet D., and Voet J. D. (2011) *Biochemistry*, 4th edition, Hoboken: Wiley.

REVIEW QUESTIONS

9.1 Theme: Amino Acids

A alanine
C cysteine
E glutamate
F phenylalanine
G glycine
M methionine
N asparagine
P proline
R arginine
V valine
W tryptophan

From the list of amino acids above, choose the amino acid that best matches each of the descriptions below.

1. Can be phosphorylated
2. Has a strongly acidic side chain
3. Has a strongly basic side chain
4. Is an imino acid, not an amino acid, and therefore imposes greater constraints on the shape of the polypeptide chain
5. Two can form a disulphide bond

9.2 Theme: Terms Used to Describe Proteins

A α helix
B β sheet
C denaturation
D disulfide bond
E domain
F hydrophobic effect
G phosphorylation
H post-translational modification
I primary structure
J salt bridge
K subunit
L van der Waals

From the above list of compounds, select the one described by each of the descriptions below.

1. A covalent bond between the side chains of two cysteine residues
2. A protein secondary structure where the backbone coils in a right-handed helix with hydrogen bonds between the amide and carboxyl groups of the peptide bonds with the hydrogen bonds running parallel with the direction of the helix
3. A separately folded region of a single polypeptide chain
4. Loss of all of the higher levels of structure with accompanying loss of biological activity of a protein
5. The tendency for hydrophobic molecules to cluster together away from water

9.3 Theme: Specific Binding Partners

A A specific base sequence on DNA
B β-galactoside sugars
C Calcium ions
D Connexin 43
E Glucose
F NFAT
G Valium

Many of the functions of proteins depend on their ability to bind very specifically to other molecules. From the list above choose a ligand to which each of the proteins below binds with high specificity.

1. Calmodulin
2. Catabolite activator protein (CAP)
3. Connexin 43
4. Glucocorticoid receptor

THOUGHT QUESTION

We have stated that hydrophobic solutes will dissolve in nonpolar solvents but will not dissolve in water. How is it that amino acids with hydrophobic side chains are able to dissolve in the aqueous cytosol, ready to be picked up by their tRNAs?

10

INTRACELLULAR PROTEIN TRAFFICKING

Eukaryotic cells contain organelles (page 43). This compartmentation allows the spatial and temporal separation of different processes so that synthesis and degradation of molecules are separated and materials destined for secretion are packaged separately from those taken up by the cell. For this to be possible each type of organelle must have a specialized set of proteins. Proteins are synthesized by ribosomes so there must be systems to move proteins to their correct destination. In this chapter we will discuss some of the ways in which a newly synthesized protein is precisely and actively moved (**translocated**) to the correct cellular compartment.

THREE MODES OF INTRACELLULAR PROTEIN TRANSPORT

Newly synthesized proteins must be delivered to their appropriate site of function within the cell. The mechanisms and machinery that eukaryotic cells use to accomplish this are highly conserved from yeast to humans. There are three possible ways by which the cell accomplishes the task (Fig. 10.1). *First*, the protein may fold into its final form as it is synthesized and then move through an aqueous medium to its final destination, remaining folded all the way. Delivery of proteins to the nucleus follows this scheme; the proteins are synthesized on cytosolic ribosomes and pass through nuclear pores into the nucleoplasm by a process called **gated transport**. In the *second* form of transport, **transmembrane translocation**, unfolded polypeptide chains are threaded across one or more membranes to reach their final destination. Proteins destined for the interior of peroxisomes and mitochondria (and chloroplasts in plants) are synthesized on cytosolic ribosomes and may fold up completely or partially into their final form before being transported to and across the appropriate membrane. Transport into mitochondria involves the protein being unfolded and threaded through a hole in a transporter; import into peroxisomes is less well understood. Proteins that are synthesized by ribosomes on the rough endoplasmic reticulum are threaded across the membrane and into the interior of that organelle while they are being synthesized—some may remain in the lumen of the endoplasmic reticulum while others are transported further. Lastly, in the *third* form of transport, small closed bags made of membrane, **vesicles**, carry newly synthesized protein from the endoplasmic reticulum to the Golgi apparatus and between other compartments in the process of **vesicular trafficking**.

The final destination of a protein is determined by sections of the protein itself that act as **sorting signals**. When a protein is first made on the ribosome, it is simply a stretch of polypeptide. The initial sorting decisions must therefore be made on the basis of particular amino acid sequences called **targeting sequences**. For proteins synthesized on rough endoplasmic reticulum-bound ribosomes, additional sorting signals such as sugars and phosphate groups may be added by enzymes. In general, sorting operates by proteins containing a specific sorting signal binding to a receptor protein, which in turn binds to

Cell Biology: A Short Course, Third Edition. Stephen R. Bolsover, Jeremy S. Hyams, Elizabeth A. Shephard and Hugh A. White.

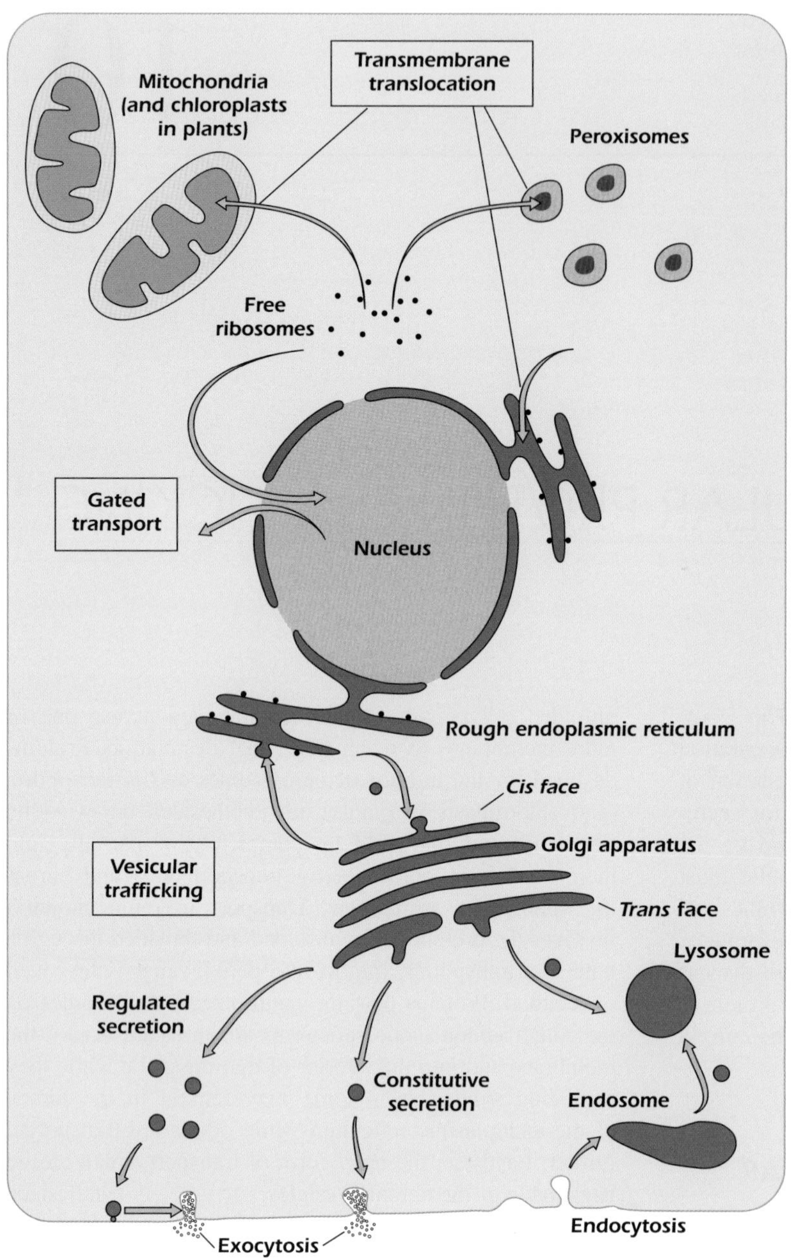

Figure 10.1. The three modes of intracellular protein transport.

translocation machinery situated in the membrane of the appropriate compartment. Proteins without sorting signals, such as hemoglobin, are made on cytosolic ribosomes and remain in the cytosol.

Targeting Sequences

Targeting sequences (also known as **localization sequences**) usually comprise a length of 3–80 amino acids that are recognized by specific receptors that guide the protein to the correct site and make contact with the appropriate translocation machinery. Once the protein has been imported into the new location the targeting sequence may be removed by enzymes that break the peptide bond between the targeting sequence and the rest of the protein. Some targeting sequences have been characterized better than others. The targeting sequence encoding import into the endoplasmic reticulum consists of about 20 mostly hydrophobic amino acids at the N terminus of the protein called the **signal sequence**. The import signal for mitochondria is a stretch of 20–80 amino acids in which positively charged side chains stick out on one side of the helix and hydrophobic side chains stick out on the other, a so-called amphipathic helix. A cluster of about five positively charged amino acids located within the protein sequence targets a protein to the nucleus, while the best-known peroxisomal targeting sequence is the C-terminal tripeptide Ser-Lys-Leu-COOH.

Retention

Another class of sorting signal does not activate the transport of a protein out of its present location but is rather a signal to the cell that the protein has reached its final destination and should not be moved. For example, proteins with the motif Lys-Asp-Glu-Leu-COOH (KDEL) at their C terminus are retained within the endoplasmic reticulum.

TRANSPORT TO AND FROM THE NUCLEUS

In contrast to the situation in prokaryotes, RNA transcription and protein synthesis are separated in space and time in eukaryotes. Exchange of material between the nucleus and the cytoplasm is essential for the basic functioning of these cells and must be tightly controlled. RNA and ribosomal subunits that are assembled in the nucleus have to enter the cytoplasm where they are required for protein synthesis. On the other hand, proteins such as histones and transcription factors must enter the nucleus to carry out their functions. The nuclear pore complex mediates trafficking between the nucleus and the cytoplasm. The pore is made from a large number of proteins, and we refer to this type of structure as a multiprotein complex. Transport through the nuclear pore complex is mediated by signals and requires both energy and transporter proteins.

The Nuclear Pore Complex

The nuclear pore complex is embedded in the double membrane of the nuclear envelope (Fig. 10.2). It is very large with a relative molecular mass (Mr) of 125 million. The basic structure of the nuclear pore complex, which has been mainly elucidated by electron microscopy and computer modeling, is very similar in all eukaryotic cells. There is still much to be found out about this structure and the way it functions. It consists of over 30 different components, called nucleoporins, which form eight identical subunits arranged in a circle. There are other proteins associated with the nucleoporins. Seen from the side the nuclear pore complex comprises three rings each made of the eight identical subunits. The central ring subunits have a transmembrane part that anchors the pore complex in the nuclear membrane and also project unstructured polypeptide chains inward to fill the hole of the pore. Small molecules can slip between these central tangled polypeptides but larger molecules are blocked and must be transported.

Gated Transport Through the Nuclear Pore

In general, a protein has to display a distinct signal for it to be transported through the nuclear pore. Proteins with a nuclear localization signal are transported in, while proteins with a nuclear export signal are transported out. Mobile transporter proteins, usually mediating either export or import, recognize the appropriate targeting sequence and then interact with the nuclear pore. The polypeptides tangled in the lumen of the pore have many glycine-phenylalanine repeats (GF repeats). Transporter proteins can bind these GF repeats and appear to hop from one to another through the pore, taking the protein to be transported with them. The precise mechanism by which large molecules move through the nuclear pore is not yet fully understood. The pore must expand or distort to allow movement of large assemblies such as ribosomal subunits. A typical nucleus will have 2000 to 4000 nuclear pores. Each pore can transport up to 500 macromolecules per second and allows transport in both directions at the same time. These complex structures vanish when the nucleus disassembles during mitosis and reform when the nuclei reform.

Example 10.1 Holding Calcium Ions in the Endoplasmic Reticulum

One of the functions of the smooth endoplasmic reticulum is to hold calcium ions ready for release into the cytosol when the cell is stimulated. A protein called calreticulin (short for calcium-binding protein of the endoplasmic reticulum) helps hold the calcium ions. Its primary structure is

(NH_2)MLLSVPLLLGLLGLAVAEPAVYFKEQFLDGDGW
TSRWIESKHKSDFGKFVLSSGKFYGDEEKDKGLQTSQD
ARFYALSASFEPFSNKGQTLVVQFTVKHEQNIDCGGGYV
KLFPNSLDQTDMHGDSEYNIMFGPDICGPGTKKVHVIF
NYKGKNVLINKDIRCKDDEFTHLYTLIVRPDNTYEVKI
DNSQVESGSLEDDWDFLPPKKIKDPDASKPEDWDERAKI
DDPTDSKPEDWDKPEHIPDPDAKKPEDWDEEMDGEWE
PPVIQNPEYKGEWKPRQIDNPDYKGTWIHPEIDNPEYSP
DPSIYAYDNFGVLGLDLWQVKSGTIFDNFLITNDEAYA
EEFGNETWGVTKAAEKQMKDKQDEEQRLKEEEEDKKR
KEEEEAEDKEDDEDKDEDEEDEEDKEEDEEEDVPGQAK
DEL(COOH)

The first 17 amino acids include 14 hydrophobic ones, shown in black: this is the signal sequence that triggers translocation of the protein into the endoplasmic reticulum. The last four amino acids, KDEL, ensure that it remains there. In between these two sorting signals is the functional core of the protein.

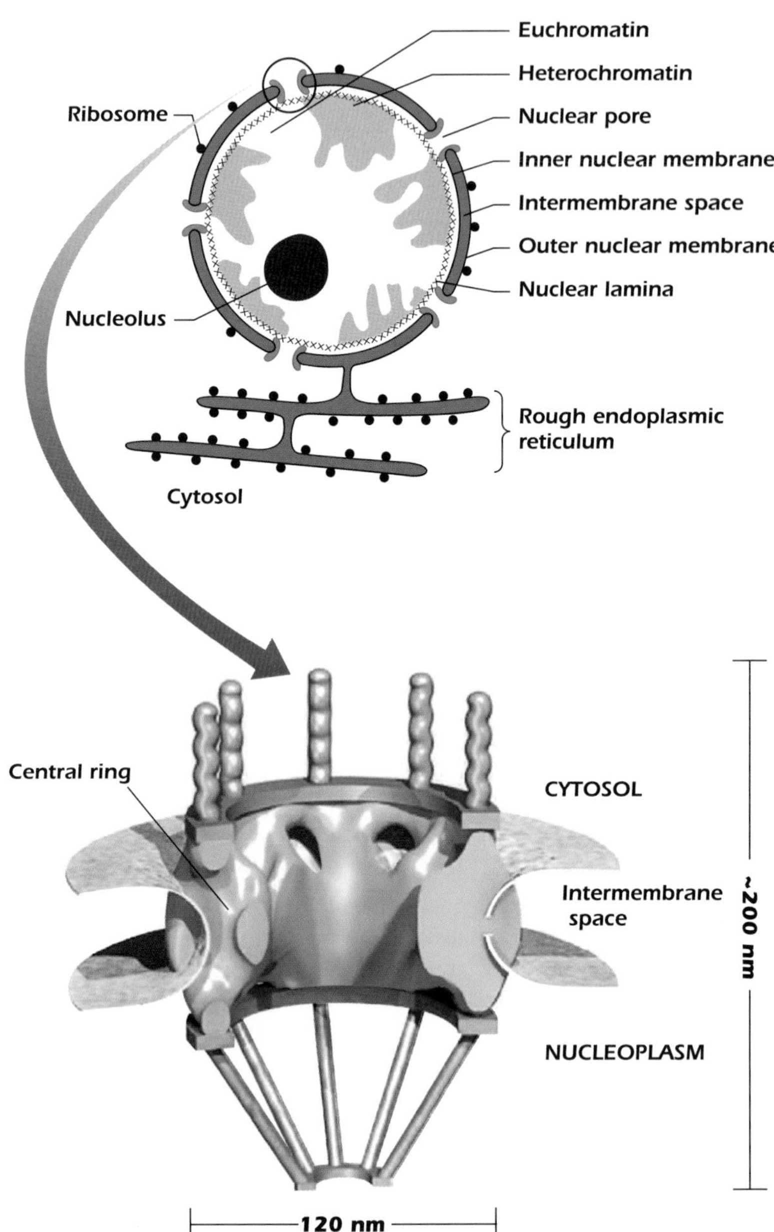

Figure 10.2. The nuclear pore.

The discovery of the role played by the GTPase **Ran** in nuclear transport has given valuable insight into how directionality of transport into and out of the nucleus is achieved. GTP hydrolysis by Ran provides the energy for transport. We will now describe this process.

GTPases and the GDP/GTP Cycle

GTPases form a family of proteins that are often involved when cells need to control complex processes (Fig. 10.3). They all share the ability to hydrolyze the nucleotide GTP but otherwise differ markedly in the processes they control and their mode of operation. We have already met three GTPases in the context of protein synthesis: IF2, EF-tu, and EF-G (pages 128–129). Once a GTPase has hydrolyzed GTP to GDP and inorganic phosphate, it adopts an inactive shape and is unable to activate its target process. In contrast, if the protein expels the GDP and binds a molecule of GTP, it then adopts its active form. Many scientists describe these proteins as switches: they are "on" with GTP bound and "off" with GDP bound. The cycle between the GDP-bound and GTP-bound state is regulated by effector proteins. **GTPase activating proteins** or **GAPs** speed up the rate at which GTPases hydrolyze GTP, and hence the rate at which they inactivate, while **guanine nucleotide exchange factors**, or **GEFs**, assist in the exchange of GDP for GTP and therefore help GTPases adopt their active configuration.

Medical Relevance 10.1 Blocking Calcineurin—How Immunosuppressants Work

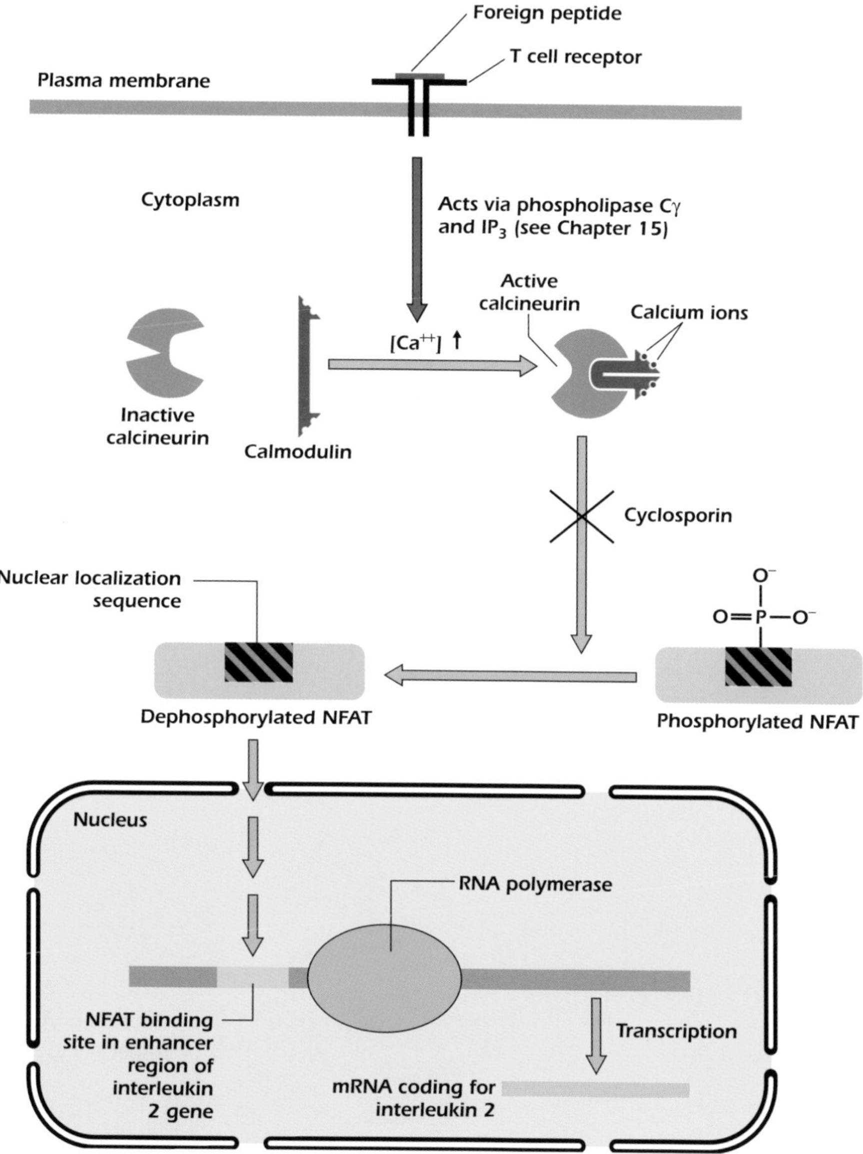

The drug cyclosporin A is invaluable in modern medicine because it suppresses the immune response that would otherwise cause the rejection of transplanted organs. It does this by blocking a critical stage in the activation of T lymphocytes, one of the cell types in the immune system. T lymphocytes signal to other cells of the immune system by synthesizing and releasing the protein interleukin 2. Transcription of the interleukin 2 gene is activated by a transcription factor called NFAT.

NFAT has a sorting signal that would normally direct it to the nucleus, but in unstimulated cells this is masked by a phosphate group, so NFAT remains in the cytoplasm and interleukin 2 is not made. However, when major histocompatability complex proteins present foreign peptides to the T lymphocyte, the concentration of calcium ions in the cytosol increases (an increase of calcium concentration is a common feature of cell stimulation to be described in Chapter 15). Calcium activates a phosphatase called **calcineurin**, which removes the phosphate group from many substrates including NFAT. NFAT then moves to the nucleus and activates interleukin 2 transcription. The released interleukin 2 activates other immune system cells that attack the foreign body. Cyclosporin blocks this process by inhibiting calcineurin, so that even though calcium rises in the cytoplasm of the T lymphocyte, NFAT remains phosphorylated and does not move to the nucleus.

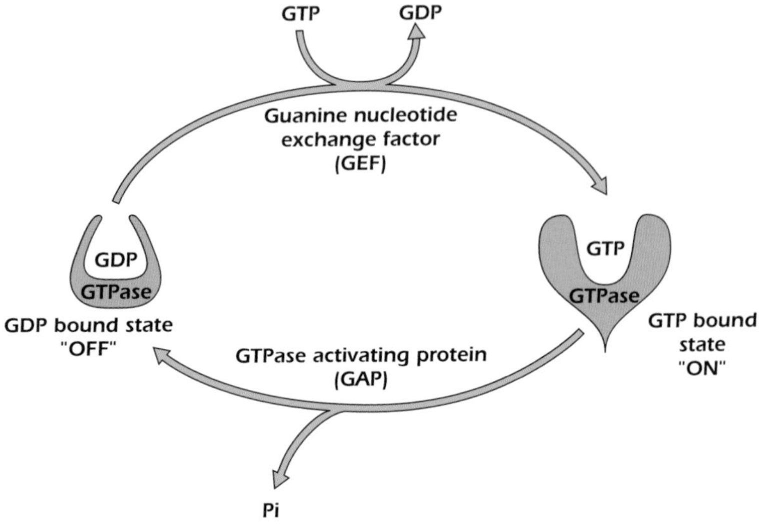

Figure 10.3. The GDP/GTP cycle of a GTPase.

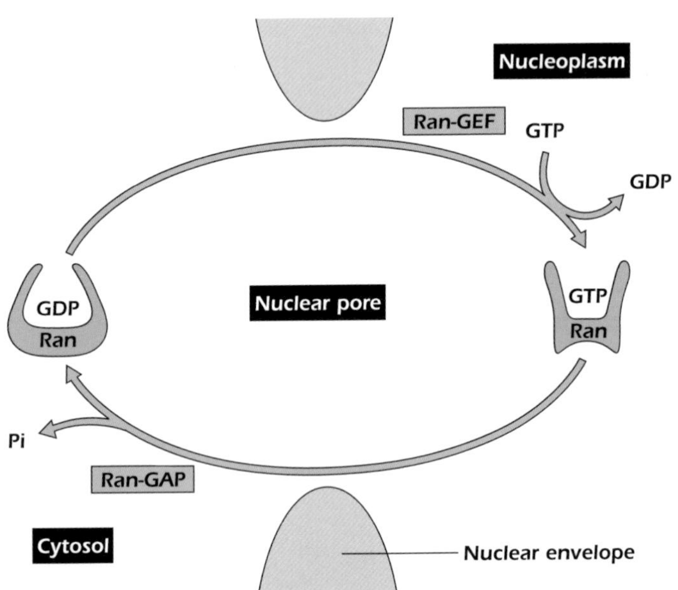

Figure 10.4. Ran GEF and GAP are localized to the nucleoplasm and cytosol, respectively.

GTPases in Nuclear Transport

In the case of nuclear pore transport, the GEFs that operate on Ran are found in the nucleus and seem to be associated with chromatin, while Ran GAPs are attached to the cytosolic face of the nuclear pore complex (Fig. 10.4). Thus nucleoplasmic Ran is predominantly in the GTP bound state (Ran:GTP), while most cytosolic Ran has GDP bound (Ran:GDP).

Figure 10.5 shows how Ran regulates import of proteins into the nucleus. An import transporter binds the nuclear localization sequence on the protein. As long as the transporter remains on the cytosolic side of the nuclear envelope, its cargo will remain bound. However, once the transporter finds itself on the nucleoplasmic side, Ran in its active, GTP-bound state binds and causes the cargo to be released. Now, as long as the transporter remains on the nucleoplasmic side, it will have Ran:GTP bound and will be unable to bind cargo. However, once the transporter finds itself on the cytoplasmic side, Ran GAPs will cause Ran to hydrolyze its bound GTP to GDP and the Ran:GDP will dissociate from the transporter, which is then able to bind more cargo.

The same principle is used when a protein is to be exported from the nucleus (Fig. 10.6). In this case the export receptor can only bind proteins with an export sequence when it is associated with Ran:GTP. As long as the transporter remains on the nucleoplasmic side of the nuclear envelope, its cargo will remain bound. However, once the transporter finds itself on the cytoplasmic side Ran GAPs will cause Ran to hydrolyze its bound GTP to GDP and both

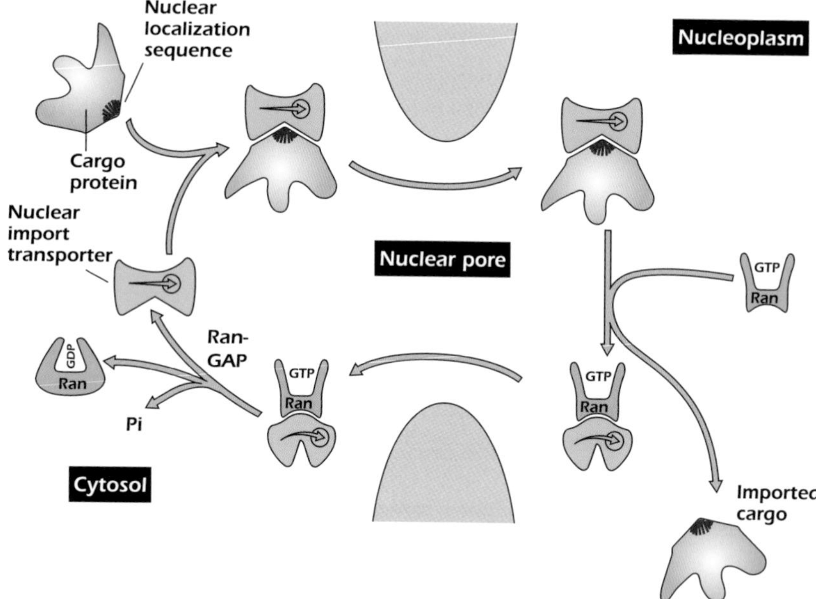

Figure 10.5. Nuclear import.

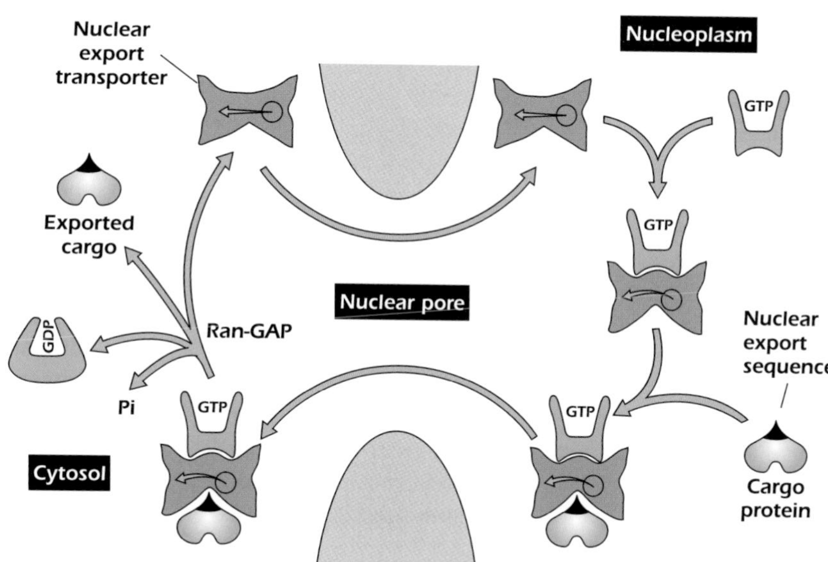

Figure 10.6. Nuclear export.

Ran:GDP and the cargo will dissociate from the transporter, which is then unable to bind more cargo until it moves back into the nucleoplasm where Ran:GTP is available.

Although Ran can drive both nuclear import and nuclear export, in each case it is the active GTP-bound form of the protein that binds to the transporter. The GDP-bound form is inactive and cannot bind the transporter.

Some proteins move back and forth between the cytosol and the nucleus by successively revealing and masking nuclear localization sequences. For example, the glucocorticoid receptor (page 94) only reveals its nuclear localization sequence when it has bound glucocorticoid.

TRANSPORT ACROSS MEMBRANES

Transport to Mitochondria

Mitochondria have their own DNA and manufacture a small number of their own proteins on their own ribosomes. However, the majority of mitochondrial proteins are coded for by nuclear genes. They are synthesized on cytoplasmic ribosomes and only imported into the mitochondrion post-translationally. For example, proteins destined for the mitochondrial matrix carry a targeting sequence at their N terminus and are recognized by a receptor protein in the outer membrane. This mitochondrial receptor makes contact with translocation complexes, which unfold the protein and moves it across both outer and inner membrane simultaneously. Once the protein is translocated, the targeting sequence is cleaved off and the protein refolded.

Chaperones and Protein Folding

A correctly addressed protein may fail to be targeted to an organelle if it folds too soon into its final three-dimensional shape. For example, movement of proteins into the mitochondrial matrix requires that a protein must move through channels through the outer and inner membranes. These channels are just wide enough to allow an unfolded polypeptide to pass through. Our cells have proteins called **chaperones**, which as the name indicates, look after young proteins. Chaperones use energy derived from the hydrolysis of ATP to keep newly synthesized proteins destined for the mitochondrial matrix in an unfolded state. As soon as the protein moves through the channels and into the matrix, the matrix targeting sequence is cleaved. The protein now folds into its correct shape. Some small proteins can fold without help. Larger proteins are helped to fold in the mitochondrial matrix by a chaperone protein called chaperonin, which provides a surface on which another protein can fold. Chaperones themselves do not change shape when helping another protein to fold.

Certain stresses that cells can experience, such as excessive heat, can cause proteins to denature (page 151). The cell responds by making a class of chaperone called **heat-shock proteins** in large amounts. The heat-shock proteins bind to misfolded proteins, usually to a hydrophobic region exposed by denaturation, and help the protein to refold. The heat-shock proteins are not themselves changed, but instead form a platform on which the denatured protein can refold itself. Heat-shock proteins are found in all cell compartments and also in bacteria.

Transport to Peroxisomes

Most organelles that are bound by a single membrane have their proteins made at the rough endoplasmic reticulum and transported to them in vesicles (Fig. 10.1). Peroxisomes (page 49) are an exception: their proteins are synthesized on free ribosomes in the cytosol and then transported to their final destination. Peroxisomal targeting sequences on the protein bind to peroxisome import receptors in the cytosol. The complex of cargo and receptor docks onto the peroxisomal membrane and then crosses the membrane to enter the peroxisome. Here, the protein cargo is released and the import receptor is shuttled back into the cytosol.

Synthesis on the Rough Endoplasmic Reticulum

Proteins destined for import into the endoplasmic reticulum have a mainly hydrophobic signal sequence of about 20 amino acids at the amino terminal. Their synthesis starts on free ribosomes. When the growing polypeptide chain is about 20 amino acids long, the endoplasmic reticulum signal sequence is recognized by a **signal recognition particle** that is made up of a small RNA molecule and several proteins (Fig. 10.7). When this particle binds to the signal sequence it stops protein synthesis from continuing. The complex of ribosome and signal recognition particle encounters a specific receptor called the signal recognition particle receptor ("docking protein") on the endoplasmic reticulum. This interaction directs the polypeptide chain to a **protein translocator**. Once this has occurred the signal recognition particle and its receptor are no longer required and are released. Protein synthesis now continues; and, as the polypeptide continues to grow, it threads its way through the membrane via the protein translocator,

Medical Relevance 10.2 How Protein Mistargeting Can Give You Kidney Stones

Primary hyperoxaluria type 1 is a rare genetic disease in which calcium oxalate "stones" accumulate in the kidney. Healthy people convert the metabolite glyoxylate to the useful amino acid glycine by the enzyme alanine glyoxylate aminotransferase (AGT) (page 184). AGT is located in peroxisomes in liver cells. If glyoxylate cannot be converted to glycine, it is instead oxidized to oxalate and excreted by the kidney, where it tends to precipitate as hard lumps of insoluble calcium oxalate. Two-thirds of patients with primary hyperoxaluria type 1 have a mutant form of AGT that simply fails to work. However, the other third have an AGT with a single amino acid change (G170R) that works reasonably well, at least in the test tube. However, this amino acid change is enough to make the mitochondrial import system believe that AGT is a mitochondrial protein and import it inappropriately, so that no AGT is available to be transported to the peroxisomes. For the clinician, the mistargeting of AGT in primary hyperoxaluria type 1 poses an unusual problem, namely, how to explain to a patient that the way to cure their kidney stones is to have a liver transplant!

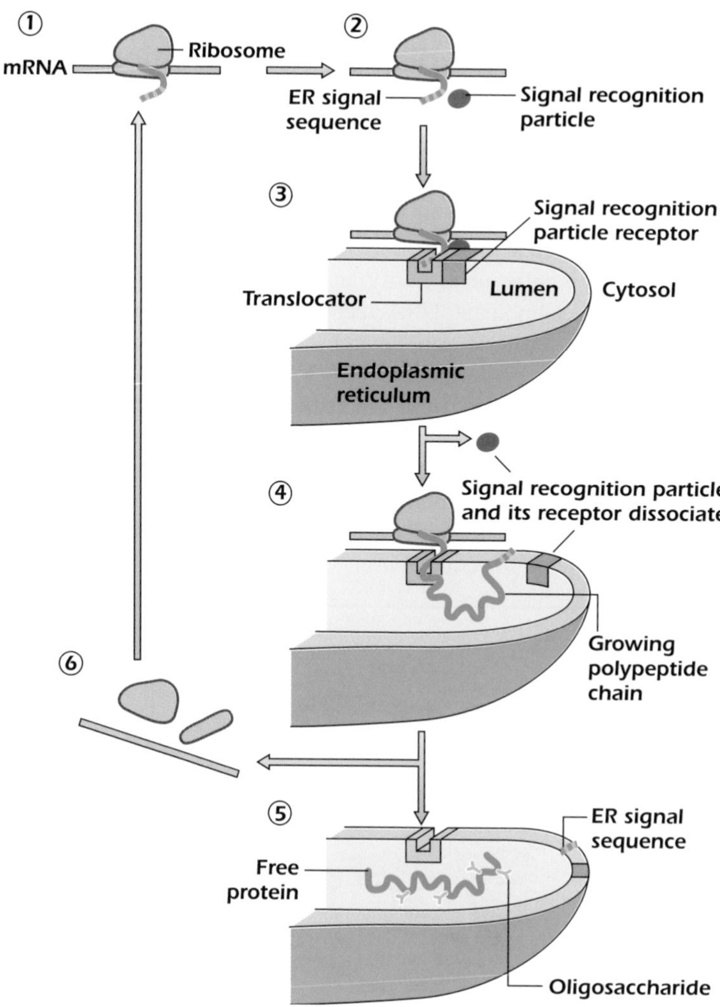

Figure 10.7. Transport of a growing protein across the membrane of the endoplasmic reticulum.

which acts as a channel allowing hydrophilic stretches of polypeptide chain to cross. Once the polypeptide chain has entered the lumen of the endoplasmic reticulum, the signal sequences may be cleaved off by an enzyme called signal peptidase. Some proteins do not undergo this step but instead retain their signal sequences.

The platelet-derived growth factor receptor is an example of an integral membrane protein. It contains a stretch of 22 hydrophobic amino acids that spans the plasma membrane (Fig. 9.6 on page 144). The first part of the polypeptide to be synthesized is an endoplasmic reticulum signal sequence, so the polypeptide begins to be threaded into the lumen of the endoplasmic reticulum. This section will become the extracellular domain of the receptor. When the stretch of hydrophobic residues is synthesized, it is threaded into the translocator in the normal way but cannot leave at the other end because the amino acid residues do not associate with water. As synthesis continues, therefore, the newest length of polypeptide bulges into the cytosol. Once synthesis stops, this section is left as the cytosolic domain.

If a protein contains more than one hydrophobic stretch, then synthesis of the second stretch reinitiates translocation across the membrane, so that the protein ends up crossing the membrane more than once.

Glycosylation: The Endoplasmic Reticulum and Golgi System

Most polypeptides synthesized on the rough endoplasmic reticulum are glycosylated, that is, they have sugar residues added to them, as soon as the growing polypeptide chain enters the lumen of the endoplasmic reticulum. In the process of N-glycosylation a premade oligosaccharide composed of two N-acetyl glucosamines (Fig. 2.11 on page 27), then nine mannoses, and then three glucoses is added to an asparagine residue by the enzyme oligosaccharide transferase. The three glucose residues are subsequently removed, marking the protein as ready for export from the endoplasmic reticulum to the Golgi apparatus.

Example 10.2 Cyclists and Glycosylation

Erythropoietin is a protein hormone produced by the kidney that stimulates the formation of red blood cells in the bone marrow. The hormone has important medical applications in the treatment of anaemias caused by chronic kidney disease or appearing as a side effect of cancer therapy. It is difficult to purify and medical use had to wait until recombinant DNA methods were developed (Chapter 7). Bacteria can be transfected so that they express human erythropoietin cDNA, but they do not add the oligosaccharides that are critical in the protein's ability to bind to its receptor. Fortunately mammalian expression systems for the human gene were developed and these produce recombinant glycosylated erythropoietin that is active. Recombinant erythropoietin is now widely available for clinical use.

Soon erythropoietin found other uses among athletes taking part in endurance sports such as cycling as it boosted the oxygen-carrying capacity of the blood (although taking erythropoietin is not without risks). As recombinant erythropoetin has the same amino acid sequence as the regular endogenous human protein, its presence is extremely hard to prove. However, the mammalian cells used to make commercial erythropoietin are not human and there are small differences in the exact glycosylation that do not affect the function. Work is being undertaken to develop tests that can discriminate between human erythropoietin and that made in non-human cells.

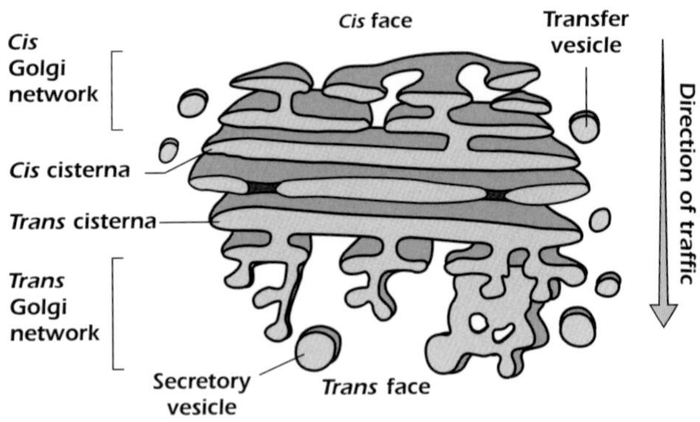

Figure 10.8. The Golgi apparatus.

The Golgi apparatus (Fig. 10.8) is a stack of flattened membraneous bags called **cisternae**. Each cisterna is characterized by a central flattened region where the luminal space, as well as the gap between adjacent cisternae, is uniform. The margin of each cisterna is often dilated and is often fenestrated (i.e., has holes through it) as well. Small, spherical vesicles are always found in association with the Golgi apparatus, especially with the edges of the ***cis* face**. These are referred to as transfer vesicles; some of them carry proteins from the rough endoplasmic reticulum to the Golgi stacks, while others transfer proteins between the stacks, that is, from *cis* to middle and from middle to *trans*. As proteins move through the Golgi apparatus, the oligosaccharides already attached to them are modified, and additional oligosaccharides can be added. As well as having important functions once the protein has reached its final destination, glycosylations play an important role in sorting decisions at the ***trans* Golgi** network.

VESICULAR TRAFFICKING BETWEEN INTRACELLULAR COMPARTMENTS

Most of the single-membrane organelles of the eukaryotic cell pass material between themselves by vesicular traffic, in vesicles that bud off from one compartment to fuse with another (Fig. 10.9). In this way the cargo proteins are never in contact with the cytosol. Two main directions of traffic can be identified (Fig. 10.1). The exocytotic pathway runs from the endoplasmic reticulum through the Golgi apparatus to the plasma membrane. The endocytotic pathway runs from the plasma membrane to the lysosome. This is the route by which extracellular macromolecules can be taken up and processed. If vesicles are to be moved over long distances, they are transported along cytoskeletal highways (page 287).

IN DEPTH 10.1 TRAFFICKING MOVIES

Chimeric proteins that contain green fluorescent protein (page 115) and appropriate sorting signals will be processed by the cells machinery just as proteins coded for on its own genes. Thus protein trafficking can be viewed by fluorescence microscopy in live cells that have been transfected with DNA coding for such chimeras. Movies illustrating many aspects of intracellular trafficking are available on the Internet.

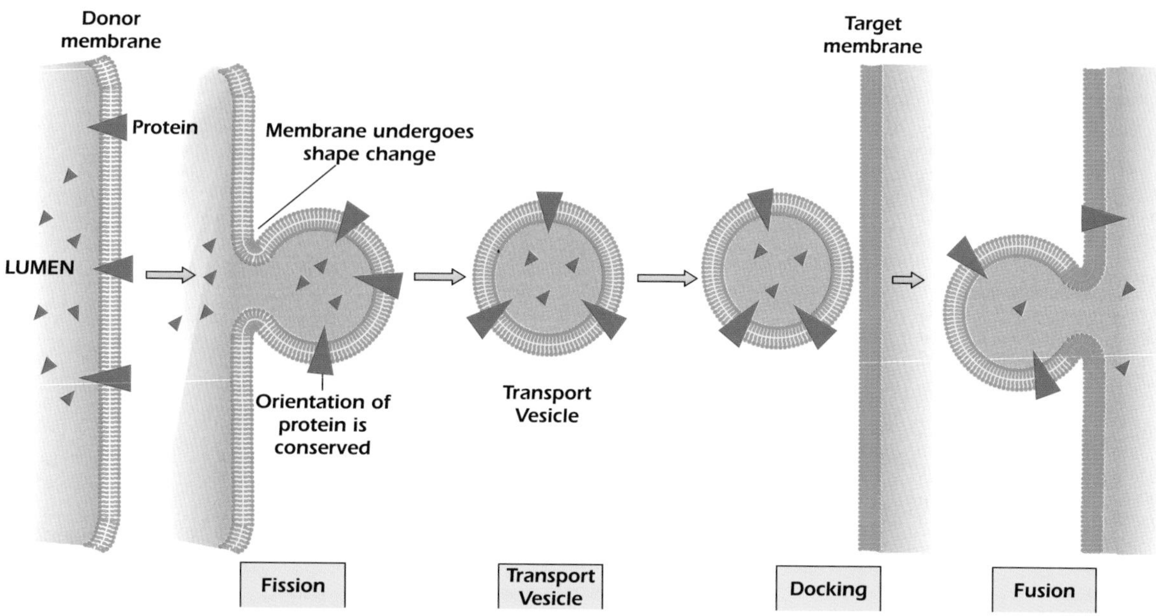

Figure 10.9. Vesicle fission and fusion.

The Principle of Fission and Fusion

Figure 10.9 illustrates how budding of a vesicle from one organelle, followed by fusion with a second membrane, can transport both soluble proteins and integral membrane proteins to the new compartment. The process retains the "sidedness" of the membrane and the compartment it encloses: the side of an integral membrane protein that faced the lumen of the first compartment ends up facing the lumen of the second compartment, while soluble proteins in the lumen of the first compartment do not enter the cytosol but end up in the lumen of the second compartment, or in the extracellular medium if the target membrane is the plasma membrane.

Vesicle Formation

Vesicle formation is the process during which cargo in the endoplasmic reticulum or Golgi lumen is captured and the lipid membrane is shaped, with the help of cytosolic proteins, into a bud which is then pinched off in a process called **fission**. The ordered assembly of cytosolic proteins into a coat over the surface of the newly forming vesicle is responsible for forcing the membrane into a curved shape (Fig. 10.10). There are two types of coats that serve this function: **coatomer** coats and **clathrin** coats. The coat must be shed before fusion of the vesicle with its target membrane can occur.

Coatomer-Coated Vesicles

Transport along the default pathway uses coatomer-coated vesicles. This is the mechanism used in trafficking between the endoplasmic reticulum and Golgi, between the individual Golgi stacks, and in budding of constitutive secretory vesicles from the *trans* Golgi. The coatomer coat consists of seven different proteins that assemble into a complex. The current model for coatomer coat formation at the Golgi is that a guanine nucleotide exchange factor (GEF) in the donor membrane exchanges GDP for GTP in the GTPase Arf. This causes Arf to adopt its active configuration, in which an α helix at the N terminus is exposed and embeds into the top of the donor membrane (Fig. 10.10A). As well as anchoring Arf to the membrane, this also increases the area of the cytosolic face of the membrane, causing it to buckle outwards to form the bud. Membrane-bound Arf is the initiation site for coatomer assembly and coatomer-coated vesicle formation. The coat is only shed when the vesicle is docking to its target membrane. An Arf-GAP in the target membrane causes Arf to hydrolyse its GTP, and the resulting conformational change causes Arf to retract its hydrophobic N terminus and become cytosolic and this causes the coat to be shed.

Clathrin-Coated Vesicles

Clathrin-coated vesicles mediate selective transport. They are, for example, the means by which protein in the *trans* Golgi that bears mannose-6-phosphate is collected into a vesicle bound for the lysosome. Clathrin-coated vesicles carry proteins and lipids from the plasma membrane to endosomes, and operate in other places where selective transport is required.

Figure 10.10B illustrates how clathrin generates a vesicle. The process starts when the cargo of interest binds

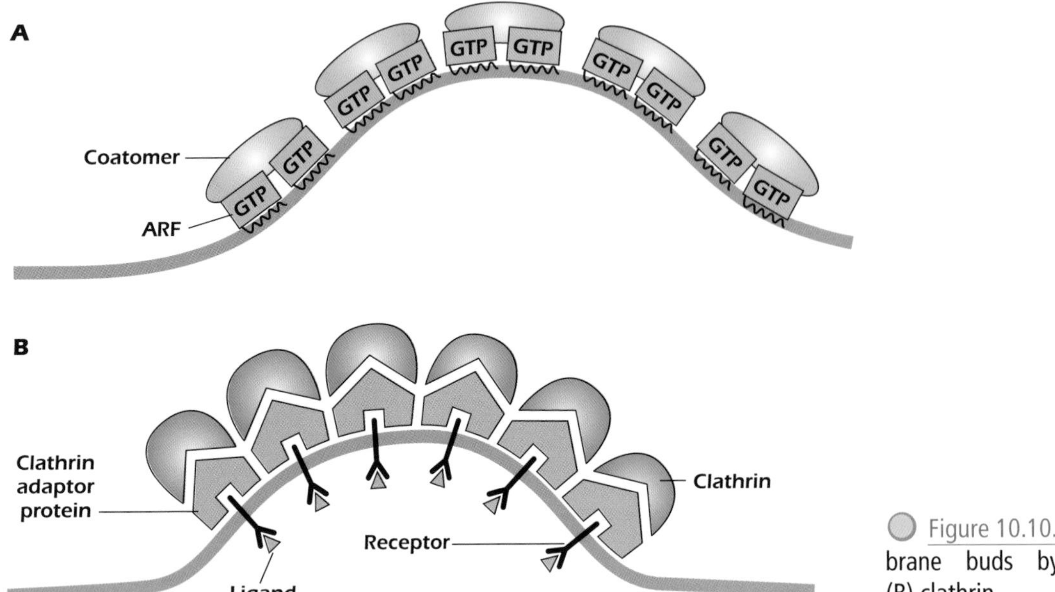

Figure 10.10. Generation of membrane buds by (A) coatamers and (B) clathrin.

to integral proteins of the donor membrane that are selective receptors for that cargo. **Clathrin adaptor proteins** then bind to cargo-loaded receptors and begin to associate, forming a complex. Lastly, clathrin molecules bind to this complex, forming the coat and bending the membrane into the bud shape

Even though clathrin can force the membrane into a bud shape, it cannot force the bud to leave as an independent vesicle. One of the best-studied membrane fission events is endocytosis. Here a GTPase called **dynamin** forms a ring around the neck of a budding vesicle. GTP hydrolysis then causes a change in dynamin's shape that mechanically pinches the vesicle off from its membrane of origin. Unlike coatamer, clathrin coats dissociate as soon as the vesicle is formed, leaving the vesicle ready to fuse with the target membrane.

Trans Golgi Network and Protein Secretion

At its *trans* face the Golgi apparatus breaks up into a complex system of tubes and sheets called the *trans* Golgi network (Fig. 10.8). Although there is some final processing of proteins in the *trans* Golgi network, most of the proteins reaching this point have received all the modifications necessary to make them fully functional and to specify their final destination. Rather, the *trans* Golgi network is the place where proteins are sorted into the appropriate vesicles and sent down one of three major pathways: **constitutive secretion**, **regulated secretion**, or transport to lysosomes.

Vesicles for constitutive or regulated secretion, though functionally different, look very much alike and are directed to the cell surface (Fig. 10.1). When the vesicle membrane comes into contact with the plasma membrane, it fuses with it. At the point of fusion the membrane is broken through, the contents of the vesicle are expelled to the extracellular space, and the vesicle membrane becomes a part of the plasma membrane. This process, by which the contents of a vesicle are delivered to the plasma membrane and following membrane fusion and breakthrough are released to the outside, is called **exocytosis**.

The difference between the constitutive and regulated pathways of secretion by exocytosis is that the former is always "on" (vesicles containing secretory proteins are presented for exocytosis continuously) while the regulated pathway is an intermittent one in which the vesicles containing the substance to be secreted accumulate in the cytoplasm until they receive a specific signal, usually an increase in the concentration of calcium ions in the surrounding cytosol, whereupon exocytosis proceeds rapidly. After secretion, vesicular membrane proteins are retrieved from the plasma membrane by **endocytosis** and transported to the **endosomes** (Fig. 10.1). From the endosomes vesicles are targeted to the lysosomes, back to the Golgi, or into the pool of regulated secretory vesicles. In a cell that is not growing in size, the amount of membrane area added to the plasma membrane by exocytosis is balanced over a period of minutes by endocytosis of the same area of plasma membrane.

Notice that in exocytosis the membrane of the vesicle becomes incorporated into the plasma membrane; consequently the integral proteins and lipids of the vesicle membrane become the integral proteins and lipids of the plasma membrane. This is the principal, if not the only way, that integral proteins made on the rough endoplasmic reticulum are added to the plasma membrane.

Major histocompatability complex (MHC) proteins are a group of integral plasma membrane proteins that reach

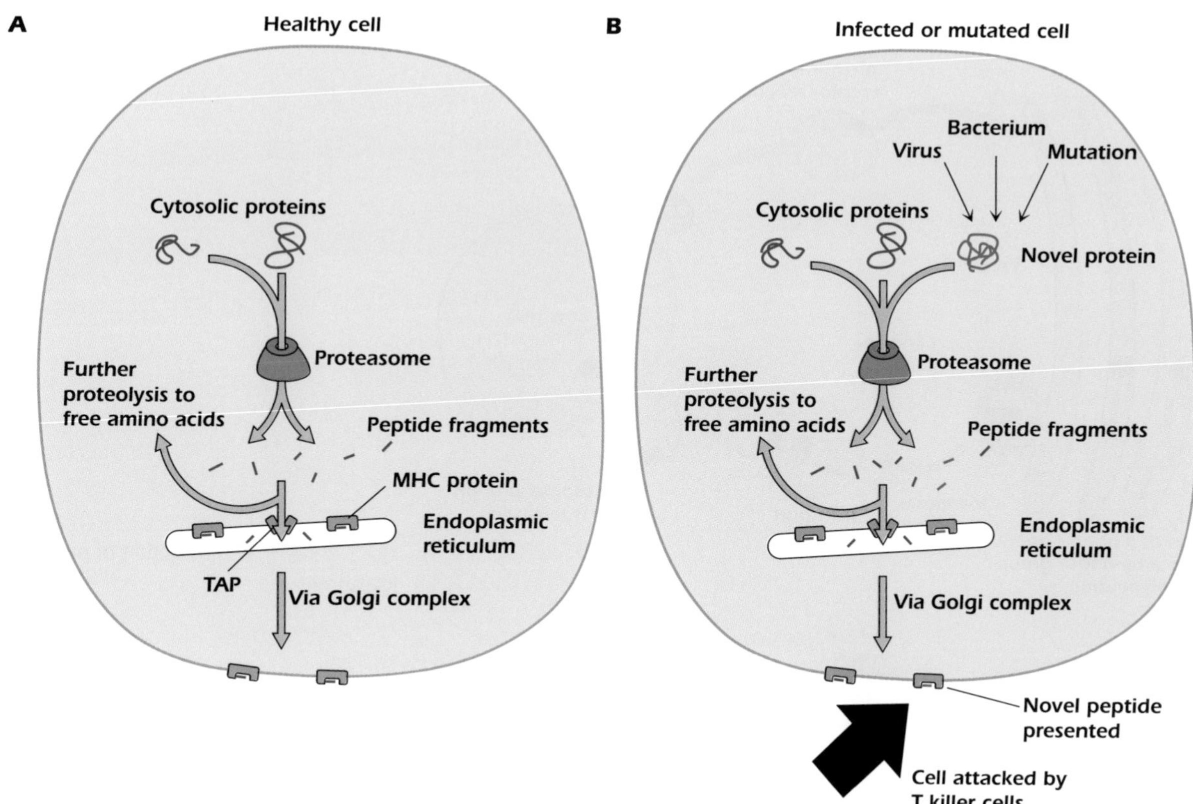

Figure 10.11. Presentation of peptides by MHC proteins.

their final location by this route. Found only in vertebrates, the function of these proteins is to present short lengths of peptide to patrolling white blood cells. Figure 10.11A shows how this works. All the time, a selection of cytosolic proteins is being degraded into peptide fragments by the proteasome (page 133). These fragments are captured by a carrier located on the endoplasmic reticulum membrane called the TAP (for Transporter associated with Antigen Processing) and moved into the lumen, where they bind to a pocket in the MHC protein. From there the complex is processed together and passes via the Golgi complex to the plasma membrane.

One class of immune system cells called **T killer** patrols the body and kills any cell that is presenting a novel peptide (Fig. 10.11B). In this way, cells infected by viruses or bacteria, or cells that have undergone somatic mutation, are detected and killed. This process is described further in Chapter 19.

Targeting Proteins to the Lysosome

One of the best-understood examples of sorting in the *trans* Golgi network is lysosomal targeting (Fig. 10.12). Proteins that are destined for the lysosome are synthesized on the rough endoplasmic reticulum and therefore, like all proteins synthesized here, have a mannose containing oligosaccharide added. Because they do not have an endoplasmic reticulum retention signal such as KDEL, they are transported to the Golgi apparatus. There proteins destined for the lysosome are modified by phosphorylation of some of their mannose residues to form mannose-6-phosphate (Fig. 2.11 on page 27). Once the proteins reach the *trans* Golgi network, specific receptors for mannose-6-phosphate recognize this sorting signal and cause the proteins to be packaged into vesicles that are transported to the lysosome, with which they fuse. In the low pH (5) environment of the lysosome, the lysosomal protein can no longer bind to its receptor. The phosphate group is removed by a phosphatase. Vesicles containing the receptor bud off from the lysosome and deliver the mannose-6-phosphate receptors back to the *trans* Golgi network.

The function of the lysosome is to degrade unwanted materials. To carry out this function, an inactive or primary lysosome fuses with a vesicle containing the material to be digested. This makes a secondary lysosome. The vesicles with which primary lysosomes fuse may be bringing materials in from outside the cell or they may be vesicles made by condensing a membrane around worn out or unneeded organelles in the cells own cytoplasm. The latter are sometimes called autophagic vacuoles. Some materials may not be digestible and remain in the lysosome for the lifetime of the cell. These small dense remnant lysosomes are called residual bodies.

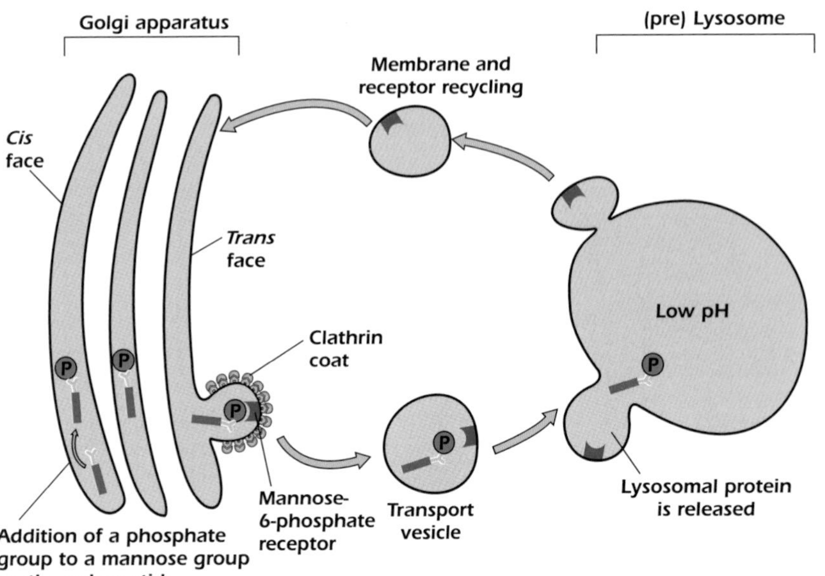

Figure 10.12. Targeting of protein to the lysosome.

Medical Relevance 10.3 Failure of the Lysosome Targeting Signal

One severe lysosomal storage disease (page 50) is different in that rather than a single enzyme, a whole group of hydrolytic enzymes is missing from the lysosomes. This is known as "I-cell disease" because of the large inclusions found in affected cells that can be easily seen with the light microscope. These inclusions are huge lysosomes swollen with undegraded materials. More curiously the enzymes missing from the lysosomes are found instead in the extracellular medium, including the blood plasma! This mass failure of targetting occurs because the enzyme that adds the mannose containing oligosaccharide to proteins in the Golgi complex is missing or defective. This means that all the proteins that should go to the lysosome travel instead along the default pathway and are exocytosed.

Fusion

Membrane fusion is the process by which a vesicle membrane incorporates its components into the target membrane and releases its cargo into the lumen of the organelle or, in the case of secretion, into the extracellular medium (Fig. 10.13). Different steps in membrane fusion are distinguishable. First, the vesicle and the target membrane mutually identify each other. Then, proteins from both membranes interact with one another to form stable complexes and bring the two membranes into close apposition, resulting in the docking of the vesicle to the target membrane. Finally, considerable energy needs to be supplied to force the membranes to fuse, since the low-energy organization—in which the hydrophobic tails of the phospholipids are kept away from water while the hydrophilic head groups are in an aqueous medium—must be disrupted, even if only briefly, as the vesicle and target membranes distort and then fuse.

Each type of vesicle must only dock with and fuse with the correct target membrane, otherwise the protein constituents of all the different organelles would become mixed with each other and with the plasma membrane. Our understanding of the molecular processes leading to membrane fusion is only just beginning to take shape, but our current understanding is that two types of proteins,

Answer to Thought Question: The nuclear pore seems to contain a forest of polypeptide chains for the import transporter protein to swing through rather like Tarzan on jungle vines. This process does not itself require energy expenditure. However if the transporter is then to carry a second molecule of cargo into the nucleus it must first return out again to the cytosol, and for each cycle of transporter in/transporter out one GTP molecule is hydrolysed, releasing energy. It is this release of energy that drives the accumulation of cargo into the nucleus. As we will see in Chapter 12 the amount of energy released by GTP hydrolysis is the same as that released by the hydrolysis of ATP.

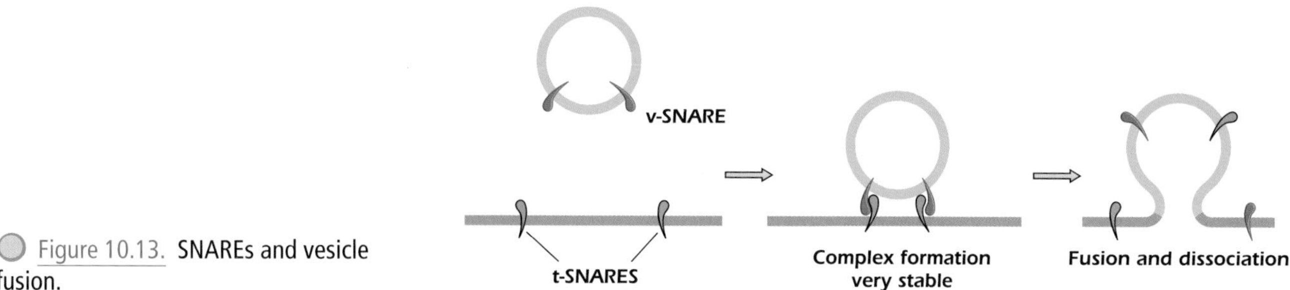

Figure 10.13. SNAREs and vesicle fusion.

called **SNARES** and **Rab family** GTPases work together to achieve this. SNARES located on the vesicles (v-SNARES) and on the target membranes (t-SNARES) interact to form a stable complex that holds the vesicle very close to the target membrane. Not all vSNARES can interact with all tSNARES, so SNARES provide a first level of specificity. The 50 or more members of the Rab family identified in mammalian cells play a different role to SNARES. Each seems to be found at one particular site where it regulates one specific transport event, thus controlling which vesicle fuses with which target. For example, the recycling of the mannose-6-phosphate receptor back from the lysosome to the *trans* Golgi network (Fig. 10.12) requires Rab9. GTP hydrolysis by Rabs is thought to provide energy for membrane fusion.

Example 10.3 SNARES, Food Poisoning and Face-Lifts

Botulism, food poisoning caused by a toxin released from the anaerobic bacterium *Clostridium botulinum*, is fortunately rare. Botulinum toxin comprises a number of enzymes that specifically destroy those SNARE proteins required for regulated exocytosis in nerve cells. Without these proteins regulated exocytosis cannot occur, so the nerve cells cannot tell muscle cells to contract. This causes paralysis—most critically, paralysis of the muscles that drive breathing. Death in victims of botulism results from respiratory failure.

Low concentrations of botulinum toxin (or "Botox") can be injected close to muscles to paralyze them. For example, in a "chemical face-lift," botulinum toxin is used to paralyze facial muscles, producing an effect variously described as "youthful" and "zombie-like."

SUMMARY

1. The basic mechanisms of intracellular protein trafficking are similar in all eukaryotic cells, from yeast cells to human nerve cells.
2. The final destination of a protein is defined by sorting signals that are recognized by specific receptors. The polypeptide chain itself contains targeting sequences while glycosylation and phosphorylation can add additional sorting signals.
3. Some sorting signals activate translocation of a protein to a new location, while others such as the endoplasmic reticulum retention signal KDEL cause the protein to be retained at its present location.
4. Nuclear proteins are synthesized on free ribosomes and carried through the nuclear pore by Ran-mediated gated transport. Other proteins with a nuclear export signal are carried the other way, again by a Ran-mediated process.
5. Peroxisomal proteins, together with that majority of mitochondrial and chloroplast proteins not coded for by mitochondrial or chloroplast genes, are synthesized on free ribosomes and then transported across the membrane(s) of the target organelle.
6. Proteins with an endoplasmic reticulum signal sequence are synthesized on the rough endoplasmic reticulum. The growing polypeptide chain is fed across the membrane as it is synthesized. The signal sequence may then be cleaved off.
7. The default pathway for proteins synthesized on the rough endoplasmic reticulum is to pass through the Golgi apparatus and be secreted from the cell via the constitutive pathway.

8. In vertebrates, peptide fragments of cytosolic proteins are presented at the cell surface on major histocompatability complex proteins for inspection by patrolling T cells.

9. Glycosylation in the endoplasmic reticulum and then in the Golgi apparatus has two functions: to produce the final, functional form of the protein and to add further sorting signals such as the lysosomal sorting signal mannose-6-phosphate.

10. Vesicles shuttle between most of the single-membrane cellular organelles, the exception being peroxisomes. Budding and then fission of vesicles from the donor membrane can be driven by coatamer family proteins or by clathrin plus dynamin.

11. A tight association of vesicular and target SNARE proteins forces vesicles into close association with the target membrane. Fusion requires the action of a Rab family GTPase.

FURTHER READING

Alberts, B., Johnson A., Lewis, J., Raff, M., Roberts, K., and Walter, P. (2008) *Molecular Biology of the Cell*, 5th edition, Garland Science, New York.

Karp, G. C. (2008) *Cell and Molecular Biology: Concepts and Experiments*, 5th edition, John Wily & Sons, New York.

REVIEW QUESTIONS

10.1 Theme: The Three Modes of Intracellular Transport

A Gated transport
B Remains in cytosol, no trafficking required
C Transmembrane translocation
D Vesicular trafficking

From the above list of transport modes, select the one experienced by each of the proteins below subsequent to its complete synthesis.

1. β globin
2. Catalase (required inside peroxisomes)
3. Glucocorticoid hormone receptor.
4. NFAT (Nuclear Factor of Activated T cells)
5. Platelet-derived growth factor receptor
6. Pyruvate dehydrogenase (required in the mitochondrial matrix)

10.2 Theme: Trafficking Processes

A Endocytosis
B Exocytosis
C Traffic through the nuclear pore
D Transport between Golgi cisternae
E Transport into mitochondria
F Transport into peroxisomes
G Transport into the endoplasmic reticulum

From the list of trafficking processes above, select a process with which each of the proteins below is associated.

1. Arf
2. Dynamin
3. Rab
4. Ran
5. Signal recognition particle
6. TAP (transporter associated with antigen processing)

10.3 Theme: GTPases

A ADP
B ATP
C Coatamer
D GDP
E GMP
F GTP
G GTPase activating protein, GAP
H Guanine nucleotide exchange factor, GEF
I Nuclear import transporter

We describe the cycle of operation of a GTPase below. For each step, identify the molecule associated with that step.

1. GTPases have a binding site for a nucleotide. Identify the nucleotide present in the pocket when the GTPase is in its off state (e.g., Arf in the state that cannot associate with membranes).
2. GTPases are activated when the nucleotide in the binding pocket is replaced by a nucleotide present at higher concentration in the cytosol. Give the general name for the protein partner that catalyzes this switch.
3. GTPases turn off when the nucleotide in the binding pocket is hydrolysed. Identify the product of this hydrolysis.
4. Hydrolysis of the nucleotide in the binding pocket is activated by a protein partner. Give the general name for the protein partner that accelerates the hydrolysis.

THOUGHT QUESTION

Proteins with a nuclear localization sequence accumulate in the nucleus, that is, they move from a region where they are at a low concentration to a region where they are at a higher concentration. As we will see in Chapter 12, a cell needs to expend energy to move solute up a concentration gradient. Yet we have stated that movement of the nuclear import transporter, with its associated cargo protein, through the nuclear pore involves repeated binding and unbinding of the transporter protein to glycine-phenylalanine repeats on polypeptides within the pore, a process that does not involve the chemical conversion of ATP or other compound. Where does the energy to drive nuclear import come from?

11

HOW PROTEINS WORK

The three-dimensional structures of proteins create binding sites for other molecules. Reversible binding of one molecule to another is central to most of the biological roles of proteins, whether the protein is a connexin that binds a connexin on another cell (page 46) or a transcription factor that binds to DNA. One special class of proteins, enzymes, have sites that not only bind another molecule but then catalyze a chemical reaction involving that molecule.

HOW PROTEINS BIND OTHER MOLECULES

Proteins can bind other protein molecules, DNA or RNA, polysaccharides, lipids, and a very large number of other small molecules and inorganic ions and can even bind dissolved gases such as oxygen, nitrogen, and nitric oxide. **Binding sites** are usually very specific for a particular ligand, although the degree of specificity can vary widely. Usually the binding is reversible so that there is an equilibrium between the free and bound ligand.

A binding site is usually a cleft or pocket in the surface of the protein molecule made up of amino acid side chains appropriately positioned to make specific interactions with the ligand. All of the forces that stabilize tertiary structures of proteins are also used in ligand–protein interaction: hydrogen bonds, electrostatic interactions, the hydrophobic effect, and van der Waals forces all have their roles. Even covalent bonds may be formed in a few cases—some enzymes form a transient covalent bond with the substrate as part of the mechanism used to effect the reaction.

DYNAMIC PROTEIN STRUCTURES

It is easy to get the impression that protein structures are fixed and immobile. In fact proteins are always flexing and changing their structure slightly around their lowest energy state. A good term for this is "breathing." Many proteins have two low-energy states in which they spend most of their time, like a sleeper who, though twisting and turning throughout the night, nevertheless spends most time lying on their back or side. An example is the **glucose carrier** (Fig. 11.1). This is a transmembrane protein that forms a tube through the membrane. It is stable in one of two configurations. In one the tube is open to the cytosol; in the other the tube is open to the extracellular medium. By switching between the two states, the glucose carrier carries glucose into and out of the cell.

Allosteric Effects

The glucose carrier is able to bind a ligand—the glucose molecule—in either of its low-energy conformations. In contrast, the *lac* repressor (page 87) can only bind its ligand, the operator region of the *lac* operon, in one conformation. On its own the protein predominantly adopts this conformation so transcription is prevented as it binds to the DNA. When the *lac* repressor binds allolactose (a signal that lactose is abundant), it is locked into a second, inactive form that cannot bind to the DNA. Transcription is no longer repressed, although the cAMP–CAP complex is additionally required if transcription is to proceed at a

Cell Biology: A Short Course, Third Edition. Stephen R. Bolsover, Jeremy S. Hyams, Elizabeth A. Shephard and Hugh A. White.

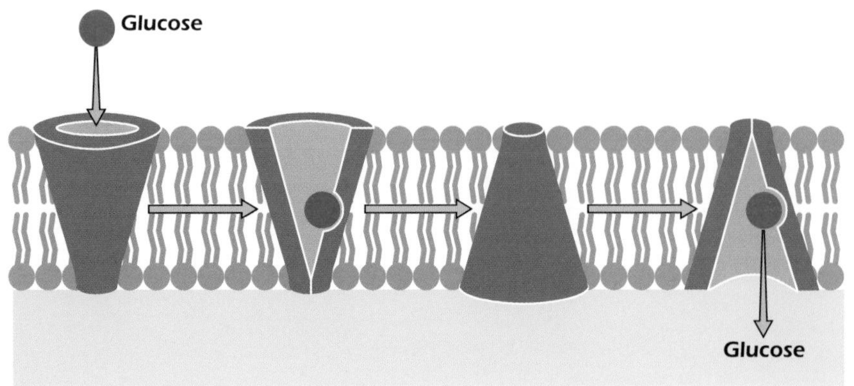

Figure 11.1. The glucose carrier switches easily between two shapes.

Medical Relevance 11.1 Oxygen for the Fetus

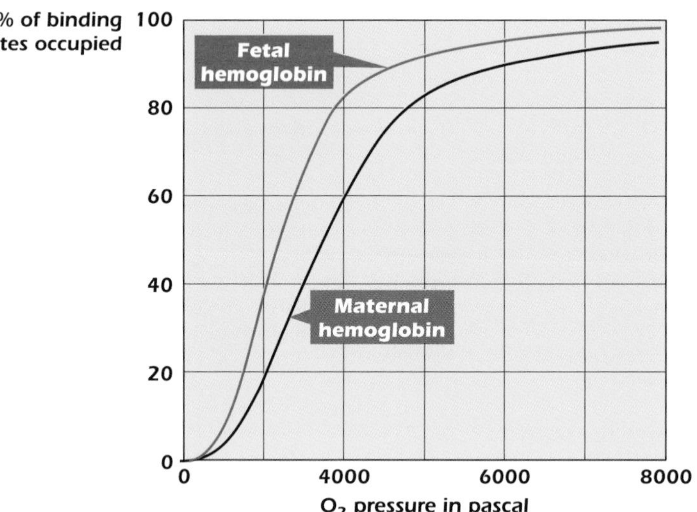

A developing fetus requires oxygen which must pass from the maternal blood to the blood in the fetus. Evolution has ensured that this occurs by using a slightly different hemoglobin in the fetal red blood cells, with a higher affinity for oxygen as shown in the graph. Adult hemoglobin is a tetramer of 2α subunits and two β subunits. Fetal hemoglobin is similar but the β subunits are replaced by γ subunits (page 77). Just before birth transcription of the gene for the γ chain is switched off and the gene for the β chain is switched on. Gradually the newborn's red blood cells with fetal hemoglobin are replaced with those filled with adult hemoglobin.

high rate. This type of interaction, in which the binding of a ligand at one place affects the ability of the protein to bind another ligand at another location, is called **allosteric** and is usually a property of proteins with a quaternary structure (i.e., with multiple subunits).

Hemoglobin (Fig. 9.14 on page 152) is an example of a protein where allosteric effects play an important role. Each heme prosthetic group, one on each of the four subunits, can bind an oxygen molecule. We can get an idea of what one subunit on its own can do by looking at myoglobin (Fig. 11.2A), a related molecule that moves oxygen within the cytoplasm. Myoglobin has just one polypeptide chain and one heme. The red line in Figure 11.2B shows the oxygen-binding curve for myoglobin. Starting from zero oxygen, the first small increase in oxygen concentration produces a large amount of binding to myoglobin; the next increase in oxygen produces a slightly smaller amount of binding, and so on, until myoglobin is fully loaded with

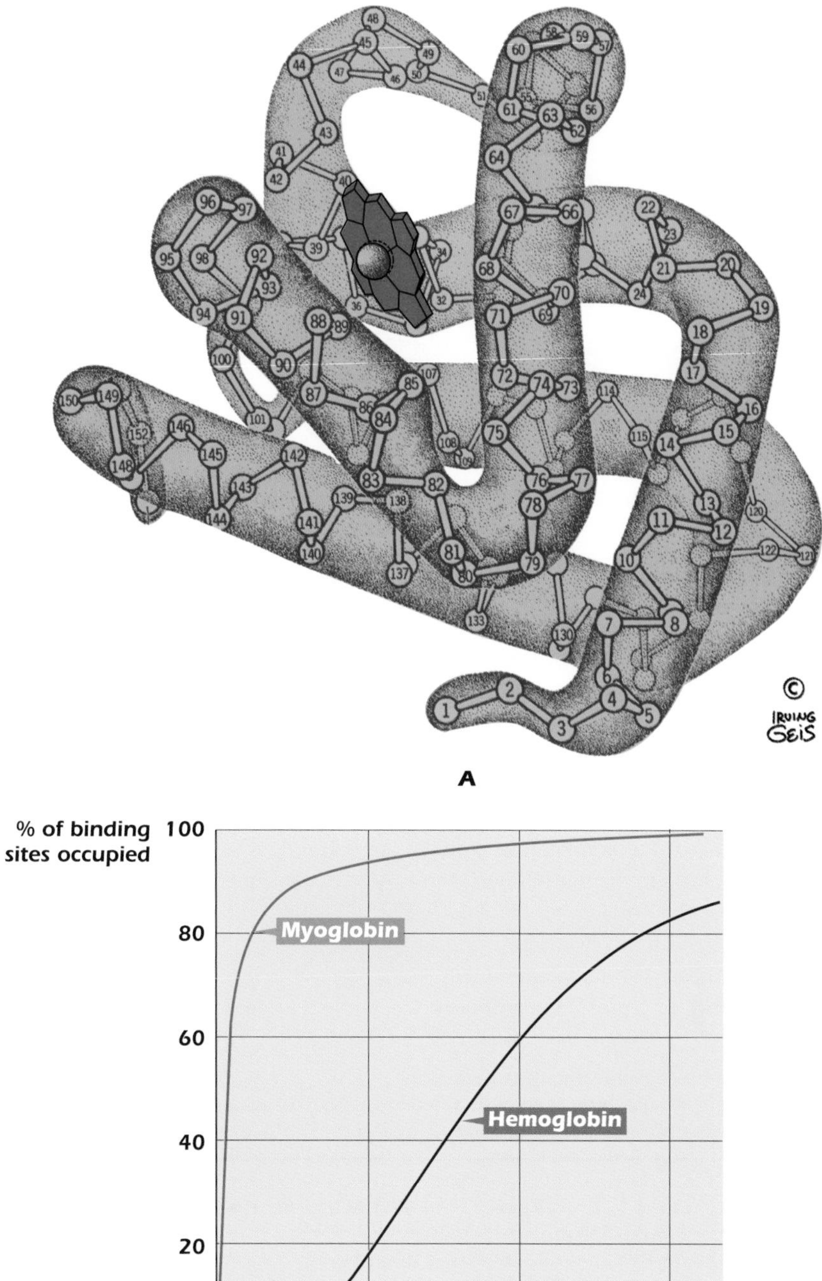

Figure 11.2. (A) The monomeric oxygen-carrying protein myoglobin. Illustration: Irving Geis. Rights owned by Howard Hughes Medical Institute. Reproduction by permission only. (B) Oxygen binding of myoglobin (in red) and hemoglobin (in purple) as oxygen pressure increases.

oxygen. A curve of this shape is known as hyperbolic. The purple line in Figure 11.2B shows the oxygen-binding curve for hemoglobin. Starting from zero oxygen, the first small increase in oxygen concentration produces hardly any binding to hemoglobin. The next increase in oxygen produces much more binding so the curve gets steeper before leveling off again as the hemoglobin becomes fully loaded. This behavior is called cooperative and the curve is described as "sigmoid." The explanation for this behavior is that the hemoglobin subunits can exist in one of two

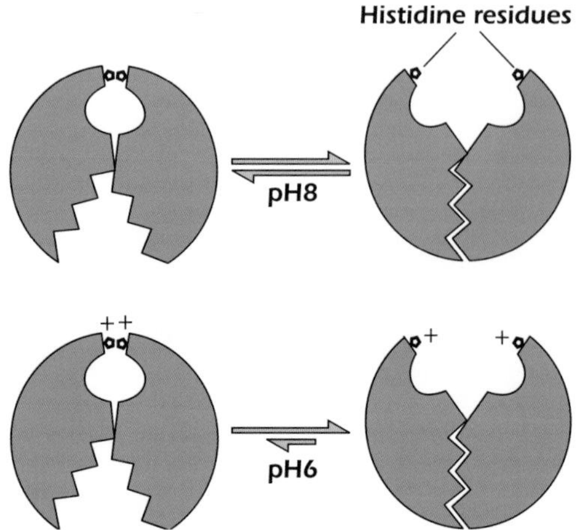

Figure 11.3. A pH change that alters the charge on histidine will alter the balance of forces within a protein. At pH 6, the structure on the right will predominate.

states, only one of which has a high affinity for oxygen. The way that the four subunits fit together means that they all must be in one form or the other. When oxygen concentration is low, most of the hemoglobin molecules have their subunits in the low-affinity form. As oxygen increases, it begins to bind to the hemoglobin—little at first as most of the hemoglobin is in the low-affinity form and only a little in the high-affinity form. As oxygen binds, more molecules switch to the high-affinity form as the low- and high-affinity forms are in equilibrium. Eventually virtually all of the molecules have made the switch to the high-affinity form. This produces the curve shown in Figure 11.2B. This cooperative oxygen binding makes hemoglobin an effective transporter as it will load up with oxygen in the lungs but will release it readily in the capillaries of the tissues where the oxygen concentration is low. Myoglobin would release little of its bound oxygen at the oxygen concentrations typical of respiring tissues.

Some enzymes show cooperative behavior caused by an allosteric effect that causes binding of one substrate molecule to make it easier for the other substrate to bind. The degree of cooperativity can be altered by the binding of other molecules (called effectors) that act to switch the enzyme on or off.

Chemical Changes That Shift the Preferred Shape of a Protein

Proteins can change conformation as a result of environmental changes, by binding a ligand or by having a particular group attached covalently to them. Anything that changes the pattern of electrostatic interactions within a protein will alter the relative energy of its states. If a protein contains histidine residues, merely changing the pH will do this (Fig. 11.3). In solutions with pH greater than 7 most of the histidine residues in a protein will be uncharged (page 138). In solutions with a pH less than 7, most of the histidine residues will bear a positive charge. A protein conformation that had two histidine residues close together would therefore be stable in alkaline conditions, but in acid conditions the two residues would each bear a positive charge and repel, destabilizing the conformation.

Another mechanism that is used to change the conformation of proteins is the addition of a negatively charged phosphate group by a class of enzymes called **protein kinases**. Proteins can be phosphorylated on serine, threonine, tyrosine, aspartate, glutamate, or histidine. The calcium ATPase (Fig. 11.4) is a transmembrane protein that forms a tube that can be open to the cytosol or to the extracellular medium. In its resting state the tube is open to

IN DEPTH 11.1 WHAT TO MEASURE IN AN ENZYME ASSAY

In an enzyme assay (page 182) we measure the appearance of product as a function of time. In principle we could do this by starting identical reactions in a series of test tubes and stopping the reaction in each test tube at a different time after the start and then measuring the amount of product in each tube.

However, if either the substrate or the product has a property that can be measured while the reaction is proceeding, then the whole assay can be performed in one test tube. Many enzyme assays are done by using changes in the absorption of light when the substrate is converted to product. For example, lactate dehydrogenase catalyzes the conversion of pyruvate to lactate (page 31) and the progress of the reaction can be conveniently measured by following the absorbance of 340 nm light: NADH absorbs at this wavelength but NAD^+ does not. Other optical properties can be used. In their original work Michaelis and Menten studied an enzyme that breaks the disaccharide sucrose down into glucose and fructose. The mixture of glucose plus fructose rotated polarized light differently from sucrose, and they used this property to follow the course of the reaction. If there is no convenient optical property, then others may be available; for example, one can monitor the progress of a reaction in which the polysaccharide glycogen (page 26) is hydrolyzed by digestive enzymes by measuring the resulting fall in viscosity.

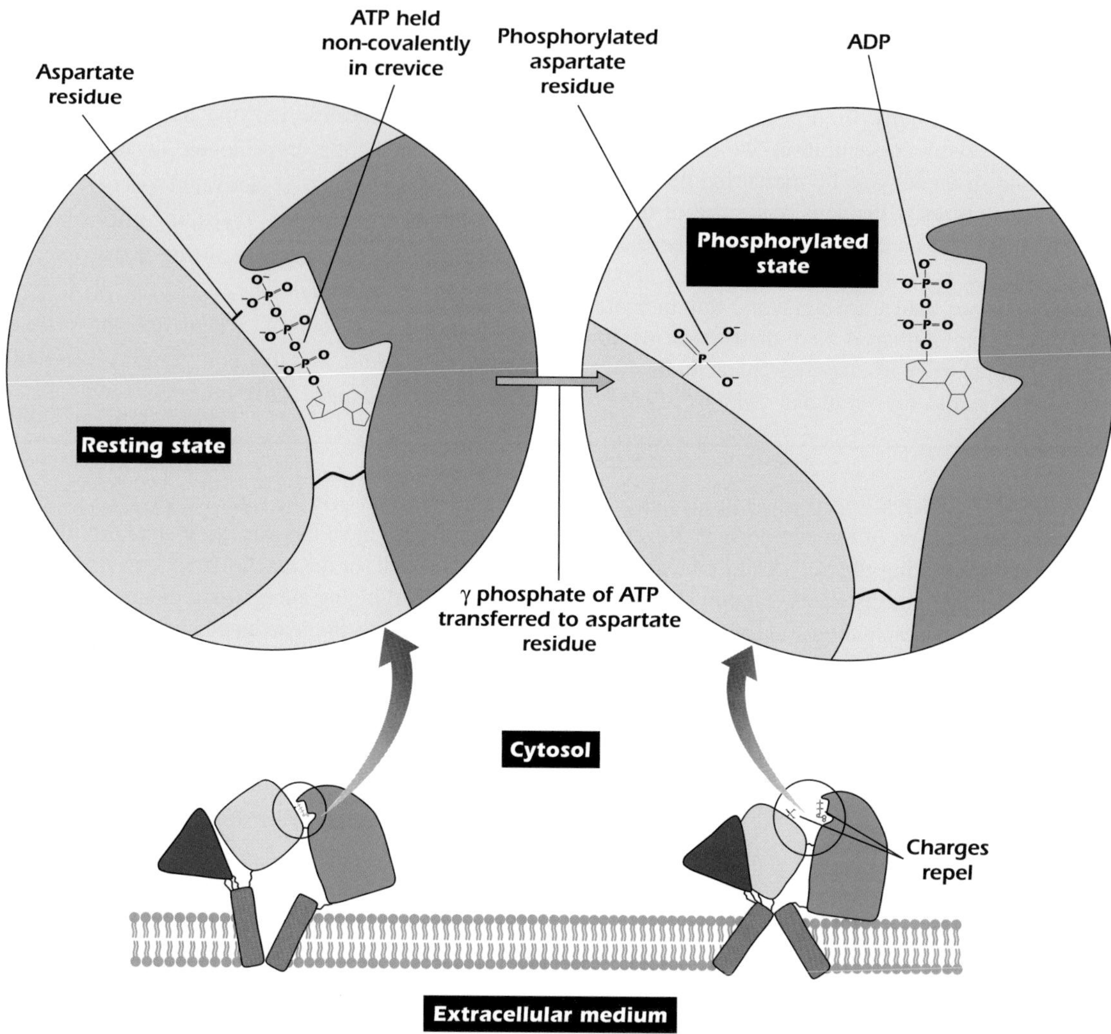

Figure 11.4. Phosphorylation changes the charge pattern, and hence the balance of forces within the calcium ATPase, forcing a change of shape.

the cytosol, while ATP is held by noncovalent interactions in a crevice in a cytosolic domain. Phosphorylation transfers the γ phosphate of ATP to a nearby aspartate residue. Repulsion between the phosphoaspartate and the remaining ADP means that the lowest energy state of the protein is now one in which the tube is open to the extracellular medium. This mechanism is used to force calcium ions out of the cytoplasm into the extracellular medium. The calcium ATPase is described in more detail on page 237.

The movements produced in a protein by phosphorylation are small, on the order of nanometers or less. When these are repeated many times and amplified by lever systems, they can produce movements of micrometers or even meters. The beating of a flagellum (page 286) or the kicking of a leg are both produced by phosphorylation-induced protein shape changes.

ENZYMES ARE PROTEIN CATALYSTS

Life depends on complex networks of chemical reactions. These are mediated by enzymes. Enzymes are catalysts of enormous power and high specificity. Consider a lump of sugar. It is combustible but quite difficult to set alight. A chemical catalyst would speed up its combustion, and we would end up with heat, a little light, carbon dioxide, and water. Swallowed and digested, the sucrose is broken down in many steps to carbon dioxide and water by the action of at least 22 different enzymes, and the energy released is used to drive other reactions in the body.

The term **substrate** (S) is used for a ligand that binds to an enzyme (E) and that is then transformed to the product

Medical Relevance 11.2 Measurement of Enzyme Concentrations in Blood

Because the initial velocity of a reaction is a linear function of the enzyme concentration, we can measure the concentration of an enzyme by measuring the rate of the reaction it catalyzes. Clinicians make use of this to measure the amount of an enzyme present in the blood. If enzymes that are normally found within cells appear in the blood in more than trace amounts, this indicates that cells are being damaged and their contents are leaking out. The presence of enzymes characteristic of a particular organ tells the clinician which tissue is affected. The digestive enzyme trypsin is stored in an inactive form called trypsinogen in the cells of the pancreas. Trypsinogen is activated once it is secreted into the lumen of the gut. Trypsinogen can be detected in the blood by converting it to the active enzyme and measuring its activity. The presence of trypsinogen in the blood of a fetus is a sign that the cells of the pancreas are dying, and this is in turn an early indicator of cystic fibrosis (page 330).

(P): $E + S \rightleftharpoons E + P$. A catalyst speeds the rate of reaction but does not alter the position of the equilibrium. Reaction speed is increased by lowering the **activation energy** barrier between the reactants and products. Figure 11.5 illustrates this effect. The top graph shows the total energy of the system as the reaction proceeds, the horizontal axis representing the progress of the reaction so that at the left we have the reactants, on the right we have the products, and in the middle we have the transition state that the system has to pass through for the reaction to occur. Because the transition state is of higher energy than the reactants, the reaction will only occur if the reactants have additional energy (e.g., as thermal motion) equal to the activation energy. The bottom graph shows the same reaction in the presence of a catalyst. Although the energy of the reactants and products has not changed, the presence of the catalyst has reduced the energy of the transition state and has therefore reduced the activation energy. A greater proportion of all the reactant molecules in solution will possess this smaller amount of extra energy, so more can undergo the reaction: the catalyst has speeded up the rate at which the reaction occurs. Each enzyme has evolved to reduce the activation energy of a particular reaction. Clearly the binding of the substrate is as central to catalysis as it is to specificity. The **active site** contains precisely positioned amino acid side chains that promote the reaction. Daniel Koshland proposed that the enzyme's active site was shaped not so much to fit the substrate but to fit a molecule that was halfway between substrate and product, and therefore to reduce the activation energy of the reaction by stabilizing, and hence reducing the energy of, this transition state. He proposed that the enzyme might actually change shape when it bound the substrate to promote catalysis. This is called induced fit and many enzymes have been shown to do this.

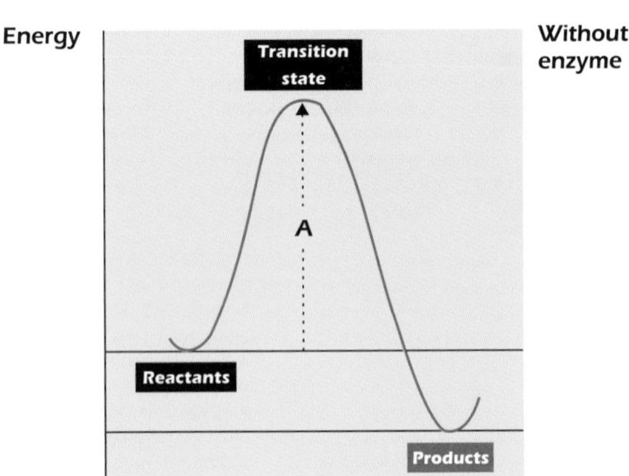

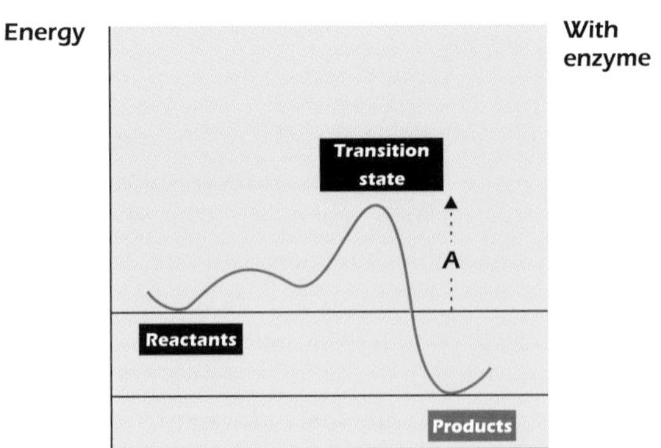

Figure 11.5. Catalysts act by reducing the activation energy (A) of a reaction.

At a basic level, a reaction carried out by an enzyme can be expressed as

$$E + S \rightleftharpoons ES \rightleftharpoons EP \rightleftharpoons E + P$$

where ES and EP represent the enzyme with bound substrate and product respectively. Binding of the substrate produces the enzyme-substrate complex ES; the catalytic function of the protein then converts the substrate to product, still bound to the enzyme in the complex EP. Finally the product dissociates from the enzyme. Binding is specific, often highly so. The enzyme β-galactosidase (page 85) is moderately specific and will split not only lactose

IN DEPTH 11.2 RAPID REACTION TECHNIQUES

Enzyme measurements are usually carried out with low concentrations of the enzyme because higher concentrations would give initial velocities too fast to measure. Initial velocity measurements are always made long after the enzyme-substrate complex has formed. It would be very interesting to be able to observe the actual formation of enzyme-substrate complexes and indeed, more generally, to observe the formation of protein-ligand complexes. These reactions are rapid, occurring on a millisecond time scale or less.

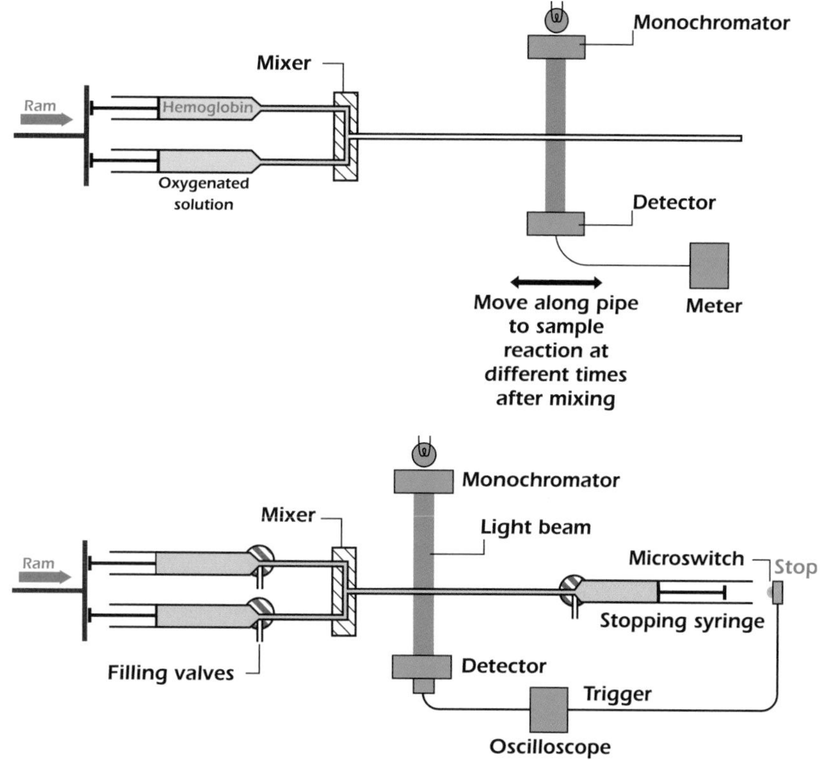

Hamilton Hartridge and Francis Roughton devised the method illustrated in the upper diagram for measuring the rate at which hemoglobin combined with oxygen. A ram pushes two syringes, one containing hemoglobin and one containing oxygenated solution. The solutions mix and pass down a tube, so that the further one looks to the right in the tube, the more time the oxygen has had to bind to the hemoglobin. The association of oxygen with hemoglobin can be monitored by an optical change, the same color change that causes arterial blood to be bright red while venous blood is a dark bluish red. Simply moving an optical detector from left to right along the tube showed the concentration of oxygen-hemoglobin complex after different reaction times. However, all the time the measurements are being made, the hemoglobin solution is running out of the pipe at the right-hand end, so that one can only use this approach when large quantities of the protein are available.

The lower figure shows a less wasteful method called stopped flow. Here two solutions are mixed and passed into an observation chamber and from it into a syringe. The plunger of this "stopping syringe" moves back until it hits a stopping plate and the flow stops. Observations are made on the solution as it ages in the observation chamber, using a high-speed recording device that is triggered by the stopping syringe as it hits the plate. Stopped flow allows observations of reactions down to about 0.1 ms after mixing. For example, the rate at which calmodulin (page 149) binds calcium can be measured by filling one syringe with calmodulin, the other with a calcium solution. In this case the reaction can be followed because one tyrosine residue on calmodulin (Tyr138) increases its fluorescence when calmodulin binds calcium.

but also any other disaccharide that has a glycosidic bond to β-galactose. By contrast each of the aminoacyl tRNA synthases (page 124) acts with absolute specificity on its specific substrates, one and only one type of amino acid and one and only one tRNA. In general, the specificity of an enzyme is conferred by the shape of the active site and by particular amino acid side chains that interact with the substrate. Chemical engineers use catalysts made of many materials, and within cells there are catalysts called ribozymes that are made of RNA (In Depth 8.2 on page 130). However, only proteins, with their enormous repertoire of different shapes, can produce catalysts for each of the enormous range of reactions occurring in cells.

The **catalytic rate constant k_{cat}** (also known as the **turnover number**) of an enzyme gives us an idea of the enormous catalytic power of most enzymes. It is defined as the number of molecules of substrate converted to product per molecule of enzyme per unit time (equally it is moles of substrate converted per mole of enzyme per unit time). Many enzymes have k_{cat} values around 1000 to 10,000 per second. The reciprocal of k_{cat} is the time taken for a single event. Thus if k_{cat} is 10,000 s^{-1}, one substrate molecule will be converted every tenth of a millisecond. Some enzymes achieve much higher rates. Catalase, an enzyme found in peroxisomes, has a k_{cat} of 4×10^7 s^{-1}, and so it takes only 25 ns to split a molecule of hydrogen peroxide into oxygen and water.

The Initial Velocity of an Enzyme Reaction

The basic experiment in the study of enzyme behavior is measurement of the appearance of product as a function of time (Fig. 11.6). This is often called an enzyme assay. The product appears most rapidly at the very beginning of the reaction. As the reaction progresses, the rate at which the product appears slows down and eventually becomes zero when the system has reached equilibrium. All reactions are in principle reversible; thus the observed overall rate of the reaction is actually the difference between the rate at which the product is being formed and the rate at which the product is being broken down again in the reverse reaction. In many enzyme reactions the equilibrium lies strongly toward the product.

We can simplify the analysis of enzyme reactions if we consider only the start of the reaction, when there is no product present and so the back reaction can be neglected. The rate of reaction at time zero (the **initial velocity v_0**, sometimes called the initial rate) is found by plotting a graph of product concentration as a function of time and measuring the slope at time zero (Fig. 11.6). In practice the slope is measured over the first 5% of the total reaction. Initial velocity, v_0, is conveniently expressed as the rate at which the concentration of the product increases, that is, moles per liter per second.

Enzymes must be assayed under controlled conditions because temperature, pH, and other factors alter their activity. Most are such highly effective catalysts that they must be assayed under conditions where the concentration of the enzyme is always very much less than the concentration of the substrate. Otherwise the reaction would be over in a fraction of a second.

Enzymes are studied for many reasons. An understanding of how they achieve their catalytic excellence and specificity is of fundamental interest and has many practical applications as we increasingly use them in industrial processes and even seek to design enzymes for particular tasks (page 114). Measurements of enzyme concentrations and properties allow us to study processes within cells and organisms. The starting point of any study of an enzyme is to determine its activity, measured as k_{cat}, and how tightly

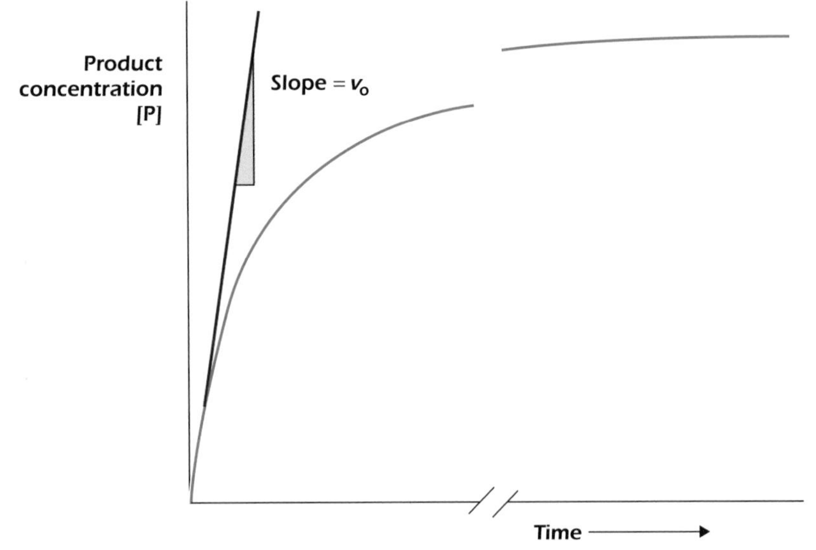

Figure 11.6. Definition of v_0, the initial velocity of a chemical reaction.

it can bind its substrate—its substrate affinity. Substrate affinity is shown by another constant called the Michaelis constant K_M. We will now explain how these constants are measured.

Effect of Substrate Concentration on Initial Velocity

Let us consider a series of experiments designed to see how the initial velocity of an enzyme reaction varies with the concentration of substrate (which is always much greater than that of the enzyme). Each of the smaller graphs in Figure 11.7 shows the result of one of these experiments. As we increase the substrate concentration, we find that at first the velocity increases with each increase in substrate concentration but that, as the substrate concentration becomes larger, the increases in rate produced get smaller and smaller. We have already met the name of a curve of this type: it is a hyperbola. The initial velocity approaches a maximum value that is never exceeded. Reactions that show this sort of dependence on substrate concentration are said to show **saturation kinetics**.

How can we explain this? The reaction sequence can be simplified to

$$E + S \rightleftarrows ES \rightarrow E + P$$

(we are using initial velocities so we can ignore any back reaction). The enzyme and substrate must collide in solution, and the substrate must bind at the enzyme's active site to form the ES complex. The chemical reaction then takes place within the ES complex, and finally the product is released. In experiments with higher and higher substrate concentrations, there is ever more ES present, and this increasing ES gives an increasing rate of product release. At very high substrate concentrations virtually all of the enzyme is present as ES, and the observed rate is limited by the $ES \rightarrow E + P$ step. Thus, as the substrate concentration increases, the reaction rate levels off as it approaches a **maximal velocity**, called $\boldsymbol{V_m}$ or $\boldsymbol{V_{max}}$. We can define V_m as the limiting initial velocity obtained as the substrate concentration approaches infinity. It is the product of the catalytic rate constant k_{cat} and the amount of enzyme present, that is, $V_m = k_{cat}[E_{total}]$. Having defined V_m, we now define the **Michaelis constant, $\boldsymbol{K_M}$** as that substrate concentration that gives an initial velocity equal to half V_m.

The plot of v_0 against [S] gives a hyperbolic curve that is described by the equation

$$v_o = \frac{V_m[S]}{K_M + [S]}$$

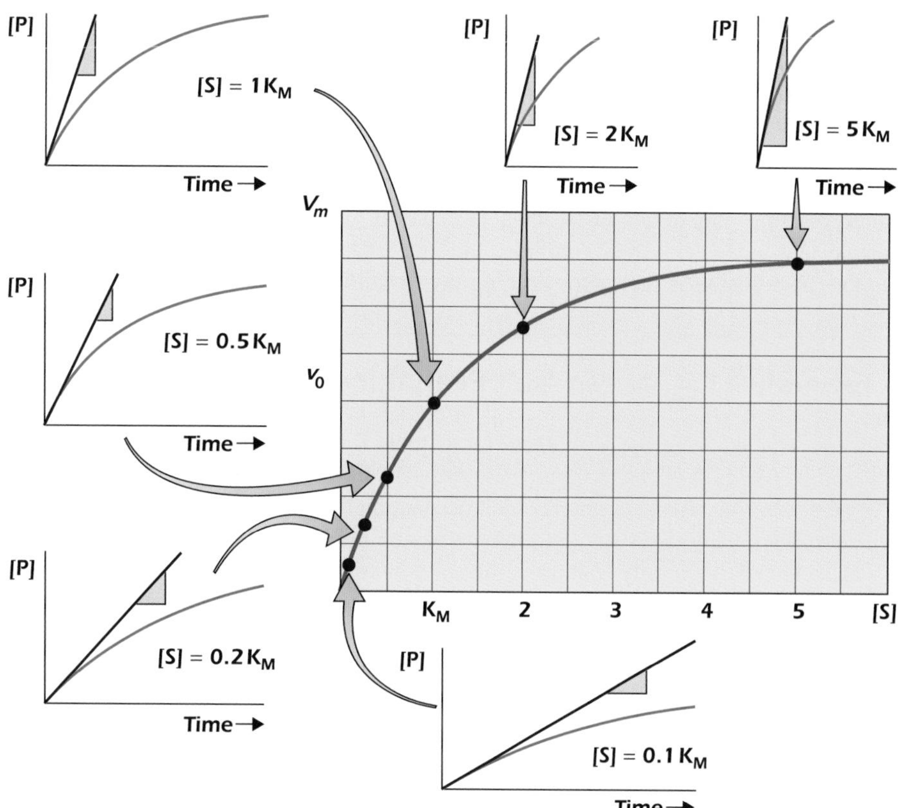

Figure 11.7. v_0 measured in a number of reaction tubes (with [E] constant and always less than [S]) forms a hyperbolic curve when plotted as a function of substrate concentration.

where [S] is the substrate concentration. This is the **Michaelis-Menten equation**, named after Maud Menten and Leonor Michaelis who propounded a general theory of enzyme action in 1913. The Michaelis constant K_M is the substrate concentration that gives an initial velocity numerically equal to half of V_m. Another way of looking at this is to say that if the enzyme is saturated with substrate, then the rate of reaction will be V_m, while at a substrate concentration giving an initial velocity of $\frac{1}{2}V_m$ the enzyme is half saturated with substrate. A small value of K_M means that the enzyme has a high affinity for the substrate.

The Effect of Enzyme Concentration

As the concentration of an enzyme increases, so does the initial velocity, and this relationship is linear: if you double the enzyme concentration, you double the initial velocity. This fact means that we can measure how much of an enzyme is present by measuring the rate of the reaction it catalyzes.

The Specificity Constant

The Michaelis constant K_M reflects the affinity of an enzyme for its substrate; k_{cat} reflects the catalytic ability of an enzyme. The ratio of these, k_{cat}/K_M, is the **specificity constant**, which is a measure of how good the enzyme is at its job. A high specificity constant means that a reaction goes fast (k_{cat} is big) and the enzyme does not need a high concentration of substrate (K_M is small). When an enzyme has relatively low specificity, that is, it can work on a number of different substrates, the substrate that has the largest specificity constant is the preferred substrate for the enzyme.

A reaction cannot go faster than the rate at which enzyme and substrate actually collide. In some enzymes this rate of collision is the factor limiting the overall rate, and such enzymes are said to be diffusion limited and are considered to have reached perfection of biological design. Diffusion-limited values of k_{cat}/K_M can be calculated to be in the range 10^8–10^{10} liter per mole per second. Table 11.1 shows k_{cat}, K_M, and k_{cat}/K_M for a number of enzymes.

COFACTORS AND PROSTHETIC GROUPS

Enzymes that need to perform reactions that are outside the repertoire of the 20 amino acid side chains recruit other chemical species to help them do the job. The aminotransferases provide a good example. These enzymes are central to amino acid metabolism (page 215) and catalyze the interconversion of amino acids and oxo-acids by moving an amino group (Fig. 11.8). Pyridoxal phosphate (derived from vitamin B_6) is bound by the protein. The amino acid donor passes its amino group to the pyridoxal phosphate and leaves as an oxo-acid. The oxo-acid substrate then binds, accepts the amino group from the pyridoxal phosphate, and leaves having been converted to an amino acid. Pyridoxal phosphate ends up exactly as it started and remains bound to the enzyme ready for another cycle. These helper chemicals are called **cofactors** when, like pyridoxal phosphate, they are not tightly bound to the enzyme. We already know the term for a molecule that is very tightly bound to a protein and helps it perform its job—it is a prosthetic group (page 150). These terms overlap and usage varies.

The prosthetic group of hemoglobin and myoglobin, heme, (Fig. 9.15 on page 152) binds oxygen. It is a versatile molecule that is also found in electron transfer proteins colored red or brown and therefore called **cytochromes** from the Latin for cell and color. The iron in a reduced cytochrome is Fe^{2+} while in an oxidized cytochrome it is Fe^{3+}. Cytochromes found in the inner mitochondrial membrane play important roles in the electron transport chain (page 196).

Example 11.1 Speed Isn't Everything

In this chapter we emphasize enzymes as tools for performing reactions that are required in the cell. For this type of enzyme, the faster the reaction occurs, the better, although in many cases an "optimized" enzyme must be controlled so that it only operates at maximum rate when lots of product is needed.

For other enzymes, slowness is a virtue. We will meet trimeric G proteins in Chapter 15. Like the GTPases that we have already met (EF-tu, Ran, and Rab), these are active when they have GTP bound, and switch to an inactive state when they hydrolyze the GTP to GDP. While they are active, they turn on their target processes. In these enzymes, therefore, a slow rate of reaction allows them to act as timer switches. They are activated when GDP is ejected and GTP bound. They then remain active until they hydrolyze the GTP. For the typical trimeric G protein listed in Table 11.1, k_{cat} is 0.02 per second, so the "timer" is set to 50 s (the reciprocal of 0.02 s^{-1}).

IN DEPTH 11.3 DETERMINATION OF V_m AND K_M

The initial rate of an enzyme reaction, v_0 (at constant enzyme concentration) depends on V_m and K_M as well as on the substrate concentration, [S], according to the Michaelis-Menten equation.

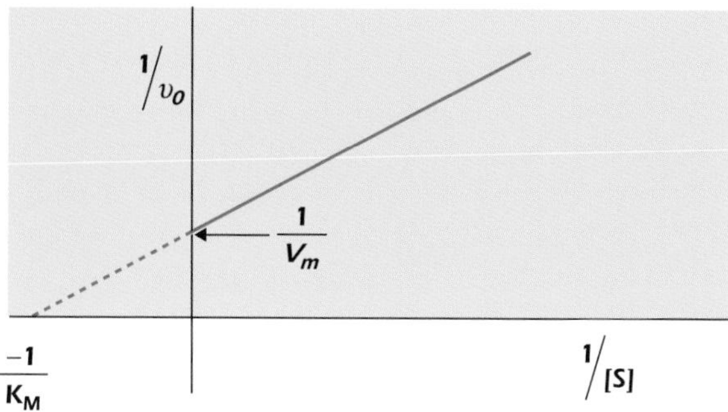

$$v_o = \frac{V_m[S]}{K_M + [S]}$$

Thus the information necessary to determine V_m and K_M is contained within the data such as those shown in Figure 11.7. It is not immediately obvious how one would go about this: the curve tends slowly towards V_m as [S] increases so that guessing V_m by eye will give a very poor value. Nowadays there are many computer programs available that will fit the Michaelis-Menten equation (an equation of a hyperbola) to the raw data and calculate the constants. Before this, biologists used mathematical transformations of the Michaelis-Menten equation that yielded straight-line relationships. The best known is called the Lineweaver-Burke plot after its inventors. It simply inverts the equation to get

$$\frac{1}{v_o} = \frac{K_M + S}{V_m[S]} = \frac{K_M}{V_m[S]} + \frac{[S]}{V_m[S]}$$

then rearranges to

$$\frac{1}{v_o} = \frac{K_M}{V_m} \times \frac{1}{[S]} + \frac{1}{V_m}$$

This is the equation of a straight line ($y = mx + c$) where y is $1/v_0$ and x is 1/[S]. One simply plots $1/v_0$ on the y axis and 1/[S] on the x axis to get a straight line where the y intercept is $1/V_m$, the x intercept is $-1/K_M$ and the slope is K_M/V_m.

IN DEPTH 11.4 CYTOCHROMES P450 HAVE MANY ROLES

The cytochromes P450 that we have met earlier (pages 95, 113, 127), are found in the membranes of the endoplasmic reticulum and mitochondria. There are at least 50 different cytochrome P450s in humans. Some of these proteins carry out very specific metabolic reactions in the process of converting cholesterol into the different steroid hormones our body utilizes. Other cytochromes P450 are less specific in their reactions. Their role is to detoxify the thousands of foreign chemicals to which we are exposed on a daily basis. However, all cytochromes P450 share one property: they all insert a single atom of oxygen, derived from molecular oxygen, into a substrate. Subsequent rearrangements often result in the oxygen being in the form of a hydroxyl group in the final reaction product. This transformation converts a hydrophobic molecule into one that is more hydrophilic. After hydroxylation, other molecules are coupled to the hydroxyl group to make the foreign chemical even more hydrophilic and hence allow it to be rapidly excreted in the urine and feces. Unfortunately, in very rare cases the addition of an oxygen atom into a molecule can increase its reactivity and, instead of being rendered harmless by the action of a cytochrome P450, it is transformed into a very dangerous chemical that can damage DNA. If the damage causes mutation or strand breakage, then cancer can result. For example polycyclic aromatic hydrocarbons, found in tobacco smoke and vehicle exhaust gases, and the chemical aflatoxin, produced by a mold that grows on peanuts, are converted into chemical carcinogens by the action of certain cytochromes P450.

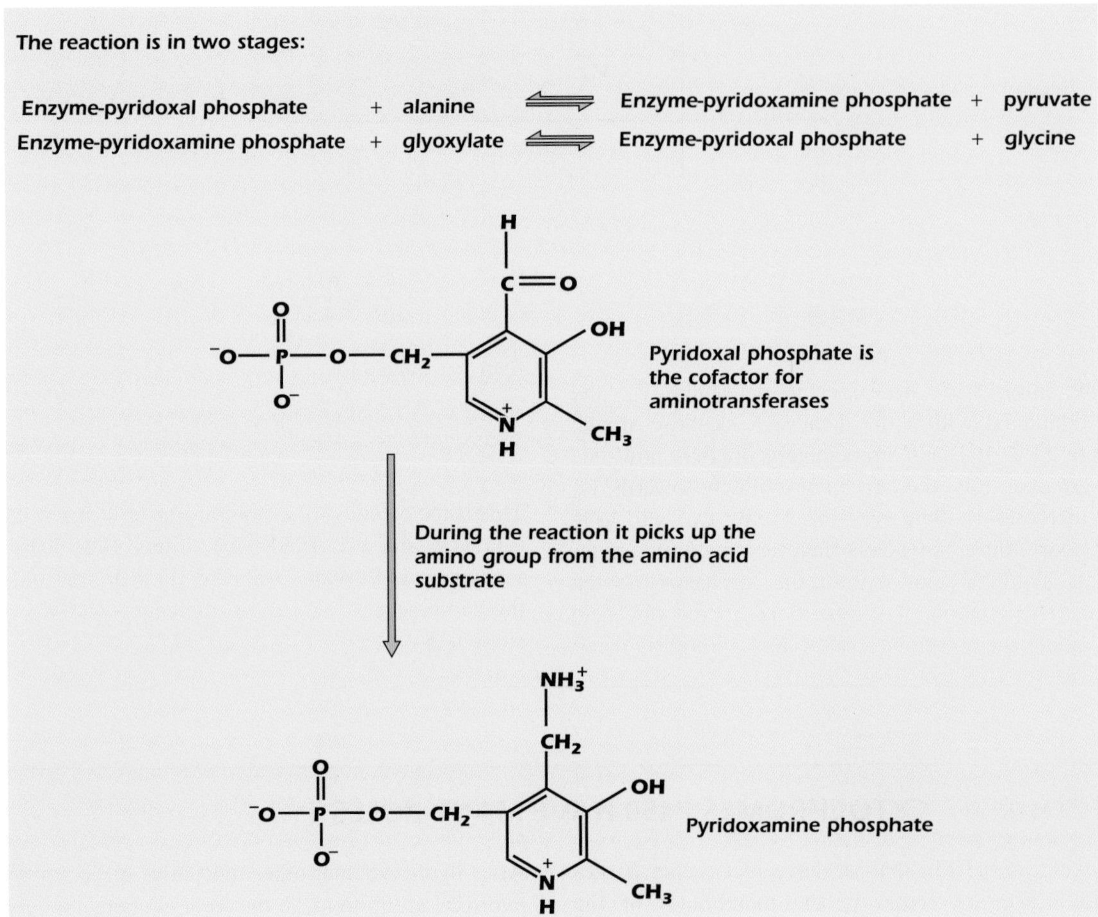

Figure 11.8. Aminotransferases use a cofactor that participates in the reaction but ends up unchanged.

ENZYMES CAN BE REGULATED

In this chapter we have discussed the catalytic role of enzymes in isolation. In fact enzymes are like other proteins in showing the complex behavior described in Chapter 9 such as multiple states, multiple binding sites, quaternary structure, and phosphorylation. For instance, some important enzymes have quaternary structures and show cooperativity between the active sites. Such enzymes will not follow the Michaelis-Menten equation, but instead the curve of initial velocity against substrate concentration will have the same S-shaped or sigmoid curve seen in the binding of oxygen to hemoglobin (Fig. 11.2).

A good example of an allosteric enzyme is to be found in the processes that allow us to use sugars as metabolic fuels. The pathway called **glycolysis** (page 209) converts

TABLE 11.1. Values of k_{cat}/K_M for a Number of Enzymes

Enzyme	k_{cat} (s^{-1})	K_M (mol liter^{-1})	k_{cat}/K_M (liter mol^{-1}s^{-1})
Trimeric G protein (page 253)	0.02	6×10^{-7}	3.3×10^4
Lysozyme (page 144)	0.5	6×10^{-6}	8.3×10^3
Aminoacyl tRNA synthase (page 124)	7.6	9×10^{-4}	8.4×10^3
Intestinal ribonuclease (page 143)	7.9×10^2	7.9×10^{-3}	1×10^5
Catalase (page 49)	4×10^7	1.1	4×10^7

Medical Relevance 11.3 Drugs and Enzymes

Many drugs act by inhibiting specific enzymes. Penicillin, for instance, works by inhibiting one of the enzymes involved in making bacterial cell walls. The target enzyme cross links the chains of sugars and peptides. If the enzyme is inactive the cells burst osmotically in their normal medium.

Enzyme inhibitors fall into two general categories: those that work by forming a covalent bond with an essential group in the enzyme, and those that bind reversibly.

The effect of an irreversible inhibitor is simply to destroy the enzyme's activity. Only synthesis of new enzyme molecules can remove the inhibition. Any unaffected molecules show normal k_{cat} and K_M.

Reversible inhibitors have more short-lived effects as the inhibitor can dissociate and be removed from the body. An inhibitor that binds reversibly at the active site so the substrate cannot bind is said to be competitive. Such inhibition becomes less effective as the substrate concentration rises. V_{max} does not change as high enough concentrations of the substrate will still saturate the enzyme but it takes a higher substrate concentration to half saturate the enzyme so K_M increases (think about this). Other types of reversible inhibitor bind elsewhere on the enzyme and may alter V_{max} or K_M or both.

Prostaglandins are a family of paracrine transmitters (page 271) that are derived from arachidonic acid (found in phospholipids—page 34). Prostaglandins have a wide range of effects including acting to stimulate inflammation, control ion transport across membranes and alter synaptic transmission—many of these effects relate to pain. Prostaglandin synthase carries out a two-step reaction to oxidize and cyclize arachidonic acid to prostaglandin H_2. This membrane-bound enzyme is irreversibly inhibited by aspirin (acetylsalicylic acid) which transfers an acetyl group to a serine in the enzyme, blocking substrate access.

glucose and fructose to pyruvate with a net production of two ATP molecules. The pyruvate is then passed to the mitochondria, which use it to produce more ATP.

The first committed step in this pathway is the addition of a second phosphate group to monophosphorylated fructose. The γ phosphate of ATP is transferred to the number 1 carbon of fructose-6-phosphate. Before this step, sugars can be converted to glycogen or metabolized in other ways, but after this step the sugar is committed to be broken down to pyruvate. The step is catalyzed by the enzyme **phosphofructokinase** (Fig. 11.9). ATP serves as an allosteric inhibitor of this enzyme so that it operates slowly when

Answer to Thought Question: Enzymes are proteins that bind their substrate(s) in such a way that the chemical reaction is catalyzed. pH affects the charge on amino acid side chains on the protein (such as aspartic and glutamic acids, histidine, lysine and arginine). If a side chain is required for binding and/or catalysis with a particular charge, then a change to a pH where this group is uncharged will reduce binding and/or prevent catalysis. The binding site is generated by the tertiary structure of the protein and this too depends on side chains having the correct charges. Change in pH may alter the structure so that the binding site no longer binds the substrate. Extremes of pH will destroy all of the tertiary structure and denature the protein.

As one increases the temperature one increases the rate of any chemical reaction. Enzyme reactions show this effect, but, as temperature increases, the rate of the reaction slows down again. The higher the temperature the more thermal motion the protein has. As the temperature increases the tertiary structure opens up and the polypeptide unfolds. Finally the protein is denatured. Substrate binding and catalysis require precise interactions with the protein. Even small changes will interfere with these. Enzymes have evolved to be stable at the temperatures the organism normally experiences.

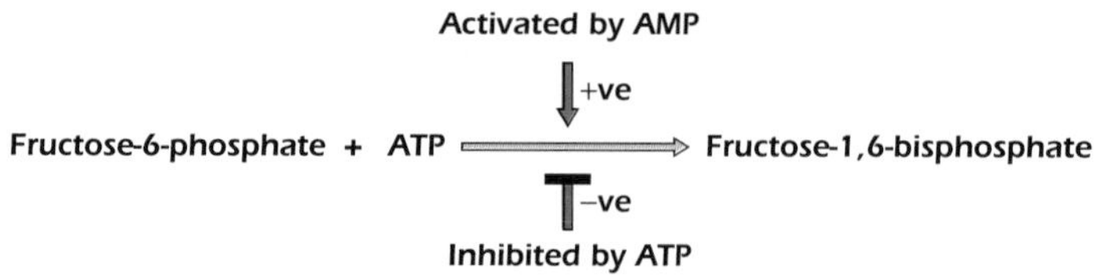

Activated by AMP

+ve

Fructose-6-phosphate + ATP ⟶ Fructose-1,6-bisphosphate

−ve

Inhibited by ATP

Figure 11.9. Phosphofructokinase is regulated by the binding of ATP or AMP at a regulatory site that is separate from the active site. Binding of ATP inhibits while binding of AMP activates.

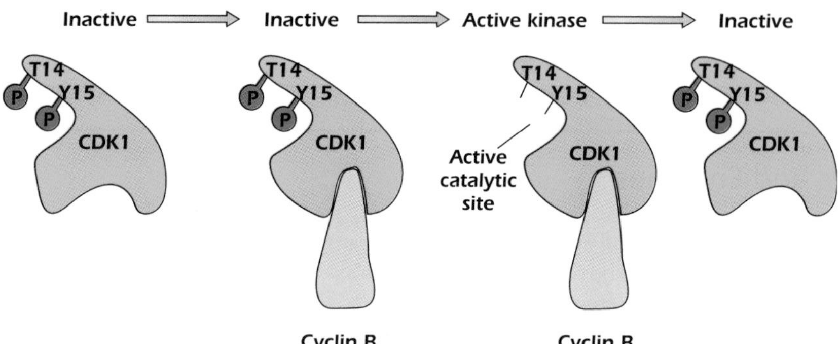

Figure 11.10. Control of Cdk1 by cyclin B and by phosphorylation.

ATP concentrations are high and there is no need for more ATP to be produced. When there is a drain on ATP, levels of AMP increase (page 192). AMP competes with ATP for the regulatory site and causes the enzyme to switch to the active, high-affinity conformation. Thus as the ATP concentration is reduced and AMP levels increase, the pathway is more and more activated until it is turned on to full speed. We very often see this type of control system, where the product of a pathway or process feeds back to allosterically inhibit an enzyme at the start of that pathway. They are an example of negative feedback (page 223).

It is not only enzymes involved in metabolism that are regulated by ligand binding. **Cdkl**, the enzyme that triggers mitosis (page 304), is an allosterically regulated enzyme. On its own, in solution, it spends the vast majority of time in an inactive conformation. In order to assume the active conformation, Cdkl must bind a protein ligand called cyclin B (Fig. 11.10). In addition to this allosteric regulation, Cdkl is also regulated by phosphorylation: for most of the time the catalytic site is blocked by two phosphate groups within the catalytic domain, at threonine 14 and tyrosine 15. In order to become an active enzyme, Cdkl must lose these two phosphates while cyclin B is bound. This complex control means that Cdkl can act as a checkpoint. If the phosphate groups have been removed from T14 and Y15, AND cyclin B is present at a high enough concentration, THEN it is safe to proceed into mitosis.

Cdkl is an example of an enzyme that is turned off by phosphorylation. Other enzymes are turned on by phosphorylation. Examples include RNA polymerase II (page 93) and glycogen phosphorylase (page 225).

SUMMARY

1. Most protein functions arise from their ability to bind to other molecules in a reversible yet specific fashion. Shape changes, triggered by the binding of other molecules, mediate protein movements and function. Allostery is a special case where a shape change induced by a ligand at one site changes the affinity at another site.
2. Enzymes are highly specific biological catalysts. The turnover number or catalytic rate constant, k_{cat}, is the maximum number of substrate molecules that can be converted to product per molecule of enzyme per unit time.
3. The initial velocity (i.e., the rate at the start when product is absent) of many enzymes shows a hyperbolic dependence on substrate concentration. At high substrate concentrations the initial velocity approaches a limiting value V_m, as the enzyme is saturated with substrate. The substrate concentration that gives an initial velocity equal to half V_m, is the Michaelis constant K_M. This indicates the enzyme's affinity for the substrate.
4. The initial velocity is related to the substrate concentration by the Michaelis-Menten equation:

$$v_o = \frac{V_m[S]}{K_M + [S]}$$

5. The initial velocity is directly proportional to the enzyme concentration.

6. Some enzymes use cofactors to carry out reactions that require different properties from those of the side chains of the 20 amino acids found in proteins.
7. Enzyme activity within cells is modulated by a variety of methods that include phosphorylation and allosteric effects.

FURTHER READING

Cornish-Bowden, A. (2004) *Fundamentals of Enzyme Kinetics*, 3rd edition, Portland Press, London.
Fersht, A. (1999) *Structure and Mechanism in Protein Science*. W. H. Freeman, New York.
Voet, D., and Voet, J. D. (2011) *Biochemistry*, 4th edition, John Wiley & Sons, Hoboken.

REVIEW QUESTIONS

11.1 Theme: Changing the Preferred Shape of a Protein

This question relates to a theoretical protein that can exist in two forms, open and closed, as in the diagram below.

```
T. . . .S. . . .E. . . . . . . . . . . . .H. . . .K. . . .T
                    open form

. . . . . .E. . . .S. . . .T
.
. . . . . .H. . . .K. . . .T
        closed form
```

Adoption of the closed form brings a glutamate, serine and threonine close to, respectively, a histidine, lysine and second threonine as shown.

A The change favors the open configuration
B The change favors the closed configuration
C The change has no effect on the equilibrium between the two configurations

For each of the perturbations described below, choose the appropriate effect on the protein from the list above.

1. A fall of pH from 7.5 to 6.5
2. Phosphorylation of the serine
3. Phosphorylation of both threonines

11.2 Theme: Enzyme Kinetics

A Equilibrium constant for an enzyme reaction

B That substrate concentration that gives an initial velocity equal to half the maximum velocity (V_m)

C The maximum initial velocity possible when an enzyme is fully saturated with its substrate

D The Michaelis-Menten equation that shows the relationship between initial velocity and substrate concentration in an enzyme reaction

E The number of moles of product formed per mole of enzyme per unit time

F The rate of an enzyme reaction when it is half-saturated with its substrate

G The rate of product formation at the start of an enzyme reaction

H The time taken for a single catalytic event

Match the definition, given above, with the term it defines from the list below.

1. k_{cat} (catalytic rate constant)
2. K_M (Michaelis constant)
3. v_0 (initial velocity)
4. V_m (maximum velocity)

11.3 Theme: Enzymes

A A hyperbolic curve obtained when v_0 is plotted against substrate concentration shows...

B A sigmoid curve obtained when v_0 is plotted against substrate concentration shows...

C A substrate with K_M of 5×10^{-3} mol L^{-1} binds to the enzyme more...

D A substrate with K_M of 5×10^{-5} mol L^{-1} binds to the enzyme more...

E If you double the amount of an enzyme (keeping other conditions constant) you will...

F The catalytic rate constant...

G When an enzyme can work on two substrates the best substrate is the one that...

For each of the phrases in the list below, there is a phrase in the list above that can be combined with it to create a single true statement. Choose the appropriate sentence start above for each of the sentence terminations below.

1. ... can be determined if both V_{max} and the total enzyme concentration are known.
2. ... gives the highest specificity constant (ratio of k_{cat} over K_M).
3. ... increase V_m by twofold.
4. ... that the enzyme binds its substrate cooperatively, that it is an allosteric enzyme.
5. ... weakly than one with $K_M = 5 \times 10^{-4}$ mol L^{-1}.

THOUGHT QUESTION

Why are enzymes sensitive to pH and to temperature?

12

ENERGY TRADING WITHIN THE CELL

In the nonliving world complex things degrade naturally to simpler things: gradients of temperature or concentration disappear, chemical reactions approach equilibrium, and uniformity triumphs. Living things do not appear to follow these trends. Cells are complex and divide to make other complex cells: a fertilized egg differentiates to make a whole complex organism. Living things must obey the laws of **thermodynamics**. The escape from the behavior of nonliving systems is allowed because living systems take matter and energy from the environment and use it to grow, to reproduce, and to repair themselves.

A chemical reaction that has reached equilibrium can do no work. A good definition of death is the state at which all of the chemical reactions in a cell/organism have reached equilibrium. In a living organism the concentrations of metabolites are often very far from the equilibrium concentrations and yet are more or less constant: this is said to be a **steady state**. Cells can do this because they are open systems taking energy and matter from their environment.

We can use an analogy with the world of economics. It is unlikely that people would spontaneously repair our houses, or feed us, or give us this book, but we can drive these otherwise unlikely processes by spending money. In a similar way, cells can drive otherwise unlikely processes by using up one of four **energy currencies** that are then replaced using energy taken from the outside world.

 CELLULAR ENERGY CURRENCIES

The scientific way of saying that a process will proceed (although a catalyst may be necessary to achieve a reasonable reaction speed) is to say that the change in **Gibbs free energy**, expressed as ΔG (Δ is the Greek uppercase delta used to denote a change), is negative. One reaction that we will meet again is the hydrolysis of glucose-6-phosphate to yield glucose and a phosphate ion:

$$\text{Glucose-6-phosphate} + H_2O \rightarrow \text{Glucose} + HPO_4^{2-} \quad \Delta G = -19\ \text{kJmol}^{-1}$$

Since this reaction has a negative ΔG, it can proceed, releasing 19 kilojoules (kJ) of energy for every mole of glucose-6-phosphate hydrolyzed. Another reaction we will meet again is the hydrolysis of nucleotides, the building blocks of DNA and RNA (Chapter 4). Simply losing the terminal phosphate from the nucleotide ATP releases 30 kJmol^{-1}:

$$\text{Adenosine triphosphate} + H_2O \rightarrow \text{Adenosine diphosphate} + HPO_4^{2-}\ \Delta G = -30\ \text{kJmol}^{-1}$$

The reverse of these reactions will of course not proceed. For instance, cells need to phosphorylate glucose to make glucose-6-phosphate but cannot use the reaction

Cell Biology: A Short Course, Third Edition. Stephen R. Bolsover, Jeremy S. Hyams, Elizabeth A. Shephard and Hugh A. White.

$$\text{Glucose} + \text{HPO}_4^{2-} \rightarrow \text{Glucose-6-phosphate} + \text{H}_2\text{0} \quad \Delta G = +19 \text{ kJmol}^{-1}$$

The reaction will not proceed because it has a positive ΔG. Crucially, though, an unfavorable (positive ΔG) reaction can occur if it is tightly coupled to a second reaction that has a negative free-energy change (negative ΔG) so the overall change for the reactions put together is negative. Thus cells phosphorylate glucose by carrying out the following reaction, catalyzed by the enzyme **hexokinase**:

$$\text{Glucose} + \text{ATP} \rightarrow \text{Glucose-6-phosphate} + \text{ADP} + \text{H}^+ \quad \Delta G = -11 \text{ kJmol}^{-1}$$

Adenosine triphosphate, or ATP, has given up the energy of its hydrolysis to drive an otherwise energetically unfavorable reaction forward. We call ATP a cellular currency to draw an analogy with money in human society. Just as we can spend money to cause someone to do something they would not otherwise do, the cell can spend its energy currency to cause processes that would otherwise not occur. However, the analogy is not exact because energy currencies are not hoarded. There is a continuous turnover of ATP to ADP and back again. ATP is therefore not an energy store but simply a way of linking reactions. It can be thought of as a truck that carries metabolic energy to where it is needed and then returns empty to be refilled. The number of trucks is small but the amount moved can be large. An average person hydrolyzes about 50 kg of ATP per day but makes exactly the same amount from ADP and inorganic phosphate. We will see how this happens in this chapter. The cell has a number of energy currencies of which four—NADH, ATP, the hydrogen ion gradient across the mitochondrial membrane, and the sodium gradient across the plasma membrane—are the most important. We will now discuss each of these in turn.

Reduced Nicotinamide Adenine Dinucleotide (NADH)

This, the most energy rich of the four currencies, is shown in Figure 12.1. NADH is a strong reducing agent (page 31). It will readily react to allow two hydrogen atoms to be added to molecules, in the general reaction $\text{NADH} + \text{H}^+ + \text{X} \rightarrow \text{NAD}^+ + \text{H}_2\text{X}$. We will later see NADH acting to reduce complex cell chemicals like pyruvate and acetoacetate. However when NADH is acting as an energy currency it simply passes its two hydrogen atoms to oxygen, making water. This releases a lot of energy: every mole of NADH that is used in this way releases 206 kJ of energy.

Nucleoside Triphosphates (ATP plus GTP, CTP, TTP, and UTP)

Adenosine triphosphate, the second most energy rich of the four currencies, is shown in Figure 12.2. In earlier chapters we met many chemical processes in the cell that are driven by ATP hydrolysis. When one mole of ATP is hydrolyzed, 30 kJ of energy are released. The γ phosphate is easily transferred between nucleotides in reactions such as this one:

$$\text{ATP} + \text{GDP} \rightleftarrows \text{ADP} + \text{GTP}$$

So as far as energy is concerned, we can regard GTP, CTP, TTP, and UTP as equivalent to the most commonly used nucleotide energy currency, ATP. Another easy, reversible reaction is this one:

$$2\ \text{ADP} \rightleftarrows \text{AMP} + \text{ATP}$$

in which one ADP transfers its β phosphate to another, so that it itself is left as AMP while converting the other ADP to ATP. This conversion is at equilibrium in cells and is important in maintaining the supply of ATP when ATP is used in reactions.

NADH and ATP take part in so many reactions within the cell that they are often called **coenzymes**, meaning molecules that act as second substrates for many enzymes as they do their particular jobs. The term is easy to confuse with cofactor (page 184). A cofactor is a chemical

Medical Relevance 12.1 NAD$^+$, Pellagra, and Chronic Fatigue Syndrome

A small fraction of our total body NAD$^+$ is lost from the body each day. New NAD$^+$ is synthesized from the vitamin niacin, and can also be synthesized from the essential amino acid tryptophan. If we don't get enough niacin or tryptophan, we develop the disease called pellagra. In pellagra parts of our bodies that use a lot of energy, such as the brain, begin to fail. Pellagra killed about 100,000 Americans in the first half of the twentieth century before Joseph Goldberger of the U.S. Public Heath Service showed that it could be prevented and cured by a varied diet. Niacin is now added to all flour, and pellagra is almost unknown in Western countries.

Since chronic fatigue syndrome is manifested as a lack of energy, some people have suggested that NAD$^+$ itself, given as a dietary supplement, might reenergize the patients. There is no scientific evidence for this idea, but nevertheless many chronic fatigue syndrome sufferers buy and take NAD$^+$ in the hope that it may help.

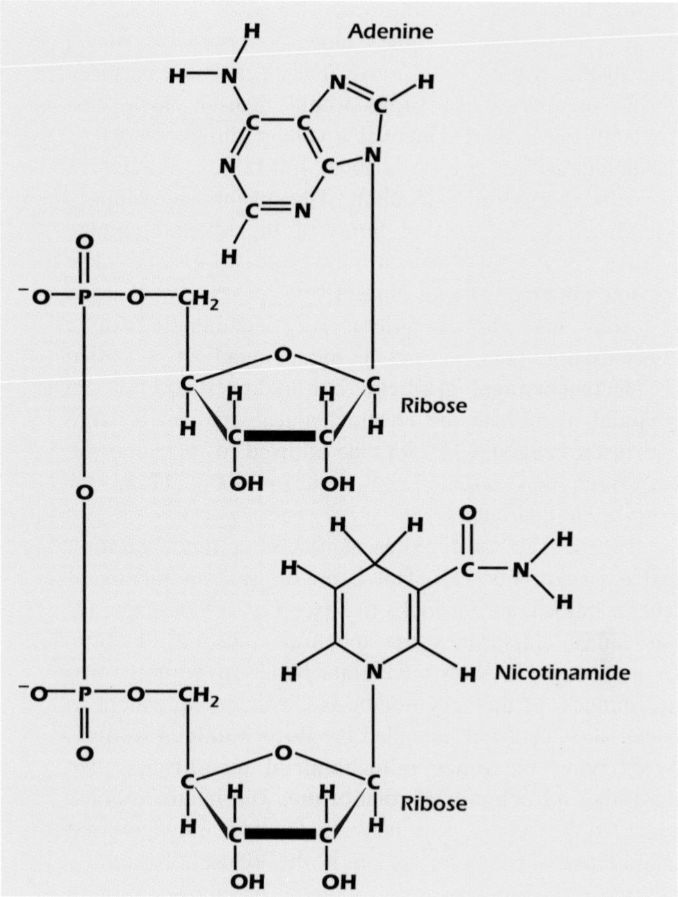

Figure 12.1. Reduced nicotinamide adenine dinucleotide (NADH) is a strong reducing agent and energy currency.

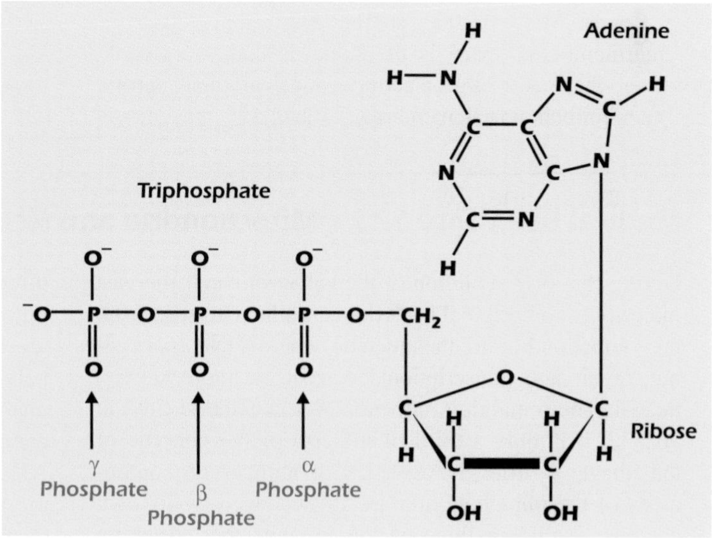

Figure 12.2. Adenosine triphosphate, an energy currency.

species that is loosely associated with an enzyme, helps it carry out its function, and, although it may undergo reactions, ends up in the same state that it began. The concept of coenzymes is very different. Coenzymes are bona fide substrates that are converted to products (e.g., NAD^+ and ADP) by the enzyme.

The Hydrogen Ion Gradient Across the Mitochondrial Membrane

The endosymbiotic theory states that mitochondria are derived from bacteria that evolved to live in eukaryotic cells (page 10). The bacterial cytosol is usually about

0.6 pH units more alkaline than the world outside; that is, H^+ ions are four times more concentrated outside than inside. If they could move freely across the bacterial plasma membrane, H^+ ions would rush in down this gradient. Furthermore, there is a voltage difference across the membrane: the inside is about 140 mV more negative than the extracellular medium. Transmembrane voltages are always referred to in terms of the internal voltage relative to that outside: in this case, −140 mV. The transmembrane voltage attracts the positively charged H^+ ions into the bacterium. Any combination of a concentration gradient and a voltage gradient is called an **electrochemical gradient**. For hydrogen ions at the bacterial membrane the electrochemical gradient is large and inward. Should H^+ ions be allowed to rush into the bacterium, they would release energy: about 17 kJ for every mole that enters.

Figure 12.3 is a representation of a mitochondrion inside a eukaryotic cell. This is not an accurate picture of what a mitochondrion looks like (see Fig. 3.6 on page 48) but rather emphasizes the topology and the function. In the center is shown the mitochondrion with its two membranes. In the very middle is a volume equivalent to bacterial cytosol that is called the **mitochondrial matrix**. Next comes the **inner mitochondrial membrane**, then the **outer mitochondrial membrane**. The intermembrane space is the small space between the two mitochondrial membranes. The green region is the cytosol, bound by the plasma membrane. An integral membrane protein of the outer mitochondrial membrane, called porin, forms a hole or channel that lets through all ions and molecules of $Mr \leq 10,000$, so the ionic composition of the intermembrane space is the same as that of cytosol. However, there is a large electrochemical gradient across the inner mitochondrial membrane. When H^+ ions move in from the **intermembrane space** to the mitochondrial matrix, they release 17 kJ mol^{-1} of energy, exactly as in the mitochondrion's proposed bacterial ancestors.

The Sodium Gradient Across the Plasma Membrane

Unlike bacteria, most eukaryotic cells do not have an H^+ electrochemical gradient across their plasma membranes. Rather, it is sodium ions that are more concentrated outside the cell than inside (Fig. 12.3). Typically, the sodium concentration in the cytosol is about 10 mmol $liter^{-1}$ while the concentration in the extracellular medium is about 150 mmol $liter^{-1}$. This chemical gradient is supplemented by a voltage gradient. The cytosol is between 70 and 90 mV more negative than the extracellular medium, that is, the transmembrane voltage of the plasma membrane is between −70 and −90 mV. There is therefore a large inward electrochemical gradient for sodium ions. If sodium ions are allowed to rush down this gradient, they release energy: approximately 15 kJ for every mole of Na^+ entering the cytosol.

ENERGY CURRENCIES ARE INTERCONVERTIBLE

A company that buys raw materials in the United States and Mexico, spending dollars and pesos, and then sells products in Europe and Japan, receiving euros and yen, simply converts from euros and yen to dollars and pesos to pay its bills. In the same way, cells convert from the energy currency in which they are in credit to the energy currency they are using up.

Medical Relevance 12.2 Mitochondria and Neurodegenerative Diseases

During the long evolution of the eukaryotic cell the vast majority of the original bacterial genes have moved from the mitochondria to the nucleus where DNA packaging, repair and transcriptional control are more sophisticated. The remaining mitochondrial genome is tiny: at 16.5 kb it is only a third of the size of the genome of the phage λ virus. Nevertheless a number of components of the mitochondrion are still encoded by its own genome, including three of the subunits that make up complex IV of the electron transport chain (described below).

Mature nerve cells do not divide. Nevertheless because they are among the most energetically active in the body they need to make new mitochondria throughout their life. This requires replication of the mitochondrial DNA, during which mutations can build up. Evidence is growing that as normal aging proceeds nerve cells contain more and more mitochondria that are unable to synthesise critical components, particularly complex IV, and which therefore are unable to generate ATP. The remaining good mitochondria may be enough to support most functions. However, the cells cannot increase their metabolism sufficiently at times of high demand, leading to slow and error-prone data processing—that is, thought—and finally to death of the nerve cell by apoptosis (page 308). Some workers believe this process underlies diseases such as Parkinson's and Alzheimer's.

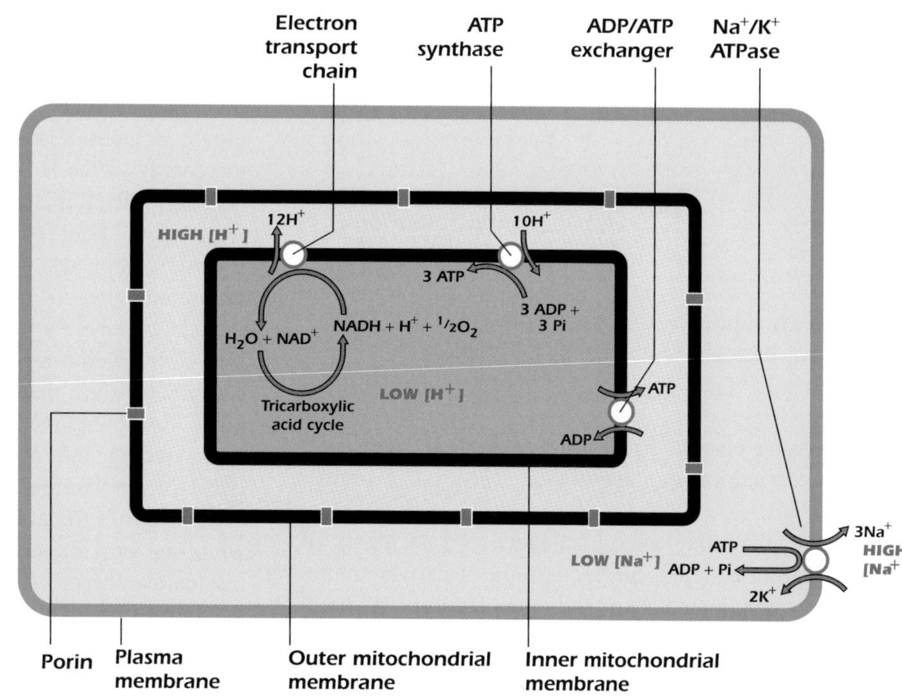

Figure 12.3. Diagram of a cell showing sites where the energy currencies are interconverted.

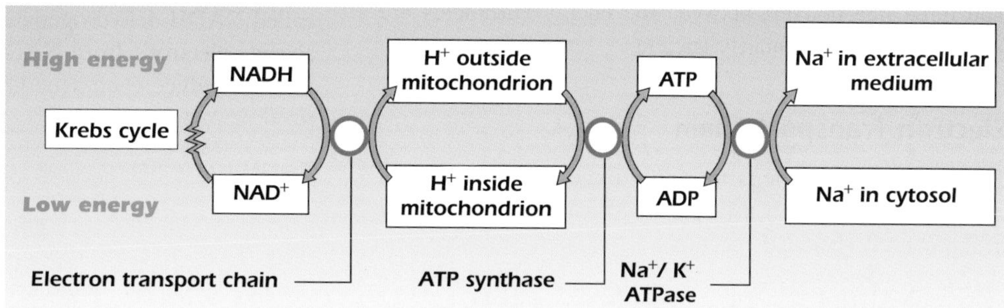

Figure 12.4. Energy flow between the currencies in a normal animal cell.

Exchange Mechanisms Convert Between the Four Energy Currencies

The cell has mechanisms that transfer energy between the four currencies. The conversions are summarized in Figure 12.4. In a typical animal cell oxidation of fuel molecules in the mitochondria (by the Krebs cycle, page 208) tops up the supply of NADH. The cell then converts this energy currency into the other three. All the interconversions are reversible. Figure 12.3 shows where each conversion mechanism is located.

Consider what happens if a few sodium ions move out of the extracellular medium and into the cytosol of a eukaryotic cell, for example, when a nerve cell transmits the electrical signal called an action potential (Chapter 14). The sodium gradient has been slightly depleted: the cell holds less of this energy currency than it did before. However, the cell still has plenty of energy in the form of ATP that it can convert into energy as a sodium gradient. It does this using the sodium/potassium ATPase. This protein is located in the plasma membrane. Its function is to move $3Na^+$ ions out of the cell and to move $2K^+$ ions into the cell. For this to happen ATP is hydrolyzed to ADP thus giving up energy that is used to push Na^+ out of the cytosol to its higher energy state in the extracellular medium.

The cell has now used some ATP, but it still holds plenty of energy in another currency—the H^+ gradient across the inner mitochondrial membrane. The enzyme ATP synthase is located in the inner mitochondrial membrane. The protein interconverts the two energy currencies: as H^+ ions move into the mitochondrion they give up energy that is then used by ATP synthase to make ATP from ADP and inorganic phosphate.

The energy of the H^+ gradient is now depleted. However, the cell still has plenty of energy in its NADH account. The electron transport chain allows interconversion of energy as NADH to energy as H^+ gradient. NADH is used to reduce molecular oxygen to

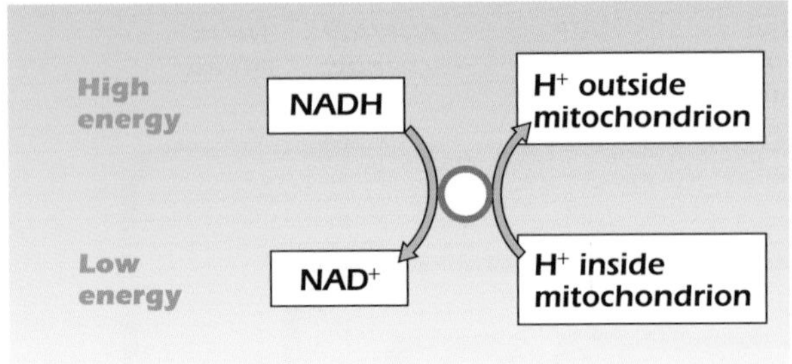

Figure 12.5. Currency conversion: the electron transport chain converts between NADH and the H^+ gradient.

water, releasing energy that is used to push H^+ ions out of the mitochondrion to their higher energy state in the cytosol.

Each of the three energy conversion systems is a protein structure called a **carrier** because it carries solute across the membrane. The sodium/potassium ATPase carries sodium and potassium ions, while both ATP synthase and the electron transport chain carry H^+ ions. There are many carriers in the cell with a wide variety of functions, some of which we will meet later in this book. The three that convert between the energy currencies are vital and are evolutionarily ancient.

Electron Transport Chain

The reason we breathe is to supply oxygen to the electron transport chain, which uses the oxygen to oxidize NADH in the overall chemical reaction:

$$NADH + H^+ + 1/2\ O_2 \rightleftharpoons NAD^+ + H_2O$$

This reaction releases energy, 206 kJ for every mole of NADH used. The energy is used to carry about $12H^+$ ions from their low-energy state in the mitochondrial matrix to their high-energy state in the cytosol. Figure 12.5 summarizes the reaction in terms of energy currencies. The circle symbolizes the linkage between the energy released in the conversion of NADH to NAD^+ and the energy used to drive H^+ out of the mitochondrial matrix. If the electron transport chain simply allowed NADH to reduce oxygen to water, then the reaction's energy would be released as heat. Instead, the enzymatic function of the electron transport chain is tightly coupled to its function as a carrier that moves H^+ ions. The energy of NADH is thus converted to the energy of the H^+ gradient.

The electron transport chain has six components (Fig. 12.6). Four of these are large integral membrane protein complexes called complexes I, II, III and IV, each containing many polypeptide chains. The complexes include members of the cytochrome family that contain iron atoms within a prosthetic heme group, and which can accept electrons that reduce their Fe^{3+} to Fe^{2+}. The other two components are the mobile electron carriers coenzyme Q and cytochrome *c*. The electron transport chain is very similar in all organisms. In eukaryotes the complexes are found in the inner mitochondrial membrane, while in prokaryotes the complexes are found in the plasma membrane. The electron transport process has been the subject of intensive investigation for decades and this has led to some rather complex nomenclature. Some aspects of the process remain unclear.

Figure 12.7 shows the reactions performed by each of the protein complexes. At the top, **Complex I** (sometimes called NADH dehydrogenase or NADH-Q oxidoreductase) accepts electrons from NADH, thus oxidizing it, and uses these to reduce coenzyme Q. The reaction can be summarized as

$$NADH + Q + 5H^+_{matrix} \rightarrow NAD^+ + QH_2 + 4H^+_{intermembrane}$$

Coenzyme Q carries both hydrogen ions and electrons (Fig. 12.8) It is a type of chemical called a quinone (hence "Q"). All quinones are fairly hydrophobic, and the hydrophobicity of coenzyme Q is increased by a long hydrocarbon tail.

For each NADH oxidized, complex I moves four hydrogen ions outward from the matrix to the mitochondrial intermembrane space. The transport of the hydrogen ions is accomplished using a coenzyme Q integral to the complex that collects hydrogen ions from the matrix then releases them to the intermembrane space. The machinery that produces this directionality is not yet clear.

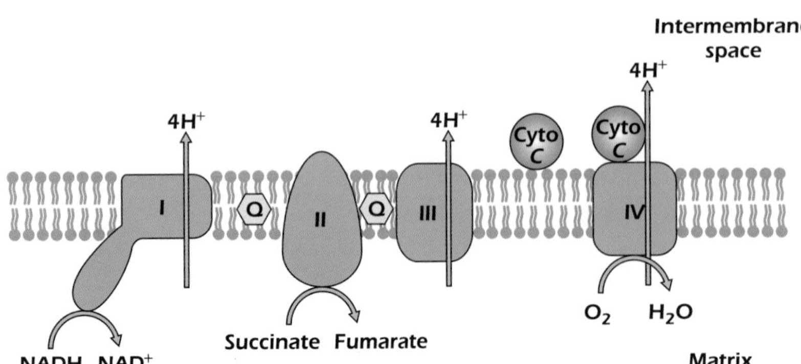

Figure 12.6. The electron transport chain comprises four large multimolecular complexes in the inner mitochondrial membrane, three of which are hydrogen ion carriers.

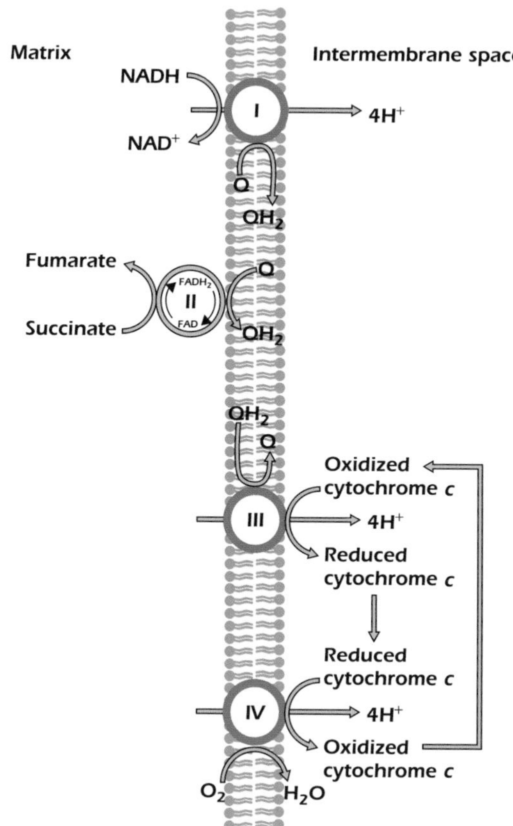

Figure 12.7. Overview of the operation of the electron transport chain.

Complex II (also known as succinate-Q reductase complex or succinate dehydrogenase) is the only one of the four complexes that is not a hydrogen ion pump. As we will see in Chapter 13 (page 208) complex II is part of the Krebs cycle: it is the enzyme that oxidizes succinate to fumarate using another coenzyme called FAD, which becomes $FADH_2$. Complex II passes its electrons to mobile coenzyme Q. Because hydrogen ions are not moved across the membrane in this process one gets less ATP formed from the electrons on $FADH_2$ than from those on NADH.

Complex III (Q-cytochrome *c* oxidoreductase or cytochrome reductase) accepts electrons from reduced coenzyme Q (QH_2) and uses them to reduce the second mobile electron carrier cytochrome *c*, a small soluble protein located in the intermembrane space that associates loosely with the inner mitochondrial membrane. Cytochrome *c*'s relative molecular mass of 12,270 means that it is just too big to pass through the channel porin in the outer membrane. Unlike coenzyme Q, cytochrome *c* can only carry one electron, so complex III marks a change from two electrons being carried to one. For each coenzyme Q fully oxidised, complex III moves four hydrogen ions outward from the matrix to the mitochondrial intermembrane space.

Cytochrome *c* moves the electrons to **complex IV (cytochrome *c* oxidase)**. This protein complex reduces oxygen to water and moves four hydrogen ions from the matrix to the intermembrane space. It contains two types of heme and copper ions. Oxygen binds to Fe^{2+} in a heme just as it binds to the heme in myoglobin or hemoglobin. The overall reaction is

$$4\ \text{Cyto}\ c_{\text{reduced}} + 8\text{H}^+_{\text{matrix}} + \text{O}_2 \rightarrow 4\ \text{Cyto}\ c_{\text{oxidized}} + 2\ \text{H}_2\text{O} + 4\text{H}^+_{\text{intermembrane}}$$

IN DEPTH 12.1 BROWN FAT

Triacylglycerols are stored within specialized fat cells in the body. Most fat cells are composed of a droplet of lipid surrounded by a thin layer of cytoplasm with a nucleus and a few mitochondria. The resulting tissue is white in color and simply releases or stores fatty acids in response to the needs of the organism. This is the kind of fat that is typically found around our kidneys and under the skin.

A second kind of fat is found in babies. Brown fat cells not only have stored triacylglycerols but are also rich in mitochondria, the cytochromes of the mitochondria giving the brown color. Brown fat is a heat-generating tissue. A channel selective for H^+ called thermogenin is found in the inner mitochondrial membrane. As fast as the electron transport chain pushes H^+ ions out of the mitochondrial matrix, they flow through thermogenin back down their electrochemical gradient into the matrix. In other cells this flux would only occur through ATP synthase and would be tightly coupled to the production of ATP from ADP. The presence of thermogenin uncouples the phosphorylation of ADP from the flow of electrons to oxygen so that the electron transport chain can work flat-out even though there is no ADP available, as long as the cell contains triacylglycerols, which can be oxidized to regenerate NADH (see β oxidation, page 214). This generates a lot of heat and helps the infant maintain body temperature. Large blocks of brown fat are found in animals that hibernate.

A similar uncoupling mechanism is used by some plants to generate heat. Some arum lilies rely on carrion-eating flies for pollination. Uncoupled mitochondria at the base of the flower generate sufficient heat to evaporate the evil-smelling odorants used to attract the flies.

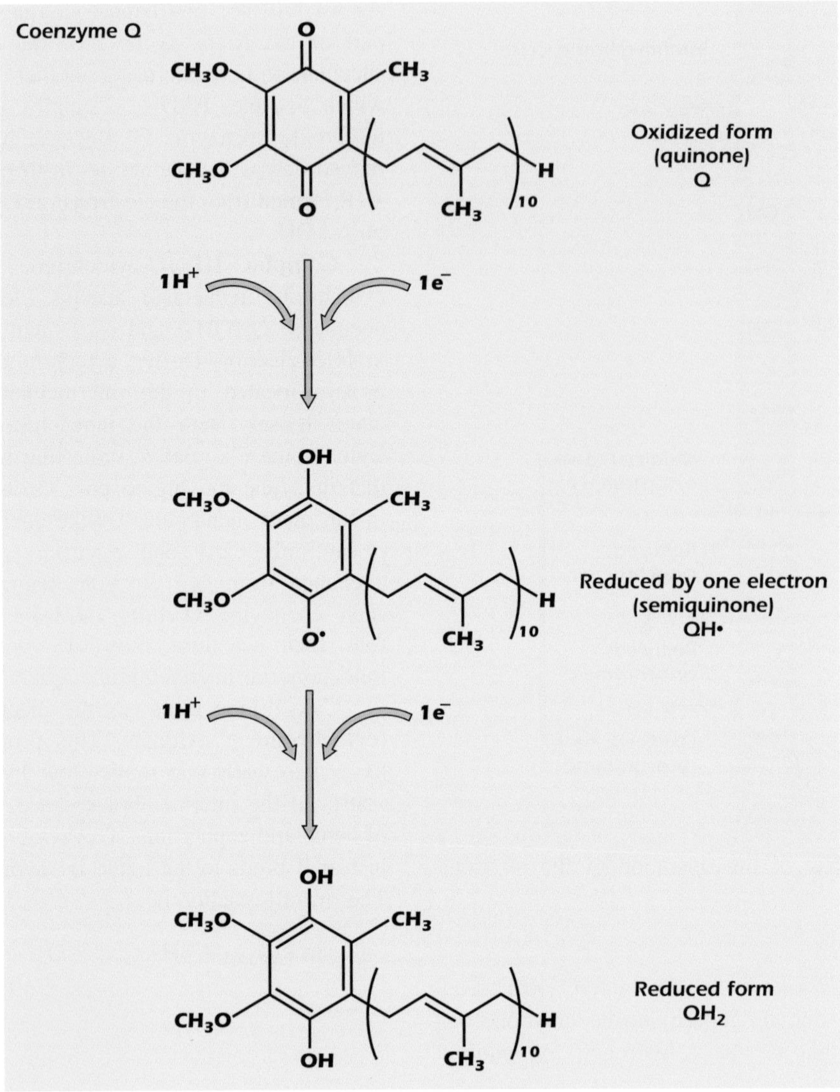

Coenzyme Q
Oxidized form (quinone) Q
1H+
1e−
Reduced by one electron (semiquinone) QH•
1H+
1e−
Reduced form QH2

Figure 12.8. Coenzyme Q carries two hydrogen atoms.

ATP Synthase

The energy stored in the hydrogen ion gradient can be used to drive the synthesis of ATP from ADP and inorganic phosphate. The conversion is accomplished by a marvelous "nanomachine" called ATP synthase. At its most basic, the operation is a simple one and is summarized in Figure 12.9. Once again, the circle symbolizes the linkage between the enzyme and carrier functions. **ATP synthase** is a protein complex that is also known as mitochondrial ATPase (because it can carry out the reverse reaction, hydrolyzing ATP and driving H^+ out of the mitochondria) or F_1F_0 ATPase (because it can be prepared in two fragments F_1 and F_0), while others call it complex V.

In most mitochondria ATP synthase is the only route by which the large electrochemical gradient for H^+ generated by the electron transport chain can be relieved. If the supply

High energy
H+ outside mitochondrion
ATP
Low energy
H+ inside mitochondrion
ADP

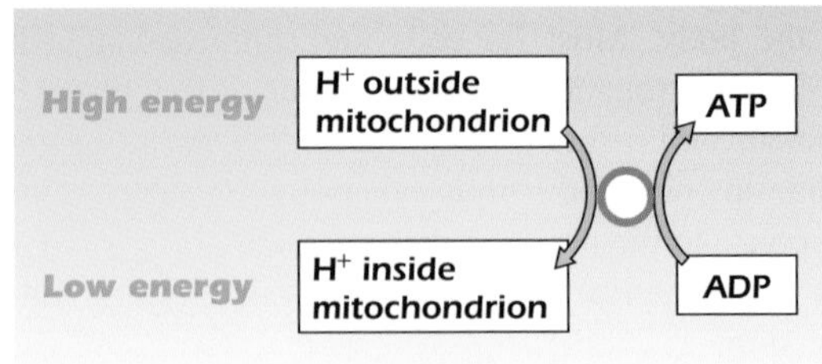

Figure 12.9. Currency conversion: ATP synthase interconverts the H^+ gradient and ATP.

of ADP runs out, then ATP synthase will stop moving H^+ in. If this happens, the electron transport chain will stop too because it cannot keep pushing H^+ out against a greater and greater electrochemical gradient. Another way of stopping ATP synthase is to block its operation with the antibiotic

oligomycin; again, if this is done, the electron transport chain will stop too.

The coupling between ATP synthesis and electron transport can be broken if hydrogen ions can be carried back across the membrane by an alternative route. Weak organic acids such as 2,4-dinitrophenol can do this. The dinitrophenol is uncharged in the intermembrane space and in this form is sufficiently hydrophobic to diffuse into the membrane. If it enters the matrix, it will tend to lose its hydrogen ion as the hydrogen ion concentration is much lower. The result is movement of hydrogen ions down their concentration gradient. Such molecules are called "uncouplers." In the presence of an uncoupler electron transport (and therefore oxidation of fuels) will carry on regardless of the availability of ADP or a working ATP synthase.

Sodium/Potassium ATPase

This is a single protein in the plasma membrane, with a carrier action that is tightly linked to an enzymatic one. Under normal conditions, it hydrolyzes ATP. The energy released drives sodium ions up their electrochemical gradient out of the cell. The Na^+/K^+ ATPase also moves potassium ions the other way, into the cytosol. For every ATP hydrolyzed, three Na^+ ions are moved out and two K^+ ions are moved in. Figure 12.10 summarizes the reaction in terms of energy currencies.

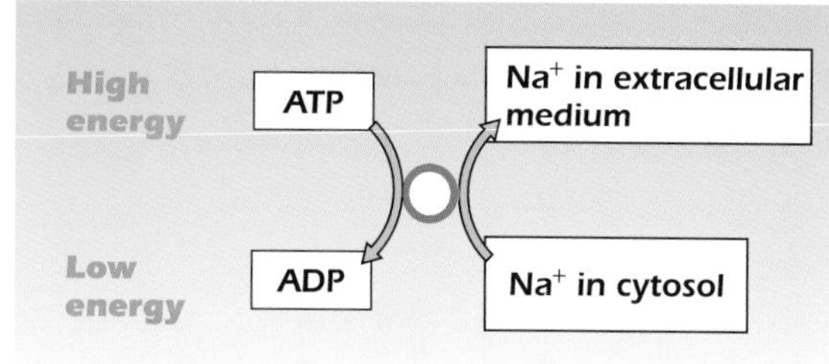

Figure 12.10. Currency conversion: the sodium-potassium ATPase interconverts ATP and the Na^+ gradient.

Example 12.1 Chemicals That Interfere with Energy Conversion Are Highly Toxic

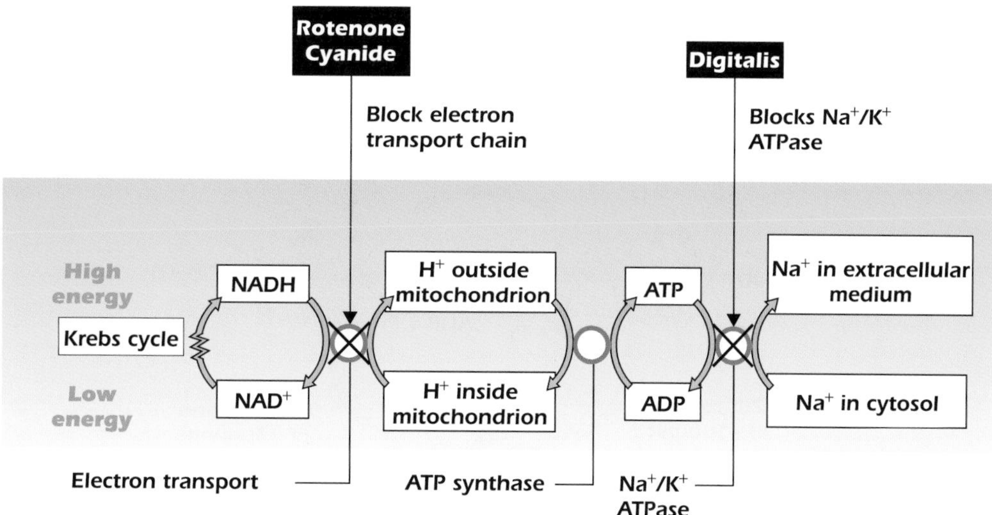

The electron transport chain, ATP synthase, and the sodium/potassium ATPase together run the energy currency market and are vital to the cell. Chemicals that interfere with them are very toxic. Rotenone, the most widely used rat poison, blocks complex I of the electron transport chain while cyanide blocks complex IV. Digitalis, from foxgloves, blocks the Na^+/K^+ ATPase.

Answer to Thought Question: One complete revolution of the γ subunit occurs when one H^+ has moved in across each of the c subunits that make up the rotor embedded in the inner mitochondrial membrane. One complete revolution generates three ATP. To estimate the stoichiometry, therefore, one can count the number of c subunits in the ring. An isoform with 10 c subunits would be expected to have the 10 H^+ to 3 ATP stoichiometry shown in Figure 12.3. An isoform with 14 c subunits would be expected to have a stoichiometry of 14 H^+ to 3 ATP.

However, like man-made rotary motors, ATP synthase can sometimes misfire or spin unproductively. The stoichiometry it actually achieves may therefore not be the same as that estimated by counting the c subunits.

IN DEPTH 12.2 ATP SYNTHASE: ROTARY MOTOR AND SYNTHESIS MACHINE

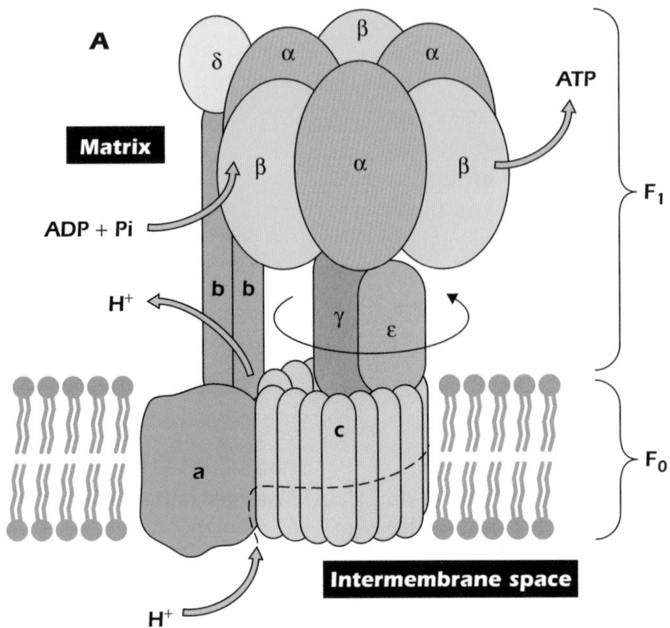

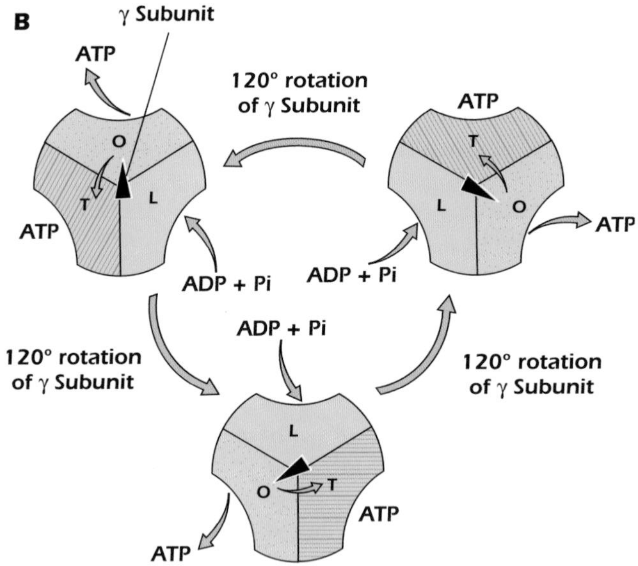

Mitochondrial ATP synthesis is carried out by ATP synthase which is a nanoscale rotating machine driven by the flow of H^+ ions from the intermembrane space to the matrix. ATP synthase has a membrane embedded part, called F_0, and a bead-like head, called F_1, that sticks into the matrix and connects to F_0 with a stalk. F_1 particles can be seen with the electron microscope.

The main part of the head consists of six subunits of two types: α and β in alternation. The β subunits carry out ATP synthesis. The γ subunit is in the center of the head and joins the head to the part buried in the membrane. ε and δ subunits complete the head. The F_0 part in the membrane has a circle of 10 to 14 subunits denoted "c" (the number depends on the species). In the center of the ring of c subunits is the γ subunit. Making contact with the outside of the circular assembly of c subunits is the a subunit that also connects with a dimer of b subunits that connect to δ and through it to the head. This is shown in Part A of the diagram.

We have seen that the splitting of ATP can drive a conformational change in the protein that has bound it and that this conformational change can produce movement (page 179). The converse is also true: motion can produce a conformational change and this can drive a reaction. H^+ flow causes the ring of c subunits to rotate and these drive rotation of the γ subunit. For each H^+ moving into the mitochrondrial matrix, the ring rotates by one c subunit. The rest of the head is stationary, stopped from rotating by the b subunit dimer and the δ subunit. The γ subunit is not symmetrical and makes different interactions with each of the β subunits in the head as it rotates. This means that each of the three β subunits changes from one to another of three conformations. These conformations are called T, O and L. As the γ subunit rotates it converts a single β subunit from T to O, then from O to L, and finally from L to T. As there are three β subunits they will be in one of the three different conformations at any time. The T conformation binds ADP and Pi very tightly and catalyses conversion to ATP but it does not release it. As γ rotates the T conformation changes to O that can release ATP. Another 120-degree rotation converts O to L that now binds ADP and Pi ready for conversion to T and another cycle. A full rotation of the γ subunit drives release of 3 ATP. Part B of the diagram summarizes the cycle. Only the γ and β subunits are shown. The γ subunit rotates counterclockwise and the β subunits are stationary.

The 1997 Nobel Prize in Chemistry was awarded to Paul Boyer and John Walker for their elucidation of this mechanism.

Example 12.2 The Fuel That Makes Bacteria Swim

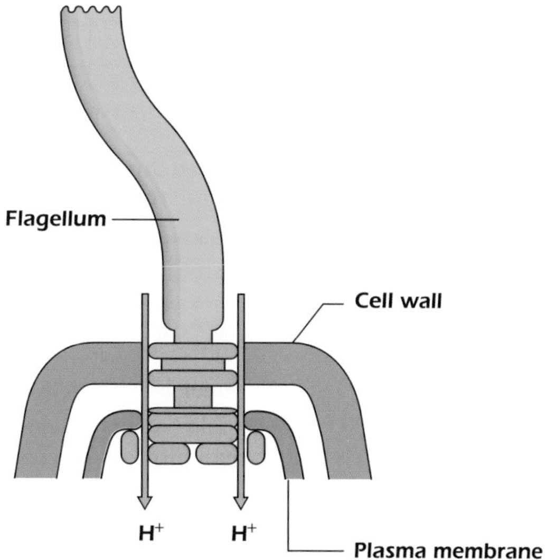

Although by no means all bacteria are motile, some, such as *Escherichia coli* and *Salmonella typhimurium*, can swim. Their small size makes the process very different from swimming as we experience it. If you, the reader, dive into a swimming pool, your momentum will carry you some distance because your mass is great relative to the density of the water. For a bacterium whose mass is extremely small, water is a highly viscous medium. A bacterium swimming in water experiences roughly what a human swimmer in molasses might feel! Despite this, motile bacteria can achieve speeds of up 100 μms^{-1} by the use of a structure called the bacterial flagellum. This consists of a rigid helical filament, 20–40 nm in diameter and up to 10 μm long, composed of a single protein, flagellin. The bacterial flagellum operates like a boat's propeller, pushing the bacterium along as it is turned by the motor at its base.

The flagellar motor consists of a series of rings that allow the motor to rotate within the complex layers of membranes and cell wall that make up the bacterial cell surface. Like ATP synthase, the flagellar motor allows H^+ ions to flow in across the plasma membrane. The energy released as the H^+ ions flow down their electrochemical gradient is used to turn the rotor of the motor and hence the flagellum at up to 100 Hz (= 100 times per second).

IN DEPTH 12.3 CAN IT HAPPEN? THE CONCEPT OF FREE ENERGY

We know from experience that some chemical reactions give out heat (the burning of organic matter for instance) while others absorb heat (so heat is used to make the reaction happen). The study of how energy affects matter and particularly how it affects chemical reactions is **thermodynamics**. A vast amount of research in the late nineteenth and early twentieth centuries led to the development of this complex field. To make a very long story short, a process results in a change in two parameters: heat and **entropy**. Heat we are familiar with and understand as the random motion of molecules, but entropy may be unfamiliar. Entropy is the degree of disorder in a system: Entropy always tends to increase. The melting of ice is favorable because it results in an increase in entropy as the ordered, crystalline ice becomes more random water.

A physical process can absorb or emit heat and/or result in an increase or decrease in entropy. If we wish to know whether a process can occur or not, we need to consider both. The concept of free energy was formulated by the American physical chemist J. Willard Gibbs in 1878. It is now called the Gibbs free energy and given the symbol G in his honor. The change in free energy in a process, ΔG is

$$\Delta G = \Delta H - T\Delta S$$

where ΔH is the heat change (called enthalpy) in the process, T is the absolute temperature in degrees Kelvin, and ΔS is the change in entropy. If ΔG for a process is negative, it will proceed, although the rate may be very slow in the absence of a catalyst. If it is not negative, the process cannot occur.

Unfortunately, we can only measure changes and not absolute values for G, H, and S in chemical processes. This difficulty is avoided by defining **standard states** so we can make comparisons. A free-energy change in a process under standard conditions is denoted ΔG°. Biochemists use a slightly different standard state, which is in water at pH 7, and this is shown as $\Delta G^{\circ\prime}$.

The free-energy change is related to the equilibrium constant for a reaction. Consider the reaction

$$A + B \rightleftarrows C + D$$
$$K_{eq} = ([C][D])/([A][B])$$

and

$$\Delta G = \Delta G^{\circ\prime} + R\ T\ \ln([C][D])/([A][B])$$

where R is the gas constant and T the absolute temperature. At equilibrium $\Delta G = 0$ so

$$\Delta\ G^{\circ\prime} = -R\ T\ \ln\ K'_{eq}$$

where K'_{eq} is the equilibrium constant under standard conditions.

Conditions in the cell are not usually the standard ones, but if we know $\Delta G^{\circ\prime}$ for a reaction and the relevant concentrations, we can calculate ΔG for the reaction. For this book we have calculated ΔG values under human cellular conditions by using the following reasonable concentration values:

[ATP] = 6 mmol liter^{-1}
[ADP] = 0.6 mmol liter^{-1}
[AMP] = 0.2 mmol liter^{-1}
[Pi] = 4.5 mmol liter^{-1}
Glucose = 250 μmol liter^{-1}
Glucose-6-phosphate = 10 mmol liter^{-1}
$[NAD^+]/[NADH] = 4$ in the mitochondrial matrix
$[O_2]$ = 21 μmol liter^{-1}
Temperature 37°C

For example, we have used the hexokinase reaction (page 192) to illustrate coupling a reaction with a negative free-energy change with one which has a positive free energy change.

$$\text{Glucose} + \text{ATP} \rightarrow \text{Glucose-6-phosphate} + \text{ADP} + H^+$$

Under biochemical standard conditions this reaction has a $\Delta G^{\circ\prime}$ of $-36\ \text{kJmol}^{-1}$. However, inside cells the concentration of glucose-6-phosphate is much higher than the concentration of unphosphorylated glucose; when this and other concentration effects are taken into account, the calculated overall free-energy change is $-11\ \text{kJmol}^{-1}$. The reaction will certainly still proceed but is not as strongly favored as the $\Delta G^{\circ\prime}$ would suggest.

The ΔG values we have given for the four energy currencies are those that apply in an aerobic animal cell, as shown in Figure 12.3, where each of the conversion reactions has a negative ΔG, as it must if it is to proceed as shown. Under conditions where one or more of the conversion reactions is operating in the opposite directions, the concentrations, and therefore ΔG values, will of course be different.

IN DEPTH 12.4 PHOTOSYNTHESIS

The free energy needed to support life comes mainly from the energy of sunlight. Photosynthetic bacteria and plants have evolved systems that use the energy of light to make ATP and produce NADPH. Plants use ATP and NADPH to fix CO_2 and convert it to sugars. The light reactions that produce NADPH and ATP also oxidise water to oxygen. All life on Earth is virtually completely dependent upon photosynthesis.

Fixation of CO_2 is carried out by the enzyme ribulose 1,5-bisphosphate carboxylase/oxygenase (known as "rubisco"). It is probably the most abundant protein on Earth! The five-carbon sugar ribulose 1,5-bisphosphate is combined with a CO_2 and splits into two three-carbon molecules of 3-phosphoglycerate. These are first phosphorylated using ATP and then reduced using NADPH (page 214) to glyceraldehyde 3-phosphate some of which is converted to dihydroxyacetone phosphate. These two are combined by the action of an aldolase to make fructose 6-phosphate. A series of reactions similar to those of the pentose phosphate pathway (page 214) regenerate ribulose 1,5-bisphosphate. The reactions that fix CO_2 are known as the **Calvin Cycle** (or dark reactions). More than 10^{10} tons of carbon are fixed every year.

The ATP and NADPH necessary to drive **carbon fixation** are generated using the energy in light. When a molecule absorbs a photon of light the electrons in the molecule jump to a higher energy level. In most molecules the electrons very rapidly drop to their ground state and the energy of the photon is lost as heat. In photosynthesis the pigments are so arranged that the excited electron is trapped by another molecule.

Plants have chloroplasts that carry out the light and dark reactions of photosynthesis. These organelles, like mitochondria, are thought to be the result of endosymbiosis (page 10). They have an inner and an outer membrane and their own DNA that codes for some of their proteins. The outer membrane is permeable but the inner membrane is impermeable. The mitochondrial inner membrane is folded into cristae to increase the area available for the electron transport chain and ATP synthase. Chloroplasts carry this further and have stacks of discs made up of membrane enclosing an inner space. These are called **thylakoids** and the centers are the thylakoid space. The protein-pigment complexes that carry out the light reactions are embedded in the thylakoid membranes. Their activity results in hydrogen ions being moved into the thylakoid space. The thylakoid membranes also have an ATP synthase that uses the flow of hydrogen ions from the thylakoid space to drive the synthesis of ATP. It is very similar to the mitochondrial ATP synthase (page 200).

Chlorophyll a is the main light-absorbing pigment but other chlorophylls and molecules called carotenoids are present to extend the wavelength range of the light absorbed. Chlorophylls have alternating double and single bonds that make them very efficient at absorbing light in the visible part of the spectrum, and magnesium ions at their center. There are two different photosystems involved in moving hydrogen ions into the thylakoid space, splitting water to release oxygen, and reducing $NADP^+$ to NADPH. Both photosystems have a special pair of chlorophyll molecules with associated "antenna" chlorophylls that relay the energy of absorbed photons to the special pair. When this happens an high-energy electron is abstracted by a quinone and is passed through a series of protein-bound carriers rather like the complexes involved in the electron transport chain. An electron is returned to the reaction center but at the lower, ground state energy level.

ADP/ATP Exchanger

ATP synthase makes ATP inside the mitochondrion. For the ATP to be available to the rest of the cell, there needs to be a mechanism to enable it to leave the mitochondrion for use in the cytosol. This job is performed by another carrier, the **ADP/ATP exchanger**. This protein has no enzymatic action; it simply moves ADP in one direction across the mitochondrial inner membrane and ATP in the opposite direction. In most eukaryotic cells the carrier operates in the direction shown in Figure 12.3. Carriers such as the Na^+/K^+ ATPase together with many synthetic processes use up ATP in the cytosol, producing ADP. ADP enters the mitochondria by the ADP/ATP exchanger and is reconverted to ATP by ATP synthase. ATP then leaves the mitochondrion with the help of the ADP/ATP exchanger.

All Carriers Can Change Direction

In a normal animal cell, the primary source of energy is the Krebs cycle (page 208). This regenerates NADH from NAD^+, making at the same time a small amount of ATP. Because the NADH currency is always being topped up, while the others are being used, the direction of operation of the energy conversion systems is usually that shown in Figure 12.3. However, all the carriers are reversible.

Yeast cells in a wine barrel, or muscle cells in the leg of a sprinter, are anaerobic; there is no oxygen available. In this situation cells can make NADH, but the electron transport chain cannot drive H^+ out of the mitochondria because there is no molecular oxygen waiting to be reduced by NADH. Instead, the cell's energy needs are met by anaerobic glycolysis (page 211), which makes ATP.

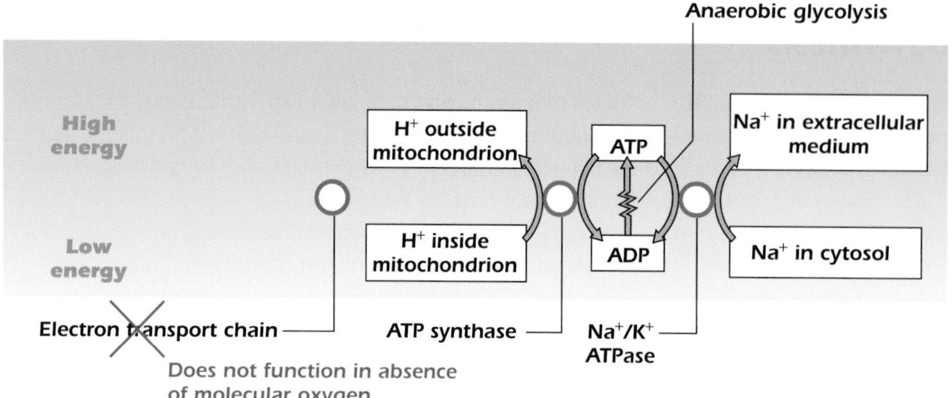

Figure 12.11. Energy flow between the currencies in an anaerobic cell.

Figure 12.11 shows how the cell maintains the amounts of energy currencies. Any drain on the mitochondrial H^+ gradient is counteracted by ATP synthase running in the opposite direction from that shown in Figure 12.3. ATP is hydrolyzed and H^+ ions are pushed out of the mitochondrion. The ADP/ATP exchanger also reverses its direction. ATP is regenerated by anaerobic glycolysis in the cytosol and is used up by ATP synthase in the mitochondrial matrix.

Healthy cells maintain themselves in a steady state in which none of the energy currencies are allowed to run down. The direction in which energy moves between the four currencies depends on the primary source of energy for that cell.

SUMMARY

1. Reactions with a positive Gibbs free-energy change (ΔG) can be caused to happen in a cell by linking them with a second reaction with a larger, negative ΔG. The second reaction drives the first.
2. The majority of such reactions in the cell are driven by one of four energy currencies: NADH, ATP, the hydrogen ion gradient across the inner mitochondrial membrane, and the sodium gradient across the plasma membrane.
3. The electron transport chain in the mitochondrial inner membrane converts between energy as NADH and energy in the hydrogen ion gradient.
4. ATP synthase in the mitochondrial inner membrane converts between energy in the hydrogen ion gradient and energy as ATP.
5. The sodium/potassium ATPase in the plasma membrane converts between energy in ATP and energy in the sodium gradient.
6. All carriers can change direction.
7. In a healthy cell, none of the energy currencies are allowed to run down. The direction of energy exchange between the four currencies depends on the primary source of energy available to the cell.

FURTHER READING

Voet, D., and Voet, J. D. (2011) *Biochemistry*, 4th edition, John Wiley & Sons, Hoboken.

REVIEW QUESTIONS

12.1 Theme: Cell Spaces and Regions in Energy Trading

A The mitochondrial matrix
B The inner mitochondrial membrane
C The intermembrane space of the mitochondrion
D The outer mitochondrial membrane
E The cytosol
F The plasma membrane
G The extracellular medium

From the above list of locations, select the location where each of the following proteins, processes or conditions is found in a healthy cell.

1. ADP/ATP exchanger
2. ATP synthase
3. Coenzyme Q
4. Cytochrome *c*
5. $[Na^+] > 100$ mmole liter^{-1}
6. porin
7. sodium/potassium ATPase
8. The electron transport chain

12.2 Theme: The Electron Transport Chain and ATP Synthase

A Complex I
B Complex II
C Complex III
D Complex IV
E ATP synthase

From the above list of proteins, select the protein described by each of the following statements,which apply to a human cell under well-oxygenated conditions.

1. Is not a carrier
2. Oxidises coenzyme Q
3. Oxidises NADH
4. Oxidises succinate
5. Oxidises the reduced form of cytochrome *c*
6. Reduces molecular oxygen to water
7. Will reverse direction if the H^+ electrochemical gradient across the membrane is dissipated after application of an uncoupler

12.3 Theme: Energy Currencies

A ATP
B GTP
C NADH
D Hydrogen ion electrochemical gradient
E Sodium ion electrochemical gradient
F UTP

From the above list of energy currencies, choose the currency corresponding to each of the following descriptions.

1. Contains a pyrimidine residue
2. Generated by the action of an integral membrane protein of the plasma membrane
3. Has no energy content under anaerobic conditions
4. In a well-oxygenated cell, this is the most energy rich of the energy currencies
5. This currency is directly depleted by the action of uncouplers such as 2,4 dinitrophenol

THOUGHT QUESTION

In Figure 12.3 we show the stoichiometry of ATP synthase as 3 ATP made per 10 H^+ moved. However this stoichiometry is different in different species and for different isoforms of ATP synthase. Based on a reading of In Depth 12.2, how would you estimate the stoichiometry of a particular isoform of ATP synthase given the structure of the isoform?

13

METABOLISM

In Chapter 12 we described the energy currencies NADH and ATP. In this chapter we will describe the chemical pathways that regenerate these currencies when their levels are depleted. We will then consider some other important chemical pathways, some that operate in all cells, others that are found only in certain types of organism or in specialized biochemical centers like the liver. Figure 13.1 is an overview of the main metabolic pathways within a cell.

All the processes that occur within a living cell are ultimately driven by energy taken from the outside world. Green plants and some bacteria take energy directly from sunlight. Other organisms take compounds made using sunlight and break them down to release energy, a process called **catabolism**. The most common way of breaking down these food compounds is to oxidize them, that is, to burn them but in a controlled way. The energy trapped in energy currencies can then be used for the building, repair, and maintenance processes termed **anabolism**. The collective term for all of the reactions going on inside a cell is **metabolism**. All metabolic reactions share some general features:

- They are catalyzed by enzymes.
- They are universal in that all organisms show remarkable similarity in the main pathways.
- They involve relatively few types of chemical reaction.
- They are controlled, often by modulation of key regulatory enzymes.
- They are compartmentalized within cells. In eukaryotes different sets of metabolic reactions are carried out in different organelles. In animals and plants this compartmentalization is carried further, so that in some cases different reactions take place in different body organs. Prokaryotes too show compartmentalization of a simpler sort—some processes are associated with regions on the inner face of the plasma membrane.
- They usually involve coenzymes, molecules that are second substrates in a number of different reactions.
- The pathways that break particular molecules down are different from those used to synthesize them. This allows them to be controlled separately.

Molecules that are common second substrates for a number of reactions are called coenzymes. These are either energy currencies (such as NADH and ATP) or carriers of chemical groups such as coenzyme A, which carries organic acids such as acetate and succinate. ATP is used to drive other reactions by transfer of a phosphate group. The resultant ADP can be rephosphorylated back to ATP. The energy to do this can be derived from NADH passing its electrons to the electron transport chain, which converts the NADH back to its oxidized form NAD^+. The total amount of ATP + ADP in a cell is relatively constant, but the proportion that is ATP can vary. Similarly the amount of $NADH + NAD^+$ is constant, but the proportion present as NADH can vary.

Cell Biology: A Short Course, Third Edition. Stephen R. Bolsover, Jeremy S. Hyams, Elizabeth A. Shephard and Hugh A. White.

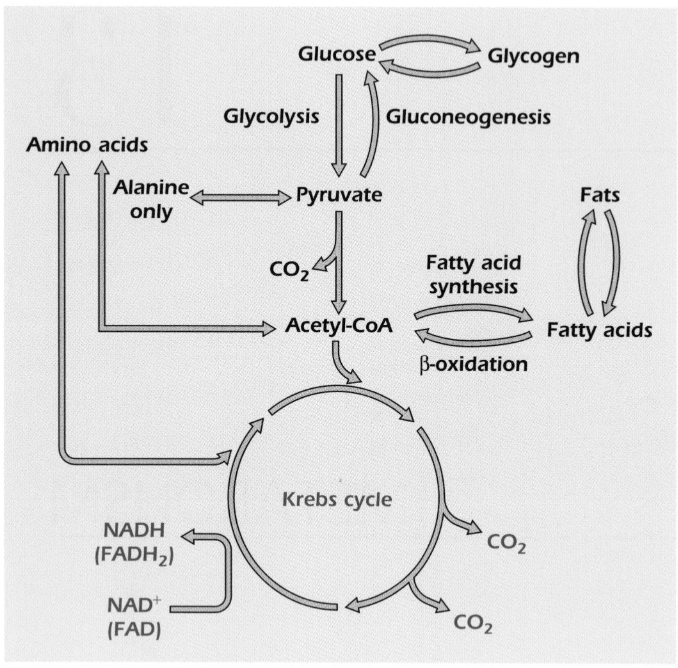

Figure 13.1. Overview of metabolism.

THE KREBS CYCLE: THE CENTRAL SWITCHING YARD OF METABOLISM

The fuels we take in in the diet are mainly fats, proteins, and carbohydrates. At the very center of metabolism is a cycle of reactions that takes place in the mitochondrial matrix. The Krebs cycle is named after its discoverer, Hans Krebs, and is also known as the **tricarboxylic acid (TCA) cycle** or the citric acid cycle. The foods we eat are converted to the two-carbon unit acetate, CH_3COO^-. The acetate is not free but is carried by a coenzyme called coenzyme A. Acetate bound to coenzyme A—acetyl-CoA for short—is then fed into the Krebs cycle and may be completely oxidized to carbon dioxide and water. In the process the energy currency NADH is produced. The Krebs cycle is central to carbohydrate, fat, and amino acid metabolism.

We shall first describe the Krebs cycle and then look at the other pathways that interact with it. The reactions are shown in Figure 13.2.

1. A molecule of acetyl-CoA enters the cycle, and the acetate (two carbons) is combined with the four-carbon molecule oxaloacetate, making citrate (as carboxyl groups are ionized at the pH in the cell, we normally speak of them as ions; so we say "citrate" rather than citric acid).
2. Citrate is rearranged to isocitrate.
3. In the first oxidation step isocitrate is oxidized to 2-oxoglutarate (sometimes called α-ketoglutarate). A carbon is lost as CO_2 and NAD^+ is reduced to NADH.
4. A second oxidation converts 2-oxoglutarate to succinate: a second carbon leaves as CO_2 and again NAD^+ is reduced to NADH. In the process the product is attached to CoA. This reaction is catalyzed by oxoglutarate dehydrogenase.
5. The bond between succinate and CoA is now broken and the energy released used to drive the phosphorylation of a GDP to GTP. γ-phosphate groups can be swapped between nucleotides, so this GTP can be used to regenerate ATP from ADP.
6. Succinate is oxidized to fumarate. The oxidant in this case is flavin adenine dinucleotide (FAD) and not NAD^+ so an $FADH_2$ is produced. $FADH_2$, the reduced form of FAD, does not carry as much energy as NADH but like NADH is used to drive H^+ up its electrochemical gradient out of the mitochondrial matrix (page 197). The enzyme involved is succinate dehydrogenase, which is actually part of the electron transport chain.
7. Water is added to the double bond in fumarate, making malate.
8. Malate is oxidized to oxaloacetate in a reaction catalyzed by malate dehydrogenase. One NAD^+ is reduced to NADH. The starting compound oxaloacetate has been regenerated and is ready to accept another acetyl group to start another turn of the cycle.

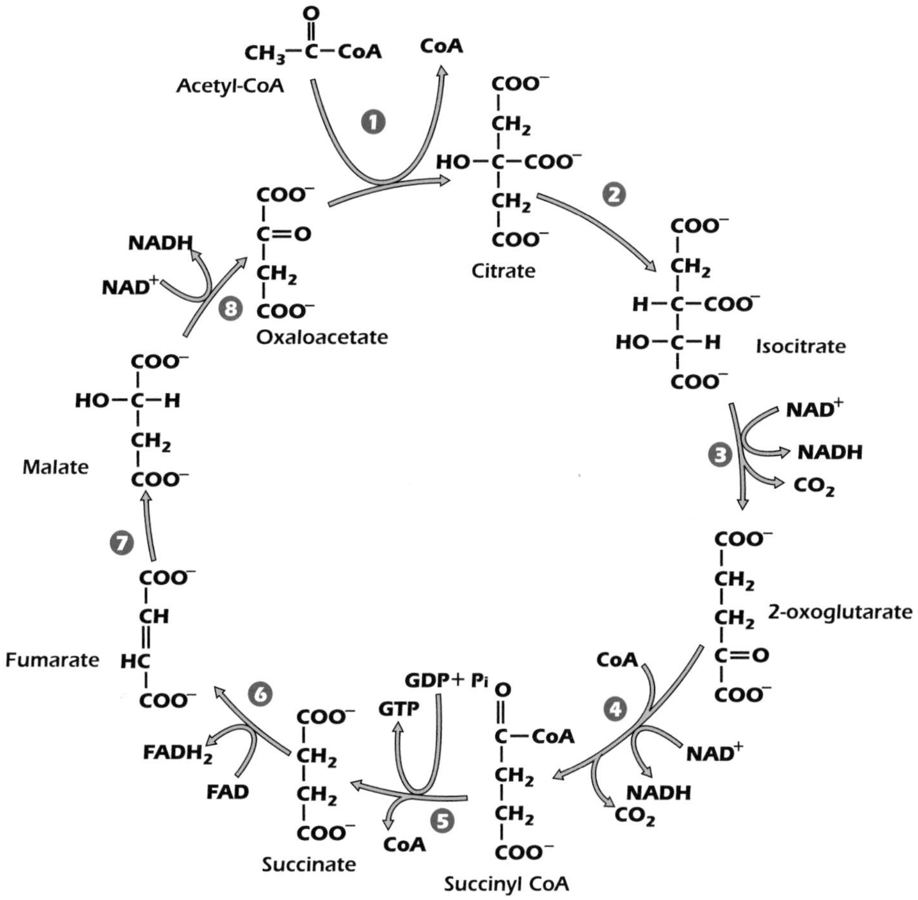

Figure 13.2. The Krebs cycle.

The reactions of the Krebs cycle can be summarized as

$$CH_3CO\text{–}CoA + (3NAD^+ + FAD) + GDP + Pi + 3H_2O \rightarrow CoA\text{–}H + (3NADH + FADH_2) + GTP + 2CO_2 + 3H^+$$

where Pi represents an inorganic phosphate ion.

FROM GLUCOSE TO PYRUVATE: GLYCOLYSIS

Glucose is an important fuel for most organisms, and **glycolysis** is the main pathway that enables it to be used. The word glycolysis simply means the breakdown of glucose. It is an ancient pathway that can function without oxygen and indeed is thought to have evolved before there was much oxygen in the atmosphere. The pathway is present in almost all cells and takes place in the cytosol. It is shown in Figure 13.3.

- Free glucose is phosphorylated on carbon 6 to produce glucose-6-phosphate. As we discussed earlier (page 192), this reaction, catalyzed by hexokinase, is driven by the free energy available from ATP.
- Glucose-6-phosphate is isomerized to fructose-6-phosphate.
- Fructose-6-phosphate is phosphorylated to produce fructose-1,6-bisphosphate. Again ATP is used and the reaction is carried out by phosphofructokinase. This reaction commits the sugar to being broken down and used to provide energy rather than used for other purposes. Phosphofructokinase is regulated allosterically, being inhibited by ATP (page 187).
- The fructose-1,6-bisphosphate is now split into two halves. Each of the halves has a phosphate attached. The two products are dihydroxyacetone phosphate and glyceraldehyde-3-phosphate. The enzyme is aldolase.
- Triose phosphate isomerase interconverts dihydroxyacetone phosphate and glyceraldehyde-3-phosphate. This allows both halves of the original glucose to be used.

The glucose-6-phosphate has now been converted into two molecules of glyceraldehyde-3-phosphate. Each of the following reactions occurs twice for each molecule of glucose fed into the pathway.

- Glyceraldehyde-3-phosphate is oxidized in a reaction that also attaches one inorganic phosphate ion to give

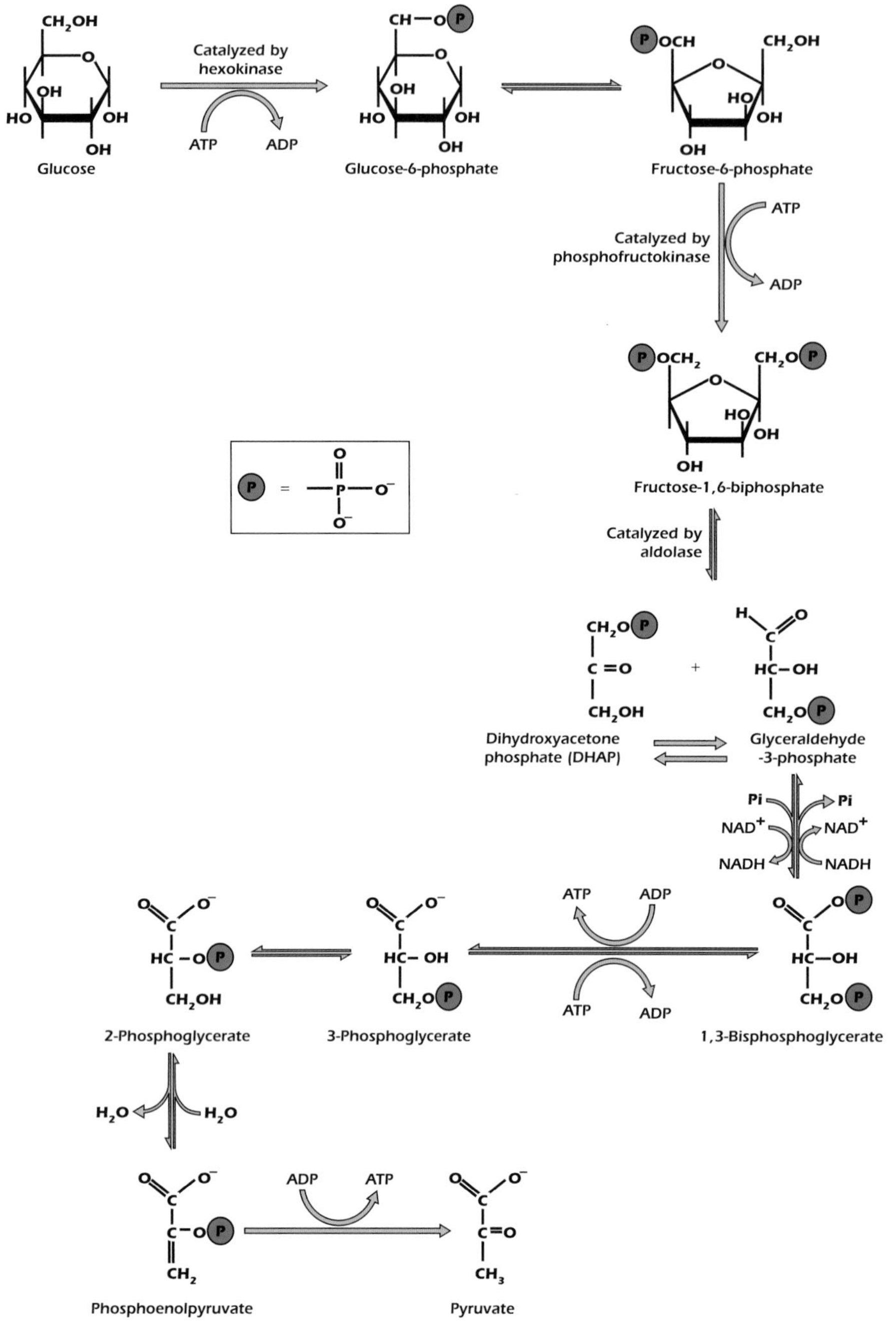

Figure 13.3. Glycolysis breaks glucose down into pyruvate.

1,3-bisphosphoglycerate. NAD^+ is the oxidant and a NADH is produced. The enzyme is glyceraldehyde-3-phosphate dehydrogenase.

- One of the phosphate groups from bisphosphoglycerate is transferred to ADP. This is a process called substrate-level phosphorylation. ATP has been made from ADP by a single enzyme, without any involvement of ATP synthase. We are left with 3-phosphoglycerate.
- After a rearrangement to 2-phosphoglycerate, water is removed leaving phosphoenolpyruvate.
- In another substrate-level phosphorylation phosphoenolpyruvate transfers its phosphate group to ADP, leaving pyruvate.

Overall glycolysis has used up two ATP molecules but has produced four: a net gain of two ATP per glucose.

Figure 13.4 illustrates the various ways that the cell can use pyruvate. If it is to be used to make fatty acids or to be oxidized in the Krebs cycle, it is carried into the mitochondrial matrix. Here it is oxidized and decarboxylated by the complex enzyme pyruvate dehydrogenase (Fig. 13.4A).

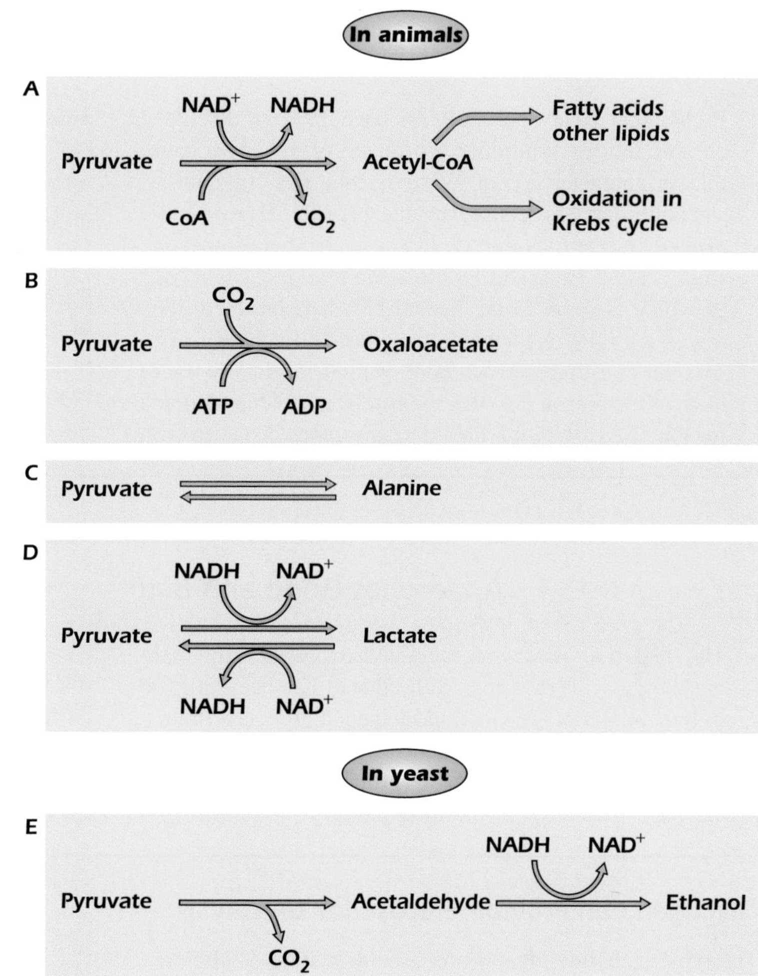

Figure 13.4. Pyruvate can be used in a number of ways.

NAD^+ is used as the oxidant, producing NADH. Coenzyme A is also added. The product is therefore acetyl-CoA, which may enter the Krebs cycle or can be converted to fatty acids or some other molecules.

The Krebs cycle is limited by the availability of oxaloacetate. This may run low as some biosynthetic pathways use one or the other component of the Krebs cycle as their starting material. If this happens new oxaloacetate is made from pyruvate (Fig. 13.4B). Pyruvate may also be converted to the amino acid alanine in a **transamination** reaction (Fig. 13.4C; Fig. 11.8 on page 186).

Glycolysis Without Oxygen

The leg muscles of a sprinter cannot be supplied with oxygen rapidly enough to supply the mitochondria. Muscle cells therefore need to make ATP by a method that does not require oxygen. As we have seen, glycolysis itself produces two ATPs per glucose. Can the cell let the pyruvate pile up and use it when oxygen becomes available again? No, because glycolysis as far as pyruvate also converts one NAD^+ to NADH. If this was all that happened, the cell would quickly convert all its NAD^+ to NADH, and glycolysis would stop as the cell does not have oxygen available for the mitochondria to oxidize the NADH back to NAD^+. To solve this problem cells reduce pyruvate to lactate (Fig. 13.4D) and in doing so regenerate the NAD^+ needed to allow glycolysis to carry on. This reaction (and its reverse) is catalyzed by lactate dehydrogenase. The buildup of lactic acid in poorly oxygenated muscles is thought to cause the pain of cramp. When we stop using the muscle, the blood can supply more oxygen to the muscle and the need for lactate production abates. The blood carries the lactate to the liver where it is reoxidized to pyruvate. Red blood cells lack mitochondria and are entirely dependent on glycolysis for their energy needs.

Some microorganisms—particularly yeasts—regenerate their NAD^+ in a different way. Pyruvate is first decarboxylated to acetaldehyde (also called ethanal) and then reduced to ethanol by alcohol dehydrogenase, which uses NADH and regenerates NAD^+. One molecule of CO_2 is also produced (Fig. 13.4E).

Medical Relevance 13.1 Sleeping Sickness

In Medical Relevance 12.1 on page 192 we discussed chronic fatigue syndrome, a problem of unknown cause that is characteristic of Western countries. In contrast sleeping sickness, caused by the parasite *Trypanosoma brucei*, is a serious disease in Africa. Trypanosomes are single-celled eukaryotes with a complex life cycle, part of which is spent in the human bloodstream. Here they reveal a prodigious appetite for glucose. Each tiny parasite can consume its own body weight in glucose every hour. The reason for the parasite's insatiable demand for glucose is twofold. First, trypanosomes dispense with mitochondrial ATP production and rely on anaerobic glycolysis to generate their ATP, so they only make two ATPs per glucose rather than the 30 or so that they could if their mitochondria were still working. Second, trypanosomes do not have any glycogen or fat stores and are therefore entirely dependent on a continuous supply of glucose in the host blood. The enormous consumption of glucose by the parasites leaves little for the host, who is overcome by extreme langor such that even keeping the eyes open is an unsurmountable effort.

Example 13.1 Anaerobes Good and Bad

The fact that yeasts regenerate their NAD^+ by making the gas CO_2 together with ethanol has been utilized in breadmaking, brewing, and winemaking since prehistoric times. Other microorganisms are like our muscles and generate lactate under anaerobic conditions. These too have their uses in the food industry: yogurts, many cheeses, sauerkraut, and dill pickles all rely on lactic acid released in anaerobic glycolysis. On the minus side, some food-spoiling bacteria can only function when there is no oxygen. These obligate anaerobes include the deadly *Clostridium botulinum* (Example 10.3 on page 171).

Glycogen Can Provide Glucose for Glycolysis

The polysaccharide glycogen is used as a store of glucose, particularly in liver and muscle cells. We saw in Chapter 2 how the glycosidic bond can be hydrolyzed with the broken ends of the bond being sealed with groups from a water molecule, so that a hydrogen atom is added to one side of the broken bond and a hydroxyl group is added to the other (page 34). The enzyme **glycogen phosphorylase** specifically breaks the $\alpha(1 \rightarrow 4)$ glycosidic bond in glycogen but seals the broken ends with groups from inorganic phosphate, so that a hydrogen atom is added to one side of the broken bond and a phosphate group is added to the freed glucose monomer (Fig. 13.5). The resulting glucose-1-phosphate is readily converted to glucose-6-phosphate for glycolysis. Breaking up glycogen this way is more energy efficient than simply hydrolyzing it (as happens in the intestine) since the ATP that would otherwise be required to make glucose-6-phosphate from free glucose is saved. The periodic $\alpha(1 \rightarrow 6)$ links that attach side arms to the glycogen chain are broken by other enzymes.

Medical Relevance 13.2 Inherited Muscle Cramps

We all experience muscle cramps as an occasional result of unusual exertion. However some people routinely experience serious cramps after any strenuous exercise. They suffer from McArdle's disease, an inherited deficiency of the muscle isoform of glycogen phosphorylase. Because their muscles are unable to use stored glycogen, the ATP runs out in these cells during exercise. Although there is probably some muscle damage because the glycogen granules get unusually large, the condition is more of an inconvenience than a serious debility. Liver cells express a different isoform of glycogen phosphorylase and are unaffected.

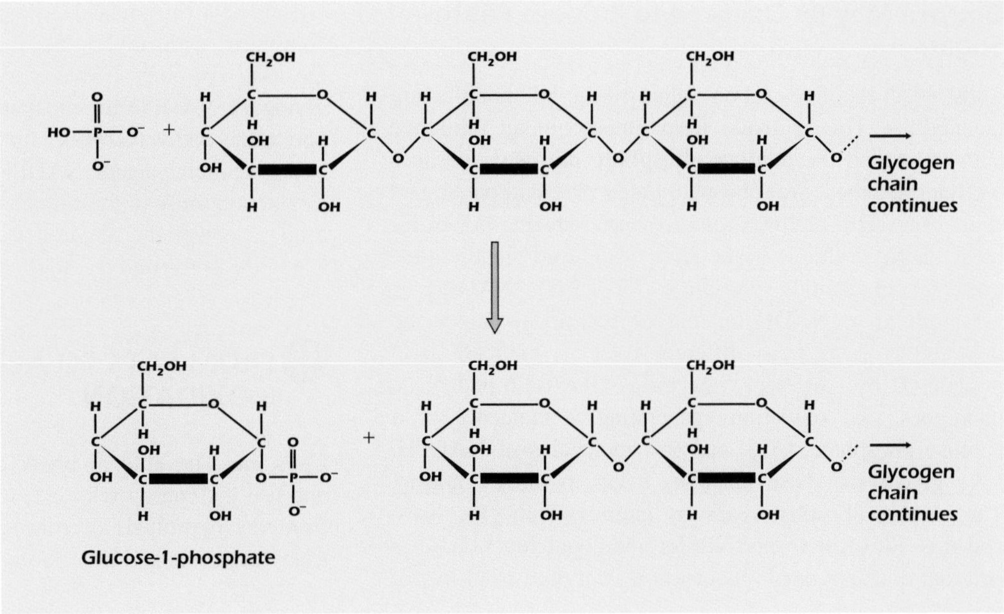

Glucose-1-phosphate

Glycogen chain continues

Figure 13.5. Glycogen phosphorylase cleaves a glucose monomer off glycogen and phosphorylates it.

One of the many important roles of the liver is to maintain the level of glucose in the blood—this is the most important of the circulating fuels and is the primary fuel for red blood cells and the brain. Glycogen stores in the liver can provide glucose when none is available from the gut. The glucose carrier (page 175) allows glucose to enter and leave the liver cells but cannot transport phosphorylated sugars, so glucose-6-phosphate must be converted to free glucose for transport out into the extracellular medium and from there to the blood. To do this, liver has the enzyme glucose-6-phosphate phosphatase, which removes the phosphate group. Muscle also has stores of glycogen, but these are for its own use: it does not have glucose-6-phosphate phosphatase and so cannot release glucose into the extracellular medium.

Medical Relevance 13.3 Red Blood Cells and Glucose-6-phosphate Dehydrogenase Deficiency

Hemoglobin is very effective at keeping its bound oxygen as O_2 but every so often a superoxide anion is released, or circulating nitrite is reduced to nitric oxide, leaving the heme iron oxidized to Fe^{3+}. When this happens the heme cannot bind any more oxygen until it is reduced back to Fe^{2+}. Most cells of our body have plenty of reducing power that originates in the NADH produced by Krebs cycle enzymes in the mitochondria. But mammalian red blood cells live dangerously. They experience a more oxidizing environment than most body cells, yet lack mitochondria. The only source of reducing power to reduce the Fe^{3+} back to Fe^{2+} is the pentose phosphate pathway, described on the next page.

Some apparently healthy men develop serious and sometimes fatal hemolytic anemia when they take some types of antimalarial drugs such as pamaquine. People suffering from this drug-induced anemia are found to have a variant form of the pentose phosphate pathway enzyme glucose-6-phosphate dehydrogenase that is less active than the normal form.

Most of the time individuals who have inherited defective glucose-6-phosphate dehydrogenase have enough activity to produce sufficient NADPH for their needs. Anything that leads to increased levels of reactive oxygen species ("oxidative stress") will use up the limited NADPH production capacity of the red blood cells. This results in damage of membrane proteins and to cross-linking of hemoglobin molecules via disulphide bonds so that the red blood cells become distorted. The result is loss of the red blood cells, sometimes to a disastrous level. The antimalarial drugs that cause this hemolytic anemia appear to cause the generation of reactive oxygen species and therefore cause anemia in individuals who have inherited the enzyme defect. Glucose-6-phosphate dehydrogenase deficiency is the most common human enzyme deficiency, affecting some 400 million people. It is X-linked so it affects only men, and is most common in people originating from parts of the Mediterranean and Africa.

Glucose May Be Oxidized to Produce Pentose Sugars

Cells need the five-carbon sugar ribose to manufacture nucleotides. This is made from glucose in an oxidative pathway called the **pentose phosphate pathway**, because its intermediates are phosphorylated five-carbon sugars (Fig. 13.6). This pathway also provides reducing power for biosynthetic reactions in the form of reduced nicotinamide adenine dinucleotide phosphate (NADPH). NADPH is very similar to NADH but has an additional phosphate; this allows it to have different roles in the cell as it is bound by different enzymes. Glucose-6-phosphate undergoes two oxidations generating a molecule of a pentose phosphate, CO_2, and two molecules of NADPH. Like glycolysis, these reactions occur in the cytoplasm. The pentose phosphate pathway interacts with glycolysis, and this allows it to perform an additional function. In a cell that does not need lots of ribose, but does need lots of NADPH for biosynthesis, the pentose is recycled by being combined with glycolytic intermediates. Going around the loop six times converts a glucose entirely to 6 CO_2, giving 12 NADPH for biosynthesis. Although the pentose phosphate pathway breaks up sugar to give CO_2 and the strong reducing agent NADPH, it is not used for energy production: only in the mitochondria, where NADH can be used to power the electron transport chain, can reducing agents be converted to other energy currencies.

FROM FATS TO ACETYL-CoA: β OXIDATION

Cells must be able to break down fatty acids (page 33) and use the energy released, whether the fatty acids come from the hydrolysis of triacylglycerols in fat droplets or triacylglycerols and phospholipids in food. Before they can

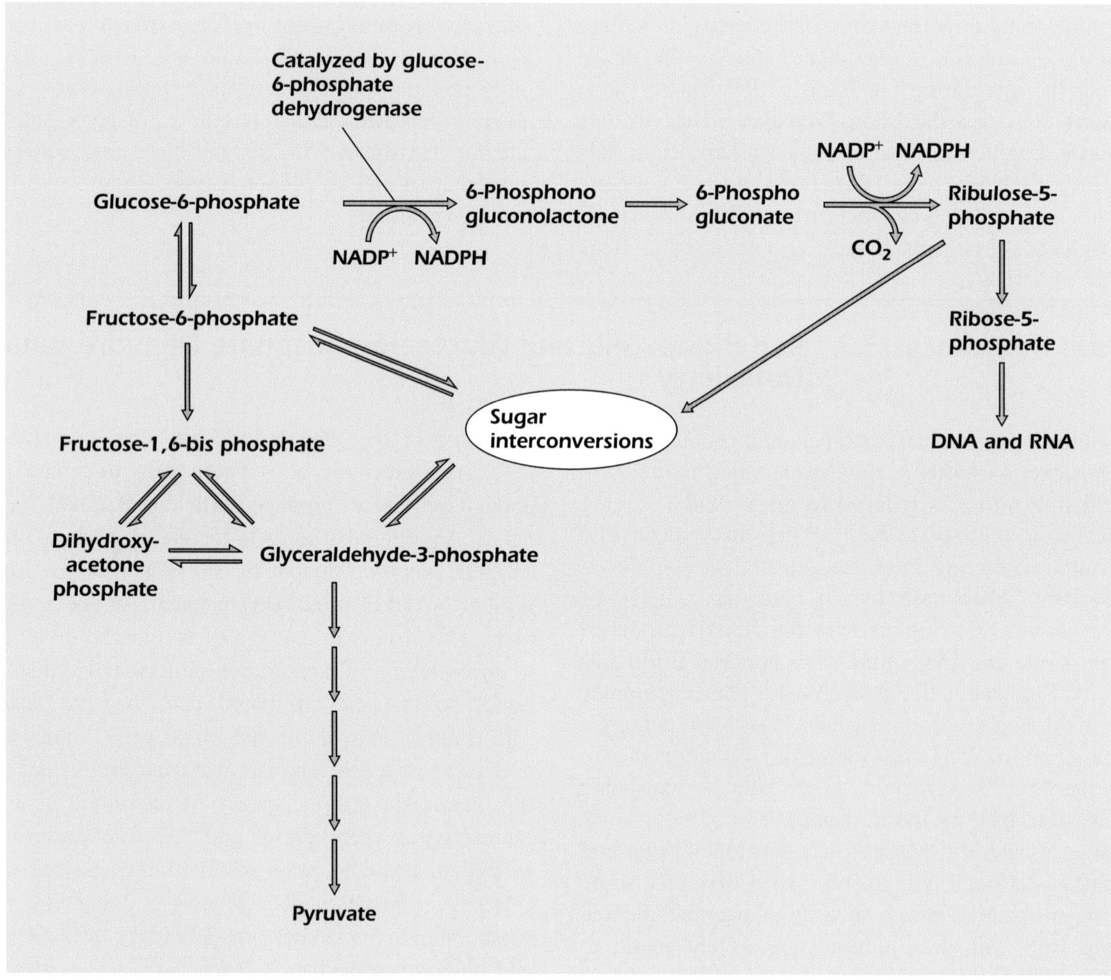

Figure 13.6. The reactions of the pentose phosphate pathway are shown in gray with the reactions of glycolysis in green.

be broken down, the fatty acids are coupled to coenzyme A to give an acyl-CoA in a reaction driven by the conversion of ATP to AMP. A spiral of reactions in the mitochondrial matrix called β oxidation (Fig. 13.7) then oxidizes the fatty acyl-CoA. (It is called **β oxidation** as oxidation is at the "β" carbon; see In Depth 2.1 on page 33) Each turn of the spiral shortens the fatty acyl chain by two carbons, releasing an acetyl-CoA and generating an NADH and an $FADH_2$ for each acetyl-CoA. The acetyl-CoA may be oxidized in the Krebs cycle or converted to other molecules.

During fasting, fat cells supply fuels for other parts of the body. There is a problem as triacylglycerols are insoluble in water, as are long-chain fatty acids. Fat cells break down their fat stores and release free fatty acids and these are bound by **albumen** in the blood. Tissues such as muscle and liver take up the bound fatty acids and oxidize them to acetyl-CoA. The acetyl-CoA is used in the Krebs cycle in most tissues but in the liver much of it is converted to soluble, circulating fuels called **ketone bodies** (Fig. 13.8; the word **ketone** means any chemical containing a carbon atom with single bonds to two other carbons and a double bond to an oxygen). The fundamental ketone body is acetoacetate, which the liver synthesizes from acetyl-CoA. Acetoacetate is then reduced to 3-hydroxybutyrate. These two molecules are important circulating fuels in mammals. Heart muscle, for instance, prefers ketone bodies to glucose as a fuel source. Ketone bodies are a normal part of everyday metabolism but they become increasingly important during starvation.

AMINO ACIDS AS ANOTHER SOURCE OF METABOLIC ENERGY

Protein forms a considerable part of the animal diet, even in vegetarians. It is broken down to free amino acids during digestion. These amino acids can be used for the biosynthesis of new proteins in the cell, but those in excess of this need can serve as metabolic fuels. To begin this process, the amino groups are removed in a process called transamination by aminotransferases such as alanine glyoxylate aminotransferase (page 184). The resulting carbon skeletons are then converted to intermediates in the Krebs cycle or to acetyl-CoA. Since there are 20 amino acids, there are many different pathways, but the overall effect is that the amino groups are passed to oxaloacetate or 2-oxoglutarate to form aspartate and glutamate, respectively. The amino groups on glutamate and aspartate are then converted to urea for excretion.

It is interesting to note that of the three major energy sources in the diet—carbohydrate, fat, and protein—only protein is not used as an energy store in our bodies. We make specific proteins when we need them but never simply as a way of storing amino acids. Of course, if one eats a lot of protein, one puts on weight: once the amino acids have been converted to carbon skeletons or to acetyl-CoA, these can then be used to make glucose and hence glycogen or to make fat.

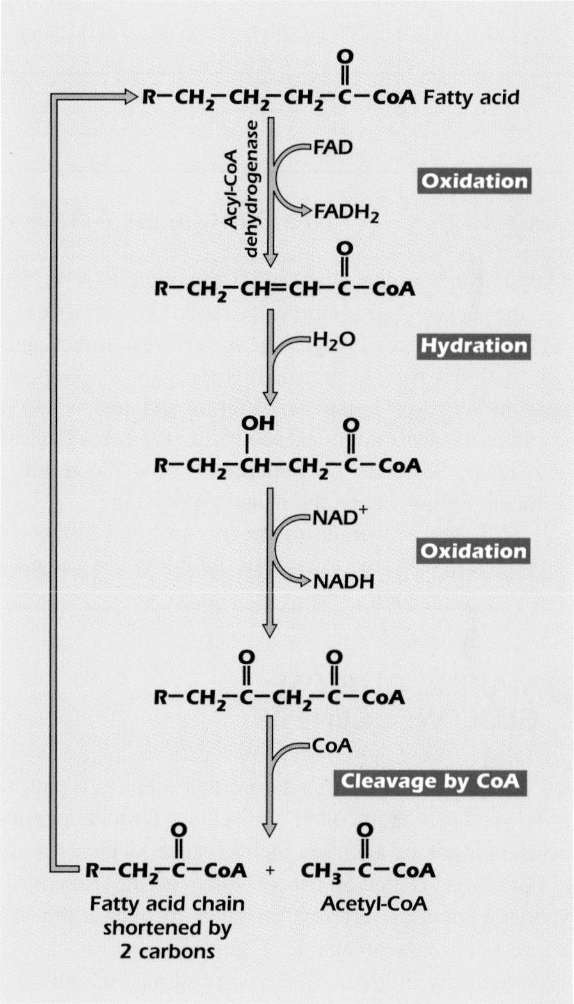

Figure 13.7. β oxidation of fatty acids produces acetyl-CoA.

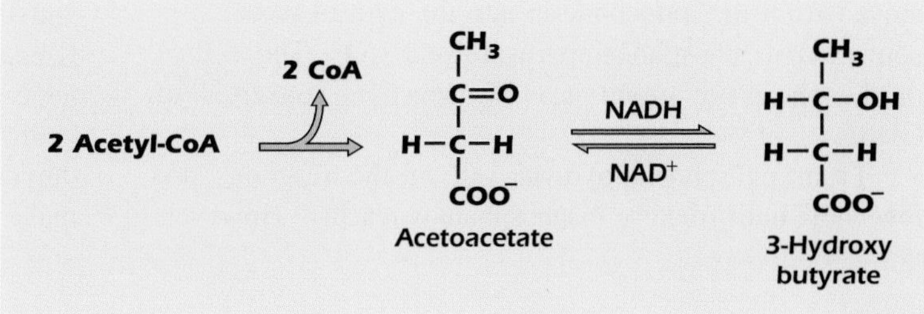

Figure 13.8. Formation of ketone bodies from acetyl-CoA.

Medical Relevance 13.4 Deficiency of Acyl-CoA Dehydrogenase and Sudden Infant Death Syndrome

The first step in β-oxidation of fatty acids is carried out by acyl-CoA dehydrogenase. There are different forms of this enzyme favoring different lengths of the fatty acyl chain. A relatively common genetic defect is the lack of active medium-chain acyl-CoA dehydrogenase. Babies who have inherited this condition develop symptoms of lethargy, vomiting and sometimes coma after fasting. Their blood glucose is very low. As glucose in the blood drops after feeding, the liver should increase ketone body formation from the acetyl-CoA resulting from fatty acid oxidation, but it cannot because β oxidation is blocked. This defect is responsible for about 10% of cases of sudden infant death syndrome. Gluconeogenesis slows and muscle (also unable to carry out fatty acid oxidation because of the enzyme defect) uses up glucose. When the deficiency is detected the treatment is simple: frequent feeding to avoid periods of fasting.

Example 13.2 The Dangerous Ackee Fruit

An unusual amino acid called hypoglycin A is found in the ackee fruit. Originally from West Africa, the ackee fruit is widely grown in the West Indies and is the national fruit of Jamaica. If the fruit is eaten when unripe it causes Jamaican vomiting sickness. The other symptoms are coma and convulsions, often followed by death. Sufferers have extremely low blood glucose concentrations, hence the name hypoglycin.

All these symptoms result from the fact that hypoglycin A is a potent inhibitor of acyl-CoA dehydrogenases (Fig. 13.7). β oxidation of fats fails, so the large energy reserves in body fat are useless to our cells. The only fuel cells can take from the blood and use is glucose, which is rapidly exhausted.

As the ackee fruit matures, the toxin moves from the fleshy part of the fruit, the aril, to the seeds. The arilli are now safe to eat, while the seeds are left alone to generate more ackee trees.

MAKING GLUCOSE: GLUCONEOGENESIS

Such is the importance of glucose that there is a pathway for its synthesis from other molecules. **Gluconeogenesis** enables animals to maintain their glucose levels even during starvation. It makes use of some of the enzymes of glycolysis but uses different enzymes to bypass the steps that are not freely reversible. Figure 13.9 shows the glycolytic pathway in green. The new reactions that allow the entire pathway to run in reverse are shown in gray.

New step 1: Pyruvate is carboxylated to make oxaloacetate (a reaction that also serves to top up levels of oxaloacetate for the Krebs cycle). The oxaloacetate is moved from the mitochondria into the cytosol where it is converted to phosphoenolpyruvate and CO_2. The source of the phosphate group is GTP, which is converted to GDP.

From phosphoenolpyruvate all of the reactions are reversible until fructose bisphosphate is reached. However, the reaction catalyzed by phosphofructokinase is not reversible and this step is avoided by:

New step 2: One of the phosphoester bonds on fructose bisphosphate is hydrolyzed by a phosphatase.

The interconversion of fructose-6-phosphate and glucose-6-phosphate is easily reversible. The final production of free glucose is accomplished by:

New step 3: The other phosphoester bond is hydrolyzed by glucose-6-phosphate phosphatase.

Gluconeogenesis is an expensive process: the conversion of two pyruvates to a glucose molecule uses four ATP, two GTP, and two NADH.

Different compounds can be fed into the gluconeogenesis pathway as appropriate. Glutamate is converted to 2-oxoglutarate (page 184), fed into the Krebs cycle, and tapped off as oxaloacetate to feed gluconeogenesis. Lactate and alanine are converted to pyruvate and hence oxaloacetate. Glycerol, released from lipids by hydrolysis, is phosphorylated and then oxidized to generate dihydroxyacetone phosphate. However, animals cannot make glucose from fats.

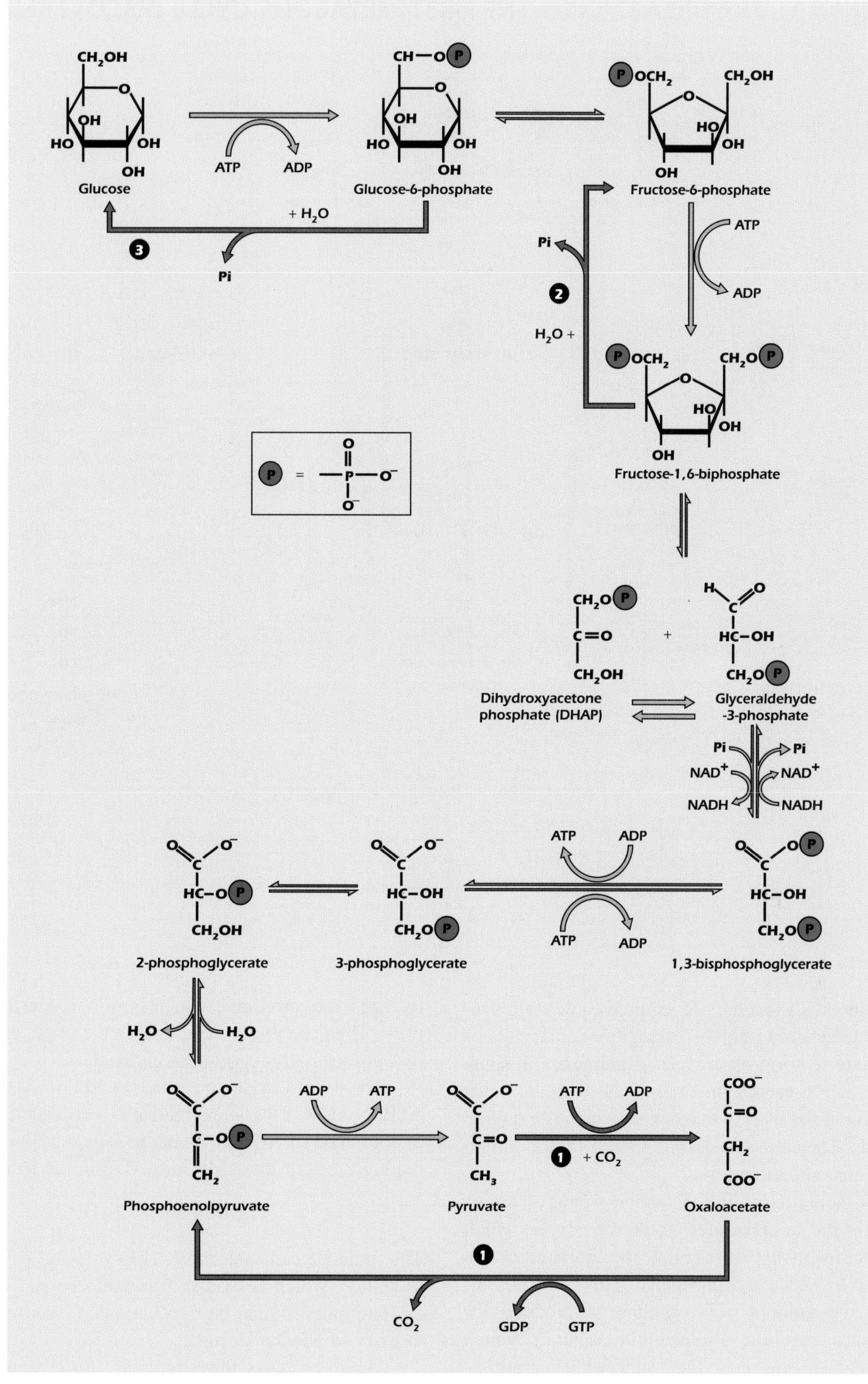

Figure 13.9. Gluconeogenesis allows glucose to be made from pyruvate.

IN DEPTH 13.1 THE UREA CYCLE—THE FIRST METABOLIC CYCLE DISCOVERED

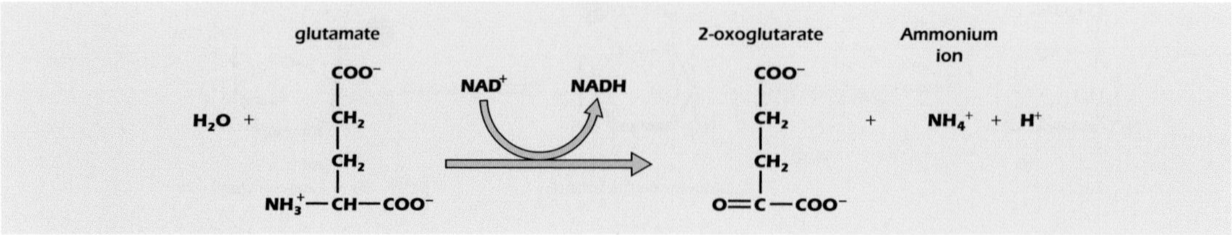

Dietary amino acids that are not required for protein synthesis have their amino groups passed to oxaloacetate or 2-oxoglutarate to form aspartate and glutamate, respectively. The amino groups on each of these amino acids are then used in the liver to make urea, which is excreted in the urine. The first step is the release of the amino group as a free ammonium ion.

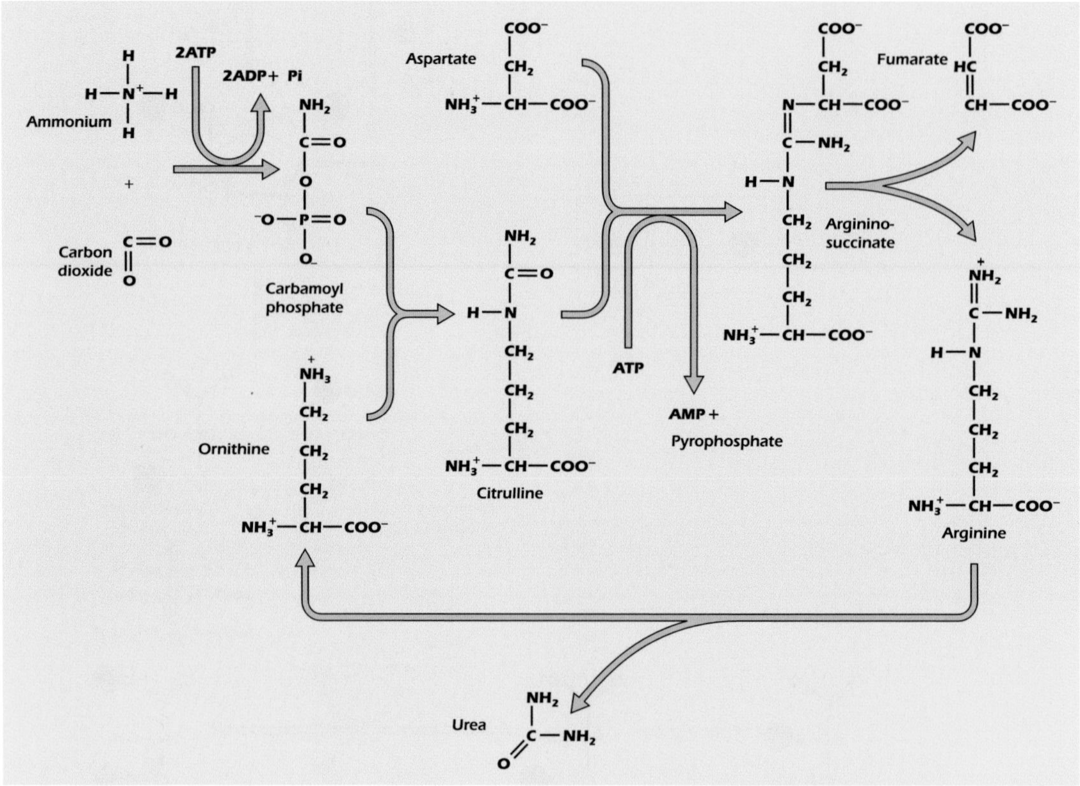

This ammonium ion, plus the amino groups on aspartate, is then used to make urea in the urea cycle, which operates mainly in the liver. First, the ammonium ion is converted to carbamoyl phosphate. This is then combined with the α-amino acid ornithine to produce citrulline. In turn, citrulline is joined with aspartate, which carries a second nitrogen into the cycle, producing argininosuccinate. Argininosuccinate is now cleaved to release fumarate and arginine. Urea is removed from arginine to regenerate ornithine, which is now ready for another cycle. Urea can then be excreted in the urine, eliminating two nitrogen atoms per urea molecule.

The overall cycle converts two ATP to ADP and one ATP to AMP. This is energetically equivalent to a total of four ATP to ADP conversions, since a fourth ATP is required to turn the AMP into ADP in the reaction:

$$\text{ATP} + \text{AMP} \rightleftarrows 2\text{ADP}$$

The fumarate can enter the Krebs cycle and be converted to malate, which is oxidized to oxaloacetate, which can (among other things) be transaminated back to aspartate to carry in another nitrogen.

Ornithine plays the same role in the urea cycle as oxaloacetate in the Krebs cycle: it accepts the incoming molecule, undergoes a series of interconversions, and is regenerated, allowing the cycle to begin again. The availability of ornithine determines the rate at which the cycle can operate. Arginine is a protein amino acid and is normally present in the diet as a source of ornithine. Vice versa, the urea cycle can make arginine so this is not normally considered to be an essential amino acid for adults. Arginine is, however, essential in the diet of growing children as the net protein synthesis of growth needs more of it than can be supplied from the urea cycle without draining too much and so impeding the cycle.

Since the urea cycle only occurs in the liver, liver failure results in a buildup of ammonium ions in the blood. This is toxic to nerve cells, producing first mental confusion and finally coma and death.

MAKING GLYCOGEN: GLYCOGENESIS

Glucose is stored as the polymer glycogen. Glucose polymerization, as a stand-alone reaction, has a positive ΔG and will not occur. The synthesis of glycogen is therefore driven by the hydrolysis of nucleoside triphosphates—not only ATP, but also UTP. If phosphorylated glucose is not available, then it is made by hexokinase using ATP (page 192). Glucose-1-phosphate then reacts with UTP to make UDP-glucose (Fig. 13.10). Glycogen synthase then transfers glucose from the UDP to the growing glycogen chain (Fig. 13.11). Other enzymes insert the $\alpha(1 \rightarrow 6)$ branches at intervals.

MAKING FATTY ACIDS, GLYCERIDES AND CHOLESTEROL

All cells need fatty acids for membrane lipids. Fat cells store large amounts of **fat** (triacylglycerols) in

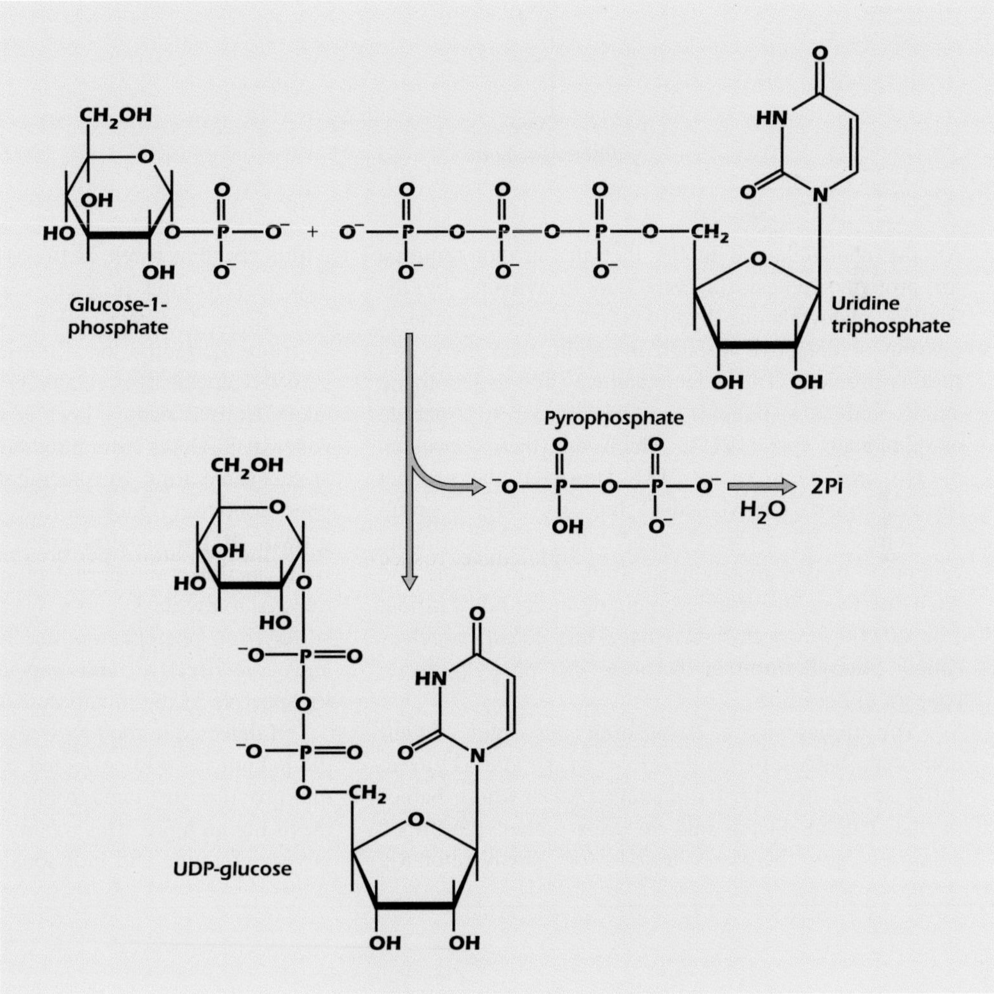

Figure 13.10. Uridine diphosphate glucose is synthesized from UTP and glucose-1-phosphate.

Medical Relevance 13.5 Phenylketonuria

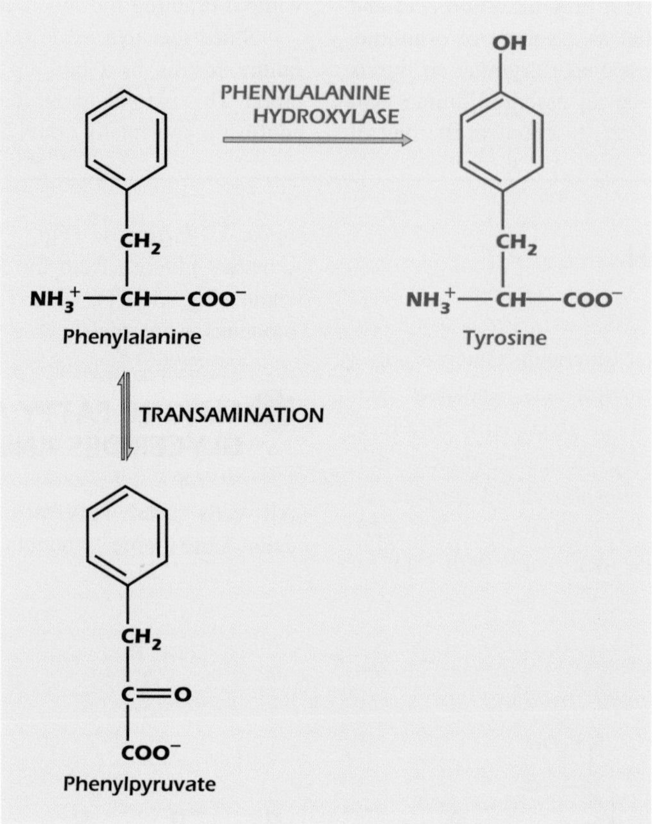

PHENYLALANINE HYDROXYLASE

Phenylalanine

Tyrosine

TRANSAMINATION

Phenylpyruvate

All the phenylalanine in the diet that is not required for protein synthesis is converted to tyrosine by the enzyme phenylalanine hydroxylase. If this enzyme is missing or defective through mutation, then there is a serious problem. The default pathway for metabolizing amino acids is to transfer the NH_3^+ group to aspartate or glutamate (page 215), which are then processed by the urea cycle (page 218). However, the product formed when phenylalanine has its NH_3^+ group transferred away, a phenylketone called phenylpyruvate, cannot be further metabolized. Both phenylalanine and phenylpyruvate therefore accumulate in the body. About 1 in 20,000 newborn babies have this defect, which is called **phenylketonuria** because the phenylpyruvate appears in the urine.

The disease is a devastating one, and, if left untreated, babies of only a few weeks old begin to suffer severe neurological damage. All newborn babies in the United States and in most other developed countries are therefore tested for phenylketonuria within the first week of life. This represents the largest genetic screening program carried out by the medical profession. The test, often called the heel-prick test or more correctly the Guthrie test after the scientist who developed it, is relatively simple. A drop of blood taken from a baby's heel is dried onto a small filter paper disk. Disks from hundreds of infants can be tested at the same time by placing their disks onto an agar plate containing bacteria that require phenylalanine for growth. If the bacteria grow, then the baby is at risk from phenylketonuria and blood from the baby will be retested a few days later to ensure the infant does indeed suffer from phenylketonuria. The Guthrie test is simple and cost-effective and can test for phenylketonuria irrespective of the mutation that has caused the problem.

If detected in the first weeks of life, the prognosis for the patient is good. Affected infants are fed a strict diet that provides just enough phenylalanine for protein synthesis but no more. This treatment, carried on to maturity, is very effective and individuals develop normally.

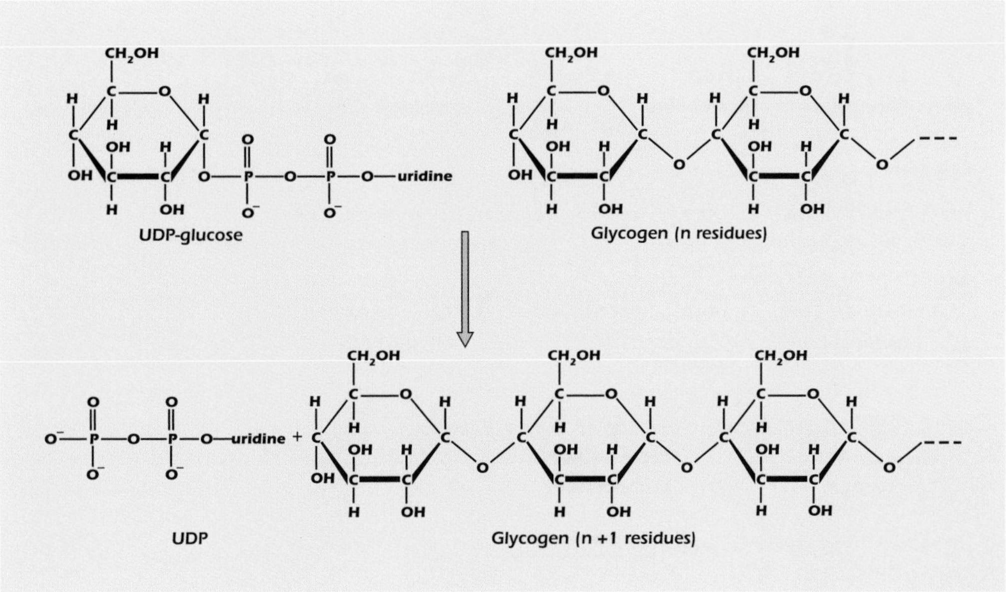

Figure 13.11. Glycogen is synthesized from UDP-glucose monomers.

times of plenty, using fats delivered to them from the gut or from the liver (delivered in soluble complexes with proteins). Triacylglycerols are made from free fatty acids and glycerol.

Fatty acids can be made from glucose and from amino acids. The basic machinery (fatty acid synthase) is a multienzyme complex (in bacteria) or a multidomain protein (in eukaryotes) that uses as a starting point the two-carbon unit of acetyl-CoA. The growing fatty acid chain is not released: it swivels from enzyme to enzyme or domain to domain in the array, adding two carbons for each complete cycle until the limiting length of 16 carbons is reached; the product, palmitic acid, is then released. Although the reactions look similar (Fig. 13.12), the process is not a reversal of β oxidation (page 214). It uses entirely different enzymes, takes place in the cytosol rather than in the mitochondria, and is separately regulated. Like much of biosynthesis it is reductive, and the reducing power comes not from NADH but from the closely related dinucleotide NADPH. The liver is an important site for fatty acid synthesis.

Initially acetyl-CoA is carboxylated to malonyl-CoA. From here on, however, fatty acid synthesis does not use free coenzyme A to carry the growing chain but instead uses a protein called acyl carrier protein (ACP). The malonyl residue is transferred to ACP from malonyl-CoA. This condenses with a molecule of acetyl-ACP (made from an acetyl-CoA) to give a four-carbon molecule with the release of ACP and CO_2. The four-carbon acetoacetyl-ACP is next reduced to hydroxybutyryl-ACP. The next enzyme (molecule or domain) removes water, leaving a double bond, which is again reduced to give butyryl-ACP. Another malonyl-ACP is condensed with this, and the cycle continues. Finally a chain 16 carbons long has been made (palmitic acid). At this point it is hydrolyzed from the ACP. Overall 14 NADPH molecules, 1 acetyl-CoA, and 7 malonyl-CoA molecules have been used to make palmitic acid. Palmitic acid is then used by enzymes on the endoplasmic reticulum that extend the chains and that can introduce double bonds. Mammals cannot, however, synthesize all the different kinds of fatty acids that they need for their membranes and must obtain essential fatty acids in food (page 39).

The main use of fatty acids is to make glycerides, both triacylglycerols for storage in fat globules and phospholipids for membranes. The process uses glycerol phosphate, which is usually generated by reduction of dihydroxyacetone phosphate. The fatty acid is then swapped in, replacing the phosphate group, which leaves as inorganic phosphate.

Cholesterol is made in the liver using a pathway that starts by combining acetyl-CoA and acetoacetate to make hydroxyglutaryl-CoA. This is then reduced to a compound called mevalonic acid. This reduction is the first committed step in cholesterol biosynthesis; the enzyme responsible is the target of both cellular controls and of the cholesterol-lowering drugs called statins.

SYNTHESIS OF AMINO ACIDS

Nitrogen is an important constituent of proteins and nucleic acids and many other molecules important in cells. Although nitrogen gas is plentiful, making up 80% of the atmosphere, it is inert. It is chemically a very difficult task to break the triple bond and reduce nitrogen gas to ammonia that can be used for incorporation into biomolecules. Chemical fertilizers are made using the Haber process, which fixes nitrogen by the use of pressures of around 300 atmospheres and temperatures of 500°C. Some nitrogen is fixed naturally by lightning,

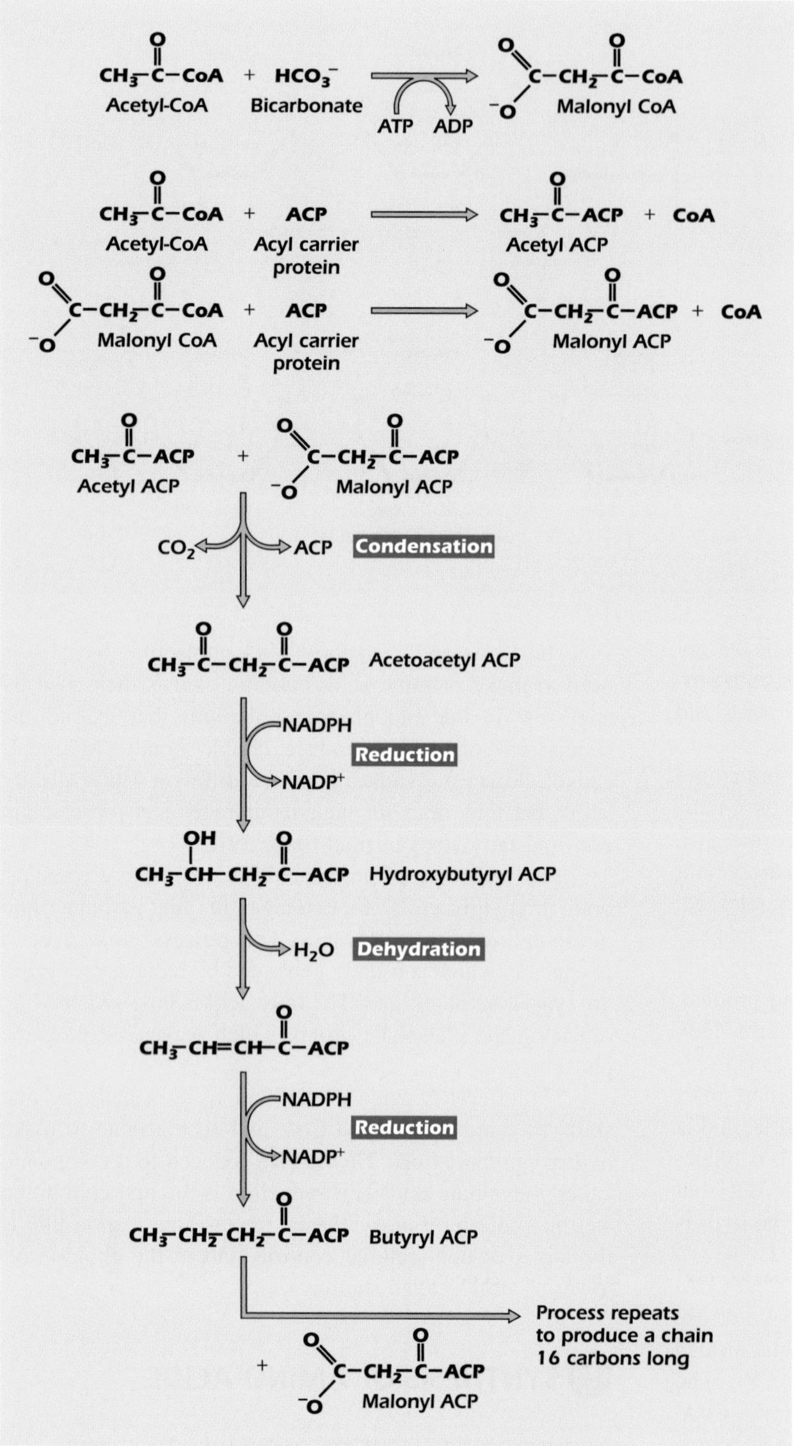

Figure 13.12. Synthesis of fatty acids.

but most is fixed by various types of prokaryotes, which possess the complex enzyme nitrogenase. Some of these nitrogen-fixing organisms are free living while others form symbiotic relationships with plants: the root nodules of legumes are a good example of plants generating a special environment for their nitrogen-fixing symbionts.

Nitrogenase consists of two protein complexes: a reductase and an iron-molybdenum protein. Nitrogen gas is bound to the iron-molybdenum cofactor where it is reduced to ammonia. Ammonia in water forms ammonium ions NH_4^+, which can be incorporated into the amino acid glutamine and from there into other amino acids and other molecules.

IN DEPTH 13.2 EATING WELL, GETTING FAT

Dihydroxyacetone phosphate

NADH + H$^+$ NAD$^+$

Glycerol-3-phosphate dehydrogenase present in most cells

Glycerol phosphate

ADP

ATP

Glycerol kinase found in liver only

Glycerol phosphate, the building block from which lipids and phospholipids are made, is obtained in most cells by reducing dihydroxyacetone phosphate. This in turn is generated in the glycolytic pathway. This means that adipose cells can only make fats when glucose is abundant—when we are eating lots of sugar or other carbohydrate. In contrast, liver cells have the enzyme glycerol kinase, which phosphorylates glycerol to make glycerol phosphate directly so that some lipids and phospholipids can be made even during times of fasting.

Given ammonium ions, plants and bacteria can synthesize all 20 amino acids. Animals are more limited and must obtain some amino acids from their diet. Amino transferases (page 184) allow animals to move amino groups from an amino acid to an oxo-acid to generate a new amino acid, but there are some carbon skeletons that cannot be synthesized. Adult humans require histidine, isoleucine, leucine, lysine, methionine, phenylalanine, threonine, tryptophan, and valine in their diet—these are the **essential amino acids**. Adult humans are in nitrogen balance—we excrete the same amount of nitrogen as we take in. Growing infants, however, have a net uptake of nitrogen: they take in more nitrogen than they excrete. Growth clearly demands more amino acids, and in this case the limited ability we have to synthesize arginine is insufficient and so it must be present in the diet as well.

Other molecules, such as the nucleic acid bases, are synthesized from amino acid starting materials.

CONTROL OF ENERGY PRODUCTION

Feedback and Feedforward

We have seen how an energy currency that runs low is topped up by conversion from another currency. This is not enough to ensure a constant energy supply, however. Therefore the cell has more mechanisms that ensure that the supply of cellular energy is accelerated or slowed as appropriate. These mechanisms are of two types: **feedforward** and **feedback**. We will introduce the terms by analogy with real money. Consider a bank teller. During the day people deposit checks but draw out cash to spend. As time passes the stock of banknotes and change in the till gets low. The teller signals to the supervisor, who opens the bank vault, takes out more cash, and refills the teller's till. This is an example of **negative feedback**. In general negative feedback is said to occur when a change in some parameter activates a mechanism that reverses the change in that parameter. We have already met an analogous negative feedback system in the control of tryptophan biosynthesis (page 90). A downward change in the concentration of tryptophan in the bacterial cell activates the mechanism that causes the cell to make more tryptophan. **Positive feedback** is less common in both biology and banking. It is said to occur when a change in some parameter activates a mechanism that accelerates the change. In banking this occurs when a rumor starts that a bank is about to fail. The lower a bank's reserves of money get, the more its depositors rush to take their money out before it is too late. Biological examples of positive feedback are unusual. Quorum sensing in luminescent bacteria (Example 6.1 on page 89) and the action potential (page 241) are two examples of positive feedback that we describe in this book.

What about feedforward? The bank is especially busy at lunchtime, with lots of cash withdrawn between 12:30 and 2:00. During this time, everyone is rushed off their feet. The supervisors do not wait until tellers signal that they are short of cash but instead open the vaults at 12 noon and bring out enough cash to see the tellers through the lunchtime rush. They are preparing for a future drain on cash by stocking up the tills before the drain occurs—this is feedforward. Feedforward happens in biological systems too, as we will see in this section.

Negative Feedback Control of Glycolysis

Phosphofructokinase catalyzes the first irreversible step in glycolysis after the paths from glucose and glycogen converge. The enzyme is allosterically regulated by ATP (Fig. 13.13). When ATP concentrations are high, ATP binds to regulatory sites on phosphofructokinase and locks it into an inactive (low affinity for fructose phosphate) conformation. When the concentration of ATP is reduced, levels of AMP increase (because ADP is converted to ATP and AMP; see page 192). AMP will compete with ATP so that an increasing number of phosphofructokinase molecules come to have AMP in the regulatory sites, which causes the enzyme to switch to the active, high-affinity conformation. Fructose-1,6-bisphosphate is produced, feeding the glycolytic pathway that in turn feeds the mitochondria with pyruvate for the production of ATP. This process is negative feedback because changes in the concentration of ATP act, through its allosteric action on phosphofructokinase, to reverse the change in ATP concentration.

Feedforward Control in Muscle Cells

When a signal goes out from our brains to the muscles in our legs to tell them to start working, it causes the endoplasmic reticulum to release calcium ions into the cytosol. Calcium is acting as an intracellular messenger, a topic we will cover in more detail in Chapter 15. The increase of calcium activates several processes. It causes the muscle cell to contract, using the energy released by ATP hydrolysis to do mechanical work (page 289). At the same time, other calcium ions pass through a channel into the mitochondrial matrix, attracted by the large negative voltage of the mitochondrion interior. Once there, calcium activates three key enzymes: pyruvate dehydrogenase, oxoglutarate dehydrogenase, and malate dehydrogenase (pages 208, 210). The

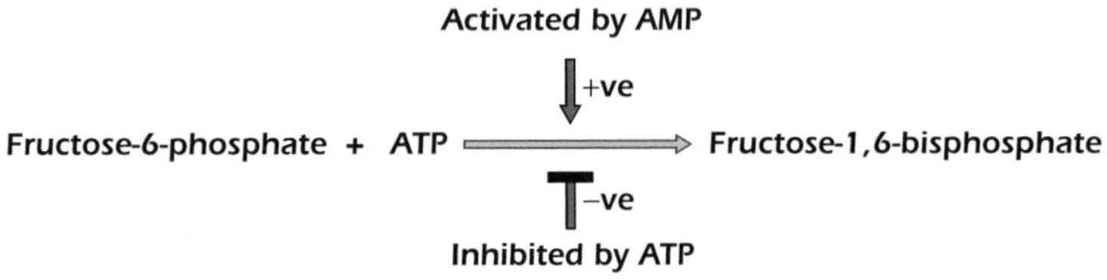

Activated by AMP

+ve

Fructose-6-phosphate + ATP ⟹ Fructose-1,6-bisphosphate

−ve

Inhibited by ATP

Figure 13.13. Phosphofructokinase is regulated by the binding of ATP or AMP at a regulatory site that is separate from the active site.

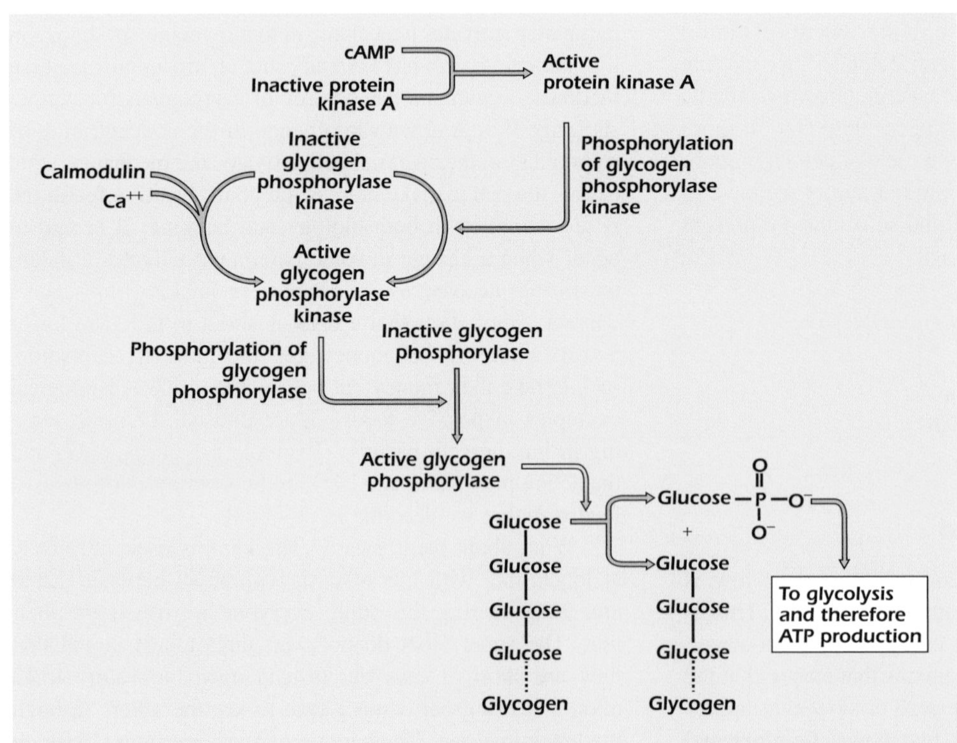

Figure 13.14. Calcium and cyclic AMP both activate glycogen breakdown in muscle and liver.

cell has not waited until the ATP concentration starts to decrease before activating the Krebs cycle in the mitochondria, so this is feedforward control. Meanwhile, in the cytosol, calcium ions bind to the protein calmodulin (page 149), which in turn activates the enzyme **glycogen phosphorylase kinase** (Fig. 13.14). Glycogen phosphorylase kinase is only able to phosphorylate one target, glycogen phosphorylase. Glycogen phosphorylase is activated by phosphorylation (of a serine residue) and proceeds to break down glycogen to release glucose-1-phosphate, which is fed into the glycolytic pathway. The cell has not waited until glucose concentration falls before activating glycogen breakdown, so this is feedforward control.

In fact, muscles can begin to break down glycogen even before the message goes out from the brain to tell them to contract. When the brain realizes that we are in a dangerous situation and might be going to have to run, it causes the release of the hormone **adrenaline** from the adrenal glands above the kidneys. Adrenaline binds to an integral membrane protein of the skeletal muscle cells called the **β-adrenergic receptor**. This causes the production of the intracellular messenger cyclic AMP (cAMP) within the cytosol of the muscle cell. This topic is dealt with in more detail in Chapter 15. cAMP then activates the serine-threonine kinase **cAMP-dependent protein kinase**, which is given the short name of **protein kinase A** (Fig. 13.14). Protein kinase A phosphorylates target proteins on serine and threonine residues. Glycogen phosphorylase kinase, the enzyme that phosphorylates glycogen phosphorylase, is itself phosphorylated by protein kinase A, and is then active even when cytosolic calcium is low.

Thus even before we know for sure that we need to run, the muscles are breaking down glycogen and making the glucose they will use if running becomes necessary.

SUMMARY

1. Metabolism is the collective term for all of the reactions going on inside a cell. These reactions are divided into catabolic—those that break down chemical compounds to provide energy—and anabolic—those that build up complex molecules from simpler ones. Catabolic reactions are oxidative and anabolic reactions are reductive.
2. The Krebs cycle is at the center of the cell's metabolism. It can act to oxidize two carbon units derived from carbohydrates, fats, or amino acids and acts as a central switching yard for the molecules in metabolism.
3. Glycolysis converts glucose to pyruvate. If the pyruvate is reduced to lactate, glycolysis can continue in the absence of oxygen as this reduction regenerates the NAD^+ needed for glycolysis.
4. Glycogen can act as a reserve of glucose for glycolysis.
5. Glucose-6-phosphate can be converted to a pentose sugar for nucleotide manufacture with production of NADPH for biosynthesis.
6. Fats (triacylglycerols) are concentrated fuel stores. Their fatty acid components are oxidized to two-carbon units by β oxidation.
7. Amino groups must be removed before excess dietary amino acids can be used as fuels. This is done by transferring them to make aspartate or glutamate, and thence to urea for excretion.
8. Gluconeogenesis allows the synthesis of glucose from noncarbohydrate precursors. However, mammals cannot make glucose from fatty acids.
9. Biosynthetic pathways for molecules follow different routes from the catabolic pathways. Good examples are fatty acid synthesis and breakdown (β oxidation) and glycogen synthesis and breakdown.
10. Metabolic reactions are controlled by feedforward and feedback mechanisms, which make use of allosteric control and covalent modification of key enzymes.

Answer to Thought Question: Each turn of the Krebs cycle generates two carbon dioxide molecules for each acetyl-CoA fed in: one turn of the cycle regenerates oxaloacetate for the next cycle. If oxaloacetate is removed for gluconeogenesis the cycle will soon stop as it runs out of oxaloacetate. (The oxaloacetate is regenerated from pyruvate using pyruvate carboxylase.) Note that the carbon dioxides released in *one* turn are not the two carbons that entered from acetyl-CoA. Subsequent turns of the cycle will release these.

FURTHER READING

Bender, D. A. (2007) *An Introduction to Nutrition and Metabolism*, 4th edition, CRC Press, London.

Buchanan, B. B., Gruissem, W., and Jones, R. L. (2000) *Biochemistry and Molecular Biology of Plants*. American Society of Plant Physiologists, Rockville, Maryland.

Devlin, T. M. (2010) *Textbook of Biochemistry with Clinical Correlations*, 7th edition, John Wiley & Sons, Hoboken.

Voet, D., and Voet, J. D. 2011. *Biochemistry*, 4th edition, John Wiley & Sons, Hoboken.

REVIEW QUESTIONS

13.1 Theme: Reactions and Pathways

A acetoacetate
B acetyl-CoA
C fructose-1,6-bisphosphate
D glucose-1-phosphate
E glucose-6-phosphate
F NADH
G NADPH
H oxalacetate
I stearic acid

Choose, from the list above, one of the *product(s)* of the reactions, pathways and enzymes listed below.

1. fatty acid synthesis
2. glucose-6-phosphate dehydrogenase
3. gluconeogenesis
4. lactate dehydrogenase
5. phosphofructokinase

13.2 Theme: Pathways and Enzymes

A anabolic metabolism
B β oxidation
C fatty acid synthesis
D gluconeogenesis
E glycogen synthesis
F glycolysis
G pyruvate dehydrogenase
H the Krebs cycle
I the pentose phosphate pathway

Match the term in the list above with the appropriate description in the list below.

1. Converts fatty acids to acetyl-CoA
2. Converts pyruvate to acetyl-CoA
3. The only way the red blood cell has to make ATP
4. The source of 5-carbon sugars and a source of NADPH for biosynthesis
5. Uses UDP-glucose

13.3 Theme: Metabolism

A Basic amino acids are . . .
B Essential amino acids are . . .
C Feedforward control works to provide glucose-6-phosphate for muscle contraction by . . .
D It is necessary to convert pyruvate to lactate when there is little oxygen available because . . .
E Ketone bodies are . . .
F Pyruvate carboxylase makes . . .
G The enzymes of the Krebs cycle are . . .
H The main difference between β oxidation of fatty acids and the synthesis of fatty acids is . . .
I When one is starving, blood glucose is maintained by . . .

For each of the phrases in the list below, there is a phrase in the list above that can be combined with it to create a single true statement. Choose the appropriate sentence start above for each of the sentence terminations below.

1. . . . amino acids that an organism cannot make and so must be present in the diet.
2. . . . amino acids with a side chain that can be protonated.
3. . . . located mainly in the mitochondrial matrix with one in the inner mitochondrial membrane.
4. . . . oxalacetate. This is used to top up oxalacetate for the Krebs cycle and is an important step in gluconeogenesis.
5. . . . the reaction regenerates the NAD^+ used up during glycolysis (during the oxidation of glyceraldehyde-3-phosphate to 1,3-bisphosphoglycerate).

THOUGHT QUESTION

We have stated that animals cannot make glucose from fats. Why not? β oxidation of fatty acids generates acetyl-CoA, and acetyl-CoA can be fed into the Krebs cycle, one of whose intermediates is oxaloacetate. Why therefore cannot animals make glucose from fats? It appears simple enough to divert oxalacetate from the Krebs cycle for gluconeogenesis.

14

IONS AND VOLTAGES

We described in Chapter 2 how membranes are composed of phospholipids arranged so that their hydrophobic tails are directed toward the center of the membrane, while the polar hydrophilic head groups face out. Membranes are a barrier to the movement of many solutes. In particular, small hydrophilic solutes such as ions and sugars cannot pass through membranes easily because, to do so, they would have to lose the cloud of water molecules that forms their hydration shell (page 18). Two consequences follow from the fact that membranes are barriers. First, the composition of the liquid on one side of a membrane can be different from the composition of the liquid on the other side. Indeed, by allowing cells to retain proteins, sugars, ATP, and many other solutes, the barrier property of the plasma membrane makes life possible. Table 14.1 shows how five important ions have different concentrations in cytosol and extracellular medium. Second, the cell must make proteins called channels and carriers whose job it is to help hydrophilic solutes across the membrane. This chapter describes how cells make use of this barrier property of membranes.

THE POTASSIUM GRADIENT AND THE RESTING VOLTAGE

Ions are electrically charged. This fact has two consequences for membranes. First, the movement of ions across a membrane will tend to change the voltage across that membrane. If positive ions leave the cytosol, they will leave the cytosol with a negative voltage, and vice versa. Second, a voltage across a membrane will exert a force on all the ions present. If the cytosol has a negative voltage, then positive ions such as sodium and potassium will be attracted in from the extracellular medium. We will begin to address the question of how ions and voltages interact by considering the effect of potassium movements on the voltage across the plasma membrane.

Potassium Channels Make the Plasma Membrane Permeable to Potassium Ions

Potassium channels (Fig. 14.1) are found in the plasma membrane of almost all cells. They are tubes that link the cytosol with the extracellular medium. Potassium ions, which cannot pass through the lipid bilayer of the plasma membrane, pass through potassium channels easily. Other ions cannot go through. The precise shape of the tube, and the position of charged amino acid side chains within the tube, blocks their movement. The channels are selective for potassium.

We saw earlier that the sodium/potassium (Na^+/K^+) ATPase (page 199) uses the energy of ATP hydrolysis to drive sodium ions out of the cell and, at the same time, bring potassium ions into the cell. This ensures that potassium is much more concentrated in the cytosol than outside—typically 140 mmol $liter^{-1}$ in the cytosol but only 5 mmol $liter^{-1}$ in the extracellular medium. There is an apparent paradox here. If potassium can pass through the potassium channel, why is this ion much more concentrated inside the cell than outside? Why doesn't all

Cell Biology: A Short Course, Third Edition. Stephen R. Bolsover, Jeremy S. Hyams, Elizabeth A. Shephard and Hugh A. White.

TABLE 14.1. Typical Concentrations for Five Important Ions in Mammalian Cytosol and Extracellular Medium

Ion	Cytosol	Extracellular Medium
Sodium Na^+	10 mmole liter^{-1}	150 mmole liter^{-1}
Potassium K^+	140 mmole liter^{-1}	5 mmole liter^{-1}
Calcium Ca^{2+}	100 nmole liter^{-1}	1 mmole liter^{-1}
Chloride Cl^-	5 mmole liter^{-1}	100 mmole liter^{-1}
Hydrogen ions H^+ (really H_3O^+)	60 nmole liter^{-1} or pH 7.2	40 nmole liter^{-1} or pH 7.4

Note: The unit n for nano (10^{-9}) is one million times smaller than unit m for milli (10^{-3}).

IN DEPTH 14.1 MEASURING THE TRANSMEMBRANE VOLTAGE

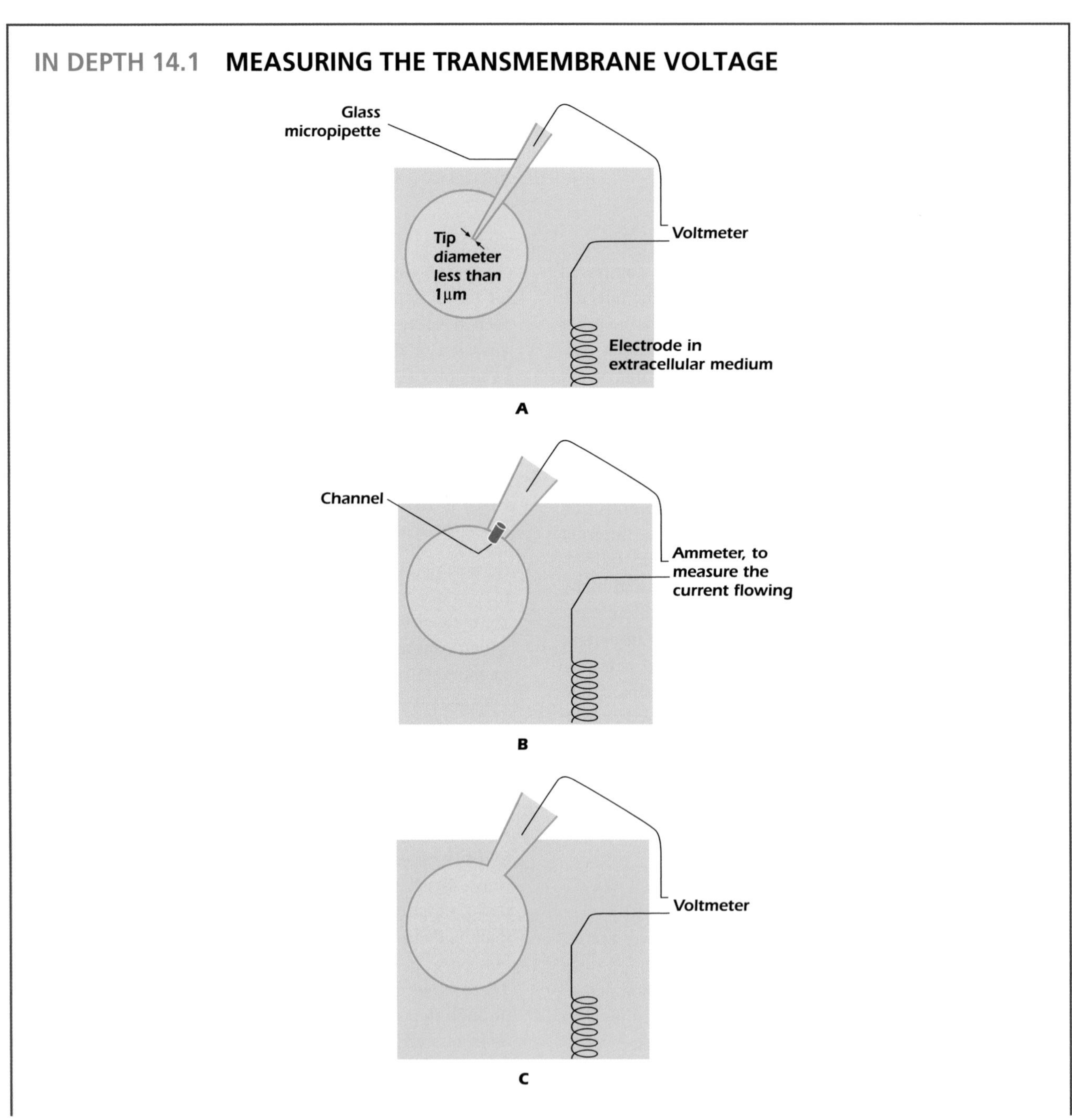

In 1949 Gilbert Ling and Ralph Gerard discovered that when a fine glass micropipette filled with an electrically conducting solution impaled a cell (panel A in the diagram), the plasma membrane sealed to the glass, so that the transmembrane voltage was not discharged. The voltage difference between a wire inserted into the micropipette and an electrode in the extracellular medium could then be measured. By passing current through the micropipette, the transmembrane voltage could be altered.

Twenty-five years later, Erwin Neher and Bert Sakmann showed that the micropipette did not have to impale the cell. If it just touched the cell, a slight suction caused the plasma membrane to seal to the glass (panel B). The technique, called cell-attached patch clamping, can measure currents through the few channels present in the tiny patch of membrane within the pipette.

Stronger suction bursts the membrane within the pipette (panel C). The transmembrane voltage can now be measured. Alternatively, current can be passed through the micropipette to change the transmembrane voltage—this is the whole-cell patch clamp technique. In 1991, Neher and Sakmann received the Nobel prize for medicine.

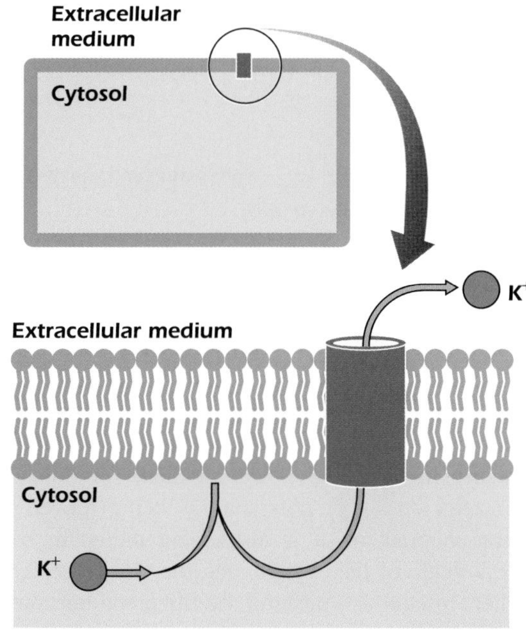

Figure 14.1. The positively charged potassium ion cannot cross the lipid bilayer, but passes easily through a water-filled tube in a potassium channel.

the potassium rush out? To explain why, we must think about the effects of ion movement on transmembrane voltage.

Concentration Gradients and Electrical Voltage Can Balance

Figure 14.2A shows a glial cell. Many glial cells express only potassium channels in their plasma membranes. For these cells, the cytosol is about −90 mV relative to the extracellular medium. Potassium ions are acted upon by two forces. They would leave the cell under the influence of the concentration gradient, but are pulled in by the negative voltage of the cytosol. For every ion present on both sides of a membrane, it is possible to calculate the transmembrane voltage that will exactly balance the concentration gradient. This voltage is the **equilibrium voltage** for that ion at that membrane. When cytosolic and extracellular potassium concentrations are 140 mmole liter^{-1} and 5 mmole liter^{-1} respectively, the potassium equilibrium voltage at human body temperature is −90 mV. Thus for the glial cell with a membrane voltage of −90 mV, the forces on potassium ions exactly balance and the cell neither gains nor loses potassium. This in turn means that the cytosol is neither gaining nor losing charge, so the membrane voltage does not change. The condition shown, in which the voltage is equal to the equilibrium voltage of potassium, is a stable one. We can see that the condition is stable by thinking about what would happen if the voltage was for some reason perturbed to, for example, −80 mV. At this new voltage the electrical force pulling the potassium ions in is not strong enough to oppose the concentration force favoring potassium loss. Potassium ions would leave and as they did so they would carry their positive charge out, so the cytosol voltage would move in a negative direction. This would continue until the concentration and voltage forces again balance, which would be when the voltage has returned to the potassium equilibrium voltage. In a similar but opposite way, if the membrane voltage were artificially perturbed to a value that is more negative than the potassium equilibrium voltage, potassium ions would move in until the resulting movement of charge has returned the membrane voltage to the potassium equilibrium voltage. In general, a cell whose plasma membrane is permeable to one ion only will have a voltage equal to the equilibrium voltage of that ion.

In other cells the situation is more complicated. In nerve cells, for instance (Fig. 14.3), the voltage of an unstimulated cell is −70 mV (Fig. 14.2B). This is because these cells have in addition a second type of channel called the **voltage-gated sodium channel** that allows sodium ions (and only sodium ions) to pass. As its name suggests, opening of the voltage-gated sodium channel is controlled by the membrane voltage. At −70 mV,

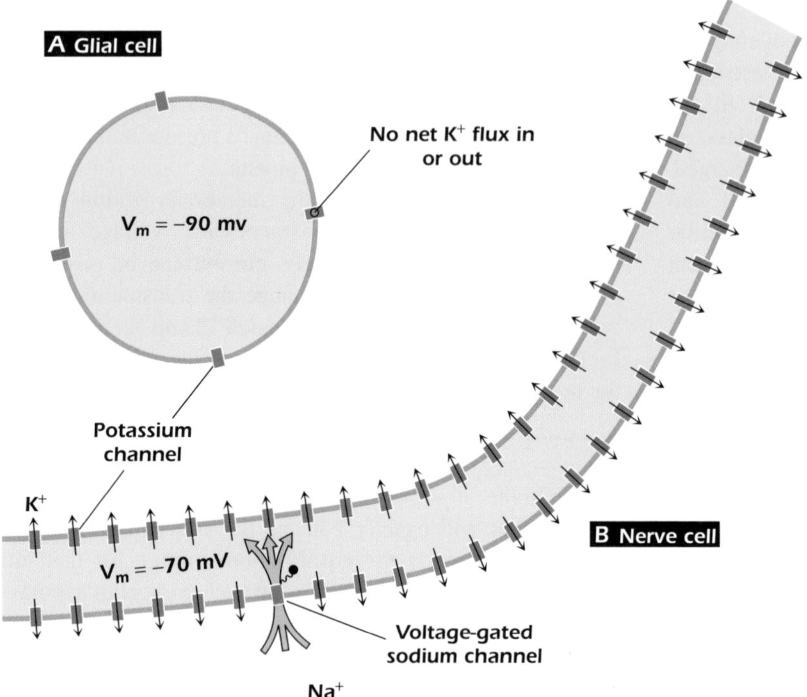

Figure 14.2. The resting voltage of (A) glial and (B) nerve cells.

the vast majority of voltage-gated sodium channels are shut: only one in every 4,000 are open at any one time. Nevertheless the few channels that are open allow sodium ions to move in, so that in an unstimulated nerve cell the membrane settles down to a **steady state** voltage in which the inward current through voltage-gated sodium channels is balanced by an outward current of potassium. Potassium flows because the cytosol is no longer negative enough to hold these ions in against their outward concentration gradient. Even in nerve cells, though, the membrane is more permeable to potassium than to any other ion, so the voltage of the unstimulated cell does not deviate very far from the potassium equilibrium voltage. However, in nerve cells and in all other cells in which the plasma membrane has a significant permeability to sodium, the sodium/potassium ATPase has to work constantly to maintain the concentration gradients across the plasma membrane. This is an example of a steady state maintained by the constant expenditure of energy (page 191). As we will see, the voltage across the nerve cell membrane changes dramatically when the cell is stimulated and transmits the electrical signals for which it is specialized. The term **resting voltage** is used for the voltage across the plasma membrane of the unstimulated cell. We also talk about the resting voltage of cells where the membrane voltage never changes, so we say that the resting voltage of a glial cell is −90 mV.

The negative resting voltage characteristic of most cells turns the action of the Na^+/K^+ ATPase into an energetically asymmetrical one. Consider one conversion cycle: one molecule of ATP is hydrolyzed, three sodium ions are pushed out of the cell, and two potassium ions move in. Very little of the energy of ATP hydrolysis is used up in moving the two potassium ions into the cell because the electrochemical gradient for this ion is close to zero. Although potassium is being moved up a concentration gradient, it is also being pulled in by the negative voltage of the cytosol, and the two forces cancel each other. In contrast, pushing the three sodium ions out of the cell requires more energy than would be required if the cytosol were at the same voltage as the extracellular medium. The sodium ions are positively charged, so they are attracted by the negative voltage inside the cell, which combines with the concentration gradient to form a large inward electrochemical gradient. Most of the energy released by ATP hydrolysis is used to push the three sodium ions up this large electrochemical gradient out of the cell. The presence of the potassium channels, and the resting voltage that they set up, means that almost all the energy of ATP hydrolysis by the Na^+/K^+ ATPase is stored in the sodium gradient, while potassium ions are close to equilibrium.

THE CHLORIDE GRADIENT

Chloride ions (Fig. 14.4) are at a lower concentration in the cytosol than in the extracellular medium. Typically, their

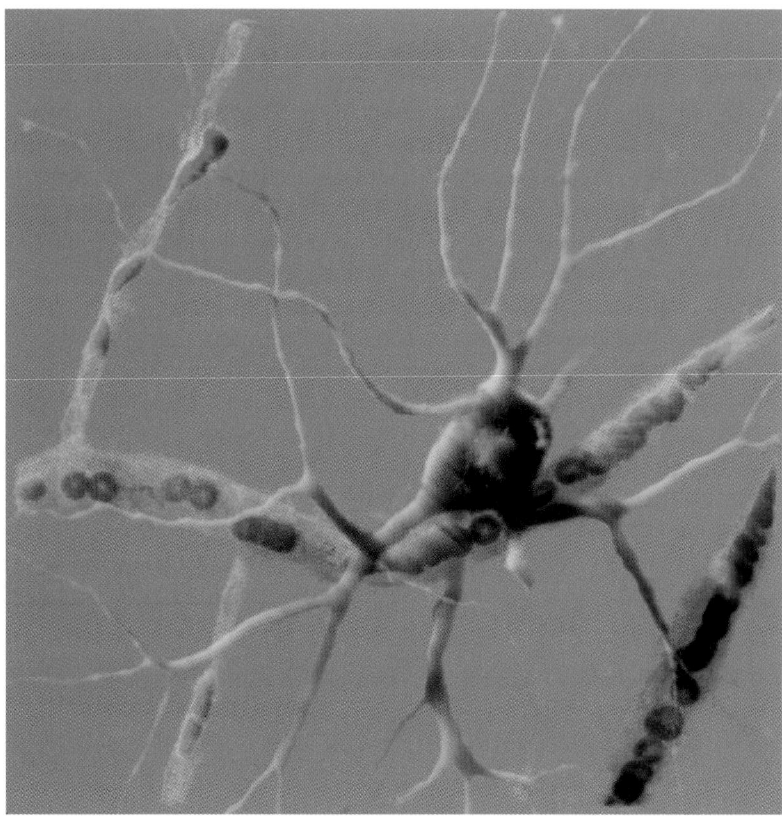

Figure 14.3. A nerve cell from the retina viewed with its nutritive capillaries. Red blood cells are an example of cells whose membrane voltage varies little, while nerve cells are specialized to transmit electrical signals over long distances. Image by Professor David Becker, University College London; used with permission.

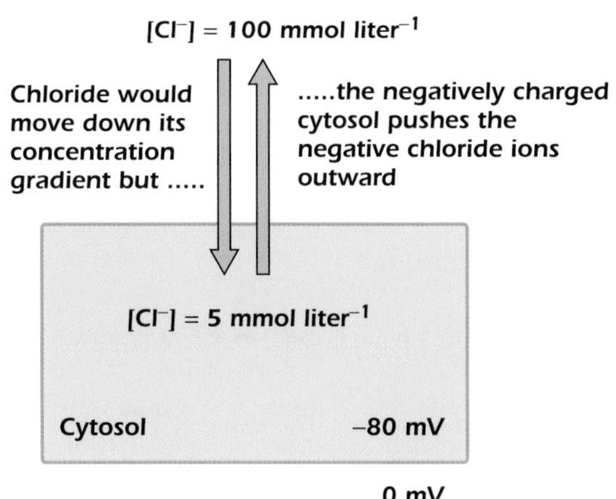

Figure 14.4. Chloride is close to equilibrium across most plasma membranes.

concentration in the cytosol is 5 mmol liter^{-1} compared with 100 mmol liter^{-1} in the extracellular medium. This is because of the resting voltage set up by the potassium channels. Chloride ions are repelled by the negative voltage of the cytosol. They leave the cell until their tendency to re-enter down their concentration gradient exactly matches their tendency to be repelled by the negative voltage of the cytosol.

GENERAL PROPERTIES OF CHANNELS

Channels are integral membrane proteins that form water-filled tubes through the membrane. We have already introduced three: the connexon (page 46), porin (page 194), and potassium channels. Channels that, like potassium channels, are selective for particular ions can set up transmembrane voltages. The connexons are much less selective than potassium channels. They form tubes 1.5 nm in diameter through which any solute of Mr $<$ 1000 can pass. Connexons are not always open. They open only when they connect with a second connexon on another cell, forming a tube through which solutes can pass from the cytosol of one cell to the cytosol of the other. Channels that are sometimes open and sometimes shut are said to be **gated**. When a connexon contacts another on another cell, its gate opens and solute can pass through; at other times the gate is shut. The usefulness of gating is obvious: if the connexons not contacting others were open, many solutes, including ATP and sodium, would leak out into the extracellular medium and exhaust the cell's energy currencies.

IN DEPTH 14.2 THE NERNST EQUATION

An ion that can pass across a membrane is acted on by two forces. The first derives from the concentration gradient. The ion tends to diffuse from a region where it is at high concentration to one where it is at low concentration. The second force derives from the transmembrane voltage. In the case of positively charged ions such as Na^+ and K^+, the ions tend to move toward a negative voltage. Negatively charged ions such as Cl^- tend to move toward a positive voltage. For each ion there is a value of the transmembrane voltage for which these forces balance, and the ion will not move. The ion is said to be at equilibrium, and this value of the transmembrane voltage is called the equilibrium voltage for that ion at that membrane.

Now consider the electrical force. The definition of a volt means that one coulomb of charge moving across a membrane with a transmembrane voltage of V volts gains V joules of free energy. However, we are working in moles, not coulombs. One mole of ions has a charge of zF coulombs, where z is the charge on the ion. For Na^+ and K^+ z is 1; for Ca^{2+} z is 2; and for Cl^- z is -1. The term F is a number that relates the coulomb to the mole. It has the value 96,500. Therefore one mole of ions I moving inward gains by virtue of the transmembrane voltage free energy equal to

$$\text{zFV joules}$$

Ion I at concentration $[I]_{outside}$

Ion I at concentration $[I]_{inside}$

Cytosol at voltage V

When the forces balance, then ions that move in will neither gain nor lose energy. This way of describing equilibrium is useful because it allows us to set equivalent the effects of the two very different gradients, concentration and voltage. For concentration, the free energy possessed by a mole of ions I by virtue of its concentration is

$$G = G^{o\prime} + RT\ \log_e[I] \text{ joules}$$

where $G^{o\prime}$ is the standard free energy, R is the gas constant (8.3 J mol^{-1} $degree^{-1}$), and T is the absolute temperature.

A mole of I passing in therefore moves from a region where it had a free energy of

$$G_{outside} = G^{o\prime} + RT\ \log_e[I_{outside}] \text{ joules}$$

to one where its free energy is

$$G_{inside} = G^{o\prime} + RT\ \log_e[I_{inside}] \text{ joules}$$

One mole of ions I moving inward therefore gains by virtue of the concentration gradient free energy equal to

$$RT\ \log_e[I_{inside}] - RT\ \log_e[I_{outside}] \text{ joules}$$

This does not mean that an ion always gains energy from the transmembrane voltage when it moves inward, as the term zFV can just as easily be negative as positive.

When the effects of concentration and voltage just balance, one mole of ions moving inward neither gains nor loses free energy. Hence, at equilibrium

$$RT\ \log_e[I_{inside}] - RT\ \log_e[I_{outside}] + zFV_{eq} = 0$$

This can be simplified to

$$V_{eq} = \left(\frac{RT}{F}\right)\left(\frac{1}{z}\right)\log_e\left(\frac{[I_{outside}]}{[I_{inside}]}\right) \text{ volts}$$

This is the **Nernst equation**. The value of $\left(\frac{RT}{F}\right)$ is 0.025 at a room temperatue of 22°C. At human body temperature $\left(\frac{RT}{F}\right)$ is 0.027.

"In" and "out" can refer to any two solutions separated by a membrane. At the plasma membrane *in* is the cytosol and *out* is the extracellular medium, but when considering equilibria across the inner mitochondrial membrane, *in* is the mitochondrial matrix and *out* is the intermembrane space.

Porin in the outer mitochondrial membrane plays an important role in energy conversion. It forms a very large diameter tube that allows all solutes of Mr $\leq$ 10,000 to pass and seems to spend a large fraction of time open under most circumstances. This is why the outer mitochondrial membrane is permeable to most solutes and ions. Appendix 1 at the end of this book lists all the different types of channels described in this book. It represents only a small fraction of the total number known.

Example 14.1 Cytochrome *c*—Vital but Deadly

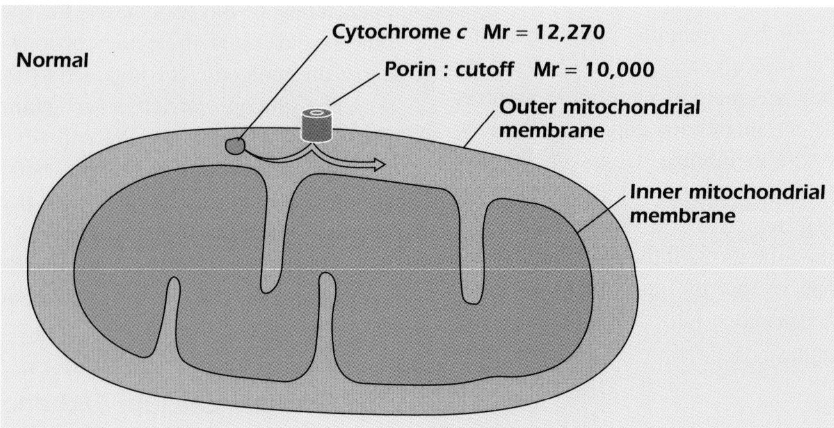

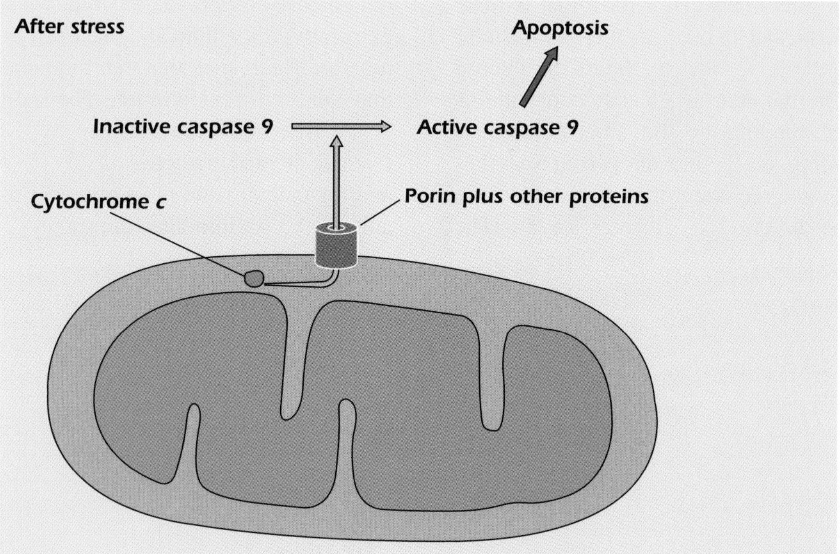

We have described how the electron carrier cytochrome *c* resides in the intermembrane space between the outer and inner mitochondrial membranes and helps the electron transport chain to convert energy as NADH to energy as the hydrogen ion electrochemical gradient across the mitochondrial inner membrane (page 196). Although cytochrome *c* is a soluble protein of relative molecular mass 12,270, it cannot escape from the intermembrane space into the cytosol because porin, the channel of the outer mitochondrial membrane, only allows solutes of Mr $\leq$ 10,000 to pass. Although cytochrome *c* is essential for mitochondrial function, it has another, deadly role. If cytochrome *c* comes into contact with a class of cytosolic enzymes called caspases, it activates them, turning on the process of cell suicide called apoptosis (page 308). Under certain conditions, porin can associate with other proteins to form a channel of larger diameter; when this happens, cytochrome *c* can leak out and the cell dies by apoptosis. This process seems to occur in hearts during heart attacks, and in the brain during a stroke; therefore a considerable research effort is aimed at preventing this from occurring.

GENERAL PROPERTIES OF CARRIERS

We have already met the three carriers that interconvert the four energy currencies of the cell (page 195). Carriers are like channels in that they are integral membrane proteins that allow solute to cross the membrane and, like channels, they form a tube across the membrane. However, there is a critical difference. In carriers the tube is never open all the way through; it is always closed at one or other end. Solutes can move into the tube through the open end. When the carrier changes shape so that the end that was closed is now open, the solute can move into the solution on the other side of the membrane.

The Glucose Carrier

One of the simplest carriers is the glucose carrier (Fig. 14.5). It switches freely between a form that is open to the cytosol and a form that is open to the extracellular medium. Inside the tube is a site to which a glucose molecule can bind. On the left, a glucose molecule is entering the tube and binding to the site. Sometimes glucose leaves the binding site before the carrier switches shape. In Figure 14.5 we see the other possibility: the carrier switches shape before the glucose leaves. The binding site is now open to the cytosol, and the glucose can escape into the cytosol. It has been carried across the plasma membrane. Unlike channels, carriers never form open tubes all the way across the membrane. Instead, they bind one or more molecules or ions, then change shape to carry the molecules or ions across the membrane.

The glucose carrier is very simple, whereas other carriers are more complex. We will next consider two carriers, the sodium/calcium exchanger and the calcium ATPase, which do much the same job—they push calcium ions up their concentration gradient out of the cell—but take their energy from different currencies. The sodium/calcium exchanger uses the sodium concentration gradient while the calcium ATPase uses ATP.

The Sodium/Calcium Exchanger

Figure 14.5 shows that the sodium/calcium exchanger, like the glucose carrier, can exist in two shapes, one open to the extracellular medium and one open to the cytosol. Inside the tube are three sites that can bind sodium ions and one site that can bind a calcium ion. The sodium/calcium exchanger is not free to switch between its two shapes at any time. Instead, it only switches if all the sodium sites are filled and the calcium site is empty, or if the calcium site is filled and all the sodium sites are empty.

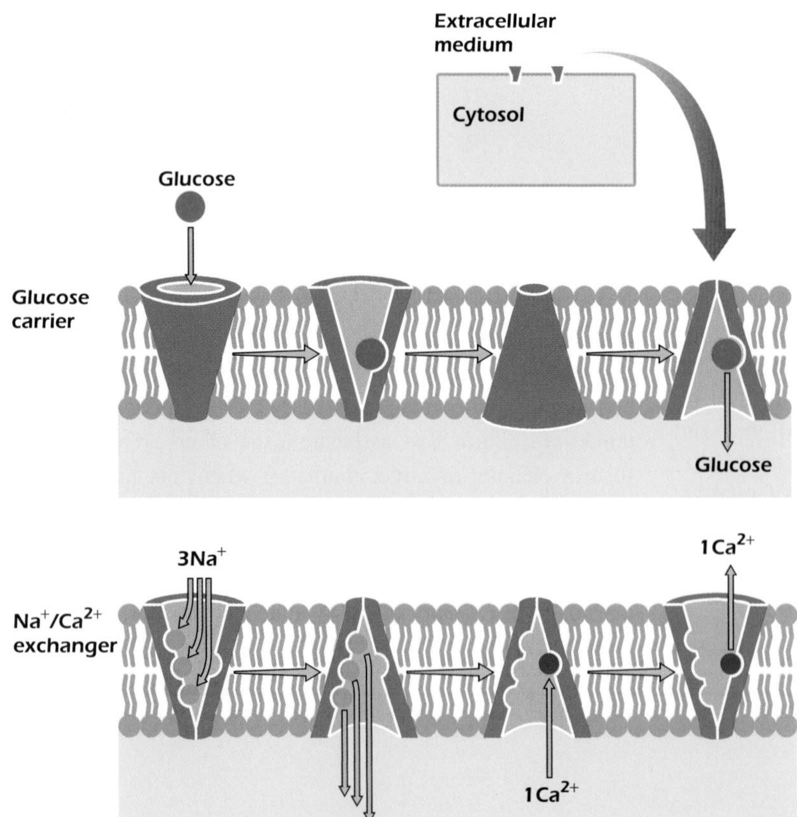

Figure 14.5. The glucose carrier and the sodium/calcium exchanger are sometimes open to the cytosol and sometimes open to the extracellular medium.

Example 14.2 The Glucose Carrier Is Essential

The cells of our bodies are bathed in a glucose-rich solution. However, glucose cannot cross the plasma membrane by simple diffusion because it is strongly hydrophilic: it can only get in via the glucose carrier. Some cells have glucose carriers in their membranes all the time, but others such as muscle and fat cells translocate the glucose carrier to the plasma membrane only in the presence of insulin. Insulin-dependent diabetics cannot produce their own insulin, and unless they inject synthetic insulin, their muscles and fat cells cannot take up glucose and therefore run out of energy, even though the concentration of glucose in the blood becomes very high. This is why muscular weakness is one symptom of diabetes.

On the left side of the figure, the carrier is open to the extracellular medium. It can switch its shape so that it is open to the cytosol if one of two things happen. It could, as shown, bind three sodium ions (keeping the calcium site empty) or it could bind one calcium and keep the sodium sites vacant. This second option does not often happen, since sodium is at high concentration in the extracellular medium, and one or more of the sodium sites are usually occupied. Therefore nearly all the switches from open-to-outside to open-to-inside are of the type shown. Once the tube has opened to the low-sodium environment of the cytosol, the sodium ions tend to leave.

Once the carrier is open to the cytosol, it can switch back to the open-to-outside form by binding either one calcium or three sodium ions. Since sodium is scarce in the cytosol, the latter event is unlikely. More frequently a calcium ion will bind and will be carried out. The carrier is now ready to bind sodium again and perform another cycle.

The overall effect of one cycle is to carry three sodium ions into the cell down their electrochemical gradient, and one calcium ion out of the cell up its electrochemical gradient. A simple rule about when the carrier can switch shape has produced a machine that uses the energy currency of the sodium gradient to do work in pushing calcium ions out of the cell. Figure 14.6 shows a simple way of representing what is happening. The circle represents one cycle of operation: from open to extracellular medium back to open to extracellular medium again.

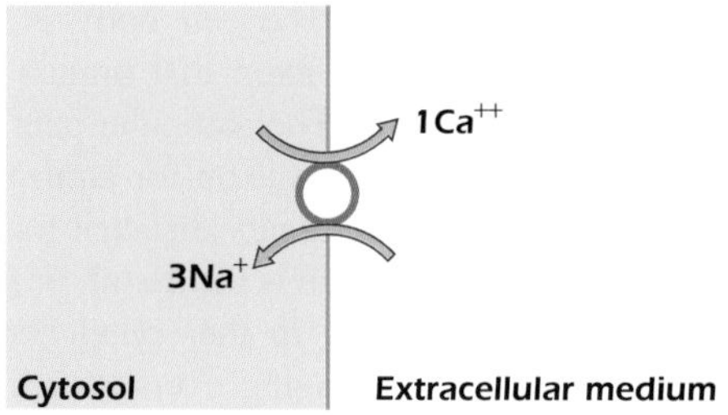

Figure 14.6. Action of the sodium/calcium exchanger.

Carriers with an Enzymatic Action: The Calcium ATPase

The electron transport chain, ATP synthase, and the Na^+/K^+ ATPase are carriers with an additional level of complexity in that they carry out an enzymatic action as well as their carrier function. Figure 14.7 illustrates another carrier with a linked enzymatic function: the calcium (Ca^{2+}) ATPase. Like the sodium/calcium exchanger this is found in the plasma membrane. The transmembrane part of this carrier forms a tube that can be open to the cytosol or to the extracellular medium. Two carboxyl groups in the tube can form a binding site for a calcium ion. Other domains of the protein lie in the cytosol and can hydrolyze ATP. Changes in the shape of the cytosolic region are transmitted to the transmembrane region and force it between the open-to-cytosol and open-to-outside shapes. Figure 14.7 shows our present understanding of how this might happen. (1) is the relaxed shape of the protein. ATP is held by noncovalent interactions in a crevice in one of the cytosolic domains. One of the carboxyl groups in the tube is protonated at this time. (2) A calcium ion moving in from the cytosol pushes the tube open and binds to both carboxyl groups, displacing an H^+ ion as it does so. The distortion in protein shape caused by the calcium ion pushing the tube open is transmitted to the cytosolic region and brings the γ phosphate of ATP close to an aspartate residue on a neighboring domain. (3) An intrinsic kinase activity of the two domains transfers the γ phosphate group from ATP to the aspartate residue. This means that a phosphate group, with two negative charges, is sitting very close to ADP, with three negative charges. (4) The repulsion between the phosphorylated aspartate and the ADP forces the two domains apart. The shape change is transmitted to the transmembrane region, which is forced into a wide open-to-outside shape. The calcium ion, which can no longer bind to the carboxyl groups on both sides of the tube, is held only weakly and tends to escape into the extracellular medium. An H^+ ion moves in to protonate one of the carboxyl groups. (5) A third cytosolic domain, which has phosphatase catalytic ability, swings in and dephosphorylates the aspartate residue. Meanwhile, ATP

Medical Relevance 14.1 Poisoned Hearts Are Stronger

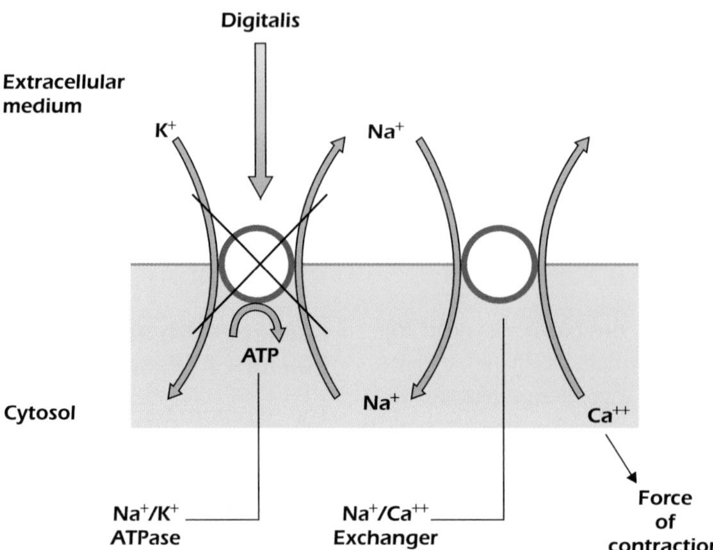

Digitalis is used to treat heart failure. Digitalis inhibits the sodium/potassium ATPase and is extremely toxic. Nevertheless, a small dose, which inhibits the sodium/potassium pump just a little, causes the heart muscle to beat more strongly. The reason is that inhibiting the sodium/potassium pump just a little causes a small increase of cytosolic sodium concentration. Because the sodium/calcium exchanger has three binding sites for sodium, its activity is extremely sensitive to sodium concentration, and even a small increase of cytosolic sodium reduces its activity significantly. The calcium concentration in the cytosol therefore rises. The mechanical motor that drives heart contraction (page 289) is controlled by calcium, so that a small increase of cytosolic calcium makes the heart beat more strongly.

moves in and displaces ADP at the binding crevice. Once the aspartate has been dephosphorylated, the protein is no longer held in the open-to-outside shape by electrical repulsion and relaxes back to the resting shape shown in (1). It is now ready to accept another calcium ion from the cytosol and repeat the process. Figure 14.8 summarizes what is happening. The circle represents one cycle of operation. One molecule of ATP is hydrolyzed to ADP and inorganic phosphate, one calcium ion moves out of the cell, and one H^+ moves in. The energy released by ATP hydrolysis has been used to drive one calcium ion up its electrochemical gradient and out of the cell.

All the cells of our bodies have one of these two calcium pumps, the sodium/calcium exchanger or the calcium ATPase, and many have both. Because of the action of these carriers, the calcium concentration in the cytosol is much less than the concentration in the extracellular medium—usually about 100 nmol liter^{-1} compared with 1 mmol liter^{-1}. Because the resting voltage is attracting the positively charged calcium ions inward, the overall result is a large electrochemical gradient favoring calcium entry into cells.

ELECTRICAL SIGNALING

For many types of cell the voltage across the plasma membrane never changes. However nerve and muscle cells use changes of transmembrane voltage as a method of rapid signaling over long distances.

The Pain Receptor Nerve Cell

We will illustrate the operation of nerve cells using pain receptors. Figure 14.9 shows the cell responsible for the sensation of pain in a finger. The cell body is close to the spinal cord and extends an **axon** that branches to the spinal cord and out to the body. The particular pain receptor illustrated in Figure 14.9 sends its axon for almost a meter to the tip of a finger, an extraordinary distance for a single cell. The axon terminal in the skin is the **distal**, or far away, terminal and the axon terminal in the spinal cord is the **proximal** one. Potentially damaging events, such as high temperatures, are detected at the finger, and the message is passed on to another nerve cell (a pain relay cell) in

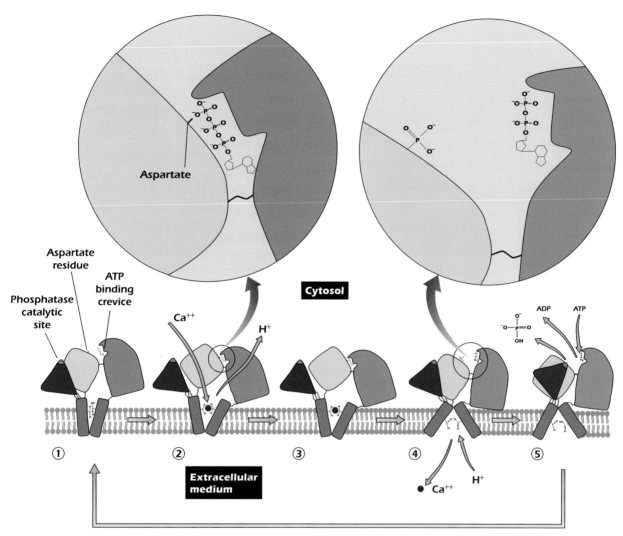

Figure 14.7. The calcium ATPase undergoes a cycle of phosphorylation and dephosphorylation. These drive shape changes that in turn push calcium ions out of the cell.

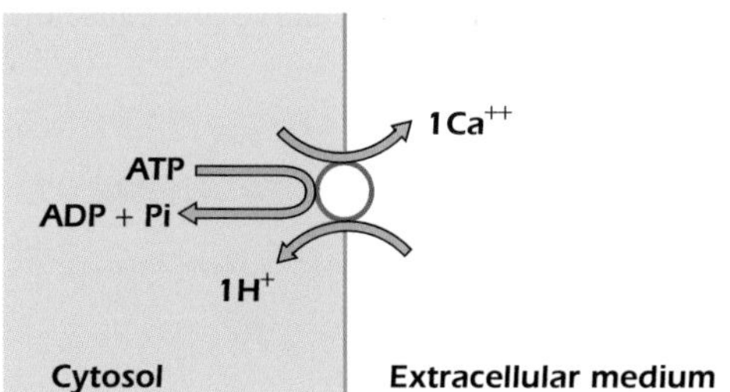

Figure 14.8. Action of the calcium ATPase.

the spinal cord. We will now explain how this function is performed.

The plasma membrane of the distal terminal contains an ion channel called TRPV that is closed at 37°C but spends a greater and greater fraction of time open at higher temperatures (Fig. 14.10). The channel is a nonselective cation channel, that is, it allows sodium, potassium, and calcium ions to pass. Since even in nerve cells potassium ions are close to equilibrium, the dominant flux through open TRPV channels is an inward movement of sodium and calcium, both of which have a large inward electrochemical gradient.

Figure 14.11A shows what happens when the finger is passed through the warm air from a hand dryer. The warmth causes some of the TRPV channels to open, and sodium and calcium ions move in. This inward current causes the transmembrane voltage in the distal axon terminal to **depolarize**. We use the word depolarization to mean any positive shift in the transmembrane voltage, whatever its size or cause. Because the cytosol is less negative than it was, and hence pulls potassium ions inward less strongly, the outward flux through potassium channels increases. The membrane reaches a new steady state in which the inward current through TRPV channels is balanced

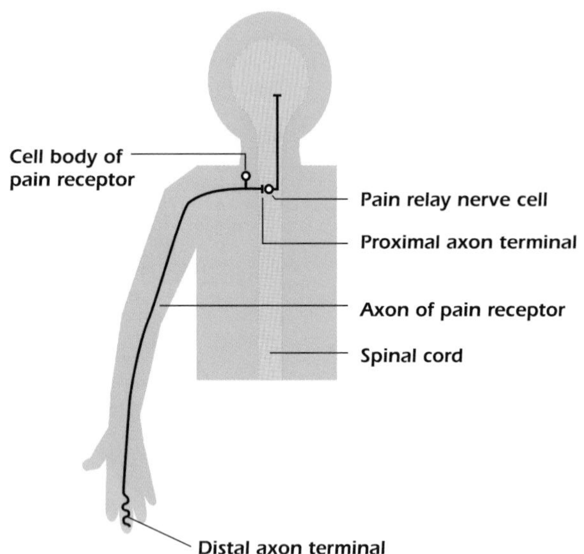

Figure 14.9. The connections of a pain receptor nerve cell.

by the increased outward current through potassium channels.

Figure 14.11B shows what happens when the finger is passed through the hot air from a hair dryer. More TRPV channels open, so the distal axon terminal depolarizes further. However, at a voltage of −40 mV, a new phenomenon appears: a massive but temporary depolarization to +30 mV. This is the **action potential**. Cells capable of generating action potentials are said to be **electrically excitable** and include nerve cells and muscle cells. To explain the action potential we must return to the voltage-gated sodium channel and describe it in more detail.

The Voltage-Gated Sodium Channel

Figure 14.12 shows how the voltage-gated sodium channel works. On the left is the shape of the protein when the transmembrane voltage is −70 mV. The protein forms a tube, but it is not open all the way through the membrane. Positively charged arginine and lysine side chains are attracted by the negative cytosol, favoring a state in which the tube is gated closed. If the membrane is depolarized, the positive side chains are pulled inwards more weakly and the channel can switch to the open state. Sodium ions can now pass through the channel. A protein domain in the cytosol called the inactivation plug is constantly jiggling about at the end of a flexible link and can bind to the inside of the open channel. The resulting blockage is **inactivation**, and it occurs after about one millisecond. As long as the plasma membrane remains depolarized, the voltage-gated sodium channel will remain inactivated. When the plasma membrane is repolarized, the positive charges are attracted back toward the inside of the cell, squeezing the plug out.

In summary:

1. When the transmembrane voltage is −70 mV, the voltage-gated sodium channel is gated shut.
2. When the plasma membrane is depolarized, the channel opens rapidly and inactivates after about 1 ms.
3. After the channel has gone through this cycle, it must spend at least 1 ms with the transmembrane voltage at the resting voltage before it can be opened by a second depolarization.

The voltage-gated sodium channel goes through this process whether or not sodium ions are present. To understand the action potential, we must now think about the

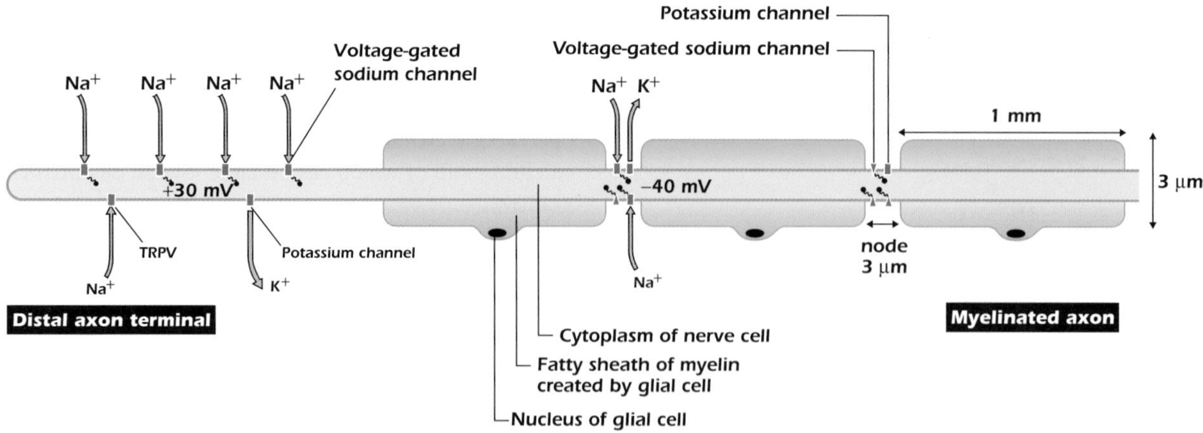

Figure 14.10. One type of pain receptor has a distal terminal in the skin and is connected to the central nervous system by a myelinated nerve fiber.

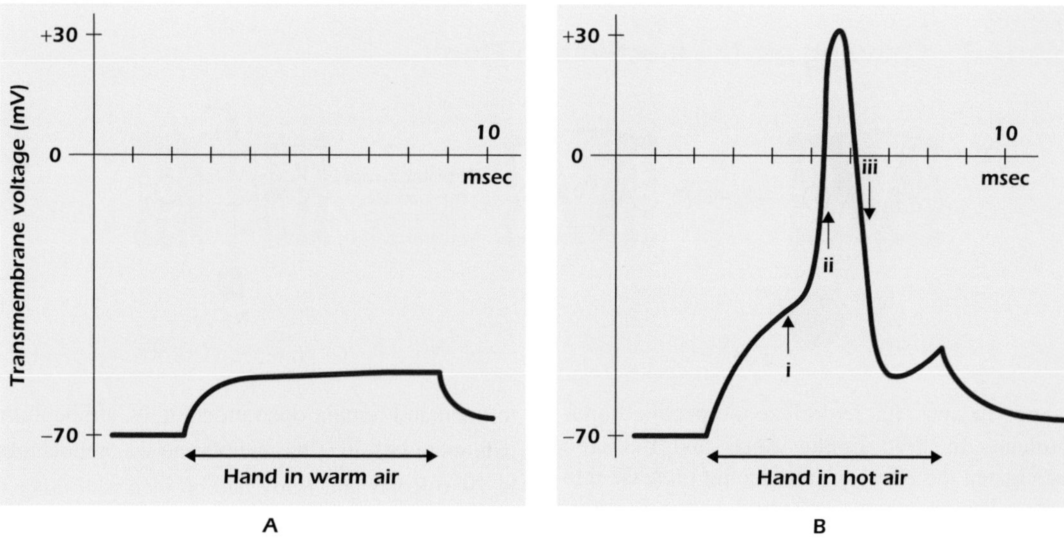

Figure 14.11. Electrical events in a pain receptor nerve cell.

effect that movements of sodium ions through this channel have on the transmembrane voltage.

The Sodium Action Potential

Even the slight depolarization in the warm air opened some voltage-gated sodium channels in the distal terminal of the pain receptor, but the additional inward current that they passed was neutralized by the additional outward potassium current, so the membrane voltage stabilized at a new depolarized value. In contrast when the membrane is depolarized past the **threshold voltage**, which for this nerve cell is about −40 mV, sufficient voltage-gated sodium channels open such that the inward current through them (plus the current through the open TRPV channels) is greater than the outward current through the potassium channels. This is the situation at time (i) in Figure 14.11B. This state is not a stable one because the cytosol is gaining charge and is therefore becoming more positive. More voltage-gated sodium channels therefore pop into the open state, letting more sodium ions pour inward, depolarizing the membrane further. Very rapidly the cell reaches state (ii), where all the voltage-gated sodium channels are open and the membrane is rapidly depolarizing.

There are so many voltage-gated sodium channels in the membrane that the voltage would approach the sodium equilibrium voltage (about +70 mV) but for the fact that the channels inactivate, stopping the sodium influx. By time (iii) almost all the voltage-gated sodium channels are inactivated, so that the only channels that can carry current are the potassium channels. Each of these is now carrying a large outward current since the membrane voltage is much more positive than the potassium equilibrium voltage. The membrane is rapidly **repolarizing**. Once the transmembrane voltage has returned to negative values, the voltage-gated sodium channels will begin to recover to the closed (but ready-to-open) state, and the cell would soon be ready to generate a second action potential if the TRPV channels were still open.

This figure shows a critical feature of an action potential: it is all or nothing. The warm air did not depolarize the pain receptor sufficiently to evoke an action potential. The hot air depolarized the cell sufficiently to start the process, and the action potential then took

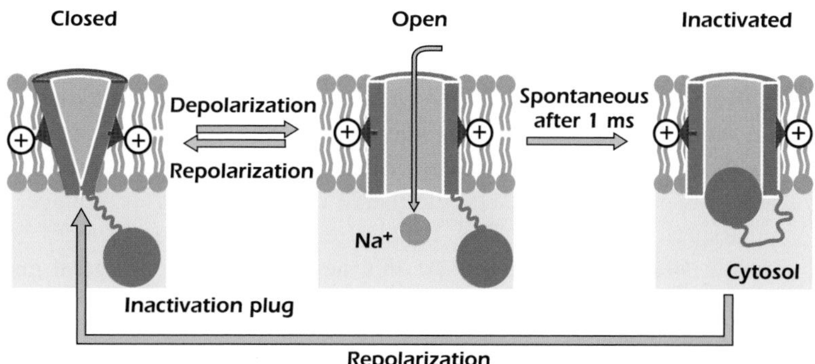

Figure 14.12. The voltage-gated sodium channel.

Example 14.3 Chewing off the Inactivation Plug

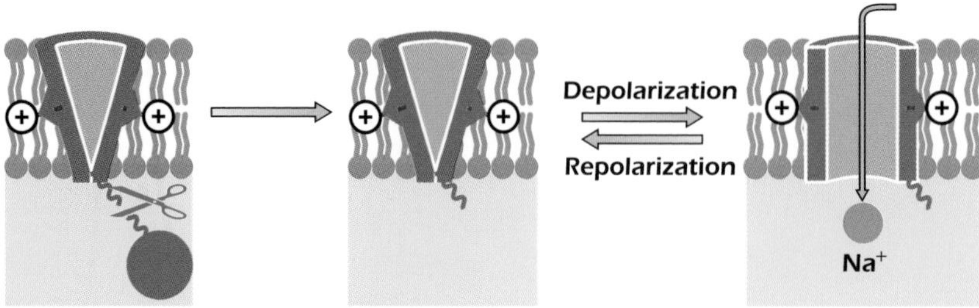

Proteases are enzymes that hydrolyze the peptide bonds within proteins. In 1976 Emilio Rojas and Bernardo Rudy investigated the effect of introducing protease into squid axons. The membrane of squid axons contains voltage-gated sodium channels that, like ours, normally inactivate about 1 ms after they are opened by a depolarization. However, after introduction of the protease, depolarization caused the voltage-gated sodium channels to open and remain open indefinitely, although the channels would close if the membrane was repolarized. Back in 1976 Rojas and Rudy had no idea why this should be so. Now we can understand the result—the protease cuts the linker between the main part of the voltage-gated sodium channel and the inactivation plug. The plug then floats off and is not available to block the open channel.

off in an explosive, self-amplifying way until all the voltage-gated sodium channels were open and the plasma membrane had depolarized greatly. In all excitable cells there is a threshold for initiating an action potential; in the pain receptor it is about −40 mV. Depolarizations to below the threshold elicit nothing, but depolarizations to voltages more positive than the threshold elicit the complete action potential. At the heart of the action potential is a positive feedback loop. Depolarization causes voltage-gated sodium channels to open, and open sodium channels cause depolarization.

The Strength of a Signal Is Coded by Action Potential Frequency

At the distal axon terminal of the pain receptor, the intensity of the stimulus is represented by the amount of depolarization produced. The hotter the air blowing on the skin, the more positive the transmembrane voltage of the terminal becomes. Other sensory cells behave in the same way. For instance, in Chapter 15 we will meet the scent-sensitive nerve cells in the nose that detect smell chemicals. The higher the concentration of the smell chemical, the more these cells depolarize. However, the signal that passes towards the brain is composed of action potentials. The strength of the stimulus cannot be encoded in the amplitude of the action potentials, because they are all-or-nothing discrete events. Rather, the strength of the stimulus is encoded by the frequency of the action potentials.

Figure 14.13 shows how this happens. In Figure 14.13A the stimulus is too weak to elicit an action potential. The graph on the right shows that away from the terminal, the membrane voltage of the pain receptor does not deviate from the resting voltage.

Example 14.4 Peppers and Pain

Capsaicin, the active ingredient in chili peppers, activates one isoform of TRPV channels and so causes the hot, burning sensation in the mouth. In contrast sanshool, the active ingredient in Szechuan peppers, activates pain receptors and other nerve cells by closing potassium channels💻. This depolarizes the cells to a new steady state voltage. At this new voltage, the outward current of potassium once again has the same amplitude as the inward current through voltage-gated sodium channels. However, since there are fewer open potassium channels, the new steady state voltage must lie further away from the potassium equilibrium voltage such that each potassium channel carries a larger outward current. The depolarization caused is great enough to reach the threshold for action potential generation.

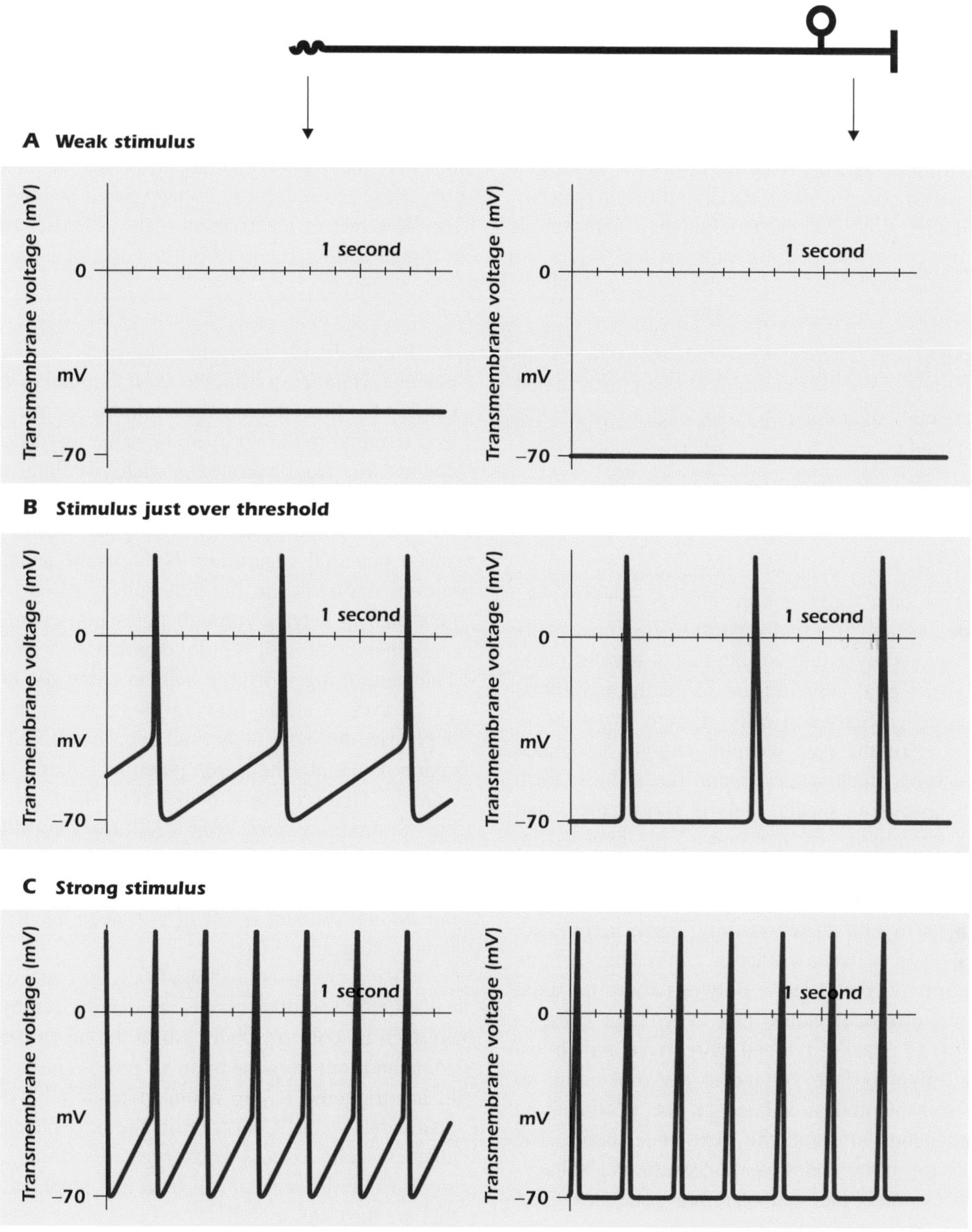

Figure 14.13. Frequency coding in the nervous system.

Figure 14.13B represents what happens when the stimulus is over threshold, but not by much. After each action potential, the membrane voltage at the distal terminal depolarizes slowly to threshold; the cell therefore fires action potentials at low frequency. The graph on the right shows that the action potentials, but not the TRPV-mediated depolarization, propagate up the axon. Figure 14.13C shows what happens when the stimulus is strong. Because more TRPV channels are open, carrying in positive charge, the depolarization after each action potential is more rapid. Therefore the time between successive action potentials is short and the frequency of action potentials is greater.

Amplitude modulation (AM) and frequency modulation (FM) are familiar terms from radio. Here we are seeing exactly the same two coding strategies in the nervous system. AM is used in the distal terminals of the pain receptor. The strength of stimulus is coded for by the amplitude of

Medical Relevance 14.2 A Sodium "Channelopathy"

Epilepsy, in which nerve cells in the brain fire action potentials in an uncontrolled fashion, is not uncommon in young children but often clears up as the child gets older. However, one form of childhood-onset epilepsy, termed GEFS+ for "generalized epilepsy with febrile seizures that extend beyond six years of age," does not improve with age. Robyn Wallace and coworkers in Australia showed that this condition could be caused by a mutation that slows the rate at which the inactivation plug moves to block the voltage-gated sodium channel. The slow rate of inactivation makes it easier for stimuli to trigger action potentials in the nerve cells in the brain.

the depolarization. FM is used in almost all axons, including those of the pain receptor. The strength of stimulus is coded for by the frequency of action potentials, each of which has the same amplitude.

Myelination and Rapid Action Potential Transmission

The axon of the pain receptor transmits its signal to the central nervous system so rapidly because for most of its length it is insulated by a fatty sheath called **myelin** made by glial cells (the glial cells outside the brain and spinal cord are often called **Schwann cells**). Figure 14.10 shows part of the axon of the pain receptor and its associated glial cells, each of which wraps around the axon to form an electrically insulating sheath. Only at short gaps called nodes is the membrane of the nerve cell exposed to extracellular medium. Note that the vertical and horizontal scales of this diagram are completely different. The axon together with its myelin sheath is only 3 μm across, but the section of axon between nodes is 1 mm long, a large distance by normal cell standards. Between nodes the axon is an insulated electrical cable: just as voltage changes at one end of an insulated metal wire have an almost instantaneous effect on the voltage at the other end, so a change in transmembrane voltage at one node has an almost instantaneous effect at the next node along. The plasma membrane at the nodes has potassium and voltage-gated sodium channels and can therefore generate action potentials.

Figure 14.10 shows the moment at which the action potential is jumping from the axon terminal to the first node. Current has flowed along the axon cytosol from the depolarized terminal to the node, in the same way as current flows down a wire from a point at a positive voltage to a negative one. Current flowing into the node has depolarized it, and the membrane at the node has now reached its threshold for action potential generation. Now, at the node itself, more current is flowing in through voltage-gated sodium channels than is leaving through potassium channels, and the node membrane will rapidly depolarize to about +30 mV. The same thing will now happen along the next millimeter of axon. Current flowing along the axon cytosol will depolarize the next node, and soon it too will generate an action potential. The action potential is jumping from node to node. This process is named after the Latin word meaning "to jump," *saltere*, so it is **saltatory conduction**. For a pain fiber conducting at 15 m s^{-1} the action potential takes 67 μs to jump each 1 mm distance. The fastest axons in our bodies conduct at 60 m s^{-1}, with each node-to-node jump taking only 17 μs.

TRPV channels are found only in pain receptor nerve cells. Other sensory neurons are depolarized by other stimuli such as cold or touch, while the nerve cells that send commands out from the brain, or process information within it, are triggered to fire action potentials by other mechanisms that we will describe in the next chapter. However all these cells express voltage-gated sodium channels and so all share the mechanism of action potential propagation that we have described.

Example 14.5 Local Anesthetic, Overall Well-being

Nerve conduction is exploited when a dentist uses local anesthetic. A patient feels pain from the drill because pain receptors in the tooth are depolarized and transmit action potentials toward the brain. The site at which they are being depolarized is inside the tooth and therefore inaccessible to drugs until the drill has made a hole. However, the axons of the pain receptors run through the gum. Local anesthetics injected into the gum close to the axon bind to the nerve cell membrane at the node and prevent the opening of voltage-gated sodium channels. Drilling into the tooth still depolarizes the pain receptor membrane, and action potentials begin their journey toward the brain, but they cannot pass the injection site because the nodes there cannot generate an action potential. The patient therefore feels no pain.

The combination of axon plus myelin sheath is a nerve fiber. The nerve fiber shown in Figure 14.10 has an overall diameter of 3 μm. The human body contains myelinated nerve fibers with diameters ranging from 1 μm, conducting at 6 m s^{-1}, to 10 μm, conducting at 60 m s^{-1}. Some nerve cells have axons that are not myelinated and generally conduct more slowly than myelinated fibers. The axons of scent-sensitive nerve cells (page 258) are an example. Getting scent information from the nose to the brain as fast as possible is apparently not a significant advantage, given that a scent can take many seconds to waft from its source to our nose.

SUMMARY

1. Channels are membrane proteins with a central water-filled hole through which hydrophilic solutes, including ions, can pass from one side of the membrane to the other. Changes in protein structure may act to gate the channel but are not required for movement of the solute from one side to the other.
2. The presence of potassium channels in the plasma membrane, and the resting voltage that they set up, means that almost all the energy of ATP hydrolysis by the Na^+/K^+ ATPase is stored in the sodium gradient, while potassium ions are close to equilibrium.
3. The resting voltage repels chloride ions from the cell interior.
4. Carriers, like channels, form a tube across the membrane, but the tube is always closed at one end. Solutes can move into the tube through the open end. When the carrier changes shape, so that the end that was closed is open, the solute can leave to the solution on the other side of the membrane.
5. The glucose carrier is present in the plasma membrane of all human cells.
6. The sodium/calcium exchanger is a carrier that uses the energy of the sodium gradient to push calcium ions out of the cell.
7. The electron transport chain, ATP synthase, the sodium/potassium ATPase, and the calcium ATPase are both carriers and enzymes, the two actions being tightly linked.
8. The voltage-gated sodium channel is shut at the resting voltage but opens upon depolarization. After about 1 millisecond the channel inactivates.
9. While the voltage gated sodium channel is open, sodium ions pour into the cell down their electrochemical gradient. The mutual effect of current through the channel on transmembrane voltage, and of transmembrane voltage on current through the channel, constitutes a positive feedback system. If the membrane is initially depolarized to threshold, the positive feedback of the sodium current system ensures that depolarization continues in an all-or-nothing fashion. Repolarization occurs when the voltage-gated sodium channel inactivates. The entire cycle of depolarization and repolarization is called an action potential.
10. Action potentials are all-or-nothing events. The strength of the signal transmitted by a nerve cell is encoded not by the amplitude of the action potentials but by their frequency.
11. Long nerve cell processes called axons transmit action potentials at speeds up to 60 m per second. They can transmit the signal at such high rates because myelin, a fatty sheath, insulates the 1 mm distances between nodes.
12. A number of axons are not myelinated; these generally conduct action potentials more slowly than do myelinated axons and are found where high conduction speed is not required.

Medical Relevance 14.3 Solute Movement in the Kidney

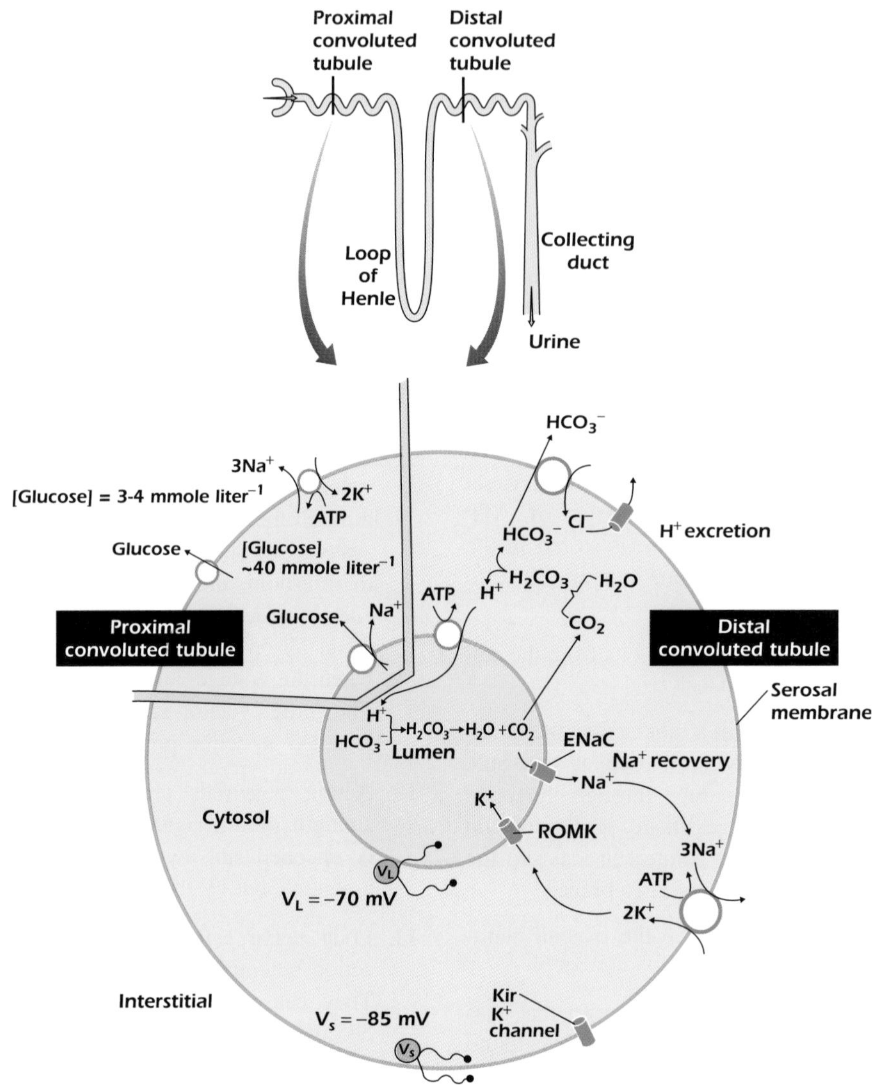

We give here a brief introduction to some of the many channels and carriers, and many complex control mechanisms, that operate in the kidney.

The operational unit of the kidney is the kidney tubule, shown in the upper part of the diagram. The liquid in the tubule is formed by filtering blood and initially has a composition similar to blood plasma. Solutes and water are then added and removed as the liquid passes down the tubule to be finally lost as urine.

One of the functions of the proximal convoluted tubule is to recover the glucose from the tubule fluid. The lower part of the diagram shows how this is achieved. The plasma membrane facing the interstitial fluid (this is called the **serosal** membrane) contains the glucose carrier we have met before (page 236). This will always carry glucose from areas where it is at high concentration to areas where the concentration is lower. In contrast the plasma membrane facing the tubule lumen (this is the luminal membrane) contains an unusual glucose carrier which carries sodium and glucose in. The two movements are locked together: if sodium moves, glucose moves, and vice versa. Sodium is at a low concentration in the cytosol due to the action of the sodium/potassium ATPase (page 199). The sodium, glucose cotransporter allows sodium to move down its electrochemical gradient into the cell and at the same time pulls glucose into the cells up its concentration gradient. The result of the glucose inward pumping is that the glucose concentration in the cytosol of kidney tubule cells is much greater than in normal cells (In Depth 12.3 on page 202) at about 40 mmole liter^{-1}. The glucose can then flow down its concentration gradient, using the regular glucose carrier in the serosal membrane, into the interstitial fluid and hence back into the blood.

Carriers have a maximum rate at which they can work—the relationship between flow rate and concentration follows the same Michaelis-Menten relationship that governs enzymes (page 184). The carriers in the kidney can recover almost all the glucose if the concentration in the initially filtered plasma is in the normal 3–4 mmole liter^{-1} range, but if it increases above 12 mmole liter^{-1}, as happens in the disease diabetes mellitus, then even working at maximum rate the carriers cannot recapture all the glucose and some is lost in the urine, which is therefore sweet, giving the disease its name.

Another function of the kidney is to excrete H^+. Metabolism of the sulfur-containing amino acids methionine and cysteine acquired in the diet creates sulfuric acid, which ionizes to sulfate and H^+. The diagram shows the processes in the distal convoluted tubule by which the H^+ is excreted. At the heart of the process is the H^+ ATPase which pushes H^+ into the lumen. This carrier is a close relative of mitochondrial ATP synthase (In Depth 12.2 on page 200). Like ATP synthase it uses a rotor system to couple the ATP $\rightleftarrows$ ADP + Pi reaction to the movement of H^+ across the membrane. However its job is the reverse one, to hydrolyse ATP and use the energy hydrolysis to drive H^+ up its electrochemical gradient.

The energetics of the ATPase are finely balanced, so that even a small fall in the energy released upon ATP hydrolysis will mean that it cannot push H^+ out of the cells. Thus any condition that reduces [ATP] or increases either [ADP] or [Pi] in kidney cells will tend to cause acid excretion to fail, leading to a dangerous increase of blood acidity. This condition, called renal tubular acidosis, can happen if the oxygen supply to the kidney is compromised, for example in sickle cell anemia (page 64). It can also be simply a consequence of kidney cell stress during the body's response to any disease.

When the adrenal gland secretes the hormone aldosterone, distal convoluted tubule cells express the ENaC sodium channel (Medical Relevance 6.2 on page 96). The result is that sodium that would otherwise be lost in the urine is recovered. The diagram shows how this is achieved. The luminal membrane already contains channels selective for potassium ions, called ROMK. When ENaC appears, sodium moves in down its electrochemical gradient, because unlike the voltage-gated sodium channel (page 240), ENaC spends about 30% of its time open, whatever the membrane voltage. Because this membrane now has a significant permeability to sodium, the voltage across it deviates significantly from the potassium equilibrium voltage; the value moves to about −70 mV. Potassium ions therefore move from the cells into the lumen.

The sodium ions are now pumped out of the cell into the interstitial fluid by sodium/potassium ATPases located on the serosal membrane. This brings potassium into the cells. The serosal membrane is highly permeable to potassium because of the presence of a second type of potassium channel called Kir. However, the serosal membrane is not particularly permeable to other ions, so this membrane voltage has a value close to the potassium equilibrium voltage. There is therefore almost no electrochemical gradient for potassium at this membrane. Rather, the potassium that enters the cells in the sodium/potassium ATPase leaves via ROMK into the lumen.

Patients whose kidneys are excreting too much potassium and resorbing too much sodium are often given the drug amiloride (sold as Midamor), which blocks ENaC. With ENaC blocked, sodium can no longer move into the cells from the lumen and is instead lost in the urine. The luminal membrane is now permeable almost exclusively to potassium, so that the transmembrane voltage across the membrane facing the lumen shifts to a value close to the potassium equilibrium voltage and potassium stops moving out from the cells into the lumen.

Patients with a genetic lesion in ROMK lose excessive amounts of sodium in their urine (Example 9.1 on page 141). However, a similar syndrome is seen in genetically normal individuals whose body tissues become too acid (acidosis). The primary cause of the acidosis can be a respiratory problem that prevents the patient from breathing out all the carbon dioxide that their cells make, or can be metabolic, as in diabetic acidosis (Medical Relevance 2.1 on page 22). Whatever the primary cause, a relatively slight acidification of kidney cells markedly reduces the time ROMK spends open. The molecular basis of this phenomenon is still under investigation, but part of the explanation lies in a series of four histidine residues at the carboxy terminus of each of the four identical protein subunits that make up the working channel. The histidine side chain has a pK_a of 7 so that an increasing fraction of the sixteen histidines in the complete channel complex are protonated, and therefore carry a positive charge, as the pH falls. These charges disrupt the quaternary structure of the complex and the channel closes so that potassium cannot move from the kidney cells into the tubule lumen. The result is that acidotic patients usually also have a high potassium concentration in their blood plasma (hyperkalemia). If untreated the effects of the high potassium on the heart in particular can be rapidly lethal.

FURTHER READING

Ashcroft, F. M. (2000) *Ion Channels and Disease*, Academic Press, San Diego, CA.

Levitan, I. B., and Kaczmarek, L. K. (2002) *The Neuron*, 3rd edition, Oxford University Press, New York.

REVIEW QUESTIONS

14.1 Theme: Cytosolic and Extracellular Concentrations of Important Ions

A $\geq 10^{-1}$ moles liter^{-1}
B 5×10^{-2} moles liter^{-1}
C 10^{-2} moles liter^{-1}
D 5×10^{-3} moles liter^{-1}
E 10^{-3} moles liter^{-1}
F 10^{-4} moles liter^{-1}
G 10^{-5} moles liter^{-1}
H 10^{-6} moles liter^{-1}
I $\leq 10^{-7}$ moles liter^{-1}

From the above list of concentrations, choose the one that most closely approximates the concentration described in the statements below. Answer for a general human cell.

1. cytosolic calcium
2. extracellular calcium
3. cytosolic chloride
4. extracellular chloride
5. cytosolic H^+ (H_3O^+)
6. extracellular H^+ (H_3O^+)
7. cytosolic potassium
8. extracellular potassium
9. cytosolic sodium
10. extracellular sodium

Answer to Thought Question: The resting voltage will depolarize, that is, become more positive. There will be two effects on voltage-gated sodium channels: they will be more likely to open but there will also be more channels in a permanently inactivated state.

Under normal conditions, with extracellular potassium at 5 mM and cytosolic potassium at 140 mM the resting voltage of cardiac muscle cells is about −80 mV. This is the voltage at which the outward current through the many open potassium channels is equal and opposite to the inward current through the few open voltage-gated sodium channels. In turn what allows an outward potassium current is the fact that the voltage is significantly more positive (about 10 mV) than the potassium equilibrium voltage of −90 mV.

If the extracellular potassium increases then the potassium equilibrium voltage will move positive. For example, when the extracellular potassium is 10 mM then the potassium equilibrium voltage will be −71 mV. In order for there still to be an outward potassium current to balance the inward sodium current, the resting voltage must be about 10 mV more positive than the potassium equilibrium voltage, so that now means about −61 mV. Thus the increase of extracellular potassium depolarizes the muscle cell.

Initially, the effect of the depolarization will be to increase the probability that voltage-gated sodium channels open, and therefore to increase the frequency of action potentials in the heart; this is the condition of a rapid heart beat or tachycardia. However, recovery of voltage-gated sodium channels from the inactivated back to the closed state requires that the membrane voltage be negative. If the resting voltage is −60 mV rather than −80 mV, many voltage-gated sodium channels will not recover but will remain in the inactivated state. If too few recover to the closed state, the cardiac muscle cells become electrically inexcitable: they do not generate action potentials at all and do not contract, a situation that is rapidly lethal.

14.2 Theme: Pathways for Solute Movement Across the Plasma Membrane

A calcium ATPase
B connexon
C glucose carrier
D potassium channel
E sodium/calcium exchanger
F TRPV
G voltage-gated sodium channel

From the above list of channels and carriers, select the protein corresponding to each of the descriptions below.

1. An enzyme that carries out a hydrolytic reaction
2. A protein with no enzymic activity that carries one solute in one direction and a second solute in the opposite direction
3. A channel that can pass glucose
4. A protein whose expression in almost all human cells is responsible for the fact that their cytosol is at a negative voltage with respect to the extracellular medium
5. A channel expressed in pain receptor neurons which opens at damagingly high temperatures
6. A channel expressed in almost all neurons, even those that are not sensitive to painful stimuli, but which is not expressed in most non-neuronal human cells (it is not found, for example, in liver or blood cells)

14.3 Theme: Ion Fluxes in a Nerve Cell

A Cl^-
B H^+ (H_3O^+)
C K^+
D Na^+

From the above list of ions, select the ion corresponding to each of the descriptions below.

1. Although gradients of this ion play a crucial role at other cell membranes, the concentration gradient of this ion across the plasma membrane is small: the concentration in the cytosol is the same, within a factor of two, as the concentration in the extracellular medium.
2. In a resting nerve cell, this ion is constantly leaking into the cell, and must be removed by the action of an ATP-consuming carrier.
3. In a resting nerve cell, this ion is constantly leaking out of the cell, and must be pumped back in by the action of an ATP-consuming carrier.
4. The more a nerve cell is depolarized, the greater the electrochemical gradient favoring entry of this ion into the cell.
5. In a resting nerve cell, this ion is at equilibrium across the plasma membrane.

THOUGHT QUESTION

In kidney failure, the concentration of potassium in the extracellular medium can rise from its normal value of 5 mM to 10 mM or even higher. What will be the effect of this on the resting voltage of heart muscle cells? What will be the consequent effect on voltage-gated sodium channels in the heart?

15

INTRACELLULAR SIGNALING

The behavior of cells is not constant. Cells need to be able to alter their behavior in response to internal changes or to external events, and the internal signaling mechanisms that allow them to do so are varied and complex. We will begin by continuing the story begun in Chapter 14 and explain how in nerve cells calcium ions carry a signal from the plasma membrane to vesicles deep within the cytosol. The rest of the chapter will describe some other, very different, methods of intracellular signaling.

CALCIUM

Calcium ions are present at a very low concentration (about 100 nmol liter^{-1}) in the cytosol of a resting cell. An enormous number of processes in many types of cells are activated when the concentration of calcium rises. Calcium can move into the cytosol from two sources: the extracellular medium or the endoplasmic reticulum.

Calcium Can Enter from the Extracellular Medium

In Chapter 14, we saw how heating the hand causes action potentials to travel from the hand to the spinal cord along the axons of pain receptors. When the action potential reaches the proximal axon terminal, the amino acid glutamate is released onto the surface of another nerve cell, the pain relay cell (Fig. 15.1). Glutamate is a transmitter that stimulates the pain relay nerve cell, so that the message that the finger is being damaged is passed toward the brain.

The proximal axon terminal is unmyelinated, so that the plasma membrane is exposed to the extracellular medium. The membrane contains not only potassium channels and voltage-gated sodium channels, but also voltage-gated calcium channels. Voltage-gated calcium channels are close relatives of voltage-gated sodium channels, but are selective for calcium. Like voltage-gated sodium channels, they open and then inactivate in response to depolarization, but the inactivation process is considerably slower. They are closed in a resting cell, but they open when an action potential travels in from the skin and depolarizes the plasma membrane of the proximal axon terminal. Calcium ions pour in, increasing their concentration in the cytosol by 10 times, from the normal concentration of 100 nmol liter^{-1} to 1 μmol liter^{-1}. At the proximal axon terminal the cytosol of the nerve cell contains regulated exocytotic vesicles (page 168). In the case of the pain receptor, these vesicles are filled with sodium glutamate. In response to the increased concentration of cytosolic calcium the regulated exocytotic vesicles move to the plasma membrane and fuse with it, releasing their contents into the extracellular medium. The glutamate then diffuses across the gap to the pain relay cell, stimulating it (we will describe how in Chapter 16). As the action potential in the axon terminal of the pain receptor cell is over in 1 millisecond, the voltage-gated calcium channels do not have time to inactivate. They simply return to the ready-to-open state and can be

Cell Biology: A Short Course, Third Edition. Stephen R. Bolsover, Jeremy S. Hyams, Elizabeth A. Shephard and Hugh A. White.

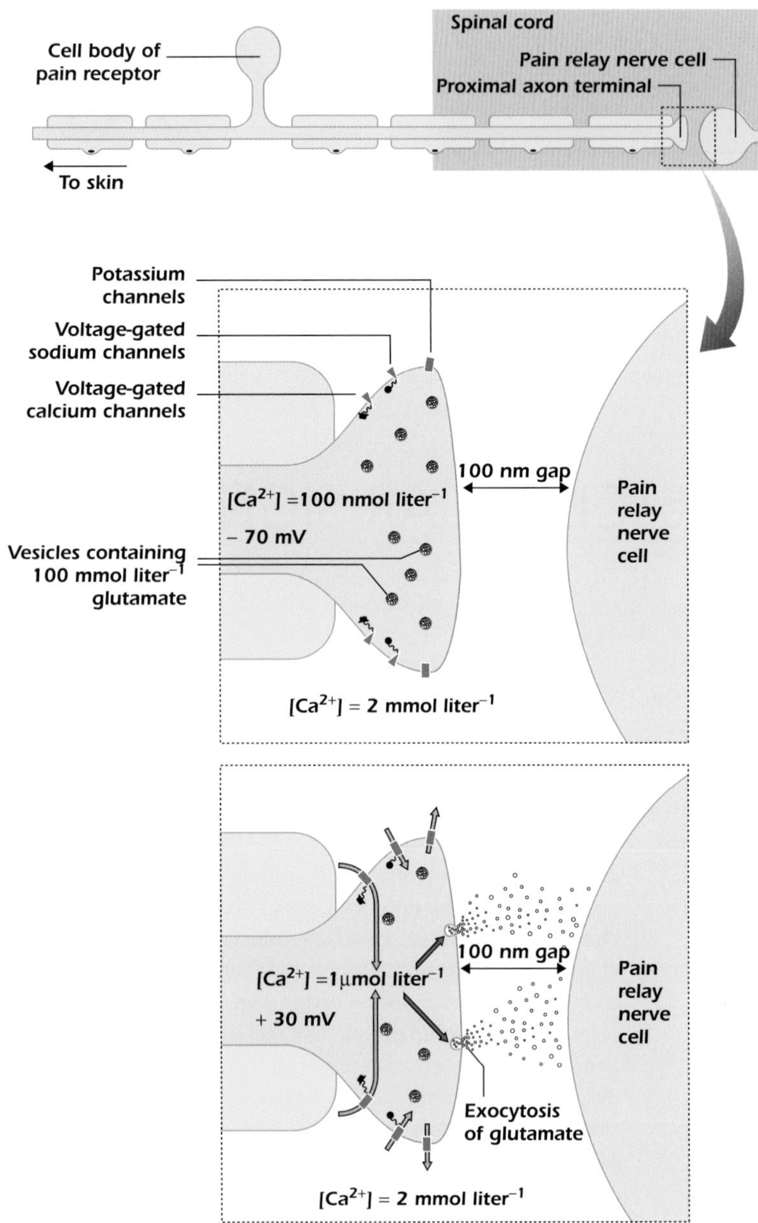

Figure 15.1. Calcium ions entering the cytosol from the extracellular medium activate regulated exocytosis in the proximal axon terminal of a pain receptor nerve cell.

reopened immediately by the next action potential that arrives.

In the process that we have described, calcium ions act as a link between the depolarization of the plasma membrane and the regulated exocytotic vesicles within the cytosol.

Calcium ions are **intracellular messengers**. Regulated exocytotic vesicles that are triggered by an increase of cytosolic calcium concentration were first discovered in nerve cells but are now known to be a feature of almost all cells.

Before we continue with the topic of intracellular messengers, we will review two general points raised by our description of the pain receptor proximal axon terminal. The first point is that different nerve cells release different transmitters. Glutamate is in the exocytotic vesicles of many nerve cells, but other nerve cells release other transmitters. The second point concerns nomenclature. Many nerve cells, like the pain receptor, have their axon terminals close to a second cell and release their transmitter onto it. In such cases the complete unit of axon terminal, gap, and the part of the second cell that receives the transmitter is called a **synapse**. The part of the axon terminal that releases transmitter is the **presynaptic terminal**, and the cell upon which the transmitter is released is the **postsynaptic cell**. Many nerve cells do not come close enough to a specific second cell to form a synapse but simply release transmitter from their axon terminals into the extracellular medium.

Example 15.1 Visualizing Calcium Signals

Calcium signals in cells can be visualized by using dyes that fluoresce brightly when they bind calcium. The images in the figure show a cell body of a pain receptor cell that has been filled with a calcium indicator dye. In image (A), at rest, the cell is dim. Image (B) was acquired after 100 ms of depolarization. In this the edges of the cell are bright because calcium ions have entered through voltage-gated calcium channels. The bright object in the lower part of the cell is the nucleus.

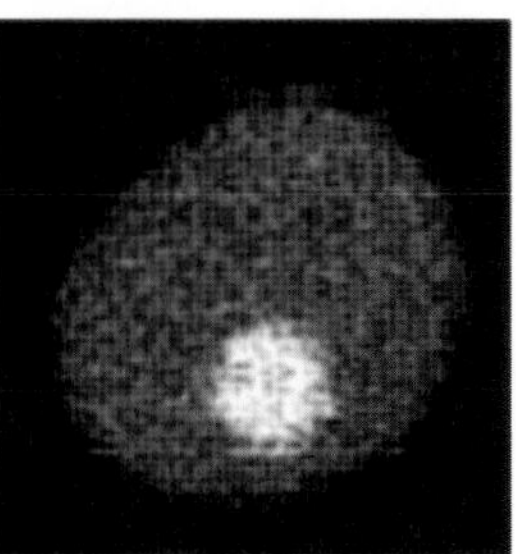

A

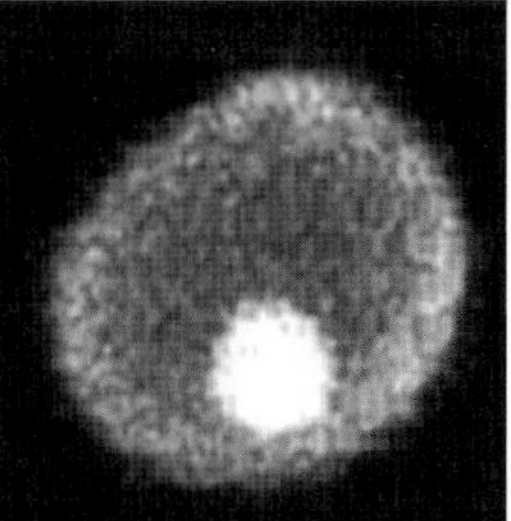

B

Calcium Can Be Released from the Endoplasmic Reticulum

Cells may show an increase of cytosolic calcium not because of an action potential but because of the appearance of a transmitter or other chemical in the extracellular medium. The presence of the chemical is detected by integral membrane proteins, each one a receptor that recognizes a particular chemical with high affinity. These receptors then participate in a more general mechanism, the end result of which is the release of calcium ions from the smooth endoplasmic reticulum into the cytosol. This mechanism will be illustrated with the particular example of blood platelets. We will then discuss how much of the mechanism is general to a wider range of cells.

Platelets are common in the blood. They are small fragments of cells and contain no nucleus, but they do have a plasma membrane and some endoplasmic reticulum. Blood platelets use the release of calcium from the endoplasmic reticulum as one step in the mechanism of blood clotting (Figs. 15.2 and 15.3). Two new mechanisms are involved in calcium release from the endoplasmic reticulum. The **inositol trisphosphate-gated calcium channel** (Fig. 15.3), like the voltage-gated calcium channel, allows only calcium ions to pass. Most of the time its gate is shut, and no ions flow. It is not opened by a change in the voltage across the endoplasmic reticulum membrane. Instead, when the intracellular solute **inositol trisphosphate (IP_3)** binds to the cytosolic face of the channel, the channel changes to an open shape.

The second new mechanism that we must describe is the one that makes IP_3 (Fig. 15.2). If a blood vessel is cut, cytosol from damaged cells at the edge of the cut can leak into the blood. The appearance of solutes that normally are found only inside cells is a sure sign that damage has occurred. Adenosine diphosphate (ADP) is one such solute, and it acts to stimulate platelets, causing them to begin a blood clot to help plug the damaged vessel. The plasma membrane of the platelet contains a protein receptor that binds ADP, that is, ADP is its ligand. When the ADP has bound, the receptor, which is free to move in the plasma membrane, becomes a guanine nucleotide exchange factor (page 160) for a **trimeric G protein** called G_q. Like the GTPases Ran, Arf, and Rab that we have met earlier, trimeric G proteins are GTPases that activate target proteins when they have GTP bound, but turn themselves

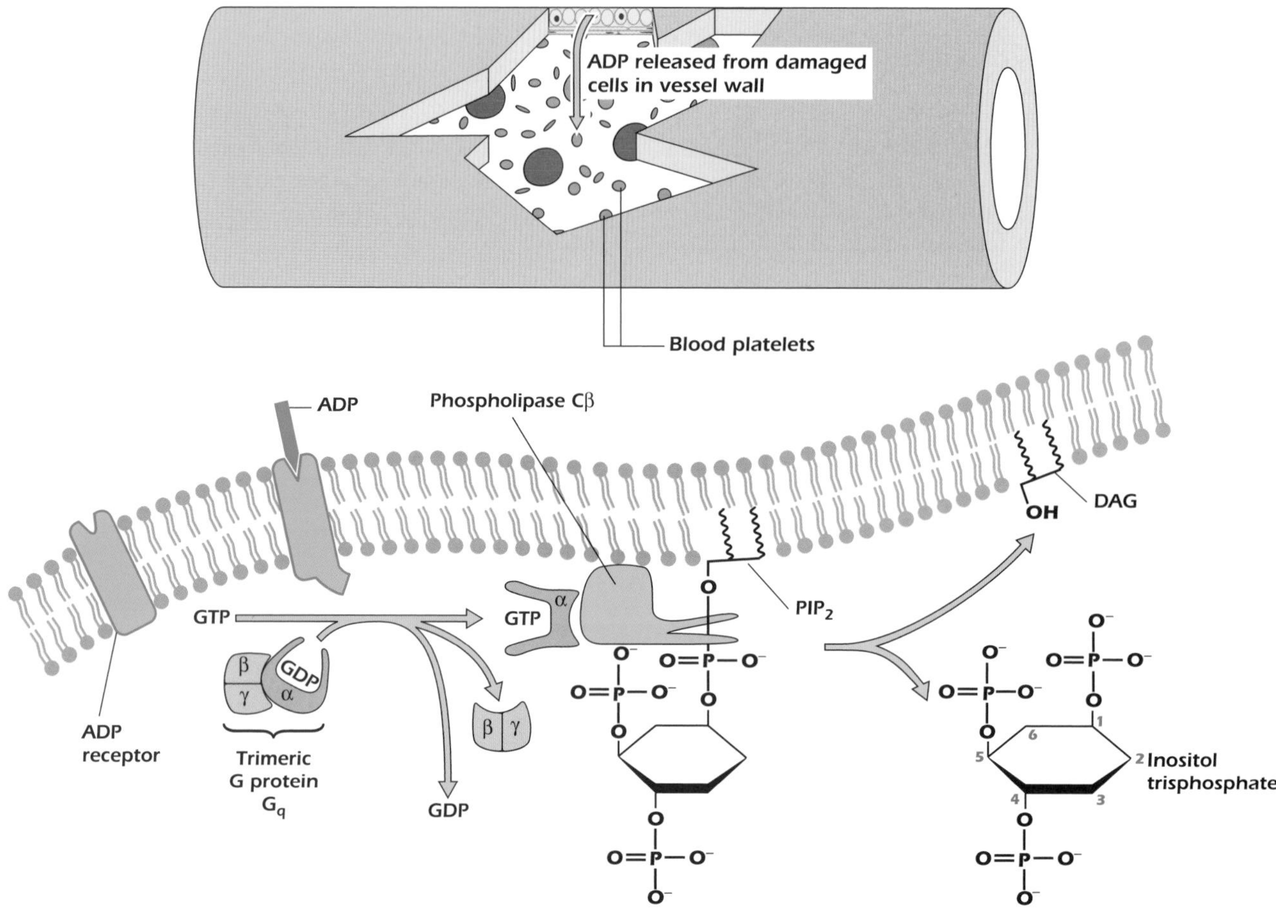

Figure 15.2. ADP from damaged cells activates G_q and hence phospholipase Cβ in platelets.

off by hydrolyzing the GTP to GDP. Trimeric G proteins have a slight additional complication in that, as the name indicates, they are composed of three subunits. The α subunit is homologous to the GTPases we have met before, while the β and γ subunits dissociate from the α subunit when it has GTP bound and only reassociate when the GTP has been hydrolyzed. GTP-loaded G_q activates the β isoform of an enzyme called phosphoinositide phospholipase C, which we will call **phospholipase C** for short or just **PLC**. (The capital C refers to the bond hydrolyzed by the enzyme; A, B, and D phospholipases also exist, but we will not meet them in this book). PLC specifically hydrolyzes **phosphatidylinositol bisphosphate (PIP_2)**, a phospholipid in the plasma membrane that has phosphorylated inositol as its polar head group. Hydrolysis releases the IP_3 to diffuse freely in the cytosol, leaving behind the glycerol backbone with its attached acyl groups: this residual lipid is called **diacylglycerol**, or **DAG**. As it diffuses through the cytosol, IP_3 reaches the endoplasmic reticulum and binds to the inositol trisphosphate-gated calcium channels. The inositol trisphosphate-gated calcium channels open, and calcium ions pour out of the endoplasmic reticulum into the cytosol. The increase of calcium concentration causes the platelet to change shape (Fig. 15.4) and to become very sticky, so platelets begin to clump together in a clot.

Phosphatidylinositol bisphosphate, G_q, PLCβ, and the inositol trisphosphate-gated calcium channel are found in almost all eukaryotic cells, but the distribution of the ADP receptor is more restricted. Only cells that express the ADP receptor will show an increase of cytosolic calcium concentration in response to ADP. Other cells respond to other chemicals. Each produces a receptor specific for that chemical, which then activates G_q. Over 100 such receptors are known. We will meet two more in the next chapter.

Processes Activated by Cytosolic Calcium Are Extremely Diverse

The targets activated by an increase of cytosolic calcium concentration differ between different cells. An increase of calcium is a crude signal that says "do it" but contains no information about what the cell should do. This

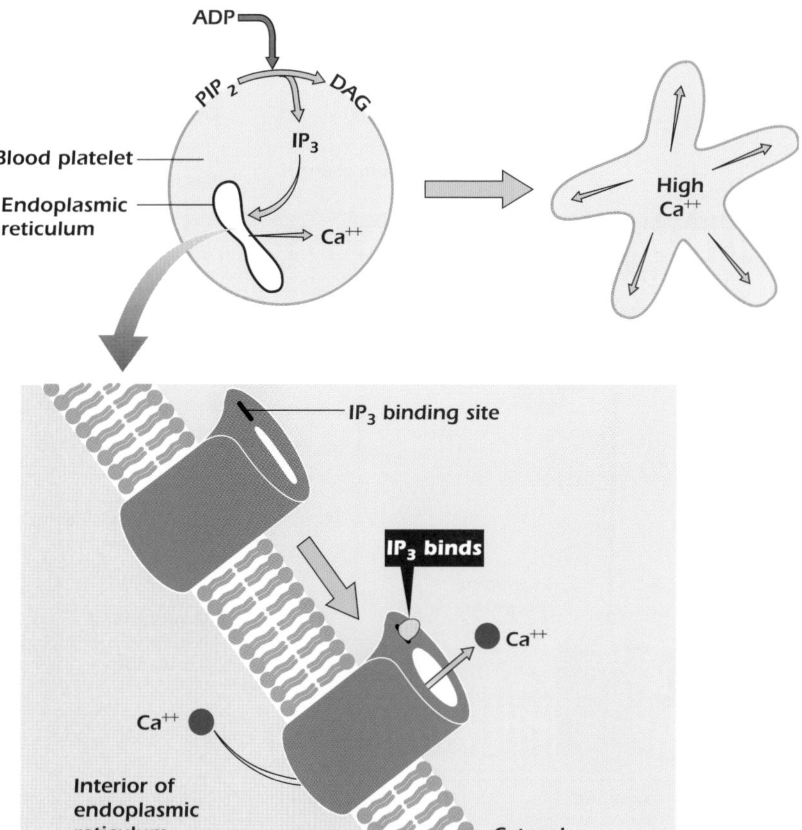

Figure 15.3. The inositol trisphosphate-gated calcium channel is a calcium-selective channel in the membrane of the endoplasmic reticulum. An increase of cytosolic calcium concentration in platelets makes them sticky, initiating blood clotting.

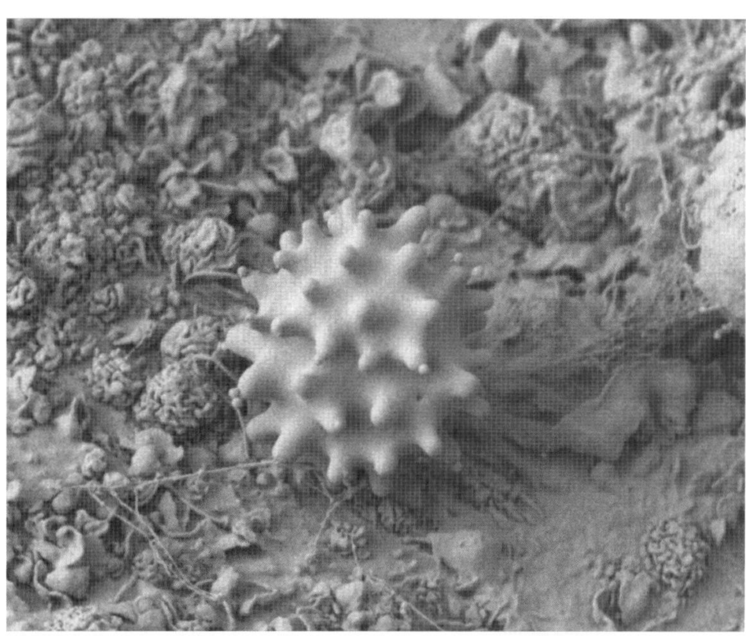

Figure 15.4. Scanning electron micrograph of a blood platelet on the damaged inner surface of a blood vessel. The platelet has activated, extending short processes called pseudopodia, and is ready to initiate blood clotting. Image by Mark Turmaine, Department of Cell and Developmental Biology, University College London. Reproduced by permission.

depends on what the cell was designed to do. Cells and cell regions designed for regulated exocytosis (e.g., salivary gland cells and axon terminals) exocytose when cytosolic calcium increases. Cells designed to contract (e.g., muscle cells) contract when calcium increases, and so on. In each case, the calcium ions bind to a calcium-binding protein then the calcium ion–calcium-binding protein complex activates the target process.

IN DEPTH 15.1 RYANODINE RECEPTORS

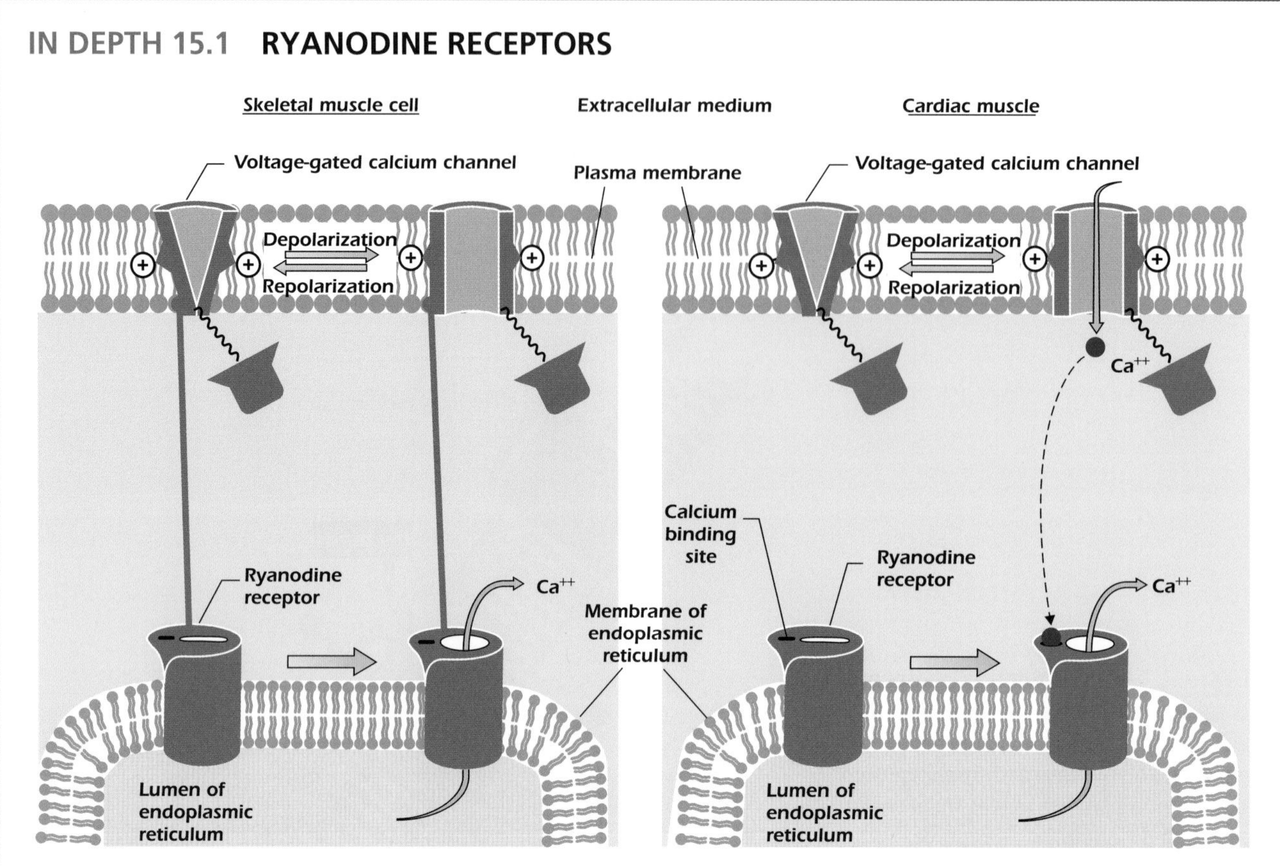

Many cells contain a second type of calcium channel in the membrane of the endoplasmic reticulum. This channel was initially distinguished from the IP_3 receptor by the fact that it bound the plant toxin ryanodine, so it was called the ryanodine receptor. Ryanodine receptors were first identified in skeletal muscle. In these cells they are physically linked to voltage-gated calcium channels in the plasma membrane (left-hand side of diagram). When the voltage-gated calcium channels switch to their open configuration, they induce the ryanodine receptors below them to switch into an open configuration, allowing calcium to flow out of the smooth endoplasmic reticulum into the cytosol.

No such physical linkage between ryanodine receptors and plasma membrane channels is found in other cell types. Instead, ryanodine receptors are opened when the concentration of calcium ions in the cytosol rises above a critical level. For example, in heart muscle (right-hand side of diagram) depolarization causes voltage-gated calcium channels in the plasma membrane to open. The calcium ions that enter through this channel bind to the cytosolic aspect of the ryanodine receptor, causing it to open too, allowing calcium to flow out of the smooth endoplasmic reticulum into the cytosol.

In skeletal muscle cells (page 13) a transmitter released from nerve axon terminals leads to the escape of calcium from the endoplasmic reticulum (Fig. 15.5). This activates several processes. First, calcium ions bind to a protein called troponin that is attached to the cytoskeleton. This causes the cytoskeleton to contract, using the energy released by ATP hydrolysis to do mechanical work (page 289). Second, calcium passes down its electrochemical gradient from the cytosol into the mitochondrial matrix through a calcium channel. Once there, calcium stimulates the mitochondria to increase the production of NADH and ATP (page 224). Lastly, the calcium binds to the protein calmodulin, which in turn activates glycogen phosphorylase kinase and hence glycogen breakdown as part of the feedforward control of energy metabolism (page 225).

The same simple intracellular messenger, calcium, has many different actions inside the skeletal muscle cell. Under its influence the cytoskeleton begins to use ATP, glycogen phosphorylase releases more glucose, and the

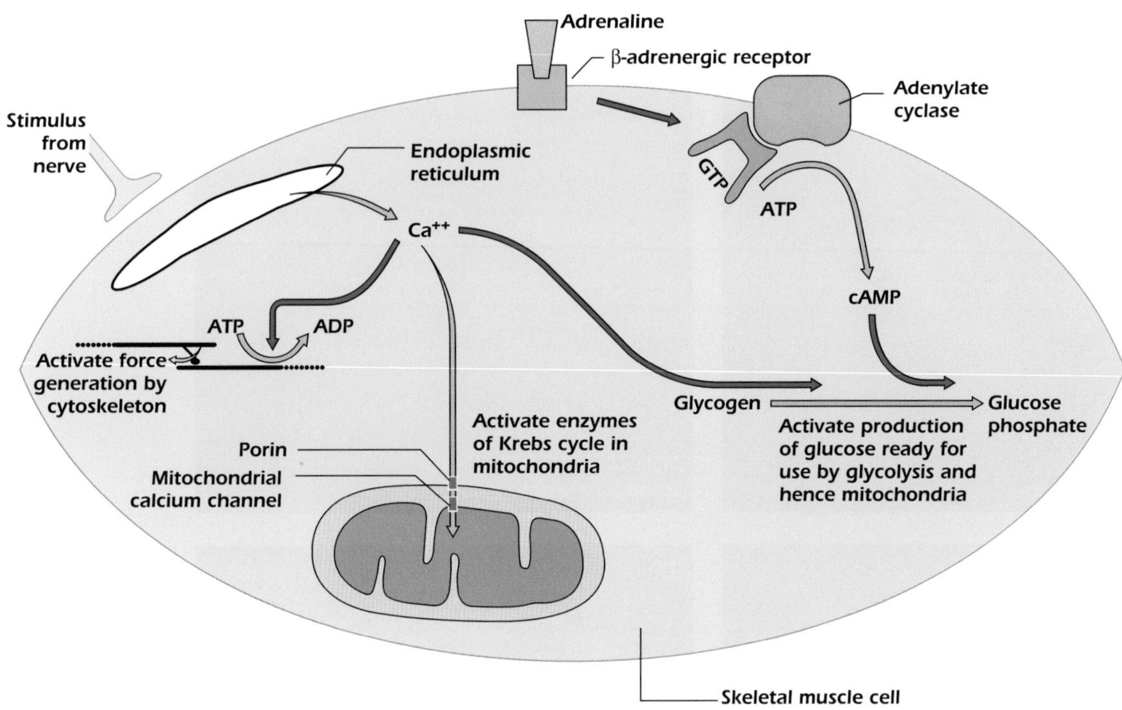

Figure 15.5. Calcium and cyclic AMP activate distinct but overlapping sets of target processes in skeletal muscle cells.

Example 15.2 Training for Anaerobic Glycolysis

Muscle cells that are contracted for long periods, cutting off the blood supply—such as those in the arms of rock climbers—have a much enhanced capacity for anaerobic glycolysis compared to those that are used briefly and then relaxed (such as those in the arms of baseball pitchers). This occurs because the calcium signals that activate muscle contraction also act to increase the transcription of the genes coding for glycolysis (via calcineurin and NFAT; see Medical Relevance 10.1 on page 161).

mitochondria produce more ATP. The skeletal muscle cell is an example of how diverse mechanisms inside a cell can be integrated by the action of intracellular messengers.

Figure 15.6 shows an experiment that used fluorescent microscopy to follow the uptake of calcium by mitochondria. The cell body of a living pain receptor nerve cell was stained with a dye to reveal the mitochondria (Fig. 15.6A) and with Hoechst 33342 to reveal the nucleus (Fig. 15.6D). It was also stained with the calcium indicator dye Rhod-2, which concentrates inside mitochondria. In the unstimulated cell the Rhod-2 emits little light, indicating that mitochondrial calcium is low (Fig 15.6B). However, when the plasma membrane was depolarized for 1 second, opening voltage-gated calcium channels, calcium flowed in from the extracellular medium to the cytosol, and from there into the mitochondria. This caused the Rhod-2 to fluoresce brightly (Fig. 15.6C).

Return of Calcium to Resting Levels

As soon as a stimulus, be it depolarization or extracellular chemical, disappears, cytosolic calcium concentration falls again. Calcium ions are pumped up their electrochemical gradients from the cytosol into the extracellular medium or into the endoplasmic reticulum by two carriers. The Ca^{2+} ATPase (page 237) uses the energy released by ATP hydrolysis to move calcium ions up their electrochemical gradient out of the cytosol, while the Na^{+}/Ca^{2+} exchanger (page 236) uses the energy released by sodium ion movement into the cytosol to do the same job.

CYCLIC ADENOSINE MONOPHOSPHATE

We have already met the nucleotide cyclic adenosine monophosphate, or cAMP (Fig. 15.7), in the context of

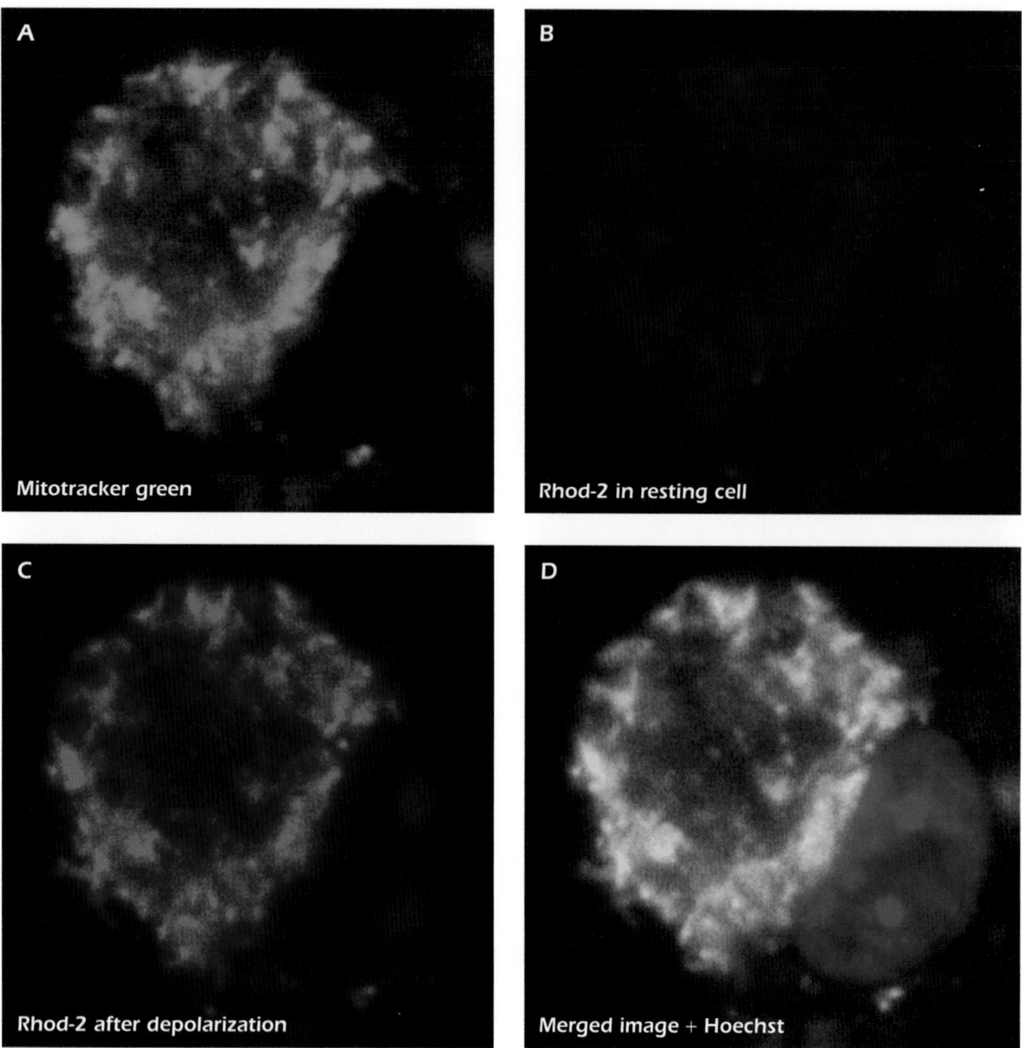

Figure 15.6. Nerve cell mitochondria take up calcium from the cytosol. Experiment of William Coatesworth and Stephen Bolsover. First published in *Cell Calcium* **39**, 217 (2006).

the regulation of the *lac* operon (page 88). (It is worth mentioning that cAMP is always pronounced "cy-clic A-M-P," with five syllables.) In eukaryotes, cAMP is also important and acts as an intracellular messenger in a great many cells, including the scent-sensitive nerve cells in our nose (Fig. 15.8). These cells have their cell bodies in the skin of the air passages in the nose. Each cell sends an axon into the brain, and shorter processes, called dendrites, into the mucus lining the air passages. Scent-sensitive nerve cells are stimulated by scents in the air. Particular chemicals in the air stimulate these cells because the cells have protein receptors that specifically bind the scent (Fig. 15.9). When the scent binds, the receptor becomes a guanine nucleotide exchange factor for a trimeric G protein called $\mathbf{G_s}$. GTP-loaded G_s activates the enzyme **adenylate cyclase** that converts ATP to cAMP. The next stage in the detection of scents is a channel in the plasma membrane called the **cAMP-gated channel**. Like the inositol trisphosphate-gated calcium channel, the cAMP-gated channel is opened by a cytosolic solute, in this case cAMP. When the channel is open, it allows sodium and potassium ions to pass through. The electrochemical gradient pushing sodium ions into the cell is much greater than the electrochemical gradient pushing potassium ions out of the cell. Thus, when the cAMP-gated channel opens, sodium ions pour in, carrying their positive charge and depolarizing the plasma membrane. The plasma membrane also contains voltage-gated sodium channels. Thus, when enough cAMP-gated channels open, the transmembrane voltage reaches threshold for generation of sodium action potentials. These then propagate along the axon to the brain, and the person becomes aware of the particular scent for which that scent-sensitive nerve cell was specific.

G_s and adenylate cyclase are found in many cells, but only scent-sensitive nerve cells are sensitive to scents, because only they have specific scent receptors in their plasma membranes. Other cells that use cAMP as an intracellular messenger are sensitive to other specific chemicals because each makes a receptor that binds

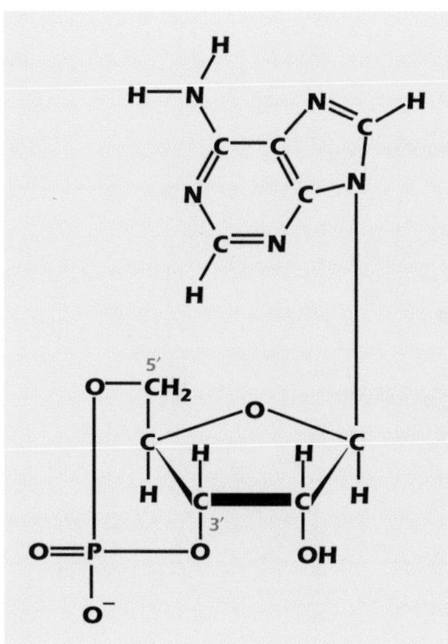

Figure 15.7. Cyclic adenosine monophosphate, also called cyclic AMP or just cAMP.

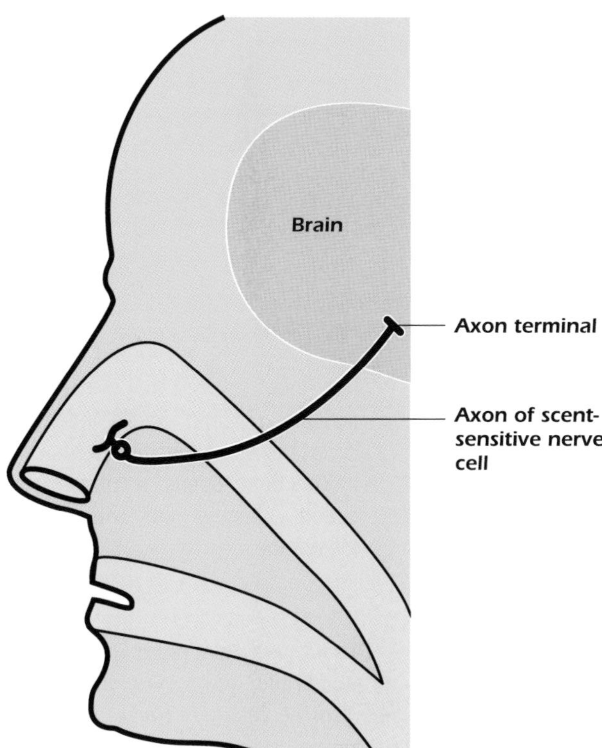

Figure 15.8. Scent-sensitive nerve cells send axons to the brain.

the chemical, whatever it may be, and then activates G_S. Once the stimulating chemical has gone, the cAMP concentration returns to resting levels. The enzyme **cAMP phosphodiesterase** hydrolyzes cAMP to AMP, which is inactive at cAMP-gated channels and other cAMP-binding proteins.

Most of the symptoms of the deadly disease cholera are caused by a toxin released by the gut bacterium *Vibrio cholera*. The toxin is an enzyme that enters the cytosol of the cells lining the gut and attaches an ADP ribosyl group (compare Example 8.2 on page 131) to the catalytic domain of G_S, preventing it from hydrolyzing GTP. G_S is therefore locked in the active state, and activates adenylate cyclase nonstop. The cAMP concentration in the cytosol then shoots up. Ion channels in the plasma membrane are opened by the increase of cAMP, allowing ions to leak from the cells into the gut contents. If untreated, this loss of ions and of the water that accompanies them leads to death from dehydration.

CYCLIC GUANOSINE MONOPHOSPHATE

Another nucleotide that acts as an intracellular messenger is cGMP. In light-sensitive nerve cells called photoreceptors, cGMP plays a role like that of cAMP in scent-sensitive nerve cells. In the dark, the enzyme guanylate cyclase makes cGMP from GTP. cGMP binds to and opens a channel that allows sodium and potassium ions to pass, so that in the dark the photoreceptor is depolarized to about −40 mV because of the constant influx of sodium ions through the cGMP-gated channel. In the light the concentration of cGMP in the cytosol falls, and the transmembrane voltage changes to the more typical resting voltage of −70 mV. The changing transmembrane voltage is transmitted to other nerve cells, making us aware of the pattern of light and dark.

MULTIPLE MESSENGERS

Many cells use more than one intracellular messenger at once. Skeletal muscle cells are a good example (Fig. 15.5). In the excitement before a race, the runner's adrenal glands release adrenaline into the blood. This binds to a receptor on skeletal muscle cells called the β-adrenergic receptor. The complex of adrenaline plus β-adrenergic receptor can now activate G_S and hence adenylate cyclase. This in turn generates cAMP, which activates **protein kinase A** (page 225) which in turn phosphorylates glycogen phosphorylase kinase and turns on the latter enzyme even when cytosolic calcium is low (Fig. 13.14 on page 224). The end result is that even before the runner begins to run, the muscles break down glycogen and make the glucose phosphate (page 212) they will need once the race begins.

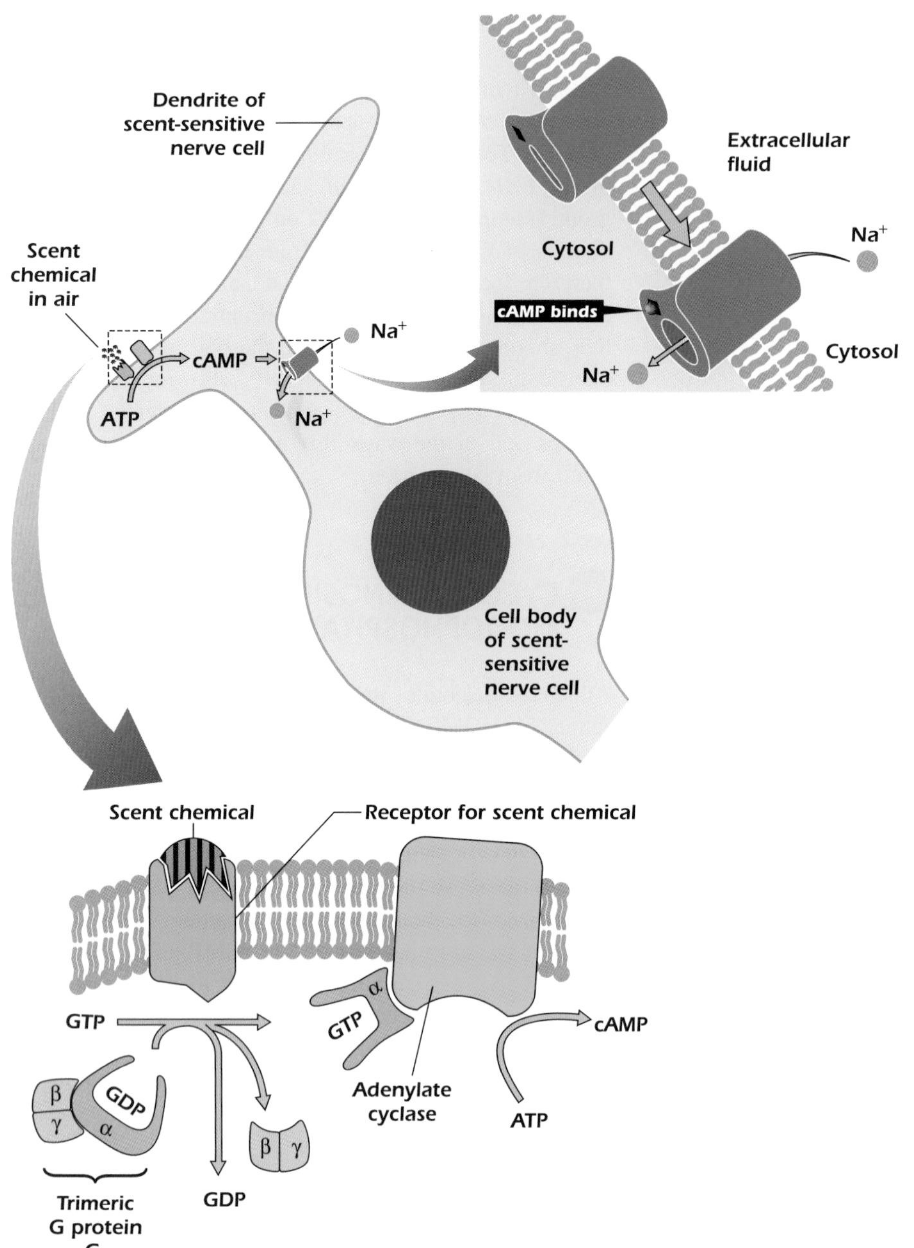

Figure 15.9. Scent chemicals activate G_s and hence adenylate cyclase in scent-sensitive nerve cells. cAMP then opens a nonselective cation channel in the plasma membrane.

BIOCHEMICAL SIGNALING

Receptor Tyrosine Kinases and the MAP Kinase Cascade

When platelets are stimulated, they not only help the blood to clot, but also trigger the damaged blood vessel to repair itself. They do this by regulated exocytosis of a protein called **platelet-derived growth factor (PDGF)**. The plasma membranes of the cells of the blood vessel contain receptors for this protein (Fig. 15.10). PDGF receptors (Fig. 9.6 on page 144) comprise an extracellular domain, which binds PDGF, a single polypeptide chain that crosses the plasma membrane, and a cytosolic domain that can phosphorylate tyrosine residues (page 142). The catalytic activity of the monomeric protein is low. PDGF can bind to two receptor molecules, drawing them close to each other, causing a shape change that dramatically increases the tyrosine kinase activity. One of the first proteins to be phosphorylated is the receptor itself.

A number of cytoplasmic proteins have a domain called **SH2** that is just the right shape to stick to phosphorylated tyrosine. At the bottom of a deep pocket in the protein surface is a positively charged arginine. Although proteins can be phosphorylated on amino acids other than tyrosine, only

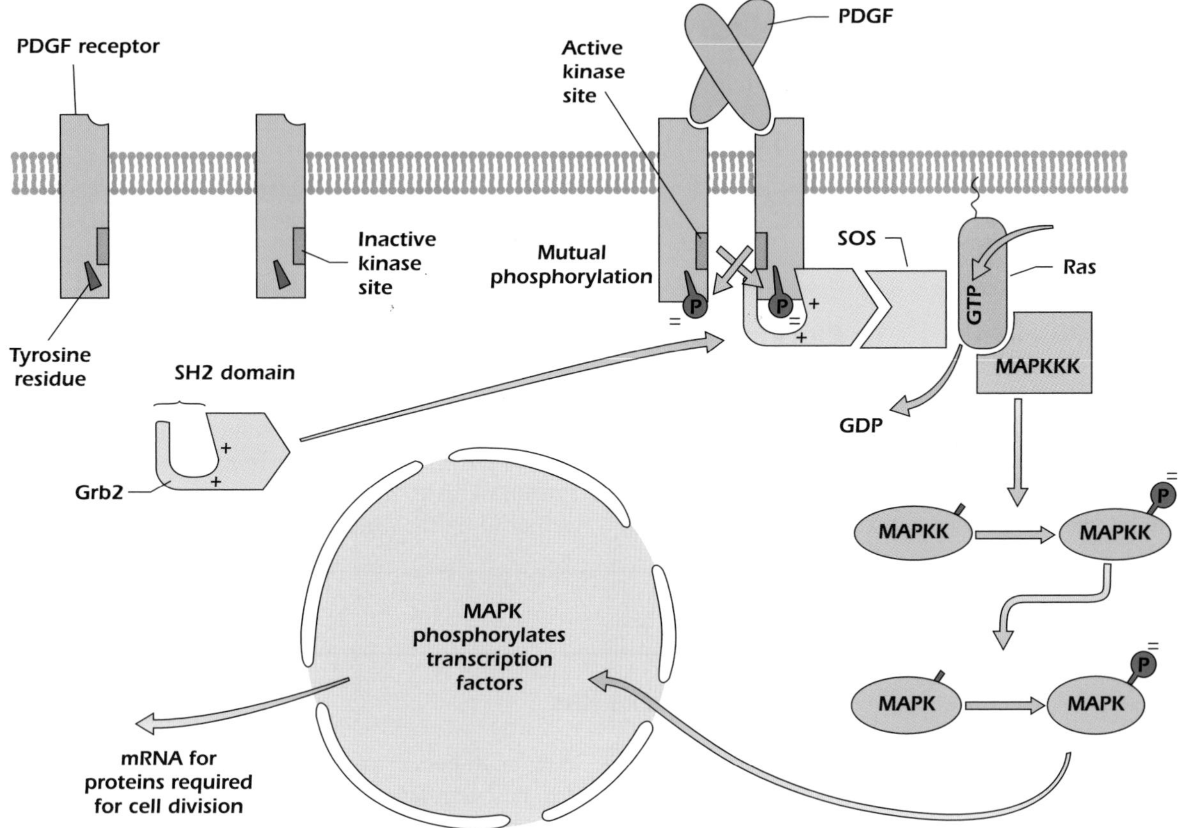

Figure 15.10. The PDGF receptor, like other growth factor receptors, activates the GTPase Ras and therefore the MAP kinase cascade.

tyrosine is long enough to insert down the pocket so that the negative phosphate can stick to the positive arginine. Proteins with SH2 domains therefore stick to dimerized PDGF receptors. In contrast, SH2 domains do not stick to solitary PDGF receptors whose tyrosines do not carry the negatively charged phosphate group.

One protein that has an SH2 domain is **growth factor receptor binding protein number 2 (Grb2)**. Grb2 has no catalytic function but recruits a second protein called **SOS**, and SOS is a guanine nucleotide exchange factor for a GTPase called **Ras**, allowing it to discard GDP and bind GTP. Active Ras turns on the first of a series of protein kinases, each of which phosphorylates and hence activates the next, culminating in a kinase called **mitogen associated protein kinase** (**MAP kinase** or **MAPK**). The kinase that phosphorylates MAP kinase is **MAP kinase kinase (MAPKK)** while the one that phosphorylates MAPKK is **MAP kinase kinase kinase (MAPKKK)**. MAPKKK is activated by Ras.

The word **mitogen** means a chemical that tends to cause mitosis, reflecting the fact that MAP kinase is activated by transmitters such as PDGF that turn on cell division (page 277). It phosphorylates many targets on serine and threonine residues. When MAP kinase is phosphorylated, it moves to the nucleus and phosophorylates transcription factors that in turn stimulate the transcription of the genes for **cyclin D** (page 304) and other proteins that are required for DNA synthesis and cell division.

Platelet-derived growth factor is one of many growth factors. All work in much the same way: their receptors are tyrosine kinases that are triggered to dimerize and phosphorylate their partners on tyrosine when the growth factor binds. The general term for this type of receptor is **receptor tyrosine kinase**. Phosphorylated tyrosine then recruits proteins with SH2 domains including Grb2, which in turn allows activation of Ras and the MAP kinase pathway, leading to DNA synthesis and cell division. In fact the situation is more complicated than a simple "SH2 domains bind phosphotyrosine" rule would suggest. Different SH2 domain proteins have specific requirements for the amino acids flanking the phosphotyrosine (Fig. 15.11). The PDGF receptor can recruit Grb2 because one of its phosphorylated tyrosines has an asparagine two amino acids away in the C terminus direction. Grb2 will only bind to phosphotyrosines that meet this criterion.

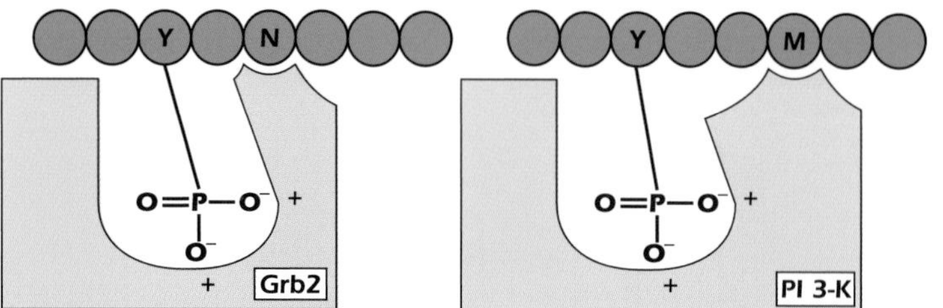

Figure 15.11. Amino acid residues adjacent to phosphotyrosine recruit specific subtypes of SH2 domain.

Medical Relevance 15.1 Blocking Growth Factor Receptors

Active growth factor receptors cause cell division by activating the MAP kinase pathway, and keep cells alive by activating PKB (page 309). Turning off growth factor receptors therefore tends both to stop cells dividing and to kill them. The drugs cetuximab and panitumumab are effective in slowing down the progression of bowel cancer because they prevent a growth factor, called epidermal growth factor (EGF), from binding to its receptor tyrosine kinase. This slows cell division and promotes cell death in the cancer cells.

Medical Relevance 15.2 Zarnestra Blocks Ras

Like other GTPases, Ras turns itself off by hydrolyzing its bound GTP. Mutant forms of Ras without GTPase activity are therefore **constitutively active**: they are always in the "on" state and will therefore be activating the pathway that terminates in MAP kinase and cell division at all times, even in the absence of growth factor. Such mutant forms of Ras are found in about 20% of all human cancers. Zarnestra (R115777) prevents the post-translational addition of a long hydrophobic group to Ras and hence keeps the active Ras away from MAPKKK. MAPKKK therefore remains inactive and does not promote cell division. Zarnestra is in late stage clinical trial for treatment of leukaemias and other cancers.

Growth Factors Can Trigger a Calcium Signal

A second protein that contains an SH2 domain and that is therefore recruited to phosphorylated receptor tyrosine kinases is another isoform of phosphoinositide phospholipase C, **PLCγ**. Binding to the phosphorylated tyrosine holds PLCγ at the growth factor receptor long enough for it itself to be phosphorylated (Fig. 15.12), and this activates its enzymatic action. Active PLCγ hydrolyzes PIP_2 into diacylglycerol and inositol trisphosphate, and in turn inositol trisphosphate triggers the release of calcium from the smooth endoplasmic reticulum.

Protein Kinase B and the Glucose Carrier: How Insulin Works

Figure 15.13 shows the **insulin receptor**. Like other receptor tyrosine kinases, the insulin receptor has an extracellular domain that can bind the transmitter—in this case the protein insulin, a single polypeptide chain that crosses the plasma membrane, and a cytosolic domain with tyrosine kinase activity. Unlike growth factor receptors, the insulin receptor exists as a dimer even in the absence of its ligand, insulin. When insulin binds, the shape and orientation of the individual insulin receptors change a little, and this allows each receptor to phosphorylate its partner upon tyrosine. An associated protein called the **insulin receptor substrate number 1 (IRS-1)** is also phosphorylated on tyrosine.

In particular, nine of the tyrosines on IRS-1 that become phosphorylated have a methionine three amino acids away in the C terminus direction. This is the preferred sequence for another SH2 domain protein, **phosphoinositide 3-kinase (PI 3-kinase)** (Fig. 15.11). Binding to the phosphorylated tyrosines holds PI 3-kinase close to the insulin receptor long enough for it to become phosphorylated, and this activates its enzymatic action. PI 3-kinase operates on the same substrate (PIP_2) that PLC does, but instead of hydrolyzing it, PI 3-kinase adds another phosphate group to the inositol head group, generating the intensely charged lipid **phosphatidylinositol trisphosphate (PIP_3)** (Fig. 15.13). In contrast, none of the phosphorylated tyrosines in the insulin receptor itself, or on IRS-1, is good at recruiting PLCγ, so insulin does not evoke a calcium signal in cells.

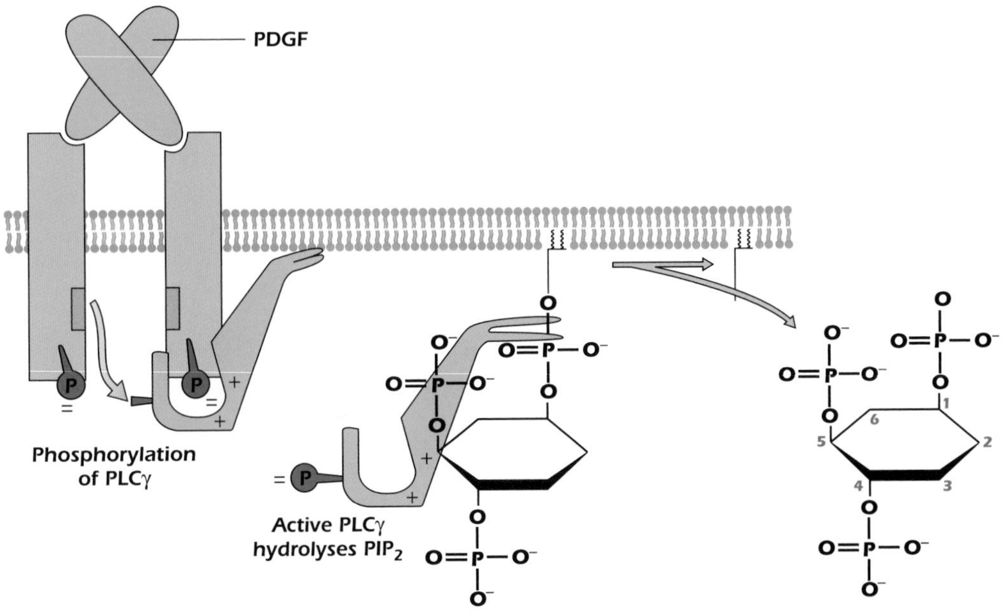

Figure 15.12. The PDGF receptor, like other growth factor receptors, phosphorylates and hence activates phospholipase Cγ.

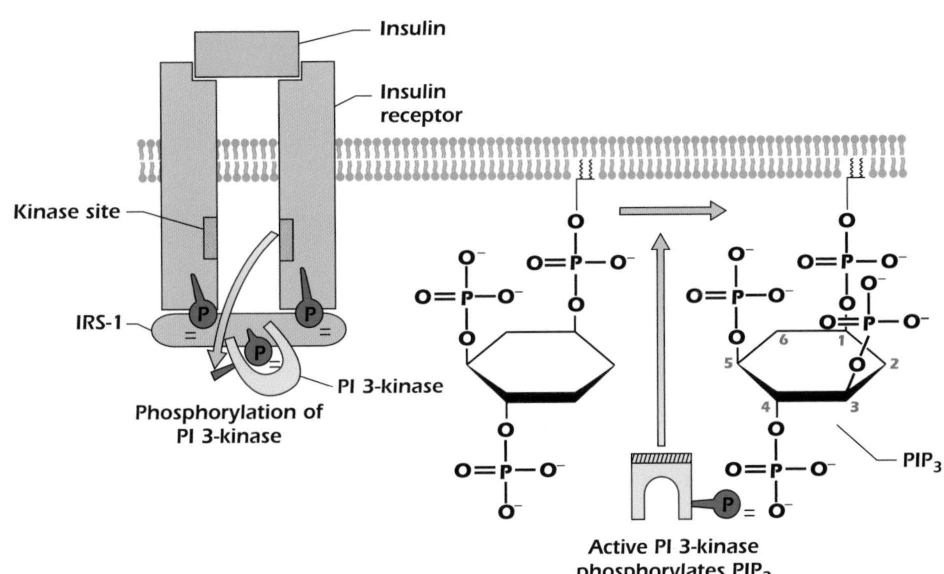

Figure 15.13. The insulin receptor phosphorylates and hence activates PI 3-kinase.

Just as phosphotyrosine is bound by a wide variety of proteins that contain SH2 domains, so highly phosphorylated inositols, such as those in PIP_3, are bound by a domain called the **PH domain** found on many proteins. Most important among these is **protein kinase B (PKB)** (Fig. 15.14). PKB, which phosphorylates targets on serine and threonine residues, is itself activated by phosphorylation, but the kinase that does this is located at the plasma membrane and therefore only gets a chance to phosphorylate PKB when PKB is held at the plasma membrane through its binding to PIP_3.

In many cell types, particularly fat cells and muscle cells, the final stage in trafficking (page 166) of the glucose carrier (page 236) from the Golgi apparatus to the plasma membrane requires active PKB. At the same time, glucose carriers are being endocytosed and are only returned to the plasma membrane if PKB is active. When we eat a large meal, insulin concentrations in the blood increase. Activation of the insulin receptor therefore causes an increase in the activation of PKB and hence a translocation of glucose carriers to the plasma membrane. This allows muscle and fat cells to take up large amounts of glucose from the extracellular medium. Muscle cells convert the glucose to glycogen (page 219); fat cells convert the glucose to fat (page 221). The action of PKB on the protein BAX, shown in Figure 15.14, will be described in Chapter 18.

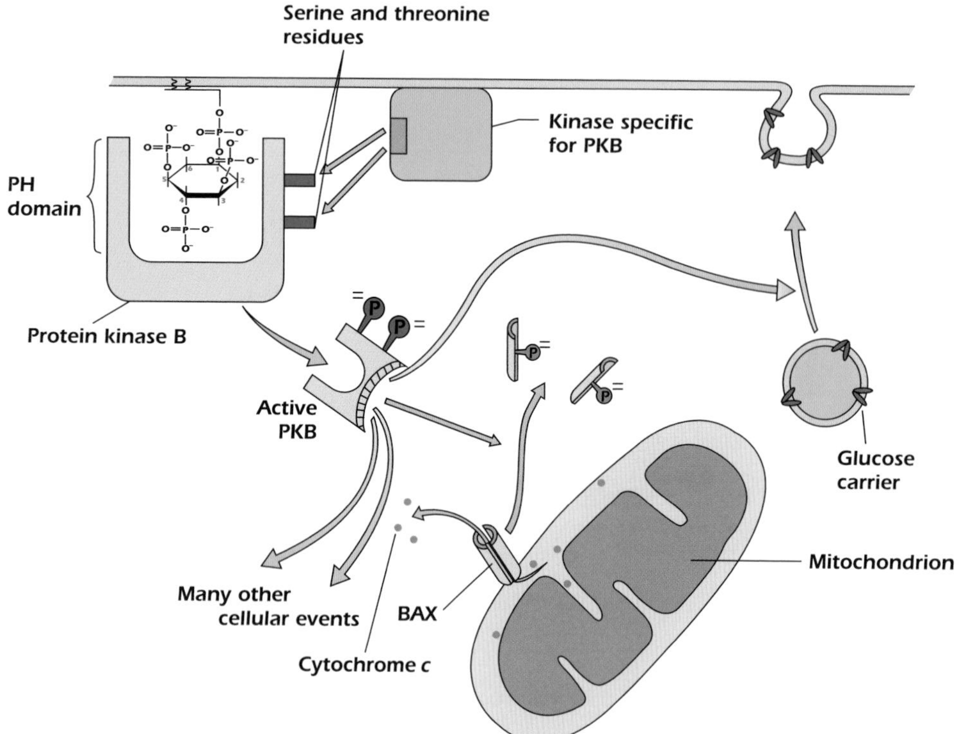

Figure 15.14. PIP3 recruits protein kinase B (PKB) to the plasma membrane, where it is activated. Active PKB has many effects including exocytosis of vesicles containing glucose carriers. The action of PKB on the protein BAX will be described in Chapter 18.

Cytokine Receptors

Some of the most complicated interactions between cells occur in the immune system of vertebrates, described in Chapter 19. The complex cell-cell interaction is mediated in part by a large family of protein transmitters called **cytokines**. A large and disparate collection of cell surface receptors bind cytokines and activate downstream processes. Some are G protein coupled receptors related to the ADP receptor and the β adrenergic receptor. Others are receptor tyrosine kinases related to the PDGF and insulin receptors. However many cytokines act through a distinct family of receptors known as the type 1 cytokine receptor family. A type 1 cytokine receptor, such as the receptor for interleukin II (page 161), comprises a number of integral membrane proteins, each of which crosses the plasma membrane only once (Figure 15.15). The extracellular aspects of these proteins form a binding site for the cytokine. Associated with the cytosolic part of the receptor are proteins called **JAK**. JAKs are tyrosine kinases, but unlike the receptor tyrosine kinases that we have met before JAKs are not integral membrane proteins but are peripheral membrane proteins (page 43) that remain at the membrane only because they associate with the cytosolic aspect of the cytokine receptor.

When the extracellular aspect of the receptor binds its cytokine ligand the cytosolic aspect changes shape. This brings the two JAKs into proximity, allowing them to phosphorylate each other on their own tyrosine residues. Phosphorylation greatly increases the catalytic activity of the JAKs, which can now phosphorylate tyrosines on the other components of the complex. These phosphotyrosines can activate many of the signaling pathways that we have already discussed in the context of receptor tyrosine kinases. For example, active interleukin II receptors can signal via Grb2 to Ras, and via phosphoinositide 3-kinase to protein kinase B. However type 1 cytokine receptors have a uniquely direct signaling route to the nucleus. This operates through a family of transcription factors called **STATs** (for Signal Transducers and Activators of Transcription) which in unstimulated cells are found mainly in the cytosol. Stats have SH2 domains and are recruited to activated type 1 cytokine receptors, where they are themselves phosphorylated by the active JAKs. This induces STATs to dimerize, with the SH2 domain of one STAT binding the phosphotyrosine on its partner. Dimerized STATs translocate to the nucleus and turn on the transcription of genes that promote cell division such as cyclin D (Chapter 18).

CROSSTALK—SIGNALING PATHWAYS OR SIGNALING WEBS?

Cell biologists often talk of signaling pathways in which a transmitter activates a chain of events culminating in an

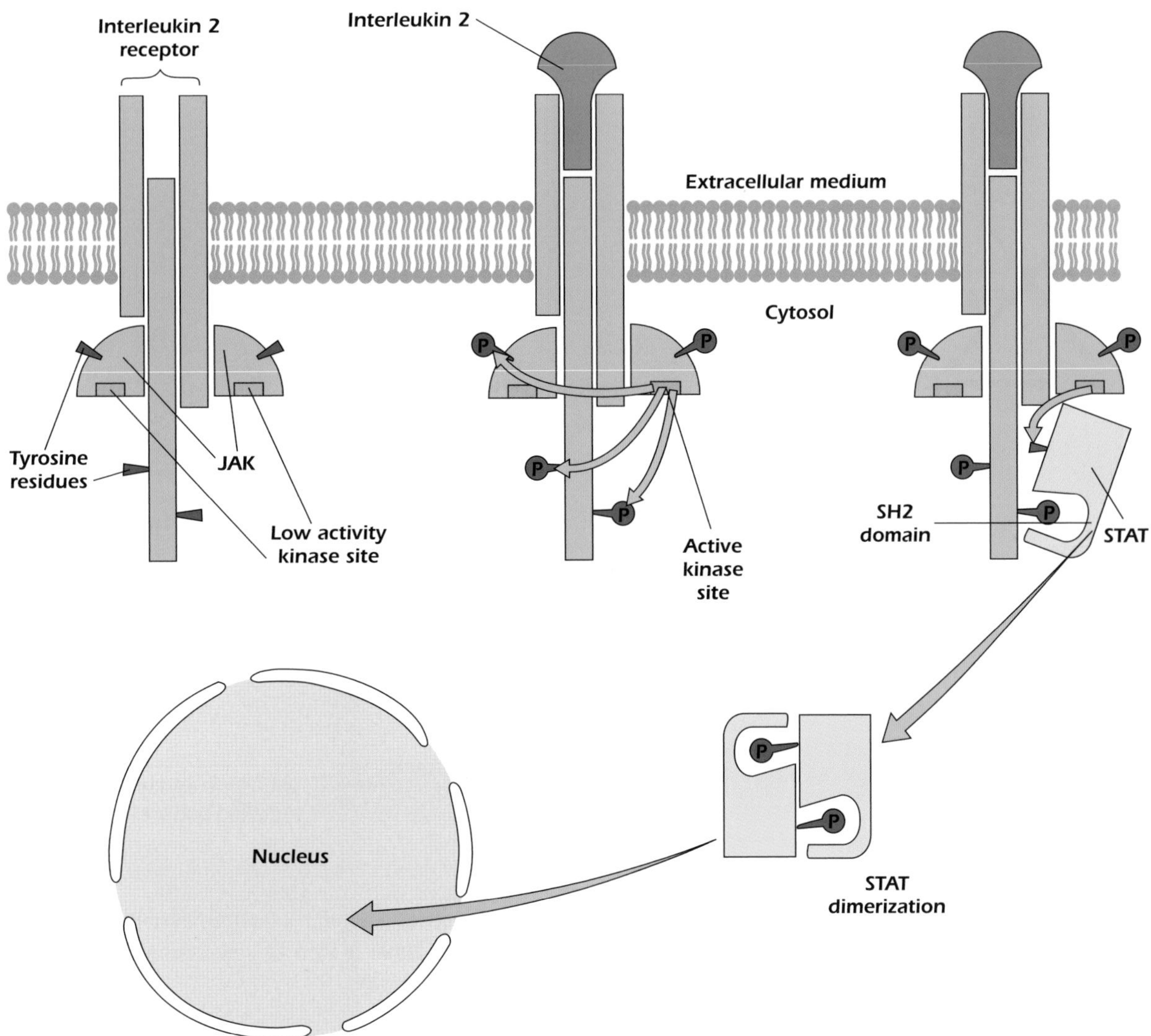

Figure 15.15. Signaling from type 1 cytokine receptors.

effect. The vertical arrows in Figure 15.16 show four signaling pathways that we have discussed in this chapter. We have already seen one example of how these pathways can interact: growth factor receptors can trigger a calcium signal by activating phospholipase Cγ. In fact, the pathways can interact in many ways and at many levels. The red arrows in Figure 15.16 show some of the most important. SOS is recruited to phosphorylated IRS-1 and can in turn activate Ras (arrow 1 in Fig. 15.16). PI 3-kinase can bind to phosphotyrosine on growth factor receptors via its SH2 domain and be activated by phosphorylation so that growth factors, as well as insulin, will activate protein kinase B (arrow 2 in Fig. 15.16). In Chapter 18 we will see that this process is literally vital for our cells—if it stops, they die (page 309). A second route by which growth factors can activate PI 3-kinase is shown as arrow 3 in Figure 15.16: active Ras can activate PI 3-kinase.

Many of the actions of cAMP occur through the phosphorylation of other proteins by protein kinase A. In particular, cAMP can activate transcription of many eukaryotic genes (arrow 4 in Fig. 15.16), but it does so by causing the phosphorylation of transcription factors, and hence activating them. This is in sharp contrast to the cAMP-CAP system in *Escherichia coli* (page 88).

Calcium ions can activate a number of protein kinases, notably **protein kinase C (PKC)** and **calcium-calmodulin-activated protein kinase**. These phosphorylate target proteins on serine and threonine residues and recognise many of the same targets as protein kinase A, hence many of the downstream events are the same (arrows 5 in Fig. 15.16). Paradoxically,

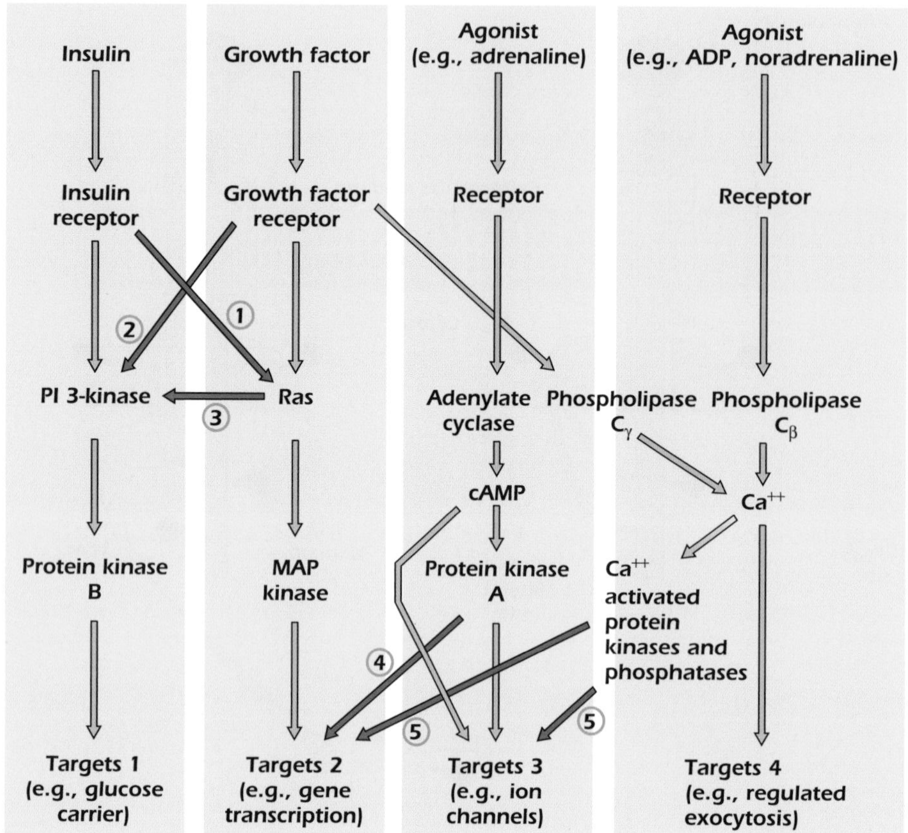

Figure 15.16. Interactions of signaling pathways.

calcium-calmodulin also activates the phosphatase calcineurin, which dephosphorylates and hence activates the transcription factor NFAT (page 161). The more we know about intracellular signaling systems, the more they come to resemble a web of interactions rather than a set of discrete pathways.

Answer to Thought Question: The outward voltage force is opposed by a large inward concentration force. Even when the inside of the cell is at +50 mV, the concentration gradient is so large as to push the calcium ions in.

We can calculate when the forces would balance by using the Nernst equation. Extracellular calcium is about 1 mmole liter^{-1} (page 230). In a resting cell, cytosolic calcium is about 100 nmole liter^{-1} (page 230) so that there is a ten-thousandfold concentration difference favouring calcium entry. However, we have stated that during an action potential calcium in axon terminals can increase to 1 μmole liter^{-1}, that is, the concentration gradient is now only a thousandfold. It would be more appropriate to find the equilibrium voltage under those conditions.

The general statement of the Nernst equation is

$$V_{eq} = \left(\frac{RT}{F}\right)\left(\frac{1}{z}\right)\log_e\left(\frac{[I_{outside}]}{[I_{inside}]}\right) \text{ volts}$$

For this calculation, $z = 2$, $[Ca_{outside}] = 10^{-3}$ mole liter^{-1}, $[Ca_{inside}] = 10^{-6}$ mole liter^{-1}, and if we work at human body temperature RT/F has the value 0.027. Thus under these conditions the equilibrium voltage for calcium is +0.093 volts or +93 mV. This means that even when the cytosolic calcium concentration has increased to 1 μmole liter^{-1}, the net electrochemical gradient for calcium will be inward unless the membrane depolarizes all the way to +93 mV, which will never happen under physiological conditions.

SUMMARY

1. The majority of transmitters are released from cells by exocytosis induced by an increase of cytosolic calcium concentration.
2. Many nerve cells have their axon terminals close to a second cell, and they release their transmitter onto it. The complete unit of axon terminal, gap, and the part of the cell that receives the transmitter is called a synapse.
3. An intracellular messenger is an intracellular solute whose concentration changes in response to cell stimulation; it activates or modulates a variety of cellular processes. The most important intracellular messengers are calcium ions, cyclic adenosine monophosphate (cAMP), and cyclic guanosine monophosphate (cGMP).
4. The increased cytosolic calcium may be derived from two sources: the extracellular medium and the smooth endoplasmic reticulum.
5. Phospholipases C cleave the hydrophilic head group inositol trisphosphate (IP_3) from the lipid phosphatidylinositol bisphosphate (PIP_2). IP_3 diffuses through the cytosol and opens a calcium channel in the membrane of the smooth endoplasmic reticulum. Calcium then floods out, increasing the cytosolic calcium concentration.
6. IP_3 can be generated by the action of the β isoform of phospholipase C (PLCβ). Binding of extracellular chemical to a cell surface receptor activates the trimeric G protein G_q, which in turn activates PLCβ.
7. A different set of receptors activates G_s and hence adenylate cyclase, which makes cyclic AMP from ATP. Many of the actions of cAMP are mediated by the serine-threonine kinase cAMP–dependent protein kinase (protein kinase A).
8. Many receptor tyrosine kinases are caused to dimerize when their ligand binds. This allows each partner to phosphorylate the other upon tyrosine; this in turn recruits proteins containing SH2 domains.
9. Grb2 is an SH2-domain-containing protein that serves to bring together SOS and Ras, allowing activation of Ras and therefore activation of the MAP kinase pathway. This culminates in the transcription of genes necessary for DNA synthesis and cell division.
10. Phospholipase Cγ (PLCγ) and phosphoinositide 3-kinase (PI3K), two enzymes that act on the membrane lipid phosphatidylinositol bisphosphate (PIP_2), both have SH2 domains and are both recruited to, and phosphorylated by, receptor tyrosine kinases: this activates them. PLCγ generates IP_3 and therefore initiates a calcium signal, while PI3K generates PIP_3.
11. The serine-threonine kinase protein kinase B is recruited to the plasma membrane by PIP_3; this allows it itself to be phosphorylated, which activates it.
12. In many cells including fat and skeletal muscle, glucose carriers only appear at the plasma membrane when PKB is kept active through the activation of the insulin receptor.
13. Type 1 cytokine receptors become tyrosine phosphorylated when their agonist binds, but unlike receptor tyrosine kinases the kinase is an independent protein called JAK.
14. Among the SH2 domain proteins recruited to activated type 1 cytokine receptors are STATs. When phosphorylated by JAK, STATs are active transcription factors.

FURTHER READING

Marks, F., Klingsmüller, U., and Müller-Decker, K. (2009) *Cellular Signal Processing*, Garland, New York.

REVIEW QUESTIONS

15.1 Theme: Processes Downstream of Receptor Tyrosine Kinases

A JAK
B phosphoinositide 3-kinase
C phospholipase Cβ
D phospholipase Cγ
E protein kinase B
F SH2
G SOS
H STAT

From the above list of proteins or protein domains, select the one corresponding to each of the descriptions below.

1. A domain comprising a pocket with a positively charged arginine at the base; proteins with this domain are recruited to phosphorylated tyrosine residues
2. A guanine nucleotide exchange factor for Ras
3. A hydrolytic enzyme that is activated when phosphorylated by receptor tyrosine kinases
4. A kinase that is activated when phosphorylated by receptor tyrosine kinases
5. An enzyme that is activated by Ras:GTP

15.2 Theme: Proteins Activated by Nucleotides

A calcium ATPase
B cAMP gated channel
C cGMP gated channel
D G_q
E G_s
F IP_3 receptor
G protein kinase A
H protein kinase B
I protein kinase C
J Ras

From the above list of proteins, select the one corresponding to each of the descriptions below.

1. A protein kinase activated by cAMP
2. A protein responsible for generating electrical signals in photoreceptors
3. A protein that activates a phospholipase C when in the GTP bound state
4. A protein that activates adenylate cyclase when in the GTP bound state
5. A protein that activates MAP kinase kinase kinase when in the GTP bound state

15.3 Theme: Inositol Compounds

A inositol
B inositol bisphosphate IP_2
C inositol trisphosphate IP_3
D inositol tetrakisphosphate IP_4
E inositol hexakisphosphate IP_6
F phosphatidylinositol bisphosphate PIP_2
G phosphatidylinositol trisphosphate PIP_3

From the above list of inositol compounds, select the one described by each of the descriptions below.

1. A ligand that binds to and opens a calcium channel
2. A lipid that recruits protein kinase B to the plasma membrane
3. A substrate for phosphoinositide 3-kinase (PI3K)
4. A substrate for phospholipases C (PLC)
5. The product of phosphoinositide 3-kinase (PI3K)
6. The product of phospholipases C (PLC)

THOUGHT QUESTION

When an action potential invades an axon terminal, the membrane voltage depolarizes to voltages in the range +30 mV to +50 mV and voltage-gated calcium channels open. We have stated that calcium ions flow into the cell through the calcium channels. How can this happen—why are they not pushed out of the cytosol by the positive voltage of the cell interior? If possible, use a calculation to support your answer.

16

INTERCELLULAR COMMUNICATION

The millions of cells that make up a multicellular organism can work together only because they continually exchange the chemical messages called **transmitters**. Here we describe how systems of cells use transmitters to cooperate for the good of the organism. Most of the many known transmitters are found in all animals, and probably evolved with ancestral multicellular organisms more than a billion years ago.

CLASSIFYING TRANSMITTERS AND RECEPTORS

Transmitter mechanisms can be classified in two ways. The first depends on the location and action of the receptor on the target cell. There are three types of receptors: ionotropic cell surface receptors, metabotropic cell surface receptors, and intracellular receptors.

Ionotropic Cell Surface Receptors

Ionotropic cell surface receptors are channels that open when a specific chemical binds to the extracellular face of the channel protein. The **ionotropic glutamate receptor** found on nerve cells (Fig. 16.1) is one example. The channel is closed in the absence of the amino acid glutamate in the extracellular medium. When glutamate binds, the channel opens and allows sodium and potassium ions to pass through. The electrochemical gradient pushing sodium ions into the cell is much greater than that pushing potassium ions out of the cell, so when the channel opens sodium ions pour in, carrying positive charge and depolarizing the plasma membrane. This mechanism is similar to that of two channels we met in Chapter 15, the inositol trisphosphate-gated calcium channel and the cAMP-gated channel. There is, however, a major difference: those two channels were opened by cytosolic solutes, while ionotropic cell surface receptors are opened by extracellular solutes.

Metabotropic Cell Surface Receptors

Metabotropic cell surface receptors are linked to enzymes. We have already met a number of metabotropic cell surface receptors in Chapter 15. When the ADP receptor binds extracellular ADP, it activates G_q and hence phospholipase Cβ. The receptor for smell chemicals and the β-adrenergic receptor are linked to G_s and hence adenylate cyclase, so ligand binding increases cytosolic cAMP. Type 1 cytokine receptors are linked to JAK tyrosine kinase. Receptor tyrosine kinases are themselves protein kinases activated when their ligand binds.

The α-adrenergic receptor (Fig. 16.2) is another receptor that causes cytosolic calcium concentration to increase. The α- and β-adrenergic receptors are distinct proteins that bind the same transmitters, adrenaline and noradrenaline, but signal to different trimeric G proteins and therefore different downstream targets. To simplify the issue somewhat, we can say that noradrenaline acts mainly on α receptors and adrenaline acts mainly on β receptors. Because the α and β receptors are distinct proteins, it is possible to design drugs (α and β blockers) that interfere with one or the other.

Cell Biology: A Short Course, Third Edition. Stephen R. Bolsover, Jeremy S. Hyams, Elizabeth A. Shephard and Hugh A. White.

Example 16.1 A Toxic Glutamate Analog

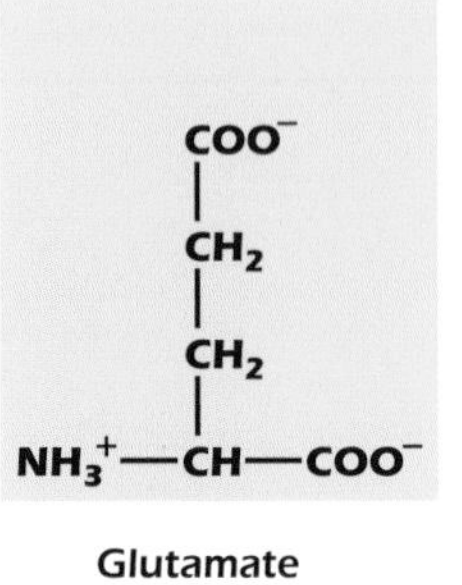

Glutamate

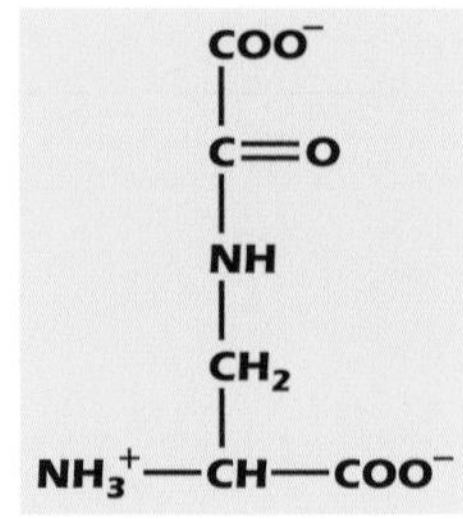

β-N-oxalyl-L-α-β-diaminopropionic acid

The grasspea is a protein-rich crop that has been cultivated since ancient times and is still an important source of calories and protein in India, Africa, and China. The peas contain the amino acid β-N-oxalyl-L-α-β-diaminopropionic acid. If untreated peas are eaten, the toxin binds to and opens the glutamate receptor on nerve cells. The resulting long-lasting depolarization damages and finally kills the nerve cells. Boiling the peas during cooking destroys the toxin; but in times of famine, when fuel is scarce, many people are poisoned, and the resulting brain damage is irreversible.

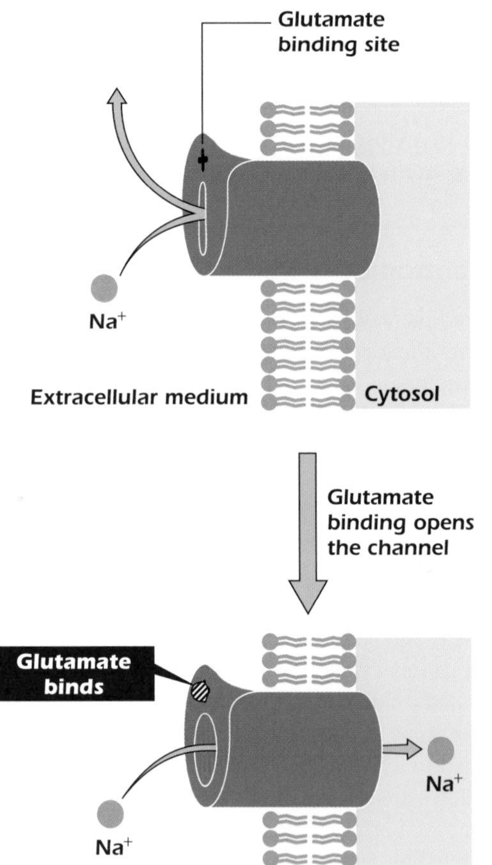

Figure 16.1. The ionotropic glutamate receptor is an ion channel that opens when glutamate in the extracellular medium binds.

Intracellular Receptors

Intracellular receptors lie within the cell (in the cytosol or in the nucleus) and bind transmitters that diffuse through the plasma membrane. They always exert their effects by activating enzymes. The receptors for nitric oxide and steroid hormones are two examples.

Nitric oxide, or NO, is a transmitter in many tissues. It is not stored ready to be released but is made at the time it is needed. NO diffuses easily through the plasma membrane and binds to various cytosolic proteins that are NO receptors. One particularly important NO receptor is the enzyme **guanylate cyclase**, which in the presence of NO converts the nucleotide GTP to the intracellular messenger cyclic guanosine monophosphate, or cGMP.

Steroid hormones have intracellular receptors such as the glucocorticoid receptor (page 94). In the absence of hormone this receptor remains in the cytosol and is inactive because it is bound to an inhibitor protein. However, when the glucocorticoid hormone binds to its receptor, the inhibitor protein is displaced. The complex of the glucocorticoid receptor with its attached hormone now moves into the nucleus. Here two molecules of the complex bind to a 15-bp sequence known as the hormone response element (HRE), which lies upstream of the TATA box (page 94). The HRE is a transcriptional enhancer sequence. The binding of the glucocorticoid hormone receptor to the HRE stimulates transcription.

Classification by Transmitter Lifetime

The second way of classifying transmitter mechanisms depends on their lifetime in the extracellular medium.

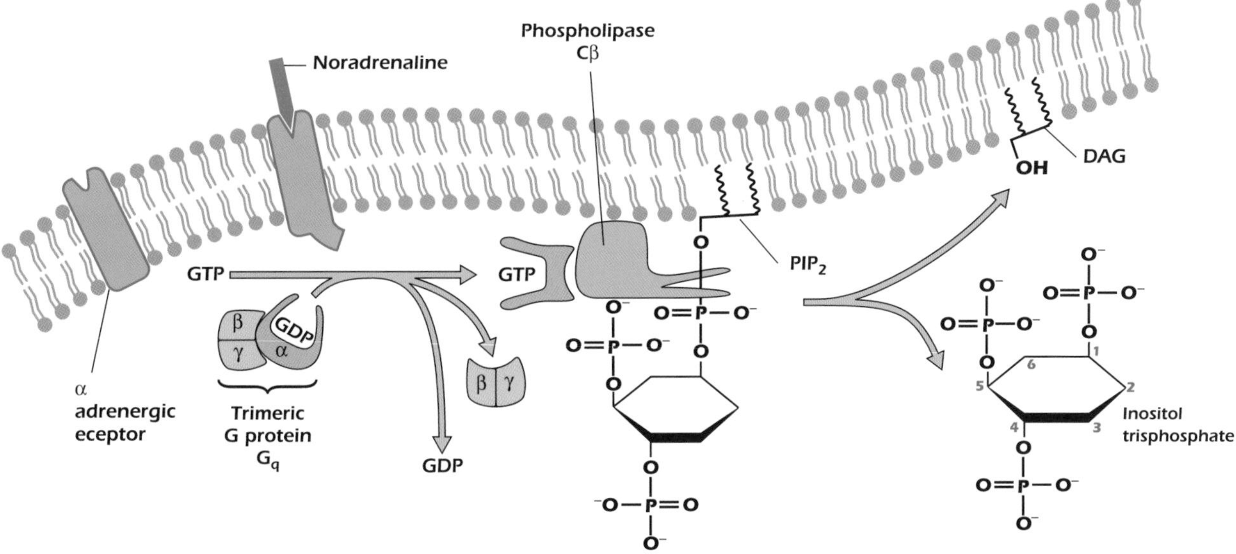

Figure 16.2. Noradrenaline activates G_q and hence phospholipase Cβ in many cells including smooth muscle.

A transmitter that is rapidly broken down or taken up into cells acts only near its release site. One that is broken down slowly can diffuse a long way and may act on cells a long distance away. The shortest lived transmitters of all are those released at synapses (page 252), where the distance from release site to receptor is only 100 nm.

At the other extreme are transmitters that last many minutes or, sometimes, even longer. **Hormones** are long-lived transmitters that are released into the blood and travel around the body before being broken down. Most are released by specialized groups of secretory cells that form a structure called an endocrine gland. **Paracrine transmitters** also last many minutes before being broken down, but they are released into specific tissues rather than into the blood, and only diffuse within the tissue before they are destroyed.

RAPID COMMUNICATION: FROM NERVE CELLS TO THEIR TARGETS

We have already introduced the synapse between the pain receptor and the pain relay nerve cell (page 251). An action potential in the axon terminal of the pain receptor raises cytosolic calcium from 100 nmol $liter^{-1}$ to 1 μmol $liter^{-1}$ and evokes release of the transmitter glutamate. Free glutamate survives only a few milliseconds in the extracellular medium and is quickly taken back up into cells by a carrier protein. However, its target—the pain relay cell—is only 100 nm away so it has time enough to bind to the ionotropic glutamate receptors present in the plasma membrane of the relay cell. These open, evoking a small depolarization (Fig. 16.3). However, the depolarization is not enough to take the transmembrane voltage to threshold which in these cells is about −40 mV. As soon as glutamate is removed from the extracellular medium, the transmembrane voltage returns to the resting level. In the axon of the relay cell, some distance from the synapse, the transmembrane voltage does not change, and no message passes on to the brain. The subject does not feel pain.

A subject does feel pain when they move their finger into the hot air from a hair dryer (Fig. 16.4). This is because the hot air heats a large area and causes many pain receptors to fire action potentials. Many of the pain receptors synapse onto one relay cell, which therefore receives many doses of glutamate. Because of this **spatial summation**, enough glutamate receptor channels open to depolarize the relay cell to threshold. Once threshold is reached the mechanism we described in Chapter 14, driven by sodium influx through voltage-gated sodium channels, takes over and an action potential travels rapidly up the axon of the relay cell to the brain so that the subject becomes aware of the painful event.

An intense stimulus to a small area is also painful. When the subject is jabbed with a needle, only one pain receptor is activated, but that receptor is intensely stimulated and fires a rapid barrage of action potentials, each of which causes the release of glutamate and an extra depolarization of the pain relay cell (Fig. 16.5). Soon the transmembrane voltage of the relay cell reaches threshold, and an action potential travels along its axon toward the brain and the subject feels pain. Such **temporal summation** only occurs if the presynaptic action potentials are frequent enough to ensure that the depolarizations produced in the postsynaptic cell add.

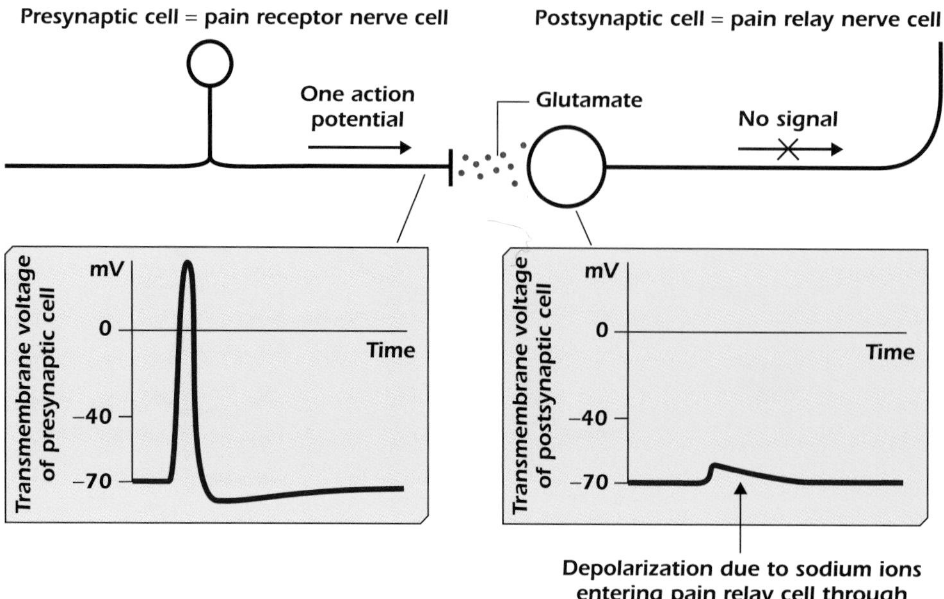

Figure 16.3. Opening of ionotropic glutamate receptors depolarizes the postsynaptic cell.

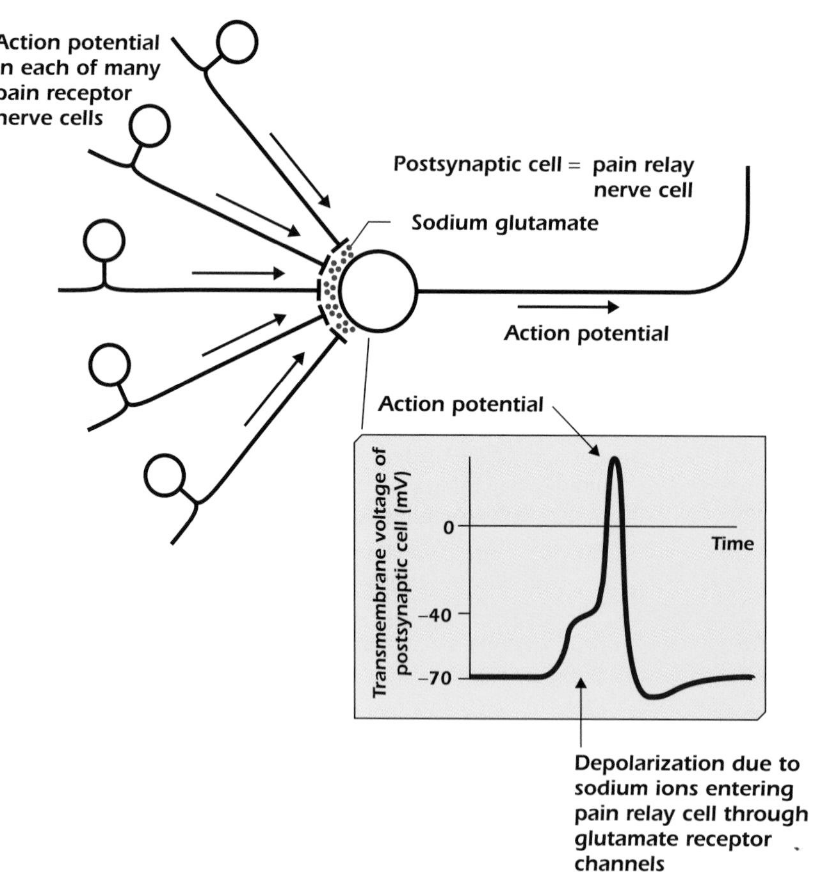

Figure 16.4. Spatial summation at a synapse.

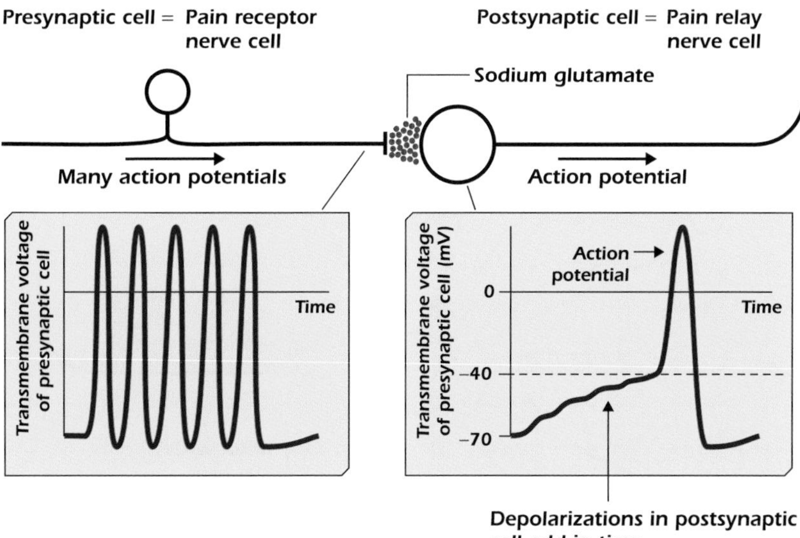

Figure 16.5. Temporal summation at a synapse.

Inhibitory Transmission: Chloride-Permeable Ionotropic Receptors

The pain relay system is powerfully modulated by the brain. One component of this is mediated through the effects of another amino acid, γ-amino butyric acid (GABA) (page 32). Ionotropic GABA receptors are selective for chloride ions. Figure 16.6 illustrates a single nerve cell bearing both glutamate and GABA receptors.

At point A on the graph an action potential in the GABA-secreting axon releases GABA onto the surface of the postsynaptic cell, causing the GABA receptor channels to open. Although chloride ions could now move into or out of the cell, they do not, because their tendency to travel into the cell down their concentration gradient is balanced by the repulsive effect of the negative voltage of the cytosol; chloride ions are at equilibrium (page 232). The opening of GABA channels therefore causes no ion movements and therefore does not alter the transmembrane voltage.

At point B, action potentials occur simultaneously in six glutamate-secreting axons. In this example, this activity provides enough glutamate to depolarize the postsynaptic cell to threshold, and the postsynaptic nerve cell fires an action potential.

At point C, action potentials occur simultaneously in six glutamate-secreting axons and also in the GABA-secreting axon. The same number of glutamate receptor channels open as before, and about the same number of sodium ions flow into the postsynaptic nerve cell, depolarizing it. However, as soon as the transmembrane voltage of the postsynaptic cell deviates from the resting value, chloride ions start to enter through the GABA receptor channels because the cytosol is no longer negative enough to prevent them from entering the cell down their concentration gradient. The inward movement of negatively charged chloride ions neutralizes some of the positive charge carried in by sodium ions moving in through the glutamate receptor channels. The postsynaptic nerve cell therefore does not depolarize as much and does not reach threshold. No action potential is generated in the postsynaptic nerve cell.

Glutamate and GABA are respectively the most important excitatory and inhibitory transmitters throughout the nervous system, and drugs that affect their operation have dramatic effects on neural processing. For example, antianxiety drugs such as Valium act on the GABA receptor and increase the chance of its channel opening. Nerve cells exposed to the drug are less likely to depolarize to threshold. Valium therefore reduces action potential activity in the brain, calming the patient.

How Nerve Cells Control the Body

We can use the gastrocnemius muscle to illustrate many of the concepts in this chapter (Fig. 16.7). This is the calf muscle at the back of the lower leg. When it contracts, it pulls on the Achilles tendon so that the toes push down on the ground. Most of the bulk of the muscle is made up of one type of cell, skeletal muscle cells. The plasma membrane of the skeletal muscle cell contains voltage-gated sodium channels, so like neurons they can generate action potentials.

All skeletal muscle cells are relaxed until they receive a command to contract from nerve cells called **motoneurons**. Motoneuron cell bodies are in the spinal cord while the myelinated axons run to the muscle. To press the foot down, action potentials travel from the spinal cord down the motoneuron axon to its terminal, opening voltage-gated calcium channels in the plasma membrane, and transmitter is released into the synaptic cleft. Motoneurons are relatively

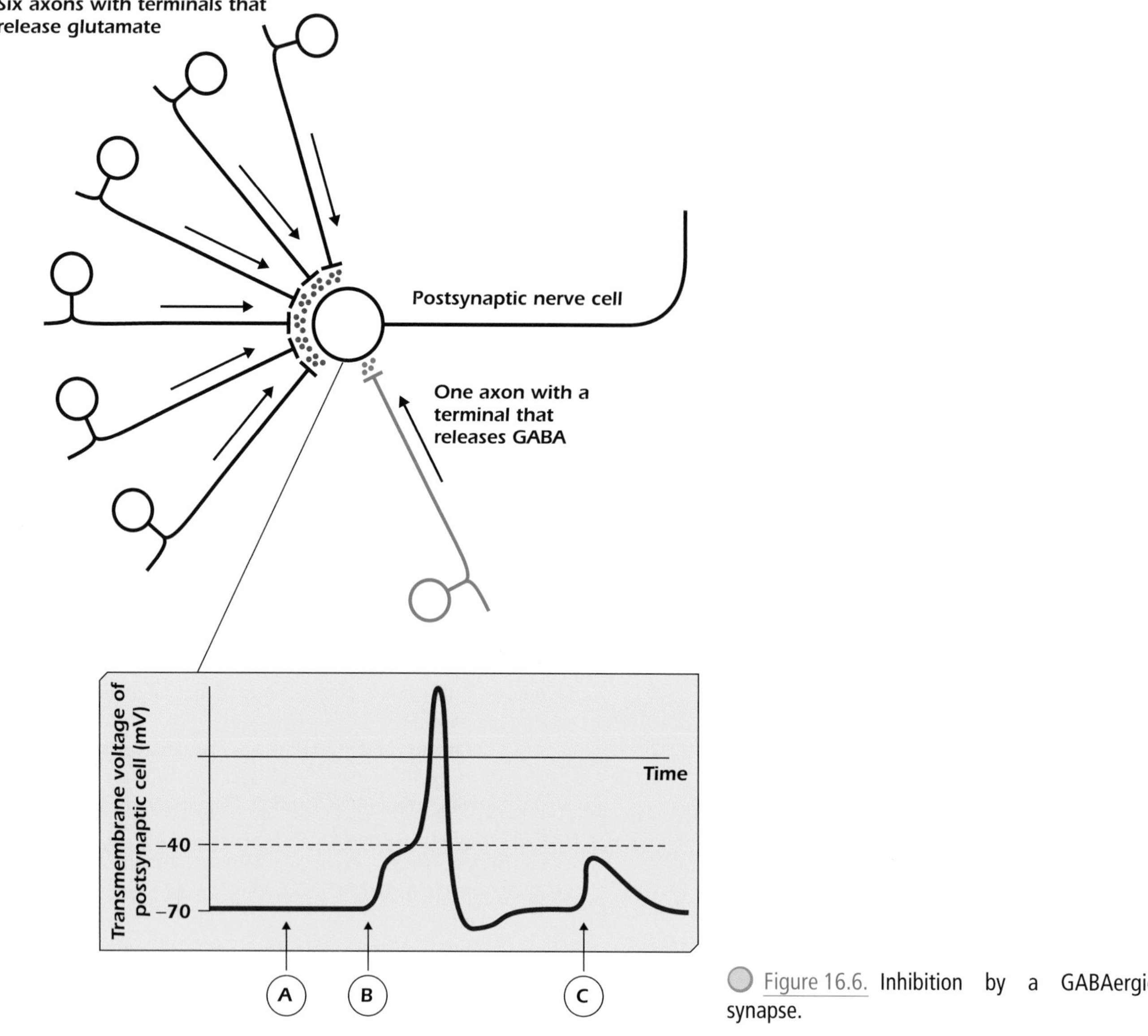

Figure 16.6. Inhibition by a GABAergic synapse.

unusual in using the chemical **acetylcholine** as their transmitter. The plasma membrane of the skeletal muscle cell contains ionotropic receptors for acetylcholine called **nicotinic acetylcholine receptors**, so named because the drug nicotine binds to them. Like ionotropic glutamate receptors, nicotinic acetylcholine receptors allow both sodium and potassium ions to pass, and cause a depolarization in the skeletal muscle cell. Indeed, just one action potential in a motoneuron elicits a depolarization so large that it depolarizes the skeletal muscle cell to threshold, which is about −50 mV in these cells. The resulting action potential in the muscle cell in turn causes release of calcium from the endoplasmic reticulum, and the increase of calcium concentration in the cytosol of the muscle cell causes it to contract (page 289).

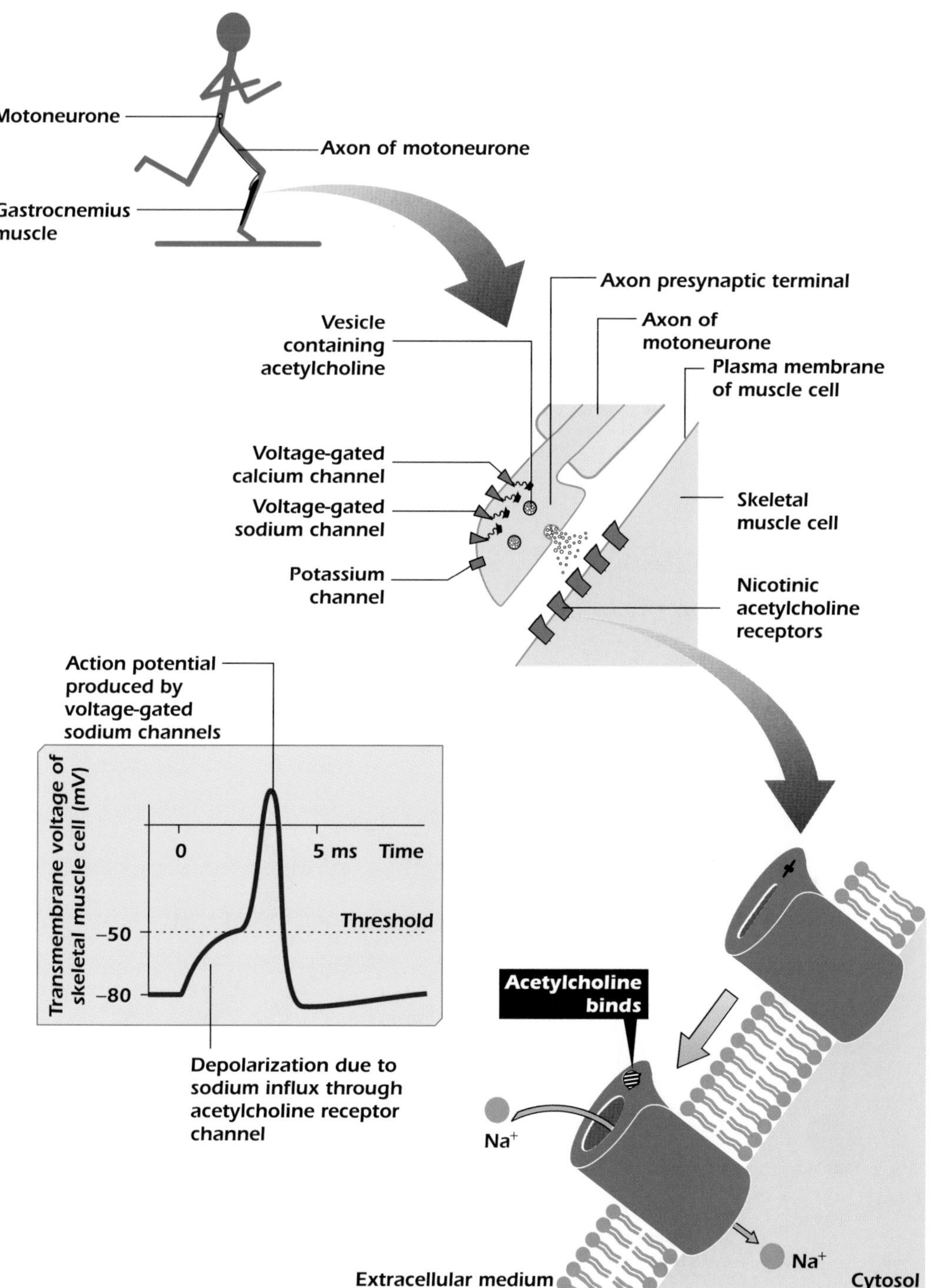

Figure 16.7. Motoneurons release the transmitter acetylcholine that binds to nicotinic receptors on skeletal muscle cells. The plasma membrane of the muscle cell is depolarized to threshold and fires an action potential.

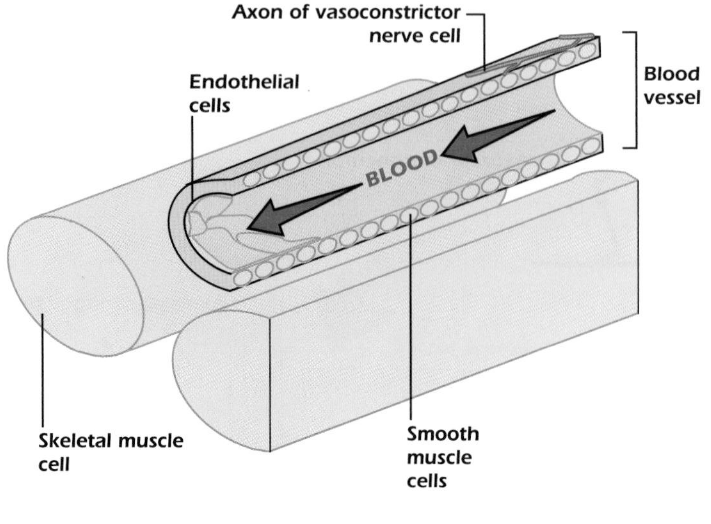

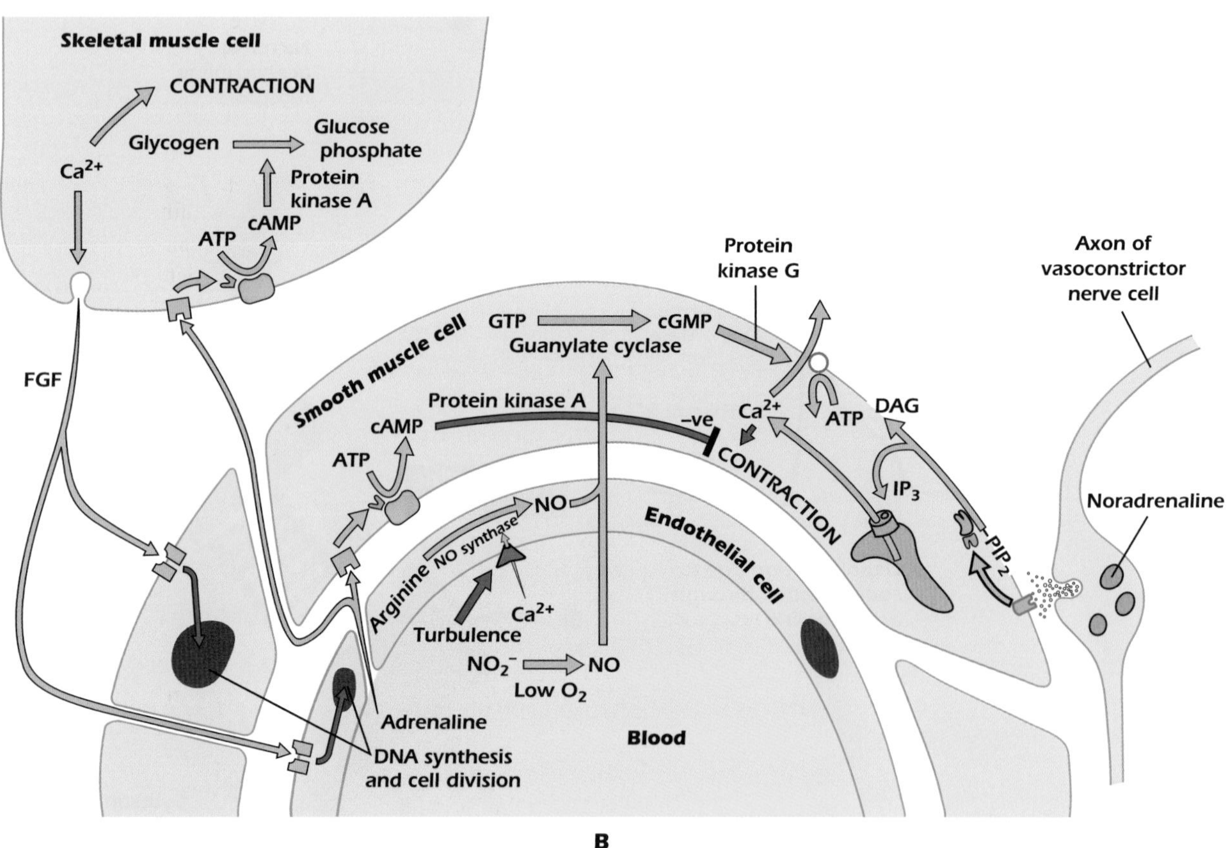

Figure 16.8. Transmitters regulate the blood supply to muscles.

PARACRINE TRANSMITTERS AND THE CONTROL OF MUSCLE BLOOD SUPPLY

While transmission between nerve cells, and between nerve cells and skeletal muscle cells, is extremely rapid, communication between the various cells within a tissue takes seconds and minutes rather than milliseconds, and operates through paracrine transmitters. We can illustrate this by considering the control of blood flow to the muscle (Fig. 16.8). The inner surface of the blood vessel is a thin layer of epithelium called **endothelium**. Wrapped around these are muscle cells of a different type: smooth muscle cells. Both endothelial cells and smooth muscle cells are much smaller than skeletal muscle cells. A small blood vessel may be only as large as a single skeletal muscle cell. The flow of blood is controlled by the relaxation and contraction of the smooth muscle cells, which in turn is regulated both by local factors and by signals originating from outside the tissue.

Exercise increases the blood flow to the muscles. A number of local mechanisms operate to cause this, including a simple action of deoxygenated hemoglobin in generating the transmitter nitric oxide. Blood contains up to 1 μmole liter^{-1} of the nitrite ion NO_2^-. When tissues take oxygen from hemoglobin, the now deoxygenated hemoglobin can reduce nitrite to nitric oxide:

$$NO_2{}^- + H^+ + \text{Hemoglobin:}Fe^{2+} \rightarrow NO + OH^- + \text{Hemoglobin:}Fe^{3+}$$

The nitric oxide easily passes through the plasma membranes of both endothelial and smooth muscle cells and reaches its receptor within the smooth muscle cells. Here it activates guanylate cyclase, causing an increase of cGMP concentration. (Meanwhile, as described in Medical Relevance 13.3 on page 213, the Fe^{3+} in the hemoglobin must be reduced back to Fe^{2+} to restore hemoglobin function.)

Just as cAMP exerts many of its effects through cAMP-dependent protein kinase, called protein kinase A for short, so cGMP exerts many of its effects through another serine-threonine kinase called cGMP-dependent protein kinase or **protein kinase G**. One of the targets of protein kinase G is calcium ATPase (page 237). When this is phosphorylated by protein kinase G it works harder, reducing the concentration of calcium ions in the cytosol. This has the effect of relaxing the smooth muscle cells. The blood vessel therefore dilates, delivering more oxygen to the active tissue. Nitric oxide lasts for only about 4 seconds before being broken down. It is therefore a paracrine transmitter, able to diffuse through and relax all the smooth muscle cells of the blood vessel, but without lasting long enough to pass into more remote tissues.

Nitric oxide plays an important role in preventing turbulent flow in blood vessels all over the body. Physical stress of the plasma membranes of endothelial cells opens **stretch activated channels** that allow calcium ions to flow into the cells. Inside the endothelial cells is the enzyme **NO synthase** that makes nitric oxide, and this enzyme is activated by calcium. The nitric oxide then diffuses to the smooth muscle cells and relaxes them. The blood vessel therefore dilates, allowing a gentler flow rate to carry the same overall amount of blood.

Superimposed on these local mechanisms are controls from outside the tissue itself. The nervous system can act to divert blood from organs that are not in heavy use to areas that need it, such as active muscles. The most important nerve cells in this regard are those called **vasoconstrictors** because they cause blood vessels to contract. Action potentials in the axons of these cells cause the exocytosis of the transmitter noradrenaline onto the surface of the smooth muscle cells. The smooth muscle cells have α-adrenergic receptors in their plasma membranes. Binding of noradrenaline to α-adrenergic receptors activates PLCβ, which generates IP_3, which in turn releases calcium from the endoplasmic reticulum into the cytosol. The increase of cytosolic calcium concentration causes the smooth muscle cells to contract, constricting the blood vessel and reducing the flow.

The Blood Supply Is Also Under Hormonal Control

The hormone adrenaline is chemically related to noradrenaline but is more stable, lasting a minute or so in the extracellular medium before being broken down. It is released from an endocrine gland (the adrenal gland) during times of stress and spreads around the body in the blood. Adrenaline that diffuses to the skeletal muscle cells stimulates them to begin breaking down glycogen to make glucose-6-phosphate (page 225). The smooth muscle cells of blood vessels within skeletal muscles also have β-adrenergic receptors connected to adenylate cyclase. However, they do not contain glycogen, and cAMP has another effect in these cells. When cAMP rises, it activates cAMP-dependent protein kinase. This in turn phosphorylates proteins that relax the smooth muscle cell. The action of adrenaline is therefore to increase the blood supply to all the muscles of the body in preparation for flight or fight. If we are very frightened and too much adrenaline is released, so much blood is diverted from the brain to the muscles that we faint.

New Blood Vessels in Growing Muscle

All the phenomena we have discussed so far occur within minutes. However, when a muscle is repeatedly exercised over many days, it becomes stronger: the individual skeletal muscle cells enlarge. This is because high cytosolic calcium acts via NFAT (Medical Relevance 10.1 on page 161) to stimulate the transcription of genes coding for structural proteins. Furthermore, new blood vessels sprout and grow into the enlarging muscle. A growth factor called FGF is released by stimulated muscle. The receptor for FGF, like that for PDGF (page 260), is found on endothelial and smooth muscle cells and like the PDGF receptor is a tyrosine kinase that signals via Ras and MAP kinase to trigger cell division and hence the growth of new blood vessels. (FGF stands for fibroblast growth factor, but FGF is effective on a vast range of cell types including both endothelial and smooth muscle cells).

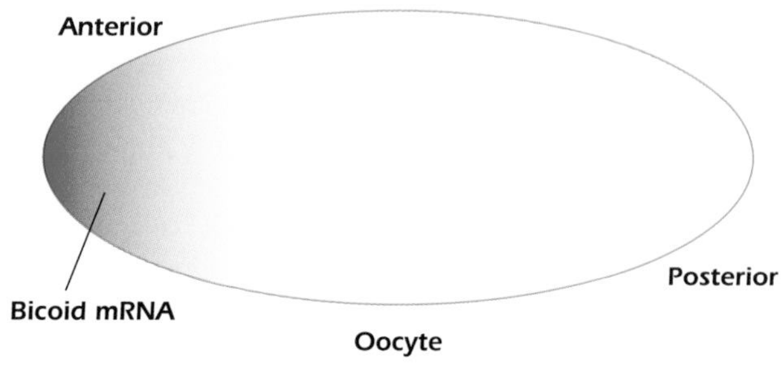

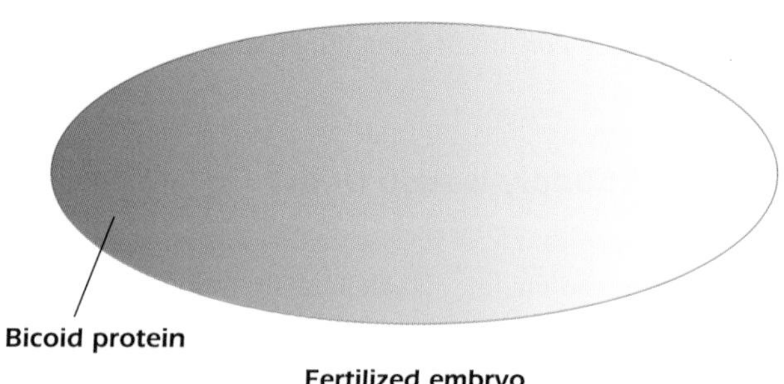

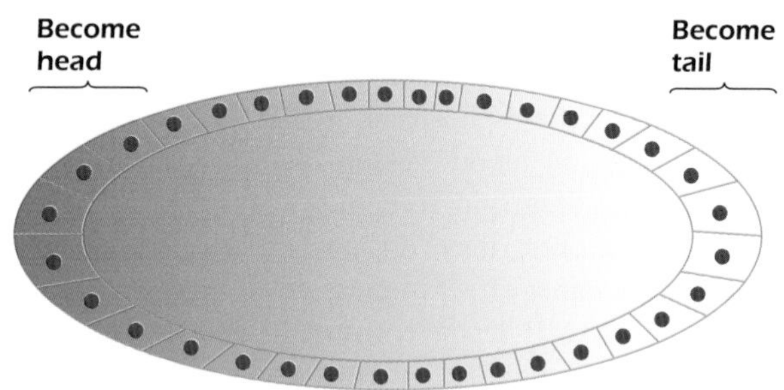

Figure 16.9. Bicoid signaling in the *Drosophila* embryo.

In our study of the gastrocnemius muscle and its blood supply, we have seen examples of all types of transmitter mechanisms. Acetylcholine acts as a synaptic transmitter at the axon terminal of the motoneuron. Adrenaline is a hormone. The other transmitters are paracrine. The nicotinic acetylcholine receptor is an ionotropic cell surface receptor. The nitric oxide receptor is intracellular. The other receptors are metabotropic cell surface receptors. There is a wide variety of timescales of action. The acetylcholine released from the axon terminal of the motoneuron causes a contraction of the skeletal muscle cell within 5 milliseconds, by which time the acetylcholine has already been destroyed by extracellular enzymes. Adrenaline lasts 1 minute and dilates blood vessels for all this time. FGF lasts 10 minutes, but its effects are much longer lasting. FGF triggers the synthesis of proteins, which then act to cause cell proliferation that lasts for days. A similar pattern of intercellular communication, using some of the same transmitters and receptors but also many others, is found in every tissue of the body.

CHEMOTAXIS

Motile cells are also affected by extracellular chemical cues, and one of their responses can be to orient their movement by the concentration gradient of the chemical. For example, peptides beginning with formyl methionine are a signal to the immune system that bacteria are close (Example 8.1 on page 128). Neutrophils and other immune system cells are activated by N-formyl methionine peptides and crawl up the concentration gradient until they meet the bacterium, whereupon they will kill it and eat (phagocytose) it.

SIGNALING DURING DEVELOPMENT

Following fertilization, a new embryo divides to generate a steadily increasing number of cells. These soon adopt particular traits so that as development proceeds choices made by cells lead them to adopt very specific identities such as motoneuron, skeletal muscle cell, or red blood cell. A focus of ongoing research and debate is the extent to which these decisions are reversible. With the single exception of the lymphocyte lineage described in Chapter 19, all the body's nucleated cells contain a complete genome and therefore

Example 16.2 Viagra

Just as cAMP phosphodiesterase hydrolyzes cAMP to AMP and hence terminates its action as an intracellular messenger, so cGMP is inactivated by cGMP phosphodiesterase. There are a number of isoforms of this enzyme in different human tissues. The drug sildenafil, sold as Viagra, inhibits the form of the enzyme found in the penis. If cGMP is not being made, this has little effect on blood flow to the region. However, when cGMP is made in response to a local production of NO, its concentration in blood vessel smooth muscle increases much more than would otherwise occur because its hydrolysis to GMP is blocked. This in turn causes a greater activation of protein kinase G, a greater activation of the calcium ATPase, a lower cytosolic calcium concentration, a greater relaxation of blood vessel smooth muscle, and therefore greater blood flow.

Example 16.3 Nitroglycerine Relieves Angina

The discovery in 1987 that nitric oxide was a transmitter explained why nitroglycerine (more familiar as an explosive than as a medicine) relieved angina pectoris. Angina is a pain felt in an overworked heart. Nitroglycerine spreads throughout the body via the bloodstream and slowly breaks down, releasing nitric oxide that then dilates blood vessels. The heart no longer has to work so hard to drive the blood around the body.

could in principle generate the RNA and protein required to build any type of cell. Much research at present is aimed at persuading cells in the adult body to revert into more or less undifferentiated stem cells (page 13) that can be used clinically to rebuild damaged tissue. Nevertheless, outside the laboratory the choices cells make are usually not reversed: neurons remain neurons, and muscle cells remain muscle cells. The cues that influence cell fate are of two types: those that are intrinsic to the cells themselves and those that arise elsewhere in the body.

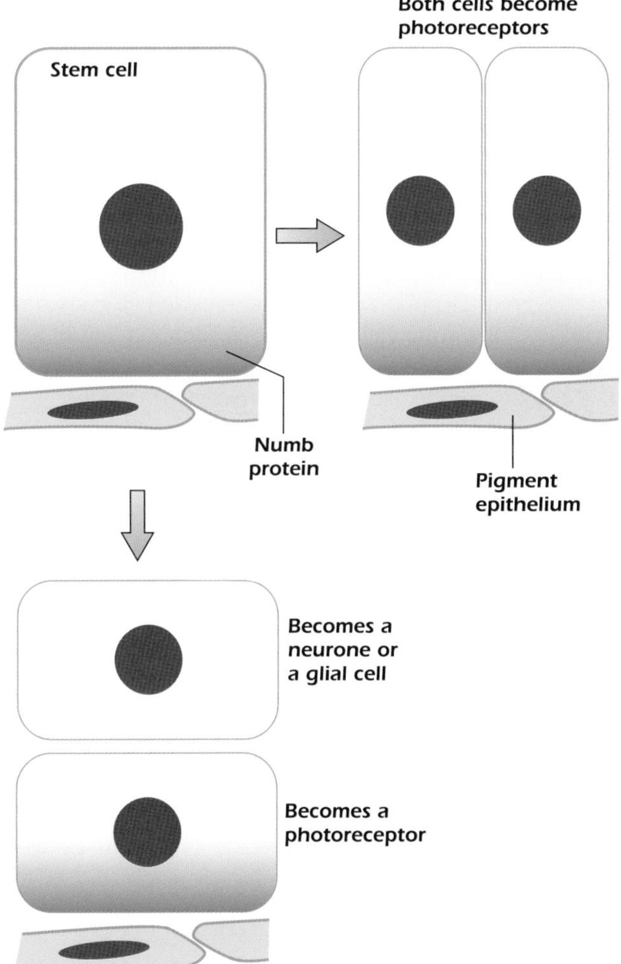

Figure 16.10. Numb signaling in the vertebrate retina.

Intrinsic Cues

Intrinsic cues are in large part responsible for the ability of cells to retain their identity. For example, cells that make the decision to become muscle express the transcription factor **MyoD**, which then acts at the enhancer regions of a number of genes, including the gene for muscle myosin. MyoD also acts on the enhancer of the *MyoD* gene, causing more MyoD to be synthesized. This positive feedback (page 223) ensures that once a cell has chosen to become a muscle cell, it remains one.

The best-characterized example of an intrinsic factor controlling a choice of cell fate is in the insect egg. The mRNA for the transcription factor **bicoid** is concentrated at one end of the ovoid egg cell (Fig. 16.9). After fertilization bicoid mRNA is translated, resulting in a spatial gradient of bicoid protein within the single cell. As the egg divides to form the multicellular early embryo, cells inherit different amounts of bicoid protein. Those that receive the most bicoid become, because of the action of bicoid at particular enhancer regions on DNA, the embryo's head. Cells receiving less and no bicoid become, respectively, mid-body and tail.

A similar mechanism operates during the development of the mammalian retina (Fig. 16.10). Within the stem cells of the retina of newborn rats the protein **numb** is found only along the surface that faces the pigment epithelium. If the stem cell divides symmetrically, so that both progeny cells inherit some of the numb, then both become photoreceptor cells. However if the stem cell divides asymmetrically, only the progeny cell that inherits the numb protein becomes a photoreceptor. The other becomes either a nerve cell or a glial cell. Figure 16.11 shows an image of dividing retinal stem cells.

Inductive Signaling

External signals are used throughout the body to allow one cell or group of cells to induce a specific direction of development in neighboring cells. In this chapter we have already seen an example in exercising muscle, where FGF released by active muscle triggers cell division in nearby endothelial and smooth muscle cells, leading to the creation of new blood vessels. The development of the motoneuron to skeletal muscle synapse yields a second example. During development the elongating axon of the motoneuron

Figure 16.11. Cell division in the retina. The micrograph shows part of the developing retina. The tissue was fixed and then stained with propidium iodide, a dye that, like Hoechst, stains DNA, but fluoresces red. Four stem cells in the field of view have divided and are in telophase/cytokinesis, with the chromosomes still condensed and visible as independent structures (page 299). Three of the cell divisions, generating the daughter cells indicated by yellow arrows, were symmetrical divisions in the plane of the retina, but one, generating the daughter cells indicated by the green arrows, was an asymmetric one at 90 degrees to the plane of the retina. Image by Professor David Becker, University College London; used with permission.

secretes the protein agrin. When the axon gets close to a skeletal muscle cell, the agrin binds to and activates muscle-specific kinase, a receptor tyrosine kinase of the skeletal muscle cell. This then phosphorylates tyrosine residues within its own cytosolic domains, leading to the recruitment of proteins containing SH2 domains and, in turn, their phosphorylation. The end result is that nicotinic acetylcholine receptors are recruited to the site of active muscle specific kinases, creating the specialized postsynaptic zone.

Once synaptic transmission is set up, further changes are triggered. Depolarization of the muscle cell causes an increase of cytosolic calcium ion concentration that in turn activates the phosphatase calcineurin. Dephosphorylation of the transcription factor NFAT (page 161) allows it to enter the nuclei of skeletal muscle cells and turn on the transcription of genes required for muscle cell growth and differentiation, such as muscle myosin.

Answer to Thought Question: The cytosol that was squeezed out contained almost all the bicoid mRNA. Very little bicoid protein was therefore made in the developing embryo, which therefore failed to develop a head.

SUMMARY

1. Transmitter mechanisms can be classified in two ways. One depends on the location and action of the receptor on the target cell; the other on their lifetime and the extracellular medium in which the transmitter is found.
2. Receptors can be divided into ionotropic cell surface receptors, metabotropic cell surface receptors, and intracellular receptors. Ionotropic cell surface receptors are ion channels that open in response to ligand binding. The effect upon the target cell is electrical. Metabotropic cell surface receptors are linked to enzymes. The effect upon the target cell is mediated through a biochemical process. Intracellular receptors lie within the target cell and bind transmitters that are able to cross the plasma membrane by simple diffusion.
3. Synaptic transmitters are extremely short lived. Paracrine transmitters and hormones have a longer lifetime and are found in specific tissues and in the blood respectively.
4. Synapses between nerve cells do not generally mediate a one-to-one transmission of the action potential. The presynaptic signal must show summation in time or space to elicit a postsynaptic action potential.
5. Stimulation of ionotropic cell surface receptors that pass chloride ions makes it more difficult for a nerve cell to be depolarized to threshold.
6. The gastrocnemius muscle provides examples of all types of intercellular signaling operating in concert to fit the operation of the tissue to the requirements of the organism.
7. Gradients of extracellular chemical can cause chemotaxis, the directed movement of cells.
8. During development, the fate of cells can be determined by cues intrinsic to the cells themselves, such as the amount of a particular protein inherited from the progenitor cell, or by extrinsic cues such as transmitters released from neighboring cells.

FURTHER READING

Levitan, I. B., and Kaczmarek, L. K. (2002) *The Neuron*, 3rd edition, Oxford University Press, New York.

Marks, F., Klingsmüller, U., and Müller-Decker, K. (2009) *Cellular Signal Processing*. Garland, New York.

Wolpert, L., et al. (2007) *Principles of Development*, 3rd edition, Oxford University Press, Oxford.

REVIEW QUESTIONS

16.1 Theme: Receptors

A α-adrenergic receptor
B β-adrenergic receptor
C glucocorticoid receptor
D guanylate cyclase
E interleukin 2 receptor
F muscle-specific kinase
G nicotinic acetylcholine receptor

From the above list of receptors, select the receptor that best fits each of the descriptions below.

1. A receptor that signals to the trimeric G protein G_q
2. A receptor that signals to the trimeric G protein G_s
3. A receptor tyrosine kinase
4. An intracellular receptor that is always cytosolic
5. An intracellular receptor that moves to the nucleus upon binding transmitter
6. An ionotropic receptor

16.2 Theme: Transmitters

A adrenaline
B GABA
C glutamate
D glycogen
E lysine
F noradrenaline
G numb
H Valium

From the above list of chemical species, select the chemical that best fits each of the descriptions below.

1. A hormone
2. A paracrine transmitter
3. A transmitter that acts to cause release of calcium from the endoplasmic reticulum of many cells including smooth muscle
4. A transmitter that is a γ-amino acid
5. A transmitter that is an α-amino acid
6. An excitatory synaptic transmitter
7. An inhibitory synaptic transmitter

16.3 Theme: Synapses

A A depolarization that does not reach the threshold for action potential generation
B A depolarization that exceeds the threshold for action potential generation
C A long-lasting (more than one second in duration) switch to a membrane voltage more positive than 0 mV
D A reduction in action potential frequency, or the complete cessation of action potentials
E No voltage change

From the above list of electrical changes in postsynaptic cells, choose the change that would be evoked by each of the stimuli below.

1. A burst of activity in a presynaptic GABAergic neuron for a postsynaptic cell that is receiving steady excitatory input from a number of presynaptic glutaminergic neurons
2. A rapid burst of action potentials in a presynaptic pain receptor neuron caused by a painful stimulus such as a pinprick
3. A single action potential in a motoneuron
4. A single action potential in a presynaptic GABAergic neuron, at a time when no other synapses onto the postsynaptic cell are active
5. A single action potential in a presynaptic glutaminergic neuron, at a time when no other synapses onto the postsynaptic cell are active

THOUGHT QUESTION

In 1986 Hans Frohnhöfer and Christiane Nüsslein-Volhard reported an influential experiment in which a newly laid *Drosophila* egg was pricked at the anterior end and a drop of cytosol, representing about 5% of the total cell volume, squeezed out. The egg was allowed to develop into an embryo. What do you think Frohnhöfer and Nüsslein-Volhard observed?

17

MECHANICAL MOLECULES

Eukaryotic cells contain a dense network of filaments called the **cytoskeleton**. The cytoskeleton is multifunctional: it determines cell shape and cell locomotion, it is responsible for moving vesicles and organelles from place to place within the cell, and it plays a key role in cell division. The cytoskeleton is composed of three distinct cytoplasmic filament networks: microtubules, microfilaments, and intermediate filaments (Fig. 17.1; see also Fig. 1.2 on page 3). Although the individual filaments making up the cytoskeleton are below the limit of resolution of the light microscope, they, and the cytoskeleton as a whole, can be readily observed within the cell by using fluorescence microscopy (e.g., In Depth 1.1 on page 6). The term cytoskeleton implies a rigid set of bones within the cell but nothing could be further from the truth. All three filament systems are highly dynamic and able to rapidly alter their organization in response to the needs of the cell.

MICROTUBULES

Microtubules possess a combination of physical properties that allows them to participate in multiple cellular functions. They can form bundles of rigid fibers that make excellent structural scaffolds; these serve an important role in the determination of cell shape and also provide tracks for the directed movement of other cellular components. Microtubules have an inherent structural polarity that supports the two-way traffic of organelles and small vesicles. Movement is powered by enzymes that interact with the microtubule surface. Microtubules can be rapidly formed and broken down, a property that allows the cell to respond to subtle environmental changes. Finally, they play a role in one of the most exquisite and precise of all movements within the cell, the segregation of chromosomes at mitosis and meiosis (Chapter 18).

Animal cells contain a network of several thousand microtubules, of variable length but with a constant diameter of 25 nm. All the cell's microtubules can be traced back to a single structure called the **centrosome**, which is tightly attached to the surface of the nucleus at the cell center (Fig. 1.2 on page 3). The centrosome is the **microtubule organizing center** of the cell and consists of amorphous material enclosing a pair of **centrioles** (Fig. 17.2). Centrioles have a characteristic ninefold symmetry that we will meet again in cilia and flagella. Figure 17.3 shows immunofluorescence images of two fibroblasts; in each, the green microtubules radiate out from the microtubule organizing center.

Microtubules are composed of the protein **tubulin** that consists of two subunits designated α and β. These have been highly conserved throughout evolution, and the α- and β-tubulins present in the cells of complex eukaryotes such as humans are much the same as those in a simple eukaryote such as a yeast. In the human genome there are about five α-tubulin genes and roughly the same number for β-tubulin. There is a third member of the tubulin superfamily, γ-tubulin, which does not itself contribute to microtubule structure but which is found at the centrosome and plays a role in initiating microtubule assembly. α-tubulin/β-tubulin

Cell Biology: A Short Course, Third Edition. Stephen R. Bolsover, Jeremy S. Hyams, Elizabeth A. Shephard and Hugh A. White.

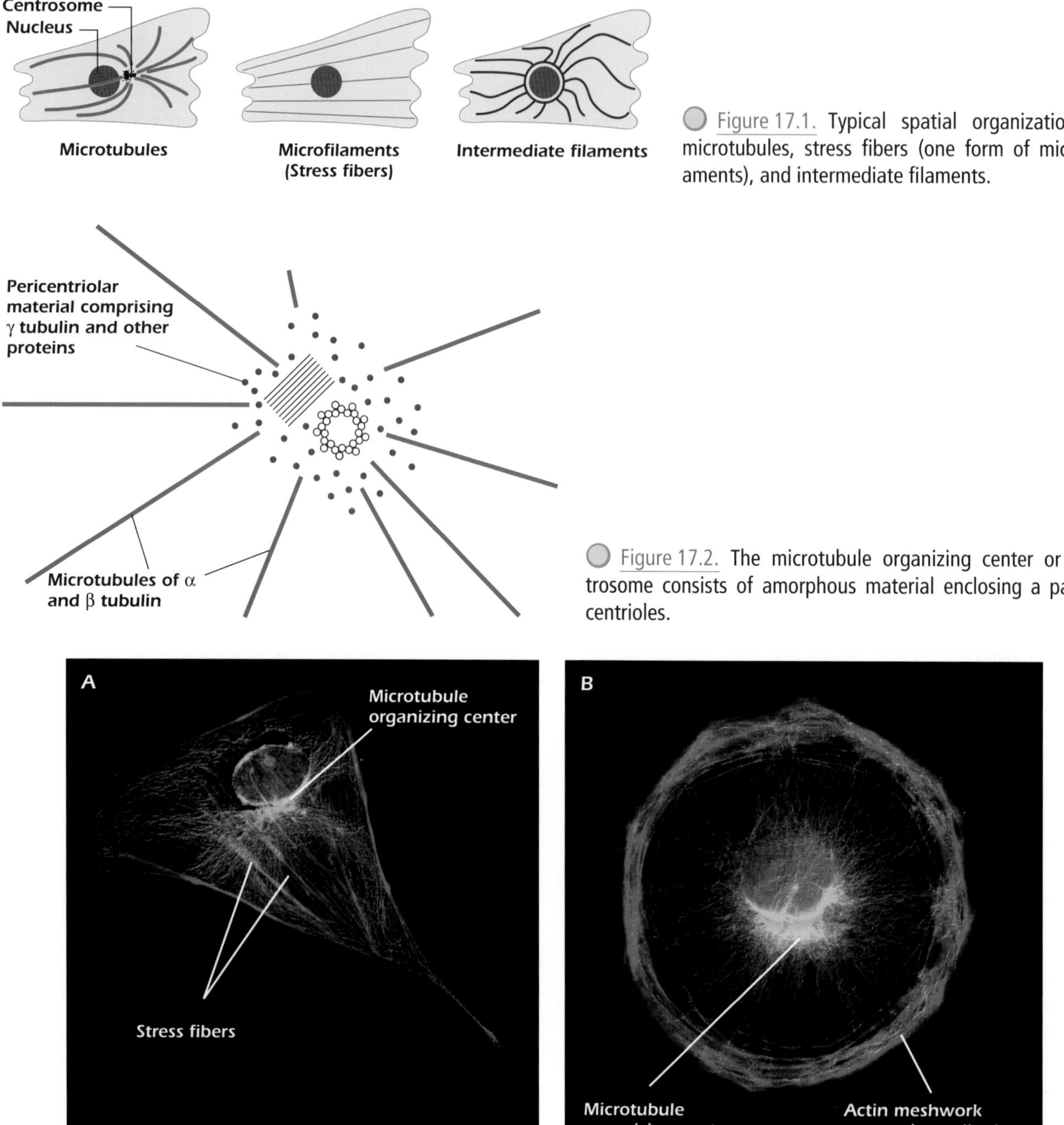

Figure 17.1. Typical spatial organization of microtubules, stress fibers (one form of microfilaments), and intermediate filaments.

Figure 17.2. The microtubule organizing center or centrosome consists of amorphous material enclosing a pair of centrioles.

Figure 17.3. Microfilaments and microtubules in fibroblasts grown in culture. The green shows microtubules radiating out from the microtubule organizing center while the red shows actin. (A) In this flattened cell the actin is organized as stress fibers. (B) In this rounded cell the actin is organized as a loose meshwork under the plasma membrane. The blue color is Hoechst staining and shows the nucleus. Images by Professor David Becker, University College London; used with permission.

dimers assemble into chains called **protofilaments**, 13 of which make up the microtubule wall (Fig. 17.4). Within each protofilament the tubulin dimers are arranged in a "head-to-tail" manner, α/β, α/β, and so on. This gives the microtubule a built-in molecular polarity that is reflected in the way it grows. Tubulin subunits are added to, and lost from, one end much more rapidly than the other. By convention, the fast growing end is referred to as the 'plus' (+) end and the slow growing end as the 'minus' (−) end.

Figure 17.4. Microtubule structure.

Figure 17.5. Microtubules show dynamic instability.

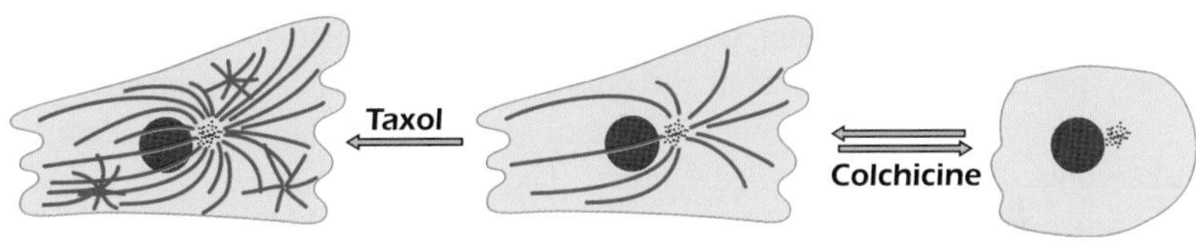

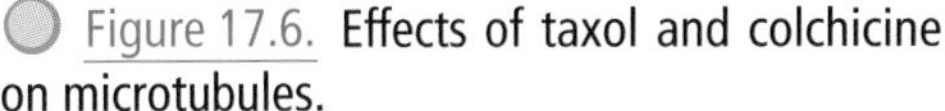

Figure 17.6. Effects of taxol and colchicine on microtubules.

In cells, the minus end of each microtubule is normally embedded in the centrosome so that only the plus ends are free to grow or shrink. This process is surprisingly complex. Individual microtubules undergo periods of slow growth followed by rapid shrinkage, sometimes disappearing completely. This phenomenon is referred to as **dynamic instability**. By chance, the growing ends of certain microtubules may be captured by sites at the plasma membrane and stabilized, so that they are protected from shrinkage. Their further growth influences the shape of the cell (Fig. 17.5). Groups of microtubules having a common orientation make an excellent structural framework. Because microtubules are dynamic, the framework can be continually remodeled as the needs of the cell change.

One of the most important tools in establishing microtubule function in cells has been the plant toxin colchicine. Extracted from the corms of the autumn crocus, *Colchicum autumnale*, colchicine has been used since Roman times as a treatment for gout. Cells exposed to colchicine lose their shape, stop dividing and the movement of organelles within the cytoplasm ceases. When the drug is washed away, microtubules reassemble from the centrosome and normal functions are resumed (Fig. 17.6). Taxol, another drug originally obtained from the bark of the Pacific yew, *Taxus brevifolia*, has the opposite effect, causing large numbers of very stable microtubules to form in the cell; this effect is difficult to reverse. Nowadays, taxol is synthesized chemically and is a widely used anti cancer drug because it very effectively blocks cell division.

MICROTUBULE-BASED MOTILITY

Cells move for a variety of reasons. Human spermatozoa in their millions swim frantically toward an ovum; the soil amoeba, *Acanthamoeba* (said to be the most abundant eukaryote on Earth), crawls over and between soil particles, engulfing bacteria and small organic particles as it does so. Cells in the early human embryo show similar crawling movement as they reorganize to form tissues and organs; the invasive properties of some cancer cells is due to their reverting to this highly motile embryonic state. Of course, not all cells show these obvious forms of motility, but careful observation of even the most sedentary eukaryotic cells often reveals a remarkable repertoire of intracellular movements. Both microtubules and microfilaments play important roles in cell motility. We will describe microtubule-based motility first; microfilament-based motility will be covered later.

Cilia and Flagella

Cilia and flagella appeared very early in the evolution of eukaryotic cells and have remained essentially unchanged to the present day. The terms **cilium** (meaning an eyelash) and **flagellum** (meaning a whip) are often used arbitrarily. Generally, cilia are shorter than flagella ($<10\ \mu m$ compared to $>40\ \mu m$) and are present on the surface of the cell in much greater numbers (ciliated cells often have hundreds of cilia but flagellated cells usually have a single flagellum). The real difference, however, lies in the nature of their movement. Cilia row like oars. The movement is biphasic, consisting of an effective stroke in which the cilium is held rigid and bends only at its base ($1 \rightarrow 4$, Fig 17.7) and a recovery stroke ($4 \rightarrow 8 \rightarrow 1$) in which the bend formed at the base passes out to the tip. Flagella wriggle like eels. They generate waves that pass along their length, usually from base to tip at constant amplitude (Fig. 17.7). Thus the movement of water by a flagellum is parallel to its axis while a cilium moves water perpendicular to its axis and, hence, parallel to the surface of the cell.

Cilia are such a conspicuous feature of some unicellular eukaryotes that they are called ciliates. The swimming of a paramecium (Fig. 17.8), for example, is generated by the coordinated motion of several thousand cilia on the cell surface. Cilia also play a number of important roles in the human body. The respiratory tract, for example, is lined with about 0.5 m^2 of ciliated epithelium, bearing in total something like 10^{12} cilia (Fig. 1.8 on page 9). The beating of these cilia moves a belt of mucus containing inhaled particles and microorganisms away from the lungs. This activity is paralyzed by cigarette smoke, with the result that mucus accumulates in the smoker's lung, causing the typical cough.

Despite their different patterns of beating, cilia and flagella are indistinguishable structurally (Fig. 17.8). Both contain a ring of nine paired microtubules (the outer doublets) surrounding two central single microtubules (the central pair). The overall structure is the **9 + 2 axoneme**. The axoneme is enclosed by an extension of the plasma membrane. Attached to the nine outer doublet microtubules are projections, or arms, composed of the motor protein **dynein**. Dynein is an ATPase that converts the energy released by ATP hydrolysis into the mechanical work of ciliary and flagellar beating. Using ATP produced by mitochondria near the base of the cilium or flagellum as fuel, the dynein arms push on the adjacent outer doublets, forcing a sliding movement to occur between them. Because the arms are activated in a strict sequence both around and along the axoneme and because the amount of sliding is restricted by the radial spokes and interdoublet links, sliding is converted into bending. Cilia and flagella are full of cytosol all the way to their tips, and use the ATP in that cytosol to generate force all the way along their length.

Bacterial flagella (Example 12.2 on page 201) use a fundamentally different mechanism. Like the propellor of a

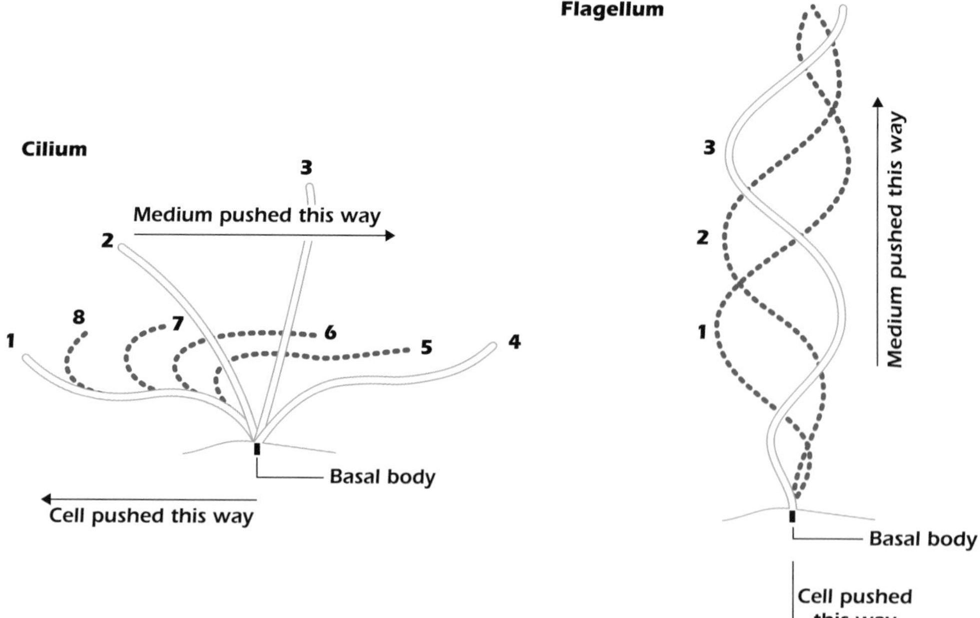

Figure 17.7. How cilia and flagella bend.

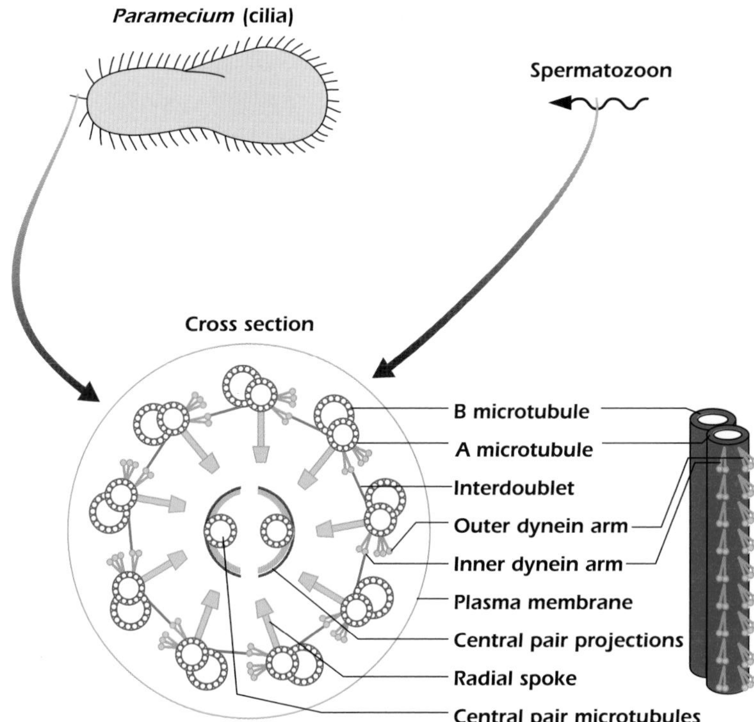

Figure 17.8. Cilia and flagella have identical structures.

boat, the motion of the bacterial flagellum is entirely driven by the rotary motor at its base. The bacterial flagellum is made of one protein, flagellin, that has no similarity to tubulin or dynein.

Intracellular Transport

The beating of cilia and flagella is an obvious manifestation of movement generated by a microtubule-based structure having a defined and geometric shape. However, motility is a general property of microtubules within cells. It is seen particularly clearly in some specialized cell types such as the pigment cells called **chromatophores** in the skin of fish and amphibia. The inward and outward movement of pigment granules along radial arrays of microtubules underlies the remarkable color changes such animals are able to display (Fig. 17.9). But this is just an exaggerated example of a process that occurs less spectacularly in all cells. For instance, nerve cell axons extend up to 1 m from the cell body. Organelles, small vesicles, and even mRNA are transported in both directions in a phenomenon referred to as **axonal transport** (Fig. 17.10). This is subdivided into outward or **anterograde** transport and inward or **retrograde** transport. Both are dependent upon microtubules that are abundant in nerve cells.

The two forms of axonal transport are dependent upon different molecular motors. Dynein, sometimes referred to as **cytoplasmic dynein** to distinguish it from its relative in cilia and flagella, moves vesicles and organelles in the retrograde direction while another protein, **kinesin**, is the motor for movement along microtubules in the anterograde direction. Their directionality is specified by the polarity of the microtubules. Thus, dynein moves along a microtubule in a plus-to-minus direction, while kinesin works in the

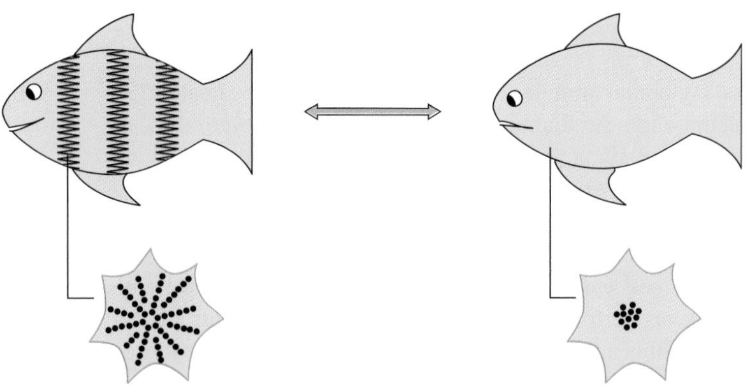

Figure 17.9. Pigment migration in fish chromatophores.

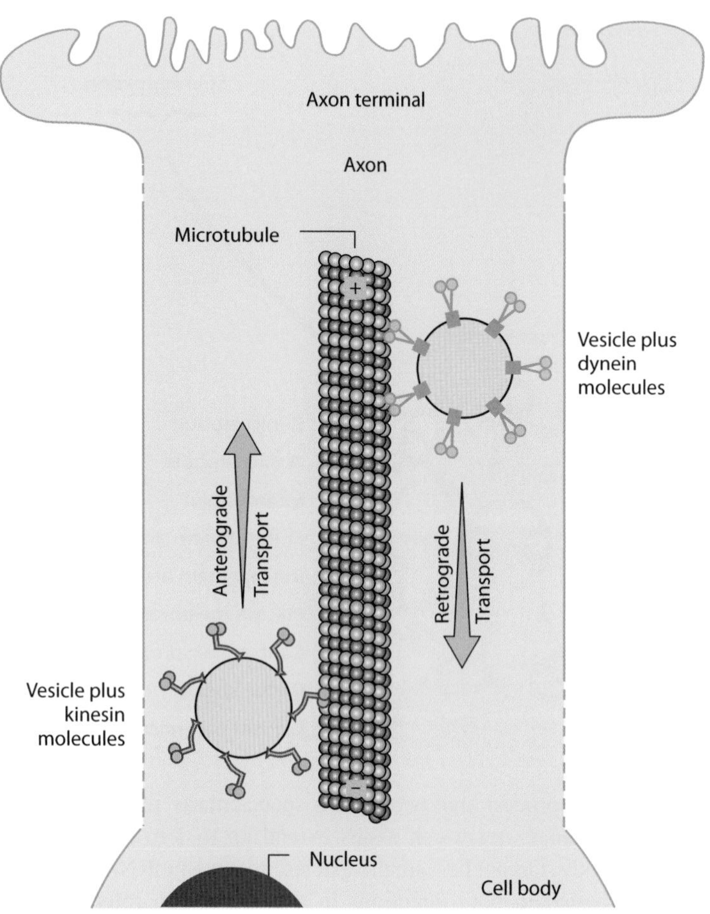

Figure 17.10. Axonal transport.

opposite direction. Both proteins consist of a tail that binds to the cargo to be transported and two (sometimes three for dynein) globular heads that interact with the surface of the microtubule, generating movement. Specificity is imparted to this process by having multiple dyneins and kinesins in a single cell, each responsible for transporting a specific type of cargo. This multiplicity of dyneins and kinesins is created by the presence of multiple genes which are alternatively spliced to generate further variants (page 92).

MICROFILAMENTS

Microfilaments are fine fibers, about 7 nm in diameter, that are made up of subunits of the protein **actin**. Because it is a globular protein, the actin monomer is designated G-actin while the filament that forms from it is referred to as F-actin. Each actin filament is composed of two chains of actin monomers twisted around one another like two strands of beads (Fig. 17.11). In animal cells, actin is particularly associated with the cell periphery. When they are grown in plastic cell culture dishes, nonmotile cells have two main types of microfilaments: bundles of actin filaments called **stress fibers** run across the cell and help to anchor it to the dish (Figs. 17.1, 17.3A) while under the plasma membrane can be seen a loose meshwork of filaments that give the edges of the cell structural strength (Fig. 17.3B). In actively moving cells the stress fibers disappear and actin becomes concentrated at the leading edge. Projections from the cell surface such as microvilli (page 12) are maintained by rigid bundles of actin filaments.

The equilibrium between G- and F-actin is affected by many factors including **actin-binding proteins** (Fig. 17.11). Filament growth is prevented by **profilin**, which binds G-actin monomers and prevents their polymerization, and by capping proteins that bind to F-actin filament ends. In contrast, filament nucleation proteins such as the Arp2/3 complex (Arp stands for **actin-related protein**) act as a base on which new filaments can form. Cross-linking proteins stick to two existing filaments, forming a mechanically strong lattice. Of these **villin**, found in microvilli, generates parallel bundles, while related proteins bind crisscrossing filaments together to form a viscous, three-dimensional cytoplasmic gel. Cells are anchored to the extracellular matrix through transmembrane proteins such as **integrins**. These are dimeric proteins that have an extracellular domain that binds to collagen and other extracellular matrix proteins and an intracellular domain that attaches to actin microfilaments (Fig. 17.12).

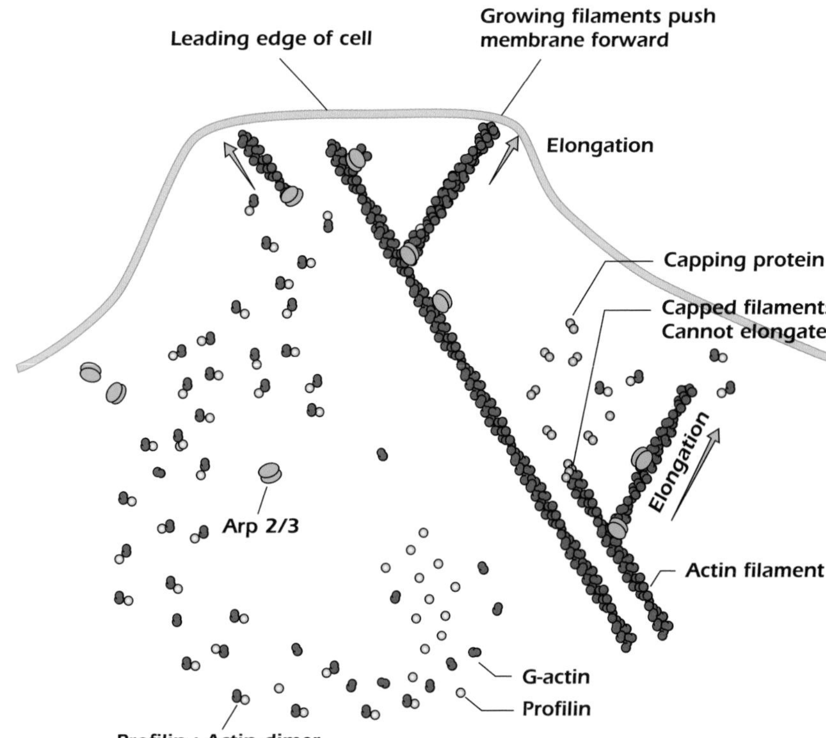

Figure 17.11. Actin polymerization is regulated by actin-binding proteins.

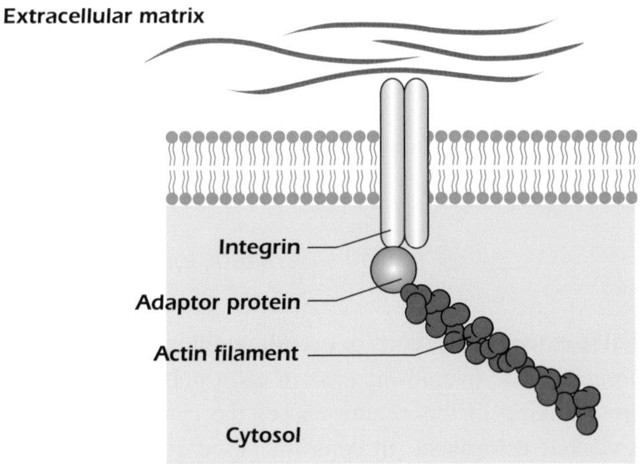

Figure 17.12. Integrins anchor the actin cytoskeleton to the extracellular matrix.

Muscle Contraction

Striated muscle, the kind of muscle found attached to bones or in our heart, has that appearance because passing along the cell one encounters in turn regions of parallel microfilaments (which are relatively transparent), then regions of **thick filaments** of a second protein called **myosin** (Fig. 17.13). The complete repeating unit is a **sarcomere** and is delineated by the **Z disc**, which holds the microfilaments in a regular pattern, so that striated muscle has an almost crystalline appearance in transverse section (Fig. 1.7 on page 8). The myosin molecule has a distinctive structure, consisting of a tail and two globular heads. The thick filament is formed from a large number of myosin molecules arranged in a tail-to-tail manner. This design means that the thick filament is a bipolar structure with myosin heads at both ends. In a resting striated muscle, the myosin heads are unable to operate on the actin microfilaments. When the cytosolic calcium concentration increases, calcium binds to the protein **troponin**, which changes shape to allow myosin access to the actin. The myosin heads now crawl along the actin filaments or **thin filaments** and generate sliding that is powered by the hydrolysis of ATP. Because of the geometry of the system, the two Z discs are pulled toward each other and the cell shortens.

Several types of myosin are found in non-muscle cells. One of them, **myosin II**, is very similar to muscle myosin but does not assemble into filaments to the same extent, probably because the levels of force required within non-muscle cells is relatively small. The primary role of myosin II is in cell division (page 299). **Myosin V** is also two-headed and is responsible for carrying cargo (vesicles or organelles) along actin filaments. Unlike the microtubule-associated motors dynein and kinesin, which can make long journeys, myosin V can only make short excursions along an actin filament before falling off.

Cell Locomotion

For a cell such as an amoeba or white blood cell to crawl across a surface it must generate traction. Less than 1%

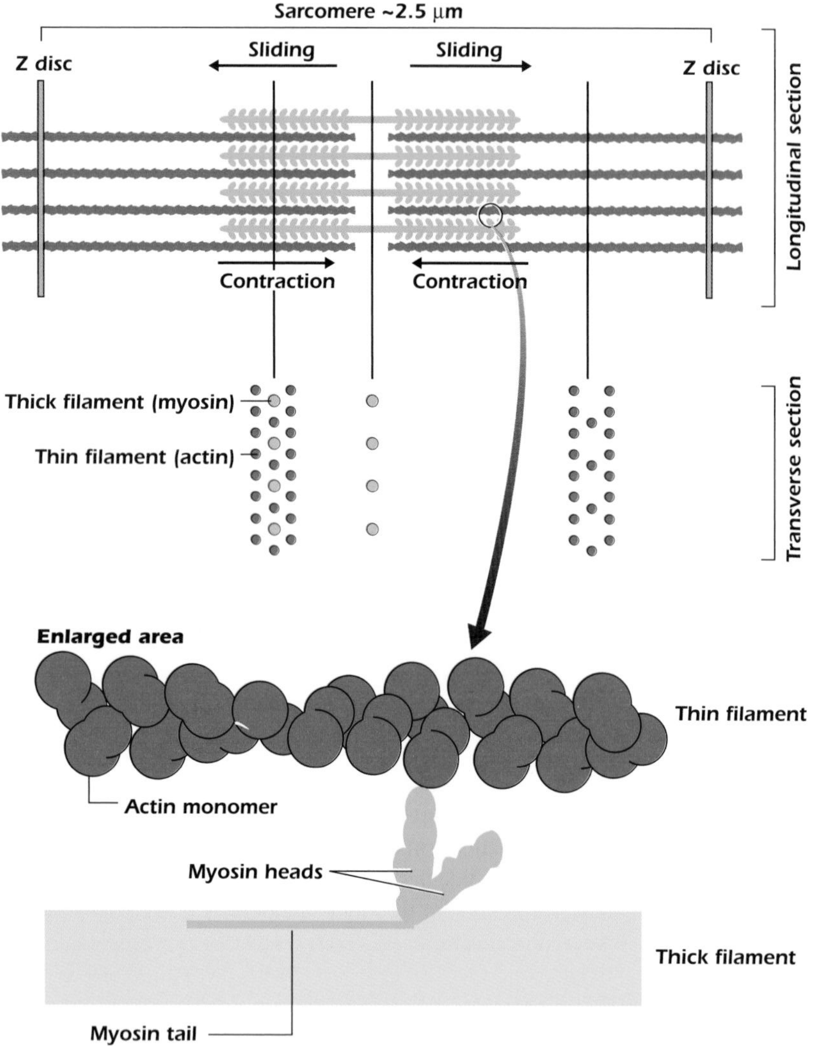

Figure 17.13. Muscle contraction.

of the ventral surface of a crawling cell is in contact with the surface over which it is moving, so that the cell does not slither along on its belly but rather "tiptoes" delicately across the surface. The transient points of attachment with the surface are called **focal contacts** (Fig. 17.14). At the leading edge actin polymerization (Fig. 17.11) causes the cell to extend projections or **pseudopodia** in the direction of progress. The overall increase in actin polymerization and cross-linking in this region causes the cytoplasm to form a viscous **ectoplasm**. In contrast, raised cytosolic calcium concentrations in the tail activate **actin-severing proteins** such as **gelsolin**. Calcium also activates myosin II, and

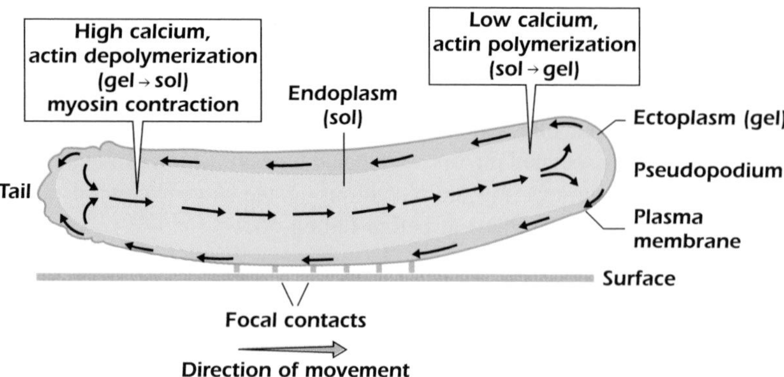

Figure 17.14. Amoeboid movement resembles the progression of a military tank.

Medical Relevance 17.1 Some Bacteria Highjack the Cytoskeleton for Their Own Purposes

We are all familiar with the ways that bacteria and viruses spread from person to person by various forms of contact, but how does a bacterium spread through the cells of a human body? A number of bacteria, including the important pathogens *Listeria* (responsible for sepsis and meningitis in immunocompromised patients and for infections of the fetus during pregnancy) and *Shigella* (which causes dysentery), avoid contact with the human body's antibodies and white blood cells by remaining hidden within the cytoplasm of our cells as they spread through our tissues. The bacteria use actin to power their journey from cell to cell. *Listeria* has a protein called ActA at one end of its rod-shaped body. ActA activates the Arp2/3 complex to promote the formation of actin filaments. The force generated by the extending microfilaments pushes the bacterium through the cytoplasm in the opposite direction. Movement is random, but occasionally the bacterium will bump into the plasma membrane, causing the formation of a finger-like projection from the cell surface. The membrane of the projection fuses with the membrane of the neighboring cell, transferring the bacterium in a membrane sac from which it quickly escapes to repeat the procedure.

the resulting contraction drives the **endoplasm**, which is now a fluid "sol", forward and displaces the existing actin cytoskeleton backward. Since actin is connected via the focal contacts to the surface, the backward movement of the ectoplasm drives the cell forward, operating rather like the tracks of a military tank.

INTERMEDIATE FILAMENTS

Intermediate filaments were so named because their diameter, 10 nm, lies between that of the thin and thick muscle filaments. They are the most stable of the cytoskeletal filament systems. Although intermediate filaments from different mammalian tissues look much the same in the electron microscope, they are in fact composed of different protein monomers (and can therefore be distinguished by immunofluorescence; see In Depth 1.1 on page 6). We have already met the lamins, which form filaments supporting the nuclear envelope (page 47). Nerve cells contain **neurofilaments**, muscle cells contain **desmin** filaments, fibroblasts contain **vimentin** filaments, and epithelial cells contain filaments composed of **keratin**, the protein that gives our skin its protective coating, forms our hair and fingernails, and makes the horns and hooves of our domestic animals and pets. These different proteins can generate a common structure because all share the same basic design (Fig. 17.15). This consists of a central, α-helical rod and a nonhelical head and tail. Most of the variation between intermediate filament proteins is in the head and tail, and these regions probably confer subtly different properties on different intermediate filament classes. The basic building block of intermediate filaments is a tetramer of pairs of subunit proteins joined by their central region to form coiled coils. Intermediate filaments tend to form wavy bundles that extend from the nucleus to the cell surface (Fig. 17.1). It appears that the nucleus is suspended by intermediate filaments stretching to the plasma membrane, rather like a sailor in a hammock.

Medical Relevance 17.2 Protected by the Dead

The epidermis of the skin is made up of a layer of living cells called keratinocytes covered by a protective layer of their dead bodies. Dead keratinocytes form a good protective layer because while alive they generate a dense internal cytoskeleton of the intermediate filament keratin, with adjacent cells being linked by desmosomes. When the cells die, the keratin fibers remain because intermediate filaments are stable. Since the intermediate filaments were joined by desmosomes, the resulting protective fibers do not stop at the edge of the now-dead cell, but are strongly connected with the fibers in the next cell, and the next, and so on, forming an extremely strong network of fibers.

Keratin mutants, collectively referred to as keratinopathies, predominantly give rise to blistering diseases of the skin. Mutations in the major skin keratins (K5 and K14), for example, give rise to epidermolysis bullosa simplex. In mild cases blistering is confined to the hands and feet although in more severe forms it is widespread. Keratinocytes cultured from these patients show disorganized keratin filaments. This results in the epidermis becoming fragile and easily damaged.

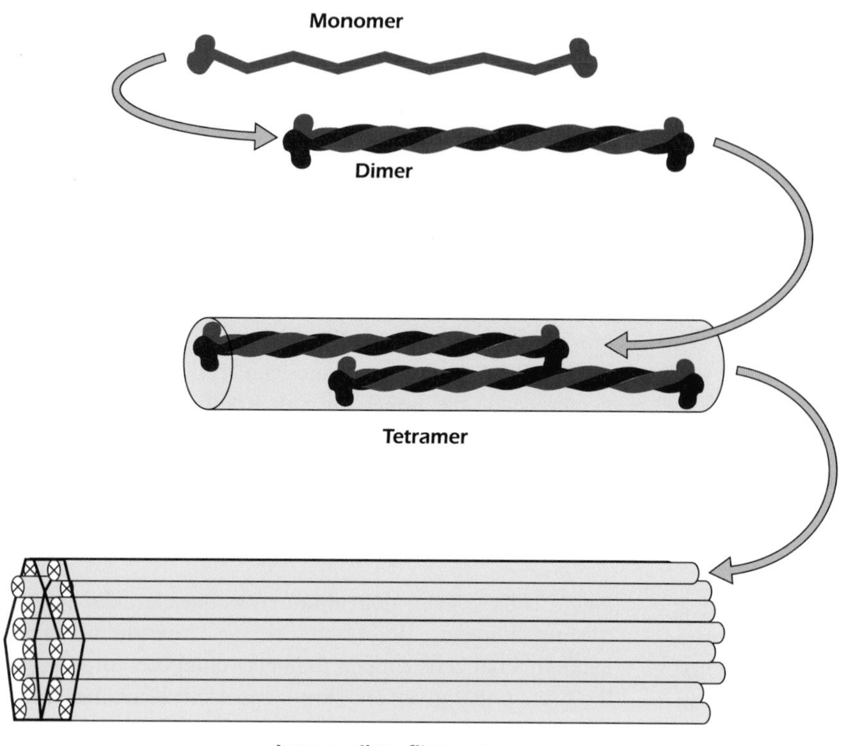

Figure 17.15. Intermediate filaments are formed from rod-shaped monomers.

Anchoring Cell Junctions

The cells that form tissues in multicellular organisms are attached together via anchoring junctions (Fig. 17.16). Integral membrane proteins called **cell adhesion molecules**, of which **cadherin** is an example, extend out from each cell and bind tightly together, while their cytosolic domains attach to the cytoskeleton. There are two basic types of anchoring junction. In **adherens junctions** the cell adhesion molecules are linked to actin microfilaments by linking proteins such as catenin. In **desmosomes** the cell adhesion molecules are linked to intermediate filaments. Tissues that need to be mechanically strong, such as the epithelial cells of the gut (page 13) and cardiac muscle, have many anchoring junctions linking the cytoskeletons of the individual cells. Anchoring junctions are one of the three types of cell junctions, the others being tight junctions (page 45) and gap junctions (page 46).

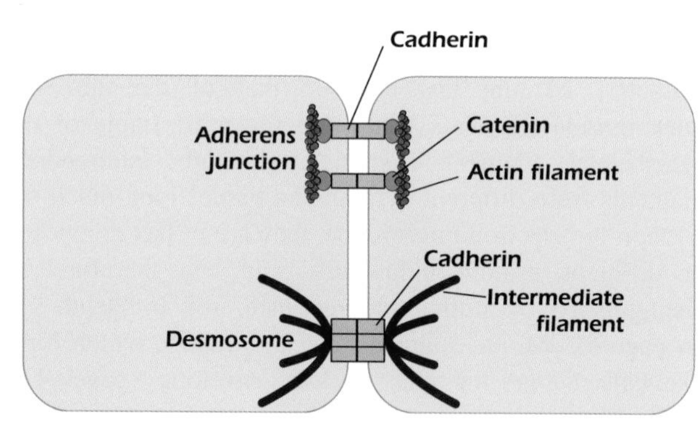

Figure 17.16. Anchoring junctions attach the cytoskeletons of adjacent cells.

Answer to Thought Question: If many cells in a tissue are killed, the remainder will not be able to find partners with which to form adherens junctions. This frees the catenin, which then (if other factors are also present) promotes cell division, regenerating the tissue. Thus the dual role of catenin is part of the mechanism that allows the cells of an animal to divide when more cells are needed, but to stop dividing when there is no more space to be filled. Another aspect of this "contact inhibition" is described on page 306.

IN DEPTH 17.1 SPECIAL PROPERTIES OF PLANT CELLS

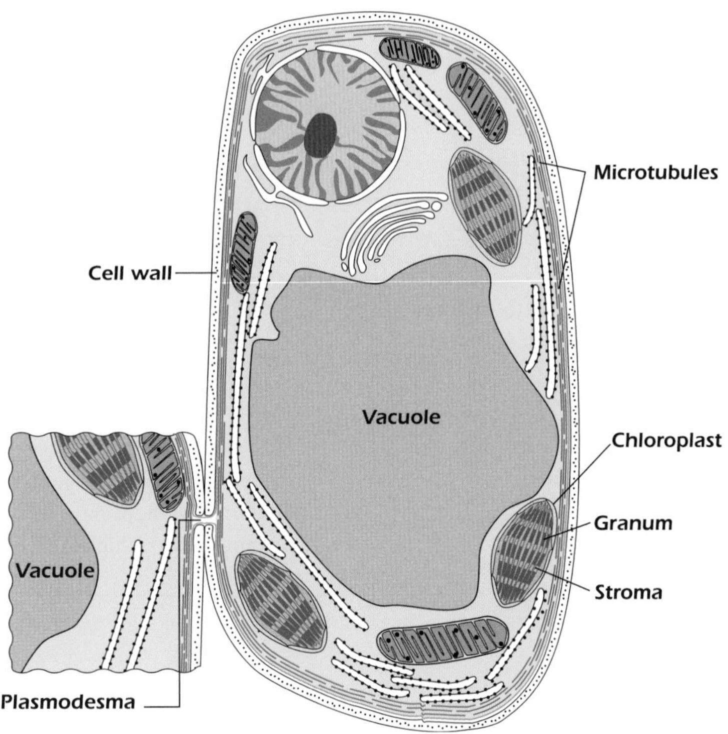

Among eukaryotic cells the most striking difference is between those of animals and plants. Plants have evolved a sedentary lifestyle and a mode of nutrition that demands that they support a leaf canopy. Rather than use an internal cytoskeleton for structural support, plant cells enclose themselves within a rigid cell wall composed of the polysaccharide cellulose (page 26) plus other components, most notably the polyphenolic compound lignin. Since the plasma membranes of adjacent plant cells are separated by the cellulose cell wall the cells cannot form the gap, tight, and anchoring junctions (pages 44, 292) seen in animals. However, **plasmodesmata** fulfill a similar role to animal gap junctions, allowing cytosolic solutes to pass from one cell to another.

Plant cells have actin filaments, intermediate filaments, and microtubules. Surprisingly, they lack centrosomes although this does not prevent them organizing a complex microtubule network. This lies immediately under the plasma membrane and directs the deposition of cellulose molecules in the cell wall on the other side of the membrane.

Plant cells frequently contain one or more vacuoles that can occupy up to 75% of the cell volume. Vacuoles accumulate a high concentration of sugars and other soluble compounds. Water enters the vacuole to dilute these sugars, generating hydrostatic pressure that is counterbalanced by the rigid cell wall. In this way the cells of the plant become stiff or turgid, in the same way that when an inner tube is inflated inside a bicycle tire the combination becomes stiff. Vacuoles are often pigmented, and the spectacular colors of petals and fruit reflect the presence of compounds such as the purple anthocyanins in the vacuole. Cells of photosynthetic plant tissues contain an unique organelle, the **chloroplast**, that houses the light-harvesting and carbon-capturing systems of photosynthesis (page 203). Chloroplasts contain internal membranes called thylakoids which form stacks called grana. The thylakoids contain the proteins and other molecules responsible for light capture. The dark reactions of photosynthesis (page 203), on the other hand, takes place in the chloroplast matrix, the **stroma**, which also contains the DNA and ribosomes.

Unlike animal cells, plant cells are naturally **totipotent**; a single cell can give rise to a complete, mature plant. When a gardener takes cuttings from a plant they take advantage of the fact that cells differentiated to form shoots and leaves can dedifferentiate and then differentiate into roots.

SUMMARY

1. The cytoskeleton is made up of microtubules, microfilaments, and intermediate filaments.
2. Microtubules are composed of equal amounts of α- and β-tubulin. In animal cells the minus (−) ends of microtubules are stabilized at the centrosome while the plus (+) ends show dynamic instability.
3. Cilia and flagella contain a 9 + 2 axoneme of microtubules plus the motor protein dynein.
4. Dynein is present elsewhere in cells where it transports cargo along microtubules in a retrograde (towards the minus end) manner. Kinesin moves cargo in the opposite, anterograde/plus end direction.
5. Microfilaments are composed of actin.
6. Actin-binding proteins control actin polymerization and the organization of actin filaments into networks.
7. The motor protein myosin is present in all cell types but is particularly prominent in muscle cells. It operates on actin microfilaments when cytosolic calcium concentration rises.
8. Cell locomotion is driven by spatially distinct zones of actin polymerization and depolymerization, aided by myosin.
9. The proteins that form intermediate filaments differ in different tissues. Intermediate filaments are more stable than microtubules and microfilaments and serve a structural role.
10. Anchoring cell junctions attach the cytoskeletons of adjacent cells together. They are divided into adherens junctions, which link to actin, and desmosomes, which link to intermediate filaments.

FURTHER READING

Gruenheid, S., and Finlay, B. B. (2003) Microbial pathogenesis and cytoskeletal function. *Nature* **422**, 775–781.

Lane, E. B., and McLean, W. H. I. (2004) Keratins and skin disorders. *Journal of Pathology* **204**, 355–366.

The Myosin Home Page, www.mrc-lmb.cam.ac.uk/myosin/myosin.html.

Pollard, T. D. (2003) The cytoskeleton, cell motility and the reductionist agenda. *Nature* **422**, 741–745.

REVIEW QUESTIONS

17.1 Theme: Cytoskeletal Structures

A actin
B intermediate filament proteins
C tubulin

For each of the structures listed below, choose from the list above the relevant cytoskeletal protein that forms or supports the structure.

1. cilia
2. fingernails
3. flagella
4. microfilaments
5. microtubules
6. microvilli
7. stress fibers

17.2 Theme: Proteins of the Cytoskeleton

A actin
B β-tubulin
C γ-tubulin
D keratin
E kinesin
F myosin
G profilin

From the above list, select the protein corresponding to each of the descriptions below.

1. A building block of intermediate filaments
2. A building block of microfilaments
3. A building block of microtubules
4. A molecular motor that acts on microfilaments
5. A molecular motor that acts on microtubules
6. A protein that binds actin and helps to prevent its polymerization
7. A protein that is concentrated at the centrosome

17.3 Theme: Fueling Movement

A actin
B dynamin
C dynein
D gelsolin
E myosin
F tubulin

From the above list of proteins, select the one that best fits each of the descriptions below.

1. Amoeboid locomotion is powered by ATP hydrolysis performed by this ATPase
2. The contraction of muscle is powered by ATP hydrolysis performed by this ATPase
3. The rowing motion of a cilium is powered by ATP hydrolysis performed by this ATPase
4. The transport of vesicles and organelles from the tips of nerve cell axons to the cell body is powered by ATP hydrolysis performed by this ATPase
5. The wriggling motion of sperm tails is powered by ATP hydrolysis performed by this ATPase

THOUGHT QUESTION

Catenin has another function that at first glance appears to be completely unconnected to its cytoskeletal role. Free catenin is a transcription factor that upregulates proteins required for cell division. What might be the advantage to the organism of this dual role?

18

CELL CYCLE AND THE CONTROL OF CELL NUMBER IN EUKARYOTES

Cells arise by the division of an existing cell. The life of a cell from the time it is generated by the division of its parent cell to the time it in turn divides is the **cell division cycle** or just the **cell cycle** for short. The duration of the cell cycle varies from 2 to 3 hours in a single-celled organism like the budding yeast *Saccharomyces cerevisiae* to around 24 hours in a human cell grown in a culture dish. During this period the cell doubles in mass, duplicates its genome and organelles, and partitions these between two new progeny cells. These events have to be carried out with great precision and in the correct order, and eukaryotic cells have established sophisticated control processes to ensure that the cell cycle proceeds with the required accuracy. In humans the cells of some tissues, such as the skin, the lining of the gut (see Chapter 1), and the bone marrow continue to divide throughout life. Others, such as the light-sensitive cells of the eye and skeletal muscle cells, show almost no replacement. The latter, laid down in infancy, must last a lifetime. Not only must cell division be precisely controlled in terms of which cells divide when, the whole process has to know when to stop. A human is bigger than a mouse and smaller than an elephant. Allowing for some minor differences in basic design, this is because humans are made up of more cells than a mouse (and, hence, are the result of more cell cycles) and less than an elephant. How are these differences achieved? Why don't humans grow to be the size of elephants or whales? What has become increasingly clear over recent years is that cells contain what have become known as cell cycle control genes. The proper functioning of such genes not only determines how big we are; it also prevents cell division going out of control and leading to cancer.

Under the microscope it is possible to distinguish two elements of the cell cycle. **Interphase** occupies about 90% of the cell cycle and is a period of synthesis and growth, during which the cell roughly doubles in mass but without displaying obvious morphological changes. Once interphase is complete, the cell enters **mitosis**, which is a brief period of profound structural changes. The focal point of mitosis is the behavior of the chromosomes, and it is this that we will deal with first.

 STAGES OF MITOSIS

Mitosis is designed to produce two progeny cells, each containing an identical set of the parental cell's chromosomes. To achieve this, the chromosomes execute a precisely choreographed sequence of movements that were first described well over a century ago. Classically, mitosis is divided into five stages, each of which is characterized by changes in the appearance of the chromosomes and their organization with respect to a new cellular structure called the **mitotic spindle** that is responsible for their segregation. Physical division (**cytokinesis**) then follows. The stages of mitosis are shown diagrammatically in Figure 18.1.

Our genome is encoded in 23 chromosomes and is made up of 3×10^9 base pairs of DNA. Human cells contain 46 chromosomes, 23 inherited from each parent. Cells that contain two complete sets of chromosomes are said to be **diploid** whereas those containing a single set are **haploid**. For simplicity Figure 18.1 shows only two chromosomes.

Cell Biology: A Short Course, Third Edition. Stephen R. Bolsover, Jeremy S. Hyams, Elizabeth A. Shephard and Hugh A. White.

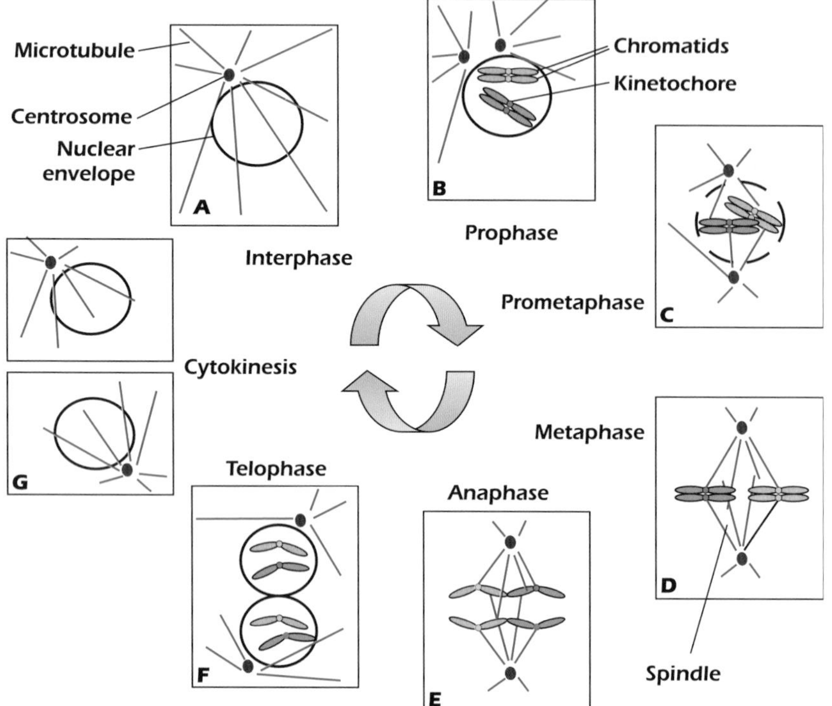

Figure 18.1. Stages of mitosis. One pair of chromosomes is shown: one chromosome originating from the father (gray), and the matching one from the mother (red).

Prophase: The first evidence of mitosis in most cells is the compaction of the threads of chromatin that existed through interphase into chromosomes that are visible in the light microscope (Fig. 18.2). As the chromosomes condense, each can be seen to composed of two **chromatids**. This is the visible effect of the DNA molecules having been replicated in interphase (page 68). Chromosome condensation reduces the chance of long DNA molecules becoming tangled and broken. Each chromosome has a constriction called the **kinetochore**, a structure that forms around a region rich in satellite DNA (page 78) called the **centromere**. The kinetochore is the point of attachment of the chromosome to the spindle. At the same time as the chromosomes are condensing within the nucleus, the centrosomes (page 283), which lie on the cytoplasmic side of the nuclear envelope, begin to separate to establish the mitotic spindle.

Prometaphase: At the breakdown of the nuclear envelope, the chromosomes become free to interact with the forming spindle. Microtubule assembly from the centrosomes is random and dynamic. The growing ends of individual microtubules make chance contact with, and are captured by, the kinetochores. Because of the random nature of these events, the kinetochores of chromatid pairs are initially associated with different numbers of microtubules, and the forces acting upon each chromosome are unbalanced. Initially, therefore, the spindle is highly unstable and chromosomes make frequent excursions toward and away from the poles. Gradually, a balance of forces is established and the chromosomes become aligned at the equator, with the kinetochores of each member of a chromatid pair oriented toward opposite poles.

Metaphase: Metaphase is the most stable period of mitosis. The system can be regarded as being at steady state with the chromosomes lined up rather like athletes at the start of a race. The metaphase spindle consists of two major groups of microtubules: one group connects the chromatids to the poles, while the second group extends from each pole toward the other, forming a zone of overlap at the spindle equator.

Example 18.1 Taxol Stops Mitosis

During cell division, cytoskeletal microtubules break down and the tubulin monomers reform as the mitotic spindle. The drug Taxol stabilizes existing microtubules, making it impossible for the cell to form spindles and therefore preventing cell division. For this reason Taxol, usually referred to by its generic name paclitaxel, is a valuable anticancer drug.

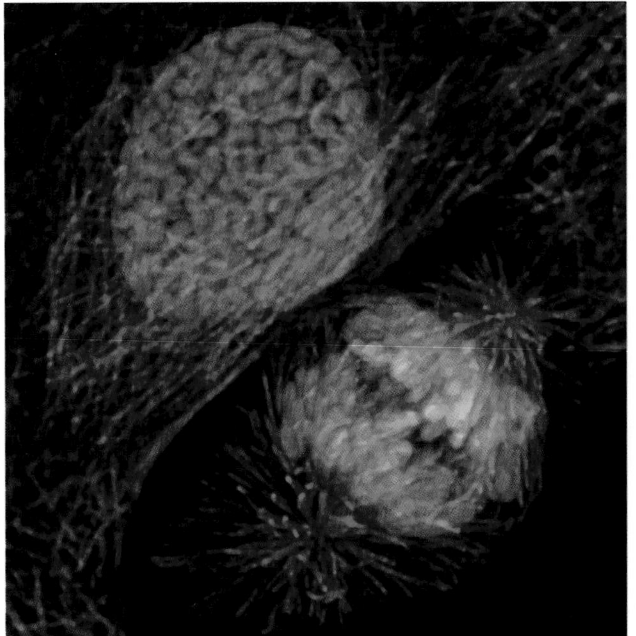

Figure 18.2. Mitosis in cultured breast cancer cells. The upper cell is in prophase. The chromosomes (shown orange) have compacted and are visible as independent structures. The microtubule cytoskeleton (shown green) has not yet reorganized into the mitotic spindle, and microtubules have not invaded the nucleus because the nuclear envelope (not visible in this image) is still intact. The lower cell is in anaphase. The microtubule cytoskeleton has been reorganized into the mitotic spindle. The two sets of chromosomes are being pulled to opposite spindle poles. Image by Professor David Becker, University College London; used with permission.

Anaphase: The trigger for the separation of the paired chromatids and the start of their journey to the spindle poles is the destruction of the protein **cohesin**, which acts as the glue holding the pairs of chromatids together (Fig. 18.3). In **anaphase A** the microtubules holding the chromosomes shorten, pulling the chromosomes to the spindle poles. The chromosomes move as a "V" with the kinetochores, the point at which the force for chromosome movement is applied, leading the way (Fig. 18.2). By contrast, in **anaphase B** the microtubules that overlap at the spindle equator lengthen and slide over one another, extending the distance between the poles. Compared to other forms of cell motility, the movement of chromosomes at anaphase is extremely slow, less than 1μm per minute. To put this in context, Los Angeles and San Francisco are moving apart by roughly the same rate as the result of plate tectonics!

Telophase: This stage sees the reversal of many of the events of prophase; the chromosomes decondense, the spindle disassembles, the nuclear envelope reforms, the Golgi apparatus and endoplasmic reticulum reform, and the nucleolus reappears. Each progeny nucleus now contains one complete copy of the parental cell genome.

Cytokinesis: During the last stages of telophase, the cell itself divides. In animal cells, a **cleavage furrow** made of actin and its motor protein myosin II (page 289) constricts the middle of the cell. The positioning of the cleavage furrow is crucial, and signals from the poles of the spindle ensure that it forms at the correct place on the cell cortex. Constriction between the two daughter nuclei is hence assured.

MEIOSIS AND FERTILIZATION

In sexually reproducing organisms, the **germ cells** that give rise to the eggs and sperm arise by a different type of cell division from the **somatic cells** that make up most of the body. In mitotic cell division DNA synthesis is followed by sister chromosome segregation so that the two progeny end up with an identical set of the parental cell's chromosomes. To produce germ cells, a single round of DNA synthesis is followed by two cell divisions known as **meiosis I and II**. The result is that an individual sperm or egg cell is haploid; that is, it contains only one copy of each chromosome. Fusion of an egg and sperm at fertilization restores the diploid state.

Meiosis

Each meiotic division involves the formation of a meiotic spindle and the same sequence of prophase, prometaphase, metaphase, anaphase, and telophase that we saw earlier in mitosis. The steps are illustrated in Figure 18.4. Two

Example 18.2 Counting Chromosomes

The DNA molecules of an interphase cell can only be observed using an electron microscope (page 3). However at mitosis, when they condense, individual chromosomes can be made out easily using a standard light microscope (Fig. 4.6 on page 59). Mitotic cells can then be screened for chromosome abnormalities such as the presence of three copies of chromosome 21, a condition that causes Down's syndrome, and the "Philadelphia chromosome," a rearrangement between chromosomes 9 and 21 (which results in a form of leukemia).

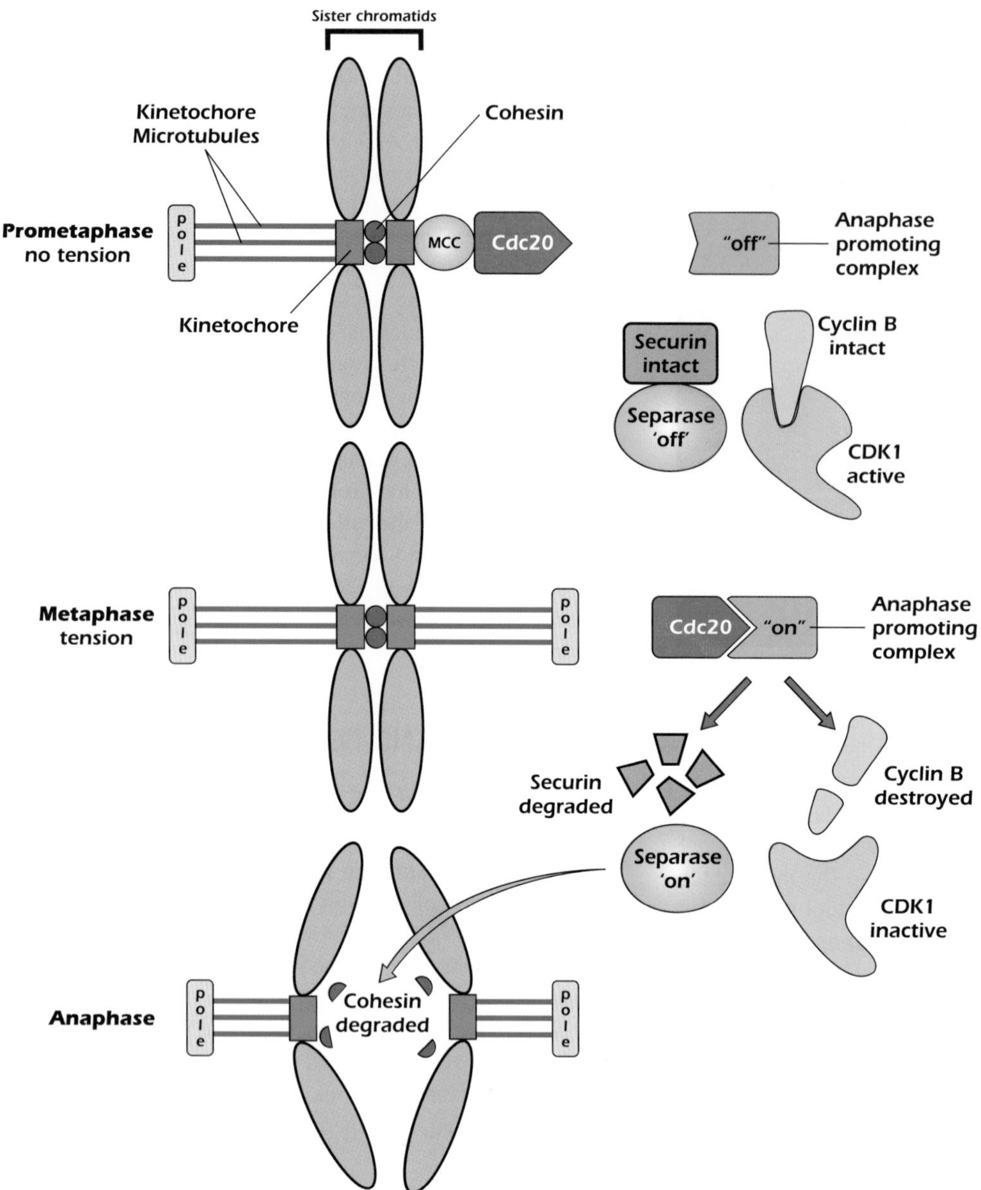

Figure 18.3. Activation of the anaphase promoting complex and the breakdown of cohesin allows cells to pass the spindle assembly checkpoint.

chromosomes are shown; the maternal chromosome is shown in gray while its paternal **homologue** (the matching chromosome from the father) is red. Because DNA synthesis has occurred in the preceding interphase, each chromosome enters meiosis as two chromatids. Mitosis is usually a rather brief process, but **meiosis** is often extended and in different organisms can last for months or even years. Most of this is occupied by a lengthy prophase of meiosis I, **prophase I**, during which the duplicated homologous chromosomes become closely aligned to form **bivalents** which can be seen to be connected at specific points along their length. We discuss these chiasmata later (page 302). As the cell progresses to metaphase I, the maternal/paternal bivalents line up along the metaphase plate. At anaphase I, the homologous pairs separate but, unlike the situation in mitosis, the paired chromatids remain attached and journey together to the pole. The two progeny nuclei formed during telophase I almost immediately enter meiosis II. Prophase II is often so brief as to be undetectable. Metaphase II and anaphase II resemble their mitotic counterparts, the chromatids finally separating to give haploid gametes (sperm or eggs). Since in Figure 18.4 we show only one pair of chromosomes, the final panel in the figure shows each gamete containing one chromosome. Although meiosis in male and female animals follows roughly the same lines, there are some important distinctions. In males, meiosis produces four equal-sized haploid cells called spermatids, each subsequently developing into a spermatozoon. In females, both meiotic divisions are asymmetric, resulting in one large cell that survives, the oocyte, and three small cells, called polar bodies, that are discarded.

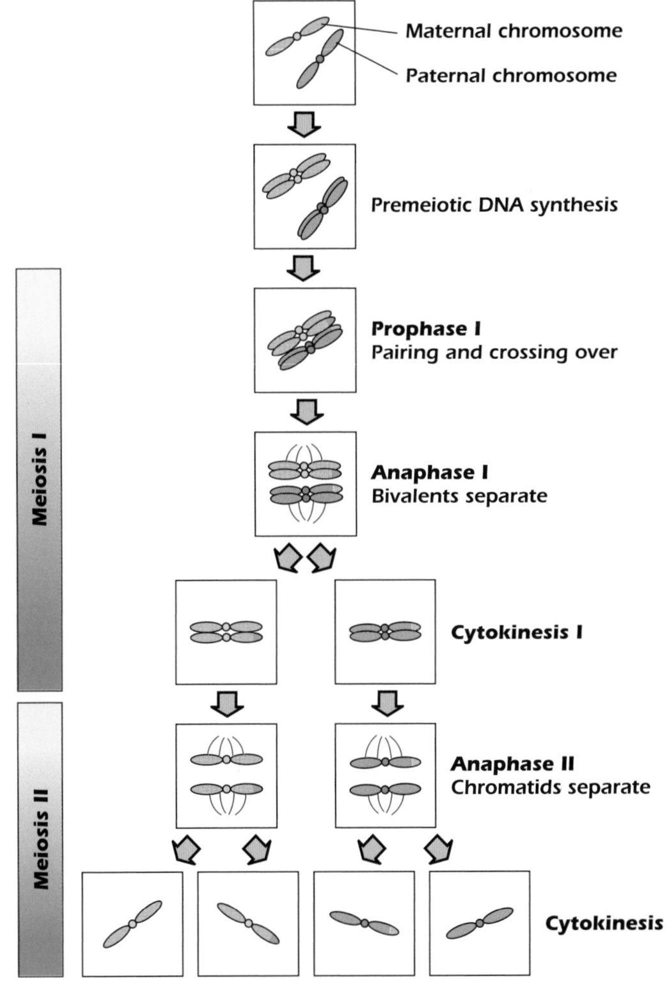

Figure 18.4. Stages of meiosis.

Fertilization and Inheritance

At fertilization the sperm cell fuses with the much larger egg cell. The nuclei that originated in sperm and egg, now called pronuclei, move together and eventually fuse to restore the normal somatic cell chromosome number. This diploid cell then divides many times by mitosis to generate the multicellular organism. Many of the attributes that make each human being distinct are determined by the pattern of genes we inherit. The simplest pattern of inheritance is seen when one form of a gene fails to make a functional protein. Brown eye color is an example. If you inherit two functional *BEY* (brown eye) genes, located on chromosome 15, then you make brown pigment and your eyes will be brown. If you inherit a chromosome from one parent with a defective *BEY* gene, and a chromosome from the other parent with a functional *BEY* gene, then you can make enough of the BEY protein to make brown pigment and your eyes will be brown. Brown eye color is said to be **dominant**.

If both of your *BEY* genes are defective, a different gene, on chromosome 19, becomes important. The *GEY* (green eye) gene codes for a protein that makes green pigment. If you have a functional *BEY* gene, then the *GEY* gene is largely irrelevant, because the brown pigment drowns out the more subtle green pigment. However, if you have two nonfunctional *BEY* genes and at least one functional *GEY* gene, you have green eyes. If both *BEY* genes and both *GEY* genes are nonfunctional, you don't make brown pigment or green pigment and your eyes are blue. Green eye color is said to be **recessive** to brown because an individual can only have green eyes if they lack any brown eye genes. Similarly, blue eye color is recessive to brown and green because an individual can only have blue eyes if he or she lacks any functional brown and green eye genes.

Clearly most of us are not unhappy with green or blue eyes. However, when a more vital protein doesn't work, the results are much less benign. Phenylketonuria is a recessive genetic disease that leads to severe mental retardation (Medical Relevance 13.5 on page 220). If an individual inherits a chromosome from one parent with a defective phenylalanine hydroxylase gene and a chromosome from the other parent with a functional phenylalanine hydroxylase gene, then they can make enough phenylalanine hydroxylase to convert phenylalanine to tyrosine and are perfectly normal.

Medical Relevance 18.1 Down's Syndrome

Cell division by mitosis is a remarkably error-free process. The cell cycle checkpoints ensure that all of the cells of our body contain the correct number of chromosomes. Meiosis on the other hand is more prone to mistakes, particularly in women. Most chromosome abnormalities in humans result from meiotic nondisjunction in the oocyte or egg. The two copies of a particular chromosome fail to segregate at meiosis, resulting in an egg with 24 chromosomes instead of the normal 23.

In most cases, fertilized embryos that contain abnormal chromosome numbers are eliminated by spontaneous abortion, often at an early stage (most first trimester spontaneous abortions are associated with chromosome abnormalities). Some, however, survive, particularly when the additional chromosome is small. The most familiar example is Down's syndrome. First described by the British physician John Langdon Down in 1866, Down's syndrome individuals are easily recognised by their characteristically flattened facial features. Most also exhibit some degree of impaired physical and cognitive development. It was not until a century after Down's first description of the condition that affected individuals were shown to be trisomic for (possess an additional copy of) chromosome 21. This is the smallest human chromosome other than the tiny, male-specific Y, and Down's syndrome is now often referred to as Trisomy 21.

The likelihood of giving birth to a Down's syndrome baby increases markedly with maternal age. The explanation for this probably lies in the fact that all of the eggs that a woman will produce throughout her lifetime are laid down some months before her own birth. Women who give birth relatively late in their normal reproductive cycle release eggs that may have been in place for more than 40 years. Why having an extra copy of chromosome 21 should have such severe consequences is not entirely clear. Chromosome 21 contains relatively few genes but perhaps an extra copy of just one or two of these is enough to disturb the normal developmental process. Other chromosome-number abnormalities such as Edward's syndrome (Trisomy 18) and Patau syndrome (Trisomy 13) are more severe than Down's and rarely survive the gestation process.

However, when a baby inherits a nonfunctional phenylalanine hydroxylase gene from each parent, then they have the potentially disastrous condition called phenylketonuria. People with one functional gene and one nonfunctional one are called **carriers**. If two carriers of the same defective gene have a child, that child has a one-in-four chance of getting a defective gene from both parents and exhibiting the recessive condition, be it nonbrown eyes or phenylketonuria.

More unusually, a defective gene may cause a problem even if the individual has one good copy, that is, the defective gene is dominant over the normal one. One example is familial Creutzfeldt-Jacob disease. As we described earlier (Medical Relevance 9.2 on page 153), the brain damage that occurs in Creutzfeldt-Jacob disease results when an alternatively folded form of PrP protein called PrP^{sc} causes all the PrP protein around it to fold up in the alternative way. Sufferers from familial Creutzfeldt-Jacob disease have a mutant form of the *PrP* gene that generates protein that spontaneously folds into the PrP^{sc} form. The PrP^{sc} then triggers the normal protein, generated by the second, normal copy of the gene, to fold into the PrP^{sc} shape. Thus one copy of the defective gene is sufficient to cause the disease.

Crossing Over and Linkage

Prophase I lasts such a long time because the chromosomes tie themselves in knots and then untangle (using topoisomerase II; see In Depth 4.1 on page 57). Figure 18.5 shows what happens. At the top we see the paternal and maternal chromosomes, each composed of two chromatids, lined up side by side. As before, we show the chromosome that originated from the mother in gray and the paternal one in red. During prophase 1 the chromosomes are cut and resealed at points called **chiasmata** (singular **chiasma**) so that lengths of paternal chromosome are transferred to a maternal one and vice versa. This process is **crossing over**. The rest of meiosis I proceeds, followed by meiosis II, and the end result is that some gametes contain chromosomes that are neither completely paternal nor completely maternal but are a **recombination** of the two. The phenomenon is called **homologous recombination**.

The biological advantage of sexual reproduction is that it allows organisms to possess a random selection of the genes from their ancestors. Those individuals with a complement of genes that makes them better suited to their environment tend to do better, allowing evolution by natural selection of the individuals posessing the better genes. Without crossing over this could not happen: those genes that are located on the same chromosome would remain **linked** down the generations, greatly reducing the number of gene permutations possible at each generation. Crossing over allows a child to inherit, for example, his grandmother's green eyes without also inheriting her defective voltage-gated sodium channel

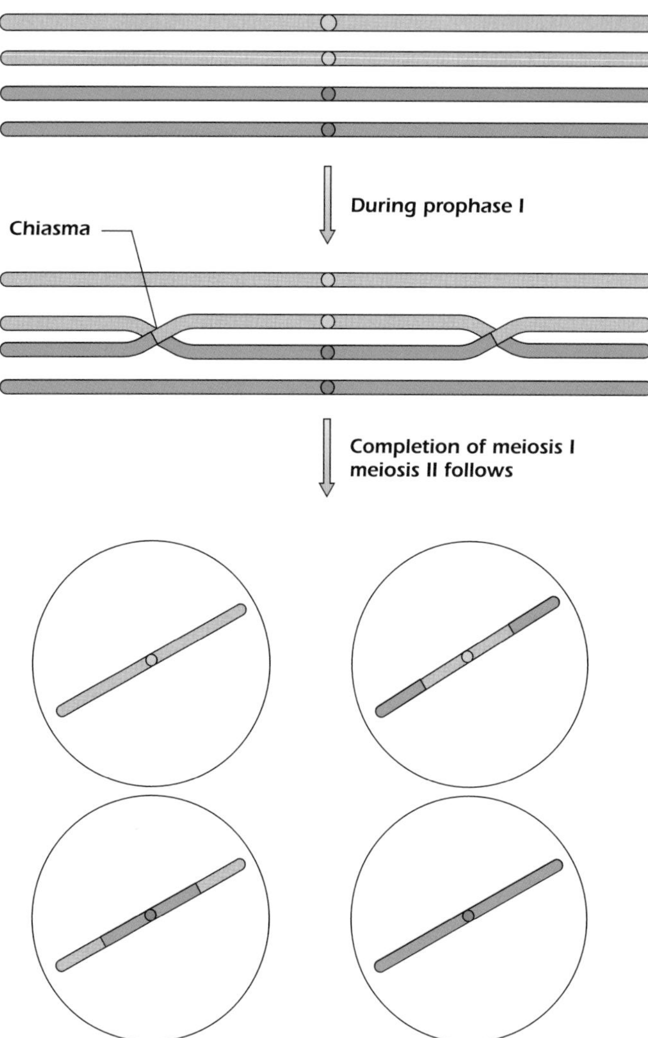

Figure 18.5. Chiasmata allow the crossing over of genetic material during prophase I of meiosis.

gene (Medical Relevance 14.2 on page 244), although both genes are on chromosome 19. Even with crossing over, genes on the same chromosome are inherited together more than they would be if they were on different chromosomes. The closer the genes, the less likely it is that a chiasma will form between them, and therefore the greater the probability that they will be inherited together. This phenomenon is used to help identify the genes responsible for specific diseases such as cystic fibrosis (Chapter 20).

CONTROL OF THE CELL DIVISION CYCLE

The segregation of chromosomes at mitosis is the culmination of a sequence of biochemical events that occur during interphase. The most important of these is the duplication of the genome in **S phase** (the S stands for DNA synthesis). In this cell cycle shorthand, cell division, mitosis and cytokinesis, is referred to as **M phase**. S phase and M phase must occur only once per cell cycle and in the correct order. To make sure this is the case, S phase and M phase do not follow immediately after one another but are separated by gaps that allow the cell to check that everything is in order before going on to the next stage. The gap between M phase and S phase is **G1 (gap 1)**, and the gap between S phase and M phase is **G2 (gap 2)**. These four phases, Gl, S, G2, and M, make up the classic cell cycle clock (Fig. 18.6). Nondividing or **quiescent** cells are said to be in the **G0 (gap 0)** phase of the cell cycle. Cells can only enter G0 from G1. G0 cells can remain viable for months or even years, and most of the cells in the human body are in fact in this nondividing state. Cells are said to be terminally differentiated if they are unable to return to the cell division cycle. Nerve cells are one such example. In contrast other differentiated cells, such as glial cells (page 12), can return to the cycle if they receive the correct signals from their neighbors.

In dividing cells three stop/go switches ensure the cell's orderly progression around its division cycle (Fig. 18.6). These are at entry to S phase, sometimes called the **G1/S**

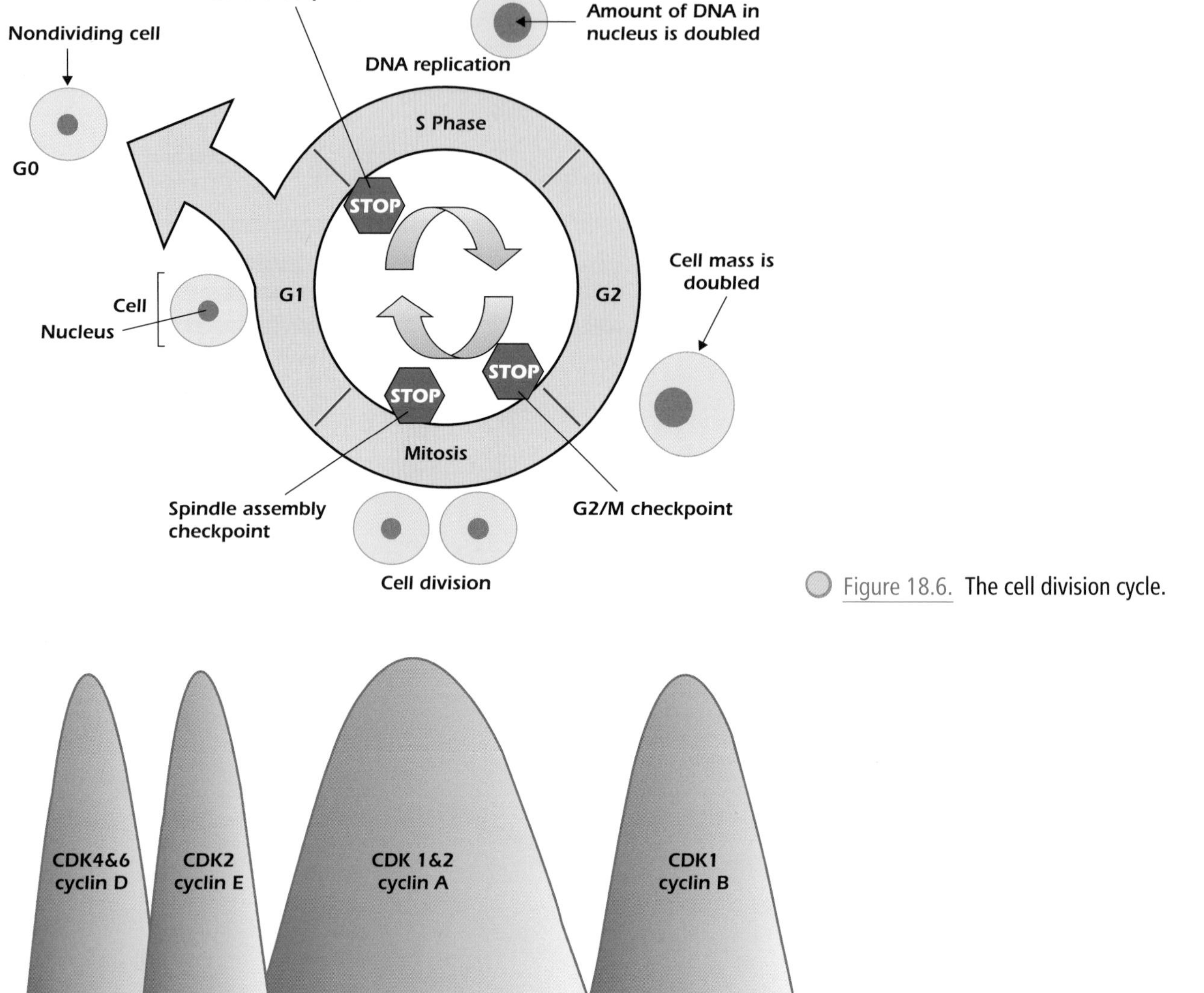

Figure 18.6. The cell division cycle.

Figure 18.7. Cdk activities through the cell cycle.

transition; entry to mitosis, the **G2/M transition**; and late in mitosis, the metaphase-anaphase transition controlled by the **spindle assembly checkpoint**. Each switch is controlled by the activation or inactivation of a unique class of cell cycle enzymes, the **cyclin-dependent protein kinases** or **CDKs**. The amounts of each CDK remain constant, but each is active only when associated with a regulatory subunit whose concentrations increase and decrease in phase with the cell cycle, hence their name, the **cyclins**. Alone, the CDK is both inactive and directionless but cyclin binding both activates the enzyme and targets it to its substrate. In simple eukaryotes such as the fission yeast, *Schizosaccharomyces pombe*, a single CDK-cyclin combination, **CDK1-cyclin B**, drives both S phase and mitosis. At G1/S the activity of the enzyme is weak but is sufficient to modify a small number of proteins at origins of DNA replication and initiate DNA synthesis (see Chapter 5). At G2/M enzyme activity is strong and a larger repertoire of substrates is phosphorylated. These induce chromosome condensation and spindle formation for mitosis together with the fragmentation of the membranes of the Golgi apparatus and endoplasmic reticulum to shut off endocytosis and secretion while mitosis proceeds.

Fission yeast is unusual in that G2/M is the major cell cycle control point. In multicellular organisms such as humans, where cells must respond to a greater variety of extracellular and intracellular signals than yeast, there are multiple waves of CDK activity (Fig. 18.7). **CDK4-cyclin D** and **CDK6-cyclin D** control early events in G1 while

Medical Relevance 18.2 Why Marrying Your Cousin is Dangerous

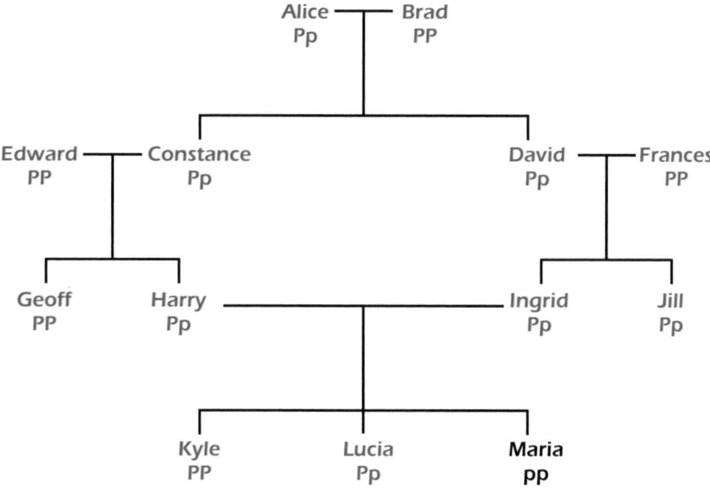

Marriage between first cousins is illegal in 30 states of the United States. The scientific rationale behind these laws is that cousin marriage provides an easy mechanism for a single carrier of a recessive genetic disease to generate great-grandchildren with the disease. The topic has recently surfaced in Britain, where a government minister has argued that the practice of cousin marriage in the Pakistani immigrant population results in an eightfold higher incidence of autosomal recessive metabolic errors compared to the general population. The accompanying figure shows how this can happen in an imaginary family tree. Harry and Ingrid are first cousins: they share a grandmother, Alice, who was a carrier for phenylketonuria. We indicate this in the letters below Alice's name: P indicates the gene coding for a functional phenylalanine hydroxylase while p indicates the gene coding for a defective protein. Both Alice's children, Constance and David, inherited her defective gene and were therefore carriers. Of their four children, Harry, Ingrid, and Jill inherited the defective gene and are carriers. The children of Harry and Ingrid therefore each had a one-in-four chance of inheriting one defective gene from Harry plus one defective gene from Ingrid. One of their children, Maria, does so and has phenylketonuria.

CDK2-cyclin E controls G1/S itself. A wave of **CDK1-cyclin A** and **CDK2-cyclin A** ensures passage through S phase and G2 while **CDK1-cyclin B** controls G2/M just as it does in yeast. Figure 18.8 shows how CDK1 is regulated at G2/M. To ensure that CDK1 remains inactive through interphase, even after cyclin binding, it is itself phosphorylated by another protein kinase called **Wee1** (because yeast mutants, with unphosphorylated CDK1, divide too early, while they are still "wee"—Scottish for small!). Wee1 adds two phosphates to CDK1, one on a threonine residue at amino acid number 14, the other on a tyrosine residue at amino acid number 15. These two amino acids lie within the ATP-binding site of CDK1, and the presence of two phosphates prevents ATP binding, the first step of the phosphorylation reaction. The phosphates are removed at G2/M by a second enzyme, the protein phosphatase **Cdc25**. This leads to a burst of CDK1 activity and entry to M phase. Once mitosis is initiated, it is essential that CDKl be deactivated for the reasons outlined above. This is achieved by the destruction of cyclin B and the rephosphorylation of CDKl by Wee1.

The decision to begin DNA synthesis is arguably an even more critical decision than entry into M phase because once S phase begins, the cell is committed to eventually dividing. Cells that replicate their DNA but then fail to undergo mitosis in a reasonable time commit suicide by apoptosis (page 308) rather than hang around in a **polyploid** state. The key target of the CDK4-cyclin D and CDK6-cyclin D enzymes is the protein **Rb** (Fig. 18.9). Rb interacts with **E2F**, a transcription factor that is required to produce the enzymes required for DNA replication during S phase. Among them is DNA polymerase alpha, the main polymerase involved in DNA replication. Rb prevents this function by interacting with E2F. Phosphorylation of Rb by CDK4-cyclin D and CDK6-cyclin D causes Rb to release E2F and allows it to get on with its job. Mutant Rb is always phosphorylated and hence cannot regulate E2F. The normal controls on DNA replication are removed, and cells become cancerous.

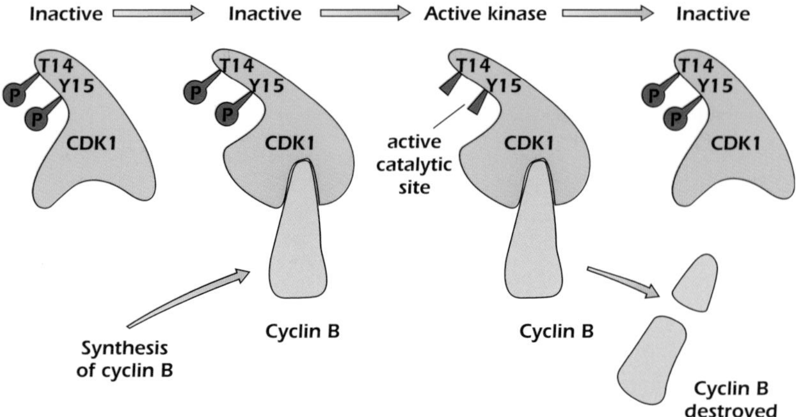

Figure 18.8. How Cdk1 is controlled by cyclin B and phosphorylation in multicellular organisms.

Checkpoints Tell the Cell Cycle When To Stop and When To Go

The cell cycle has built-in safety devices to ensure that each cell division results in a perfect copy of the parental cell's genome being passed on to its progeny. These were termed **checkpoints** by Leland Hartwell, who shared with Paul Nurse and Tim Hunt the 2001 Nobel Prize for Physiology or Medicine for their pioneering work on the molecules that control the cell cycle. A checkpoint pathway detects errors and passes a message on down the line to stop the cell cycle until the problem has been attended to. Checkpoints are like the air bags on a car; they serve no purpose whatsoever for most of the journeys that you take but can save your life in the event of a crash. Similarly, checkpoint proteins have no function during normal cell growth but are activated only in response to trauma. The best-characterized cell cycle checkpoints are those that respond to DNA damage or the failure to complete DNA replication. Both situations result in the activation of the protein kinase **ATM**. In response to DNA damage, for example by ultraviolet light (page 73), ATM activates two more protein kinases, Chk1 and Cds1 (Fig. 18.10). In the case of incomplete DNA replication only Chk1 is activated. When active, these kinases prevent cells passing both the G2/M and the G1/S control points. To prevent entry into mitosis, Chk1 and Cds1 phosphorylate Cdc25. The phosphorylated form of Cdc25 is inactive, and without active Cdc25 cells are unable to activate CDK1-cyclin B and enter M phase. This buys time for the cell to repair the damage to its DNA or to complete S phase. To prevent entry into S phase, Chk1 and Cds1 phosphorylate the transcription factor **p53** (Fig. 18.9). p53 is produced all the time in cells but is broken down rapidly, so its concentration is usually low. However, phosphorylated p53 is not destroyed so its concentration increases. In turn, the systems that repair DNA are activated (page 73) and expression of a number of new proteins is induced. One of these is $p21^{CIP1}$, a member of a class of proteins called **CKIs** for **cyclin dependent kinase inhibitors**. These bind to and inactivate CDKs; in the case of p21 the target is CDK4-cyclin D. p21 binding therefore arrests the cell in G1, preventing it from entering another round of S phase while the previous one is incomplete (Fig. 18.9). There are many CKIs mediating a number of pathways, all with the same outcome: stopping cell division. One of these pathways mediates **contact inhibition** (Fig 18.9). Cells in a culture dish, or at the edge of a wound, divide until they touch each other. When they contact neighboring cells, they stop dividing because contact causes production of two CKIs called $p16^{INK4a}$ and $p27^{KIP1}$. These inhibit the G1 Cdks and therefore prevent DNA synthesis. Loss of contact inhibition is one of the first changes seen in the transformation of normal cells into cancer cells. p53 also upregulates the transcription of microRNAs (In Depth 6.1 on page 94) that act to reduce expression of proteins required for cell cycle progression, including CDK4 and Cdc25.

Ending the Cycle

Once mitosis is initiated, it is essential that CDK1 be deactivated. Failure to do so would result in cells trying to reinitiate mitosis while the previous M phase was still in progress. Figure 18.3 shows how this is achieved. A group of proteins called the mitotic checkpoint complex (MCC) binds to kinetochores and in turn binds a protein called Cdc20. When all the kinetochores are connected to the mitotic spindle the MCC dissociates, releasing Cdc20 which can now activate a collection of enzymes called the anaphase promoting complex. This in turn modifies proteins in such a way that they are sent to the proteasome (page 133) and destroyed. Among the proteins so destroyed is cyclin B. A second target is securin, an inhibitor of an enzyme named separase whose function is to destroy the cohesin complex that keeps sister kinetochores attached. Thus at one and the same time anaphase begins and CDK1 is rendered inactive through the destruction of its cyclin partner.

Activation of the anaphase promoting complex is the critical step that allows cells to pass the spindle assembly checkpoint. The MCC is sensitive to tension, and only

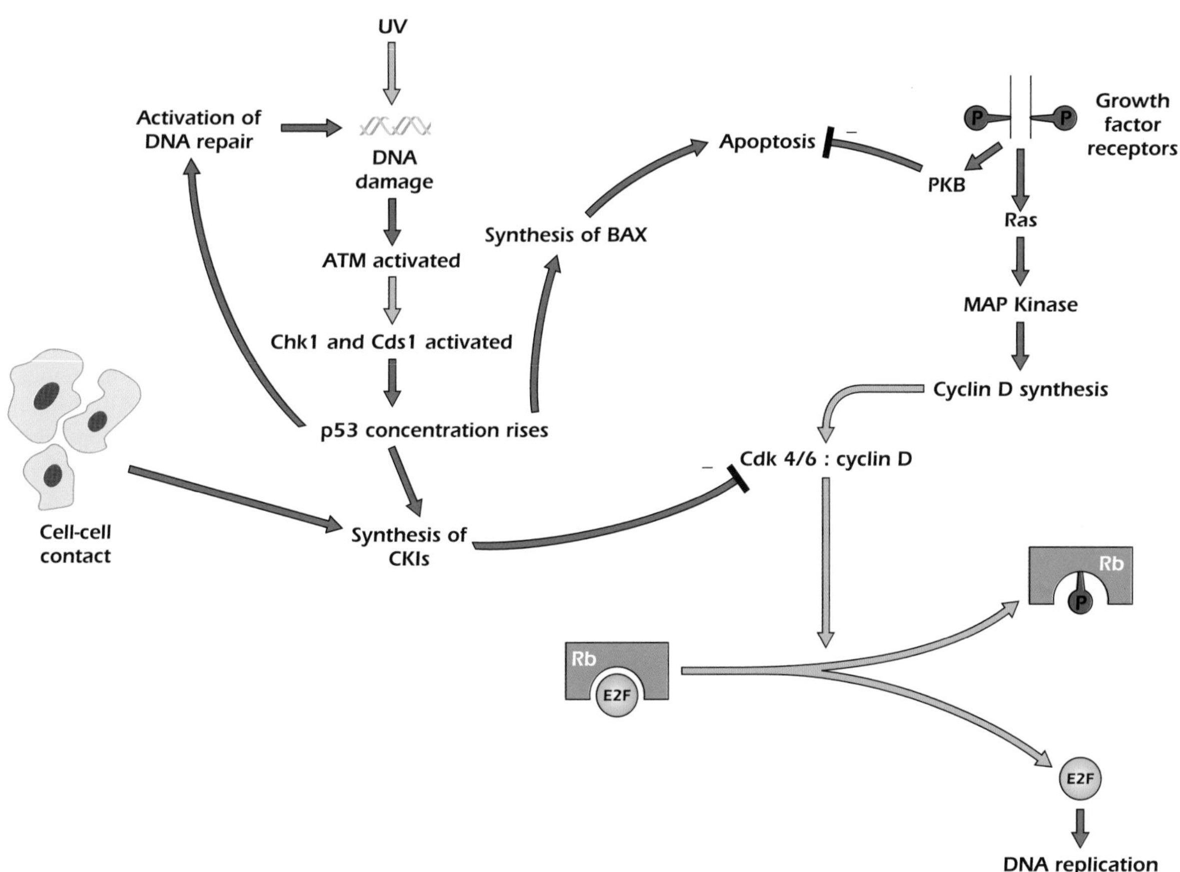

Figure 18.9. Retinoblastoma protein Rb sequesters E2F, the critical transcription factor controlling entry into S phase.

dissociates from the kinetochores when the spindle fibers pull the chromatids in opposite directions. Even a single chromatid not under tension sequesters enough Cdc20 to prevent activation of the complex. Anaphase can only begin when all the chromatids are correctly lined up on the metaphase plate.

The Cell Cycle and Cancer

Cancer occurs when cells evade the mechanisms that normally control the cell division cycle. To become cancerous cells must accumulate somatic mutations in a number of key genes including p53 and Rb. Loss of division control results in the formation of a **tumor** that can be either benign, meaning it remains in one place and is relatively easy for the surgeon to remove, or malignant, able to invade other tissues and establish secondary cancers (Fig. 18.11). Death from cancer is almost always due to such metastases (from metastasis; the spread of a disease from one organ to another) and not the original tumor.

There are many types of cancer that are broadly named after the tissues in which they occur. Carcinomas are cancers of the epithelial layers of the skin, gut and breast; sarcomas are cancers of muscle, bone and connective tissue; lymphomas are cancers of the immune system, and leukemias are cancers of blood-forming tissues such as the bone marrow.

Medical Relevance 18.3 Retinoblastoma

Retinoblastoma is a cancer of the retina of the eye that generally occurs in children below the age of five. It is usually caused by two independent somatic mutations in a photoreceptor cell lineage within one eye, each of which damages or destroys an *Rb* gene inherited from one parent. *Rb* encodes the Rb protein that sequesters E2F. A rarer form, familial retinoblastoma, occurs in children who inherit one defective *Rb* gene. Every cell in the body therefore has only one working *Rb* gene, and a single somatic mutation that destroys this gene can generate a tumor. Such children also have an increased risk of other, non-optical cancers.

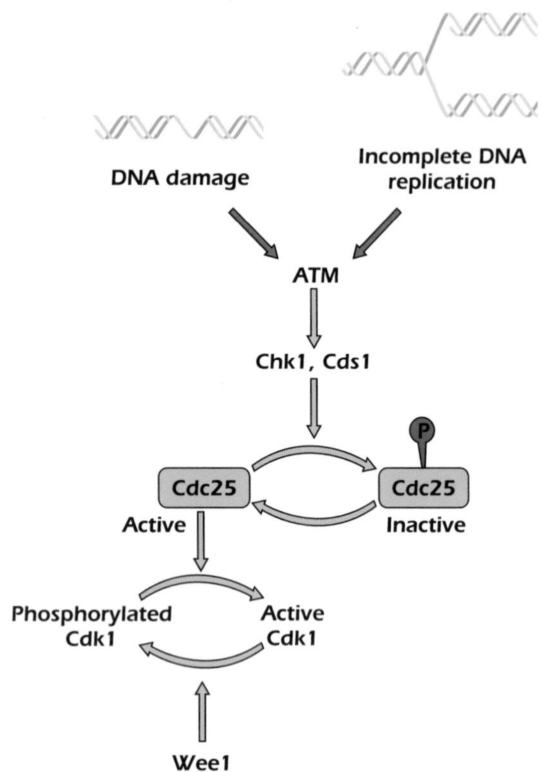

Figure 18.10. ATM activation stops the cell cycle.

Carcinogenesis (the formation of a cancer, also known as oncogenesis or tumorigenesis) is a complex process that may last several decades. In a few cases the genetic changes leading to the development of the disease are well understood. Colon cancer is a good example. The first step is mutation of the gene adenomatous polyposis coli. This is referred to as a "gatekeeper" gene as colon cancer cannot develop without this crucial step. Mutation of adenomatous polyposis coli alone leads, at worst, to the formation of a small, benign growth that is often asymptomatic. Further progress requires mutations in a number of oncogenes (cancer genes) including Ras and PI 3-kinase (pages 261, 262). Despite these multiple genetic changes the resulting tumor remains benign, forming only growths called adenomatous polyps that gave the gatekeeper gene its name. The crucial step in transforming a benign tumor into a malignant one is the mutation of *p53* and therefore the elimination of the cell cycle checkpoints. *p53*, which is mutated in more than 50% of all human cancers, is called a tumor suppressor gene. The loss of even one functional *p53* gene results in a large reduction of p53 protein synthesis and permits the development of an aggressive, fast-dividing tumor.

Recruiting Blood Vessels

Metastasis cannot occur without angiogenesis, the formation of a network of blood vessels that supplies the developing tumor with oxygen. Blood vessels arise from the proliferation of vascular endothelial cells. These cells are inhibited from division for long periods. However, when blood supply has to be reestablished, for example, following a wound or during the menstrual cycle, division is stimulated. Cancer cells secrete numerous angiogenic molecules including FGF, which binds to its receptor on the endothelial cell surface triggering the transcription of a number of genes that together stimulate endothelial cell growth. Activated endothelial cells secrete metalloproteinases which degrade the surrounding extracellular matrix, allowing the cells to migrate into the tumor where they form hollow tubes that will eventually become new blood vessels. Considerable effort has been devoted to the development of drugs that inhibit angiogenesis since choking off the blood supply to a tumor should inhibit its growth and prevent metastasis. One such drug, the monoclonal antibody Avastin, was approved as a treatment for metastatic colon cancer in 2004, and other drugs have targeted cancers of other tissues.

APOPTOSIS

An adult human is made up of about 30 trillion (3×10^{13}) cells, all of which originate from a single fertilized egg. If this first cell divides into two, the two progeny cells into four, and so on, it would take only about 45 rounds of division to produce the number of cells required to make an adult human. Cell division does not stop when we have finished growing. In fact, when one counts up all the cell divisions of active stem cells (page 13) that occur in the adult body, one finds that 3×10^{13} new cells are created every two weeks. The reason that we do not double in size every two weeks is because the proliferation of cells is balanced by cell death. Cells die for two quite different reasons. One is accidental, the result of mechanical trauma or exposure to some kind of toxic agent, and often referred to as **necrosis**. The other type of death is deliberate, the result of a built-in suicide mechanism known as **apoptosis** or **programmed cell death**. The two types of cell death are quite distinct. In cells that are injured, ATP concentrations fall so low that the Na^+/K^+ ATPase can no longer operate, and therefore ion concentrations are no longer controlled. This causes the cells to swell and then burst. The cell contents then leak out, causing the surrounding tissues to become inflamed. Cells that die by suicide on the other hand shrink, and their cell contents are packaged into small membrane-bound packets called blebs. The nuclear DNA becomes chopped up into small fragments, each of which becomes enclosed in a portion of the nuclear envelope. The dying cell modifies its plasma membrane, signaling to macrophages (page 12), which respond by engulfing the blebs and the remaining cell fragments and by secreting cytokines that inhibit inflammation. The changes that occur

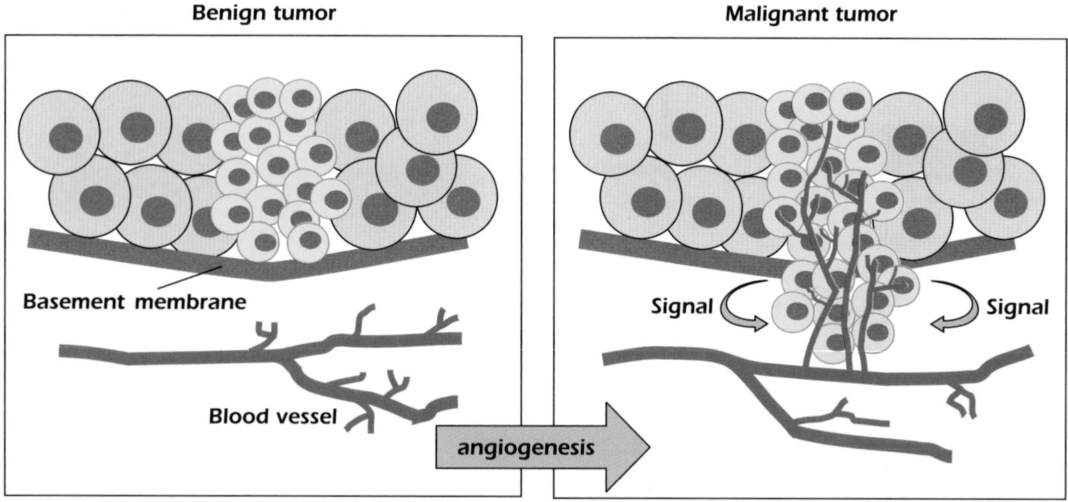

Figure 18.11. Recruitment of blood vessels is essential for tumor growth.

during apoptosis are the result of hydrolysis of cellular proteins by a family of proteases called **caspases** (short for cysteine-containing, cleaving at aspartate). All the cells of our body contain caspases, but they are normally locked in an inactive form by an integral inhibitory domain of the protein. Proteolysis cleaves the inhibitory domain off, releasing the active caspases. The advantage to the cell of this strategy is that no protein sythesis is required to activate the apoptotic pathway—all the components are already present. Thus, for example, if a virus infects a cell and takes over all protein synthesis, the cell can still commit suicide and hence prevent viral replication. Figure 18.12 summarizes the complex control systems that regulate the decision to die or survive. Cells activate apoptosis in response to three types of event.

Instructed Death: Death Domain Receptors

Cells are instructed to die when a ligand binds to one of a family of **death domain** receptors. This occurs, for example, in the case described above: if a cell is infected by a virus, white blood cells recognize viral proteins on the cell surface and activate Fas, a death domain receptor on the surface of the unlucky cell. On binding its ligand, a death domain receptor causes caspase 8 to activate. In turn, caspase 8 can hydrolyze and hence activate the **effector caspases** that begin the processes of cell destruction.

Default Death: Absence of Growth Factors

Cells that are not required by the organism die. To make sure that this occurs, death is the default option for the cells of a multicellular organism—only if a cell receives growth factors from other cells will it survive. We have already described the first part of this pathway (page 262). Active receptor tyrosine kinases activate phosphatidylinositol 3-kinase, which generates the highly charged membrane lipid phosphoinositide trisphosphate. This causes protein kinase B to visit the membrane, where it is itself phosphorylated and hence activated. Protein kinase B phosphorylates **BAX**, one of a family of **bcl-2 family proteins**. Phosphorylated

Example 18.3 Sunburn, Cell Death, and Skin Cancer

Without knowing it, we have all observed apoptosis in action. All of us at some time or other have been out in the sun without proper protection. The ultraviolet (UV) light causes damage to the DNA of the skin cells and activates p53. If the DNA damage is minor, p53 simply arrests the cell cycle until the DNA repair machinery has had time to repair the lesions. However, if the damage was more severe, the skin cells activate the apoptoptic pathway so that the sunburned skin dies and sloughs off.

The reason why this is a long-term danger is that in areas of skin where sunburn has caused cell death there has been strong selection for any cells in which an earlier somatic mutation has inactivated the p53 system. The new skin becomes enriched in these mutant cells, so that any further oncogenic mutations will generate not a benign tumor but a malignant one.

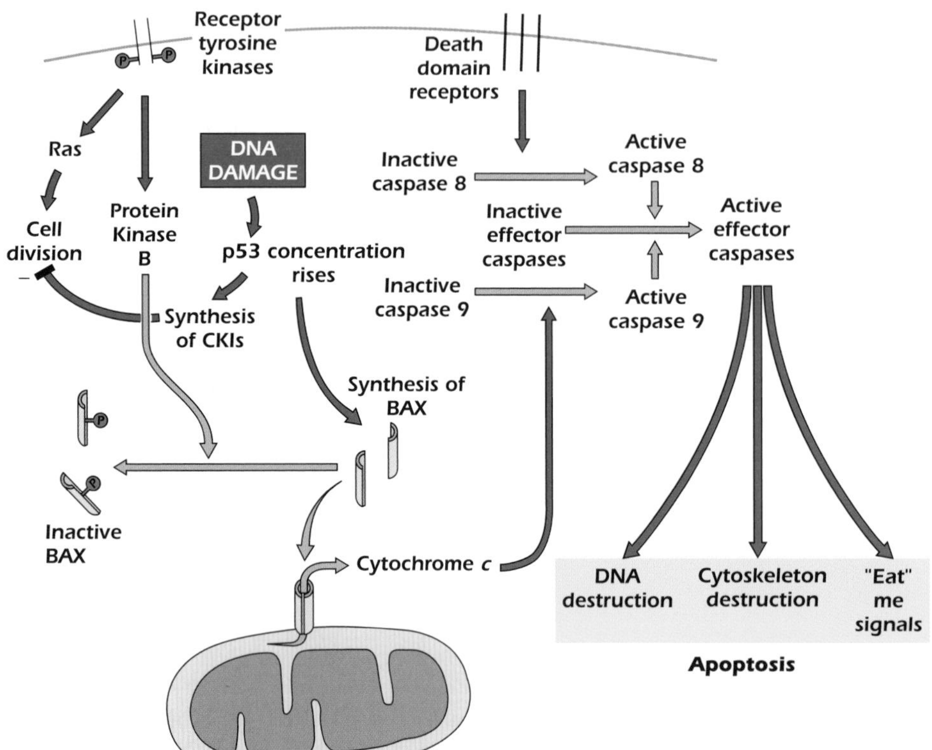

Figure 18.12. Pathways controlling apoptosis.

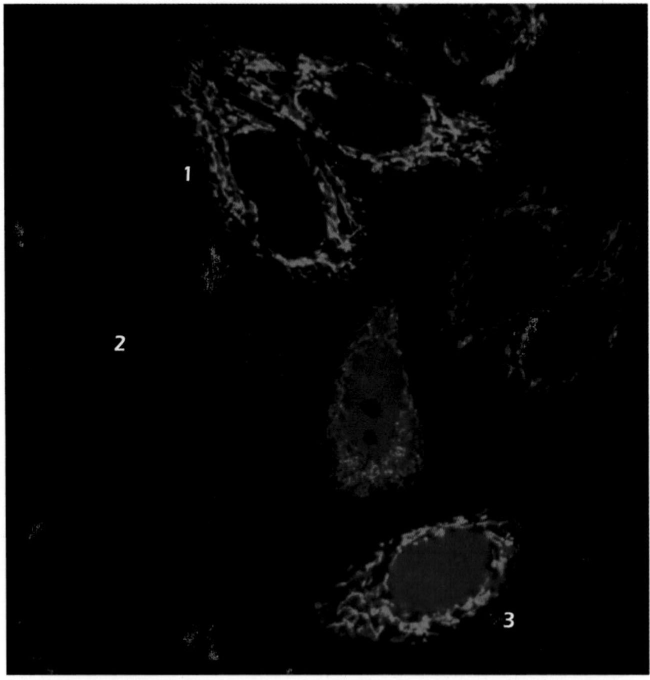

Figure 18.13. A green fluorescent protein chimera revels cytochrome *c* translocation in cells. Live human cells (HeLa) were transfected to express a chimaera of cytochrome *c* and GFP then treated with staurosporine, a drug that evokes apoptotic cell death. Data of Choon Hong Tan and Professor Michael Duchen, University College London; used with permission.

BAX is inactive. However, if PKB is ever allowed to stop working, BAX loses its phosphates and migrates to the outer mitochondrial membrane, where it dimerizes to form a channel big enough to allow cytochrome *c* to escape into the cytosol. Although cytochrome *c* plays a vital role in allowing mitochondria to generate the H^+ gradient energy currency (page 196), once in the cytosol it is deadly. It activates another caspase, number 9, which then hydrolyzes and hence activates the effector caspases. Figure 18.13 shows live cells that have been transfected to express a chimera of

Example 18.4 Neurotrophin Trafficking

During fetal development motoneurons die unless they succeed in growing their axons all the way to an appropriate muscle. Those that do find the target are bathed in **neurotrophin** 3, a growth factor released by the muscle cells. Neurotrophin 3 binds to its receptor tyrosine kinase, Trk C, on the nerve cell surface. However, the activated kinase then has to activate protein kinase B in the cell body in order to prevent apoptosis, and the cell body may be many millimeters away. To achieve this, the neurotrophin-receptor complex is endocytosed using the clathrin mechanism (page 167). The endocytotic vesicles are then transported back to the cell body by dynein (page 287). Once in the cell body, protein kinase B is activated and apoptosis prevented.

The nerve that contains motoneurons also contains pain receptor axons. Instead of Trk C these express a related receptor, Trk A, which requires a different neurotrophin (number 1): Trk A cannot be activated by neurotrophin 3. However, both motoneurons and pain receptors express a second, unrelated receptor called the p75 neurotrophin receptor, which binds both neurotrophin 1 and neurotrophin 3. p75 is a death domain receptor. The overall result is that when an axon arrives at a target, it will automatically receive a signal to die. However, if it has arrived at the correct location, it will receive a countermanding signal to survive.

cytochrome *c* and green fluorescent protein. Fluorescence imaging reveals the location of the chimera. The cells have been treated with a drug that evokes apoptotic cell death through activation of BAX. The appearance of some cells (e.g., cell 1) has not yet changed: the cytochrome *c* chimera is localized to mitochondria, as it is in control cells. In some cells (e.g., cell 2) all the cytochrome *c* has been lost from the mitochondria and has distributed evenly throughout the cell, including the nucleus. Cell 3 is intermediate: the signal from the nucleus shows that some cytochrome *c* has been lost from mitochondria, but the mitochondria can still be seen as bright green organelles. With time, all the cells will come to look like cell 2, and will then show the other signs of apoptosis such as shrinking and blebbing.

The Sick Are Left to Die: Stress-Activated Apoptosis

If a unicellular organism is damaged, it will try to repair itself since the alternative is the death of the organism. However, if a cell of a multicellular organism is stressed or damaged, it may be more efficient to allow the cell to undergo a quick and polite suicide and replace it by cell division of a healthy neighbor. There are therefore a number of mechanisms that trigger apoptosis in response to cell stress. One mechanism operates directly on mitochondria. When mitochondria are stressed, they can spontaneously release cytochrome *c*; this appears to happen when porin, a channel of the outer mitochondrial membrane, associates with other proteins to form a channel of large enough diameter to allow cytochrome *c* to leak out. This is a major medical problem because it can occur in tissues that are unable to rebuild by cell division, for example in hearts during heart attacks and in brain nerve cells during a stroke.

A second apoptosis mechanism operates through the transcription factor p53, which we have already met in the context of cell cycle control. p53 concentrations increase when DNA is damaged. DNA repair mechanisms are activated, but so is transcription of the *BAX* gene. If the DNA is not repaired in time, BAX concentrations increase to the point where they overwhelm the protein kinase B inactivation system. Cytochrome *c* escapes and apoptosis results. Without this mechanism, cancers, which result from changes in the DNA of our body cells, would be far more abundant.

A third mechanism (not shown in Fig. 18.12) operates through a protein kinase called p38. Although p38 is related to MAP kinase, it is activated not by growth factors but rather by cell stresses such as swelling, shrinkage, or radiation. The effects of p38 and another **stress-activated protein kinase**, JNK, are complex, and we will consider only two. p38 phosphorylates p53, which protects it from the rapid breakdown that normally occurs. An increase of p53 concentration then triggers the usual consequences, including synthesis of BAX. p38 also has a direct effect on the founder member of the bcl-2 family, bcl-2 itself. Unlike BAX, bcl-2 cannot make channels in the outer mitochondrial membrane. Indeed, bcl-2 is antiapoptotic because it can bind to BAX monomers and prevent them from forming a complete channel. p38 phosphorylates bcl-2, preventing it from associating with BAX which is therefore released to trigger cytochrome *c* release.

Since caspases can self-activate by mutual proteolysis, there is a steady slow rate of activation even in a healthy cell. To prevent this from triggering apoptosis, cells produce **apoptosis inhibitor proteins**, which block caspase action. If the extent of caspase activation exceeds the capacity of the inhibitory proteins, death results.

SUMMARY

1. Haploid cells contain only one copy of each chromosome. Diploid cells contain two copies, one from the organism's father, one from the mother.
2. Mitosis comprises prophase, prometaphase, metaphase, anaphase and telophase. The end result of mitosis plus cytokinesis is two diploid cells whose chromosome complement is the same as that of the original cell before it underwent S phase.
3. Meiosis generates haploid germ cells: in vertebrates, eggs and sperm. Like mitosis, it follows an S phase but comprises two cycles of cell division, so that the end result is four cells whose chromosome complement is only half that of the original cell before it underwent S phase.
4. During meiosis I homologous chromosomes undergo recombination, a physical resplicing of homologous chromosomes that allows information on chromosomes originating from father and mother to be mixed.
5. Recessive genes are usually those that fail to make functional protein. Examples are the gene for blue eyes and the gene for phenylketonuria. The functional gene is called dominant because an individual need inherit only one copy to be able to make functional protein.
6. In a few cases a defective gene, for example, the gene for familial Creutzfeldt–Jacob disease, is dominant over a normal one.
7. The cell cycle comprises S, G1, M, and G2 phases. G2, S, and G1 phases together constitute interphase. DNA is replicated in S phase, so that each chromosome becomes a pair of identical chromatids. The cell physically divides in M phase.
8. The cell cycle has three built-in control points that act as stop/go switches. These are at G1/S, G2/M, and at the metaphase/anaphase transition of M phase.
9. Cells enter mitosis when cyclin-dependent kinase 1 is active. This in turn requires that cyclin B be present in high enough concentration and that cyclin-dependent kinase 1 be dephosphorylated by Cdc25. Once mitosis is initiated, cyclin-dependent kinase 1 is rapidly turned off through phosphorylation by Wee1 in parallel with the proteolytic destruction of cyclin B.
10. The entry of cells into S phase is a more complex decision involving cyclin dependent kinases 2, 4, and 6. The main effect of active cyclin-dependent kinase 4 is to phosphorylate Rb, causing it to release the transcription factor E2F and hence allowing the synthesis of proteins required for DNA synthesis. Important components of the decision are a raised concentration of cyclin D as a result of MAP kinase activity and a low concentration of CKIs. Cell-cell contact upregulates CKIs so that when an organ has filled the space available it stops growing.
11. Cell cycle checkpoints stop the cycle if DNA is damaged or if DNA synthesis (S phase) is incomplete. A third checkpoint ensures that all chromosomes are properly aligned on the spindle before anaphase begins.
12. p53 is continually produced in cells but is as quickly destroyed. An increase in the concentration of p53 follows DNA damage or other cell stress and has three main effects: (a) activation of DNA repair mechanisms, (b) synthesis of CKIs, preventing cell division, and (c) activation of apoptosis.
13. Cells become cancerous when oncogenes are activated or tumor suppressor genes are inactivated. Tumor formation is accompanied by the proliferation of blood vessels or angiogenesis.
14. In contrast to necrosis, which causes inflammation, apoptosis is a regulated mechanism of cell suicide that has little effect on the surrounding tissue. The final effectors of apoptosis are a family of proteases called caspases.
15. Apoptosis can be triggered in three ways: (a) binding of ligand to death domain receptors, (b) denial of growth factors, and (c) cell stress.

Answer to Thought Question: In normal cells PIP_3 is constantly destroyed by PTEN, so in order to keep protein kinase B active and hence avoid apoptotic death, cells must be constantly bathed in growth factors that act via receptor tyrosine kinases to cause the activation of PI 3-kinase. Cells in which PTEN is inactive no longer need so much growth factor, because once PIP_3 is made it persists, so that PKB in turn remains active. This allows the cells to proliferate even though the relevant growth factors are in short supply.

FURTHER READING

Chial, H. (2008) Genetic regulation of cancer. *Nature Education* **1**(1).

Michael, D., Jacobson, M. D., and McCarthy, N. (eds.) (2002) *Apoptosis: The Molecular Biology of Programmed Cell Death*, Oxford University Press, Oxford.

Mitchison, T. J., and Salmon E. D. (2001) Mitosis: A history of division. *Nature Cell Biol*. **3**, E17–E21.

Morgan, D. O. (2007) *The Cell Cycle: Principles of Control*. New Science Press, London.

National Cancer Institute (US National Institutes of Health), www.cancer.gov/.

O'Connor, C. (2008) Cell Division: Stages of Mitosis. *Nature Education* **1**(1).

O'Connor, C. (2008) Meiosis, genetic recombination, and sexual reproduction. *Nature Education* **1**(1).

Snustad, D. P., and Simmons, M. J. (2006) *Principles of Genetics*, 4th edition, John Wiley & Sons, New York.

Weinberg, R. A. (2006) *Biology of Cancer*, Garland Science, USA.

REVIEW QUESTIONS

18.1 Theme: Cell Division

A anaphase
B cytokinesis
C metaphase
D prometaphase
E prophase
F telophase

From the stages of cell division listed above, select the one in which each of the events described below occur.

1. Chromosome condensation occurs in the nucleus and spindle formation begins in the cytoplasm

2. The nuclear envelope breaks down and chromosomes become associated with the spindle

3. The chromosomes are aligned on the spindle and no longer make individual excursions toward and away from the spindle poles

4. Paired chromatids separate and begin to move toward the spindle poles

5. Chromosomes decondense and the nuclear envelope reforms

6. Physical separation into two cells

18.2 Theme: Checkpoints in the Cell Cycle

A cohesin
B cyclin A
C cyclin B
D dephosphorylation at T14 and Y15
E G1 phase
F G2 phase
G M phase
H S phase
I p53
J phosphorylation at T14 and Y15
K Rb
L recruitment to the plasma membrane
M separase

From the above list of proteins, select the one that answers each of the questions below.

1. During which period of interphase does the cell replicate its DNA?
2. For animal cells to begin DNA replication, many conditions must be met. One is that the transcription factor E2F must be released from a ligand that holds it in an inactive dimer. What is this ligand?
3. If the DNA is damaged, it must be repaired before it is replicated. DNA damage activates two kinases, Chk1 and Cds1. These phosphorylate a transcription factor that acts to upregulate cyclin dependent kinase inhibitors, CKIs. Phosphorylation of the transcription factor allows its concentration to increase in cells What is the identity of this antidivision transcription factor?
4. Once DNA is replicated and the cell is large enough, it can enter mitosis. Passing the G2/M checkpoint requires activity of cyclin dependent kinase 1, Cdk1. Cdk1 is only active when dimerized with a protein partner; what is this essential partner?
5. Cdk1 is also regulated by posttranslational modification. Which reaction must be performed in order for Cdk1 to be active?
6. During prometaphase the chromosomes line up on the metaphase plate. The sister chromatids are joined together at the kinetochores by which protein?
7. When all the kinetochores are under tension the anaphase promoting complex targets securin for destruction, allowing activation of an enzyme that digests the link between the chromatids, allowing the separation of the chromatids in anaphase. What is that link-destroying enzyme?

18.3 Theme: Life and Death

A bcl-2
B caspases
C Cdk1
D cytochrome *c*
E Fas
F p53
G protein kinase A
H protein kinase B

From the above list of compounds, select the one that best fits each of the descriptions below.

1. In animals, cells are kept alive by the activation of other cells that supply growth factors. Growth factor receptors activate PI 3-kinase, which generates PIP_3 at the plasma membrane. PIP_3 recruits a critical survival-promoting kinase to the plasma membrane; name that kinase.
2. If PIP_3 disappears from the plasma membrane and the kinase described above becomes inactive, BAX is activated, allowing a protein to escape from the mitochondria. Name the released mitochondrial protein.
3. White blood cells can kill target cells by activating this cell surface death domain receptor.
4. Cells also die if their DNA is damaged so badly that it cannot be repaired in a reasonable time. Name the transcription factor whose concentration increases after DNA damage and which upregulates the synthesis of BAX.
5. All the death initiation pathways described above converge on the activation of this family of cytosolic proteases.

THOUGHT QUESTION

Medical Relevance 15.2 describes how Ras mutations are often found in cancers. Another protein that is often mutated in cancers is PTEN, an enzyme that hydrolyses PIP_3, generating PIP_2. Mutations that inactivate PTEN are found in 75% of gliomas. Suggest why inactivating mutations of PTEN are selected for in cell lineages leading to cancer.

19

THE CELL BIOLOGY OF THE IMMUNE SYSTEM

The body of a multicellular organism is a perfect place for microorganisms and viruses to live and multiply. Animals, therefore, have evolved mechanisms to eliminate or control foreign organisms and viruses. These mechanisms include a population of cells, collectively called the immune system, that patrol the body and attack invaders. The majority of these defense mechanisms have been developed by natural selection and evolution of the inherited genome and constitute **innate immunity**. These include the N-formyl methionine receptor (page 128), coded for by an inherited gene, which allows cells of the immune system to detect the presence of bacteria. However, the vertebrate animals possess an even more sophisticated system called **adaptive immunity** in which genomic variation, natural selection, and evolution within one organism allow novel threats never before encountered by the organism, or even by the species, to be countered. The unusual genetic mechanism that underlies this system, first proposed by Frank Burnet in 1958 in his "Clonal Selection Theory," is only found in two types of immune system cells: **B cells** and **T cells**, collectively called **lymphocytes**.

 CELLS OF THE IMMUNE SYSTEM

Figure 19.1 shows how differentiation of a blood system stem cell gives rise to the major types of immune system cell plus red blood cells (erythrocytes) and platelets. On the left are the lymphoid B and T cells which, as their name suggests, are found in the lymph nodes of adults together with dendritic cells. The lymphoid cell class also includes natural killer cells. Natural killer cells do not undergo genetic variation but rather bear an array of standard receptors that allow them to recognize various threats. On the right of Figure 19.1 we show the myeloid leukocytes, of which the commonest is the neutrophil. Myeloid leukocytes are phagocytic: they engulf and digest foreign bodies, such as bacteria.

Another type of phagocytic cell is the macrophage. These are not found in the blood but rather within body tissues, where they engulf and digest not only foreign bodies but also the fragments of apoptosing cells within the tissue.

 B CELLS AND ANTIBODIES

B cells secrete antibodies. This, probably the most familiar aspect of the immune system to the layman, is nevertheless one of the most complex and remarkable. Circulating **antibodies** (also known as immunoglobulins) are soluble proteins of the blood plasma and other extracellular spaces that bind with high affinity to particular targets. Immunologists use the word **antigen** to mean any target that can be bound with high affinity by an antibody or by the corresponding receptors on T cells. Any complex molecule can be an antigen although in most cases this means a protein. One isoform, immunoglobulin G (IgG), accounts for 80% of all the immunoglobulin in the blood, and we will describe this isoform first. Figure 19.2A illustrates secreted IgG molecules. Although the basic structure is common to

Cell Biology: A Short Course, Third Edition. Stephen R. Bolsover, Jeremy S. Hyams, Elizabeth A. Shephard and Hugh A. White.

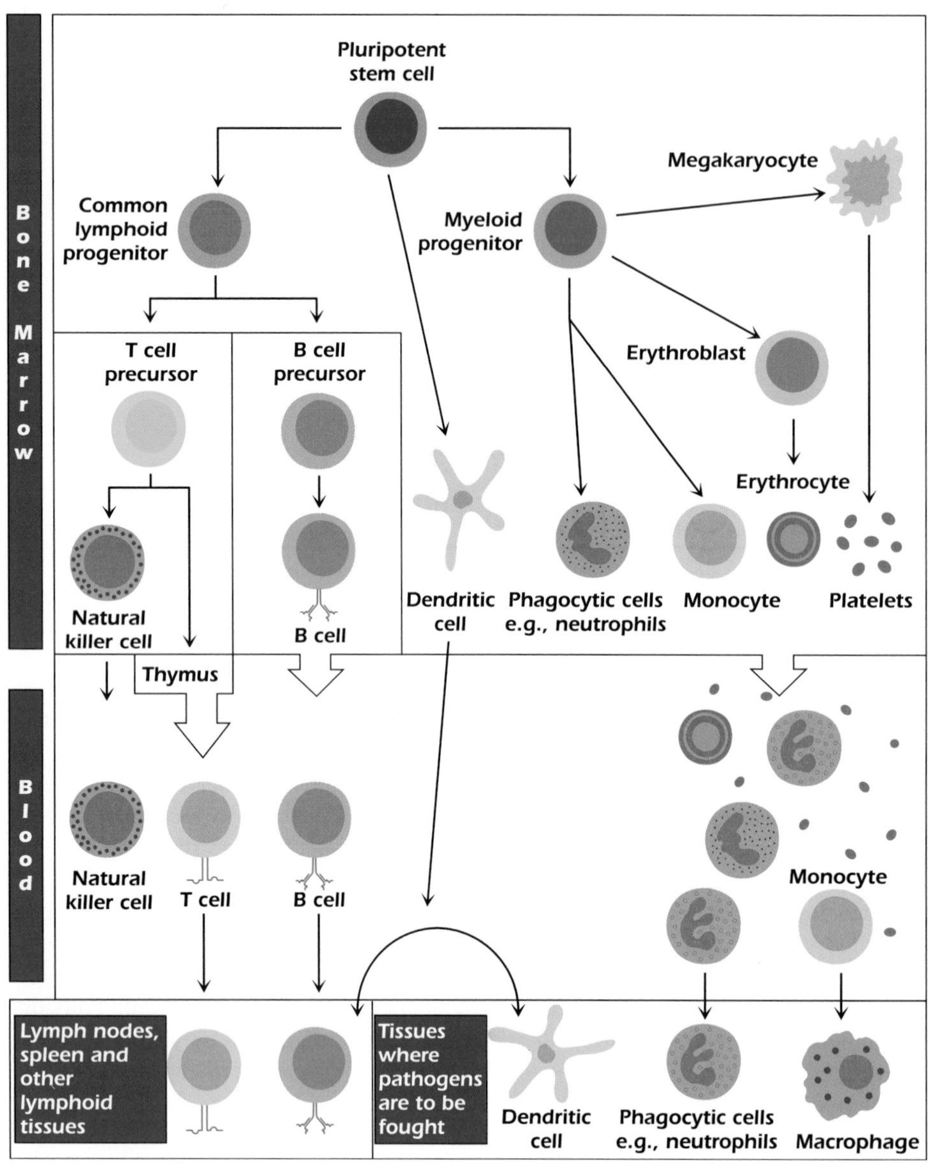

Figure 19.1. Cell types of the immune system.

all IgGs, the molecules are not identical. Experiments in the mid-twentieth century showed that treatment with proteolytic enzyme split IgG into two populations of polypeptides (Figure 19.2B). One population bound antigen but could not be easily crystallized, implying that it was not a homogeneous population of identical polypeptide chains. This population was termed the **Fab**, for antigen binding fragment. The part of the IgG molecule that becomes the Fab upon proteolysis is called the Fab region. The other population generated by proteolysis of IgG could not bind antigen but could be crystallized, indicating that it was homogeneous. This was termed the **Fc**, for crystallizable fragment. The part of the whole antibody molecule that becomes the Fc upon proteolysis is called the Fc region.

Each IgG molecule comprises four polypeptide chains, two identical short (and therefore **light**) chains and two identical chains that are longer and therefore **heavy**. The four chains are held together by disulfide bonds (page 138) and also by hydrophobic interactions to form a very stable structure. The N-terminal domain on each of the four polypeptide chains is variable, that is, it differs between IgGs within the same organism. It is this variability that made the Fabs resistant to crystallization. The variable domain of one heavy chain, and the variable domain of its associated light chain, create one antigen binding site. A single IgG therefore bears two identical antigen binding sites. Because of the molecular variability, another IgG from the same organism will contain antigen binding sites that are identical one to another but which are different from the antigen binding sites on the first IgG, and therefore bind a different antigen.

When we are exposed to a particular antigen, be it a toxin, a protein on the surface of a virus, or a protein on the surface of a bacterium, B cells secrete large amounts of

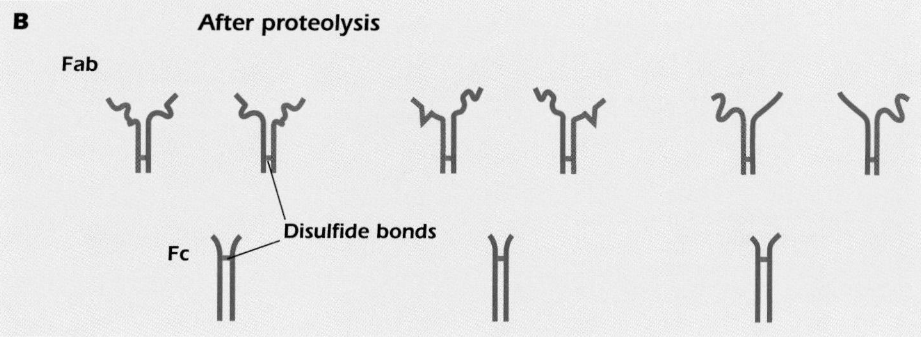

Figure 19.2. IgG structure.

those antibodies that bind to that particular antigen. Antibody binding acts in a range of ways to protect the body. First, dangerous foreign proteins are neutralized by antibody binding. For example, beauticians and doctors who work with botulinum toxin (Example 10.3 on page 171) are **immunized** so that they develop antibodies to the toxin. Now if they accidentally inject themselves with botulinum toxin, the IgG will bind to the protein, blocking its catalytic ability. Figure 19.3A shows another example—a bacterial flagellum (page 201) becomes so loaded with IgG that it cannot function as a propellor and the bacterium is rendered immobile. Second, attachment of IgG **opsonizes** target objects (Fig. 19.3B). Phagocytic cells such as neutrophils and macrophages respond to the Fc portions of the IgG sticking out from the object by engulfing and digesting the particle. Third, after IgG has bound to antigen on the surface of a bacterium (or other invading cell), the invariant Fc portion of IgG recruits an enormous battery of defense proteins collectively called **complement**. Components of complement further opsonize the particle (Fig. 19.3C) while others attract patrolling neutrophils and macrophages (Fig. 19.3D). Assembly of complement culminates in the assembly of a channel called the **membrane attack complex** within the plasma membrane of the target cell (Fig. 19.3E). Successful assembly of the membrane attack complex kills the target cell by allowing the sodium (in eukaryotes) or H^+ (in bacteria) energy currency (page 193) to run down, as well as allowing other critical solutes to leak out and in. However, most pathogenic bacteria have a strong cell wall that prevents insertion and assembly of the membrane attack complex.

Other Antibody Isoforms

Although 80% of immunoglobulin molecules are IgG, there are four other major classes: IgA, IgD, IgE and IgM. The isoforms differ in their Fc components, which confer different properties on each isoform. For example, the Fc portion of IgA is recognized by receptors on the epithelial cells of the gut and respiratory tract and is endocytosed and then directed into secretory vesicles that are exocytosed into the lumen, where they can bind to pathogens in those locations. All the immunoglobulins, of all classes, are secreted by B cells.

The Genetic Basis of Antibody Structure

Antibodies are coded for by the DNA within enormous stretches of chromosome called **loci**. The locus for the heavy chain is located on human chromosome 14 and is shown at the top of Figure 19.4. We are familiar with the concept of alternative splicing (page 76) in which a primary RNA transcript can be cut and reformed so as to code for alternative versions of a protein. This occurs in the production of antibodies, but in addition there is also cutting and splicing of the DNA itself. This process only occurs in lymphocytes and their precursors and only occurs in the DNA coding for two proteins: antibodies and, as we shall see later, the T cell receptor.

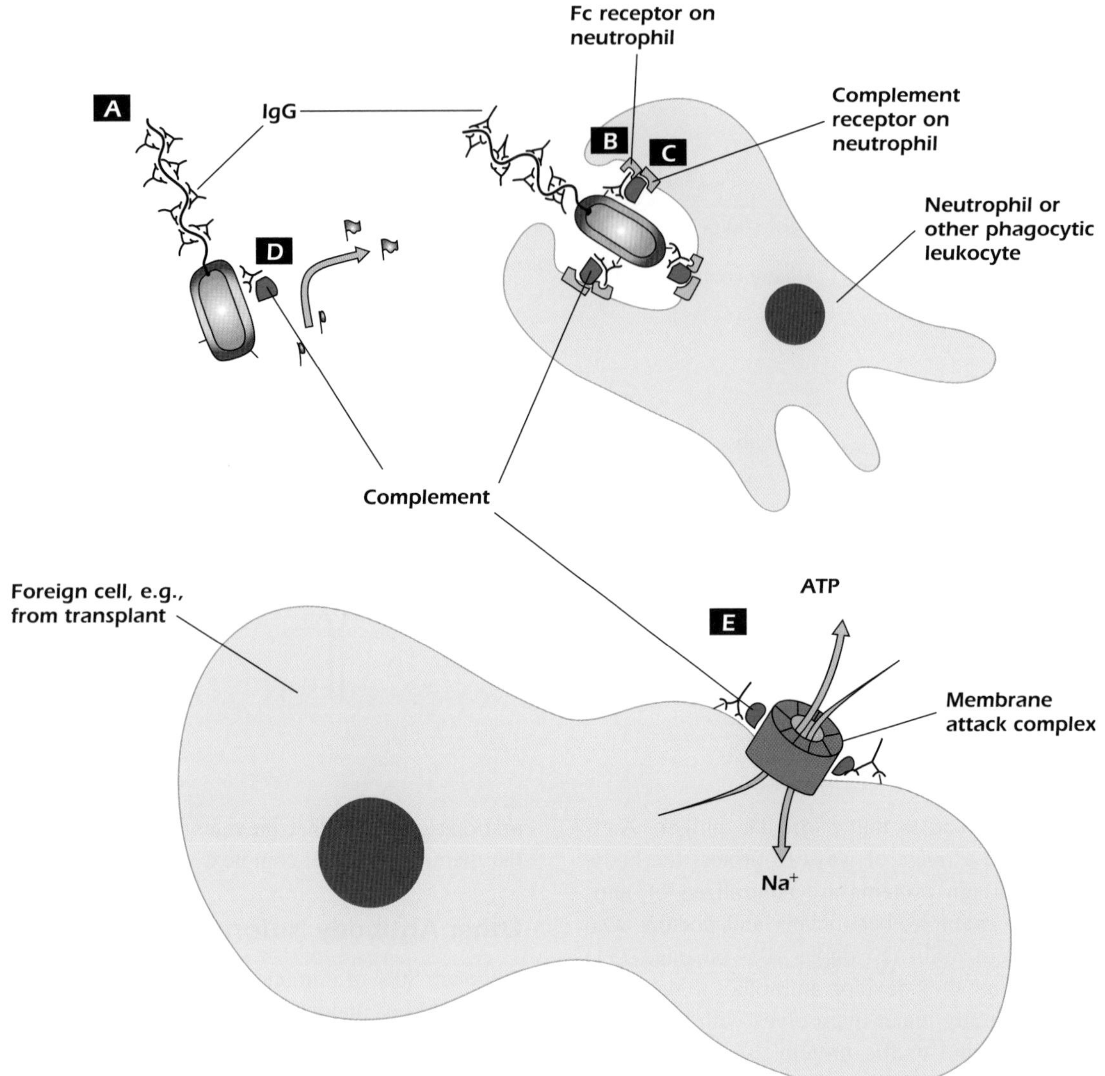

Figure 19.3. Effects of antibody binding. (A) Binding of antibody can itself neutralize a protein. (B) Fc receptors on phagocytic cells recognize antibody and initiate phagocytosis. (C) Complement receptors on phagocytic cells further stimulate phagocytosis. (D) Complement converts inactive serum components into proteins that attract neutrophils and other leukocytes. (E) If a membrane attack complex can form, the target cell is directly killed.

Close to the bottom of Figure 19.4 we show the mature messenger mRNA coding for an antibody heavy chain. This has four components. The V, D and J components code for the Fab region, and are followed by the mRNA coding for the Fc region. The germline DNA, shown at the top of the figure, contains approximately 100 copies of V, all slightly different; 23 different Ds; and 6 Js. A functional Fab can be generated from any combination of V, D and J, and each combination will generate an antibody with a different antigen binding site. Following the six copies of J come nine lengths of DNA, each of which code for an Fc. The sections are given the Greek letter corresponding to the immunoglobulin class. Thus the first two blocks of DNA, μ and δ, code for the Fc part of the heavy chain of the IgM and IgD isoforms respectively. While there are only single copies of μ and δ, there are two slightly different copies of α and four slightly different copies of γ.

In the bone marrow of the fetus, B cell precursors use the enzyme **V(J)D recombinase** to cut the DNA, eliminating large portions and splicing the cut ends together. All but one of the V sections are removed together with all but one of the D sections and a variable number of J sections. Figure 19.4 shows two out of many possibilities. The B cell on the left generates a chromosome in which all the copies of V except number 17, all the Ds except number 4, and Js 1 through 4 have been eliminated. There is a transcription

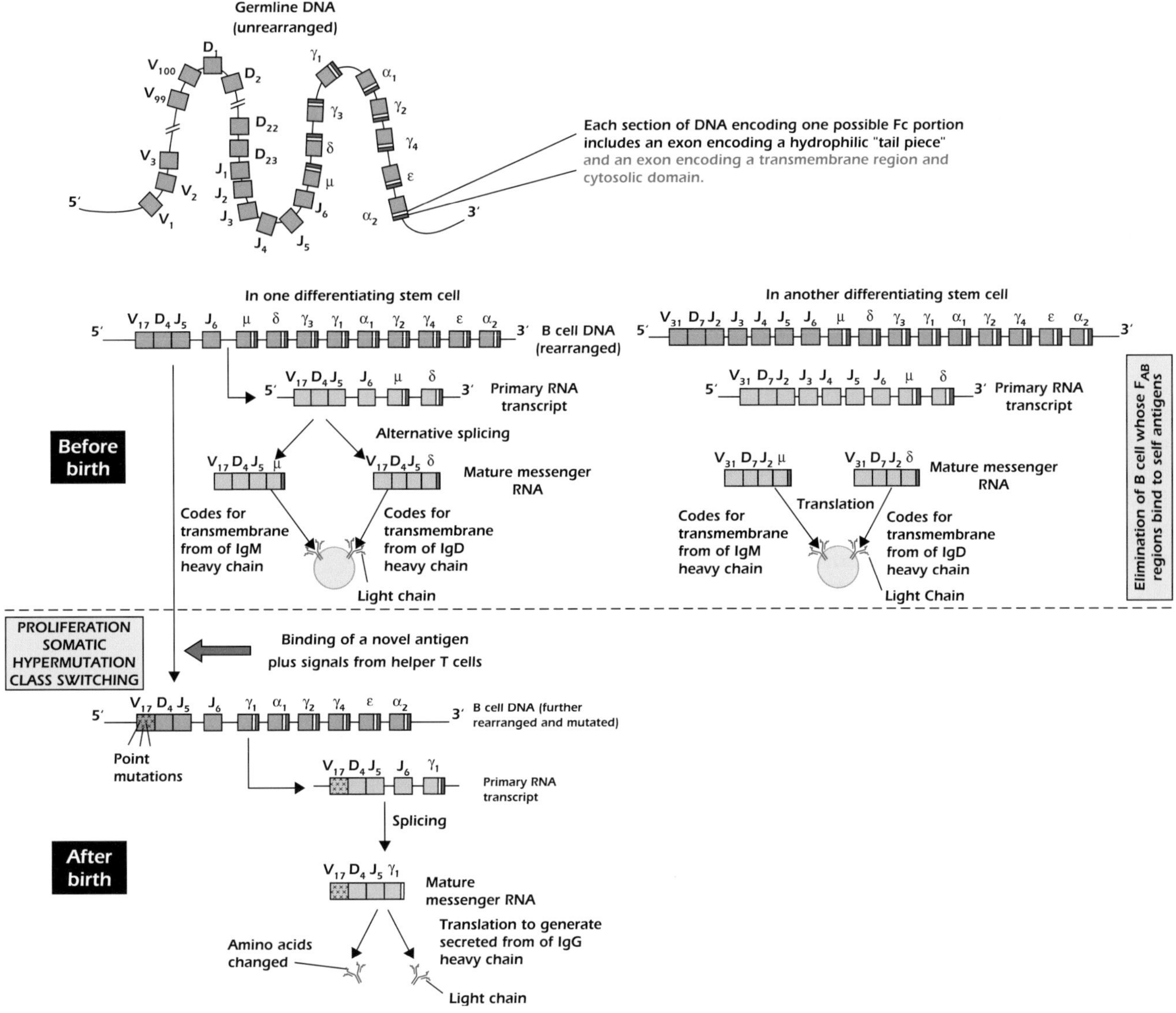

Figure 19.4. The heavy chain locus.

termination site 3′ of the δ section on the chromosome, so when the now rearranged heavy chain locus is transcribed the primary RNA transcript ends after the section coding for the D isoform of Fc.

RNA splicing now removes all J sections except the one adjacent to the D section. The transcript can be alternatively spliced to generate either one ending in μ, and therefore coding for the heavy chain of IgM, or one ending in δ, and therefore coding for the heavy chain of IgD. Alternative exons within the μ and δ regions code for transmembrane or secreted proteins. At its present stage of development the B cell generates transmembrane antibodies, with the antigen binding site facing out into the extracellular medium and a cytosolic tail that can signal to downstream processes.

Any combination of V, D and J sections is possible. For example, the B cell on the right of Figure 19.4 generates antibodies by using V_{31}, D_7 and J_2. Thus from the one heavy chain locus, $100 \times 23 \times 6 = 13{,}800$ possible different Fabs can be generated. A similar process of rearrangement generates the light chain. Simply by using the alternative lengths of DNA present in the inherited genome, B cells can generate about 300 different light chains. Since any one B cell expresses one of the 13,800 possible heavy chain Fabs, and one of the 300 possible light chains, there are over four million ($13{,}800 \times 300 = 4{,}140{,}000$) possible different antigen binding sites coded for by the genome. Within the protected environment of the womb, the embryo is unlikely to be exposed to pathogens or toxins. However, many of the generated antibodies bind endogenous proteins with high affinity. Such self-specific B cells are removed: the binding of antigen causes them either to die by apoptosis (page 308) or to enter a nonresponsive state known as anergy. The remaining live, responsive B cells are termed mature and reside in the lymph nodes and other lymphoid organs such as the spleen.

The majority of mature B cells never bind an antigen tightly enough to trigger a response. However, if a mature B cell is exposed to an antigen to which it binds avidly, a number of responses are triggered. The amplitude and exact nature of the responses are controlled by signals from other immune system cells, including cytokines (page 264). First, the stimulated B cell proliferates. From one B cell whose antibody is suited to respond to the new threat are made millions of B cells, all synthesizing the same antibody. Second, the B cell can undergo **class switching**. This involves a further cut-and-splice action on the chromosomal DNA. The μ and δ sections are cut out as are some of the alternatives that lie 3′ of them. The result is that transcription followed by translation will now produce not IgM and IgD but one of the other antibody isoforms. In the example shown in Figure 19.4, DNA rearrangement has removed the μ, δ and γ_3 sections and the B cell is now programmed to use the γ_1 section to synthesize the commonest antibody isoform, IgG. The third response to stimulation is the most remarkable. Cells appear in which the DNA sequence of the one V section remaining in their heavy chain locus, and in the V section of the light chain locus, differs from the inherited sequence. We have already met somatic mutation, but here it is occurring at more than 10,000 times the rate seen in other genes and chromosomes and is therefore called **somatic hypermutation**. The underlying cause of the mutation is the upregulation of an enzyme that deaminates cytosine in the V region to uracil. Base excision repair (page 74) normally replaces the cytosine, but the repair is not error-free and in the face of multiple deaminations a significant number of mutations build up. The one parental B cell therefore gives rise to a population of cells that generates a range of antibodies, all based on the original successful antibody but with an amino acid sequence that differs slightly from it. Those B cells whose antibody binds the antigen the most tightly are selected for, and proliferate further. In this remarkable process we see mutation, natural selection, and evolution within a single organism. During the course of the response to an invading pathogen or other threat, antibodies that bind the antigen tightly enough to be useful are modified and then selected, so that after a few days the antibodies being generated bind to the pathogen or toxin with even higher affinity than did the original, genomically encoded version. Both this process of fine tuning the antibody, and the earlier process of eliminating B cells that recognized self-antigens, were proposed in Frank Burnet's theory and described as "clonal selection," in that a selection process that removed some cells and stimulated the division of others allowed the appearance of clones of B cells, all identical, that synthesized antibody with high affinity for foreign antigens.

The last change that occurs as a result of antigen stimulation is the differentiation of **plasma cells**. In these cells, alternative splicing of the section of RNA encoding the Fc region generates antibody that is not membrane bound but rather is secreted into the extracellular medium. A large number of antigen-specific plasma cells are generated, each secreting antibodies that bind to and defeat the pathogen or other threat. Not all the stimulated B

IN DEPTH 19.1 MONOCLONAL ANTIBODIES

Antibodies made in domestic animals such as rabbits and horses have been used in research since the 1940s. Such antibodies are called **polyclonal** because even if generated in one animal they represent the output of many different B cell lineages, each secreting an antibody with a different Fab. The various antibodies will recognize different aspects of the protein or organism that was initially used to immunize the animal. Polyclonal antibodies are still very useful; almost all secondary antibodies for use in immunofluorescence (page 6) are polyclonal. However, polyclonal antibodies are by their nature variable, especially if one needs more than can be generated from one animal.

The invention of **monoclonal** antibodies by Georges Köhler and César Milstein in 1975 greatly increased the range of uses to which antibodies could be applied. An animal (usually a mouse) is immunized and generates memory B cells whose antibody binds tightly to the antigen of interest. The mouse is killed and B cells from its spleen fused with cancer cells that are capable of dividing indefinitely in cell culture. The fused cells inherit from the cancer cell the ability to divide indefinitely, and from the B cell the reorganized DNA encoding one antibody molecule. Individual clones of cells that develop from a successful fusion are grown up and tested for their ability to secrete antibody of the required type, affinity, and specificity. The product of these cells is a unique monoclonal antibody and can be produced in whatever quantities the market demands simply by expanding the culture of **hybridoma** cells. Köhler and Milstein received the 1984 Nobel Prize in Physiology and Medicine for this invention.

cells become plasma cells. Some remain as **memory B cells** expressing membrane-bound antibody. These wait for reappearance of the antigen; if it does reappear, it can be countered rapidly since these cells already contain mutated DNA encoding high-affinity antibodies. This is how vaccination works: we are injected with a killed or weakened pathogen, or proteins characteristic of a pathogen, and develop memory B cells encoding antibody specific to the protein(s). Now if the real pathogen attacks, the memory B cells proliferate and generate large numbers of plasma cells within one to three days.

T CELLS

The other large class of lymphocytes are T cells. T cells are so called because they mature in the thymus gland that lies behind the breastbone. Here self-reactive T cells are killed or suppressed, and the survivors leave to migrate to lymph nodes and other tissues. T cells fall into two groups, **CD8+** and **CD4+**. CD8+ T cells express a cell surface protein called CD8, while CD4+ cells express CD4. The functions of these two cell types are very different. CD8+ T cells kill cells of the body that are infected with viruses or other intracellular parasites. CD4+ T cells regulate the proliferation and maturation of B cells. All T cells express on their surface a protein called the **T cell receptor** (Fig. 19.5). This slightly confusing term refers to the antigen-binding receptor expressed by T cells. T cell receptors are similar in many ways to membrane-bound antibodies. Like antibodies each is the product of two lengths of DNA. The majority of T cell receptors comprise one α and one β chain, coded for by loci on human chromosomes 14 and 7 respectively. In the same way that many different antibodies are generated by rearrangement of the DNA within the heavy and light chain loci, so rearrangement of the DNA within the T cell receptor α and β loci generates clones of cells, each of which expresses a different receptor capable of binding a different antigen. However, unlike the situation with B cells, the generation of diversity stops there. Somatic hypermutation does not occur in T cells. The T cell receptor binds peptides presented by major histocompatability complex proteins on the surface of other cells (page 168).

The first place where T cells encounter peptides presented on MHC proteins is at their birthplace in the thymus. The resident cells of the thymus present peptides derived from endogenous proteins; any young T cells whose T cell receptor binds strongly to such MHC-peptide complexes die by apoptosis. This removes self-reactive T cells that would otherwise cause autoimmune disease.

The mature T cells migrate to lymph nodes where they are constantly exposed to peptides presented on **dendritic cells**. Dendritic cells are often described as professional antigen presenting cells because this is their principal function. They are found throughout the body, but are concentrated in areas such as the skin and gut where pathogen invasion is most likely. Proteins present in the extracellular medium are endocytosed, and the endosome fuses with a lysosome (Fig. 19.6). Lysosomal enzymes hydrolyze the ligand protein into short peptides. The vesicle then fuses with vesicles transporting newly synthesized major histocompatability complex protein (page 168) from the Golgi complex to the plasma membrane. The peptides bind to the major histocompatability complex protein and are therefore presented at the cell surface. Dendritic cells are constantly travelling back and forth between the peripheral location where they pick up and present antigens, and lymph nodes where they present their antigens to T cells. Any T cell whose T cell receptor binds strongly to a presented MHC-peptide complex is activated and generates calcium signals. The calcium-sensitive transcription factor NFAT (Nuclear Factor of Activated Thymocytes) is activated (Medical Relevance 10.1 on page 161) and turns on the transcription of a number of genes, one being the critical cytokine interleukin 2.

A positive feedback of cell-cell interactions and cytokine signaling now ensues. To simplify enormously, interleukin 2, through its action on its receptors (page 264) on the initially activated T cell and on its neighbors, triggers T cell proliferation. Those progeny cells that bind to the presenting dendritic cell get a further impetus to divide. The resulting family of active, antigen-specific T cells, which may number 100,000, leave the thymus and start patrolling the body. If they now encounter cells expressing the MHC-peptide complex to which they are

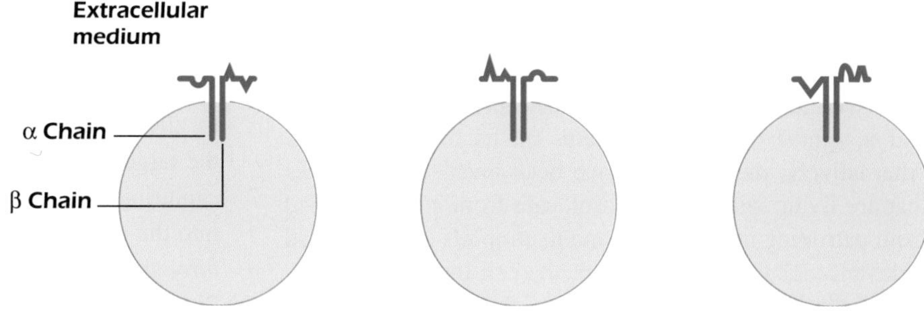

Figure 19.5. The T cell receptor.

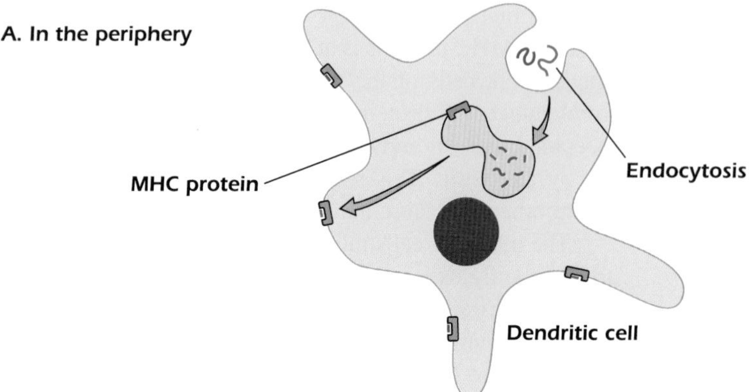

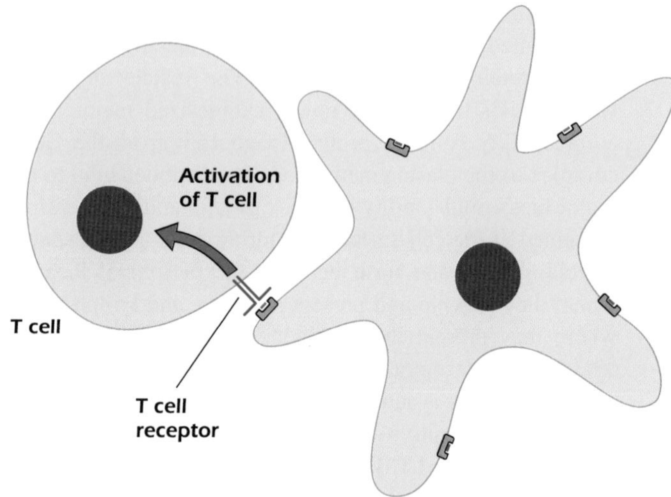

Figure 19.6. Dendritic cells present peptide fragments to T cells.

tuned, they will respond. However, two very different outcomes ensue, depending on the type of T cell and the identity of the antigen presenting cell.

The Action of CD8+ T Cells

As we described earlier (page 168) all cells of the body present on their cell surfaces a selection of peptide fragments generated by the hydrolysis of cytosolic proteins. The function of CD8+ T cells is to patrol the body and detect when novel peptides are presented on the major histocompatibility complex (Fig. 19.7). During their development in the thymus, T cells whose T cell receptors bind peptides derived from endogenous proteins are deleted or inactivated. Thus only novel peptides, not before generated in the cells of the organism, generate a response from CD8+ T cells. The most likely reason that a cell might be expressing novel peptides is that it has been infected with a virus and is synthesizing viral proteins on its own ribosomes. Alternatively, the cell may have been invaded by bacteria that are living within the cytosol, safe from antibodies and from patrolling macrophages and neutrophils (e.g., Medical Relevance 17.1 on page 291), or the cell may have undergone a somatic mutation causing it to synthesize a novel and potentially dangerous protein. When they bind to a target cell CD8+ T cells initiate apoptosis in the target cell—that is, they kill it. They use two mechanisms to achieve this lethal end.

First, CD8+ T cells express on their surface **Fas ligand** that binds to Fas on the surface of the infected cell. Fas can then activate caspase 8 and hence initiate apoptosis (page 309). Second, activated CD8+ cells release granzymes and perforin. Granzymes are a family of proteases that hydrolyze intracellular proteins. In particular, granzymes hydrolyze a bcl-2 family protein (page 309) called Bid. As synthesized at the ribosome, Bid is innocuous, but the fragment that results from hydrolysis, truncated or t-Bid, causes the release of cytochrome *c* from mitochondria which in turn activates caspase 9 (page 310), once again initiating apoptosis. The function of the perforin is to allow the granzymes to gain access to the cytosol of the target cell. The target cell endocytoses the granzymes, capturing them within the endosome, but perforin inserts into the endosome membrane and assembles into a channel through which the granzymes can escape into the cytosol. Since the function of CD8+ T cells is to kill infected or

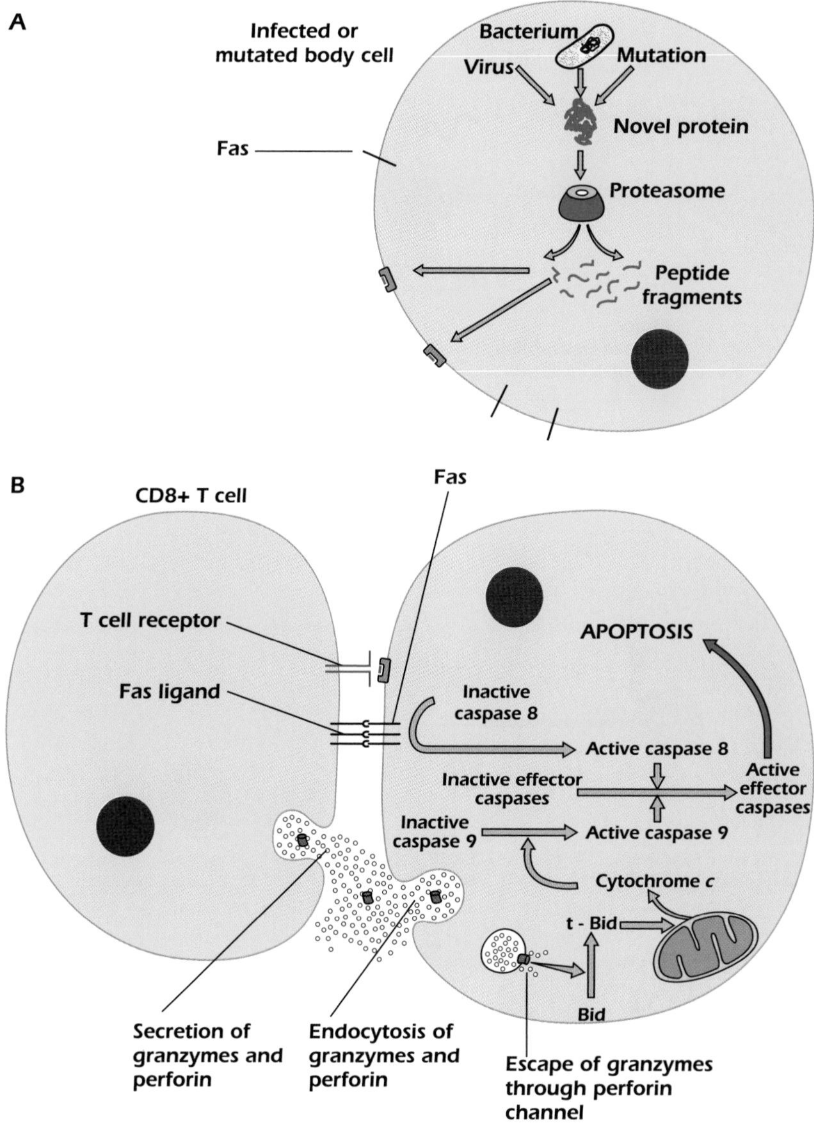

Figure 19.7. The action of CD8+ T cells.

mutant cells within the organism, they are also given the names **T killer** or **cytotoxic T cells**.

Since patrolling CD8+ cells attack and kill cells that are presenting viral peptides on their major histocompatability complex proteins, most viruses have mechanisms to turn off expression of major histocompatability complex proteins in the cells that they infect. To counter this stratagem, natural killer cells in mammals release granzymes and perforin onto any cell that they encounter that does not have major histocompatability complex protein on its surface.

The Action of CD4+ T Cells

The description on page 320 of how B cells are triggered to proliferate and secrete their antibody is too simple. If the situation were this simple, then any novel chemical that adhered to a membrane-bound antibody would activate that B cell. Yet the vast majority of novel chemicals that enter our bodies are better dealt with by the kidneys than by the immune system. Furthermore, since antibodies recognize the shape of molecules, an endogenous protein that appeared in a new tertiary structure would elicit an immune response. This does not happen; for example patients who catch Creutzfeldt-Jacob disease, and therefore begin to fold their PrP protein into an alternative shape (page 153), do not develop antibodies to PrP^{sc}. The explanation is that, as a general rule, a molecule will not activate a mature B cell unless chopping up that molecule generates peptides that a CD4+ T cell recognizes as novel. Thus to elicit antibody production a protein must not only be novel in shape, but must also be novel in amino acid sequence.

Figure 19.8 shows how this works. The basic principle is that in order to proliferate, the B cell needs signals originating from two cell surface proteins: its own specific antibody and a receptor called CD40. Figure 19.8A shows a

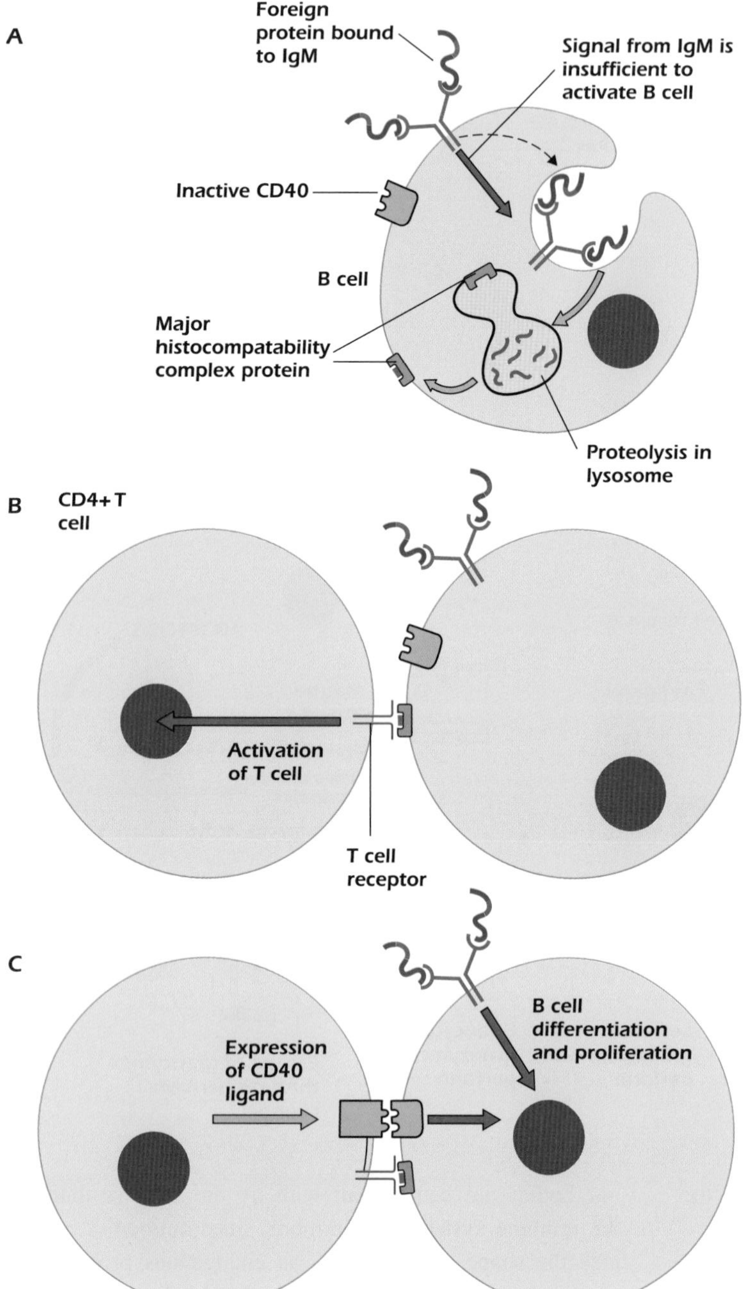

Figure 19.8. The action of CD4+ T cells.

B cell that has bound a protein on its membrane bound antibody. The antibody and its attached ligand are endocytosed and the endosome fuses with a lysosome. The resulting short peptides are recruited by newly synthesized major histocompatability complex protein and presented at the surface of the B cell. If the presented peptide is novel, then as shown in Figure 19.8B a T cell whose T cell receptor has a high affinity for that peptide will bind. Binding of the major histocompatability complex protein:peptide complex activates the T cell which allows the transcription factor NFAT (nuclear factor of activated T cells) to move to the nucleus and bind to the promoter region of various genes. Figure 19.8C shows one result of this activation: the T cell expresses a protein that acts as an activating ligand for the CD40 on the B cell. This protein is simply called CD40 ligand or CD40L. Looking now at the B cell, we see that the combination of signals from occupied antibodies and activated CD40 triggers proliferation and differentiation in the B cell. Thus B cells are only activated if two things are true: they bear an antibody that captures ligand and that ligand, when hydrolyzed, is recognized by a CD4+ T cell. Because CD4+ T cells assist B cells in mounting an antibody response to a novel protein, they are also known as **helper T cells**.

Medical Relevance 19.1 Therapeutic Monoclonal Antibodies

More than a century ago, the Nobel Prize–winning immunologist Paul Ehrlich predicted that antibodies would provide “magic bullets” to target and destroy human diseases. An excellent example of the fulfillment of Ehrich’s prediction is the anticancer drug Herceptin (also known as Trastuzumab). Herceptin is a monoclonal antibody that binds to the extracellular portion of the Her2 receptor (human epithelial growth factor receptor type 2). Normal cells with two copies of the *HER2* gene contain 20,000 to 50,000 molecules of Her2. These are part of the normal control system that tells cells when or when not to divide. In about 30% of breast cancers, the *HER2* gene becomes amplified (i.e., cells contain more than two copies of the gene). The level of the Her2 protein is elevated up to 100 times and the “divide” signal is correspondingly intensified. The resulting tumors can be identified in biopsies by immunocytochemistry, using a Her2 antibody or by FISH (fluorescence in situ hybridization) which lights up the extra copies of *HER2*. Herceptin blocks the rogue receptors in cancerous cells, shutting down the “divide” response and leading to the shrinkage of the tumor. Herceptin is particularly effective in combination with chemotherapy, for example, with the antimicrotubule drug Taxol (page 285). Herceptin has been in the news because its high cost (approximately $50,000 for a course of treatment) has restricted its availability in some countries.

Medical Relevance 19.2 Passive Immunity

Because antibodies can bind to and detoxify many animal toxins, stocks of IgG specific for various snake and scorpion venoms are kept in hospitals for injection into bite and sting victims. The protection afforded by the injection into a patient of antibodies created in another animal, or in cultures secreting monoclonal antibodies, is called passive immunity.

Another gene activated by NFAT is that coding for the cytokine interleukin 2 (page 321). Interleukin 2 secreted by CD4+ T cells is a powerful mitogen for CD4+ T cells, CD8+ T cells, and natural killer cells. Thus the number of these cells available for fighting the infection or foreign body increases dramatically. Blocking the activation of NFAT by immunosuppressant drugs (page 161) is necessary to prevent the rejection of organs following transplantation surgery.

AUTOIMMUNE DISEASE

The mechanisms that remove or inactivate self-reactive lymphocytes are not perfect. T or B cells that bind to endogenous molecules can survive the screening process and if they do, will attack the body’s own cells and cause inflammation and even cell death. A clear example is type 1 (juvenile onset) diabetes. In this disease the immune system mounts an attack on the β cells of the pancreas, whose job is to secrete insulin. In the early stages of the disease CD4+ and CD8+ T cells infiltrate the pancreatic tissue, while antibodies against β cells are found in the circulation. The result is death of the β cells through the processes we have described above. The CD8+ T cells attack the β cells by activating Fas and by releasing granzymes and perforin. Antibodies, which have been generated through the helper action of CD4+ T cells, opsonize the β cells and recruit complement. Once all the β cells are dead, the patient has an absolute requirement for injected insulin.

Autoimmune diseases are not normally present from birth. Rather, an infection, which in most individuals would have no long-lasting effect other than the production of

Answer to Thought Question: Evolution has generated an ever more complex arsenal of defenses and counterdefenses in viruses and animals. We describe in the chapter how many viruses, including cytomegalovirus, evade detection by CD8+ T cells by turning off the expression of major histocompatibility complex proteins in the infected cells. We then describe how mammals counter this viral strategy by programming natural killer cells to attack any host cell that is not expressing major histocompatibility complex protein at the cell surface.

Cytomegalovirus carries the evolution of defense and counterdefense one stage further by causing the infected cells to express a mutant major histocompatibility complex protein. Although this protein is not recognized by CD8+ T cells, it is close enough to the real thing to fool natural killer cells, which therefore leave the infected cells alone.

memory B cells, triggers the proliferation and activation of self-reactive lymphocytes. The current theory is that T and B cells whose receptors and antibodies recognize the pathogen become activated, and that on rare occasions these receptors and antibodies happen by accident to also recognize endogenous molecules.

SUMMARY

1. Defenses against infection comprise innate immunity, fully encoded by the inherited genome, and adaptive immunity that is tuned by experience. Adaptive immunity is found only in the vertebrates.
2. Most cells of the immune system fall into the myeloid or lymphoid lineages.
3. Myeloid cells, including neutrophils and macrophages, phagocytose foreign bodies and apoptosing cells.
4. Lymphoid cells comprise B and T cells plus natural killer cells. B and T cells modify their nuclear DNA to allow the generation of thousands of variants of antibodies and T cell receptors respectively. B and T cells are the only somatic cells that can rearrange their DNA.
5. B cells synthesize antibodies. Membrane-bound antibodies allow mature B cells to proliferate in response to ligands, called antigens, that bind tightly. Further differentiation generates plasma cells that produce large quantities of secreted antibodies.
6. Antibodies comprise an invariant Fc portion and a variable Fab portion. The Fab portion is encoded within long lengths of DNA called loci that are modified by DNA splicing and somatic hypermutation.
7. T cell receptors bind peptides presented on MHC proteins on both regular somatic cells and professional antigen presenting cells such as dendritic cells. DNA rearrangement and splicing of the T cell receptor loci generates individual T cells responsive to different antigens.
8. CD8+ T cells, also called cytotoxic T cells, kill somatic cells that present novel peptides.
9. CD4+ T cells, also called helper T cells, activate B cells if the peptide presented by MHC proteins on the B cell binds tightly to their own T cell receptor.
10. Although the immune system is a sophisticated and highly accurate differentiator of self and foreign, it makes mistakes. Autoimmune diseases result when the system attacks the body's own tissues.

FURTHER READING

Coico, R., and Sunshine, G. (2009) *Immunology, A Short Course*, 6th edition, Wiley-Blackwell, New Jersey.

Abbas, A., Lichtman, A. H., and Pillai, S. (2007) *Cellular and Molecular Immunology*, 6th edition, Elsevier Saunders, Edinburgh.

REVIEW QUESTIONS

19.1 Theme: Immune System Cells

A B cell
B CD4+ T cell
C CD8+ T cell
D dendritic cell
E macrophage
F natural killer cell
G neutrophil

From the above list of immune system cells, choose the cell that best fits each of the descriptions below.

1. A cell generated by a process that includes somatic hypermutation
2. A cell that attacks any body cell that does not express major histocompatability complex protein on its cell surface

3. A cell that attacks any body cell whose major histocompatability complex proteins are presenting novel peptides
4. A cell that makes antibodies
5. A cell that migrates from the tissues to lymph nodes, where it presents antigens to the lymphocytes
6. A cell that upon stimulation expresses CD40L, a protein that in turn activates B cells
7. An important phagocytic cell of body tissue which differentiates from blood monocytes
8. The commonest phagocytic cell of the blood

19.2 Theme: The Antibody Heavy Chain Locus

A V
B β
C γ
D D
E J
F L

From the above list of suggested components, choose the component of the antibody heavy chain that best fits each of the following descriptions.

1. A component that experiences somatic hypermutation
2. A component that includes a sequence encoding a transmembrane domain
3. Although chromosome 14 in every other cell of the body contains 6 of these arranged sequentially, the DNA of mature B cells contains a linear sequence comprising any number from 1 to 6
4. Genomic DNA contains more than 20, but less than 100, slightly different versions of this section arranged sequentially along the chromosome
5. The first protein coding region that would be encountered when reading along the locus from the 5′ end

19.3 Theme: T Cells and their Interaction with Other Cells

A CD4
B CD40
C CD8
D cytochrome *c*
E Fas
F granzyme
G interleukin 2
H major histocompatibility complex protein
I NFAT
J proteasome
K T cell receptor

From the above list of proteins, select the protein that best fits each of the descriptions below.

1. A death domain receptor that, when activated, causes the proteolysis and activation of caspase 8 and hence cell death by apoptosis
2. A mitogen secreted by CD4+ T cells
3. A protease secreted by CD8+ T cells
4. A protein complex that cuts cytosolic proteins into short lengths of peptide
5. A protein found in the cytosol of resting T cells which translocates to the nucleus upon cell stimulation
6. A protein whose expression allows scientists and clinicians to recognize the cell as a T killer cell, that is, a member of that population of T cells that attacks body cells infected by viruses or other pathogens
7. A receptor expressed by mature B cells that, when activated, causes differentiation and proliferation
8. A transmembrane protein that can, if the match is correct, bind to a major histocompatability complex protein that is presenting a foreign peptide
9. A transmembrane protein that presents short lengths of peptide at the cell surface

THOUGHT QUESTION

As a general rule, the genome of viruses is tiny, with the virus relying on proteins encoded by genes of the host cell for its replication and release. Yet the genome of the very common human cytomegalovirus contains a gene for a mutant form of a major histocompatibility complex protein. The mutant protein is incapable of presenting antigen to host T cells. Why should including such a gene be selected for in the evolution of the virus?

20

CASE STUDY: CYSTIC FIBROSIS

In the final chapter of the book we describe the biochemical basis of the disease cystic fibrosis (CF), the search for the *CF* gene, the prospects for CF therapy, and prenatal diagnosis. We have selected this disease as a case study to show how the combined efforts of biochemistry, genetics, molecular and cell biology, and physiology were needed to find the cause of CF and the function of the protein encoded by the *CF* gene. The basic principles of many of the techniques used in achieving this knowledge are described in Chapters 1 through 19.

 CYSTIC FIBROSIS IS A SEVERE GENETIC DISEASE

Among Caucasians (white, non-Jewish) about 1 child in 2500 is born with CF. Inheritance is simple: when both parents are carriers, there is a one-in-four chance that a pregnancy will result in a child with the disease. The disease is a distressing one. Most of its symptoms arise from faults in the way the body moves liquids, leading in particular to a buildup of inadequately hydrated, sticky mucus in various parts of the body. In the lungs this leads to difficulty in breathing, a persistent cough, and a greatly increased risk of infection. A bacterium, *Pseudomonas aerigunosa*, thrives in the lung mucus and is particularly recalcitrant to antibiotic treatment. The pancreas—which provides a digestive secretion that flows to the intestine—is also affected and may be badly damaged (which explains the disease's full name, cystic fibrosis of the pancreas). Often, there are digestive problems because the damaged pancreas is not producing enzymes. The reproductive system is also harmed and most males who survive to adolescence are infertile. Even though there are so many varied symptoms, the simple Mendelian pattern of inheritance led researchers to believe that the cause was an abnormality or absence of a single protein.

Until recently babies with CF did not survive to their first birthday. Today, the life expectancy is close to 40 years. This remarkable improvement is due to intensive therapy designed to reverse individual symptoms. Digestive enzymes are taken by mouth to replace those proteins the pancreas fails to produce. Physiotherapy—helping patients to cough up the mucus in their lungs by slapping their backs—reduces the severity of the lung disease. Nevertheless CF patients still die tragically young.

According to the Almanac of Children's Songs and Games from Switzerland, "the child will soon die whose brow tastes salty when kissed." Cystic fibrosis has long been recognized as a fatal disease of children. The link between a salty brow and a disease known for its effects on the pancreas and lungs was not, however, made until 1951. During a heat wave in New York that summer, Paul Di Sant' Agnese and coworkers noticed that babies with cystic fibrosis were more likely than others to suffer from heat prostration. This prompted them to test the sweat, and they found that it contained much more salt than the sweat of unaffected babies.

Cell Biology: A Short Course, Third Edition. Stephen R. Bolsover, Jeremy S. Hyams, Elizabeth A. Shephard and Hugh A. White.

THE FUNDAMENTAL LESION IN CYSTIC FIBROSIS LIES IN CHLORIDE TRANSPORT

Sweat glands have two regions that perform different jobs (Fig. 20.1). The secretory region deep in the skin produces a fluid that has an ionic composition similar to that of extracellular medium; that is, it is rich in sodium and chloride. If the sweat glands were simply to pour this liquid onto the surface of the skin, they would do a good job of cooling but the body would lose large amounts of sodium and chloride. A reabsorptive region closer to the surface removes ions from the sweat, leaving the water (plus a small amount of sodium chloride) to flow out of the end of the gland. CF patients produce normal volumes of sweat, but this contains lots of sodium and chloride, implying that the secretory region is working fine but that the reabsorptive region has failed. The pathways by which sodium and chloride ions are removed are distinct. Which one has failed in CF? A simple electrical test gave the answer. A normal sweat gland has a small voltage difference across the gland epithelium (Fig. 20.2). In CF patients the inside of the gland is much more negative than usual. This result tells us at once that it is the chloride transport (and not sodium transport) that has failed. In CF sweat glands the reabsorptive region has the transport systems to move sodium, but the chloride ions remain behind in the sweat, giving it a negative voltage. The sodium transport system cannot continue to move sodium ions out of the sweat in the face of this larger electrical force pulling them in, so sodium movement stops too, and sodium chloride is lost in the sweat, which then tastes salty. Every one of the symptoms suffered by CF patients is caused by a failure of chloride transport.

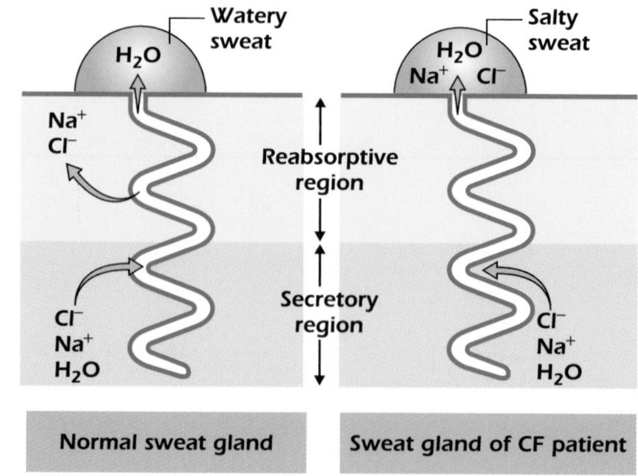

Figure 20.1. Sodium, chloride and water transport in the sweat gland.

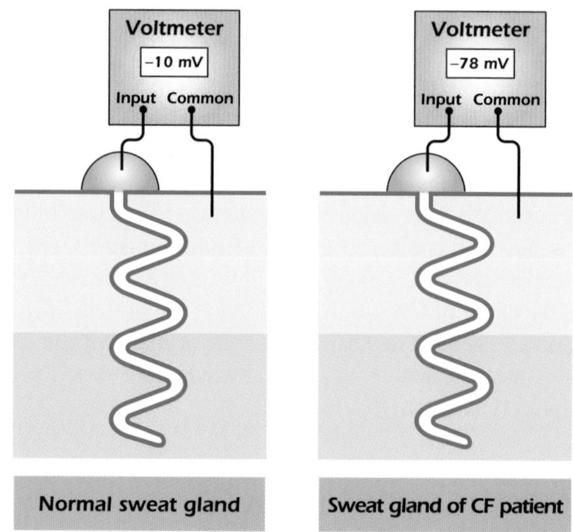

Figure 20.2. The transepithelial voltage of CF sweat glands is much more negative.

HOMING IN ON THE CF GENE

The first family studies were carried out more than 50 years ago, and family pedigrees showing CF as a classic case of recessive inheritance were published in 1946. However, there was no obvious way to identify the millions of people who carry a single copy of the defective CF gene. There were many attempts to find a test for CF. For a time it seemed that a simple stain might do the job but this was abandoned. Other tests were more eccentric. For example, there was a claim that an extract from the blood of CF patients, and perhaps even from that of their parents, who must be carriers, slowed the beating of the cilia of oysters.

Most other claims that particular gene products were peculiar to CF carriers also failed to stand up or proved to be symptoms rather than cause. For instance, trypsinogen, a precursor of the digestive enzyme trypsin, is found in the fluid bathing CF fetuses and not in others—but this proves to be because the pancreas is already dying at this time. In the 1980s there was hope that this result could be used to diagnose affected pregnancies but the approach was quickly overtaken by developments in studying the gene itself. As for most inborn diseases, the big problem was that nobody had any real idea what the faulty gene did; looking for a defect in an unknown protein is a task that requires hard work and good ideas. After much time-consuming work, in 1985 the research group led by Lap-Chee Tsui published data indicating that the gene for CF lay on chromosome 7. Researchers then used linkage measurements (page 302) to try to pinpoint the region of chromosome 7 containing the CF gene. An international scientific collaboration analyzed DNA from 200 families. The results indicated that the CF gene must lie between two marker regions on chromosome 7, named met and D7S8. The distance between the markers was 2 million base pairs. It was now possible to think about isolating the CF gene.

CLONING THE GENE FOR CF

Sheer hard work won the day when it came to cloning the CF gene. Scientists knew that the gene was on chromosome 7, somewhere between the met and D7S8 markers. Their strategy was to "walk" along the chromosome (page 117), away from each marker, and hopefully approach the CF gene from either side. In the 1980s, the only genomic vectors available were cosmids (insert size 40,000 bp) and those based on the bacteriophage lambda (insert size 20,000 bp) (page 106). So each genomic clone covered only a tiny part of chromosome 7. Clever tricks had to be devised to speed up the process and finally a series of cosmid clones, thought to possibly contain regions of the CF gene, were isolated. But how could they tell which clones were the right ones?

Since CF affects so many organs, the gene is likely to be a critical one conserved through mammalian evolution. A zoo blot (page 112) was therefore used in the hope of identifying the gene. In turn, each of the human cosmid clones was radiolabeled and used to probe a zoo blot. Researchers found three clones that hybridized to the DNA of all the test mammals. The next step was to check whether these clones contained sequences that code for mRNAs expressed in normal tissues that are affected by CF. Each clone was used to probe a northern blot (page 113) of sweat gland mRNA. One sequence passed the test—with luck, it would contain part of the CF gene. With hindsight we now know that the scientists' luck had almost run out: the one positive sequence did contain part of the gene, but only a tiny part—113 bp, less than 1% of the total gene.

Things moved swiftly from then on. The 113 bp section was used to screen cDNA libraries (page 102) made from sweat gland mRNA. This gave scientists cDNA for the entire *CF* mRNA, which could then be sequenced. Meanwhile, the cDNA was used to screen a genomic library to find the rest of the *CF* gene. The gene is 220,000 bp long and has 24 exons.

THE *CFTR* GENE CODES FOR A CHLORIDE ION CHANNEL

Once CF cDNA had been sequenced, it could be read, revealing that *CF* codes for a protein of 1480 amino acids. Although we knew that CF was caused by a defect in chloride ion transport, it was not immediately clear that the product of the CF gene was itself a chloride channel. Although hydropathy plots (In Depth 9.1 on page 140) indicated that the CF protein was a transmembrane protein, investigators had to consider the possibility that it was, for example, a cell surface receptor whose activation triggered expression of the proteins mediating chloride transport. The protein was therefore given the catch-all name "cystic fibrosis transmembrane regulator" (CFTR) and its gene was named *CFTR*. However, for once things were simple: CFTR is indeed itself the chloride ion channel in the plasma membrane (Fig. 20.3). We know this because if purified CFTR is introduced into lipid bilayers, chloride currents can be measured. In resting cells the channel mouth is blocked by a plug called the regulatory domain. When the regulatory domain is phosphorylated by cAMP-dependent protein kinase (page 225), the plug leaves the channel, allowing chloride ions to flow through.

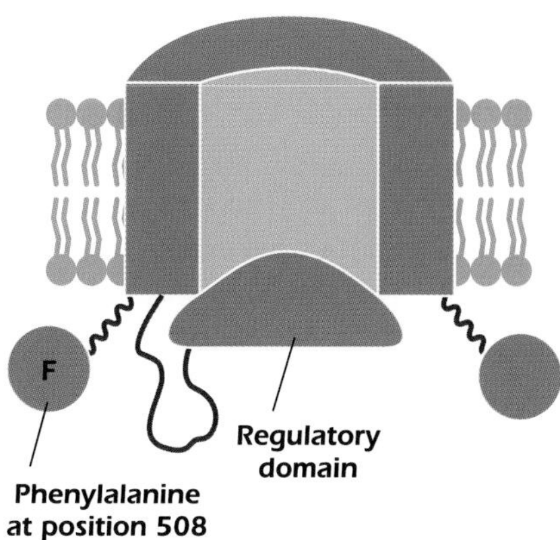

Figure 20.3. The CFTR protein forms a chloride ion channel in the plasma membrane of epithelial cells.

NOVEL THERAPIES FOR CF

It is exciting to think that we may one day treat a genetic disease by introducing a normal gene into the cells of patients. However, gene therapy is a difficult procedure and one that is not without risk to the patient. To date, it has not provided the wonder cure that was hoped for.

In 1992 the U.S. National Institutes of Health granted permission for a *CFTR* gene therapy trial. Because viruses are so good at infecting our cells, they are useful tools for transferring DNA into somatic cells. Viral vectors for gene therapy have been engineered to remove their harmful genes. The gene to be transferred into the patient is inserted into the modified viral genome. A modified adenovirus vector containing the *CFTR* cDNA was introduced into the lung cells of four patients. Each of the patients made CFTR protein, but only for a short time. Some patients showed adverse reactions to the therapy, probably due to the adenovirus itself. Subsequent trials have delivered an adenovirus containing the *CFTR* cDNA to the nasal epithelium of CF patients. However, adverse reactions to the treatment again developed.

Attempts are however being made to find alternatives to gene therapy. The goal is to improve the amount and function of the protein made from the patient's own genes. Even if the protein does not do its job very well, a little is better than no protein at all. The strategy selected for this type of treatment depends on the mutation that causes CF in the affected person. There are five classes of mutation that can cause CF. Class I, II, and III mutations cause the classic symptoms of CF and eventually lead to both lung and pancreas damage. Class I mutations cause the production of truncated, nonfunctional proteins. Some such mutations are deletions in the gene, so that the DNA coding for large parts of the protein is missing. However, some class I mutations are simple nonsense mutations, in which a base change in an amino acid codon has created one of the three stop codons UAA, UAG or UGA (page 63). Translation stops at these codons, generating a truncated protein. In Medical Relevance 8.1 on page 131 we described how antibiotics that cause the ribosome to read through a stop codon are being used to treat lysosomal storage disease. The same strategy is being used to treat CF patients. However, long-term use of these antibiotics can cause health risks. Recently, the chemical PTC124 has been shown to cause the ribosome to read through a premature stop codon, but not the normal stop codon of *CFTR*. This type of therapy holds promise for the treatment of CF, and clinical trials are now underway.

IN DEPTH 20.1 LIPID BILAYER VOLTAGE CLAMP

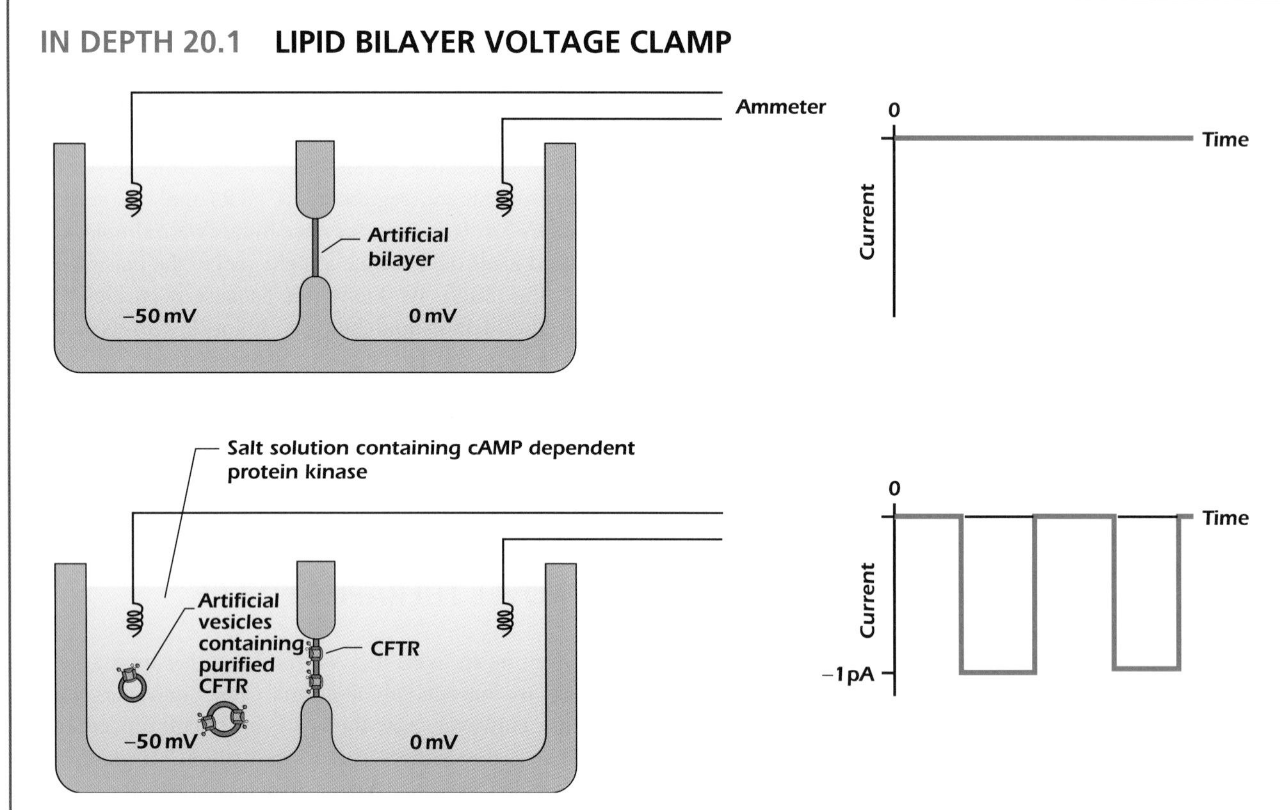

In 1992 Christine Bear and coworkers published a definitive demonstration that the CFTR is a chloride ion channel. They used a technique called the lipid bilayer voltage clamp, which is illustrated in the accompanying figure. A lipid bilayer is constructed from phospholipid and used to plug a hole between two chambers filled with salt solution. Since a lipid bilayer is a barrier to ion movement, no current is recorded when a voltage is imposed across the bilayer. However, when artificial vesicles containing integral membrane proteins are added to one bath, the vesicles fuse with the artificial membrane so that the proteins now span the barrier bilayer. When Bear and her coworkers did this with purified CFTR, they recorded currents with the direction and amplitude expected of a chloride-selective channel. In particular, no current flowed when the voltage difference across the membrane was equal to the calculated chloride equilibrium voltage (page 234). No other integral membrane proteins were present in this experiment, so the CFTR must itself be a chloride channel.

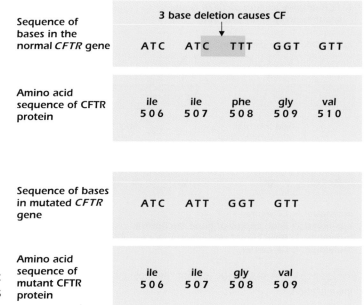

3 base deletion causes CF

Sequence of bases in the normal *CFTR* gene	ATC	ATC	TTT	GGT	GTT
Amino acid sequence of CFTR protein	ile 506	ile 507	phe 508	gly 509	val 510

Sequence of bases in mutated *CFTR* gene	ATC	ATT	GGT	GTT
Amino acid sequence of mutant CFTR protein	ile 506	ile 507	gly 508	val 509

Figure 20.4. The mutation seen in 70% of CF patients: a deletion of three nucleotides from the CFTR gene has removed a phenylalanine at position 508.

Class II mutations generate proteins with a slightly different amino acid sequence that nevertheless can function as chloride channels, albeit often at lower efficiency, but which never arrive at the plasma membrane. The changed primary structure disrupts folding and the cell detects the misfolded protein and sends it to the proteasome (page 133). This class of mutations includes the most common CF mutation, F508del (Fig. 20.4). A number of chemicals are being tested in cell systems to prevent the misfolded protein from being degraded.

In contrast, class III and IV mutations generate proteins that are delivered to the plasma membrane but which do not function correctly. Class III channels do not open when needed (for example because they lack the serine residues phosphorylated by protein kinase A), while the chloride permeability of a phosphorylated class IV mutant channel is reduced or nil. A number of chemicals are also being tested to see if they can help to restore the regulation of such proteins.

Class V mutations affect the transcription of the *CFTR* gene. Less CFTR than normal is made, but what is made is normal. In these patients the symptoms are less severe.

DIAGNOSTIC TESTS FOR CF

Because 70% of CF patients carry the same mutation, the deletion of phe at position 508, a diagnostic test was developed. The test, based on PCR (page 116), requires a few cells from the inside of the mouth and two DNA primers that will amplify a stretch of DNA that will include

Medical Relevance 20.1 Gene Therapy for Leber Congenital Amaurosis

Massive effort has gone into trying to find more effective treatments for cystic fibrosis because it is the most common severe genetic disease. However, genetic lesions that affect individual organs may well be more amenable to genetic intervention. Among these, the eye has received considerable attention, in part because the gene vectors can be injected directly to the site where they are needed, and in part because even a partial rescue of sight gives a dramatic improvement in patients' quality of life.

RPE65 is a retinal enzyme that makes 11-*cis*-retinal, the prosthetic group of rhodopsin. Babies with defective RPE65 lose all vision within the first year of life in a condition known as Leber congenital amaurosis. Recently three groups, at University College London, University of Pennsylvania, and University of Florida, carried out preliminary trials in which a vector encoding RPE65 was injected into the eye of human patients. Although these were very preliminary proof of concept tests, improvements in retinal sensitivity were seen in all three studies and one subject showed a dramatic recovery of vision. Further, more extensive, tests are planned. This example shows that gene therapy does indeed hold out the hope of treatment of severe genetic disorders, although the pace at which our basic knowledge is resulting in clinical gains is painfully slow.

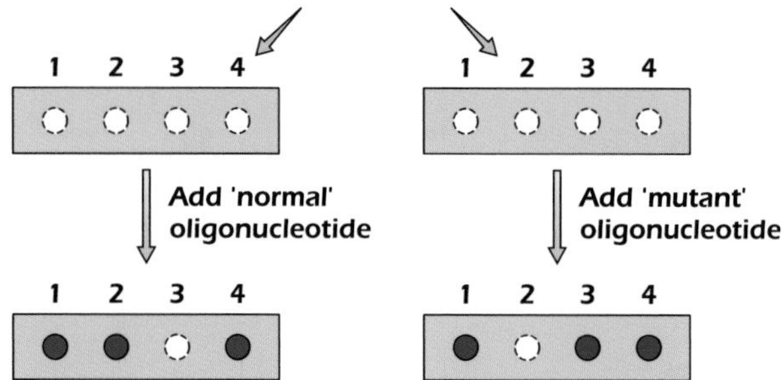
1. PCR amplify section of CFTR gene spanning codon for phe 508
2. Dot DNA onto membrane filters
1 2 3 4
1 2 3 4
Add 'normal' oligonucleotide
Add 'mutant' oligonucleotide
1 2 3 4
1 2 3 4
Individual 1: CF carrier
Individual 2: Both copies of gene are normal
Individual 3: Both copies of gene are mutant: CF patient
Individual 4: CF carrier

Figure 20.5. A diagnostic test for the most common CFTR gene mutation.

the region encoding amino acid 508 (Fig. 20.5). The PCR products are dotted onto two membrane filters. One membrane is incubated with an oligonucleotide containing the sequence encoding phe508 and the other is incubated with an oligonucleotide lacking the codon for phe508. By comparing the patterns of hybridization, carriers of CF can be detected. Other mutations that cause CF can also be tested for. However, because there are so many *CFTR* mutations, many families may require a new predictive test to be designed to match their version of the defective gene.

Prenatal Implantation Diagnosis for CF

In 1992 the first preimplantation diagnosis for CF was carried out. Three sets of parents took part in the trial. Each parent was a carrier for the phe508 mutation. Oocytes were removed from the women and fertilized with sperm from their partners. Success was reported for only one couple. Six of their oocytes developed into fertilized embryos. A single cell was tested from each embryo, using PCR, for the presence of the defective *CFTR* gene. Five embryos were characterized. Two embryos each had two copies of the normal *CFTR* gene, one on each of their chromosome 7s. Two embryos each had two copies of the phe508 deletion. One embryo tested positive for both the normal gene and the phe508 deletion and was therefore a carrier. One normal/normal embryo and the carrier embryo were implanted into the mother. A baby girl was born. Her DNA showed that both copies of her *CFTR* gene were normal. Since this initial success many more babies have been born following preimplantation diagnosis of a selected embryo. This type of treatment is now offered by a number of clinics worldwide. In 2006, 22 healthy babies were born in Europe to CF carrier parents after preimplantation diagnosis. A less expensive approach is antenatal screening during early conventional pregnancy followed by advice and parental choice on termination. In Edinburgh, UK, CF births have been reduced by two-thirds using this method.

THE FUTURE

Gene therapy strategies for CF have progressed, but much needs to be improved before this type of treatment can offer real hope for patients. The best gene therapy can offer is an alleviation of some of the symptoms. A cure for CF does not seem a likely prospect. Detection of CF carriers and prenatal diagnosis together with genetic counseling informs parents of the risks and gives them the choice of terminating an affected pregnancy.

Answer to Thought Questions: 1. It is in the Golgi apparatus that additional oligosaccharides are added to proteins destined for secretion (page 166). Sugar residues, being rich in hydroxyl groups, are strongly hydrophilic, so that a heavily glycosylated protein adsorbs lots of water and forms a hydrated gel. The proteins that form airway mucus are extremely heavily glycosylated, 70% of their total mass being carbohydrate. They therefore form the thin, watery gel that lubricates the airways and other internal spaces of the body. Barasch and coworkers suggested that an abnormal pH in the Golgi lumen, resulting from a failure to transport Cl^- as the counter ion to H^+ (in the same way that Na^+ transport fails in the sweat gland because Cl^- cannot move to neutralize the positive charge on Na^+) impairs the glycosylation of mucus proteins. The question is difficult to study because the lungs of even symptom-free CF patients contain many more bacteria than those of a non-CF subject, and bacterial enzymes digest polysaccharides. It is difficult to differentiate the primary effects of the mutation from the secondary effects resulting from bacterial invasion.

2. Northern blot hybridization (page 113) is a technique that enables us to analyze if a particular mRNA is made in a cell. We need mRNA and a gene probe. In this case mRNA was isolated from sweat gland cells and separated on an agarose gel. The RNA was transferred from the gel to a membrane. The 113 bp piece of DNA was used to make the gene probe, a radioactive copy of itself. The aim is to show that the probe is complementary in sequence to an mRNA produced by the sweat gland. If it is then a black band will show up on the autoradiogram. Even though the CFTR mRNA turned out to be much longer than 133 nucleotides it does not matter. The mRNA will hybridize (H-bond) to the 133 nucleotides in the radioactive DNA probe that are complementary in sequence to it.

SUMMARY

1. Cystic fibrosis is the most common serious single-gene inherited disease in the Western world. Many organs are affected. Sticky mucus builds up in the reproductive tract and in the lungs. The pancreas is always affected and usually fails completely.
2. Electrical measurements show that the basic problem is in chloride transport.
3. CF is a classic recessive disorder. The gene is on chromosome 7.
4. A combination of hard work and novel techniques helped isolate a small part of the *CFTR* gene. This sequence helped to isolate the normal *CFTR* cDNA from a sweat gland clone library.
5. From the cDNA it was possible to identify the gene and infer its amino acid sequence in both normal and mutated forms. The gene codes for a chloride channel protein called CFTR.
6. Hundreds of different *CFTR* mutations have now been found. Tests have been devised for prenatal diagnosis and to detect carriers of the commonest mutations.
7. The focus of current research is to generate therapies geared to the specific genetic lesion. For the most common genetic error, F508del, this aims to block the mechanisms that normally prevent misfolded proteins from being trafficked to the plasma membrane.

FURTHER READING

Bear, C. E., Li, C. H., Kartner, N., et al. (1992) Purification and functional reconstitution of the cystic fibrosis transmembrane conductance regulator (CFTR). *Cell* **68**, 809–818.

Bragonzi, A., and Conese, M. 2002. Non-viral approach toward gene therapy of cystic fibrosis lung disease. *Cuff. Gene Therapy* **2**, 295–305.

Handyside, A. H., Lesko, J. G., Tarin, J. J., et al. 1992. Birth of a normal girl after in vitro fertilization and preimplantation diagnostic testing for cystic fibrosis. *N. Engl. J Med*. **327**, 905–909.

Kerem, B., Rommens, J. M., Buchanan, J. A., et al. (1989) Identification of the cystic fibrosis gene: genetic analysis. *Science* **245**, 1073–1080.

Pearson, H (2009) Human genetics: One gene, twenty years. *Nature* **460**, 164–169.

Rich, D. P., Anderson, M. P., Gregory, R. J., et al. (1990) Expression of cystic fibrosis trans-membrane conductance regulator corrects defective chloride channel regulation in cystic fibrosis airway epithelial cells. *Nature* **347**, 358–363.

Riordan, J. R., Rommens, J. M., Kerem, B., et al. (1989) Identification of the cystic fibrosis gene: Cloning and characterization of complementary DNA. *Science* **245**, 1066–1073.

Welsh, M. J., and Smith, A. E. (1995) Cystic fibrosis. *Sci. Am*. **273**, 52–59.

REVIEW QUESTIONS

20.1 Theme: CFTR Mutations

A That subset of class I mutations in which a deletion in the gene, eliminating the DNA coding for a large part of the protein, means that a completely nonfunctional protein is generated.

B That subset of class I mutations in which a single base change within an exon results in a truncated protein that is completely nonfunctional.

C Class II mutations, that is, mutations that generate proteins with a slightly different amino acid sequence that nevertheless can function as chloride channels, albeit often at lower efficiency, but which never arrive at the plasma membrane because the cell detects a misfolding of the protein and targets it for destruction.

D That subset of class III mutations in which a full-length protein is delivered to the plasma membrane but does not open because it lacks the serine residues phosphorylated by protein kinase A.

E Class IV mutations in which the architecture of the channel is aberrant, preventing movement of chloride ions even when the protein is in what would normally be the open state.

F Class V mutations, that is, mutations in which less CFTR than normal is made, but what is made is normal.

From the above list of CFTR mutations, select the one that best fits each of the descriptions below.

1. This condition is likely to be a nonsense mutation and therefore may be treatable by drugs that cause the ribosome to read through a stop codon.
2. This condition is likely to be caused by a mutation in a promoter or enhancer region of the *CFTR* gene.
3. This condition may in the future be treatable by drugs that favor the open configuration of the CFTR channel.
4. This condition may in the future be treatable by drugs that inhibit or modify the function of the proteasome.

20.2 Theme: The CFTR as a Permeability Pathway

A The resting membrane voltage would become more negative and the electrical resistance of the membrane would increase

B The resting membrane voltage would become more negative and the electrical resistance of the membrane would become smaller

C The resting membrane voltage would become more negative but the electrical resistance of the membrane would not change

D The resting membrane voltage would become more positive (depolarize) and the electrical resistance of the membrane would increase

E The resting membrane voltage would become more positive (depolarize) but the electrical resistance of the membrane would not change

F The resting membrane voltage would become more positive (depolarize) and the electrical resistance of the membrane would become smaller

G The resting membrane voltage would not change but the electrical resistance of the membrane would increase

H The resting membrane voltage would not change but the electrical resistance of the membrane would become smaller

I The resting membrane voltage would not change, and the electrical resistance of the membrane would not change

A nerve cell is transfected with a vector encoding CFTR. A day later it is impaled with a micropipette to record the membrane voltage (see panel A, page 230) and occasional current pulses are passed to allow the electrical resistance of the plasma membrane to be measured. (This is a technique known as current clamp. Measurement of electrical properties using a micropipette filled with potassium gluconate is a technique used when the experimenter wishes to cause as little change as possible to the ionic composition of the cytosol.) Which of the possible outcomes listed above best describes what would be observed when the cell was treated with an agent, such as a membrane permeable cAMP analog, that activates protein kinase A?

20.3 Theme: A Medly of Sentences about CF

A The more negative transepithelial voltage of cystic fibrosis patients compared to controls indicated that . . .

B A zoo blot is useful to . . .

C Because CF is recessive, the disease must . . .

D CFTR protein was definitively proven to be a chloride channel by . . .

E Gene therapy for CF is very difficult because . . .

F Prenatal preimplantation diagnosis allows . . .

G The severity of CF varies because . . .

H Reading through a stop codon may help some CF patients because . . .

For each of the start phrases in the list above, there is a phrase in the list below that can be combined with it to create a single true statement. Match the appropriate sentence start with the sentence ending.

1. ... be inherited from both parents.
2. ... some of the many CF mutations have a mutation that introduces a premature stop. The truncated protein does not function properly.
3. ... identify genes that have been conserved in several species. Such genes are likely to code for important proteins.
4. ... a lipid bilayer voltage clamp experiment. CFTR protein was inserted into a lipid bilayer and chloride ions were shown to move through the channel.
5. ... the healthy gene needs to be inserted into a huge number of cells in the lungs and these cells are relatively short-lived. In addition the modified virus vectors used seem to cause side effects.

THOUGHT QUESTIONS

1. A hypothesis put forward by Barasch and coworkers in 1991 proposed that the mucus abnormalities seen in CF are in part the result of an unusual pH in the Golgi complex. What processes that occur in the Golgi complex might affect the ability of mucus proteins to form a hydrated gel?
2. The first piece of the *CFTR* gene that scientists isolated was just 113 bp in length. We now know the *CFTR* gene is much longer, 200,000 bp. Why was it possible to use the 113 bp piece of DNA to show that it corresponded to a gene that was transcribed in sweat gland cells?

APPENDIX

Channels and Carriers

All channels and carriers described in the book are listed here for reference.

CHANNELS

Name	Page for details	Location	Selective for	Opened by	Comments
Large Channels					
Connexon	46	Plasma membrane of many cells	Any solute of Mr < 1000	Contact with second connexon	Allows solutes to pass from cytosol to cytosol
Porin	194	Outer mitochondrial membrane	Any solute of Mr< 10,000	Probably spends a large fraction of time open under most circumstances	Stress can cause switch to a larger form that allows cytochrome *c* to pass
Membrane attack complex	317	Assembled from complement in plasma membrane of target cells	Any solute of Mr < ~50,000	Once assembled, probably remains open all the time	Allows critical solutes to leak out and sodium to leak in; kills the cell
Perforin	322	Released by CD8+ T cells and inserted into endosomal membrane of target cells	Any solute of Mr < ~50,000	Once assembled, probably remains open all the time	Allows granzyme to enter the target cell
BAX	309	Outer mitochondrial membrane of apoptotic cells	Any solute of Mr < ~2,000,000	Can assemble into a channel if dephosphorylated. Once assembled, probably remains open all the time	Allows cytochrome *c* and other soluble proteins of the intermembrane space to escape into the cytosol

Cell Biology: A Short Course, Third Edition. Stephen R. Bolsover, Jeremy S. Hyams, Elizabeth A. Shephard and Hugh A. White.

Name	Page for details	Location	Selective for	Opened by	Comments
H^+ channels					
Thermogenin	197	Inner mitochondrial membrane	H^+ ions	Probably spends a large fraction of time open under most circumstances	Expressed in brown fat. Uncouples electron transport from ATP synthesis
Potassium channels					
Potassium channel—includes neuronal isoforms, ROMK and Kir	229, 247	Plasma membrane of all cells	Potassium ions	Some isoforms are opened by depolarization, some are open all the time	Responsible for the resting voltage. ROMK and Kir facilitate potassium transport across epithelia. ROMK is unusually sensitive to cytosolic pH, closing when pH falls to 7.0 or below
Calcium channels					
Inositol trisphosphate-gated calcium channel	253	Membrane of endoplasmic reticulum	Calcium ions	Binding of cytosolic inositol trisphosphate	Part of the system that allows an extracellular solute to raise cytosolic calcium
Mitochondrial calcium channel	256	Inner mitochondrial membrane	Calcium ions	Not easily studied, probably always open	Allows an increase of cytosolic calcium concentration to activate mitochondria
Ryanodine receptor, standard isoforms	256	Membrane of endoplasmic reticulum	Calcium ions	Binding of cytosolic calcium	Allows a small increase of cytosolic calcium concentration to trigger a larger increase
Ryanodine receptor; skeletal muscle isoform	256	Membrane of endoplasmic reticulum	Calcium ions	Direct physical link to voltage-gated calcium channels in the plasma membrane	Allows rapid increase of cytosolic calcium concentration in response to depolarization of plasma membrane
Voltage-gated calcium channel	251	Plasma membrane of axon terminal, muscle and other cells	Calcium ions	Depolarization of plasma membrane	The inward movement of calcium ions raises the calcium concentration in the cytosol
Sodium and unselective cation channels					
ENaC (Epithelial Na Channel)	247	Plasma membrane of epthelial cells	Sodium ions	Always spends about 30% of time open	Facilitates sodium transport across the epithelium
Voltage-gated sodium channel	240	Plasma membrane of nerve and muscle cells	Sodium ions	Depolarization of plasma membrane	Produces brief action potentials
TRPV	239	Plasma membrane of pain receptor nerve cells	Sodium, potassium and calcium	High temperature and other damaging events. TRPV1 isoform is opened by capsaicin from chili peppers	Depolarizes the axon terminal allowing generation of action potentials

Name	Page for details	Location	Selective for	Opened by	Comments
Stretch activated channel	277	Plasma membrane of endothelial cells	Sodium, potassium and calcium	Membrane stretch	Allows calcium ions in, triggering downstream events including NO synthesis
Nicotinic acetylcholine receptor channel	274	Plasma membrane of skeletal muscle cells and some nerve cells	Sodium and potassium; some isoforms also pass calcium	Extracellular acetylcholine	An ionotropic cell surface receptor that causes the plasma membrane to depolarize
Ionotropic glutamate receptor	269	Plasma membrane of some nerve cells	Sodium and potassium, some isoforms also pass calcium	Extracellular glutamate	An ionotropic cell surface receptor that causes the plasma membrane to depolarize
cyclic AMP–gated channel	258	Plasma membranes of scent sensitive nerve cells	Sodium and potassium	cyclic AMP in the cytosol	Depolarizes the plasma membrane so that the cell produces action potentials
cyclic GMP–gated channel	259	Plasma membranes of photoreceptor cells	Sodium and potassium	cyclic GMP in the cytosol	Depolarizes the plasma membrane
Chloride channels					
GABA receptor	273	Plasma membrane of some nerve cells	Chloride ions	Extracellular GABA	When this channel is open, it is more difficult to depolarize the cell to threshold
CFTR	331	Plasma membrane of sweat gland cells, airway epithelial cells, and of many other cells	Chloride ions	Phosphorylation by protein kinase A	Required for several transport processes in glands

CARRIERS

Name	Page for details	Location	Mode of action	Comments
Carriers with no enzymatic action				
Glucose transporter	236	Plasma membrane of all cells	Glucose	Required by all human cells
ADP/ATP exchanger	203	Inner mitochondrial membrane	ADP / ATP	Gets ATP and ADP across inner mitochondrial membrane

Name	Page for details	Location	Mode of action	Comments
Na^+/Ca^{++} exchanger	236	Plasma membrane of many cells	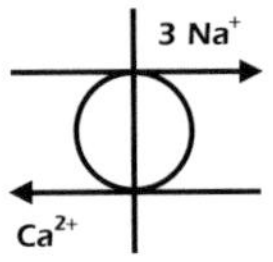 3 Na^+; Ca^{2+}	Pushes calcium ions out of the cytosol
β-galactoside permease	85	Bacterial plasma membrane	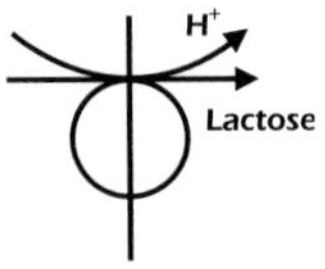 H^+; Lactose	Brings lactose into the cell. One of the products of the *lac* operon
Carriers with an enzymatic action as well as a carrier action				
ATP synthase	200	Inner mitochondrial membrane	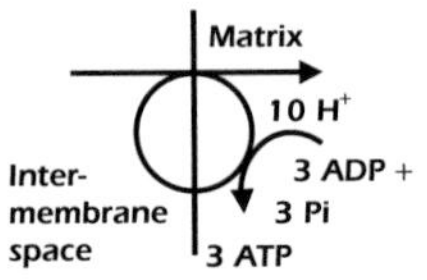 Matrix; 10 H^+; 3 ADP + 3 Pi; 3 ATP; Inter-membrane space	Converts between energy in the H^+ gradient and energy as ATP. Stoichiometry may differ from 10:3 in different species
Ca^{++} ATPase	237	Plasma membrane of many cells	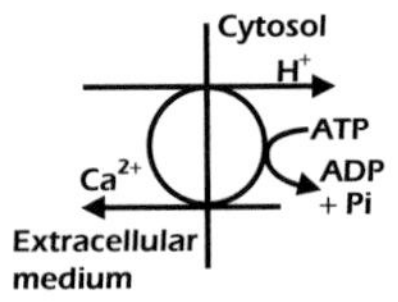 Cytosol; H^+; ATP; ADP + Pi; Ca^{2+}; Extracellular medium	Pushes calcium ions out of the cytosol. All cells have either this or the Na^+/Ca^{++} exchanger; many have both
Na^+/K^+ ATPase	199	Plasma membrane of all cells	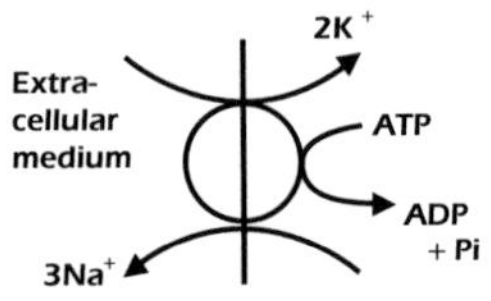 2K$^+$; Extra-cellular medium; ATP; ADP + Pi; 3Na$^+$	Converts between energy as ATP and energy in the Na^+ gradient
Electron transport chain	196	Inner mitochondrial membrane	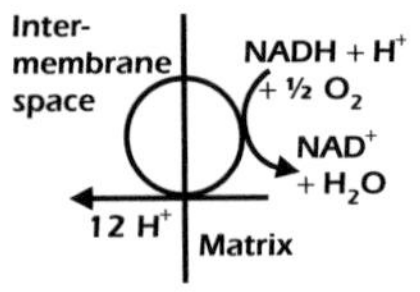 Inter-membrane space; NADH + H^+ + ½ O_2; NAD^+ + H_2O; 12 H^+; Matrix	Converts between energy as NADH and energy in the H^+ gradient

GLOSSARY

7-methyl guanosine cap: modified guanosine found at the 5′ terminus of eukaryotic mRNA. A guanosine is attached to the mRNA by a 5′-5′-phosphodiester link and is subsequently methylated on atom number 7 of the guanine.

9+2 axoneme: structure of cilium or flagellum; describes the arrangement of nine peripheral microtubules surrounding two central microtubules.

A (adenine): one of the bases present in DNA and RNA; adenine is a purine.

α-amino acid: amino acid in which the carboxyl and amino groups are attached to the same carbon.

α helix: a common secondary structure in proteins in which the polypeptide chain is coiled, each turn of the helix taking 3.6 amino acid residues. The nitrogen atom in each peptide bond forms a hydrogen bond with the oxygen four residues ahead of it in the polypeptide chain.

A site (aminoacyl site): site on a ribosome occupied by an incoming tRNA and its linked amino acid.

acceptor: in a hydrogen bond, the atom (oxygen, nitrogen or sulfur) that is not covalently bonded to the hydrogen but which nevertheless accepts a small share of the electrons.

acetone: CH_3—CO—CH_3. A chemical with a fruity smell that is produced from acetoacetate during the ketosis that can occur in diabetes.

acetylcholine:

$$CH_3\text{-}\underset{\underset{O}{\|}}{C}\text{-}O\text{-}CH_2\text{-}CH_2\text{-}\overset{\overset{CH_3}{|}}{\underset{\underset{CH_3}{|}}{N^+}}\text{-}CH_3$$

A transmitter released by various nerve cells including motoneurons.

acetylcholine receptor: integral membrane protein that binds acetylcholine. There are two types: the nicotinic acetylcholine receptor is ionotropic while the muscarinic acetylcholine receptor is metabotropic, linked via G_q to phospholipase Cβ.

acid: a molecule that readily gives H^+ to water. Most organic acids are compounds containing a carboxyl group, although thiols (—SH) are also weakly acidic.

ActA: a protein on the surface of the bacterium *Listeria* from which actin filaments polymerize.

actin: subunit protein of microfilaments. G-actin is the monomeric form while microfilaments are formed of F-actin.

actin binding proteins: proteins that bind to and modulate the function of G-actin or F-actin.

actin related protein (Arp): an actin nucleation protein. New actin filaments can grow out from an Arp base.

Cell Biology: A Short Course, Third Edition. Stephen R. Bolsover, Jeremy S. Hyams, Elizabeth A. Shephard and Hugh A. White.

actin severing protein: one of a group of enzymes that cuts actin microfilaments. Gelsolin is an example.

action potential: an explosive depolarization of the plasma membrane.

activation energy: the height of the energy barrier between the reactants and the products of a chemical reaction.

active site: the region of an enzyme where the substrate binds and the reaction occurs.

acyl group: a group having the general formula:

$$C_nH_m{-}\underset{\substack{\|\\ O}}{C}{-}$$

Acyl groups are formed when fatty acids are attached to other compounds by ester bonds.

adaptive immunity: a form of immunity found only in vertebrates in which genomic variation, natural selection, and evolution within one organism allow it to defend itself against threats never before encountered.

adenine: one of the bases present in DNA and RNA. Adenine is a purine.

adenosine: adenine linked to the sugar ribose. Adenosine is a nucleoside.

adenosine diphosphate (ADP): adenosine with two phosphates attached to the 5′ carbon of ribose. ADP is a nucleotide.

adenosine monophosphate (AMP): adenosine with one phosphate attached to the 5′ carbon of ribose. AMP is a nucleotide.

adenosine triphosphate (ATP): adenosine with three phosphates attached to the 5′ carbon of ribose. ATP is a nucleotide, a coenzyme, and one of the cell's energy currencies.

adenylate cyclase (adenyl cyclase): an enzyme that converts ATP to the intracellular messenger cyclic AMP (cAMP).

adherens junctions: type of anchoring junction in which the cell adhesion molecules are linked to actin microfilaments.

adipocytes: cells that store fats (triacylglycerols).

adipose tissue: type of fatty connective tissue.

ADP: adenosine diphosphate—adenosine with two phosphates attached to the 5′ carbon of ribose. ADP is a nucleotide.

ADP/ATP exchanger: a carrier in the inner mitochondrial membrane. ADP is moved in one direction and ATP in the other.

adrenaline: a hormone released into the blood when an individual is under stress. Adrenaline acts at β-adrenergic receptors to activate G_s and hence adenylate cyclase.

adrenergic receptor: a receptor for the related chemicals adrenaline and noradrenaline. There are two isoforms, α and β. To a first approximation, noradrenaline acts mainly on α receptors linked to G_q and therefore generates a calcium signal, while adrenaline acts mainly on β receptors linked to G_s and therefore generates a cAMP signal.

α helix: a common secondary structure in proteins in which the polypeptide chain is coiled, each turn of the helix taking 3.6 amino acid residues. The nitrogen atom in each peptide bond forms a hydrogen bond with the oxygen four residues ahead of it in the polypeptide chain.

Akt: the earlier name for protein kinase B, a protein kinase that is activated when it is itself phosphorylated; this in turn only occurs when Akt is recruited to the plasma membrane by phosphatidylinositol trisphosphate (PIP_3). Akt phosphorylates proteins on serine and threonine residues (for example, the bcl-2 family protein BAX).

albumin: a blood serum protein that binds hydrophobic molecules and particularly free fatty acids.

alkali: a strong base (that will take H^+ from water) such as sodium hydroxide or potassium hydroxide.

alkaline: (of a solution): one with a low concentration of H^+ (really H_3O^+) so that the pH is greater than 7.0.

allolactose: a disaccharide sugar. Lactose is converted to allolactose by the enzyme β-galactosidase. Allolactose is an inducer of *lac* operon transcription.

all or nothing: a phenomenon that, once initiated, proceeds to completion even if the stimulus is removed. A process that involves positive feedback, such as an action potential, will tend to be all or nothing.

allosteric; allostery: when the binding of a ligand to one site on a protein affects the binding at another site. These interactions can be between sites for the same ligand or different ligands. Molecules which are allosteric almost always have a quaternary structure.

alternative splicing: phenomenon in which a single eukaryotic primary mRNA transcript can be processed to yield a number of different processed mRNAs and can therefore generate a number of different proteins.

amino acid: a chemical that has both a carboxyl group and an amino group. In an α-amino acid, both the carboxyl and amino groups are attached to the same carbon atom. All proteins are generated using the genetically encoded palette of 19 α-amino acids plus proline.

aminoacyl site (A site): site on a ribosome occupied by an incoming tRNA and its linked amino acid.

aminoacyl tRNA: tRNA attached to an amino acid via an ester bond.

aminoacyl tRNA synthases: family of enzymes, each of which attaches an amino acid to the appropriate tRNA.

γ-amino butyric acid: a γ-amino acid that acts to open chloride channels in the plasma membrane of sensitive nerve cells. Usually called γ-amino butyric acid rather than γ-amino butyrate, even though the latter (NH_3^+—CH_2—CH_2—CH_2—COO^-) is the form in which it is found at neutral pH.

amino group: the —NH_2 group. Amino groups are often basic, accepting a proton at normal body pH to become the positively changed —NH_3^+ group.

amino terminus: the end of a peptide or polypeptide which has a free α-amino group.

amorphous: without form. Not a specifically scientific word.

AMP: adenosine monophosphate—adenosine with one phosphate attached to the 5′ carbon of ribose. AMP is a nucleotide.

amphipathic: "hating both." A molecule with a hydrophobic region and a hydrophilic region is said to be amphipathic.

amphiphilic: "loving both." A synonym for amphipathic, that is, a molecule with a hydrophobic region and a hydrophilic region.

anabolism: those metabolic reactions which build up molecules; biosynthesis.

anaerobic: without air. Obligate anaerobes are poisoned by oxygen and therefore can only function anaerobically. Other cells, such as yeast and skeletal muscle, can switch to using anaerobic glycolysis when they are denied oxygen.

anaerobic glycolysis: the partial breakdown of sugars and other cellular fuels that can be accomplished in the absence of oxygen.

anaphase: the period of mitosis or meiosis during which sister chromatids or homologous chromosome pairs separate; consists of anaphase A and anaphase B.

anaphase A: the part of anaphase in which the chromosomes move to the spindle poles.

anaphase B: the part of anaphase in which the spindle poles are separated.

anaphase I: anaphase of the first meiotic division (meiosis I).

anaphase II: anaphase of the second meiotic division (meiosis II).

anchoring junction: class of cell junction that attaches the cytoskeleton of one cell to the cytoskeleton of its neighbor, forming a physically strong connection. There are two types: desmosomes which connect to intermediate filaments, and adherens junctions which connect to actin microfilaments.

anion: a negatively charged ion, e.g., chloride (Cl^-) or phosphate (HPO_4^{2-}).

anode: a positively charged electrode, for example in a gel electrophoresis apparatus used for SDS-PAGE.

anterograde: forward movement; when applied to axonal transport it means away from the cell body.

antibiotic: a chemical that is produced by one type of organism and kills others, often by inhibiting protein synthesis. The most useful antibiotics to man are those that are selective for prokaryotes.

antibody: a protein formed by the immune system that binds to and helps eliminate another chemical. Antibodies can be extremely selective for their ligand and are useful in many aspects of cell biology such as immunofluorescence microscopy and western blotting.

anticodon: the three bases on a tRNA molecule that hydrogen bond to the codon on an mRNA molecule.

antigen: any molecule that can be bound with high affinity by an antibody or (when presented by an MHC complex protein) by a T cell receptor.

antiparallel β sheet: β sheet in which alternate parallel polypeptide chains run in opposite directions.

AP endonuclease: a DNA repair enzyme that cleaves the phosphodiester links on either side of a depurinated or depyrimidinated sugar residue. AP stands for apurinic/apyrimidinic.

apoptosis: a process in which a cell actively promotes its own destruction, as distinct from necrosis. Apoptosis is important in vertebrate development, where tissues and organs are shaped by the death of certain cell lineages.

apoptosis inhibitor protein: proteins which block the action of caspases and hence help prevent apoptosis.

aporepressor: protein that binds to an operator region and represses transcription only when it is complexed with another molecule.

aqueous: watery.

Arf: GTPase that plays a critical role in the formation of coated vesicles.

Arp: (actin related protein): an actin nucleation protein. New actin filaments can grow out from an Arp base.

A site (aminoacyl site): site on a ribosome occupied by an incoming tRNA and its linked amino acid.

assay: a term for a chemical measurement, for example, one in which the activity of an enzyme reaction is measured.

ATM: (ataxia telangiectasia mutated). A component of the checkpoint pathway that arrests the cell cycle in response to DNA damage or incomplete DNA replication. ATM is a protein kinase whose targets include two downstream protein kinases, Cds1 and Chk1.

ATP: adenosine triphosphate—adenosine with three phosphates attached to the 5′ carbon of ribose. ATP is a nucleotide, a cofactor and one of the cell's energy currencies.

ATP/ADP exchanger: a carrier in the inner mitochondrial membrane. ADP is moved in one direction and ATP in the other.

ATP synthase (ATP synthetase): a carrier of the inner mitochondrial membrane that is built around a rotary motor. Ten H^+ enter the mitochondrial matrix for every three ATP made.

autoinduction: the process of self-induction that occurs when a product of a reaction stimulates the production of more of itself. An example is the activation of the *lux* operon by the small molecule N-acyl-HSL; transcription of the *lux* operon then causes the production of more N-acyl-HSL.

autoradiography: a process which detects a radioactive molecule. For example in a Southern blot experiment, the membrane that has been hybridized to a radioactive gene probe is placed in direct contact with a sheet of X-ray film. Radioactive decay activates the silver grains on the emulsion of the X-ray film. When the film is developed, areas that have been in contact with radioactivity will show as black.

axon: the long process of a nerve cell, specialized for the rapid conduction of action potentials.

axonal transport: movement of material along microtubules within a nerve cell process; can be outward (anterograde) or inward (retrograde).

B cells: the immune system cells that present or secrete antibodies.

B-DNA: right-handed DNA double helix.

β oxidation: process by which fatty acids are broken down into individual two-carbon units coupled to CoA to form acetyl-CoA. The process, which takes place in the mitochondrial matrix, generates both NADH and $FADH_2$.

β sheet: common secondary structure in proteins, in which lengths of fully extended polypeptide run alongside each other, hydrogen bonds forming between the peptide bonds of the adjoining strands.

BAC (bacterial artificial chromosome): A cloning vector used to propagate DNAs of about 300,000 bp in bacterial cells.

bacteriophage: (sometimes shortened to phage) a virus that infects bacterial cells.

basal lamina: a synonym for the basement membrane, that is, the thin planar layer of extracellular matrix that supports epithelial cells. Some authors reserve the name basal lamina for a very thin sheet of fibres that are directly connected to the plasma membrane and the cytoskeleton by integrins and other integral membrane proteins, distinguishing this from a thicker, but still sheetlike, basement membrane. However this distinction is rather artificial and is not generally accepted.

base: there are two meanings in cell biology:

1. A chemical that will accept an H^+. Many organic bases contain amino groups, which accept H^+ to become $—NH_3{}^+$.
2. One of a group of ring-containing nitrogenous compounds that combine with a sugar to create a nucleoside; the members include adenine, guanine, hypoxanthine, cytosine, thymine, uracil and nicotinamide.

base excision repair: process that repairs DNA double helices that have lost a purine (depurination), or in which a cytosine has been deaminated to uracil (U). The entire damaged monomer is removed and a correct deoxyribonucleotide inserted in its place.

basement membrane: the thin planar layer of extracellular matrix that supports epithelial cells. Some authors use the name basal lamina for a very thin sheet of fibers that are directly connected to the plasma membrane and the cytoskeleton by integrins and other integral membrane proteins, distinguishing this from a thicker, but still sheetlike, basement membrane. However this distinction is rather artificial and is not generally accepted.

base pair: the Watson-Crick model of DNA showed that guanine in one DNA strand would fit nicely with cytosine in another strand while adenine would fit nicely with thymine. The two hydrogen-bonded bases are called a base pair. RNA can also participate in base pairing: instead of thymine it is uracil that now pairs with adenine. The rare base inosine, found in some tRNAs, can base pair with any of uracil, cytosine or adenine.

BAX: a bcl-2 family protein that dimerizes into a channel that allows the release of cytochrome *c* from mitochondria, triggering apoptosis. BAX is phosphorylated and thereby inactivated by protein kinase B.

B cells: the immune system cells that present or secrete antibodies.

bcl-2: an anti-apoptotic protein. By binding to BAX it prevents the release of cytochrome *c* from mitochondria.

bcl-2 family: a family of related proteins that regulate cytochrome c release from mitochondria.

B-DNA: right-handed DNA double helix.

β-galactosidase: an enzyme that cleaves the disaccharide lactose to produce glucose and galactose, and which also catalyses the interconversion of lactose and allolactose. β-galactosidase is a product of the *lac* operon.

bicoid: an insect transcription factor that causes developing cells to adopt an anterior fate in a dose-dependent manner. Bicoid mRNA is concentrated at one end of the unfertilized insect egg and is translated into protein following fertilization.

binary fission: the mechanism of cell division in prokaryotes, in which the two copies of the single chromosome are partitioned between the progeny cells without the obvious cell division machinery seen in eukaryote mitosis.

binding site: a region of a protein which specifically binds a ligand. A property of the protein's tertiary structure.

bioluminesence: the production of light by a living organism.

bisphosphate: of a compound, bearing two independent phosphate groups (as opposed to a diphosphate, which bears a chain of two phosphates in a line). Fructose 1,6-bisphosphate is an example.

bivalent: the structure formed when homologous chromosomes (one from the mother, one from the father) associate during prophase I of meiosis.

blastocyst: an early embryo.

Bloom's syndrome: a disease resulting from a deficiency in helicase. Affected individuals cannot repair their DNA and are susceptible to developing skin cancer and other cancers.

blue-green algae: the old name for the photosynthetic prokaryotes now known as cyanobacteria.

blunt ends: ends of a DNA molecule produced by an enzyme that cuts the two DNA strands at sites directly opposite one another.

β oxidation: process by which fatty acids are broken down into individual two-carbon units coupled to CoA to form acetyl-CoA. The process, which takes place in the mitochondrial matrix, generates both NADH and $FADH_2$.

bright-field microscopy: the most basic form of light microscopy. The specimen appears against a bright background and appears darker than the background because of the light it has absorbed or scattered.

brown fat: tissue specialized for generating heat. Thermogenin in the inner mitochondrial membrane allows H^+ to enter down its electrochemical gradient, so that the rate at which the electron transport chain works is not limited by the availability of ADP.

β sheet: common secondary structure in proteins, in which lengths of fully extended polypeptide run alongside each other, hydrogen bonds forming between the peptide bonds of the adjoining strands.

bulky lesion: distortion of the DNA helix caused by a thymine dimer.

C-terminus (carboxyl terminus): the end of a peptide or polypeptide that has a free α-carboxyl group. This end is made last on the ribosome.

Ca^{++} ATPase: a carrier that uses the energy released by ATP hydrolysis to move calcium ions up their concentration gradient out of the cytosol. Different isoforms of calcium ATPase are located at the plasma membrane and in the membrane of the endoplasmic reticulum.

cadherin: a cell adhesion molecule that helps form adherens junctions.

calcineurin: a calcium-calmodulin-activated phosphatase, that is, an enzyme that removes phosphate groups from proteins, opposing the effects of kinases. Calcineurin is inhibited by the immunosuppressant drug cyclosporin.

calcium ATPase: a carrier that uses the energy released by ATP hydrolysis to move calcium ions up their concentration gradient out of the cytosol. Different isoforms of calcium ATPase are located at the plasma membrane and in the membrane of the endoplasmic reticulum.

calcium binding protein: any protein that binds calcium. Calmodulin, troponin and calreticulin are examples found in the cytosol, attached to actin filaments in striated muscle and in the endoplasmic reticulum respectively.

calcium-calmodulin activated protein kinase: an important regulatory enzyme, activated when calcium-loaded calmodulin binds, which phosphorylates target proteins on serine and threonine residues.

calcium induced calcium release: a process in which a rise of calcium concentration in the cytosol triggers the release of more calcium from the endoplasmic reticulum. The best-understood mechanism of calcium-induced calcium release is via ryanodine receptors.

calcium pump: a carrier that moves calcium ions up their electrochemical gradient out of the cytosol into the extracellular medium or into the endoplasmic reticulum. There are two important calcium pumps: the sodium/

calcium exchanger is found on the plasma membrane, while different isoforms of the calcium ATPase are found on the plasma membrane and on the membrane of the endoplasmic reticulum.

calmodulin: a calcium binding protein found in many cells. When calmodulin binds calcium it can then activate other proteins such as the enzymes calcineurin and glycogen phosphorylase kinase.

Calvin cycle: series of reactions in the chloroplast that act to fix atmospheric carbon dioxide and build it into larger molecules.

CaM-kinase: another name for calcium-calmodulin activated protein kinase: an important regulatory enzyme, activated when calcium-loaded calmodulin binds, that phosphorylates target proteins on serine and threonine residues.

cAMP (cyclic adenosine monophosphate): A nucleotide produced from ATP by the action of the enzyme adenylate cyclase. cAMP is an intracellular messenger in many cells.

cAMP-dependent protein kinase (protein kinase A; PKA): a serine-threonine protein kinase that is activated by the intracellular messenger cyclic AMP.

cAMP gated channel: a channel found in the plasma membrane of scent-sensitive nerve cells. The channel opens when cAMP binds to its cytosolic face and allows sodium and potassium ions to pass.

cAMP phosphodiesterase: enzyme that hydrolyses cyclic AMP, producing AMP and hence turning off signaling through the cAMP system.

cap: methylated guanine added to the 5′ end of a eukaryotic mRNA.

CAP (catabolite activator protein): a protein found in prokaryotes that binds to cAMP. The CAP-cAMP complex then binds within the promoter region of some bacterial operons and helps RNA polymerase to bind to the promoter.

carbohydrates: monosaccharides and all compounds made from monosaccharide monomers.

carbon fixation: the process of reducing atmospheric carbon dioxide and building it into larger molecules.

carboxylation: introduction of a carboxyl group (—COOH).

carboxyl group: the —COOH group. Carboxyl groups give up hydrogen ions to form the deprotonated group $—COO^-$, so molecules that bear carboxyl groups are usually acids.

carboxyl terminus: the end of a peptide or polypeptide which has a free α-carboxyl group. This end is made last at the ribosome.

cardiac muscle: the form of striated muscle that is found in the heart.

carrier: there are two meanings used in this book.

1. An integral membrane protein that forms a tube through the membrane that is never open all the way through. Solutes can move into the tube through the open end. When the channel changes shape, so that the end that was closed is open, the solute can leave on the other side of the membrane.
2. A person who has one nonfunctional or mutant copy of a gene, but who shows no effects because the other copy produces sufficient functional protein.

caspase: a cysteine-containing protease that cleaves at aspartate residues. Caspases are responsible for the degradative processes that occur during apoptosis.

catabolism: those metabolic reactions that break down molecules to derive chemical energy.

catabolite activator protein (CAP): a protein that binds to cAMP. The CAP-cAMP complex then binds within the promoter region of some bacterial operons and helps RNA polymerase to bind to the promoter.

catalyst: a chemical or substance that reduces the activation energy of a reaction, allowing it to proceed more quickly. Many biological reactions would proceed at an infinitesimal rate without the aid of enzymes, which are protein catalysts.

catalytic rate constant (k_{cat}): the proportionality constant that relates the maximal initial velocity (V_m) of an enzyme catalyzed reaction to the enzyme concentration. $k_{cat} = V_m/[E]$. The units are reciprocal time. See also turnover number and maximal velocity.

cation: positively charged ion, for example Na^+, K^+ and Ca^{++}.

CD4+ T cell: T cell expressing the integral plasma membrane protein CD4. CD4+ T cells, also called helper T cells, activate B cells if the peptide presented by MHC proteins on the B cell binds tightly to their own T cell receptor.

CD8+ T cell: T cell expressing the integral plasma membrane protein CD8. CD8+ T cells, also called T killer or cytotoxic T cells, kill somatic cells that present novel peptides.

cdc: cell division cycle. The acronym is usually used to denote genes which, when mutated, cause the cell division cycle to be abnormal.

Cdc25: a protein phosphatase involved in the regulation of **Cdk1**.

Cdk1 (cyclin-dependent kinase 1): a protein kinase involved in the regulation of the G2/M transition of the cell cycle. Associates with cyclin B.

Cdk2 (cyclin-dependent kinase 2): a protein kinase involved in the regulation of the G1 phase of the cell cycle. Associates with cyclin E.

Cdk4 (cyclin-dependent kinase 4): a protein kinase involved in the regulation of the G1 phase of the cell cycle. Associates with cyclin D.

Cdk6 (cyclin-dependent kinase 6): a protein kinase involved in the regulation of the G1 phase of the cell cycle. Associates with cyclin D.

cDNA (or complementary DNA): a DNA copy of an mRNA molecule.

cDNA library: collection of bacterial cells each of which contains a different foreign cDNA molecule.

cell: fundamental unit of life. A membrane-bound collection of protein, nucleic acid and other components that is capable of self-replication using simpler building blocks.

cell adhesion molecule: an integral membrane protein responsible for sticking cells together. The extracellular domain binds a cell adhesion molecule on another cell while the intracellular domain binds to the cytoskeleton, either directly or via a linker protein.

cell center: point immediately adjacent to the nucleus of eukaryotes where the centrosome and Golgi apparatus are located.

cell cycle (cell division cycle): ordered sequence of events that must occur for successful eukaryotic cell division; consists of G1, S, G2 and M phases.

cell junctions: points of cell-cell interaction in tissues; includes tight junctions; anchoring junctions and gap junctions.

cell membrane: the membrane that surrounds the cell; also known as the plasmalemma or plasma membrane.

cell surface membrane: another name for the plasma membrane.

cellulose: the major structural polysaccharide of the plant cell wall; a $\beta 1 \rightarrow 4$ polymer of glucose.

cell wall: a rigid case that encloses plant and fungal cells and many prokaryotes. The cell wall lies outside the plasma membrane. Plant cell walls are composed of cellulose plus other polysaccharide molecules.

central dogma (of molecular biology): "DNA makes RNA makes protein"—the concept that the sequence of bases on DNA defines the sequence of bases on RNA, and the sequence of bases on RNA then defines the sequence of amino acids in protein.

centriole: the structure found at the centrosome (= microtubule organizing center) of animal cells; composed of microtubules.

centromere: the region of the chromosome at which the kinetochore (where the microtubules of the mitotic or meiotic spindle attach) is formed.

centrosome (microtubule organizing center): structure from which cytoplasmic microtubules arise.

CF (cystic fibrosis): inherited disease characterized by failure of the pancreas and by thick sticky mucus in the lungs leading to fatal lung infection unless treated. Cystic fibrosis is caused by failure to make, or properly target, functional plasma membrane chloride channels.

cGMP-dependent protein kinase (PKG; protein kinase G): a protein kinase that is activated by the intracellular messenger cyclic GMP. PKG phosphorylates proteins (for example the calcium ATPase) on serine and threonine residues.

channel: an integral membrane protein that forms a continuous water-filled hole through the membrane.

chaotropic reagents: reagents such as urea, which cause proteins to lose all their higher levels of structure and adopt random, changing conformations.

chaperone: a protein that helps other proteins to remain unfolded for correct protein targeting, or to fold into their correct three-dimensional structure.

charge

1. an excess or deficit of electrons giving a negative or positive charge respectively.
2. a transfer RNA is said to be charged when it has an amino acid attached.

charged tRNA: a tRNA attached to an amino acid.

checkpoint: a control point in the eukaryote cell cycle, where progression is allowed only if all the necessary processes have been completed.

chiasmata (singular chiasma): structures formed during crossing over between the chromatids of homologous chromosomes during meiosis; the physical manifestation of genetic recombination.

chimera: structure formed from two different parts. Chimeric proteins are generated by joining together all or part of the protein coding sections of two distinct genes. Chimeric organisms are formed by mixing two or more distinct clones of cells.

chimeric protein: proteins generated by joining together all or part of the protein coding sections of two distinct genes, for example GFP and a protein of interest.

chiral: structure whose mirror image cannot be superimposed on it. Organic molecules will be chiral if a carbon atom has four different groups attached to it and is therefore asymmetric.

chlorophyll: the major photosynthetic pigment of plants and algae.

chloroplast: the photosynthetic organelle of plant cells.

cholesterol: a hydrophobic molecule made up of four fused rings with a short hydrocarbon tail at one end and a single hydroxyl group at the other. Cholesterol is a component of eukaryotic (but not prokaryotic) membranes, and helps to keep them flexible. Steroid hormones are synthesized from cholesterol.

chromatid: a complete DNA double helix plus accessory proteins subsequent to DNA replication in eukaryotes. At mitosis, the chromosome is seen to be composed of two chromatids; these then separate to form the chromosomes of the two progeny cells.

chromatin: a complex of DNA and certain DNA-binding proteins such as histones.

chromatophore: a pigment cell in the skin of fish and amphibia.

chromosome: a single, enormously long molecule of DNA, together with its accessory proteins. Chromosomes are the units of organization of the nuclear chromatin and carry many genes. In eukaryotes chromosomes are linear; in prokaryotes they are circular.

chromosome walking: investigating a chromosome bit by overlapping bit, each bit being used to clone the next.

cilium (*plural* cilia:): locomotory appendage of some epithelial cells and protozoa.

cip1: gene encoding a cyclin-dependent kinase inhibitor protein which prevents cells entering the S phase of the cell cycle. $p21^{Cip1}$ expression is increased by cell-cell contact.

***cis* face:** the side to which material is added. Of the Golgi complex, the surface that receives vesicles from the endoplasmic reticulum.

cisternae: flattened membrane-bound sacs, for example, those that make up the Golgi apparatus.

CKI (cyclin dependent kinase inhibitor): a type of cell cycle regulatory protein. Binds to and inactivates CDKs.

class switching: the process that changes a B cell from one expressing IgD and IgM to one expressing either IgA, IgE or IgG. Class switching is a permanent change in which the μ and δ sections and a variable number of other sections are removed from the heavy chain locus.

clathrin: a protein that functions to cause vesicle budding in response to binding of specific ligand.

clathrin adaptor protein: protein that binds to specific transmembrane receptors and which in turn recruits clathrin to form a coated vesicle. The vesicle therefore contains the molecule for which the receptor is specific.

cleavage furrow: in animal cells, the structure that constricts the middle of the cell during cytokinesis.

clone: a number of genetically identical individuals.

clone library: a collection of bacterial clones where each clone contains a different foreign DNA molecule.

cloning: strictly, the creation of a number of genetically identical organisms. In molecular genetics, the term is used to mean the multiplication of particular sequences of DNA by an asexual process such as bacterial cell division.

cloning vector: a DNA molecule that carries genes of interest, can be inserted into cells, and which will then be replicated inside the cells. Cloning vectors range in size from plasmids to entire artificial chromosomes.

closed promoter complex: a structure formed when RNA polymerase binds to a promoter sequence prior to the start of transcription.

coatamer: protein complex that encapsulates one class of coated vesicle. Formation of the coatamer coat on a previously flat membrane forces the membrane into a curved shape and therefore drives vesicle formation.

coated vesicle: a cytoplasmic vesicle encapsulated by a protein coat. There are two types of coated vesicles, coated by coatamer and clathrin respectively.

codon: a sequence of three bases in an mRNA molecule that specifies a particular amino acid.

coenzyme: a molecule that acts as a second substrate for a group of enzymes. ATP/ADP, NADH/NAD^+ and acyl-coenzyme A/coenzyme A are all coenzymes.

cofactor: a nonprotein molecule or an ion necessary for the activity of a protein. Cofactors are associated tightly with the protein but can be removed. Examples are pyridoxal phosphate in aminotransferases and zinc in zinc finger proteins.

cohesin: protein that holds two chromatids together. Degradation of cohesin allows the two chromatids to separate at the start of anaphase in mitosis and meiosis.

colchicine: a plant toxin from the autumn crocus, *Colchicum autumnale*; binds to tubulin.

collagen: the major structural protein of the extracellular matrix.

columnar: taller than it is broad. Used as a description of some types of epithelial cells.

competent: in molecular genetics this refers to a bacterial culture treated with a solution such as calcium chloride so that uptake of foreign DNA is enhanced.

complement: in immunology, the noun complement refers to a collection of proteins that help antibodies neutralize pathogens and other targets.

complementary: two structures are said to be complementary when they fit into or associate with each other. The two strands of the DNA double helix are complementary, as are the anticodon and codon of transfer and messenger RNA.

complementary DNA (cDNA): a DNA copy of an mRNA molecule.

complex (as a noun): an association of molecules that is held together by noncovalent interactions, and often can be readily dissociated.

complex I, complex II, complex III, complex IV: large multimolecular complexes that make up the electron transport chain in the inner mitochondrial membrane.

condenser lens: the lens of light and electron microscopes that focuses light (or electrons) onto the specimen.

connective tissue: a tissue that contains relatively few cells within a large volume of extracellular matrix.

connexon: an integral membrane protein that can open to form a channel about 1.5 nm in diameter when it contacts a second connexon on another cell. This forms a water-filled tube that runs all the way through the plasma membrane of the first cell, across the small gap between the cells, and through the plasma membrane of the second cell, so allowing passage of solute from the cytosol of one cell to the cytosol of the other.

constitutive: operating all the time without obvious regulation. Housekeeping genes, expressed all the time, are sometimes called constitutive genes. Proteins that are secreted all the time are said to use the constitutive route.

constitutively active: a mutant or modified protein that is always in the "on" state. A constitutively active enzyme is active in the absence of its normal regulators. A constitutively active GTPase activates its downstream targets in the absence of GTP exchange factors. Paradoxically, the easiest way to generate a constitutively active GTPase is to eliminate its enzymatic activity, so that it does not hydrolyze GTP and therefore remains in the active GTP-bound state.

constitutive secretion: secretion that continues all the time, without the need for a signal such as an increase of cytosolic calcium concentration.

contact inhibition: inhibition of cell division by cell-cell contact. Contact inhibition allows cells to proliferate to fill a gap and then stop dividing.

contigs: overlapping lengths of DNA which can together be used to build up a map of the genome.

—COOH: carboxyl group. Carboxyl groups give up hydrogen ions to form the deprotonated group $—COO^-$, so molecules that bear carboxyl groups are usually acids.

cortex: the outer part of any organ or structure. For instance, the tissue that forms the outer region of the brain, and the outer regions of a cell occupied by an actin lattice, are each called cortex.

covalent bond: a strong bond between two atoms in which electrons are shared.

cristae: the name given to the folds of the inner membrane of mitochondria.

crossing over: the physical exchange of material that takes place between homologous chromosomes during recombination and is manifest in the formation of chiasmata.

crosstalk: two messenger systems show crosstalk when one messenger can produce some or all of the effects of the other.

C-terminus (carboxyl terminus): the end of a peptide or polypeptide that has a free α-carboxyl group. This end is made last on the ribosome.

cyanobacteria: the photosynthetic prokaryotes that were formerly known as blue-green algae.

cyclic adenosine monophosphate (cAMP): a nucleotide produced from ATP by the action of the enzyme adenylate cyclase. cAMP is an intracellular messenger in many cells and exerts many of its actions by activating protein kinase A.

cyclic guanosine monophosphate (cGMP): a nucleotide produced from GTP by the action of the enzyme guanylate cyclase. cGMP is an intracellular messenger in many cells and exerts many of its actions by activating protein kinase G.

cyclin: one of a family of proteins whose level oscillates (cycles) through the cell division cycle. Cyclins associate with and activate cyclin dependent kinases and hence allow progression through cell cycle control points.

cyclin dependent kinase inhibitors (CKIs): a type of cell cycle regulatory protein. Binds to and inactivates CDKs.

cyclin dependent protein kinase (CDK): one of the family of protein kinases that regulate the cell cycle. Cyclin dependent kinases are only active when bound to one of the family of cyclin proteins. For example, CDK1 associates with cyclin B and regulates the G2/M transition while

CDK2 associates with cyclin D and cyclin E and regulates the G1/S transition.

cystic fibrosis (CF): inherited disease characterized by failure of the pancreas and by thick sticky mucus in the lungs leading to fatal lung infection unless treated. Cystic fibrosis is caused by failure to make, or properly target, functional plasma membrane chloride channels.

cystine: a double amino acid formed by two cysteine molecules joined by a disulphide bond.

cytochemistry: the use of chemical compounds to stain specific cell structures and organelles.

cytochrome: proteins with a heme prosthetic group that are able to transfer electrons. Cytochromes form a critical part of the electron transport chain of mitochondria and also form part of the cytochrome P450 detoxification system in the liver.

cytochrome *c*: a soluble protein of the mitochondrial intermembrane space, often found loosely associated with the inner mitochondrial membrane. Cytochrome *c* transports electrons between components of the electron transport chain. If it is allowed to escape from mitochondria cytochrome *c* activates caspase 9 and hence triggers apoptosis.

cytokine: one of a large family of protein transmitters that are especially important in the immune system.

cytokinesis: the process by which a cell divides in two; part of the M phase of the cell division cycle.

cytology: the study of cell structure by light microscopy.

cytoplasm: the semi-viscous ground substance of the cell. All the volume outside the nucleus and inside the plasma membrane is cytoplasm.

cytoplasmic dynein: a motor protein that moves organelles along microtubules in a retrograde direction.

cytosine: one of the bases present in DNA and RNA; cytosine is a pyrimidine.

cytoskeleton: cytoplasmic filament system consisting of microtubules, microfilaments, and intermediate filaments.

cytosol: the viscous, aqueous medium in which the organelles and the cytoskeleton are bathed.

cytotoxic T cell: T cell expressing the integral plasma membrane protein CD8. Cytotoxic T cells kill somatic cells that present novel peptides.

dark reactions: reactions occuring in plants that use ATP and reducing power provided by NADPH to fix atmospheric carbon dioxide and use it to make sugar.

DAG (diacylglycerol): two acyl groups (fatty acid chains) joined by ester bonds to a glycerol backbone. Diacylglycerol is produced by the action of phospholipase C on phospholipid and helps to activate protein kinase C.

deamination: the removal of an amino group. Deamination of cytosine to form uracil is a form of DNA damage.

death domain: a domain found on proteins concerned with regulating apoptosis, such as Fas and the p75 neurotrophin receptor. When activated, death domain proteins turn on caspase 8 and hence initiate apoptosis.

denature: to cause to lose three-dimensional structure by breaking noncovalent intramolecular bonds.

dendrite: a branching cell process. The term is commonly used to name those processes of nerve cells that are too short to be called axons.

dendritic cells: professional antigen presenting cells that constantly travel back and forth between peripheral locations in the body where they pick up and present antigens, and lymph nodes where they present their antigens to T cells.

deoxyribonucleic acid (DNA): a polymer of deoxyribonucleotides. DNA specifies the inherited instructions of a cell.

deoxyribonucleotide: the building block of DNA that is made up of a nitrogenous base and the sugar deoxyribose to which a phosphate group is attached.

deoxyribose: a ribose that instead of —OH has only —H on carbon 2. Deoxyribose is the sugar used in the nucleotides that make up DNA.

depolarization: any positive shift in the transmembrane voltage, whatever its size or cause.

depth of focus: the distance towards and away from an object over which components of the object remain in clear focus.

depurination: the removal of either of the purine bases, adenine and guanine, from a DNA molecule. Depurination is a form of DNA damage.

desmin: protein that makes up the intermediate filaments in muscle cells.

desmosome: a type of anchoring junction which joins the intermediate filaments of neighboring cells. Desmosomes are common in tissues such as skin.

diabetes: the word diabetes simply refers to conditions in which a patient produces lots of urine. The only form of diabetes we discuss in this book is diabetes mellitus, in which the patient produces large volumes of urine containing sugar. Diabetes mellitus is caused by either a failure in the endocrine gland that produces insulin, or a failure of the tissues to respond adequately to insulin.

diacylglycerol (DAG): two acyl groups (fatty acid chains) joined by ester bonds to a glycerol backbone. Diacylglycerol is produced by the action of phospholipase C on phospholipid and helps to activate protein kinase C.

dideoxyribonucleotide: a man-made molecule similar to a deoxyribonucleotide but lacking a 3′ hydroxyl group on its sugar residue. Used in DNA sequencing.

differentiation: the process whereby a cell becomes specialized for a particular function.

diffusion: the movement of a substance that results from the individual small random thermal movements of its molecules.

dimeric: formed of two parts. Of a molecule, formed of two parts that are not covalently linked.

diphosphate: of a compound, bearing a chain of two phosphates in a line (as opposed to a bisphosphate, which bears two independent phosphate groups). Adenosine diphosphate is an example.

diploid: containing two sets of chromosomes; in humans, this means two sets of 23, one from the father, one from the mother. Most of the cells of the body (the somatic cells) are diploid.

dipolar bond: an unusual form of covalent bond found in trimethylamine-*N*-oxide.

dipole: a molecule that has a positive and a negative charge (partial or whole) separated by a (usually) small distance.

disaccharide: a dimer of two monosaccharides. Examples are lactose (galactose $\beta 1 \rightarrow 4$ glucose) and sucrose (glucose $\alpha 1 \rightarrow \beta 2$ fructose).

distal: far from the center.

disulfide bond (bridge): covalent bond between two sulfur atoms. In proteins disulfide bonds form by oxidation of two thiol (—SH) groups of cysteine residues. Found chiefly in extracellular proteins.

DNA (deoxyribonucleic acid): a polymer of deoxyribonucleotides, DNA specifies the inherited instructions of a cell.

DNAa: DNA binding protein that causes the two strands of the double helix to separate in the first stages of DNA replication.

DNAb: helicase that moves along a DNA strand, breaking hydrogen bonds, and in the process unwinding the helix.

DNAc: DNA binding protein that serves to bring DNAb to the DNA strands.

DNA chip: a tiny glass wafer to which cloned DNAs are attached. Also known as gene chips or microarrays.

DNA excision: process of cutting out damaged DNA prior to repair.

DNA fingerprint: the individual pattern of DNA fragments determined by the number and position of specific repeated sequences.

DNA ligase: an enzyme that joins two DNA molecules by catalyzing the formation of a phosphodiester link.

DNA polymerase: enzyme that synthesizes DNA by catalyzing the formation of a phosphodiester link. DNA is always synthesized in the 5′ to 3′ direction.

DNA repair enzymes: enzymes that detect and repair altered DNA.

DNA replication: a process in which the two strands of the double helix unwind and each acts as a template for the synthesis of a new strand of DNA.

DNA sequencing: determining the order of bases on the DNA strand.

docking protein (signal recognition particle receptor): the receptor on the endoplasmic reticulum to which the signal recognition particle binds during the process of polypeptide chain synthesis and import into the endoplasmic reticulum.

dogma: belief. The 'central dogma' of molecular biology is that 'DNA makes RNA makes protein'—the concept that the sequence of bases on DNA defines the sequence of bases on RNA, and the sequence of bases on RNA then defines the sequence of amino acids on protein.

domain: a separately folded segment of the polypeptide chain of a protein.

dominant: a gene that exerts its effect even when only one copy is present. Most dominant proteins are dominant because they produce functional protein while the recessive gene does not. However, some mutant genes are dominant because even having 50% of one's protein as the mutant form is enough to cause an effect; the gene causing familial Creutzfeldt-Jacob disease is an example.

donor: that which gives. In a hydrogen bond, the donor is the atom (oxygen, nitrogen or sulfur) to which the hydrogen is covalently bonded and which gives up some of its share of electrons to a second electron-grabbing atom.

double helix: the structure formed when two filaments wind about each other, most commonly applied to DNA, but also applicable to, for example, f-actin.

downstream: a general term meaning the direction in which things move. When applied to the DNA within and adjacent to a gene, it means lying on the side of the transcription start site that is transcribed into RNA. When applied to signaling pathways it means in the direction in which the signal travels, thus MAP kinase is downstream of Ras.

duplication: the doubling up of a particular sequence of the genetic material.

dynamic instability: term that describes the behavior of microtubules in which they switch from phases of growth to phases of shortening.

dynamin: GTPase that plays a critical role in the formation of clathrin-coated vesicles.

dynein: a motor protein. Cytoplasmic dynein moves vesicles along microtubules, while dynein arms power ciliary and flagellar beating by generating sliding between adjacent outer doublet microtubules.

E site: the site on the ribosome from which the tRNA exits after its amino acid has been transferred to the peptide chain.

E2F: transcription factor required for DNA synthesis. In quiescent cells E2F is prevented from activating transcription by being bound to RB, the product of the retinoblastoma gene. E2F is released when RB is phosphorylated by CDK4 or CDK6.

ectoplasm: viscous, gel-like outer layer of cytoplasm.

effective stroke: the part of the beat cycle of a cilium that pushes on the extracellular medium.

effector caspase: a caspase that digests cellular components in the process of apoptosis. Effector caspases are differentiated from caspases 8 and 9 which do not themselves digest cellular components but which activate effector caspases by hydrolyzing particular peptide bonds in them.

electrically excitable: able to produce action potentials.

electrochemical gradient: the free energy gradient for an ion in solution. The arithmetical sum of gradients due to concentration and voltage.

electron gun: the source of electrons in electron microscopes.

electron microscope: a microscope in which the image is formed by passage of electrons through, or scattering of electrons by, the object.

electron transport chain: the series of electron acceptor/donator molecules found in the inner mitochondrial membrane which transport electrons from the reduced carriers NADH and $FADH_2$ (thus reoxidizing them) to oxygen. The entire complex forms a carrier that uses the energy of NADH and $FADH_2$ oxidation to transport hydrogen ions up their electrochemical gradient out of the mitochondrion.

electrophoresis: a method of separating charged molecules by drawing them through a filtering gel material using an electrical field.

electrostatic bond: a strong attraction between ions or charged groups of opposite charge.

electrostatic interaction: attraction or repulsion between ions or charged groups.

elongation factors: proteins that speed up the process of protein synthesis at the ribosome. Elongation factors tu and G are GTPases.

embryonic stem cell: cells derived from an early embryo. Embryonic stem cells have the capability to divide indefinitely and to become any cell type in the body.

endocytosis: the inward budding of plasma membrane to form vesicles. Endocytosis is the process by which cells retrieve plasma membrane and take up material from their surroundings.

endogenous: belonging to the cell or organism; not introduced by a pathogen or human intervention.

endonuclease: an enzyme that digests nucleic acids by cleaving internal phosphodiester bonds; compare to exonuclease.

endoplasm: the fluid inner layer of cytoplasm that streams during cytoplasmic streaming.

endoplasmic reticulum (ER): a network (reticulum) of membrane-delimited tubes and sacs that extends from the outer membrane of the nuclear envelope to the plasma membrane. There are two types of ER: rough endoplasmic reticulum (RER) with a surface coating of ribosomes, and smooth endoplasmic reticulum (SER).

endosome: the organelle to which newly formed endocytotic vesicles are translocated and with which they fuse.

endosymbiotic theory: the proposal that some of the organelles of eukaryotic cells originated as free-living prokaryotes.

endothelial cells: epithelial cells that line blood vessels and other body cavities that do not open to the outside.

endothelium: a layer of cells that lines blood vessels and other body cavities that do not open to the outside.

energy currency: a source of energy that the cell can use to drive processes that would otherwise not occur because their ΔG is positive. Energy currencies can be coenzymes such as ATP and NADH, which give up energy on conversion to respectively ADP and NAD^+, or electrochemical ion gradients such as the hydrogen ion gradient across the mitochondrial membrane and the sodium gradient across the plasma membrane.

enhancer: a specific DNA sequence to which a protein binds to increase the rate of transcription of a gene.

enthalpy: thermodynamic description of heat.

entropy: the degree of disorder in a system.

envelope: a closed sheet enclosing a volume. The term is used to describe, among other things, the double membrane layer enclosing certain organelles, and the outer membrane of certain viruses.

enzyme: a biological catalyst. Like all catalysts, enzymes work by reducing the activation energy of the reaction.

epidermis: the protective outer cell layer of an organism.

epithelial cells: The cells that make up an epithelium.

epithelium: a sheet of cells.

equilibrium: the total balance of opposing forces. A process or object is in equilibium if the tendency to go in one direction is exactly equal to the tendency to go in the other direction. For an ion this condition is equivalent to saying that the electrochemical gradient for that ion across the membrane is zero. For a chemical reaction, equilibrium occurs when the rate of the forward reaction is equal to the rate of the reverse reaction, e.g., $2H_2O \rightleftarrows OH^- + H_3O^+$.

equilibrium voltage: the transmembrane voltage that will exactly balance the concentration gradient of a particular ion.

ER (endoplasmic reticulum): network (reticulum) of membrane channels that extends from the outer membrane of the nuclear envelope to the plasma membrane; there are two types of ER: rough endoplasmic reticulum (RER) with a surface coating of ribosomes and smooth endoplasmic reticulum (SER).

erythrocyte: red blood cell.

ES (embryonic stem) cell: a cell derived from a very early embryo. ES cells have not yet determined their developmental fate and can therefore, depending on the conditions, generate the entire range of tissue types.

E-site: the site on the ribosome from which the tRNA exits after its amino acid has been transferred to the peptide chain.

essential amino acid: an amino acid that the organism needs but cannot synthesize. Histidine, isoleucine, leucine, lysine, methionine, phenylalanine, threonine, tryptophan and valine are essential for humans.

essential fatty acid: a fatty acid that the organism needs but cannot synthesize. Linoleic and linolenic acids are essential for humans.

ester bond: the bond formed between the hydrogen of an alcohol group and the hydroxyl of a carboxyl group by the elimination of water.

euchromatin: that portion of the nuclear chromatin that is not tightly packed. Euchromatin contains genes that code for proteins that are being actively transcribed.

eukaryotic: an organism whose cells contain distinct nuclei and other organelles; includes all known organisms except prokaryotes (bacteria and cyanobacteria).

eukaryotic initiation factors (eIFs): proteins that assist the small and large ribosomal subunits of eukaryotes to assemble correctly on the mRNA at the start of protein synthesis.

exit site (E site): the site on the ribosome from which the tRNA exits after its amino acid has been transferred to the peptide chain.

exocytosis: the fusion of a vesicle with the plasma membrane. Exocytosis causes the soluble contents of the vesicle to be released to the extracellular medium, while integral membrane proteins of the vesicle become integral membrane proteins of the plama membrane.

exon: in a eukaryotic gene, exons are those parts that after RNA processing leave the nucleus. In contrast introns are spliced out before the RNA leaves the nucleus.

exonuclease: an enzyme that digests nucleic acids by cleaving phosphodiester links successively from one end of the molecule; compare to endonuclease.

expression (of a gene): the appearance of the protein for which the gene encodes.

expression vector: a cloning vector containing a promoter sequence recognised by the host cell, thus enabling a foreign DNA insert to be transcribed into mRNA.

extracellular matrix: the meshwork of filaments and fibers that surrounds and supports mammalian cells. The major protein of the extracellular matrix is collagen.

extracellular medium: the aqueous medium outside cells. For a unicellular organism the extracellular medium is the outside world. For a multicellular organism such as a human being the extracellular medium is the interstitial fluid.

extragenic DNA: DNA that can neither be identified as coding for protein or RNA nor as being promoters or enhancers regulating transcription.

Fab: the antigen binding fragment of an antibody. The term is also used to refer to the corresponding part of a complete antibody molecule.

Fas: a metabotropic cell surface receptor that activates caspase 8 and hence triggers apoptosis.

Fas ligand: an integral plasma membrane protein of CD8+ T cells that binds to and activates Fas on a target cell, initiating apoptosis.

fat: a triacylglycerol (triglyceride) that is solid at room temperature. In contrast, oils are liquid at room temperature.

fat cells (adipocytes): cells that store fats (triacylglycerols).

fatty acid: a carboxyl group attached to a long chain of carbon atoms with attached hydrogens, that is, a chemical of the general form:

$$C_nH_m-\underset{\underset{O}{\|}}{C}-OH$$

Fc: the crystalizable fragment of an antibody molecule. The fragment is crystalizable because it is the same in all antibodies of the same class (IgM, IgD, etc.). The term is also used to refer to the corresponding part of a complete antibody molecule.

feedback: a process in which the result of a process modifies the mechanisms carrying out that process to increase or decrease their rate. In negative feedback a change in some parameter activates a mechanism that reverses the change in that parameter; an example is the effect of tryptophan on expression of the *trp* operon. In positive feedback a change in some parameter activates a mechanism that accelerates the change; an example is the effect of depolarization on the opening of voltage-gated sodium channels.

feedforward: a control process in which a change in a parameter is *predicted*, and mechanisms initiated that will act to reduce the change in that parameter.

fibroblast: a cell found in connective tissue. Fibroblasts synthesize collagen and other components of the extracellular matrix.

fibroblast growth factor (FGF): a paracrine transmitter that opposes apoptosis and promotes cell division in target cells. Although named for its effect on fibroblasts, FGF triggers proliferation of many tissues and plays critical roles in determining cell fate during development.

fission: breakage into two parts. The word is used of prokaryote replication. In addition it describes the division of a single membrane-bound organelle into two and the process whereby a vesicle breaks away from a membrane.

flagellin: the protein that is the bacterial flagellum.

flagellum: a swimming appendage. In eukaryotes flagella are extensions of the cell that use a dynein/microtubule motor system. In prokaryotes flagella are extracellular proteins rotated by a motor at their base.

fmet (formyl methionine): methionine modified by the attachment of a formyl group; fmet is the first amino acid in all newly made bacterial polypeptides.

focal contact: points at which a locomoting cell makes contact with its substrate.

formyl methionine (fmet): methionine modified by the attachment of a formyl group; fmet is the first amino acid in all newly made bacterial polypeptides.

frameshift mutation: a mutation that changes the mRNA reading frame, caused by the insertion or deletion of a nucleotide.

G0 (gap 0): describes the quiescent state of cells that have left the cell division cycle.

G1 (gap 1): period of the cell division cycle that separates mitosis from the following S phase.

G1/S transition checkpoint: the checkpoint controlling entry into the S phase of the eukaryotic cell cycle. For example, DNA replication is blocked if the DNA is damaged.

G2 (gap 2): period of the cell division cycle between the completion of S phase and the start of cell division or M phase.

G2/M transition checkpoint: the checkpoint controlling entry into mitosis. Cells can proceed into mitosis only when cyclin dependent kinase 1 is active, which in turn only occurs if the DNA is undamaged and the cell is large enough.

GABA (γ-amino butyric acid): a γ-amino acid that acts to open chloride channels in the plasma membrane of sensitive nerve cells. Usually called γ-amino butyric acid rather than γ-amino butyrate, even though the latter (NH_3^+—CH_2—CH_2—CH_2—COO^-) is the form in which it is found at neutral pH.

G-actin: globular, subunit form of actin.

gamete: a sperm or egg. Gametes are haploid; that is, they contain just one set of chromosomes (23 in humans).

gamma-amino butyric acid (γ-amino butyric acid, GABA): a γ-amino acid that acts to open chloride channels in the plasma membrane of sensitive nerve cells. Usually called γ-amino butyric acid rather than γ-amino butyrate, even though the latter (NH_3^+—CH_2—CH_2—CH_2—COO^-) is the form in which it is found at neutral pH.

GAP (GTPase activating protein): a protein that speeds up the rate at which GTPases hydrolyze GTP and therefore switch from the active to the inactive state.

gap 0 (G0): describes the quiescent state of cells that have left the cell division cycle.

gap 1 (G1): period of the cell division cycle that separates mitosis from the following S phase.

gap 2 (G2): period of the cell division cycle between the completion of S phase and the start of cell division or M phase.

gap junction: type of cell junction that allows solute to pass from the cytosol of one cell to the cytosol of its neighbor without passing through the extracellular medium. Gap junctions consist of many paired connexons.

gap junction channel: the channel formed by two connexons. Connexons only open when they contact a second connexon on another cell, in this case they open and form a water-filled tube about 1.5 nm in diameter that runs all the way through the plasma membrane of the first cell, across the small gap between the cells, and through the plasma membrane of the second cell, so allowing passage of solute from the cytosol of one cell to the cytosol of the other.

gastrocnemius muscle: the muscle at the back of the shin. When it contracts the toes move down.

gated, gating: a channel is gated if it can switch to a shape in which the tube through the membrane is closed.

gated transport: transport of fully folded proteins through intracellular pores that open to allow their passage.

GEF (guanine nucleotide exchange factor): a protein that accelerates the rate at which GDP leaves a GTPase to be replaced by GTP, thus switching the GTPase from its inactive state to its active state.

GEFS (generalized epilepsy with febrile seizures): a relatively common form of childhood epilepsy, usually clearing up spontaneously. $GEFS^+$ is the less common condition in which the seizures still occur past the age of six years.

gelsolin: type of actin-binding protein that binds to and fragments actin filaments.

gene: the fundamental unit of heredity. In many cases a gene contains the information needed to code for a single polypeptide.

gene chip: a tiny glass wafer to which cloned DNAs are attached. Also known as microarrays or DNA chips.

gene family: a group of genes that share sequence similarity and usually code for proteins with a similar function.

gene probe: a cDNA or genomic DNA fragment used to detect a specific DNA sequence to which it is complementary in sequence. The probe is tagged in some way to make it easy to detect. The tag could be, for example, a radioactive isotope or a fluorescent dye.

gene therapy: a correction or alleviation of a genetic disorder by the introduction of a normal gene copy into an affected individual.

genetically modified (GM): an organism with a genome that has been modified by modern molecular techniques, usually by the addition of novel gene(s) or by swapping in new DNA to replace existing gene(s).

genetic code: the relationship between the sequence of the four bases in DNA and the amino acid sequence of proteins.

genome: one complete set of an organism's genes. In humans, the genome resides on 23 chromosomes, and each cell has two sets.

genomic DNA library: a collection of bacterial clones, each of which contains a different fragment of foreign genomic DNA.

germ cells: the cells that give rise to the eggs and sperm.

Gibbs free energy: if the change of Gibbs free energy during a reaction is negative, the reaction is favored and will proceed if the activation energy is sufficiently low.

glial cells: electrically inexcitable cells found in the nervous system.

glucocorticoid: a steroid hormone produced by the adrenal cortex that forms part of the system controlling blood sugar levels.

glucocorticoid receptor: the intracellular receptor to which glucocorticoid hormone binds.

gluconeogenesis: the synthesis of glucose from noncarbohydrate precursors such as amino acids and lactate.

glucose: a hexose monosaccharide. Glucose is the commonest sugar in the blood and is the dominant cellular fuel in animals, being used in glycolysis to generate ATP and pyruvate, the latter fuelling the Krebs cycle.

glucose carrier: a plasma membrane protein that carries glucose into or out of cells. Some cells, such as skeletal muscle cells, will only translocate glucose carriers to their membranes when protein kinase B is active.

glyceride: compound formed by attaching units to a glycerol backbone. Triacylglycerols (previously called triglycerides) and phospholipids are glycerides.

glycerol: $CH_2OH—CHOH—CH_2OH$. The backbone to which acyl groups (fatty acid chains) are attached to make triacylglycerols and phospholipids.

glycogen: a glucose polymer that can be quickly hydrolysed to yield glucose; an $\alpha 1 \rightarrow 4$ polymer of glucose with $\alpha 1 \rightarrow 6$ branches.

glycogen phosphorylase: an enzyme that releases glucose-1-phosphate monomers from glycogen. The glucose-1-phosphate is then converted to glucose-6-phosphate which can be used in respiration or dephosphorylated to glucose for release into the blood.

glycogen phosphorylase kinase (phosphorylase kinase): a kinase that is activated by the calcium-calmodulin complex and phosphorylates glycogen phosphorylase, activating the latter enzyme.

glycolysis: the breakdown of glucose to pyruvate.

glycosidic bond: a bond linking monosaccharide residues in which the carbon backbones are linked through oxygen and a water molecule is lost.

glycosylation: the addition of sugar residues to a molecule. Both proteins and lipids can be glycosylated.

glyoxylate: CHO—COO$^-$; an intermediate in various metabolic pathways.

GM (genetically modified): an organism with a genome that has been modified by modern molecular techniques, usually by the addition of novel gene(s) or by swapping in new DNA to replace existing gene(s).

Golgi apparatus: system of flattened cisternae concerned with glycosylation and other modifications of proteins.

G protein (trimeric G protein): a protein that links a class of metabotropic cell surface receptors with downstream targets. Trimeric G proteins comprise an α subunit that binds and hydrolyses GTP, and a $\beta\gamma$ subunit that dissociates from the α subunit while the latter is in the GTP-bound state. Important G proteins are G_q, which activates phospholipase Cβ, and G_s, which activates adenylate cyclase.

G_q: the isoform of trimeric G protein that activates phospholipase Cβ and therefore generates a calcium signal.

gratuitous inducer: an inducer of transcription that is not itself metabolized by the resulting enzymes.

Grb2 (growth factor receptor binding protein number 2): a linker protein that has an SH2 domain and is therefore recruited to phosphotyrosine, e.g., on receptor tyrosine kinases. Grb2 in turn recruits SOS, bringing SOS to the plasma membrane where it can act as a guanine nucleotide exchange protein (GEF) for Ras.

green fluorescent protein: a fluorescent protein made by the jellyfish *Aequorea victoria*. Unlike other colored or fluorescent proteins, it contains no prosthetic groups and therefore will fluoresce when expressed by any cell in which the gene is successfully inserted and expressed.

growth factor: a paracrine transmitter that modifies the developmental pathway of the target cell, often by causing cell division.

growth factor receptor binding protein number 2 (Grb2): a linker protein that has an SH2 domain and is therefore recruited to phosphotyrosine, e.g., on receptor tyrosine kinases. Grb2 in turn recruits SOS, bringing SOS to the plasma membrane where it can act as a guanine nucleotide exchange protein (GEF) for Ras.

G_s: the isoform of trimeric G protein that activates adenylate cyclase and therefore causes an increase of cAMP concentration.

GTPase: an enzyme that hydrolyses GTP. The name is usually restricted to that family of proteins that bind GTP and adopt a new shape that can then activate target proteins. Once they hydrolyze their bound GTP they switch back to the original form. Examples are EF-tu, Arf, Ran, dynamin, G_q and G_s.

GTPase activating protein (GAP): a protein that speeds up the rate at which GTPases hydrolyse GTP and therefore switch from the active to the inactive state.

guanine: one of the bases found in DNA and RNA. Guanine is a purine.

guanine nucleotide exchange factor (GEF): a protein that accelerates the rate at which GDP leaves a GTPase to be replaced by GTP, thus switching the GTPase from its inactive state to its active state.

guanosine: guanine linked to the sugar ribose. Guanosine is a nucleoside.

guanylate cyclase: one of a family of enzymes that generates cyclic guanosine monophosphate (cGMP) from GTP. Two isoforms of guanylate cyclase are mentioned in this book. The guanylate cyclase found in photoreceptors is constitutively active, making cGMP all the time. The guanylate cyclase found in smooth muscle cells is activated by nitric oxide.

Guthrie test: test for phenylketonuria. Newborn babies' blood is tested for the presence of phenylalanine at unusually high concentration.

hairpin loop: a loop in which a linear object folds back on itself. Used to describe the loop formed in a RNA molecule due to complementary base pairing.

haploid: containing a single copy of each chromosome, in humans, this means 23 chromosomes. Sperm and eggs are haploid while somatic cells contain one set of 23 from each parent and are referred to as being diploid.

head group: the hydrophilic group found in phospholipids. The head group is attached to the glycerol backbone by a phosphodiester link. Examples are choline and inositol.

heat shock proteins: a class of proteins found in all cells but expressed in higher amounts when cells are stressed by factors such as heat. They have roles in assisting protein folding and rescuing unfolded proteins.

heavy chain: the longer of the two polypeptide chains that form an antibody.

heel prick test (Guthrie test): test for phenylketonuria. Newborn babies' blood is tested for the presence of phenylalanine at unusually high concentration.

helicase: an enzyme that helps unwind the DNA double helix during replication.

α helix: a common secondary structure in proteins in which the polypeptide chain is coiled, each turn of the helix taking 3.6 amino acid residues. The nitrogen atom in each peptide bond forms a hydrogen bond with the oxygen four residues ahead of it in the polypeptide chain.

helix-turn-helix: a motif in protein structures that consists of a length of α helix separated from another section of α helix by a turn. The motif is found in many DNA-binding proteins.

helper T cell: T cell expressing the integral plasma membrane protein CD4. Helper T cells activate B cells if the peptide presented by MHC proteins on the B cell binds tightly to their own T cell receptor.

hemoglobin: the oxygen-carrying, iron-containing protein of the blood.

heterochromatin: that portion of the nuclear chromatin that is tightly packed. Much of the heterochromatin is repetitive DNA with no coding function.

heterozygote: an individual whose two copies of a gene differ: for example, one may be mutant. We have not used the term in this book, but have used the simpler word "carrier" for the more restricted case where an individual is a heterozygote for a recessive gene.

hexokinase: enzyme that phosphorylates glucose on the number 6 carbon.

histone: a positively charged protein that binds to negatively charged DNA and helps to fold DNA into chromatin.

histone octamer: two molecules each of histones H2A, H2B, H3 and H4, the whole forming a nucleosome.

homologous, homology: objects that are similar because they have a common ancestor.

homologous chromosomes: chromosomes carrying the same set of genes. One of a pair of homologous chromosomes is inherited from the mother, the other from the father.

homologous proteins: proteins that are similar because they have a common evolutionary origin. For example, small GTPases such as Ras, Ran, Arf and Rab are thought to have a common ancestor encoded by a gene that was then duplicated and mutated.

homologous recombination: process in which a length of DNA with ends that are homologous to a section of chromosome swaps in, replacing the existing length of DNA in the chromosome. Homologous recombination occurs naturally at chiasmata during crossing over during meiosis. It can also occur in some cells when they transfected with the appropriate exogenous DNA, and in those cases is thought to use some of the same enzymes.

homozygote: an individual whose two copies of a gene are identical. Most of us are homozygous for most of our genes.

hormone: long-lived transmitter that is released into the blood and travels around the body before being broken down.

hormone response element (HRE): a specific DNA sequence to which a steroid hormone receptor binds.

housekeeping gene: a gene that is transcribed into mRNA in nearly all the cells of a eukaryotic organism; in bacterial cells a housekeeping gene is one that is always being transcribed.

HRE (hormone response element): a specific DNA sequence to which a steroid hormone receptor binds.

hybridization: the association of unlike things. In molecular genetics, the association of two nucleic acid strands (either RNA or DNA) by complementary base pairing.

hybridoma: an immortal cell line generated by fusing two cells, one of which is a cancer cell. The term is usually used of clones generated by fusing a B cell with a cancer cell to generate immortal cells secreting monoclonal antibodies.

hydration shell: the cloud of water molecules that surrounds an ion in solution.

hydrocarbon tail: the long chain of carbon atoms with attached hydrogens found in phospholipids and triacylglycerols. The tail represents all of a molecule of fatty acid except the carboxyl group.

hydrogen bond: a relatively weak bond formed between a hydrogen atom and two electron grabbing atoms (such as nitrogen or oxygen) where the hydrogen is shared between the other atoms.

hydrogen ion gradient: an energy currency. Hydrogen ions are more concentrated outside bacteria and mitochondria than inside, and this chemical gradient is supplemented by a voltage gradient pulling hydrogen ions in. If hydrogen ions are allowed to rush in down their electrochemical gradient they release 17,000 Joules/mole.

hydrolysis: breakage of a covalent bond by the addition of water. —H is added to one side, —OH to the other.

hydropathy plot: a running average of side chain hydrophobicity along a polypeptide chain. From the hydropathy plot one can predict, for example, membrane-spanning domains in integral membrane proteins.

hydrophilic: a molecule or part of a molecule that can interact with water.

hydrophobic: a molecule or part of a molecule that will associate with other hydrophobic molecules in preference to water.

hydrophobic effect: the tendency of hydrophobic molecules or parts of molecules to cluster together away from water, such as hydrophobic amino acid residues in the center of a protein, or the fatty acid chains in lipid bilayers.

hydroxyl group: the —OH group. The term is specifically *not* used of an —OH that forms part of a carboxyl group.

hypoxanthine: a purine that is used to make the nucleotide inosine. Inosine can pair with any of uracil, cytosine or adenine.

imino acid: an organic molecule containing both carboxyl (—COOH) and imino (—NH—) groups. Proline is an imino acid although it is usually called an amino acid.

immunized: injected with an antigen, so that the immune system will respond by generating antibodies that bind the antigen.

immunofluorescence: the use of fluorescently labelled antibodies to reveal the location of specific chemicals, e.g., in fluorescence microscopy or in western blotting.

inactivation (of voltage gated channels): blockage of the open channel with a plug that is attached to the cytosolic face of the protein.

indirect immunofluorescence: a technique in which a fluorescent secondary antibody is used to label a preapplied primary antibody specific for a particular protein or subcellular structure. Also called secondary immunofluorescence.

inducible operon: an operon that is transcribed only when a specific substance is present.

initial velocity (of a reaction): the rate at which an enzyme converts substrate to product in the absence of product, that is, at the onset of a reaction.

initiation factor: a protein that assists the small and large ribosomal subunits of eukaryotes to assemble correctly on the mRNA at the start of protein synthesis.

innate immunity: a form of immunity found throughout multicellular organisms that is fully encoded by the inherited genome.

inositol: a cyclic polyalcohol, $(CHOH)_6$, that forms the head group of the phospholipid phosphatidylinositol. Phosphorylation of inositol yields inositol trisphosphate.

inositol trisphosphate (IP_3, $InsP_3$): a small (Mr = 420) phosphorylated cyclic polyalcohol that is released into the cytosol by the action of phospholipase C on the membrane lipid phosphatidylinositol bisphosphate, which acts to cause release of calcium ions from the endoplasmic reticulum.

inositol trisphosphate-gated calcium channel: a channel found in the endoplasmic reticulum of many cells. The channel opens when inositol trisphosphate binds to its cytoplasmic aspect. It allows only calcium ions to pass.

insertional mutagenesis: the insertion of additional nucleotides into a stretch of DNA.

in situ hybridization: the binding of a particular labelled sequence of RNA (or DNA) to its matching sequence in the genome as a way of searching for the location of that sequence in a tissue sample.

$InsP_3$ (inositol trisphosphate, IP_3): a small phosphorylated cyclic polyalcohol that is released into the cytosol by the action of phospholipase C on the membrane lipid phosphatidylinositol bisphosphate and which acts to cause release of calcium ions from the endoplasmic reticulum.

insulin: a hormone produced by endocrine cells in the pancreas. It activates its own receptor tyrosine kinase, which in turn acts mainly through activation of PI-3-kinase and hence protein kinase B.

insulin receptor: a receptor tyrosine kinase specific for insulin, which acts mainly through activation of PI-3-kinase and hence protein kinase B.

insulin receptor substrate number 1 (IRS-1): a protein phosphorylated on tyrosine by the insulin receptor. Once phosphorylated it recruits PI-3-kinase, which can then be phosphorylated and hence activated.

integral protein (of a membrane): class of protein that is tightly associated with a membrane, usually because its polypeptide chain crosses the membrane at least once. Integral membrane proteins can only be isolated by destroying the membrane, e.g., with detergent. In contrast, peripheral membrane proteins are more loosely associated.

integrin: dimeric proteins with an extracellular domain that binds to extracellular matrix proteins, and an intracellular domain that attaches to actin microfilaments.

intermediate filament: one of the filaments that makes up the cytoskeleton; composed of various subunit proteins.

intermembrane space: in organelles such as mitochondria, chloroplasts and nuclei that are bound by two membranes the intermembrane space is the aqueous space between the inner and outer membranes. The intermembrane space of nuclei is continuous with the lumen of the ER. The intermembrane space of mitochondria has the ionic composition of cytosol because porin in the outer mitochondrial membrane allows solutes of $M_r < 10{,}000$ to pass.

interphase: period of synthesis and growth that separates one cell division from the next; consists of the G1, S, and G2 phases of the cell division cycle.

interstitial fluid: the extracellular medium that lies between the cells of a multicellular organism.

intracellular messenger: a cytosolic solute that changes in concentration in response to external stimuli or internal events, and which acts on intracellular targets to change their behavior. Calcium ions, cyclic AMP and cyclic GMP are the three common intracellular messengers.

intracellular receptors: receptors that are not on the plasma membrane but which lie within the cell and bind transmitters that diffuse through the plasma membrane.

intron: in a eukaryotic gene, introns are those parts that are spliced out before the RNA leaves the nucleus. In contrast exons are those parts that after RNA processing leave the nucleus.

ion: a charged chemical species. A single atom that has more or less electrons than are required to exactly neutralize the charge on the nucleus is an ion (e.g., Na^+, Cl^-). A molecule with one or more charged regions is also an ion (e.g., phosphate, $HPO_4{}^{2-}$ and leucine, $NH_3{}^+$—CH(CH_2.CH(CH_3)$_2$)—COO^-).

ionotropic cell surface receptors: channels that open when a specific chemical binds to the extracellular face of the channel protein.

ionotropic glutamate receptor: a channel which opens when glutamate binds to its extracellular aspect, and which allows sodium and potassium ions to pass; some isoforms also pass calcium.

IP_3 (inositol trisphosphate, $InsP_3$): a small (Mr = 420) phosphorylated cyclic polyalcohol that is released into the cytosol by the action of phospholipase C on the membrane lipid phosphatidylinositol bisphosphate and which acts to cause release of calcium ions from the endoplasmic reticulum.

IP_3 receptor: a calcium channel in the membrane of the endoplasmic reticulum that opens when inositol trisphosphate binds to its cytosolic aspect.

IRS-1: a protein phosphorylated on tyrosine by the insulin receptor. Once phosphorylated it recruits PI-3-kinase, which can then be phosphorylated and hence activated.

isoelectric point: the pH at which a protein or other molecule has no net charge.

isoforms: related proteins that are the products of different genes or differential splicing of mRNAs from one gene.

isomers: different compounds with the same molecular formula. For example glucose and mannose are both $C_6H_{12}O_6$ and are therefore isomers.

JAK: a cytoplasmic tyrosine kinase that associates with type 1 cytokine receptors and phosphorylates targets, including the STAT transcription factors, when the cytokine ligand binds to the receptor.

JNK: a stress activated protein kinase that stimulates cell repair but can also trigger apoptosis.

K^+/Na^+ ATPase (potassium/sodium ATPase): a plasma membrane carrier. For every ATP hydrolysed three Na^+ ions are moved out of the cytosol and two K^+ ions are moved in. Usually called the sodium/potassium ATPase.

keratin: a protein that makes up the intermediate filaments in epithelial cells.

ketone: any chemical containing a carbon atom with single bonds to two other carbons and a double bond to an oxygen. Acetone (CH_3—CO—CH_3) and acetoacetate (CH_3—CO—CH_2—COO^-) are ketones.

ketone bodies: acetoacetate (CH_3—CO—CH_2—COO^-) and 3-hydroxybutyrate (CH_3—CHOH—CH_2—COO^-); circulating fuels formed from fats and used to fuel body tissues in times of starvation. Acetoacetate slowly loses carbon dioxide to form acetone.

ketosis: the overproduction of ketone bodies that is seen in extreme starvation and in diabetes.

kinase: an enzyme that phosphorylates a molecule by transferring a phosphate group from ATP to the molecule.

kinesin: the molecular motor protein responsible for movement along microtubules in the anterograde direction.

kinetochore: the point of attachment of the chromosome to the spindle. The kinetochore forms around the centromere.

K_M (Michaelis constant): the substrate concentration at which one measures an initial velocity that is half as fast as the maximal velocity (V_m) of an enzyme reaction.

knockout: a knocked out gene is one whose function has been disrupted artificially and which can no longer code for a functional protein. A knockout animal is one in which this procedure has been performed.

Krebs cycle (tricarboxylic acid cycle): the series of reactions in the mitochondrial matrix in which acetate is completely oxidized to CO_2 with the attendant reduction of NAD^+ to NADH and FAD to $FADH_2$.

lac (lactose) operon: the cluster of three bacterial genes which encode enzymes involved in metabolism of lactose.

lactose: a disaccharide comprising galactose linked to glucose by an $\beta 1 \rightarrow 4$ glycosidic bond.

lactose intolerant: unable to hydrolyse dietary lactose.

lactose (lac) operon: the cluster of three bacterial genes which encode enzymes involved in metabolism of lactose.

lagging strand: the strand of DNA that grows discontinuously during replication: in contrast to the leading strand, which is synthesised continuously.

lamins: intermediate filament proteins that make up the nuclear lamina.

leading strand: a strand of DNA that grows continuously in the 5′ to 3′ direction by the addition of deoxyribonucleotides.

leukocyte: a white blood cell.

ligand: when two molecules bind together, one (often the smaller one) is called the ligand, and the other (often the bigger one) is called the receptor.

light chain: the shorter of the two polypeptide chains that form an antibody.

linkage, linked (of genes): the physical association of genes on the same chromosome. Linked genes tend to be inherited together.

linker DNA: a stretch of DNA that separates two nucleosomes.

lipid: any cellular component that is soluble in an organic solvent such as octane. The term includes triacylglycerols, cholesterol, steroid hormones and many other materials.

lipid bilayer: two layers of lipid molecules that form a membrane.

localization sequence (targeting sequence): a stretch of polypeptide that determines the cellular compartment to which a synthesized protein is sent.

locus, loci: a position, e.g., on a chromosome. In particular the sections of DNA encoding the different polypeptides that comprise antibodies and T cell receptors are termed loci, the word gene being reserved for components within these loci.

lumen: the inside of a closed structure or tube.

lymphocyte: a class of immune system cell comprising B and T cells together with natural killer cells.

lymphoid: of the lymph nodes. B and T cells, which reside in the lymph nodes for some of their lives, are termed lymphoid cells or simply lymphocytes.

lysosome: a membrane-bound organelle containing digestive enzymes.

M phase: period of the cell division cycle during which the cell divides; consists of mitosis and cytokinesis.

macromolecule: a large molecule.

macrophage: a phagocytic housekeeping cell that engulfs and digests bacteria and dead cells.

major groove (of DNA): the wider of the two grooves along the surface of the DNA double helix.

major histocompatability complex (MHC) protein: an integral protein of the plasma membrane that presents short lengths of peptide to patrolling T cells.

mannose: a hexose monosaccharide.

mannose-6-phosphate: mannose that is phosphorylated on carbon number 6. Mannose-6-phosphate is the sorting signal that identifies lysosomal proteins.

MAP kinase (MAPK, mitogen associated protein kinase): an enzyme that phosphorylates numerous targets, including transcription factors that trigger transcription of the cyclin D gene.

MAPKK (MAPK kinase): the enzyme that phosphorylates and activates MAP kinase, it itself being phosphorylated and hence activated by MAPKK kinase.

MAPKKK (MAPKK kinase): the enzyme that phosphorylates and activates MAPK kinase. The most important isoform of MAPKKK is also called Raf. Raf is activated by the G protein Ras.

matrix: a vague term meaning a more or less closed location, often but not exclusively one that is a solid basis on which things can grow or attach. The term is used of the extracellular matrix in animals, formed of collagen and other fibers, and of the aqueous volume inside various organelles.

maximal velocity (of an enzyme catalyzed reaction): the limiting value of the initial velocity of a reaction as the substrate concentration is increased at constant enzyme concentration. Occurs when the enzyme is saturated with substrate. Written V_m or V_{max}.

meiosis: the form of cell division that produces gametes, each with half the genetic material of the cells that produce them.

meiosis I and II: the first and second meiotic divisions.

meiotic spindle: a bipolar, microtubule-based structure on which chromosome segregation occurs during meiosis I and II.

membrane: a planar sheet. Biological membranes comprise a lipid bilayer plus protein.

membrane attack complex: a channel formed in the membrane of target cells as a result of CD8+ T cell attack.

memory B cell: a B cell that has undergone the final tuning of antibody structure (somatic hypermutation plus, usually, class switching) but which expresses membrane bound antibody and is therefore able to respond, by rapid proliferation, to a repeat of the same antigen challenge. It is the presence of memory B cells (and memory T cells, which are not described in this book) that gives us immunity to a toxin or pathogen following an initial exposure.

messenger ribonucleic acid (mRNA): the RNA molecule that carries the genetic code. The order of bases on mRNA specifies the amino acid sequence of a polypeptide chain. In eukaryotes, the mRNA leaves the nucleus and is translated into protein in the cytoplasm.

metabolism: all of the reactions going on inside a cell.

metabotropic cell surface receptors: receptors that are linked to, and activate, enzymes.

metaphase: the period of mitosis or meiosis at which the chromosomes align prior to separation at anaphase.

metaphase plate: the equator of the mitotic or meiotic spindle; the point at which the chromosomes congregate at metaphase of mitosis or meiosis.

7-methyl guanosine cap: modified guanosine found at the 5′ terminus of eukaryotic mRNA. A guanosine is attached to the mRNA by a 5′ 5′-phosphodiester link and is subsequently methylated on atom number 7 of the guanine.

MHC complex proteins: a family of integral plasma membrane proteins expressed by all somatic cells whose function is to present short lengths of peptide, generated by proteolysis of cytosolic or endocytosed proteins, to patrolling T cells.

Michaelis constant (K_M): the substrate concentration at which one measures an initial velocity that is half as fast as the maximal velocity (V_m) of an enzyme reaction.

Michaelis-Menten equation: an equation that defines the effect of substrate concentration on the initial velocity of an enzyme-catalyzed reaction:

$$v_o = \frac{V_m[S]}{K_M + [S]}$$

where V_m is the limiting initial velocity obtained as the substrate concentration approaches infinity, [S] is the substrate concentration and K_M is a constant called the Michaelis constant, equal to the substrate concentration at which one measures an initial velocity that is half as fast as the maximal velocity.

microarray: a tiny glass wafer to which cloned DNAs are attached. Also known as gene chips or DNA chips.

microfilament: one of the major filaments of the cytoskeleton; also known as actin filament or F-actin; synonymous with the thin filament of striated muscle.

microorganism: any single celled organism, whether prokaryote or eukaryote. Bacteria like *E. coli*, paramecium and yeast are microorganisms.

microsatellite DNA: repetitious DNA of unknown function comprising many repeats of a unit of four or fewer base pairs.

microtubular molecular motor: protein that moves vesicles along cytoskeletal filaments; examples of microtubule-based motors are cytoplasmic dynein and kinesin.

microtubule: tubular cytoplasmic filament composed of tubulin. Tubulin is the major component of 9+2 cilia and flagella and of the mitotic and meiotic spindle.

microtubule organizing center (abbr. **MTOC**)**:** structure from which cytoplasmic microtubules arise; synonymous with the centrosome.

microvilli (sing. **microvillus**)**:** projections from the surface of epithelial cells that increase the absorptive surface; contain actin filaments.

minisatellite DNA: repetitious DNA of unknown function comprising up to 20,000 repeats of a unit of about 25 base pairs.

minor groove (of DNA): the narrower of the two grooves along the surface of the DNA double helix.

mismatch repair: a cell marks its DNA strands by methylation. This allows the cell to distinguish the template strand from a newly synthesized DNA. If an incorrect base, the mismatch, is introduced into the new strand, the repair enzymes will remove the incorrect nucleotide and insert the correct nucleotide in place using the methylated strand as the template.

missense mutation: a base change in a DNA molecule that changes a codon so that it now specifies a different amino acid.

mitochondrial inner membrane: the inner membrane of mitochondria that is elaborated into cristae. The electron transport chain and ATP synthase are integral membrane proteins of the mitochondrial inner membrane.

mitochondrial matrix: the aqueous space inside the mitochondrial inner membrane, where the enzymes of the Krebs cycle are located.

mitochondrial outer membrane: the outer membrane of mitochondria, permeable to solutes of $M_r < 10{,}000$ because of the presence of the channel porin.

mitochondrion: the cell organelle concerned with aerobic respiration.

mitogen: anything that promotes cell division. FGF and PDGF (fibroblast growth factor and platelet-derived growth factor) are potent mitogens for endothelial and smooth muscle cells.

mitogen associated protein kinase (MAP kinase, MAPK): an enzyme that phosphorylates numerous targets, including transcription factors that trigger transcription of the cyclin D gene.

mitosis: the type of cell division found in somatic cells, in which each progeny cell receives the full complement of genetic material present in the original cell.

mitotic spindle: microtubule based structure upon which chromosomes are arranged and translocated during mitosis.

mole: an amount of substance comprising 6.023×10^{23} (Avogadro's number) molecules. One mole has a mass equal to the value of the relative molecular mass expressed in grams.

monoclonal: pertaining to a single clone. A preparation of monoclonal antibody is secreted by a clone of B cell hybridoma cells and therefore comprises identical immunoglobulin molecules, all of which recognize exactly the same target.

monomer: a single unit, usually used to refer to a single building block of a larger molecule. Thus DNA is formed of nucleotide monomers. By analogy, the word is sometimes used to describe proteins that act as a single unit to distinguish them from related proteins that act as larger units, so myoglobin is said to be monomeric by comparison with hemoglobin, which has four subunits. GTPases such as Ran, Arf and Ras are sometimes called "monomeric G proteins" to distinguish them from trimeric G proteins such as G_q and G_s.

monosaccharide: a sweet-tasting chemical with many hydroxyl groups that can adopt a form in which an oxygen atom completes a ring of carbons. All the monosaccharides in this book have the general formula $C_n(H_2O)_n$ where $n = 5$ (pentoses) or 6 (hexoses).

motif: a recognizable conserved sequence of bases (in DNA) or amino acids (in a polypeptide). A motif in DNA may bind one transcription factor, for example 5′ AGAACA 3′ binds the glucocorticoid receptor. A motif in a polypeptide may, for example, fold in a particular way (for example a helix-turn-helix motif), or bind to a target (e.g., KDEL).

motoneuron: the nerve cell that carries action potentials from the spine to the muscles. It releases the transmitter acetylcholine onto the muscle cells, causing them to depolarize and hence contract.

M phase: period of the cell division cycle during which the cell divides; consists of mitosis and cytokinesis.

mRNA (messenger RNA): the RNA molecule that carries the genetic code. The order of bases on mRNA specifies the amino acid sequence of a polypeptide chain.

MTOC (microtubule organizing center): structure from which cytoplasmic microtubules arise; synonymous with the centrosome.

muscle: tissue specialized for generating contractile force.

mutation: an change in the structure of a gene or chromosome that is inherited by progeny cells or, in the case of gametes, by children.

myelin: a fatty substance that is wrapped around nerve cell axons by glial cells.

MyoD: a transcription factor of skeletal muscle that acts to increase the expression of muscle specific proteins such as myosin.

myosin: type of motor protein that moves along, or pulls on, actin filaments. The thick filaments in skeletal muscle are formed of the myosin II isoform.

myosin II: an isoform of myosin that forms the large filaments seen in striated muscle but which is also found in other cell types.

myosin V: an isoform of myosin that carries cargo along actin filaments.

N-terminus (amino terminus): the end of a peptide or polypeptide that has a free α-amino group. This end is made first on the ribosome.

NADH (nicotinamide adenine dinucleotide, reduced form): a combination of two nucleotides that is a strong reducing agent and one of the cell's energy currencies.

NADPH (nicotinamide adenine dinucleotide phosphate, reduced form): phosphorylated NADH; like NADH it is a strong reducing agent, but NADPH is not an energy currency, rather it is used in synthetic reactions in the cytoplasm.

Na^+/K^+ ATPase: a plasma membrane carrier. For every ATP hydrolyzed, three Na^+ ions are moved out of the cytosol and two K^+ ions are moved in.

necrosis: cell death that is due to damage so severe that the cell cannot maintain the level of its energy currencies and therefore falls apart—distinct from apoptosis.

negative feedback: a control system in which a change in some parameter activates a mechanism that reverses the change in that parameter; an example is the effect of tryptophan on expression of the *trp* operon.

negative regulation (of transcription): the inhibition of transcription due to the presence of a particular substance which is often the end product of a metabolic pathway.

Nernst equation: the equation that allows the equilibrium voltage of an ion across a membrane to be calculated. The general form is:

$$V_{eq} = \frac{RT}{zF} \log_e \left(\frac{[I_{outside}]}{[I_{inside}]} \right) \text{ volts}$$

where R is the gas constant (8.3 Joules per degree per mole), T is the absolute temperature, z is the charge on the ion in elementary units, and F is 96,500 coulombs in a mole of monovalent ions. The value of RT/F is 0.025 at a room temperature of 22°C, and 0.027 at human body temperature.

nerve cell: an electrically excitable cell with a long axon specialized for transmission of (usually sodium) action potentials.

nerve growth factor (NGF): the older name for neurotrophin 1.

nervous tissue: a tissue formed of nerve and glial cells that carries out electrical data processing.

neurofilament: a type of intermediate filament found in nerve cells.

neuron: other name for a nerve cell.

neurotrophin: one of a family of paracrine transmitters that act upon nerve cells to prevent apoptosis and trigger differentiation. Neurotrophin number one used to be called nerve growth factor or NGF.

NFAT: a transcription factor that only translocates to the nucleus when dephosphorylated by the calcium-calmodulin-activated phosphatase calcineurin.

NGF (nerve growth factor): the older name for neurotrophin 1.

nicotinamide: a base used in the double nucleotides NADH and NADPH.

nicotinamide adenine dinucleotide (reduced form) (NADH): a combination of two nucleotides that is a strong reducing agent and one of the cell's energy currencies.

nicotinamide adenine dinucleotide phosphate (reduced form) (NADPH): phosphorylated NADH; like NADH it is a strong reducing agent but NADPH is not an energy currency, rather it is used in synthetic reations in the cytoplasm.

nicotinic acetylcholine receptor: a channel which opens when acetylcholine binds to its extracellular aspect, and which allows sodium and potassium ions to pass; some isoforms also pass calcium.

nine plus two (9+2) axoneme: structure of cilium or flagellum; describes the arrangement of nine peripheral microtubules surrounding two central microtubules.

nitric oxide (NO): a paracrine transmitter that acts on intracellular receptors, the most important of which is guanylate cyclase.

nitrogenase: an enzyme complex used by some prokaryotes to reduce nitrogen gas to ammonia. It uses ATP and a strong reducing agent (reduced ferredoxin). It has an iron-molybdenum cofactor.

node (of a myelinated axon): the gap between adjacent lengths of myelin sheath, where the nerve cell plasma membrane directly contacts the extracellular medium. Also called node of Ranvier, after its discoverer, Louis Ranvier.

nonpolar: the covalent bonds in nonpolar molecules have electrons shared equally so that the constituent atoms do not carry a charge.

nonsense mutation: a change in the base sequence that generates a stop codon.

northern blotting: a blotting technique in which RNAs, separated by size, are probed using a single-stranded cDNA probe or an antisense RNA probe.

N-terminus (amino terminus): the end of a peptide or polypeptide that has a free α-amino group. This end is made first on the ribosome.

NTR: neurotrophin receptor. Although the Trk family are receptor tyrosine kinases for neurotrophins, the name NTR is usually reserved for the death domain receptor called p75 NTR.

nuclear envelope: double membrane system enclosing the nucleus; it contains nuclear pores and is continuous with the endoplasmic reticulum.

nuclear lamina: meshwork of intermediate filaments lining the inner face of the nuclear envelope, formed of the monomer lamin.

nuclear pores: holes running through the nuclear envelope that regulate traffic of proteins and nucleic acids between the nucleus and the cytoplasm.

nuclear pore complex: multiprotein complex that lies within and around the nuclear pore and which regulates import into and export from the nucleus in a process called gated transport.

nuclease: an enzyme that degrades nucleic acids.

nucleic acid: a polymer of nucleotides joined together by phosphodiester links. DNA and RNA are nucleic acids.

nucleoid: the region of a bacterial cell which contains the chromosome.

nucleolar organizer regions: region of one or more chromosomes at which the nucleolus is formed.

nucleolus (pl. nucleoli): the region(s) of the nucleus concerned with the production of ribosomes.

nucleoside: a purine, a pyrimidine, or nicotinamide attached to either ribose or deoxyribose.

nucleosome: the bead-like structure formed by a stretch of DNA wrapped around a histone octamer.

nucleotide: building block of nucleic acids; a nucleoside that is phosphorylated on its 5′ carbon atom.

nucleotide excision repair: process that removes a thymine dimer, together with some 30 surrounding nucleotides, from DNA. The gap is then repaired by the actions of DNA polymerase I and DNA ligase.

nucleus: the cell organelle housing the chromosomes; enclosed within a nuclear envelope.

numb: a peripheral plasma membrane protein that controls cell fate choices, particularly in the nervous system.

objective lens: the lens of a light or electron microscope that forms a magnified image of a specimen.

obligate anaerobe: an organism that is poisoned by oxygen so that it can only function anaerobically.

oil: a triacylglycerol (triglyceride) that is liquid at room temperature. In contrast, fats are solid at room temperature.

Okazaki fragment: a series of short fragments that are joined together to form the lagging strand during DNA replication.

oleic acid: $C_{17}H_{33}$—COOH; a fatty acid with a single double bond. The dominant acyl group in olive oil is derived from oleic acid.

olfactory neuron: a nerve cell found in the nose, whose dendrites are sensitive to smell chemicals.

oligonucleotide: a short fragment of DNA or RNA.

oligosaccharide: a chain of up to a hundred or so monosaccharides linked by glycosidic bonds.

oligosaccharide transferase: an enzyme that adds an oligosaccharide group onto a protein.

oocyte: a cell that undergoes meiosis to give rise to an egg.

open complex: a complex between DNA and protein which causes the two strands of DNA to separate, e.g., during DNA replication and transcription.

open promoter complex: a structure formed when the two strands of the double helix separate so that transcription can commence.

operator: the DNA sequence to which a repressor protein binds to prevent transcription from an adjacent promoter.

operon: a cluster of genes that encode proteins involved in the same metabolic pathway and which are transcribed as one length of mRNA under the control of one promoter. As far as we know, operons are only found in prokaryotes.

opsonin: proteins that, upon binding to a target, render that target attractive to phagocytic leukocytes. The process is termed opsonization.

optical isomers: two molecules that differ only in that one is the mirror image of the other.

organelle: a membrane-bound, intracellular structure such as a mitochondrion, chloroplast, lysosome, etc.

organic: in chemistry, an organic compound is one that contains carbon atoms. When applied to farming and food, organic means the avoidance of high-intensity agricultural techniques including genetic engineering.

organism: a cell or clone of cells that functions as a discrete and integrated whole to maintain and replicate itself.

origin of replication: the site on a chromosome at which DNA replication can commence.

osmolarity: the overall strength of a solution; the more solute that is dissolved in a solution, the higher the osmolarity.

osmosis: the movement of water down its concentration gradient.

outer doublet microtubules: the paired microtubules that make the "9" of the 9+2 axoneme in cilia and flagella.

oxidation: the removal of electrons from a molecule, e.g., by adding oxygen atoms, which tend to take more than their fair share of electrons in any bonds they make. Removal of hydrogen atoms from a molecule oxidizes it. (Don't confuse with deprotonation, which is the loss of H^+).

β oxidation: process by which fatty acids are broken down into individual two-carbon units coupled to CoA to form acetyl-CoA. The process, which takes place in the mitochondrial matrix, generates both NADH and $FADH_2$.

oxidizing agent: an agent that will act to lessen the share of electrons that the atoms of a molecule have. Oxidizing agents often work by adding oxygen atoms or by removing hydrogen atoms. (Don't confuse with a base, which will accept H^+, not H). Examples are oxygen itself, NAD^+, and FAD.

P site (peptidyl site): the site on the ribosome occupied by the growing polypeptide chain.

$p21^{Cip1}$: an inhibitor of cyclin dependent kinases. $p21^{Cip1}$, which is upregulated by cell-cell contact, prevents cells entering the S phase of the cell cycle.

p38: a stress-activated protein kinase related to mitogen associated protein kinase (MAP kinase) but with a very different role: p38 is activated by cell stress and stimulates both cell repair and apoptosis.

p53: a transcription factor that stimulates cell repair but also apoptosis. Cancer cells are frequently found to have mutated, nonfunctional *P53* genes.

p75 neurotrophin receptor: a death domain receptor for neurotrophins. Upon binding neurotrophins p75 activates caspase 8 and hence initiates apoptosis unless a countermanding signal to survive is present.

PAC (P1 artificial chromosome): a cloning vector, derived from the bacteriophage P1, that is used to propagate, in *E. coli*, DNAs of about 150,000 bp.

pain receptor: a nerve cell whose distal axon terminal is depolarized by potentially damaging events such as heat or stretch.

pain relay cell: this nerve cell, upon which a pain receptor synapses, carries the message on towards the brain.

paracrine transmitters: agonists released by cells into the interstitial fluid that can last many minutes and can therefore diffuse widely within the tissue before they are destroyed.

parallel β sheet: β sheet in which all the parallel polypeptide chains run in the same direction.

patch clamp: a technique in which a glass micropipette is sealed to the surface of a cell to allow electrical recording of cell properties. See In Depth 14.1 on page 230.

PCR (polymerase chain reaction): a method for making many copies of a DNA sequence where the sequence, at least at each end, is known.

PDGF (platelet derived growth factor): a paracrine transmitter that opposes apoptosis and promotes cell division in target cells such as endothelial cells and smooth muscle.

pentose: monosaccharide with five carbon atoms.

pentose phosphate pathway: pathway that makes pentoses and generates reducing power in the form of NADPH by oxidizing glucose-6-phosphate.

peptide: short linear polymer of amino acids.

peptide bond: the bond between amino acids. The bond is formed between the carboxyl group of one amino acid and the amino group of the next.

peptidyl site (P site): the site on a ribosome that is occupied by the growing polypeptide chain.

peptidyl transferase: enzyme that catalyses the formation of a peptide bond between two amino acids. *E. coli* peptidyl transferase is an example of a ribozyme.

peripheral membrane protein: class of protein that is easily detached from a cell membrane, unlike integral membrane proteins, which can only be isolated by destroying the membrane, e.g., with detergent.

peroxisome: class of cell organelles of diverse function. Peroxisomes frequently contain the enzyme catalase, which breaks down hydrogen peroxide into oxygen and water.

pH: measure of the acidity of a solution, equal to minus the logarithm to base ten of the hydrogen ion concentration in moles liter^{-1}. The smaller the pH value, the more acid the solution. A neutral solution has a pH of 7, that is, the H^+ concentration is 10^{-7} moles liter^{-1} or 100 nmoles liter^{-1}.

phage: short for bacteriophage, that is, a virus that infects bacterial cells.

phase contrast microscopy: the type of light microscopy in which differences in the refractive index of a specimen are converted into differences in contrast.

PH domain: a protein domain that binds phosphorylated inositols. The PH domain on protein kinase B is recruited to the membrane by the intensely charged lipid phosphatidylinositol trisphosphate.

phenylketonuria: an inherited disease in which the enzyme phenylalanine hydroxylase is missing or defective. Unless intake of phenylalanine is drastically curtailed phenylalanine and its transamination product phenylpyruvate build up in the body and cause brain damage. The heel prick or Guthrie test, performed on all newborn babies, tests for this condition.

phosphatase: an enzyme that removes phosphate groups from substrates. Glucose 6-phosphate phosphatase and calcineurin are phosphatases operating on a sugar and proteins respectively.

phosphate: properly, a name for the ions $H_2PO_4^-$, HPO_4^{2-} and PO_4^{3-}. The word is also very commonly used to mean the group:

```
      O⁻
      |
  −O−P=O
      |
      O⁻
```

and we have used this convention in this book. If one wants to specifically describe the group, *not* the ions, one can refer to a phosphoryl group (a term that does not include the leftmost oxygen in the diagram above).

phosphatidylinositol bisphosphate (PIP_2): a membrane lipid. PIP_2 releases inositol trisphosphate into the cytosol upon hydrolysis by phospholipase C. Alternatively, PIP_2 can be further phosphorylated to yield phosphatidylinositol trisphosphate.

phosphatidylinositol trisphosphate (PIP_3): a phospholipid whose head group is an inositol phosphorylated on its 3, 4 and 5 carbons. PIP_3 recruits a subset of proteins containing PH domains (among them protein kinase B) to the plasma membrane.

phosphodiester link: a link between two parts of a molecule in which a phosphate is attached through an oxygen atom to each of the two parts. In phospholipids, the head group is attached to glycerol by a phosphodiester bond. In DNA and RNA, successive nucleotides are joined together by phosphodiester links.

phosphofructokinase: enzyme that phosphorylates fructose-6-phosphate to generate fructose-1,6-bisphosphate.

phosphoinositide-3-kinase (PI-3-kinase): an enzyme that phosphorylates phosphatidylinositol bisphosphate on the 3 position of the inositol ring to generate the intensely charged lipid phosphatidylinositol trisphosphate (PIP_3). PIP_3 can then recruit proteins containing PH domains to the plasma membrane.

phospholipase C: an enzyme that hydrolyzes the membrane lipid phosphatidylinositol bisphosphate (PIP_2) to release inositol trisphosphate into the cytosol. The β isoform is activated by the trimeric G protein G_q, while the γ isoform is activated by tyrosine phosphorylation.

phospholipid: a glyceride lipid in which acyl groups are attached to the first and second hydroxyl group of glycerol, the third hydroxyl group being connected through a phosphodiester link to a polar, often electrically charged, head group. Phospholipids are amphipathic and are a major component of biological membranes.

phosphorylase: an enzyme that cleaves a glycosidic bond by adding phosphate.

phosphorylase kinase (glycogen phosphorylase kinase): a kinase that is activated by the calcium-calmodulin complex and which phosphorylates glycogen phosphorylase, activating the latter enzyme.

phosphorylated: having had a phosphate group added. The phosphate groups are usually substituted into a hydroxyl group to form a phosphoester bond but are sometimes substituted into an acid to form a phosphoanhydride or attached to a nitrogen atom to form a phosphoimide.

phosphorylation: addition of a phosphate group to a molecule. Usually substituted into a hydroxyl group to form a phosphoester bond. Sometimes substituted into an acid to form a phosphoanhydride, or attached to a nitrogen atom to form a phosphoimide.

phosphoryl group: The group

$$-\overset{\displaystyle O^-}{\overset{|}{\underset{\underset{\displaystyle O^-}{|}}{P}}}=O$$

which is often simply called a phosphate group.

photosynthesis: the synthesis of complex organic molecules, and oxidation of water to release oxygen, that is driven by the energy of light.

PI-3-kinase (phosphoinositide 3-kinase): an enzyme that phosphorylates phosphatidylinositol bisphosphate on the 3 position of the inositol ring to generate the intensely charged lipid phosphatidylinositol trisphosphate (PIP_3). PIP_3 can then recruit proteins containing PH domains to the plasma membrane.

PIP_2 (phosphatidylinositol bisphosphate): a membrane lipid. PIP_2 releases inositol trisphosphate into the cytosol upon hydrolysis by phospholipase C. Alternatively, PIP_2 can be further phosphorylated to yield phosphatidylinositol trisphosphate.

PIP_3 (phosphatidylinositol trisphosphate): a phospholipid whose head group is an inositol phosphorylated on its 3, 4 and 5 carbons. PIP_3 recruits a subset of proteins containing PH domains (among them protein kinase B) to the plasma membrane.

pKa: a parameter equal to $-\log_{10}K_a$ and representing the pH at which the concentration of the dissociated acid is equal to the concentration of the undissociated acid.

PKA (protein kinase A; cAMP-dependent protein kinase): a protein kinase that is activated by the intracellular messenger cyclic AMP. PKA phosphorylates proteins (for example, glycogen phosphorylase kinase) on serine and threonine residues.

PKB (protein kinase B): a protein kinase that is activated when it is itself phosphorylated; this in turn only occurs when PKB is recruited to the plasma membrane by phosphatidylinositol trisphosphate (PIP_3). PKB phosphorylates proteins on serine and threonine residues (for example, the bcl-2 family protein BAX). An older name for PKB is Akt.

PKC (protein kinase C): a protein kinase that is activated by a rise of cytosolic calcium concentration; it is also activated by diacylglycerol. PKC phosphorylates proteins on serine and threonine residues.

PKG (protein kinase G; cGMP-dependent protein kinase): a protein kinase that is activated by the intracellular messenger cyclic GMP. PKG phosphorylates proteins (for example, calcium ATPase) on serine and threonine residues.

plaque: an area of dead bacteria in a lawn of live bacteria that is caused by infection of bacterial cells by bacteriophages.

plasma cell: a differentiated B cell that generates large amounts of secreted antibody.

plasmalemma: the membrane that surrounds the cell. Also called the plasma membrane or the cell membrane.

plasma membrane: the membrane that surrounds the cell. Also called the plasmalemma or the cell membrane.

plasmid: a circular DNA molecule that is replicated independently of the host chromosome in bacterial cells.

plasmodesmata (singular; plasmodesma): a type of cell junction unique to plant cells, which provides a much bigger hole for passage of substances between the cytoplasm of the two cells than gap junctions do.

platelet: small fragment of cell that contains no nucleus, but which has a plasma membrane and some endoplasmic reticulum. Platelets are critical in the process of blood clotting, and also release platelet-derived growth factor.

platelet-derived growth factor (PDGF): a paracrine transmitter that opposes apoptosis and promotes cell division in target cells such as endothelial cells and smooth muscle.

PLC (phospholipase C): an enzyme that hydrolyses the membrane lipid phosphatidylinositol bisphosphate (PIP_2) to release inositol trisphosphate into the cytosol. The β isoform is activated by the trimeric G protein G_q, while the γ isoform is activated by tyrosine phosphorylation.

polar: having covalent bonds in which the electrons are unequally shared, so that atoms have partial charges. Polar molecules can interact with water by electrostatic interactions and by hydrogen bonding.

polarized: divided into two parts with disparate properties. More specific uses in chemistry and biology include ***polarized light*** whose electromagnetic oscillations occur in defined planes; ***polarized molecules*** where the electrons are not shared equally so that one end of the molecule has a +ve charge and another end has a −ve charge; ***electrically polarized*** membranes that separate volumes at different voltages; and ***polarized cells*** that exhibit different morphologies, and express different proteins, at different ends.

polyadenylation: the process whereby a poly-A tail is added to the 3′ end of a eukaryotic mRNA.

poly-A tail (poly adenosine tail): a string of adenine residues added to the 3′ end of a eukaryotic mRNA.

polycistronic mRNA: an mRNA that, when translated, yields more than one polypeptide.

polyclonal: pertaining to many clones. The term is usually only used to describe traditional antibodies generated by immunizing an animal and collecting its blood plasma. A polyclonal antibody is secreted by all those B cell clones that proliferated because their antibody bound the target.

polymer: a chemical composed of a long chain of identical or similar subunits.

polymerase: an enzyme that makes polymers, that is, long chains of identical or very similar subunits. DNA and RNA polymerase respectively are involved in making DNA and RNA.

polymerase chain reaction (PCR): a method for making many copies of a DNA sequence when the sequence, at least at each end, is known.

polypeptide: a polymer of more than fifty amino acids joined by peptide bonds.

polyploid: having three or more sets of chromosomes.

polyribosome (or polysome): a chain of ribosomes attached to an mRNA molecule.

polysaccharide: a chain of more than a hundred or so monosaccharides linked by glycosidic bonds.

polysome (or polyribosome): a chain of ribosomes attached to an mRNA molecule.

polyunsaturated: having several double bonds. Usually applied to fatty acids; triacylglycerols with polyunsaturated acyl (fatty acid) groups are liquid even at low temperatures.

porin: a channel found in the mitochondrial outer membrane. It is always open and allows all solutes of M_r $<10{,}000$ to pass.

positive feedback: a process in which the consequences of a change act to increase the magnitude of that change, so that a small initial change tends to get bigger and bigger.

positive regulation (of transcription): a process whereby transcription is activated in the presence of a particular substance.

postsynaptic cell: the cell upon which a nerve cell releases its transmitter at a synapse.

potassium channel: a channel in the plasma membrane of many cells that allows potassium ions to pass.

potassium/sodium ATPase (K^+/Na^+ ATPase): a plasma membrane carrier. For every ATP hydrolyzed, three Na^+ ions are moved out of the cytosol and two K^+ ions are moved in. Usually called the sodium/potassium ATPase.

pre-initiation complex: complex formed between transcription factors and RNA polymerase at the promoter of a eukaryotic gene.

presynaptic terminal: the region of an axon terminal that is specialized for exocytosis of transmitter.

primary antibody: the first antibody applied to the preparation. In many cases the primary antibody is not itself labelled, but must be revealed by applying a labelled secondary antibody.

primary endosome: an acidic cell compartment with which coated vesicles fuse.

primary immunofluorescence: technique in which the location or presence of a chemical is revealed by treating the sample with a dye-labelled antibody.

primary structure (of a protein): the sequence of amino acids held together by peptide bonds making up a polypeptide.

primase: the enzyme that synthesizes the RNA primers needed for the initiation of synthesis of the leading and lagging DNA strands.

primer: a short sequence of nucleic acid (RNA or DNA) which acts as the start point at which a polymerase can initiate synthesis of a longer nucleic acid chain.

prion: a protein that is itself an infective agent for a disease.

profilin: a type of actin-binding protein that regulates the assembly of actin filaments.

progenitor: ancestor; an individual from which others are descended. In this book we use the term "progenitor cell" to mean the cell that divides to give rise to two other cells (in mitosis) or four others (in meiosis).

progeny: children, offspring. In this book we use the term "progeny cells" to mean the products of cell division.

programmed cell death (apoptosis): a process in which a cell actively promotes its own destruction, as distinct from necrosis. Apoptosis is important in vertebrate development, where tissues and organs are shaped by the death of certain cell lineages.

projector lens: lens of light or electron microscope that carries the image to the eye; more commonly known as the "eyepiece."

prokaryotic: a type of cellular organization found in bacteria in which the cells lack a distinct nucleus and other organelles.

prometaphase: the period of mitosis or meiosis that sees the breakdown of the nuclear envelope and the attachment of the chromosomes to the mitotic spindle.

promoter: the region of DNA to which RNA polymerase binds to initiate transcription.

pronuclei: the nuclei of the egg and sperm prior to fusion.

prophase: the period of mitosis or meiosis in which the chromosomes condense.

prophase I: the first prophase of meiosis.

prophase II: the second prophase of meiosis.

prostaglandin: one of a family of paracrine transmitters that diffuse into tissue and are short-lived. They are made from the fatty acid arachidonate (20 carbons, 4 double bonds).

prosthetic group: a nonprotein molecule necessary for the activity of a protein. The concept overlaps with the concept of cofactor. The difference is just how tightly they are bound: a prosthetic group is very tightly bound and cannot be removed without at least partial unfolding of the protein. Examples are the heme groups of myoglobin, hemoglobin and the cytochromes.

proteasome: a barrel-shaped proteolytic machine characteristic of eukaryotic cells that degrades proteins that are no longer needed. Proteasomes are found in bacteria but in most prokaryotes the components are not so obviously organized into a discrete barrel-shaped structure.

protein: a polypeptide (a polymer of α-amino acids) that has a preferred way of folding.

protein engineering: designing a novel protein and then causing it to be built by altering the sequence of DNA on a vector or in the chromosome.

protein kinase: an enzyme that phosphorylates a protein by transferring a phosphate group from ATP to the molecule.

protein kinase A (PKA): a protein kinase that is activated by the intracellular messenger cAMP. Protein kinase A phosphorylates proteins (for example, glycogen phosphorylase kinase) on serine and threonine residues.

protein kinase B (PKB): a protein kinase that is activated when it is itself phosphorylated; this in turn only occurs when PKB is recruited to the plasma membrane by phosphatidylinositol trisphosphate (PIP_3). PKB phosphorylates proteins on serine and threonine residues (for example, the bcl-2 family protein BAX). An older name for PKB is Akt.

protein kinase C (PKC): a protein kinase that is activated by a rise of cytosolic calcium concentration; it is also activated by diacylglycerol. PKC phosphorylates proteins on serine and threonine residues.

protein kinase G (PKG; cGMP-dependent protein kinase): a protein kinase that is activated by the intracellular messenger cyclic GMP. PKG phosphorylates proteins (for example, calcium ATPase) on serine and threonine residues.

protein phosphatase: an enzyme that removes phosphate groups from proteins.

protein phosphorylation: the addition of a phosphate group to a protein. The addition of the charge on the phosphoryl group can markedly alter the tertiary structure and therefore function of a protein.

protein targeting: the delivery of proteins to their correct cellular location.

protein translocator: channel-like protein of the endoplasmic reticulum that allows polypeptide chains to cross the membrane as they are synthesized.

proteolysis: the hydrolysis of peptide bonds within a protein to release separate fragments.

proteome: the complete protein content of the cell.

proteomics: the study of the proteome. For example, one might compare the protein profiles of cells in two tissues.

protofilaments: the chains of subunits that make up the wall of a microtubule.

protonated: having accepted an H^+. For example the lactate ion, $CH_3CH(OH)COO^-$, will become protonated in acid solutions to become lactic acid, $CH_3CH(OH)COOH$. The word "proton" is being used as shorthand for "hydrogen atom nucleus" since, for the commonest isotope of hydrogen, the nucleus is a single proton.

proximal: close to the center.

pseudogene: a gene that has mutated such that it no longer codes for a functional protein.

pseudopodium: a projection extended by an amoeba or other crawling cell in the direction of movement.

P site (peptidyl site): the site on the ribosome occupied by the growing polypeptide chain.

purine: a nitrogenous base found in nucleotides and nucleosides. Adenine, guanine, and hypoxanthine are purines.

pyrimidine: a nitrogenous base found in nucleotides and nucleosides. Cytosine, thymine, and uracil are pyrimidines.

pyrophosphate:

$$^-O-\overset{\overset{O}{\|}}{\underset{\underset{OH}{|}}{P}}-O-\overset{\overset{O}{\|}}{\underset{\underset{O^-}{|}}{P}}-O^-$$

the pyrophosphate ion is stable if kept away from biological tissue but is rapidly hydrolyzed to two phosphate ions by intracellular and extracellular enzymes.

quaternary structure: the structure in which subunits of a protein, each of which has a tertiary structure, associate to form a more complex molecule. The subunits associate tightly, but not covalently, and may be the same or different.

quiescent: at rest. Nondividing cells are often said to be quiescent.

Rab family: family of GTPases that mediate fusion of vesicles with other membranes.

Raf: an isoform of MAPKKK, the enzyme at the top of the mitogen associated protein kinase cascade.

Ran: a GTPase that gives direction to transport through the nuclear pore.

Ras: a GTPase that activates the MAP kinase pathway and hence promotes cell division. Ras is activated by SOS, its GTP exchange factor (GEF), which is in turn recruited to the plasma membrane via receptor tyrosine kinases and the adaptor protein Grb2.

Rb: a protein that binds to the transcription factor E2F and therefore prevents transcription of proteins required for DNA synthesis. Phosphorylation of Rb by CDK4 causes it to release E2F, allowing entry to S phase. Mutations in *Rb* lead to the formation of the eye cancer called retinoblastoma.

reading frame: A reading of the genetic code in blocks of three bases; there are three possible reading frames for each mRNA, only one of which will produce the correct protein.

receptor: a protein that specifically binds a particular solute. Receptors can be transmembrane or cytosolic proteins. Particular receptor proteins perform additional functions (ion channel, enzyme, activator of endocytosis, transcription factor, etc.) that are activated by the binding of the solute.

receptor-mediated endocytosis: a process in which ligands bind to specific receptors in the plasma membrane and trigger clathrin-mediated vesicle budding.

receptor tyrosine kinase: an integral membrane protein with a binding site for transmitter on the extracellular aspect and tyrosine kinase catalytic ability on the cytosolic aspect. Binding of transmitter causes self-phosphorylation on tyrosine and hence recruitment of proteins with SH2 domains such as Grb2, phospholipase Cγ, and phosphoinositide 3-kinase, which may then themselves be phosphorylated. The PDGF receptor, the FGF receptor, the Trk family of receptors, and the insulin receptor are all receptor tyrosine kinases.

recessive: a gene is recessive if its effects are hidden when the organism posesses a second, dominant, version. Recessive genes usually code for nonfunctional proteins, so that if the individual can make the protein using the other, working gene, no effects are seen.

recombinant DNA: an artificial DNA molecule formed of DNA from two or more different sources.

recombinant plasmid: a plasmid into which a foreign DNA sequence has been inserted.

recombinant protein: a protein expressed from the foreign DNA inserted into a recombinant plasmid or other cloning vector. Recombinant proteins are often expressed in bacteria, yeast, insect, or mammalian cells.

recombination: a "cut and paste" of DNA. Recombination occurs at chiasmata during meiosis, and also in embryonic stem cells, allowing insertional mutagenesis.

recovery stroke: part of the beat cycle of a cilium in which the cilium is moved back into a position where it can push again; used in contrast to the effective stroke which generates the force.

reducing agent: a chemical that can give electrons to another and in doing so becomes oxidised. NADH is a reducing agent.

reduction: the addition of electrons to a compound, e.g., by the addition of hydrogen atoms or the removal of oxygen atoms.

refractive index: a measure of the capacity of any material to slow down the passage of light.

regulated secretion: secretion that only occurs in response to a signal, such as a rise in the cytosolic concentration of calcium ions.

release factors: proteins that occupy the A site in the ribosome when a stop codon is encountered, and which act to trigger termination of polypeptide synthesis.

repetitious DNA: a DNA sequence that is repeated many times within the genome.

replication (of DNA): the process whereby two DNA molecules are made from one.

replication fork: Y-shaped structure formed when the two strands of the double helix separate during replication.

repolarization: a movement of the membrane voltage back towards the normal resting voltage following a depolarization.

repressible operon: operon whose transcription is repressed in the presence of a particular substance; often the final product of the metabolic pathway.

RER (rough endoplasmic reticulum): portion of the endoplasmic reticulum associated with ribosomes and concerned with the synthesis of secreted proteins. Proteins destined to remain within the majority of single-membrane-bound organelles (Golgi, lysosomes, etc.) are also made on the rough ER, as are integral proteins of these organelles and of the plasma membrane.

residue: that which is left over. In chemistry, when a molecule is built into a larger molecule with the loss of some part, the part that remains and forms part of the larger molecule is called a residue. For example, an amino acid that has been built into a polypeptide, with the loss of the elements of water, is said to be an amino acid residue.

resolving power: a measure that defines the smallest object that can be distinguished using a microscope.

resting voltage: the voltage across the plasma membrane of an unstimulated cell, typically −70 to −90 mV.

restriction endonuclease: an enzyme that cleaves phosphodiester bonds within a specific sequence in a DNA molecule.

retinoblastoma: a cancer of the eye, usually caused by a mutation in the *Rb* gene.

retrograde: backward movement; when applied to axonal transport it means towards the cell body.

retrovirus: a virus whose genetic information is stored in RNA.

reverse genetics: research that begins with a gene of known sequence but unknown effect and works towards deducing its effects and therefore function.

reverse transcriptase: an enzyme of some viruses that copies RNA into DNA.

reverse transcription: the process whereby RNA is copied into DNA.

ribonuclease: an enzyme that cleaves phosphodiester links in RNA.

ribonuclease H: an enzyme that cleaves phosphodiester links in an RNA molecule that is joined to a DNA molecule by complementary base pairing.

ribonucleic acid (RNA): a polymer of ribonucleoside monophosphates. See mRNA, tRNA and rRNA.

ribonucleic acid polymerase: enzyme that synthesizes RNA.

ribonucleic acid primer: a short length of RNA, complementary in sequence to a DNA strand, that allows DNA polymerase III to attach and begin DNA synthesis.

ribonucleoside monophosphate: a nitrogenous base attached to the sugar ribose which has one phosphate group on its 5′ carbon atom;also known as a ribonucleotide.

ribose: a pentose sugar used to make the nucleotides that form RNA.

ribosomal RNA (rRNA): the RNA component of ribosomes. rRNA forms a major part of the ribosome and participates fully in the translation process.

ribosome: the ribosome has two subunits, the large and the small, each of which is made up of several different proteins and rRNA molecules. The two subunits come together to provide the platform on which tRNAs can bind to mRNA so that protein synthesis can take place.

ribosome recycling factor: a protein that assists the ribosomal subunits to dissociate after protein synthesis is complete. The dissociated subunits can then be reused in another round of protein synthesis.

ribozyme: an RNA molecule with enzyme-like catalytic activity.

RNA (ribonucleic acid): a polymer of ribonucleoside monophosphates. Cells contain three types of RNA: messenger RNA, ribosomal RNA and transfer RNA.

RNA polymerase: enzyme that synthesizes RNA.

RNA polymerase I: the enzyme that transcribes most rRNA genes in eukaryotes.

RNA polymerase II: the enzyme that transcribes mRNA genes in eukaryotes.

RNA polymerase III: the enzyme that transcribes tRNA and some other small RNAs in eukaryotes.

RNA primer: a short length of RNA, complementary in sequence to a DNA strand, that allows DNA polymerase III to attach and begin DNA synthesis.

RNA splicing: The removal of introns from an RNA molecule and the joining together of exons to form the final RNA product.

rough endoplasmic reticulum (RER): portion of the endoplasmic reticulum associated with ribosomes and concerned with the synthesis of secreted proteins. Proteins destined to remain within the majority of single-membrane-bound organelles (Golgi, lysosomes, etc.) are also made on the rough ER, as are integral proteins of these organelles and of the plasma membrane.

rRNA (ribosomal RNA): the RNA component of ribosomes. rRNA forms a major part of the ribosome and participates fully in the translation process.

ryanodine: a plant toxin that binds to the ryanodine receptor, with complex effects on gating of the channel.

ryanodine receptor: a calcium channel found in the membrane of the endoplasmic reticulum. In most cells it opens in response to a rise of calcium concentration in the cytosol. In skeletal muscle ryanodine receptors are directly linked to voltage-gated calcium channels in the plasma membrane and open when the latter open.

sigma factor (σ factor): subunit of bacterial RNA polymerase that recognizes the promoter sequence.

S phase: period of the cell division cycle during which DNA replication occurs.

S value (sedimentation coefficient): a value that describes how fast macromolecules and organelles sediment in a centrifuge.

saltatory conduction: the jumping of an action potential from node to node down a myelinated axon.

salt bridge (in protein structure): an interaction between a positively charged amino acid residue (such as arginine) and a negatively charged residue (such as aspartate).

sarcomere: contractile unit of striated muscle.

sarcoplasmic reticulum: type of smooth endoplasmic reticulum found in striated muscle; concerned with the regulation of the concentration of calcium ions.

satellite DNA: a DNA sequence that is tandemly repeated many times.

saturated (of fatty acids): containing no carbon-carbon double bonds.

saturation kinetics: are said to occur when the rate of a reaction approaches a maximum, limiting value as the concentration of reactant increases. The rate becomes limited by the availabilty of binding sites on the catalyst for the reactant.

scanning electron microscope (SEM): a type of electron microscope in which the image is formed from electrons that are reflected back from the surface of a specimen as the electron beam scans rapidly back and forth over it. The scanning electron microscope is particularly useful for providing topographical information about the surfaces of cells or tissues.

Schwann cell: glial cell of the peripheral (outside the brain and spinal cord) nervous system.

SDS-PAGE (sodium dodecyl sulphate-polyacrylamide gel electrophoresis): technique for separating proteins by their relative molecular mass.

secondary antibody: a labelled antibody that is used to reveal the location of an unlabelled primary antibody. Secondary antibodies are produced by one species of animal in response to the injection of another animal's antibodies, thus a "goat anti-rabbit" secondary will bind to any primary antibody that has been generated by a rabbit. Labelled secondary antibodies can be used to study many different questions because the specificity in the final image or western blot is provided by the primary antibody, not by the secondary.

secondary immunofluorescence: a technique in which a fluorescent secondary antibody is used to label a preapplied primary antibody specific for a particular protein or subcellular structure. Also called indirect immunofluorescence.

secondary structure: regular, repeated folding of the backbone of a polypeptide. The side chains of the amino acids have an influence but are not directly involved. There are two common types of secondary structure: the α helix and the β sheet.

secretion: the synthesis and release from the cell of a chemical.

secretory vesicles: vesicles derived from the Golgi apparatus which transport secreted proteins to the cell membrane with which they fuse.

second messenger: another name for intracellular messenger (the "first messenger" being the extracellular transmitter chemical). A cytosolic solute that changes in concentration in response to external stimuli or internal events, and which acts on intracellular targets to change their behavior.

Calcium ions, cyclic AMP and cyclic GMP are the three common intracellular messengers.

sedimentation coefficient (S value): a value that describes how fast macromolecules and organelles sediment in a centrifuge.

semi-conservative replication: the mode of replication used for DNA; both strands of the double helix serve as templates for the synthesis of new daughter strands.

serine-threonine kinase: an enzyme that phosphorylates proteins on serine or threonine residues, by transferring the γ phosphate of ATP to the amino acid side chain. With a few exceptions, protein kinases are either serine-threonine kinases (which can phosphorylate on serine or on threonine) or tyrosine kinases (which only phosphorylate on tyrosine).

serosal: the side (for example, of an epithelium) that faces the interstitial fluid.

seven-methyl guanosine cap: modified guanosine found at the 5′ terminus of eukaryotic mRNA. A guanosine is attached to the mRNA by a 5′–5′-phosphodiester link and is subsequently methylated on atom number 7 of the guanine.

SH2 domain: a domain found in a number of proteins that binds to phosphorylated tyrosine. Many proteins with SH2 domains are recruited to receptor tyrosine kinases when the latter phosphorylate themselves in response to ligand binding. Important proteins with SH2 domains are Grb2, PI 3-kinase and phospholipase Cγ.

β sheet: common secondary structure in proteins, in which lengths of fully extended polypeptide run alongside each other, hydrogen bonds forming between the peptide bonds of the adjoining strands.

Shine-Dalgarno sequence: a sequence on a bacterial mRNA molecule to which the ribosome binds.

side chain (of amino acids): the group attached to the α carbon of an amino acid.

sigma factor (σ factor): subunit of bacterial RNA polymerase that recognizes the promoter sequence.

signal peptidase: enzyme that cleaves the signal sequence from a polypeptide as it enters the lumen of the endoplasmic reticulum.

signal recognition particle: ribonucleoprotein particle that recognizes and binds to the signal sequence at the N terminus of a polypeptide.

signal recognition particle receptor: receptor on the endoplasmic reticulum to which the signal recognition particle binds during the process of polypeptide chain synthesis and import into the endoplasmic reticulum. Also called the "docking protein."

signal sequence: short stretch of amino acids found at the N terminus of polypeptides which targets them to the endoplasmic reticulum.

simple diffusion: not a specific term: we often say "simple diffusion" to emphasise that a solute movement is passive, down a concentration gradient, and does not require a carrier or channel.

single-strand binding protein: the protein that binds to the separated DNA strands to keep them in an extended form during replication, thus preventing the double helix from reforming.

site-directed mutagenesis: a technique used to change the base sequence of a DNA molecule at a specific site.

skeletal muscle cells: large multinucleate muscle cells that are attached to bone. Most cuts of meat are mainly skeletal muscle.

small nuclear RNAs (snRNAs): small RNA molecules found in the nucleus that play a role in RNA splicing.

smooth endoplasmic reticulum (SER): the portion of the endoplasmic reticulum without attached ribosomes, among its functions are the synthesis of lipids and the storage and stimulated release of calcium ions.

smooth muscle: nonstriated muscle, found in many places in the body including blood vessels and the intestine.

smooth muscle cells: small muscle cells that lack the characteristic striations seen in skeletal and cardiac muscle.

SNARES: proteins that mediate fusion of vesicles with other membranes.

snRNAs (small nuclear RNAs): small RNA molecules found in the nucleus that play a role in RNA splicing.

sodium action potential: an action potential driven by the opening of voltage-gated sodium channels and the resulting sodium influx.

sodium/calcium exchanger: a carrier in the plasma membrane. Three sodium ions move into the cell down their electrochemical gradient and one calcium is moved out up its concentration gradient.

sodium gradient: an energy currency. Sodium ions are more concentrated outside the cell than inside, and this chemical gradient is usually supplemented by a voltage gradient pulling sodium ions in. If sodium ions are allowed to enter down their electrochemical gradient they release 15,000 Joules/mole.

sodium/potassium ATPase (sodium pump): a plasma membrane carrier. For every ATP hydrolyzed, three Na^+ ions are moved out of the cytosol and two K^+ ions are moved in.

sodium pump (sodium/potassium ATPase): a plasma membrane carrier. For every ATP hydrolysed three Na^+ ions are moved out of the cytosol and two K^+ ions are moved in.

solute: a substance that is dissolved in a liquid.

somatic cells: cells that make up all the normal tissues of the human body; distinct from the germ cells that form the gametes (eggs and sperm).

somatic hypermutation: mutation of particular sections encoding the antigen binding site on antibodies. This occurs at such a high rate that each differentiating B cell gives rise to a clone whose DNA sequence, in these specific regions of the chromosome, is different from that in all other B cell clones.

somatic mutation: a mutation that occurs in the DNA of a somatic cell. Somatic mutations are not passed on to children. Rather they lead to a falling off in the performance of tissues with age, and to cancers.

sorting signal: a section of a protein that causes the cell to direct the protein to a specific cell compartment such as the nucleus or mitochondrion. Sorting signals can be lengths of peptide (targeting sequences), such as the signal sequence that targets a protein to the endoplasmic reticulum, or can be the result of post-translational modification, for example mannose-6-phosphate which targets a protein to the lysosome.

sorting vesicle: vesicle that carries proteins from one membrane compartment to another.

SOS: guanine nucleotide exchange factor for a GTPase called Ras. SOS is recruited to the plasma membrane by binding to growth factor receptor binding protein number 2 which in turn binds activated receptor tyrosine kinases.

Southern blotting: a blotting technique in which DNAs, separated by size, are probed using a single-stranded cDNA probe.

spatial summation: the phenomenon whereby although one presynaptic action potential does not generate a depolarization large enough to elicit an action potential in a postsynaptic cell, action potentials that occur more or less simultaneously in a number of presynaptic cells do depolarize the postsynaptic cell to threshold.

specificity constant: the ratio between the catalytic rate constant (k_{cat}) and the Michaelis constant (K_M) for an enzyme. It is a rate constant with dimensions of liters mol^{-1} s^{-1} and is used to compare different substrates for the same enzyme or to compare the effectiveness of one enzyme with another.

spermatid: cell formed by meiosis that differentiates to form a spermatozoon.

spermatozoon: motile male gamete.

S phase: period of the cell division cycle during which DNA replication occurs.

spindle: microtubule based structure upon which chromosomes are arranged and translocated during cell division.

spindle assembly checkpoint: the checkpoint controlling entry into anaphase. Cells can only proceed if the anaphase promoting complex is active, which in turn only occurs when all the chromatids are correctly lined up on the metaphase plate.

spliceosome: a complex of proteins and small RNA molecules involved in RNA splicing.

squamous: flat. A term used of epithelial cells.

standard state: a defined set of conditions under which to state the thermodynamics of a reaction. For example, $\Delta G^{\circ\prime}$ is determined for aqueous solutions at pH 7.0.

start signal: the start signal for protein synthesis is the codon AUG specifying the incorporation of methionine.

STAT: one of a family of transcription factors that dimerize and translocate to the nucleus when phosphorylated on tyrosine. STATs are phosphorylated by JAKs associated with activated type 1 cytokine receptors.

steady state: a state whose parameters do not change with time. A steady state may be very far from equilibrium and be maintained by the constant expenditure of energy. Examples include the constant concentration of ATP in cells and the constant 37°C temperature of the human body.

stearic acid: $C_{17}H_{35}$—COOH; a fatty acid with no double bonds, that is, it is fully saturated. The acyl group derived from stearic acid is common in animal fats and phospholipids.

stem cell: undifferentiated cells that retain the ability to divide and generate more stem cells as well as cells committed to differentiate. Some stem cells are committed to one differentiated fate, others are more or less pluripotent and are capable of creating a range of cell types.

stereo isomers: isomers (that is, different compounds with the same molecular formula) in which the atoms have the same connectivity but differ in the way the atoms are arranged in space.

steroid hormone: these transmitters act on intracellular receptors to activate transcription of particular genes. Glucocorticoids are one type of steroid hormone; the sex hormones (oestradiol, testosterone, progesterone) are another.

steroid hormone receptor: transcription factors that, upon binding their appropriate steroid hormone, move from the cytosol to the nucleus and activate transcription of particular genes.

sticky ends (of DNA): the short single-stranded ends produced by cleavage of the two strands of a DNA molecule at sites which are not opposite to one another.

stoichiometry: the numbers of molecules of reactants and products taking part in a chemical reaction.

stop codon: the codons UAA, UAG, and UGA are codons that signal protein synthesis to stop. Also known as termination codons.

stop signal: the signal, to stop protein synthesis, given to the ribosome by a stop codon on mRNA.

stratified epithelium: type of epithelium consisting of several layers, such as the skin.

stress-activated protein kinase: a kinase that stimulates cell repair but can also trigger apoptosis. p38 and JNK are stress-activated protein kinases.

stress fiber: bundle of actin filaments commonly seen in cultured, nonmotile animal cells.

stretch activated channel: a channel found in the plasma membrane of endothelial and other cells that opens when the membrane is stretched. It is permeable to sodium, potassium and calcium ions. Although well characterized it has not to date been genetically identified.

striated muscle: striped muscle; includes skeletal and cardiac muscle.

stroma: the volume within the chloroplast inner membrane but outside the thylakoids.

substrate: in normal English, a substrate is a solid base. In biology, the term is used to mean (1) a reactant in a reaction catalyzed by an enzyme, e.g., "lactose is a substrate for β-galactosidase"; (2) a base that cells grow and move on, e.g., "collagen is a good substrate for cell attachment."

sucrose: a disaccharide comprising glucose linked to fructose by an $\alpha 1 \rightarrow \beta 2$ glycosidic bond.

summation (at synapses): the additive effects of more than one presynaptic action potential upon the postsynaptic voltage.

supercoiling: the organization of a linear structure into coils at more than one spatial scale.

S value (sedimentation coefficient): a value that describes how fast macromolecules and organelles sediment in a centrifuge.

Svedberg unit (S value): a value that describes how fast macromolecules and organelles sediment in a centrifuge.

synapse: the structure formed from the axon terminal of a nerve cell and the adjacent region of the postsynaptic cell. Transmitter released by the axon terminal diffuses across the synapse gap and acts upon the postsynaptic cell.

T cell receptor: a receptor expressed by T lymphocytes, the product of DNA rearrangement within large chromosomal loci, such that receptors on different T cells have different amino acid sequences in the ligand binding domain. A particular T cell receptor is selective for, and activated by, a particular short peptide presented by a major histocompatability complex (MHC) protein.

T killer cell: T cell expressing the integral plasma membrane protein CD8. T killer cells kill somatic cells that present novel peptides.

tandem repeats: many copies of the same DNA sequence that lie side by side on the chromosome.

targeting sequence: a stretch of polypeptide that determines the cellular compartment to which a synthesized protein is sent.

TATA box: a sequence found about twenty bases upstream of the beginning of many eukaryotic genes that forms part of the promoter sequence and is involved in positioning RNA polymerase for correct initiation of transcription.

Taxol: compound obtained from the bark of the Pacific yew, *Taxus brevifolia*; binds to tubulin. Taxol is a powerful anticancer drug.

T cells: the immune system cells that respond to peptides presented by MHC proteins on the surface of other cells. T cells are divided into CD4+ T cells that help B cells respond to antibody, and CD8+ T cells that kill somatic cells expressing novel proteins.

telomere: specialized region at the ends of eukaryotic chromosomes. Telomeres are rich in minisatellite DNA.

telophase: final period of mitosis or meiosis in which the chromosomes decondense and the nuclear envelope reforms.

telophase I: telophase of the first meiotic division (meiosis I).

telophase II: telophase of the second meiotic division (meiosis II).

temporal summation: the phenomenon whereby although one presynaptic action potential does not generate a depolarization large enough to elicit an action potential in a postsynaptic cell, a series of action potentials at rapid succession in one presynaptic cell does depolarize the postsynaptic cell to threshold.

terminally differentiated: term that describes a cell that cannot return to the cell division cycle. Nerve cells are terminally differentiated; glial cells are not.

termination codon: the codons, UAA, UAG and UGA are codons that signal protein synthesis to stop. Also known as stop codons.

terminator (of transcription): a DNA sequence that, when transcribed into mRNA, causes transcription to terminate.

tertiary structure: the three dimensional folding of a polypeptide chain into a biologically active protein molecule. It usually includes regions of secondary structure. Interactions of the amino acid side chains are central in its formation.

thermodynamics: the study of how energy affects matter and chemical reactions.

threshold (voltage): the plasma membrane transmembrane voltage at which enough voltage gated sodium (or sometimes calcium) channels open to initiate an action potential.

thick filament: one of the two filaments that form the cytoskeleton of striated muscle; composed of the motor protein myosin II.

thin filament: one of the two filaments that form the cytoskeleton of striated muscle; composed of actin.

thiol group: the —S—H group.

thylakoid: folded inner membrane of the chloroplast; site of the light reaction of photosynthesis. Thylakoids are arranged in stacks called grana.

thymine: one of the four bases found in DNA. Thymine is a pyrimidine.

tight junctions: type of cell junction in which a tight seal is formed between adjacent cells occluding the extracellular space.

tissue: group of cells having a common function.

T killer cell: T cell expressing the integral plasma membrane protein CD8. T killer cells kill somatic cells that present novel peptides.

topoisomerases: enzymes that cut and rejoin DNA strands. Topoisomerase I relieves simple torsional stress by cutting one strand, allowing rotation about the phosphodiester link of the other strand, then rejoining the cut strand. Topoisomerase II cuts both strands of a double helix and passes another complete double helix through the gap, keeping hold of the ends and rejoining them when the other strand has passed through. Topoisomerases are essential during DNA replication.

totipotent: able to differentiate into (or divide to create cells that then differentiate into) any cell type of the organism. While all plant cells are totipotent, in animals only embryonic stem cells are naturally totipotent.

toxin: poison.

transamination: the enzyme-catalyzed reaction that moves an amino group from an amino acid (generating an oxo-acid) to an oxo-acid (making a new amino acid). Carried out by transaminases (aminotransferases).

transcription: synthesis of an RNA molecule from a DNA template.

transcription bubble: structure formed when two strands of DNA separate and one acts as the template for synthesis of an RNA molecule.

transcription complex: a complex of RNA polymerase and various transcription factors.

transcription factor: a protein (other than RNA polymerase) that is required for gene transcription.

***trans*-face:** the side from which material is removed. Of the Golgi apparatus, the surface from which vesicles bud to pass to the plasma membrane and to lysosomes.

transfected, transfection: a cell, prokaryotic or eukaryotic, that has been infected by a foreign DNA molecule(s) is said to be transfected. The process of DNA infection is called transfection.

transfer RNA (tRNA): the RNA molecule that carries an amino acid to an mRNA template.

transform, transformation: in addition to its common English meaning, this is used in molecular genetics to mean introduction of foreign DNA into a cell.

transgenic animal: an animal carrying a gene from another organism; the foreign gene is usually injected into the nucleus of a fertilized egg.

***trans* Golgi network:** the complex network of tubes and sheets that comprise the trans face of the Golgi apparatus. It is in the trans Golgi network that proteins made on the rough endoplasmic reticulum are sorted as to their final destination.

translation: the synthesis of a protein molecule from an mRNA template.

translocation: movement. When used of the ribosome, it means the movement, three nucleotides at a time, of the ribosome on the mRNA molecule.

transmembrane proteins: class of proteins that span the plasma membrane.

transmembrane translocation: form of protein transport in which unfolded polypeptide chains are threaded across one or more membranes as a simple polypeptide chain and then (re)folded at their final destination.

transmembrane voltage: the voltage difference between one side of a membrane and the other. It is usually stated as the voltage inside with respect to outside.

transmission electron microscope: type of microscope in which the image is formed by electrons that are transmitted through the specimen.

transmitter: a chemical that is released by one cell and which changes the behavior of another cell.

transport vesicle: membrane vesicle that transports proteins from one membrane compartment to another.

triacylglycerol (triglyceride): three acyl (that is, fatty acid) groups attached to a glycerol backbone by ester bonds. If the compound is liquid at room temperature, it is called an oil; if it is solid, it is called a fat.

tricarboxylic acid cycle (Krebs cycle): the series of reactions in the mitochondrial matrix in which acetate is completely oxidized to CO_2 with the attendant reduction of NAD^+ to NADH and FAD to $FADH_2$.

triglyceride (triacylglycerol): three acyl (that is, fatty acid) groups attached to a glycerol backbone by ester bonds. If the compound is liquid at room temperature, it is called an oil, if it is solid, it is called a fat.

trimeric: formed of three subunits.

trimeric G protein: protein with three subunits, α, β and γ, where the α subunit is a GTPase that dissociates from the $\beta\gamma$ units when it is in its GTP bound state. Both the α subunit, in its GTP bound state, and the now independent $\beta\gamma$ subunit, can activate target proteins. Examples are G_s that activates adenylate cyclase and G_q that activates phospholipase Cβ.

trimethylamine: the substance that gives rotting fish its characteristic smell. A base; when dissolved in water trimethylamine will accept an H^+. Produced by the action of intestinal bacteria on trimethylamine *N*-oxide (from dietary fish), choline (from dietary phospholipid) or other chemicals containing a trimethylamine group. Trimethylamine is oxidized in mammalian liver by the action of flavin-containing monooxygenase. A mutation in the *FMO3* gene causes the disorder trimethylaminuria. Affected individuals excrete large amounts of trimethylamine in their urine, sweat, and breath, and consequently have a fishy body odor.

The structure of trimethylamine and protonated trimethylamine are:

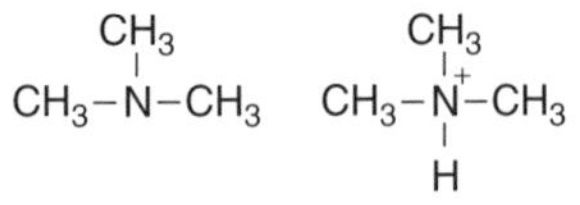

$$\begin{array}{c} CH_3 \\ | \\ CH_3{-}N{-}CH_3 \end{array} \qquad \begin{array}{c} CH_3 \\ |\;\; \\ CH_3{-}\overset{+}{N}{-}CH_3 \\ | \\ H \end{array}$$

trimethylamine N-oxide: the compound:

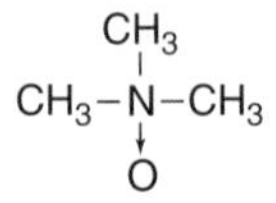

$$\begin{array}{c} CH_3 \\ | \\ CH_3{-}N{-}CH_3 \\ \downarrow \\ O \end{array}$$

The arrow indicates a so-called dipolar bond between nitrogen and oxygen. This bond can also be represented as N^+—O^-. Trimethylamine *N*-oxide is generated from trimethylamine by the action of flavin-containing monooxygenase. A mutation in the *FMO3* gene causes the disorder trimethylaminuria. Affected individuals excrete large amounts of trimethylamine in their urine, sweat, and breath, and consequently have a fishy body odor.

triphosphate: having a chain of three phosphate groups attached. Nucleoside triphosphates such as ATP are the most familiar examples.

trisphosphate: having three phosphate groups attached at three different points on the molecule. Inositol trisphosphate and phosphatidylinositol trisphosphate are examples.

Trk: a family of receptor tyrosine kinases that bind growth factors of the neurotrophin class.

tRNA (transfer RNA): the RNA molecule that carries an amino acid to an mRNA template.

troponin: a calcium-binding protein found in muscle cells.

tryptophan (trp) operon: a cluster of five bacterial genes involved in the synthesis of the amino acid tryptophan.

tubulin: subunit protein of microtubules; exists as α, β, and γ isoforms.

tumor: proliferative cell mass associated with many cancers.

turnover number: the number of moles of substrate converted to product per mole of enzyme per unit time. Another term for the catalytic rate constant, k_{cat}.

tyrosine kinase: an enzyme that phosphorylates tyrosine residues in proteins by transferring a phosphate group from ATP.

ultramicrotome: machine for cutting thin sections ($<$100 nm) for electron microscopy.

ultrastructure: fine structure of the cell and its organelles revealed by electron microscopy.

uncouple, uncoupler, uncoupled (of mitochondria): a mitochondrion is uncoupled when the tight link between the respiratory chain and ATP synthesis is broken. Since the link is the gradient of H^+ across the inner mitochondrial membrane, any chemical that allows H^+ ions to cross the inner mitochondrial membrane is an uncoupler. If the chemical is present in more than a small fraction of the cells of the body, it will kill, but expression of the protein uncoupler thermogenin in brown fat allows that tissue to generate heat.

unsaturated (of fatty acids): containing carbon-carbon double bonds.

untranslated sequence: sequence of bases in a mRNA molecule that does not code for protein. Untranslated regions are found at the 5′ and 3′ ends of an mRNA.

upstream: a general term meaning the direction from which things have come. When applied to the DNA within and adjacent to a gene, it means lying on the side of the transcription start site that is not transcribed into RNA. When applied to signaling pathways it means opposite to the direction in which the signal travels, thus the insulin receptor is upstream of protein kinase B.

uracil: one of the four bases found in RNA. Uracil is a pyrimidine.

uracil-DNA glycosidase: a DNA repair enzyme that recognises and removes uracil from DNA molecules.

urea: H_2N—(CO)—NH_2, compound made in the liver that contains lots of nitrogen but is less toxic than ammonia. Urea is chaotropic: at high concentration it reversibly denatures proteins.

urea cycle: a metabolic cycle in the liver that makes urea that is then excreted in the urine.

V(J)D recombinase: the enzyme, active only in B and T cells, that cuts and splices the DNA within loci encoding antibodies and T cell receptors.

vacuole: large membrane-bound compartment. Plant cells often contain a large vacuole filled with sugars and pigments.

Valium: an antianxiety drug that increases the chance that the GABA receptor channel will open, allowing chloride ions to pass.

van der Waals force: weak close-range attraction between atoms.

vascular tissue: blood vessels. The term is also used to describe the water transporting and support tissue of plants.

vasoconstrictor: substance that constricts blood vessels.

vasodilator: substance that dilates (widens) blood vessels.

vector: something that carries something else. The term is often used to describe a plasmid or bacteriophage that carries a foreign DNA molecule and is capable of independent replication within a bacterial cell.

vesicle: small closed bags made of membrane.

vesicular trafficking: the precisely controlled movement of vesicles between different organelles and/or the plasma membrane.

villin: type of actin-binding protein that cross-links actin filaments.

villus (plural villi): fingerlike extension of an epithelial surface that increases the surface area.

vimentin: protein that makes up the intermediate filaments in cells of mesenchymal origin such as fibroblasts.

virus: a packaged fragment of DNA or RNA that uses the synthetic machinery of a host cell to replicate its component parts.

V(J)D recombinase: the enzyme, active only in B and T cells, that cuts and splices the DNA within loci encoding antibodies and T cell receptors.

V_m: the maximal velocity of an enzyme catalyzed reaction; the limiting initial velocity obtained as the substrate concentration approaches infinity. V_m is also often used to mean transmembrane voltage.

VNTRs (variable number tandem repeats): DNA sequences that occur many times within the human genome. Each person carries a different number of these repeats.

voltage clamp: a technique in which the experimenter passes current to one side of a membrane to artificially set the value of the transmembrane voltage to a desired level.

voltage-gated calcium channel: a channel that is selective for calcium ions and which opens upon depolarization. Found in the plasma membrane of many cells.

voltage-gated sodium channel: a channel that is selective for sodium ions and which opens upon depolarization. Found in the plasma membrane of nerve and muscle cells.

Wee1: protein kinase that phosphorylates and hence inactivates CDK1.

wobble (in tRNA binding): flexibility in the base pairing between the 5′ position of the anticodon and the 3′ position of the codon.

Xeroderma pigmentosum: an inherited human disease caused by defective DNA repair enzymes. Affected individuals are sensitive to ultraviolet light and contract skin cancer when exposed to sunlight.

YAC (yeast artificial chromosome): a cloning vector used to propagate DNAs, of about 500,000 bp, in yeast cells.

Z disc: disc that is set within and at right angles to the actin microfilaments in striated muscle, holding them in a regularly spaced array.

Z-DNA: a left-handed helical form of DNA.

zinc fingers: a structural motif in some families of DNA binding proteins in which a zinc ion coordinated by cysteines and/or histidines stabilizes protruding regions which touch the edges of the base pairs exposed in the major groove of DNA.

ANSWERS TO REVIEW QUESTIONS

1.1 Theme: Dimensions in Cell Biology

General comment: 1 m = 10^3 mm = 10^6 μm = 10^9 nm.

1. E. 2,000 nm = 2 μm. Most bacteria are in the range 1 μm to 2 μm.
2. F. 20,000 nm = 20 μm. Most eukaryotic cells are in the range 5 μm to 100 μm.
3. I. 1,000,000,000 nm = 1 m. Nerve cells supplying the fingers and toes are about this length.
4. D. 250 nm. The wavelength of green light is 500 nm; light microscopes can achieve a resolution of about half this.
5. B. 0.2 nm. This allows the electron microscope to reveal structures that are invisible in the light microscope.

1.2 Theme: Types of Cell

1. C = fibroblast. Fibroblasts secrete collagen and other components of the connective tissue extracellular matrix.
2. E = glial cell. Glial cells and nerve cells are the two main types found in nervous tissue.
3. B = epithelial cell. An epithelium is a sheet of cells.
4. D = macrophage. Macrophage means "big eater" in Latin; macrophages take up unwanted material and digest it.
5. A = bacterium. Prokaryotic cells have no nuclear envelope; the genetic material is free in the cytosol.
6. F = skeletal muscle cell. Many precursor cells, each with one nucleus, fuse to form the large cells that form our muscles.
7. G = stem cell.

1.3 Theme: Some Basic Components of the Eukaryotic Cell

1. Cytosol = D.
2. Internal membranes = C.
3. Mitochondrion = F.
4. Nucleus = B.
5. Plasma membrane = A.

2.1 Theme: Types of Organic Chemicals

1. A disaccharide = F, lactose. Lactose is a disaccharide found, as the name suggests, in milk.
2. A fatty acid = H, oleic acid.
3. A nucleoside = B, adenosine.
4. A nucleotide = C, adenosine triphosphate. A nucleotide is a phosphorylated nucleoside.
5. A pentose sugar = I, ribose. Ribose, one of the building blocks of nucleic acids, is the most important pentose sugar in the body.

2.2 Theme: Chemical Groups and Bonds

General comments: molecule (i) is an acetylated sucrose molecule. You should have recognized the oxygen-

Cell Biology: A Short Course, Third Edition. Stephen R. Bolsover, Jeremy S. Hyams, Elizabeth A. Shephard and Hugh A. White.

containing ring structures as sugar residues, and hence the link between them is a glycosidic bond. On the right, attached to the sucrose by an ester bond, is an acetate residue. Singly acetylated sugars like this can easily be made but have no major role either in nature or industry. However sucrose with many acetate and butyrate residues attached by ester bonds, a molecule called sucrose acetate isobutyrate, is a food additive—it is referred to as E444 in Europe.

Molecule (ii) is a dipeptide of cysteine on the left and serine on the right, connected by a peptide bond. The side chain of serine has been phosphorylated, a modification we will meet again in Chapter 9.

Diagram (iii) shows part of an RNA molecule hydrogen bonded to a base on another nucleic acid, which could be either DNA or RNA. We will cover RNA in depth in Chapter 6.

1. amino group = D.
2. carboxyl group = G.
3. hydroxyl group = A. Note that OH groups within a carboxyl group, such as the group at G, are not called hydroxyl groups.
4. phosphate group = F. It would not be unreasonable to answer I, but since the phosphate atom at I is connected to the rest of the molecule by two bonds, we would not normally refer to the group at I as a phosphate group.
5. ester bond = C. Adding the elements of H_2O to this bond would regenerate a hydroxyl group on the sugar and a carboxyl group on the acetate, so this is an ester bond between a hydroxyl group and an organic acid.
6. glycosidic bond = B.
7. hydrogen bond = H.
8. peptide bond = E.
9. phosphodiester link = I.

2.3 Theme: Acids and Bases

1. Considering the acetate, both the protonated form CH_3COOH and the deprotonated form CH_3COO^- are present at significant concentrations = B. pH 5. When the pH is close to the pK_a of the reaction $CH_3COOH \rightleftarrows CH_3COO^- + H^+$ then both the protonated and unprotonated forms will be present.
2. Considering the ammonium, both the protonated form NH_4^+ and the unprotonated form NH_3 are present at significant concentrations = D. pH 9. When the pH is close to the pK_a of the reaction $NH_3 + H^+ \rightleftarrows NH_4^+$ then both the protonated and unprotonated forms will be present.
3. The vast majority of both the acetate and the ammonium are in their protonated forms, CH_3COOH and NH_4^+ respectively = A. pH 3. At this pH the H^+ concentration is very high, driving protons onto all potential acceptors.
4. The vast majority of both the acetate and the ammonium are in their unprotonated forms, CH_3COO^- and NH_3 respectively = E. pH 11. At this pH the H^+ concentration is very low, so that even ammonium ions give up their H^+.
5. The vast majority of both the acetate and the ammonium are in their ionic forms, CH_3COO^- and NH_4^+ respectively = C. pH 7. This pH is more alkaline than the pK_a of the reaction $CH_3COOH \rightleftarrows CH_3COO^- + H^+$, so the acetic acid gives up its H^+ to become acetate. However, the pH is more acid than the pK_a of the reaction $NH_3 + H^+ \rightleftarrows NH_4^+$, so ammonium remains in its protonated form, NH_4^+.
6. The vast majority of both the acetate and the ammonium are in their uncharged forms, CH_3COOH and NH_3 respectively = F. This situation is impossible. For acetate to be in its uncharged, protonated form CH_3COOH, the pH must be much more acid than the pK_a of the reaction $CH_3COOH \rightleftarrows CH_3COO^- + H^+$, that is, more acid than pH 4.8. However for ammonia to be in its unchanged, unprotonated form the pH must be much more alkaline than the pK_a of the reaction $NH_3 + H^+ \rightleftarrows NH_4^+$, that is, more alkaline than pH 9.2.

3.1 Theme: Membranes

1. D. Mitochondria, the nucleus, and in plant cells, chloroplasts, are surrounded by a double membrane envelope.
2. A.
3. B.
4. F.
5. A.

3.2 Theme: Organelles in Eukaryotic Cells

1. a site of protein synthesis = A, endoplasmic reticulum.
2. contains many powerful digestive enzymes = C, lysosome.
3. contains small circular chromosomes = D, mitochondrion. Only the nucleus, mitochondria, and (in plants) chloroplasts contain DNA.
4. contains the enzyme catalase = F, peroxisome.
5. filled with chromatin = E, nucleus. Chromatin, so called because it stains strongly with colored dyes, is a complex of DNA and histones.
6. made up of flattened sacks called cisternae = B, Golgi apparatus.
7. most of the cell's ATP is made here = D, mitochondrion.
8. usually found at the cell centre = B, Golgi apparatus. The cell center is a specific region, close to the nuclear envelope, where the Golgi complex and the centrosome are found.

3.3 Theme: Transport Across Membranes

1. An RNA molecule of Mr = 10,000 = C. RNA molecules are hydrophilic because their building blocks are hydrophilic molecules such as sugars and phosphate. They cannot therefore cross lipid bilayers by simple diffusion. Most are also too large to go through gap junctions. The Mr of this particular RNA was 10,000, and gap junctions do not allow molecules of Mr > 1000 to pass.
2. Inositol trisphosphate (Mr = 649) = B. Inositol trisphosphate is highly hydrophilic since it bears six negative charges. It cannot therefore cross lipid bilayers by simple diffusion. However, inositol trisphosphate is small enough to pass through gap junctions. This is thought to be an important mechanism of cell-cell signaling; we will learn some of the signaling functions of inositol trisphosphate in Chapter 15.
3. K^+ (atomic weight = 39) = B. As a small ion, K^+ is highly hydrophilic and cannot cross lipid bilayers by simple diffusion. However K^+ is readily permeable through gap junctions. Passage of electrical current, by the movement of K^+ and other ions, allows heart cells to contract in concert.
4. Nitric oxide = NO (Mr = 30) = A. NO is a small uncharged molecule and has sufficient solubility in hydrophobic solutes to be able to cross lipid bilayers by simple diffusion. In Chapter 16 we will describe how the movement of NO from one cell to its neighbor in this way is critical in allowing the blood supply to respond to the needs of the tissues.

4.1 Theme: Mutations

1. A change of a U to a A in the 6th codon of the sequence, generating the sequence 5′ACU AUC UGU AUU AUG UAA CAC CCA3′ = C, nonsense mutation. UAA is a stop codon, so translation into a polypeptide chain will terminate prematurely.
2. A change of a U to a C in the 6th codon of the sequence, generating the sequence 5′ACU AUC UGU AUU AUG CUA CAC CCA3′ = D, none of the above. CUA, like the original UUA, codes for leucine, so the encoded polypeptide sequence is unchanged. This type of mutation is known as a synonymous mutation.
3. A change of a U to a G in the 2nd codon of the sequence, generating the sequence 5′ACU AGC UGU AUU AUG UUA CAC CCA3′ = B, missense mutation. The AGC now codes for serine in place of the original isoleucine.
4. Deletion of a U in the 3rd codon, generating the sequence 5′ACU AUC UGA UUA UGU UAC ACC CA3′ = C, nonsense mutation. The deletion has generated the stop codon UGA, so translation into a polypeptide chain will terminate prematurely.
5. Deletion of an A in the 4th codon, generating the sequence 5′ACU AUC UGU UUA UGU UAC ACC CA3′ = A, frameshift mutation. From the deletion onwards, the sequence will be read using the wrong reading frame, creating a different polypeptide sequence.

4.2 Theme: Bases and Amino Acids

1. A nitrogen-rich base that is not a component of DNA = J, uracil. Uracil replaces thymine in RNA.
2. A positively charged amino acid that is found in large amounts in chromatin, where it neutralizes the negative charge on the phosphodiester links of DNA = C, arginine. Both positively charged amino acids, arginine and lysine, are present in large amounts in histones, the proteins around which DNA wraps to form the nucleosome.
3. A protein is described as having the mutation G5E. Which amino acid is present in this protein in place of the amino acid present in the normal protein? = F, glutamate. The notation G5E means that the fifth amino acid in the protein is usually glycine (G) but is glutamate (E) in the mutant.
4. The base that pairs with guanine in double stranded DNA = E, cytosine.
5. The base that pairs with thymine in double stranded DNA = A, adenine.

4.3 Theme: Structures Associated with DNA

1. A highly compacted, darkly staining substance comprising DNA and protein found at the nuclear periphery = E, heterochromatin. Heterochromatin is the form adopted by DNA that is not being transcribed into RNA. In Chapter 3 we described how heterochromatin is located at the edge of the nucleus.
2. A mass of DNA and associated proteins lying free in the cytoplasm = F, nucleoid. This is the organization found in prokaryotes.
3. A structure formed when a 146 base pair length of DNA winds around a complex of histone proteins = G, nucleosome.
4. The form adopted by those parts of chromosomes that are being transcribed into RNA = C, euchromatin.

5.1 Theme: Synthesis on a DNA Template

1. the sequence that is generated from 3′GCGAAGTCGTA 5′ during the process of transcription = F. Because the strand is being transcribed (i.e., RNA synthesis) a U is inserted into the RNA strand when an A occurs in the template strand.

2. the sequence that is generated from 3′GCGAAGTCGTA 5′ during the process of replication = G. Replication is the process of DNA synthesis. Sequence G is complementary to the template strand shown.
3. the sequence that is generated from 5′ATGCTGAAGCG3′ during the process of transcription = F. Sequence F is complementary to the template strand shown. Because the strand is being transcribed (i.e., RNA synthesis) a U is inserted into the RNA strand when an A occurs in the template strand.
4. the sequence that is generated from 5′ATGCTGAAGCG3′ during the process of replication = G. Replication is the process of DNA synthesis. Sequence G is complementary to the template strand shown.

5.2 Theme: DNA Replication

1. DNA ligase = B. DNA ligase connects adjacent deoxyribonucleotides within an otherwise complete double-stranded DNA molecule, both during normal DNA synthesis, as here, and during DNA repair.
2. DNA polymerase I = F. DNA polymerase I is responsible for building the DNA chain in those regions previously occupied by the RNA primer.
3. DNA polymerase III = E. DNA polymerase III runs nonstop to create the leading strand. It also synthesizes most of the lagging strand, but is incapable of running through the sections already occupied by RNA primers; DNA polymerase I is required in those sections.
4. exonuclease I = D. We have described how during mismatch repair exonuclease I runs in a 3′ to 5′ direction, destroying the DNA strand.
5. exonuclease III = D. We have described how during mismatch repair exonuclease VII runs in a 5′ to 3′ direction, destroying the DNA strand.
6. helicase = A. As the first step in DNA replication, helicase splits the DNA double helix into two single strands.
7. primase = C. This occurs repeatedly during synthesis of the lagging strand. As we will describe in Chapter 6, other enzymes (the RNA polymerases) synthesize RNA during transcription, but here we are concerned with DNA replication.
8. ribonuclease H = G. In prokaryotes DNA polymerase performs this function and then goes on to synthesize the complimentary DNA strand; however in eukaryotes there is a specialized stand-alone enzyme to do the job.

5.3 Theme: Regions within Eukaryotic Chromosomes

1. A section of DNA that, read in triplets of bases, encodes successive amino acids in a polypeptide chain = A. In eukaryotic genes, only the exons encode the polypeptide sequence of a protein.
2. A section of DNA with a sequence similar to a working gene but which no longer encodes a functional protein = E, pseudogene.
3. A section of DNA within a gene that is transcribed into RNA but which does not encode amino acids and which must be removed from the RNA before it leaves the nucleus = C, intron.
4. A series of identical or almost identical genes all of which are transcribed so as to generate identical or almost identical RNA products and, in the case of protein coding genes, identical proteins = G. Examples are the genes that encode ribosomal RNA, transfer RNA, and histone proteins. Note that the answer "gene family" is wrong—members of a gene family encode significantly different, albeit strongly similar, proteins.
5. A type of DNA with a presumed structural role that makes up much of the chromosome in the centromere region = F, satellite DNA.
6. An extragenic DNA sequence that is repeated more than a million times = F, satellite DNA. Note that the answer "long interspersed nuclear element" is wrong—the copy number of LINEs is in the thousands, not in the millions.

6.1 Theme: Codes within the Base Sequence

1. A stretch of DNA rich in adenine and thymine, called the TATA box = E, transcription initiation in eukaryotes. TATA binding protein, a component of transcription factor IID, binds to the TATA box on the DNA, allowing recruitment of the other transcription factors and then RNA polymerase II.
2. A stretch of DNA rich in guanine and cytosine, followed by a string of adenines = H, transcription termination in prokaryotes. The resulting RNA strand forms a hairpin loop in the GC-rich region, reducing the size of the transcription bubble, while the attachment of the following uracils to the string of adenines is weak since there are only two hydrogen bonds in each UA pair.
3. The DNA sequence GGGGCGGGGC, called the GC box = E, transcription initiation in eukaryotes. This is an alternative transcription initiator used in many eukaryotic genes. A protein called Sp1 binds to the GC box and recruits TATA binding protein, allowing recruitment of the other transcription factors and then RNA polymerase II.
4. The DNA sequence TATATT, called the −10 or Pribnow box = F, transcription initiation in prokaryotes. The −10 box is part of the prokaryotic promoter sequence, which recruits the σ factor, allowing recruitment of the other subunits of RNA polymerase.
5. The RNA motif GU. . . . AG, where the. . . . indicates a long sequence of bases = D, removal of introns in eukaryotic mRNA. All eukaryotic introns conform to the sequence GU. . . . AG, and although the process is

incompletely understood, these bases are involved in the process by which the spliceosome recognizes the length of RNA to be removed.

6. The RNA sequence AAUAAA = C, polyadenylation of eukaryotic mRNA. This sequence is found close to the 3′ end of most mRNAs and is thought to be a signal for poly-A polymerase to attach and add a string of A residues to the 3′ end of the mRNA.

6.2 Theme: The Control of Transcription

1. Catabolite activator protein = C. In the presence of cyclic AMP, catabolite activator protein binds to a regulatory site of the *lac* operon of *E. coli* and increases transcription. In turn, cAMP is only high if glucose is in low supply.
2. Glucocorticoid hormone receptor = C. In the presence of glucocorticoid hormone, the glucocorticoid hormone receptor binds to an enhancer site of various mammalian genes and increases transcription.
3. *Lac* repressor protein = B. If the *lac* repressor protein binds a β-galactoside sugar, it adopts a shape that can no longer bind to the operator region of the operon. Thus the enzymes that allow utilization of β-galactoside sugars are only made if these sugars are present.
4. Trp aporepressor protein = D. When the aporepressor protein binds tryptophan, it can then bind to the operator region of the *trp* operon and inhibit transcription. Thus the enzymes that synthesize tryptophan are only made when tryptophan is lacking.

6.3 Theme: Events that Occur after Transcription in Eukaryotes

1. A chemical modification of the 3′ end of the RNA molecule = D, polyadenylation. A long stretch of adenosine residues is added to the 3′ end of the mRNA transcript.
2. A chemical modification of the 5′ end of the RNA molecule = A, capping. The methylated guanosine is added by a 5′ to 5′ phosphodiester link, unlike the 3′ to 5′ links formed by RNA polymerase.
3. A process that allows two or more polypeptide chains of different amino acid sequence to be generated from the same mRNA transcript = E, RNA splicing. Alternative splicing allows one primary mRNA transcript to give rise to two processed mRNAs that share some exons but differ in others. It is distinct from the phenomenon of polycistronic mRNA seen in prokaryotes, where sequential lengths of mRNA on the same molecule are translated to generate completely different proteins.
4. A process that reduces, often dramatically, the length of the RNA molecule prior to its subsequent translation into protein = E, RNA splicing. Splicing removes the introns to leave only exonic, coding RNA. Answer B, digestion by nucleases, is wrong because after digestion the mRNA cannot be translated into protein.

7.1 Theme: A Mammalian Expression Plasmid

1. DNA encoding cytochrome *c* is inserted into the plasmid. Which element in the plasmid makes this possible = D, the multiple cloning site. This comprises the recognition sites of a number of restriction enzymes, allowing the experimenters to choose a site also present at the appropriate places on the DNA encoding the cytochrome *c*, cut both pieces of DNA with the same enzyme, and then assemble the recombinant plasmid.
2. To generate large amounts of the recombinant plasmid, the plasmid is grown up in bacteria. The plasmid is used to transform competent *E. coli* which are then cultured in such a way that only bacteria containing the plasmid survive. Which element of the plasmid allows the survival of host bacteria when sister bacteria are dying = E, antibiotic resistance gene. The bacteria are grown on agar containing the toxic antibiotic. Only bacteria containing the resistance gene survive.
3. The transformed bacteria divide repeatedly, producing colonies each derived from a single transformed progenitor. What element in the plasmid allows it to be copied in parallel with the host bacterium's DNA = A, origin of replication. This is the point where the host enzyme DnaA attaches, allowing formation of the open replication complex.
4. Clones containing the recombinant plasmid are grown up and then lysed, allowing purification of large amounts of recombinant plasmid. The purified plasmid is then used to transfect human cells, which synthesize the GFP:cytochrome *c* chimera. Which element in the plasmid allows the chimaeric protein to be expressed in the HeLa cells, even though it was not expressed in the bacteria = B, cytomegalovirus promoter. This is a powerful promoter used by the cytomegalovirus to drive transcription of its genes in preference to those of the host, and is very commonly used by experimenters to drive expression from plasmids introduced into mammalian cells. This promoter is not recognized by the bacterial RNA polymerase and therefore the chimeric mRNA (and hence protein) will not be made in bacterial cells.

7.2 Theme: Choosing an Oligonucleotide for a Specific Task

1. In the polymerase chain reaction: indicate the oligonucleotide that should be used together with the oligonucleotide 5′TACGGATCCCTTTGCAGGAT3′ to

amplify the double stranded DNA molecule shown at the top. = B. The oligonucleotide 5′TGCCTACTGCAGCG TCTGCA3′ can be used to copy the top strand of the sequence shown in the 5′ to 3′ direction.

2. If you wished to clone the amplified product into the EcoR1 site of a plasmid how would you modify the oligonucleotide 5′TACGGATCCCTTTGCAGGAT3′ = E. The sequence the enzyme will recognize is simply added to the 5′ end of the oligonucleotide generating 5′GAATTCTACGGATCCCTTTGCAGGAT3′.
3. An oligonucleotide that could be used to prime the synthesis of DNA from most of the mRNAs present in a tissue, in order to generate a cDNA library = A. Most eukaryotic mRNAs have at their 3′ end a poly-A tail. The primer 5′TTTTTTTTTTTTTTT3′ will hydrogen bond with the tail and can then be extended by reverse transcriptase.
4. An oligonucleotide that could be used in a Southern blot to identify carriers of sickle cell anemia. Note that this disease is caused because an A in the sequence 5′GTGCATCTGACTCCTG**A**GGAGAAGTCT3′ in the normal β globin gene is mutated to a T to generate the sequence 5′GTGCATCTGACTCCTG**T**GGAGAAGT CT3′. = F. Both strands of DNA are present on a Southern blot so the sequence 5′GTGCATCTGACTCCTG**T**G GAGAAGTCT3′ will hydrogen bond with its complementary sequence.
5. An oligonucleotide that could be used for northern blotting to detect mRNA containing the sequence 5′GUCAGCUUACGAUGGCAGUC3′ = G, 5′GACTGCCATCGTAAGCTGAC3′. The oligonucleotide must be complementary in sequence to the mRNA. Oligonucleotides are made as DNA and therefore contain Ts instead of Us.

7.3 Theme: Uses of cDNA Clones

1. Amplifying a known or partially known DNA sequence using the polymerase chain reaction = A. The strands are separated at high temperature and then the primers (which have been chosen such that once attached, the 3′ ends point towards each other) allowed to anneal to their respective strands. Thermostable DNA polymerase then generates the complementary strands so that for each cycle of operation there is a replication of the DNA between the primer two attachment points. At present polymerase chain reaction amplification of lengths of up to 4 kb are routine; some users amplify lengths of up to 8 kb.
2. Automated DNA sequencing by the dideoxy chain termination method = D. DNA polymerase will attach to the 3′ end of the annealed oligonucleotide and then run along copying the remainder of the DNA sequence until a dideoxynucleotide is incorporated, at which point replication of that molecule stops. Note that this is automated sequencing, in which detection of the various products is by fluorescence of labelled dideoxynucleotides. The oligonucleotide does not need to be radiolabelled (and both cost and safety considerations mean that a radioactive reagent would never be used when a nonradioactive one will suffice).
3. Detection of specific DNA sequences by Southern blotting, for example to differentiate DNA from two human subjects = C. The oligonucleotide will hybridize to the specific DNA sequence and can then be detected by autoradiography.
4. Investigation, by northern blotting, of the degree to which a gene of interest is transcribed in a particular tissue = C. The oligonucleotide will hybridize to the specific RNA sequence and can then be detected by autoradiography.

8.1 Theme: Translation Initiation

1. In the early stages of translation initiation in prokaryotes, the small ribosomal subunit attaches by complimentary base pairing to this sequence at the 5′ end of the mRNA, sometimes called the Shine-Dalgano sequence = H, 5′GGAGG3′.
2. In contrast in the early stages of translation initiation in eukaryotes, the small ribosomal subunit attaches to this group at the extreme 5′ end of the mRNA molecule = M, 7-methyl guanosine cap.
3. The eukaryotic small subunit then slides along the mRNA until it encounters this sequence, known as the Kozak sequence = E, 5′CCACC3′.
4. The subsequent steps are similar in prokaryotes and eukaryotes. The small subunit slides a few more bases along until it encounters this sequence, the start codon for translation = D, 5′AUG 3′.
5. Initiation factors then act to catalyze the assembly of the complete ribosome. The first tRNA, with its formyl methionine (in prokaryotes) or methionine (in eukaryotes) attached, locates in one of the three tRNA binding sites on the ribosome: state which one = C, P site.

8.2 Theme: Translation Elongation and Termination

1. Following translocation of the ribosome three bases along the mRNA, this site on the ribosome is empty and can be occupied by a charged tRNA whose anticodon is complimentary to the corresponding codon on the mRNA = A, A site.
2. Peptidyl transferase now catalyzes the formation of a peptide bond between the new amino acid and the existing polypeptide chain. Immediately following this, the polypeptide chain is only attached to the mRNA via the tRNA in which of the three sites on the ribosome = A, A site.

3. The next step is translocation, the physical movement of the ribosome three bases along the mRNA. Energy from this is provided by the hydrolysis of GTP by which enzyme, which occupies the A site on the ribosome = D, EF-G. EF-G is a GTPase, one of a large family of fundamentally similar proteins that guide and drive biological processes through the binding and hydrolysis of GTP. GTPases will be described in more detail in Chapter 10.
4. As a result of translocation the uncharged tRNA, which gave up its amino acid during the formation of the peptide bond, moves to this site on the ribosome, from where it is released = B, E site.
5. When translocation brings a stop codon UGA, UAA or UAG into the position facing the A site on the ribosome, the A site becomes occupied not by a normal charged tRNA molecule but rather by this molecule = I, release factor 1 or 2.
6. Finally the ribosome splits into two subunits as a result of energy released in GTP hydrolysis by EF-G. The released small ribosomal subunit already has one initiation factor attached, ready to accept the other initiation factors and the first charged amino acid to initiate synthesis of a new polypeptide. Name this pre-attached initiation factor = F, IF3.

8.3 Theme: The Wobble

1. Methionine = B, 5′CAU3′. If you answered H = 5′UAC3′ you forgot that nucleic acid strands pair up in an antiparallel fashion.
2. Asparagine = D, 5′GUU3′. The wobble phenomenon is operating: the G at the 5′ end of the anticodon can pair with either U or C at the 3′ end of the codon.
3. Phenylalanine = C, 5′GAA3′. Once again the wobble phenomenon is operating: the G at the 5′ end of the anticodon can pair with either U or C at the 3′ end of the codon.
4. Isoleucine = E, 5′IAU3′. The inosine at the 5′ end of the anticodon can pair with any of U, C or A at the 3′ end of the codon.

9.1 Theme: Amino Acids

1. Can be phosphorylated = E, glutamate.
2. Has a strongly acidic side chain = E, glutamate.
3. Has a strongly basic side chain = R, arginine.
4. Is an imino acid, not an amino acid, and therefore imposes greater constraints on the shape of the polypeptide chain = P, proline.
5. Two can form a disulphide bond = C, cysteine.

9.2 Theme: Terms Used to Describe Proteins

1. A covalent bond between the side chains of two cysteine residues = D, disulfide bond.
2. A protein secondary structure where the backbone coils in a right-handed helix with hydrogen bonds between the amide and carboxyl groups of the peptide bonds with the hydrogen bonds running parallel with the direction of the helix = A, α helix.
3. A separately folded region of a single polypeptide chain = E, domain.
4. Loss of all of the higher levels of structure with accompanying loss of biological activity of a protein = C, denaturation.
5. The tendency for hydrophobic molecules to cluster together away from water = F, hydrophobic effect.

9.3 Theme: Specific Binding Partners

1. Calmodulin = C, calcium ions.
2. Catabolite activator protein = A, a specific base sequence on DNA. Catabolite activator protein also binds cAMP, but does not directly bind either glucose or β-galactoside sugars. Rather, low glucose concentrations cause the concentration of cAMP to rise, and it is cAMP that activates catabolite activator protein so that it can bind to its specific base sequence on DNA.
3. Connexin 43 = D, connexin 43. Connexins on one cell bind a compatible connexin on a neighboring cell to form a gap junction channel.
4. Glucocorticoid receptor = A, a specific base sequence on DNA. The glucocorticoid receptor also binds glucocorticoid hormones, and can only adopt a shape that binds to the DNA when it has the steroid bound.

10.1 Theme: The Three Modes of Intracellular Transport

1. β globin = B. β globin is used to assemble hemoglobin, which remains in the cytosol.
2. Catalase (required inside peroxisomes) = C. All other organelles bound by a single membrane receive the majority of their proteins by vesicular trafficking, but proteins destined for peroxisomes are synthesized on cytosolic ribosomes and only then imported into peroxisomes.
3. Glucocorticoid hormone receptor = A. If the glucocorticoid hormone receptor binds its steroid hormone, it is imported into the nucleus.
4. NFAT (Nuclear Factor of Activated T cells) = A. Dephosphorylated NFAT reveals a nuclear localization sequence and is imported into the nucleus.
5. Platelet-derived growth factor receptor = D. All integral proteins of the membrane reach this location *via* the endoplasmic reticulum and Golgi apparatus.
6. Pyruvate dehydrogenase (required in the mitochondrial matrix) = C. Proteins destined for mitochondria are

synthesized on cytosolic ribosomes, and only then imported into mitochondria.

10.2 Theme: Trafficking Processes

1. Arf = D, transport between Golgi cisternae.
2. Dynamin = A, endocytosis.
3. Rab = B, exocytosis. Rab family members control fusion events, including the fusion of vesicles with the plasma membrane.
4. Ran = C, traffic through the nuclear pore.
5. Signal recognition particle = G, transport into the endoplasmic reticulum.
6. TAP (transporter associated with antigen processing) = G, transport into the endoplasmic reticulum.

10.3 Theme: GTPases

1. GTPases have a binding site for a nucleotide. Identify the nucleotide present in the pocket when the GTPase is in its off state (e.g. Arf in the state that cannot associate with membranes) = D, GDP.
2. GTPases are activated when the nucleotide in the binding pocket is replaced by a nucleotide present at higher concentration in the cytosol. Give the general name for the protein partner that catalyzes this switch = H, guanine nucleotide exchange factor, GEF.
3. GTPases turn off when the nucleotide in the binding pocket is hydrolysed. Identify the product of this hydrolysis = D, GDP.
4. Hydrolysis of the nucleotide in the binding pocket is activated by a protein partner. Give the general name for the protein partner that accelerates the hydrolysis = G, GTPase activating protein, GAP.

General comment: The term GTPase activating protein is somewhat confusing. These proteins activate the GTPase catalytic activity of the GTPase, and therefore speed the process whereby the GTPase turns itself off.

11.1 Theme: Changing the Preferred Shape of a Protein

1. A fall of pH from 7.5 to 6.5 = B, favours the closed configuration. At the lower pH the histidine will spend a greater fraction of time protonated and therefore positively charged. While protonated it will be attracted to the negative charge on the glutamate.
2. Phosphorylation of the serine = B, favours the closed configuration. The negative phosphate group will be attracted to the positive charge on the lysine.
3. Phosphorylation of both threonines = A, favours the open configuration. The two negative phosphate groups will repel.

11.2 Theme: Enzyme Kinetics

1. k_{cat} (catalytic rate constant) = E, the number of moles of product formed per mole of enzyme per unit time.
2. K_M (Michaelis constant) = B, that substrate concentration that gives an initial velocity equal to half the maximum velocity (V_m).
3. v_0 (initial velocity) = G, the rate of product formation at the start of an enzyme reaction.
4. V_m (maximum velocity) = C, the maximum initial velocity possible when an enzyme is fully saturated with its substrate.

11.3 Theme: Enzymes

1. F. The catalytic rate constant... can be determined if both V_{max} and the total enzyme concentration are known.
2. G. When an enzyme can work on two substrates the best substrate is the one that... gives the highest specifity constant (ratio of k_{cat} over K_M).
3. E. If you double the amount of an enzyme (keeping other conditions constant) you will... increase V_m by twofold.
4. B. A sigmoid curve obtained when v_0 is plotted against substrate concentration shows... that the enzyme binds its substrate cooperatively, that it is an allosteric enzyme.
5. C. A substrate with K_M of 5×10^{-3} mol L^{-1} binds to the enzyme more... weakly than one with $K_M = 5 \times 10^{-4}$ mol L^{-1}.

12.1 Theme: Cell Spaces and Regions in Energy Trading

1. ADP/ATP exchanger = B.
2. ATP synthase = B.
3. Coenzyme Q = B.
4. Cytochrome *c* = C.
5. $[Na^+] > 100$ mmole liter^{-1} = G.
6. porin = D.
7. sodium/potassium ATPase = F.
8. The electron transport chain = B.

12.2 Theme: The Electron Transport Chain and ATP Synthase

1. Is not a carrier = B.
2. Oxidises coenzyme Q = C.
3. Oxidises NADH = A.
4. Oxidises succinate = B.
5. Oxidises the reduced form of cytochrome *c* = D.
6. Reduces molecular oxygen to water = D.

7. Will reverse direction if the H^+ electrochemical gradient across the membrane is dissipated after application of an uncoupler = E.

12.3 Theme: Energy Currencies

1. Contains a pyrimidine residue = F. UTP is a nucleotide comprising the pyrimidine uracil coupled to triphosphorylated ribose. In contrast NADH, ATP and GTP all contain purine residues: adenine, adenine, and guanine respectively. NADH also contains a nicotinamide residue.
2. Generated by the action of an integral membrane protein of the plasma membrane = E. The sodium ion electrochemical gradient across the plasma membrane is created by the action of the sodium/potassium ATPase.
3. Has no energy content under anaerobic conditions = C. Under anaerobic conditions NADH cannot pass hydrogens to molecular oxygen, and so cannot be used to drive H^+ up its electrochemical gradient out of the mitochondrial matrix.
4. In a well oxygenated cell, this is the most energy rich of the energy currencies = C.
5. This currency is directly depleted by the action of uncouplers such as 2,4 dinitrophenol = D. Uncouplers allow H^+ to cross the inner mitochondrial membrane, dissipating the H^+ electrochemical gradient.

13.1 Theme: Reactions and Pathways

1. Fatty acid synthesis = I, stearic acid. Stearic is one of many fatty acids synthesized by the body. Double bonds and longer-chain fatty acids are made by other systems after fatty acid sythesis has made stearic acid.
2. Glucose-6-phosphate dehydrogenase = G, NADPH. This is the first step in the pentose phosphate pathway, generating reducing power as NADPH.
3. Gluconeogenesis = E, glucose 6-phosphate. Liver cells can then dephosphorylate this compound to generate glucose itself for export from the cell.
4. Lactate dehydrogenase = F, NADH. Lactate dehydrogenase oxidizes lactate to pyruvate, reducing NAD^+ in the process.
5. Phosphofructokinase = C, fructose-1,6-bisphosphate. This reaction commits the sugar to being broken down and used to provide energy rather than used for other purposes.

13.2 Theme: Pathways and Enzymes

1. Converts fatty acids to acetyl CoA = B, β oxidation.
2. Converts pyruvate to acetyl CoA = G, pyruvate dehydrogenase.
3. The only way the red blood cell has to make ATP = F, glycolysis.
4. The source of 5-carbon sugars and a source of NADPH for biosynthesis = I, the pentose phosphate pathway.
5. Uses UDP-glucose = E, glycogen synthesis.

13.3 Theme: Metabolism

1. B. Essential amino acids are... amino acids that an organism cannot make and so must be present in the diet.
2. A. Basic amino acids are... amino acids with a side chain that can be protonated.
3. G. The enzymes of the Krebs cycle are... located mainly in the mitochondrial matrix with one in the inner mitochondrial membrane.
4. F. Pyruvate carboxylase makes... oxalacetate. This is used to top up oxalacetate for the Krebs cycle and is an important step in gluconeogenesis.
5. D. It necessary to convert pyruvate to lactate when there is little oxygen available because... the reaction regenerates the NAD^+ used up during glycolysis (during the oxidation of glyceraldehyde-3-phosphate to 1,3-bisphosphoglycerate).

14.1 Theme: Cytosolic and Extracellular Concentrations of Important Ions

1. Cytosolic calcium = I. A typical value for cytosolic calcium is 100 nmoles liter^{-1}.
2. Extracellular calcium = E.
3. Cytosolic chloride = D.
4. Extracellular chloride = A.
5. Cytosolic H^+ (H_3O^+) = I. A typical value for cytosolic H^+ is 60 nmoles liter^{-1}, so the response "less than or equal to 100 nmoles liter^{-1}" is correct.
6. Extracellular H^+ (H_3O^+) = I. A typical value for extracellular H^+ is 40 nmoles liter^{-1}, so the response "less than or equal to 100 nmoles liter^{-1}" is correct.
7. Cytosolic potassium = A. A typical value for cytosolic potassium is 140 mmoles liter^{-1}, so the response "greater than or equal to 100 mmoles liter^{-1}" is correct.
8. Extracellular potassium = D.
9. Cytosolic sodium = C.
10. Extracellular sodium = A. A typical value for extracellular sodium is 150 mmoles liter^{-1}, so the response "greater than or equal to 100 mmoles liter^{-1}" is correct.

14.2 Theme: Pathways for Solute Movement Across the Plasma Membrane

1. An enzyme that carries out a hydrolytic reaction = A. The calcium ATPase hydrolyses ATP to ADP + Pi; this hydrolysis provides the energy that drives a calcium ion out of the cytosol.
2. A protein that carries one solute in one direction and a second solute in the opposite direction = E.

3. A channel that can pass glucose = B. The molecular weight of glucose is 180, well below the 1000 or so limit for a connexon. Note that C is not correct because the glucose carrier is not a channel.
4. A protein whose expression in almost all human cells is responsible for the fact that their cytosol is at a negative voltage with respect to the extracellular medium = D.
5. A channel expressed in pain receptor neurons which opens at damagingly high temperatures = F.
6. A channel expressed in almost all neurons, even those that are not sensitive to painful stimuli, but which is not expressed in most non-neuronal human cells (it is not found, for example, in liver or blood cells) = G.

14.3 Theme: Ion Fluxes in a Nerve Cell

1. Although gradients of this ion play a crucial role at other cell membranes, the concentration gradient of this ion across the plasma membrane is small: the concentration in the cytosol is the same, within a factor of two, as the concentration in the extracellular medium = B. H^+ has a crucial role in mitochondria, but the concentration gradient across the plasma membrane is small, with the concentration in the cytosol being about 1.5 times that in the extracellular medium.
2. In a resting nerve cell, this ion is constantly leaking into the cell, and must be removed by the action of an ATP-consuming carrier = D.
3. In a resting nerve cell, this ion is constantly leaking out of the cell, and must be pumped back in by the action of an ATP-consuming carrier = C.
4. The more a nerve cell is depolarized, the greater the electrochemical gradient favouring entry of this ion into the cell = A. When the cytosol becomes less negative, the voltage force pulling the positive ions H^+, K^+ and Na^+ will get smaller. Only a negative ion such as Cl^- will experience a greater inward electrochemical gradient when the cell is depolarized.
5. In a resting nerve cell, this ion is at equilibrium across the plasma membrane = A. Chloride is at equilibrium across the plasma membranes of most resting animal cells.

15.1 Theme: Processes Downstream of Receptor Tyrosine Kinases

1. A domain comprising a pocket with a positively charged arginine at the base. Proteins with this domain are recruited to phosphorylated tyrosine residues = F.
2. A guanine nucleotide exchange factor for Ras = G.
3. A hydrolytic enzyme that is activated when phosphorylated by receptor tyrosine kinases = D.
4. A kinase that is activated when phosphorylated by receptor tyrosine kinases = B. A kinase is an enzyme that transfers the γ phosphate of ATP to another molecule; most of the kinases described in this book are protein kinases that phosphorylate serine, threonine, or tyrosine residues, but phosphoinositide 3-kinase phosphorylates the lipid PIP_2. Note that protein kinase B is not a correct answer because this kinase is activated by being phosphorylated on serine and threonine residues; this phosphorylation is not carried out by receptor tyrosine kinases, which only phosphorylate tyrosine residues.
5. An enzyme that is activated by Ras:GTP = B. As well as activating MAPKKK, Ras can also activate PI 3-kinase, as shown in Figure 15.16.

15.2 Theme: Proteins Activated by Nucleotides

1. A protein kinase activated by cAMP = G.
2. A protein responsible for generating electrical signals in photoreceptors = C.
3. A protein that activates a phospholipase C when in the GTP bound state = D.
4. A protein that activates adenylate cyclase when in the GTP bound state = E.
5. A protein that activates MAP kinase kinase kinase when in the GTP bound state = J.

15.3 Theme: Inositol Compounds

1. A ligand that binds to and opens a calcium channel = C.
2. A lipid that recruits protein kinase B to the plasma membrane = G.
3. A substrate for phosphoinositide 3-kinase (PI3K) = F.
4. A substrate for phospholipases C (PLC) = F.
5. The product of phosphoinositide 3-kinase (PI3K) = G.
6. The product of phospholipases C (PLC) = C.

General comment: Unphosphorylated inositol, IP_2, and IP_4 play important roles in mammalian cells but are not covered in this book. IP_6 is found in plants but is relatively unimportant in animals.

16.1 Theme: Receptors

1. A receptor that signals to the trimeric G protein G_q = A.
2. A receptor that signals to the trimeric G protein G_s = B.
3. A receptor tyrosine kinase = F. Note that the interleukin 2 receptor is not a receptor tyrosine kinase; its downstream effects require the assistance of JAK tyrosine kinase which is associated with, but not part of, the receptor.
4. An intracellular receptor that is always cytosolic = D.
5. An intracellular receptor that moves to the nucleus upon binding transmitter = C.
6. An ionotropic receptor = G.

16.2 Theme: Transmitters

1. A hormone = A. Adrenaline is released by the adrenal gland and is distributed throughout the body by the blood system.
2. A paracrine transmitter = F. Noradrenaline is released by the axons of vasoconstrictor and other nerve cells and diffuses through the tissue, causing smooth muscle cell contraction and other responses.
3. A transmitter that acts to cause release of calcium from the endoplasmic reticulum of many cells including smooth muscle = F.
4. A transmitter that is a γ amino acid = B.
5. A transmitter that is an α amino acid = C.
6. An excitatory synaptic transmitter = C.
7. An inhibitory synaptic transmitter = B.

16.3 Theme: Synapses

1. A burst of activity in a presynaptic GABAergic neurone for a postsynaptic cell that is receiving steady excitatory input from a number of presynaptic glutaminergic neurones = D. Since the postsynaptic cell is already depolarized, the opening of GABA receptors will cause chloride ions to move in, carrying negative change and reducing the depolarization of the postsysnaptic cell.
2. A rapid burst of action potentials in a presynaptic pain receptor neurone caused by a painful stimulus such as a pin prick = B. Note that the stimulus was painful. For the stimulus to be perceived as painful by the subject, the pain relay cell must have been depolarized to threshold.
3. A single action potential in a motoneurone = B. The synapse between motoneurons and skeletal muscle cells is unusual in that one presynaptic action potential evokes a postsynaptic depolarization large enough to evoke an action potential.
4. A single action potential in a presynaptic GABAergic neurone, at a time when no other synapses onto the postsynaptic cell are active = E. Chloride is at equilibrium across the plasma membrane of an otherwise unperturbed cell.
5. A single action potential in a presynaptic glutaminergic neurone, at a time when no other synapses onto the postsynaptic cell are active = A.

17.1 Theme: Cytoskeletal Structures

1. Cilia = C, tubulin.
2. Fingernails = B, intermediate filament proteins. Fingernails are formed of keratin.
3. Flagella = C, tubulin.
4. Microfilaments = A, actin.
5. Microtubules = C, tubulin.
6. Microvilli = A, actin.
7. Stress fibers = A, actin.

17.2 Theme: Proteins of the Cytoskeleton

1. D. Keratin forms the intermediate filaments in skin and structures formed from skin such as hair, fingernails, horns and hooves.
2. A. Microfilaments are made of actin monomers.
3. B. Microtubules are formed from α- and β-tubulin.
4. F. Myosin is found in all cells, and in skeletal muscle cells forms part of the contractile apparatus. It is a motor molecule that interacts with actin filaments (microfilaments)
5. E. Kinesin and dynein are the two molecular motors that act on microtubules
6. G.
7. C. γ-tubulin does not form microtubules, but resides at the centrosome as part of the microtubule organizing center.

17.3 Theme: Fueling Movement

1. Amoeboid locomotion is powered by ATP hydrolysis performed by this ATPase = E, myosin.
2. The contraction of muscle is powered by ATP hydrolysis performed by this ATPase = E, myosin.
3. The rowing motion of a cilium is powered by ATP hydrolysis performed by this ATPase = C, dynein.
4. The transport of vesicles and organelles from the tips of nerve cell axons to the cell body is powered by ATP hydrolysis performed by this ATPase = C, dynein.
5. The wriggling motion of sperm tails is powered by ATP hydrolysis performed by this ATPase = C, dynein.

General comment: Do not confuse dynein, an ATPase, with dynamin. The latter uses energy released by GTP hydrolysis to power the pinching off of vesicles from budding membranes (page 168).

18.1 Theme: Mitosis

1. chromosome condensation occurs in the nucleus and spindle formation begins in the cytoplasm = E, prophase.
2. the nuclear envelope breaks down and chromosomes become associated with the spindle = D, prometaphase.
3. the chromosomes are aligned on the spindle and no longer make individual excursions towards and away from the spindle poles = C, metaphase.
4. paired chromatids separate and begin to move toward the spindle poles = A, anaphase.
5. chromosomes decondense and the nuclear envelope reforms = F, telophase.
6. physical separation into two cells = B, cytokinesis.

18.2 Theme: Checkpoints in the Cell Cycle

1. During this period of interphase, the cell replicates its DNA = H, S phase.
2. For animal cells to begin DNA replication, many conditions must be met. One is that the transcription factor E2F must be released from a ligand that holds it in an inactive dimer. What is this ligand = K, Rb.
3. If the DNA is damaged, it must be repaired before it is replicated. DNA damage activates two kinases, Chk1 and Cds1. These phosphorylate a transcription factor that acts to upregulate cyclin dependent kinase inhibitors, CKIs. Phosphorylation of the transcription factor allows its concentration to increase in cells. What is the identity of this anti-division trancription factor = I, P53.
4. Once the DNA is replicated and the cell is large enough, it can enter mitosis. Passing the G2/M checkpoint requires activity of cyclin dependent kinase 1, Cdk1. Cdk1 is only active when dimerized with a protein partner; identify this essential partner = C, cyclin B.
5. Cdk1 is also regulated by posttranslational modification. Which reaction must be performed on Cdk1 in order for it to be active = D, dephosphorylation at T14 and Y15.
6. During prometaphase the chromosomes line up on the metaphase plate. The sister chromatids are joined together at the kinetochores by which protein A, = cohesin.
7. When all the kinetochores are under tension the anaphase promoting complex targets securin for destruction, allowing activation of an enzyme that digests the link between the chromatids, allowing the separation of the chromatids in anaphase. What is that link-destroying enzyme = M, separase.

18.3 Theme: Life and Death

1. In animals, cells are kept alive by the activation of other cells that supply growth factors. Growth factor receptors activate PI 3-kinase, which generates PIP_3 at the plasma membrane. PIP_3 recruits a critical survival-promoting kinase to the plasma membrane; name that kinase = H, protein kinase B.
2. If PIP_3 dissapears from the plasma membrane and the kinase described above becomes inactive, BAX is activated, allowing a protein to escape from the mitochondria. Name the released mitochondrial protein = D, cytochrome *c*.
3. White blood cells can kill target cells by activating this cell surface death domain receptor = E, Fas. Many other mechanisms by which white blood cells kill targets are described in Chapter 19.
4. Cells also die if their DNA is damaged so badly that it cannot be repaired in a reasonable time. What is the transcription factor whose concentration increases after DNA damage and which upregulates synthesis of BAX = F, p53.
5. All the death initiation pathways described above converge on the activation of this family of cytosolic proteases = B, caspases.

19.1 Theme: Immune System Cells

1. A cell generated by a process that includes somatic hypermutation = A. Somatic hypermutation only occurs in the B cell lineage. It occurs in no other lineage, not even that leading to T cells.
2. A cell that attacks any body cell that does not express major histocompatibility complex protein on its cell surface = F.
3. A cell that attacks any body cell whose major histocompatibility complex proteins are presenting novel peptides = C.
4. A cell that makes antibodies = A.
5. A cell that migrates from the tissues to lymph nodes, where it presents antigens to the lymphocytes = D.
6. A cell that upon stimulation expresses CD40L, a protein that in turn activates B cells = B.
7. An important phagocytic cell of body tissue which differentiates from blood monocytes = E.
8. The commonest phagocytic cell of the blood = G.

19.2 Theme: The Antibody Heavy Chain Locus

1. A component that experiences somatic hypermutation = A. Only the V sections of the locus experience somatic hypermutation, in the process that generates antibodies that bind the antigen with even higher affinity than did the original, genetically encoded version.
2. A component that includes a sequence encoding a transmembrane domain = C. At all stages before conversion into a plasma cell, antibodies are integral membrane proteins. Their heavy chain crosses the membrane once in a polypeptide chain encoded by a length of DNA that is given a Greek letter corresponding to the Latin letter of the corresponding immunoglobulin class, in this case, γ for IgG.
3. Although chromosome 14 in every other cell of the body contains 6 of these arranged sequentially, the DNA of mature B cells contains a linear sequence comprising any number from 1 to 6 = E. The DNA splicing that occurs during B cell maturation leaves only one V and one D section. In contrast, between zero and five of the six J sections are removed. The primary RNA transcript contains all the remaining J sections, with all except the first remaining one being removed during RNA splicing.

4. Genomic DNA contains more than 20, but less than 100, slightly different versions of this section arranged sequentially along the chromosome = D. The best current estimate is that there are 23 D sections in the human genome.
5. The first protein coding region that would be encountered when reading along the locus from the 5′ end = A. The heavy chain locus begins with approximately 100 slightly different copies of a V section.

19.3 Theme: T Cells and Their Interaction with Other Cells

1. A death domain receptor that, when activated, causes the proteolysis and activation of caspase 8 and hence cell death by apoptosis = E.
2. A mitogen secreted by CD4+ T cells = G. CD40L on the surface of CD4+ T cells induces mitosis in B cells, but it is an integral membrane protein—it is not secreted.
3. A protease secreted by CD8+ T cells = F.
4. A protein complex that cuts cytosolic proteins into short lengths of peptide = J.
5. A protein found in the cytosol of resting T cells which translocates to the nucleus upon cell stimulation = I. NFAT is named for this behavior: "Nuclear Factor of Activated T cells."
6. A protein whose expression allows scientists and clinicians to recognise the cell as a T killer cell, that is, a member of that population of T cells that attacks body cells infected by viruses or other pathogens = C.
7. A receptor expressed by mature B cells that, when activated, causes differentiation and proliferation = B.
8. A transmembrane protein that can, if the match is correct, bind to a major histocompatability complex protein that is presenting a foreign peptide = K.
9. A transmembrane protein that presents short lengths of peptide at the cell surface = H.

20.1 Theme: CFTR Mutations

1. This condition is likely to be a nonsense mutation and therefore may be treatable by drugs that cause the ribosome to read through a stop codon = B. Only a premature stop codon can generate a truncated protein. If the stop codon was created by a base change, then there will be no frameshift problems. Thus if the ribosome can be caused to read through the stop, the protein generated will either be completely normal or (more likely) differ in one amino acid only.
2. This condition is likely to be caused by a mutation in a promoter or enhancer region of the *CFTR* gene = F. Mutations that change a promoter or enhancer sequence, and in turn decrease how effectively RNA polymerase II can bind, will decrease the rate of transcription and less mRNA will be made. Some mutations can of course change a promoter or enhancer sequence to one that makes RNA polymerase II bind more efficiently. In this case more mRNA protein will be made. Once the mRNA is made it will be translated into a normal protein.
3. This condition may in the future be treatable by drugs that favour the open configuration of the CFTR channel = D. If CFTR is present at the plasma membrane and would allow chloride ions to pass if it opened, but is aberrant in the residues that allow it to be switched into the open state, then a drug that artificially opens the channel is likely to be of use.
4. This condition may in the future be treatable by drugs that inhibit or modify the function of the proteasome = C. The proteasome is the protein complex that destroys misfolded proteins, and by inhibiting its function it may be possible to retain partially misfolded CFTR that will nevertheless function, at least to some extent, if it reaches the plasma membrane.

20.2 Theme: The CFTR as a Permeability Pathway

Answer: H. If protein kinase A is activated then it will phosphorylate CFTR, causing the channels to open. Chloride is at equilibrium across the plasma membrane of most cells, so opening CFTR channels will not cause a net flow of chloride in either direction, and will therefore not change the membrane voltage. However, the open CFTR channels are a pathway for electrical current flow, so the resistance of the plasma membrane will decrease.

20.3 Theme: A Medly of Sentences about CF

1. C. Because CF is recessive the disease must . . . be inherited from both parents.
2. H. Reading through a stop codon may help some CF patients because . . . some of the many CF mutations have a mutation that introduces a premature stop. The truncated protein does not function properly.
3. B. A zoo blot is useful to . . . identify genes that have been conserved in several species. Such genes are likely to code for important proteins.
4. D. CFTR protein was definitively proven to be a chloride channel by . . . a lipid bilayer voltage clamp experiment. CFTR protein was inserted into a lipid bilayer and chloride ions were shown to move through the channel.
5. E. Gene therapy for CF is very difficult because . . . the healthy gene needs to be inserted into a huge number of cells in the lungs and these cells are relatively short-lived. In addition the modified virus vectors used seem to cause side effects.

INDEX

Cell Biology: A Short Course, Third Edition. Stephen R. Bolsover, Jeremy S. Hyams, Elizabeth A. Shephard and Hugh A. White.